3rd International Seminar on Sciences (ISS 2016)

Sciences on Precision and Sustainable Agriculture

IOP Conference Series: Earth and Environmental Science
Volume 58

Bogor, Indonesia
4 November 2016

Editors:

Yessie Widya Sari

ISBN: 978-1-5108-3896-3
ISSN: 1755-1307

Printed from e-media with permission by:

Curran Associates, Inc.
57 Morehouse Lane
Red Hook, NY 12571

Some format issues inherent in the e-media version may also appear in this print version.

Printed by Curran Associates, Inc. (2017)

Institute of Physics
Dirac House, Temple Back
Bristol BS1 6BE UK

Phone: 44 1 17 929 7481
Fax: 44 1 17 920 0979

techtracking@iop.org

Additional copies of this publication are available from:

Curran Associates, Inc.
57 Morehouse Lane
Red Hook, NY 12571 USA
Phone: 845-758-0400
Fax: 845-758-2633
Email: curran@proceedings.com
Web: www.proceedings.com

Table of contents

Volume 58

The 3rd International Seminar On Sciences "Sciences On Precision And Sustainable Agriculture" (ISS-2016)

4 November 2016, Bogor, Indonesia

Accepted papers received: 20 March 2017
Published online: 4 April 2017

Preface

Oral Session Contributors

Posters

THE 3RD INTERNATIONAL SEMINAR ON SCIENCES
"SCIENCES ON PRECISION AND SUSTAINABLE AGRICULTURE"
(ISS-2016)

PREFACE

The International Seminar on Science 2016 conference was organized by the Faculty of Mathematics and Natural Sciences, Bogor Agricultural University in November 4th 2016. It is an annual event which purpose is to share knowledge and expertise among scientist, academicians, and industries in the area of sciences and the applications. In addition, the conference was arranged to build networks and collaborations among scientists from various institutions at a national and international level.

ISS 2016 was organized as a scientific seminar with the theme "Sciences on Precision and Sustainable Agriculture", which covers two keynote speeches, some invited talks, and some contributed presentations. The parallel presentations came from 104 papers, which were presented in the form of oral as well as poster presentations. Various scientists from several research and higher education institutions in Indonesia as well as some neighboring countries contributed to this conference.

In this occasion, we were honored to have our distinguished keynote speakers:

PROF. HIROSHI TAKAGI (Nara Institute of Science and Technology, JAPAN)

and

PROF. DR. JPM (JOHAN) SANDERS (Wageningen University, the Netherlands)

We were also pleased to have the following invited speakers:

PROF. DR. IR. H.J. (ERIK) HEERES (University of Groningen, Netherlands)

PROF. DANIEL MURDIYARSO (Bogor Agricultural University, Indonesia)

DR. BRUCE MATTHEWS (University of Hawaii at Hilo, United States)

We would like to thank all participants for their contributions especially those who submit their full papers to the editorial board. In total, 70 selected papers were published. We are also honored to have authors coming from various institutions around the world namely Netherland, Japan, Indonesia, United States, and Malaysia. Finally, we would also thank and give our highest appreciation to the reviewers for maintaining a high quality standard of the accepted manuscripts for publication.

Bogor, February 2017

Dr. Bagus Sartono
ISS 2016 Chairperson
Bogor Agricultural University

Dr. Yessie Widya Sari
ISS 2016 Proceeding Editor
Bogor Agricultural University

Peer review statement

All papers published in this volume of *IOP Conference Series: Earth and Environmental Science* have been peer reviewed through processes administered by the proceedings Editors. Reviews were conducted by expert referees to the professional and scientific standards expected of a proceedings journal published by IOP Publishing.

Clustering Module in OLAP for Horticultural Crops using SpagoBI

D Putri[1] **and I S Sitanggang**[1]

[1]Computer Science Department, Faculty of Mathematics and Natural Science, Bogor Agricultural University, Jalan Meranti Wing 20 level 5 Kampus IPB, Bogor, West Java, Indonesia

E-mail: defrianiputri@gmail.com, imas.sitanggang@ipb.ac.id

Abstract. Horticultural crops data are organized by the Ministry of Agriculture, Republic of Indonesia. The data are presented annually in a tabular form and result a large data set. This situation makes users difficult to obtain summaries of horticultural crops data. This study aims to develop a clustering module in the SOLAP system for the distribution of horticultural crops in Indonesia and to visualize the results of clustering in a map using SpagoBI. The algorithm used for clustering is K-Means. Horticultural crops data include vegetables, ornamental plants, medicinal plants, and fruits from 2000 to 2013. The clustering module displays clustering results of horticultural crops in the form of text and table on SpagoBI. This module can also visualize the distribution of horticultural crops in the form of map on the HTML page. The application is expected to be useful for users in order to easily obtain summaries of the horticultural crops distribution data and its clusters. The summaries and clusters can be beneficial for the stakeholders to determine potential areas in Indonesia for horticultural crops.

1. Introduction
Ministry of Agriculture of Republic of Indonesia (Ministry of Agriculture RI) is the institution that provides data on agricultural sector including horticultural crops. The data are presented annually for all districts in Indonesia in a tabular form which results in a large data set. The situation makes data accumulation and further processing on the data is required if users need summaries of horticultural crops data. The horticultural crops data can be accessed on the website http://aplikasi.pertanian.go.id /bdsp/index.asp. The data include vegetables, ornamental plants, medicinal plants, and fruits from 2000 to 2013.

Spatial data are geographically oriented data and have a coordinate system as a reference [1]. Spatial data cube contains spatial dimensions and measures. Online Analytical Processing (OLAP) is one of application that integrates data warehouse in order to simplify a process of information extracting from large data sets [1]. A spatial data warehouse for horticultural crops has been constructed by integrating thematic and geographically referenced data from multiple sources and it provides a spatial data cube that is useful for multidimensional spatial data analysis [1].

Currently there are several open source business intelligence and reporting tools including Pentaho and SpagoBI. In this study, SpagoBI was used to display horticultural crops distribution as clustering results. Whereas, visualization of clustering results of horticultural crops is done in form of map on a HTML page. SpagoBI is an open source, complete, and flexible business intelligence tool with a wide set of analytical engines and it can be used to find a best solution for end users, saving time, and economic resources [2]. SpagoBI has an advantage in integrating new modules and it gives the original developers the possibility to develop and to improve products [3].

In this study, a clustering module in the SOLAP system was developed to provide distribution of horticultural crops in Indonesia and to visualize clustering results of horticultural crops in a map using SpagoBI. The SOLAP system is expected to be used by users to obtain summaries of the horticultural crops data and distribution of its clusters at district level in Indonesia. The summaries and clusters can be beneficial for the stakeholders to determine potential areas in Indonesia for horticultural crops.

2. Data and Method

2.1. Data

Horticultural crops data consist of four fields including commodities, status id, location, and time. This study uses horticultural crops data from 2000 to 2013. In addition, this study utilizes the administrative boundary map of districts in Indonesia in 2011. The map was collected from Geospatial Information Agency. There are several types of horticultural crops analysed in this study that are medicinal plants, fruits, vegetables, and ornamental plants. At the initial system development, we selected three types of data for each commodity category based on the completeness of data and data with less null. Medicinal plants commodity includes ginger, galangal, and turmeric. Fruit commodity includes avocados, star fruit, and durian. Vegetable crops include beans, carrots, and chili. Ornamental plants include orchids, roses, and tuberose flower.

2.2. Analysis of spatial data warehouse horticultural crops

Spatial data analysis of horticultural crops was done to identify characteristics of the data. Spatial data are geographically oriented data and have coordinate system as a reference [1]. A spatial data warehouse is a collection of spatial data that has several characteristics such as subject oriented, integrated, time variant, and non-volatile [1]. Data analysis is useful for determining the scheme of data warehouse. In this study, the data warehouse's scheme applied is the star schema.

2.3. OLAP operations and horticultural crops clustering

OLAP is an application that integrates data warehouse in order to simplify information extracting from large data sets [1]. In this study, OLAP operations are performed based on the type of each commodity, location, status id, year, harvest area, production, and productivity. Furthermore, clustering of horticultural crops is done on OLAP results. Clustering is a process of partitioning a set of data objects into several groups [1]. Data clustering is based on the similarity between objects. Objects that have high similarity are placed on one cluster. The similarity between two objects can calculated based on distance between those objects. This study uses Euclidean distance to measure the similarity between objects. The Euclidean distance (*d*) between objects *i* and *j* is define as follows in which x_i and x_j are features respectively in object i and object j [1].

$$d(i,j) = \sqrt{(x_{i1} - x_{i1})^2 + (x_{i2} - x_{i2})^2 + \cdots + (x_{ip} - x_{ip})^2} \quad (1)$$

The quality of clustering results is measured by Sum of Square Error (SSE). SSE is calculated to obtain number of clusters that minimizes the square error. For example $p \in C_i$ is each data point in cluster i, c_i is the centroid of the cluster i, k is total number of clusters, *dist* is the distance to each cluster i [1]. The Sum of Square Error (SSE) is defined as follows [2]

$$\mathrm{SSE} = \sum_{i=1}^{k} \sum_{p \in C_i} \mathrm{dist}(p, c_i)^2 \quad (2)$$

This study applies the well-known K-Means clustering algorithm. The steps in the K-Means algorithm are as follows [1]:

1. Select k initial objects as the centroids of the clusters,

2. Insert the objects into a cluster whose objects that are most similar based on the average value of objects in a cluster,
3. Renew the centroid of clusters by calculating the average value of objects for each cluster,
4. Repeat step two and three until the objects in the clusters have not changed.

2.4. *Integrations of clustering modules using R and SpagoBI*
This study utilizes the tools R and SpagoBI. SpagoBI is utilized to analyze horticultural crops using OLAP operations whereas R is used to cluster the results of OLAP operations. The integration between R and the clustering module on SpagoBI aims to cluster the output of OLAP operations using the data mining module that is available on SpagoBI and to present clustering results of horticultural crops in the tabular format.

2.5. *Visualization of clustering results*
The visualization of the clustering results is provided in form of map on the HTML page. In the development of the visualization module, this study utilizes ggplot2, maptools, rgdal, and plyr packages that are available in R. The visualization module provides a map that plots distribution of horticultural crops in district level in Indonesia.

3. Result and Discussion

3.1. *Analysis of spatial data warehouse horticultural crops*
The data warehouse of horticultural crops was designed based on the star schema with a single fact table and several dimension tables. The fact table contains unique attributes from each dimension and measures. Measures contained in the fact table are production (in tons), productivity (in tons/tree), and harvested area (in ha). There are 4 dimensions that are created in the data warehouse including commodity, location, time, and status id. Data analysis was done based on horticultural crops category, commodity, islands, province, district, year, and status of measure values. Figure 1 represents the star scheme for data warehouse of horticulture. The red sign on figure 1 indicates the measures to be calculated when the OLAP operations are performed.

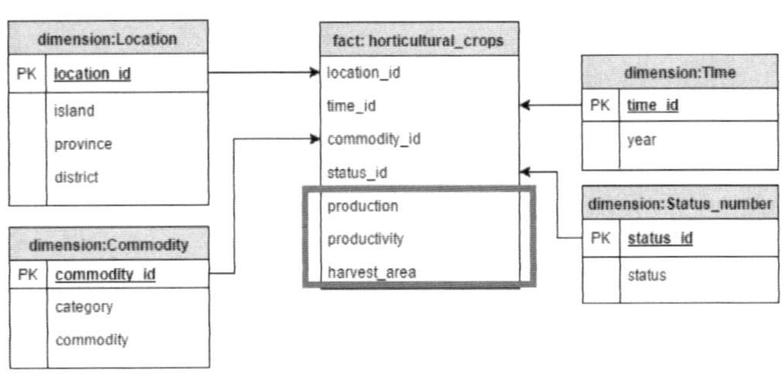

Figure 1. Star scheme for data warehouse of horticulture.

3.2. *OLAP operations and horticultural crops clustering*
OLAP operations are performed by selecting dimensions and measures on the data cube. OLAP operations on horticultural crops data include roll up, drill down, slice, and dice. The dimensions are commodity, location, status of measure values whereas the measures include harvested area (in hectares), production (in ton), and productivity (in ton/tree). The aggregation function sum is calculated for each measure in OLAP operations. Summary of horticulture data as results of OLAP operations are exported into the Microsoft Excel file on SpagoBI.

K-Means algorithm is applied to cluster datasets as the results of OLAP operations. Number of clusters (k) was chosen based on the lowest sum square error (SSE) to get the best clustering results.

Experimental results shows that the best clustering is obtained at the value k of 5. Therefore this work groups the horticulture data into 5 clusters with the label very low, low, medium, high, and very high. Table 1 shows the quality of clustering results at k of 5. Percentage of clustering quality in R is calculated by dividing *betweenss* with *totss*. The functions betweenss and totss are derived from the output component provided by the function of K-Means in R. For example, clustering results of harvested area, production, and productivity of durian fruit are presented respectively in table 2, table 3, and table 4.

Table 1. Quality of clustering results.

Number of cluster (k)	Clustering quality (%)		
	Harvested area (ha)	Production (ton)	Productivity (ton/tree)
3	91.07	81.63	92.09
4	96.72	86.68	95.39
5	98.16	87.81	96.12

Table 2. Clustering results of durian's harvested area.

Cluster	Cluster size	Centre	Minimum	Maximum.	Average	Percentage
1	12	32,321.500	23,335	44,972	32,322	2.73%
2	2	181,125.500	167,559	194,692	181,126	0.45%
3	399	208.251	0	6303	208.3	90.68%
4	19	13,199.368	7500	20,778	13,199	4.32%
5	8	63,805.625	52,895	75,813	63,806	1.82%

Table 3. Clustering results of durian's production (ton).

Cluster	Cluster size	Centre	Minimum	Maximum.	Average	Percentage
1	16	335,996.438	246,868	703,107	335,996	3.64%
2	327	3986.538	0	24,241	3987	74.32%
3	3	1,158,382.000	824,391	1,752,916	1,158,382	0.68%
4	35	130,450.314	88,980	224,386	130,450	7.95%
5	59	46,455.424	26,334	79,257	46,455	13.41%

Table 4. Clustering results of durian's productivity (ton/tree).

Cluster	Cluster size	Centre	Minimum	Maximum.	Average	Percentage
1	366	297.79	0	5725	297.79	83.18%
2	4	286,599.75	240,189	373,656	266,277	0.91%
3	7	106,352.14	79,605	130,677	106,352	1.59%
4	51	12,646.63	6993	21,131	12,647	11.59%
5	12	31,593.92	22,421	53,383	31,594	2.73%

3.3. Integration between R and data mining module on SpagoBI

This study uses the tools R version 3.2.4 and SpagoBI version 5.1.0. Clustering datasets is performed by integrating R with SpagoBI server. The integration between R and SpagoBI requires the rJava package. In rJava package, there are JRI files which are needed to execute scipts from R to Java. Furthermore, this study did the configuration on catalina.bat file that is contained in the folder bin in SpagoBI server. The configuration includes adding a path of JRI file that is available in rJava folder. The R code of Catalina.bat file configuration is given in figure 2.

```
set JAVA_OPTS="-Djava.library.path=C:\Program Files\R\R-3.2.4\library\rJava\jri\x64"
set R_HOME ="C:\Program Files\R\R-3.2.4"
set JAVA_OPTS= %JAVA_OPTS% -Xms1024m -Xmx1024m -XX:MaxPermSize=512m
```

Figure 2. Configuration on Catalina.bat file.

Integration between R and SpagoBI was implemented using the XML code template. On that template, R scripts are inserted in the SCRIPT tag. The program template is used when a new document is created on the data mining module in SpagoBI. Figure 3 shows the XML template to run R scripts on SpagoBI.

```
<?xml version="1.0" encoding="ISO-8859-15"?>
<DATA_MINING>
  <DATASETS>
    <DATASET name="data" type="file" readType="table" canUpload="true" label="Upload">
      <![CDATA[header=TRUE,sep =",",dec="."]]>
    </DATASET>
  </DATASETS>
  <SCRIPTS>
    <SCRIPT name="k-means" mode="manual" datasets="data" label="k-means clustering" libraries="fpc,RCurl,R2HTML,plyr">
      <![CDATA[

              ]]>
    </SCRIPT>
  </SCRIPTS>
  <COMMANDS>
    <COMMAND name="k-means" scriptName="k-means" action="kmeans" label="k-means cluster" mode="manual">
      <OUTPUTS>
        <OUTPUT type="text" name="LabelCluster" value="Cluster" mode="manual" label="Label Cluster"></OUTPUT>
      </OUTPUTS>
    </COMMAND>
  </COMMANDS>
</DATA_MINING>
```

Figure 3. XML template to run R scripts on SpagoBI.

A new document in the data mining module on SpagoBI is created by selecting the analytical documents feature in the data mining folder. This work selects the generic document as one of the analytical documents types. Figure 4 shows the document fields in the data mining module of SpagoBI.

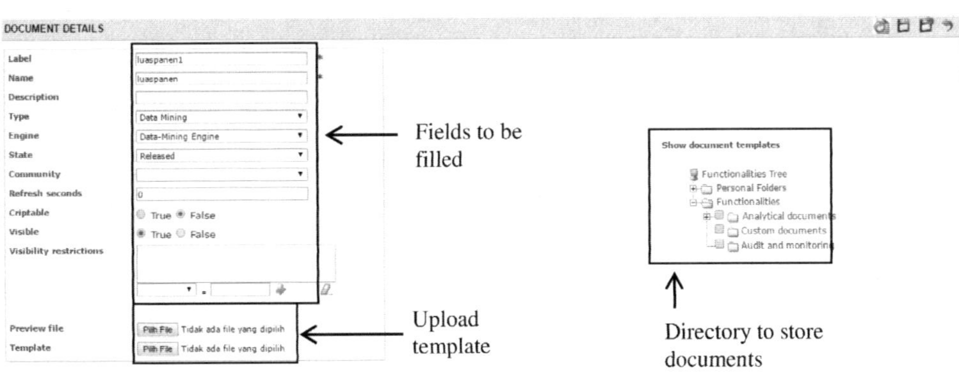

Figure 4. Document fields on the data mining module of SpagoBI.

Clustering results in SpagoBI are displayed in three types of output that are text, table, and image. The output type can be selected by adding the argument of text, html, or image on the output type in the XML template. Figure 5 shows the example of clustering results of durian commodity with the measure production that are represented in the text format.

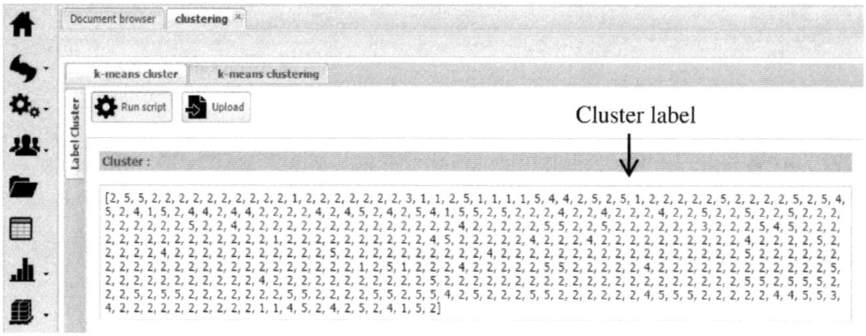

Figure 5. Example of clustering results represented in the text format.

In addition, clustering results can also be displayed in a tabular form using the data frame function. This feature requires the package R2HTML in R to display the output in tabular format on GUI of SpagoBI. Figure 6 represents the example of clustering results of durian commodity with the measure production represented in the tabular form.

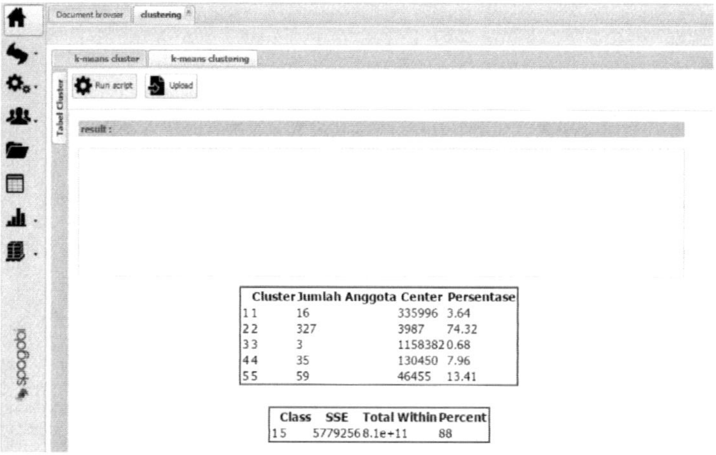

Figure 6. Example of clustering results represented in the tabular form.

3.4. Visualization of clustering results in a map

The visualization module of horticultural crops clusters was implemented using ggplot2, maptools, rgdal, and plyr packages that are available in R. The visualization module describes the distribution of horticultural crops in district level in Indonesia which is displayed on the HTML page. For example, figure 7 represents the distribution map of durian commodity based on harvested area.

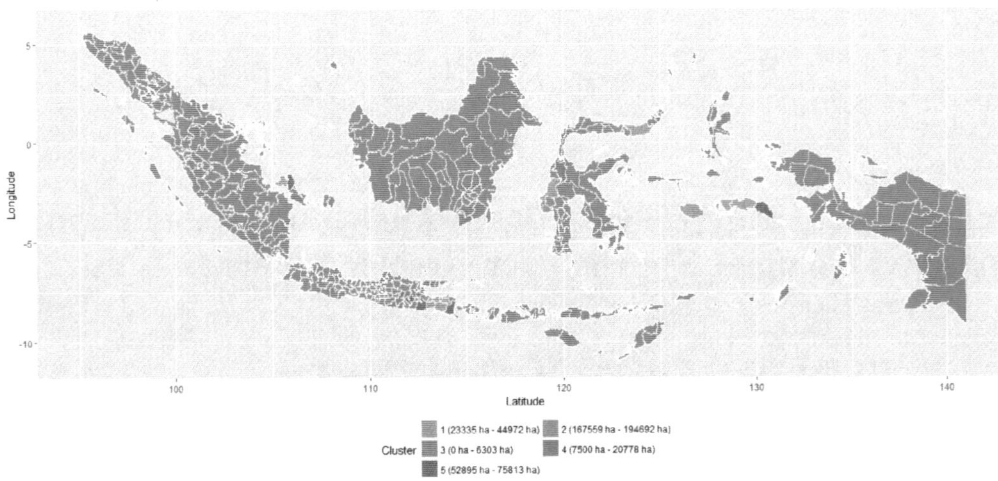

Figure 7. Distribution of durian commodity based on harvested area.

Different colors of each polygon on the distribution map in figure 7 represent different clusters. Each polygon in the map shows a district in Indonesia. Figure 7 shows that most of all districts in Indonesia are members of cluster 3 in which the harvested area in these districts is less than or equal to 6,303 ha. The main features of clustering module were tested based the black box approach. The main features include

- Load horticultural commodities dataset for clustering process
- Run the R scripts and displays the clustering results in the text format
- Run the R scripts and display the clustering results in the tabular format
- Plot clustering results on a map on the HTML page

The testing results show that all main features work properly.

4. Summary

A clustering module of horticultural crops data in the SOLAP system has successfully developed using SpagoBI integrated with R. Clustering results of horticultural crops are visualized in maps utilizing the ggplot2, maptools, rgdal, and plyr packages that are available in R. Maps are displayed on HTML pages that provide information about the distribution of horticultural crops in district level in Indonesia. Distribution of horticultural crops can be used to identify potential area for horticultural production based on harvested area (ha), production (tons), and productivity (tons/tree). System testing was carried out to evaluate the features in the visualization module. The testing results show that the clustering module in SpagoBI has successfully displayed horticulture crops clusters in the three types of output namely text, tabular and map.

References

[1] Han J, Kamber M and Pei J 2012 *Data Mining. Concept and Techniques* 3rd Edition (Amsterdam: Elsevier) pp 83-443

[2] Cazzin G 2012 *Business intelligence with SpagoBI* (Padua (IT) SpagoBI Competency Center) pp 2-59

[3] Golfarelli M 2009 Open source BI platforms: a functional and architectural comparison, *11th International Conference DaWak* (Austria. Bologna (IT): University of Bologna) pp 289-297

Application of Graph-Theory Based Algorithm for Identifying Convective Complex Systems over Greater Jakarta basins

D E Nuryanto[1,2*], E Aldrian[3], H Pawitan[1] and R Hidayat[1]

[1]Department of Geophysics and Meteorology, FMIPA Bogor Agricultural University, Kampus IPB Darmaga, Bogor 16680, West Java, Indonesia
[2]Research and Development Center, The Indonesian Agency for Meteorology Climatology and Geophysics, Jl Angkasa I No 2 Kemayoran Jakarta Pusat 10720, Indonesia
[3]Agency for The Assessment and Application of Technology, Jl MH Thamrin No 8 Jakarta 10340, Indonesia

Email : danang.eko@bmkg.go.id

Abstract. Heavy rainfalls over the Greater Jakarta (GJ) basins during January 14 - 15th, 2013 were suspected to be derived from convective complex systems (CCSs) as subsets of mesoscale convective systems (MCSs). This study, implementing a graph-theory based algorithm on black body temperature (TBB) dataset, identified CCS over GJ during heavy rainfall event on January 14 - 15th, 2013. We found that convective cloud cluster for $\leq$ 221 K of TBB more than 6 hours (about 16 hours of duration) with average of eccentricity was 0.47. The maximum area of this system was about 249,731.9 km^2 at 0300 Local Time (LT) January 15th, 2013. Ones of these convective system features is not match with MCC definition. We defined this convective system as convective cloud system that large, long lived, and very cold of cloud shield. This system is in agreement with heavy rainfall (more than 40 mm/3h) at 0200 – 0400 LT January 15th, 2013 over GJ basins.

1. Introduction

Mesoscale convective systems (MCSs) are features of complex organized thunderstorms that vary in size and continues for hours. Mesoscale convective complexes (MCCs) are a subset of large and well-organized mesoscale convective systems (MCSs) defined as convective cloud system that is large, long lived, and exhibits a quasi-circular cloud shield [1].

Identification of convective systems utilizes a particular spatial-temporal criterion that requires high-resolution (< 3 hourly) geostationary weather satellite data [1]. Past studies identifying the characteristics of the convective systems were fully manual and required both infrared (IR) images that supplied the cloud-top brightness temperature, and visible (VIS) images that provided information on cloud characteristics [2]. Recently, a fully automated method namely the Tracking of Organized Convection (TOOCAN) algorithm has been introduced for identifying and tracking convective systems in meteorological applications [3]. The existing methods for identification of convective systems in infrared datasets still need human intervention.

Graph theory is still rarely used in the atmospheric sciences for predicting details of the changing atmosphere for a period between 0 to 6 h over a small area (e.g. a city) and forecasting MCSs, specifically thunderstorms with radar and satellite data [4,5]. According to Chaudhuri and Middey

[4,5], the implementation of graph theory is suitable to express relationship among the atmospheric variables at a particular time.

This study was attempted to implement a graph-theory based algorithm on black body temperature (TBB) dataset, that were derived from the hourly infrared data of Multi-functional Transport Satellite (MTSAT)-IR imagery, for identifying convectice complex system (CCS) over the Greater Jakarta (GJ) during heavy rainfall event on January 14 - 15th, 2013. This method is called the "Grab 'em, Tag 'em, Graph 'em" (GTG) algorithm and it is a fully automated graph-theory based algorithm [6]. For the preliminary study, we used MCC criteria from Blamey and Reason [7] for identifying CCS.

2. Data and Methods

The equivalent black body temperature that was derived from hourly infrared data of Multi-functional Transport Satellite (MTSAT)-IR imagery was obtained from http://database.rish.kyoto-u.ac.jp/arch/ctop/index_e.html [8]. The satellite imagery was used to identify an objective of a convective complex system and their physical characteristics based on the parameters given by Maddox [1] and adopted by Blamey and Reason [7]. The studied area was located in the GJ basins as part of Indonesian Maritime Continent (IMC), see figure 1. The synoptic data from several weather stations of BMKG around GJ, i.e., at Serang (6.117°S, 106.133°E), Cengkareng (6.117°S, 106.650°E), Curug (6.233°S, 106.650°E), Tanjung Priok (6.100°S, 106.867°E), Kemayoran (6.183°S, 106.583°E), and Citeko (6.700°S, 106.933°E) were compared with a convective complex cloud cluster.

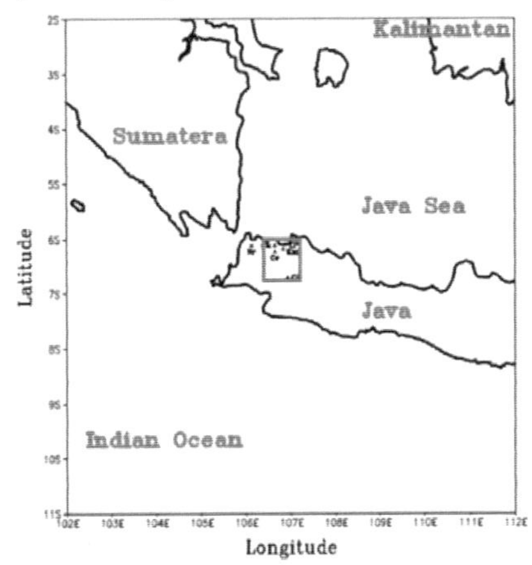

Figure 1. The study area (106.40°E – 107.20°E, 6.01°S – 6.76°S) around of the Great Jakarta (red box).

The methods for identifying the convective systems are usually based on forward or backward algorithm that executes in two stages, that is a cloud detection stage, and a tracking stage [3,9,10]. The cloud detection stage involves identifying the cloud areas of interest from a TBB data via a particular criterion of temperature and area at time increments. These criterion varies according to the particular convective systems and the location study.

Previous studies identified the temperature ranges of TBB associated with convective clouds in large convective systems to range from a 255 K temperature threshold to identify large-scale areas, to a 195 K temperature threshold to identify the convectively most active areas of the system [1,3,7,11,12,13,14]. There are some advantages in using the colder temperature threshold since it tracks the center of the system better and there was also a better representation of two systems that were in close proximity of each other [7].

In this study we implemented the "Grab 'em, Tag 'em, Graph 'em" (GTG) algorithm and is a fully automated graph-theory based algorithm [6]. The GTG algorithm works by using cloud elements

(CEs) with a particular size and temperature of TBB as the graph's vertices, defining the graph's edges as areas of overlap between CEs, and defining CEs that are correlated throughout time as cloud clusters (CCs). The advantages of graph theory implementation is to anticipate a problem relating to spontaneous splitting or merging of cloud areas though they belonged to the same system as identifying by Mathon and Laurent [15].

Figure 2 provides a schematic of the algorithm that works by first identifying areas of interest in the gridded TBB dataset to create a graph nodes and edges, then searching the graph for the type of convective systems. For a preliminary study, we used MCC criteria from Blamey and Reason [7] for identifying CEs and CCs, see given in Tabel 1. The properties of a CE in the region of interest are defined by cloud shield of 221 K, representing by graph nodes. The graph edges represented CEs in consecutive times that are correlated via the percentage of area overlap. The minimum weight of an edge occurs while the percentage overlap between two nodes is at least 66 %, and the maximum weight occurs for a 50 % overlap or in excess of 10,000 km^2 [16,17] as implemented by Whitehall et al. [6]. Within the graph construct, a cloud cluster (CC) is defined as the collection of CEs and edges. A CC can have a minimum duration of six times (6 hours), and maximum of *m* times, where *m* is number of times being considered.

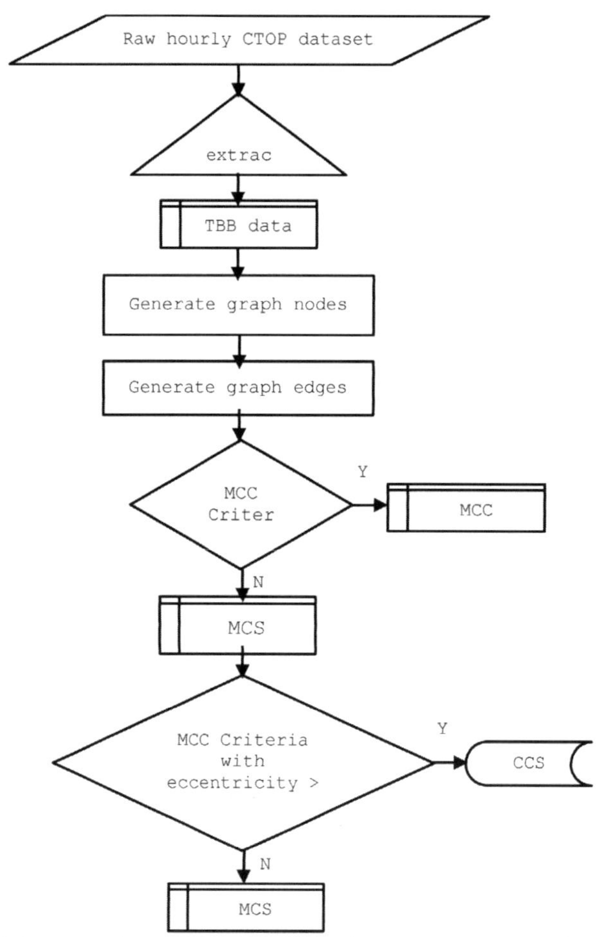

Figure 2. A schematic of the graph theory based algorithm for detection and tracking of CCS.

Table 1. Physical characteristics based on the parameters given by Maddox [1] and adopted by Blamey and Reason [7] to identify a convective complexes system in this study.

Physical characteristics	**Blamey and Reason [7] Criteria**
Cloud Top Temperature	≤ 221 K
Size	ε 50000 km^2
Shape-eccentricity	ε 0.7 at maximum extent
Duration	Size and temperature definition must be met for a period ε 6 hours
Initiation	Size and temperature thresholds are first met
Termination	Size definition are no longer satisfied

3. Results and Discussions

The area over GJ basins was examined for the 48 hour periods between January 14th, 2013 0000 UTC (0700 Local Time (LT) on January 14th to January 15th, 2013 2300 UTC (0600 LT on January 16th) using the infrared TBB data from the MTSAT-IR dataset. Table 2 shows the results of implementing a graph-theory based algorithm on MTSAT-IR dataset. We found some matches of MCC criteria for cloud size and duration for ≤ 221 K of TBB. But we did not find any match of the essential criteria of MCC (eccentricity < 0.7). The average of eccentricity was 0.47 for 16 hours of duration. The highest eccentricity from this study was about 0.69. The averages of TBB was about 208 – 214 K. The maximum area of this system was about 249,731.9 km2 at 0300 LT January 15th, 2013.

Figure 3 shows the spatial distribution of black body temperature from infrared data obtained by MTSAT-IR over GJ basins at 1600 LT, 2200 LT January 14th, 2013 through 0300 LT and 0700 LT January 15th, 2013. We described the phase of the convective systems as development stage (1600 – 2200 LT January 14th, 2013), mature stage (0300 LT January 15th, 2013) and dissipation stage (0700 LT January 15th, 2013). From this results, we defined this system as CCS defining as convective cloud system that large, long lived, and very cold of cloud shield.

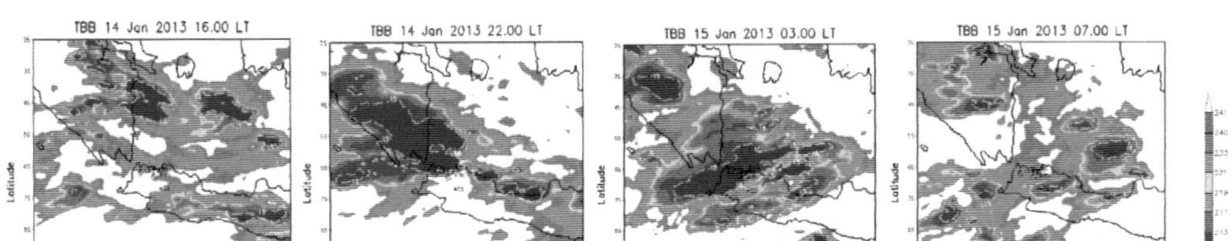

Figure 3. Horizontal distribution of black body temperature from infrared data obtained by MTSAT-IR over GJ basins respectively at 1600 LT, 2200 LT January 14th, 2013 through 0300 LT and 0700 LT January 15th, 2013.

Table 2. The results of implementing graph-theory based algorithm.

Time (LT)	Center of lat	Center of lon	Area Size (km^2)	Averages of brightness temperature (K)	Min of brightness temperature (K)	Max of brightness temperature (K)	Eccentricity
14/01/2013 16:00	-3.56	105.36	64187.5	210.8980	197	221	0.56
14/01/2013 17:00	-3.88	106.24	133185.1	211.6234	200	221	0.36
14/01/2013 18:00	-4.12	105.92	163859.4	212.0409	198	221	0.28
14/01/2013 19:00	-4.20	105.28	161730.4	210.7845	196	221	0.36
14/01/2013 20:00	-4.36	104.96	161651.5	209.1844	194	221	0.50
14/01/2013 21:00	-4.68	104.32	158655.1	208.1417	196	221	0.69
14/01/2013 22:00	-4.92	104.56	195164.6	210.2626	196	221	0.59
14/01/2013 23:00	-5.40	105.52	237351.7	211.5355	195	221	0.51
15/01/2013 00:00	-5.48	106.00	225365.9	210.2239	194	221	0.41
15/01/2013 01:00	-5.80	106.64	222921.4	208.2395	193	221	0.43
15/01/2013 02:00	-5.72	106.64	247129.7	210.2856	195	221	0.45
15/01/2013 03:00	-5.80	106.64	249731.9	212.9719	199	221	0.49
15/01/2013 04:00	-5.80	106.48	211960.6	213.8616	199	221	0.52
15/01/2013 05:00	-6.12	107.20	130030.9	213.5318	198	221	0.49
15/01/2013 06:00	-6.04	107.60	104560.9	213.4962	201	221	0.34
15/01/2013 07:00	-5.88	109.20	54803.8	214.2705	204	221	0.46

Figure 4 shows the TBB and rainfall distribution at specific sites around GJ, i.e: Cengkareng (202 K/41 mm), Citeko (224 K/1 mm), Kemayoran (200 K/19 mm), Tanjung Priok (211 K/44 mm), Curug (201 K/26 mm) and Serang (206 K/59 mm) during 0200 – 0400 LT January 15th, 2013. During this time the convective systems reach the mature stage at 0300 LT January 15th, 2013. This system in mature stage have good agreement with heavy rainfall at 0200 – 0400 LT January 15th, 2013 over GJ basins. Previous study showed that maximum rainfall was found at 0000 – 1100 LT over around the GJ [18]. Except at Kemayoran, the low temperature of clouds (TBB < 212 K) provides a significant rainfall at Cengkareng, Tanjung Priok and Serang with measured more than 40 mm/3h at such time. The higher portion of convective rainfall corresponded well to the thicker clouds colder than 210 K [19]. The TBB about 210 K also had been convinced to detect the precipitating core around the tropical deep convective activity [20].

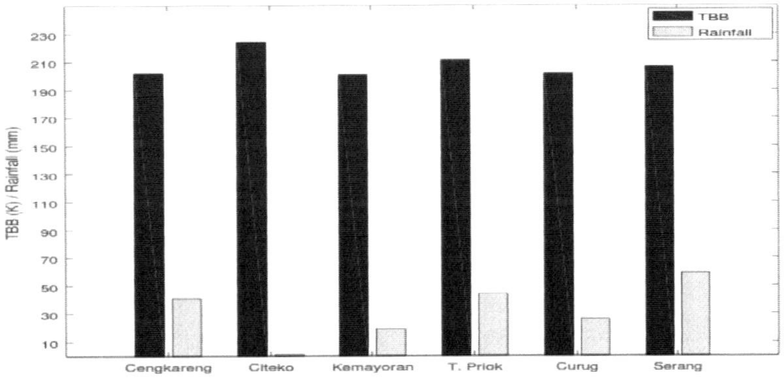

Figure 4. TBB (K) at 0400 LT January 15th, 2013 and rainfall distribution (mm/3 hour) during 0200 – 0400 LT January 15th, 2013 at specific sites around GJ, i.e: Cengkareng, Citeko, Kemayoran, Tanjung Priok, Curug and Serang.

4. Conclusions

The implementation of a graph-theory based algorithm was successfully performed on black body temperature (TBB) dataset, that identified CCS over GJ basins during heavy rainfall event on January 14 - 15th, 2013. We found convective cloud cluster for $\leq$ 221 K of TBB more than 6 hours (over 16 hours of duration) with average of eccentricity was 0.47. The maximum area of this system was about 249,731.9 km^2 at 0300 LT January 15th, 2013. The eccentricity of these convective system was not match with MCC definition. Then, we defined this convective system as convective cloud system that large, long lived, and very cold of cloud shield. This system was in good agreement with heavy rainfall (more than 40 mm/3h) at 0200 – 0400 LT January 15th, 2013 over GJ basins.

Acknowledgments

This research was supported by Department of Geophysics and Meteorlogy, Bogor Agricultural University and the Indonesian Agency for Meteorology Climatology and Geophysics through the Graduate Study Program in Appllied Climatology.

References

[1] Maddox R A 1980 *Bull. A. Met. Soc.* **61(11)** 1374 – 1387
[2] Velasco I and Fritsch JM. 1987 *J. Geophys. Res. Atmos.* **92(D8)** 9591–9613
[3] Fiolleau T and Roca R. 2013 *IEEE Trans. Geosci. Remote Sens.* **51(7)** 1–14
[4] Chaudhuri S and Middey A 2009 *Advances in Meteorol.* pp 1– 12
[5] Chaudhuri S, Middey A 2011 *Meteorol. Appl.* **18(2)** 238– 249
[6] Whitehall K, Mattmann C, Jenkins G, Rwebangira M, Demoz B, Waliser D, Kim J, Goodale C, Hart A, Ramirez P, Joyce M, Boustani M, Zimdars P, Loikith P and Lee H 2014 *Earth Sci. Inform*
[7] Blamey R C and Reason C J C 2012 *J. Climate.* **25** 753 – 766
[8] Hamada A and Nishi N *J. Appl. Meteor. Climatol.* **49** 2035 – 2049
[9] Carvalho L M and Jones C A 2001 *J. Appl. Meteorol.* **40(10)** 1683–1701
[10] Schröder M, König M and Schmetz J 2009 *J. Geophys.* Res. **114** D5
[11] Miller D and Fritsch J M 1991 *Mon. Wea. Rev.* **119** 2978 – 2992
[12] Laing A G and Fritsch J M 1993 *Mon. Wea. Rev.* **121** 2254 – 2263
[13] Laurent H, D'Amato N and Lebel T 1998 *Phys. Chem. Earth* **23(5–6)** 629–633

[14] Machado L A T, Rossow W B, Guedes R L and Walker A W 1998 *Mon. Wea. Rev.* **126(6)** 1630–1654
[15] Mathon V and Laurent H 2001 *Q. J. R. Meteorol. Soc.* **127**(572) 377–406
[16] Arnaud Y, Desbois M and Maizi J 1992 *J. Appl. Meteorol.* **31**(5) 443–453
[17] Williams M and Houze RA Jr 1987 *Mon. Wea. Rev.* **115**(22) 505–519
[18] Nuryanto D E, Pawitan H, Hidayat R and Aldrian E 2016 *Procedia Environ. Sci.* **33** 178 – 186
[19] Inoue T and Aonashi K 2000 *J. Appl. Meteorol.* **39** 2292 – 2301
[20] Chen S S and Houze RA Jr 1997 *Quart. J. Roy. Meteor. Soc.* **123** 357–388

Synthesis of activated carbon-nanomagnetite-pyrazolone(1-phenyl-3-methyl-5-pyrazolone) composite as adsorbent for Cd^{2+}

D Pratiwi[1], D Saprudin[1] and E Rohaeti[1]

[1] Department of Chemistry, Bogor Agricultural University, INDONESIA.

E-mail: dewi_pratiwi12@yahoo.co.id , dedensaprudin@yahoo.com , etirohaeti@ipb.ac.id

Abstract. Pyrazolone was impregnated onto activated carbon-nanomagnetite (AC-NM) composite and used as an adsorbent for cadmium (Cd(II)). AC-NM composite has been synthesized by hydrothermal method. The single phase of spinel magnetite and crystalinity of nanomagnetite (NM) and AC-NM composite were confirmed by X-ray diffraction (XRD). It has been found that the crystalinity of AC-NM composite was lower than NM. The synthesized activated carbon-nanomagnetite-pyrazolone (AC-NM-PZ) was characterized by Fourier Transform Infra-Red (FTIR). Moreover, the sorption capabilities for Cd(II) ions were evaluated by three adsorbent AC, AC-NM, and AC-NM-PZ. The pH range, adsorption time, and adsorbent dosage were optimized by Response Surface Method (RSM). Cd(II) ions sorption process was studied and evaluated by two adsorption isotherms and found to follow the Langmuir model. A value of 20.64 µmol/g was identified as the maximum sorption capacity of AC. AC-NM composite was found to increase the sorption capacity to be 36.73 µmol/g. The sorption capacity for AC-NM-PZ composite was 27.67 µmol/g, it was lower than KA-NM but higher than AC.

1. Introduction

Cadmium (Cd) is known to be toxic and effect on human health in a certain number. The Environmental Protection Agency provides maximum limit Cd level in drinking water is 0.005 mg/L while the World Health Organization is 0.003 mg/L. Poisoning of Cd can cause hypertension, bone fragility, malfunction of kidney, and even lung cancer [1,2]. Cd metal may enter and accumulate in the human body through foods, beverage, respiration and skin. Therefore, it is important to reduce the concentration of Cd in environment, especially the aquatic environment. One of the most common method to reduce the concentration of heavy metal is adsorption, for example using activated carbon.

Activated carbon (AC) has commonly been used as adsorbent due to the advantages, such as high surface area, micro-pore structure, and the presence of functional groups on the surface of the adsorbent. AC pore structure can be modified through surface functionalization of AC that can easily adsorb heavy metal ions. Functionalization of AC surface has been done by oxidation using nitric acid [3] and ammonium peroxydisulfate solution [4] which could increase the adsorption capacity of AC. Chemical or biological modification on the surface of AC has also been recognized as an effective way to increase adsorption capacity of AC [5]. One of the chemical modification on AC surfaces was peroxide acid which could increase the adsorption capacity of Hg(II) [6]. Therefore, the increase of adsorption capacity of AC by modification is very interesting to be explore.

Nanomagnetite (NM) immobilized on AC can increase the surface area of AC and also cause increase the adsorption capacity. Zach-Maor et al obtained the adsorption capacity of AC was 68% while NM modified AC adsorption capacity of 91% for the adsorption of Cr(VI) while the adsorption capacity to Cu(II) increased from 32% to 85% by activated carbon and activated carbon immobilized NM, respectively [7]. However, NM can form aggregates of magnetic properties [8]. Formation of aggregates causing adsorption process becomes less efficient so we need NM surface modification.

NM surface modification using specific ligands can improve the efficiency of heavy metal adsorption. One ligands have been reported on the extraction of heavy metals is EDTA. EDTA immobilized NM successfully applied to the extraction of Cu, Cd, Co, and Pb metal ions [9]. However, it cannot adsorb one heavy metal ions specifically because the extraction percentage of those metals did not differ significantly. Adsorption capacity of EDTA-immobilized NM for Cd extraction was found 0.32 μmol/g [9]. Thus its needed the development and selection of other ligands to determine the selectivity of the adsorbent.

Pyrazolone (PZ) compounds and its derivatives have been used for the extraction of lanthanide and transition metals such, La(III), Eu(III), Lu(III), Cu(II), Ni(II), Zn(II), Cd(II) , Hg (II), Pb (II), Mo (VI) [10-13]. Keto group in pyrazolone compounds can form complexes with metal ions. Todorova et al modified silica gel and kieselguhr with pyrazolone compound (1-phenyl-3-methyl-5-pyrazolone) and the largest adsorption capacity was kieselguhr composite with pyrazolone was found 0.71, 0.26, 0.58, and 0.83 mmol/g for Cu(II), Mn(II), Cd(II) and Pb(II), respectively [14]. Heavy metal adsorption using AC from apricot modified with pyrazolone also studied by Lakov et al which resulted with adsorption capacity of Cd was 0.54 mmol/g [15].

The activated carbon-nanomagnetite-pyrazolone (AC-NM-PZ) has been synthesized successfully and determined the adsorption capacity for Cd(II). In addition to the adsorption capacity, the synthesis process was observed by XRD's difractograms and FTIR's spectras.

2. Materials and Methods

2.1. Materials and Instrumentation

2.1.1. Materials. $FeCl_3.6H_2O$, sodium citrate ($Na_3C_6H_5O_7$), urea, activated carbon, pyrazolone, ethanol, and standard solution of cadmium (Cd), copper (Cu), lead (Pb) from Merck, New Jersey, USA.

2.1.2. Instrumentation. The evaluation of metal ions content was performed using FAAS AA–6800 (Shimadzu, Kyoto, Japan) and the wavelength used for monitoring Cd, Pb and Cu was 228.8; 283.3; and 213.4 nm, respectively. Adjustment of pH was carried out using pH meter HM-20S (TOA, Tokyo, Japan). Unitronic OR P shaker water batch (J. P. Selecta, Barcelona, Spain) was used for batch adsorption. Infrared spectrums were recorded on a FTIR Bruker (Tensor, Billerica, USA). XRD patterns were obtained on XRD 7000 diffractometer (Shimadzu, Kyoto, Japan).

2.2. Method

2.2.1. Synthesis and Characterization of NM, AC-NM, and the AC-NM-PZ. The NM and AC-NM were synthesized using hydrothermal method. $FeCl_3.6H_2O$ 0.2 M was added by 0.4 M sodium citrate and 0.6 M urea dissolved in 100 ml of distilled water. The mixture was added five grams of AC for AC-NM. The solution was stirred using a magnetic stirrer for 30 minutes and then put in teflon hydrothermal and heated in oven at 200 °C for 12 hours. Black solids that formed were washed with distilled water and ethanol. The sample was dried at 60 °C for 12 hours and then synthesized and characterized with FTIR and XRD.

Five grams of AC-NM was added 10 grams of PZ and then dissolved in 250 ml $NaHCO_3$ 10%. The solution was stirred using a stirrer for 20 hours. The sample was then separated from the solution with centrifugation at 3000 rpm and washed with distilled water; acetone plus distilled water in the ratio

1:1; and distilled water to pH neutral. The resulting AC-NM-PZ was dried at 60°C for 12 h and then characterized with FTIR.

2.2.2. Determination of Optimum Condition Adsorption of Cd (II) with RSM. Optimization was conducted at pH (4-8), adsorbent dosage (0.1-0.5 g) and adsorption time (5-25 min). RSM was employed to investigate the optimum condition for Cd(II) adsorption. The method was performed by inputting the combination of variables in statistics software, Minitab.v.16 English.

2.2.3. Determination of Adsorption isotherms. The adsorbent (0.5 g) was contacted with 50 ml of Cd(II) metal ion solution in various concentration (10, 20, 30, 40, and 50 mg/L) in a shaker batch under optimum condition. Unadsorbed Cd(II) ions in the solution was determined with FAAS. Adsorption isotherm of adsorbent was analyzed by using Langmuir and Freundlich isotherm models.

2.2.4. Effect of Other Coexisting Metal Ions. To investigate the selectivity of AC, AC-NM, and the AC-NM-PZ adsorbent in this study, 0.5 g of adsorbent was shaken with a mixture of metal ions (Cd, Pb, and Cu) under optimum condition using a shaker batch. Unadsorbed Cd, Pb, and Cu metal ions in the solution were evaluated by FAAS.

3. Result and Discussion

3.1. Characterization

3.1.1. Characterization of AC and AC-NM by XRD. XRD characterization is aimed to determine NM formed on the sample, the degree of crystallinity, crystal size and synthesized crystalline form. The XRD results is shown in Figure 1. Diffractograms obtained were compared with JCPDS 19-0629 2002 PCPDFWIN program version 2.3 of the International Centre for Diffraction Data. Figure 1 shows that the peaks are produced in accordance the standards of magnetite in JCPDS 19-0629 with the highest intensity at 2θ ° 35.422, 35.353 °, and 35.445° respectively for the standard magnetite, NM synthesized, and composites AC-NM. The XRD pattern showed no XRD pattern for γ-Fe2O3 is at the height of 210 and 211 [16, 17]. So that the synthesized powders and composites are magnetite with AC also did not change the crystal structure of magnetite. The difference is in the background of AC-NM, it was greater than the NM because AC amorphous. The XRD pattern of AC amorphous appears at 2θ values of 20-30 [18]. Amorphous nature of the air conditioner can also reduce the degree of crystallinity of samples that has been synthesized from 64.07% to 38.42%. Synthesized NM and AC-NM crystal size by Scherrer equation was obtained 39.6 nm and 30.3 nm, respectively. While the crystal form follows the pattern face centered cubic crystalline form.

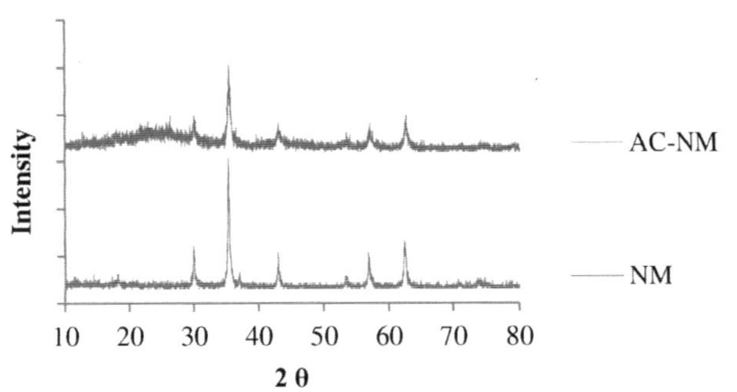

Figure 1. XRD diffractograms of NM and AC-NM.

3.1.2. Characterization of AC, NM, PZ, AC-NM, and AC-NM-PZ by FTIR. To confirm the presence of specific functional groups and additional functional groups in the materials of this study, FTIR were performed to AC, NM, PZ, AC-NM, AC-NM-PZ as shown in figure 2. AC did not indicate specific functional group absorption. It is caused AC did not functionalized well, indicated by no functional group present. The FTIR spectrum for synthesized NM indicated by a sharp peak at wave numbers 565 cm^{-1}. It is a sharp peak absorption area bond Fe-O which lies in the area absorption wave numbers <570 cm^{-1} [18, 19]. The FTIR spectrum for AC-NM composite showed no significant difference with the results of the analysis of functional groups AC. It concluded that the composite does not indicate the formation of a new bond is formed. But there is a change in the regional OH absorption wave numbers 3650-3200 cm^{-1} [16] and Fe-O with a weak peak, which indicates NM.

Pyrazolone (PZ) has several particular peaks such as carbonyl and phenyl group. The band at 1805 cm^{-1} indicates C=O (carbonyl) group of pyrazolone ring. C=C group of phenyl is shown by an absorption at 1601 cm-1. Absorption at 1151 cm^{-1} shows the vibration of the C-N. Vibration at 1027 cm^{-1} shows the C-H of ethyl group (CH_2). The band at 3056 and 3029 cm^{-1} shows the vibration of C-H group (methyl). Absorption at 3128 cm^{-1} shows the vibration of C-H aromatic groups. While vibration O-H is seen at 3422 cm^{-1}. Pavia stated that the vibration of carbonyl group located at 1820-1660 cm^{-1}, C=C aromatic vibration at 1600-1475 cm^{-1} and CN at 1350-1000 cm^{-1}, CH vibration of the aromatic hydrocarbon group at about 3000 cm^{-1}, CH vibration of the methyl group located at 3100-2500 cm^{-1} and OH vibration is located at 3600-3200 cm^{-1} [20].

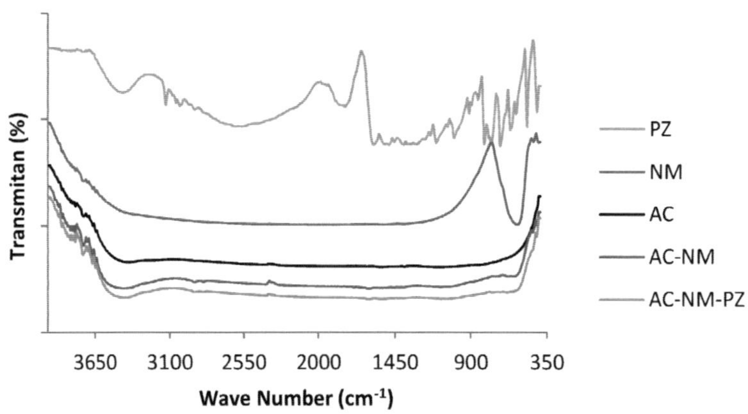

Figure 2. FTIR spectra AC, NM, PZ, AC and AC-NM-NM-PZ.

3.2. Optimum conditions Adsorption of Cd (II)

RSM was applied to develop an appropriate regression model to find out the optimal values of pH, adsorbent dosage, and adsorption time in order to enhance the adsorption efficiency. Central Composite Design (CCD) with 20 runs consisted of 8 runs for full factorial, 6 star points, 6 replicates at the centre point were successfully employed for batch experimental design. The adsorption efficiency of AC, AC-NM, and AC-NM-PZ in optimum condition is shown as the darkest green colour of contour plot shown in Figure 3(a), 3(b), 3(c) for AC, AC-NM, and AC-NM-PZ, respectively. The results are suitable with the data analysis using CCD where the optimum condition obtained were pH 6, adsorbent dosage 0.5 g, and adsorption time 25 minutes. Weights and pH optimum was accordance that conducted by Narin et al using the conventional method on Cd(II) adsorption with AC [17]. While AC-NM was obtained at pH 4, adsorbent dosage 0.5 g, adsorption time 5 minutes and AC-NM-PZ at pH 8, adsorbent dosage 0.5 g, adsorption time 25 minutes. The optimum condition was then used to determine the adsorption capacity and isotherm adsorption of the third adsorbent to Cd(II) metal ions.

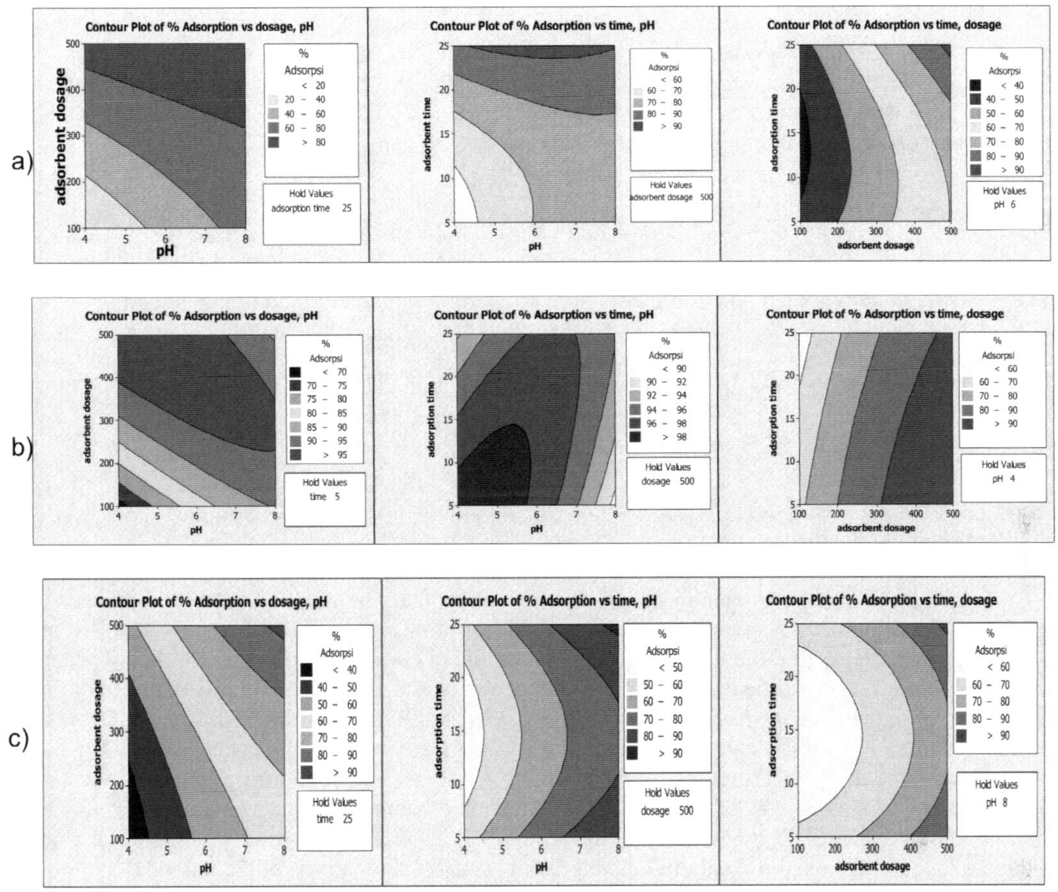

Figure 3. Contour plot of adsorption efficiency of Cd (II) by AC (a), AC-NM (b), and AC-NM-PZ (c) with various of pH, adsorption time, and amount of adsorbent.

3.3. Adsorption Isotherm

Determination of adsorption capacity and isotherm adsorption of AC, AC-NM, and AC-NM-PZ were investigated at the optimum condition with different initial concentrations of Cd(II) solutions (10-50 mg/L). Thermodynamic factors were not inserted as parameters and the determination was conducted only at room temperature. Table 1 shows the adsorption capacity and adsorption efficiency of AC, AC-NM, and AC-NM-PZ for each Cd(II) initial concentration. The efficiency adsorption of Cd (II) adsorbed by the third adsorbent decreases with the increase in the initial concentration of Cd(II) as can be seen in Table 1. While the adsorption capacity was increase.

Table 1. Adsorption efficiency and adsorption capacity of AC, AC-NM, and AC-NM-PZ.

[Cd(II)]	% Adsorption			Adsorption Capacity (μmol/g)		
(mg/L)	AC	AC-NM	AC-NM-PZ	AC	AC-NM	AC-NM-PZ
10	70.75	88.82	88.50	7.83	11.48	12.28
20	65.04	86.95	67.86	14.68	23.66	20.28
30	49.61	72.28	50.65	16.28	28.82	22.42
40	48.02	60.43	49.37	18.41	28.56	22.42
50	35.85	51.92	41.92	17.79	34.61	26.60

Isotherm adsorption of the third adsorbent was analyzed by using two isothermal equation models, Langmuir and Freundlich, showed in Fig.4a and 4b. Based on the results, the isotherm adsorption of the third adsorbent follows the Langmuir model since the R^2 value is higher than that of the Freundlich equation. This demonstrates that the third adsorbent has a homogeneous surface that based on adsorption occurs only in a single layer (monolayer) on the active side of the adsorbent.

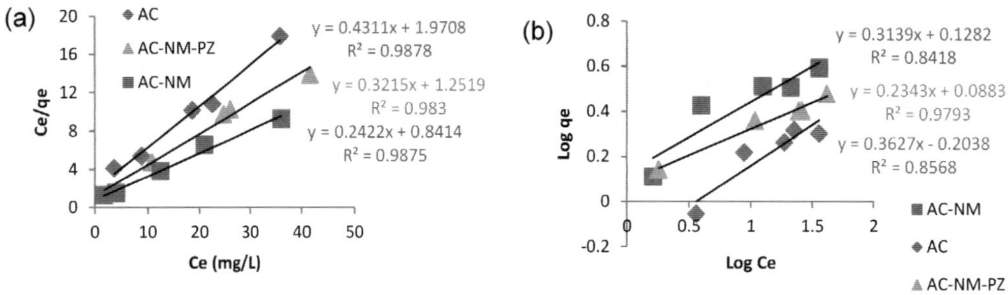

Figure 4. Isotherm adsorption Model for Cd (II) adsorption by AC, AC-NM, and AC-NM-PZ (a) Langmuir (b) Freundlich.

Cu and Pb metal ions were often in polluted waters except Cd. Therefore, Cu and Pb were used in this research to determine the selectivity of AC, AC-NM, and AC-NM-PZ in adsorbing Cd (II) in the mixture. Figure 5 shows that the three adsorbents (at optimum condition) were not selectively adsorb Cd (II) metal ion in spite of the existence of other metal ions in the mixture. The adsorption capacity for Cu (II) was higher than for the other two metal ions (Cd and Pb) by the three adsorbents. The high adsorption capacity of Cu (II) can be explained due to Cu (II) is more electronegative than other two metal ions (Cd and Pb). Therefore, Cu (II) more easily attracts the lone pair of adsorbent. Electronegativity value of Cu (II) > Pb (II) > Cd (II) according to Pauling scale estimates with electronegativity values 2.00, 1.87, and 1.69, respectively [21]. Allen and Brown states that the more electronegative metal ions, the metal ions are increasingly attracted to the adsorbent surface [22].

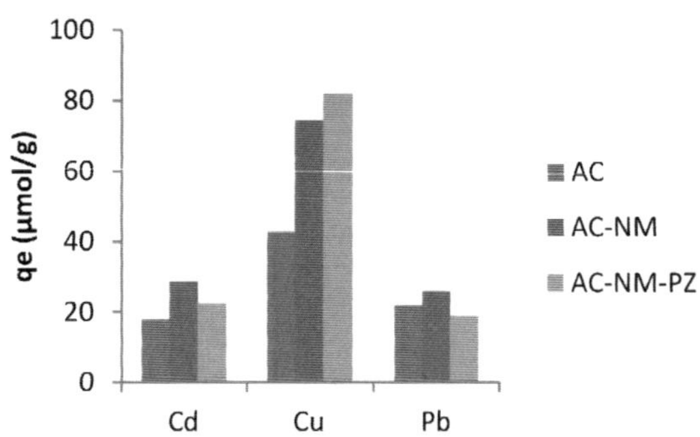

Figure 5. Effect of metal ions Cu (II) and Pb (II) on the adsorption capacity of Cd (II).

The existence of Cu (II) and Pb (II) was affects the efficiency and adsorption capacity of the third adsorbent adsorption at Cd (II) adsorption. The presence of Cu (II) and Pb (II) was reduces the adsorption capacity of the third adsorbent to Cd (II). The value adsorption capacity of Cd (II) at 50 mg/L was 17.79, 34.61, and 26.60 μmol/g for AC, AC-NM, and the AC-NM-PZ, respectively (Table 1). While the value adsorption capacity of Cd (II) in a mixture was 5.50, 9.92, and 8.58 μmol/g for AC, AC-NM, and the AC-NM-PZ, respectively. Adsorption capacity of AC decreased was ± 69% for Cd (II). While AC-NM and AC-NM-PZ decreased to ± 71% and ± 68%, respectively.

4. Conclusion
Adsorption capacity of Cd(II) on activated carbon-nanomagnetite-pyrazolone composite was lower than activated carbon-nanomagnetite composite but higher than activated carbon. The maximum adsorption capacity of activated carbon, activated carbon composite-nanomagnetite, and activated carbon composite-nanomagnetite-pirazolon were 20.64, 36.73, and 27.67 μmol/g, respectively. Isotherm model of three adsorbents followed Langmuir isotherm models. Activated carbon, activated carbon-nanomagnetit, and activated carbon-nanomagnetit-pyrazolone did not show selectivity in adsorption cadmium ions when mixed with Cu (II) and Pb (II). The results of XRD analysis showed the magnetite phase on activated carbon-nanomagnetit composite. The FTIR spectrum did not indicate any new absorption.

Acknowledgments
One of the author (DP) is thankful to the Directorate General of Higher Education for the Scholarship (BPPDN) of Lecturer Candidate in 2013.

References
[1] Walkees M P 2003*Mutation* **533** 107-120
[2] Sudarmaji, Mukono J and I P Corie 2006 *J Kesehatan Lingkungan* **2(2)**129-142
[3] Jia Y F and Thomas K M 2000 *Langmuir* **16** 1114-22
[4] Machida M, Fotoohi Y, Amamo Y, Ohba T and Kanoh H 2012 *J of Hazardous Materials* **221-222** 220-227
[5] Mahmoud ME, Abdel-Fattah TM, Osman M M and Ahmed S B 2012 *J of environ sci and health* **47** 130-141
[6] Lu X, Jiang J, Sun K, Wang J and Zhang Y 2014 *Marine Pollution Bull.* **78** 69-76
[7] Zach-Maor A, Seminat R and Shemer H 2012 *Desalination and water treatment* **31** 64-70
[8] Lyles BF, Terrot MS, Hammond P T and Gast A P 2004 *Langmuir* **20** 3028-31
[9] Jang JH and Lim H B 2010 *Microchem J.* **94** 148-158
[10] Bou-Maroun E, Chebib H, Leroy MJ F, Boos A and Goetz-Grandmont G J 2006*Separation and purification technology* **50** 220-228
[11] Oliva A, Molinari A, Zuniga F and Ponce P 2002 *Microchim acta.* **140** 201-203
[12] Hamzah B, Jalaluddin N, Wahab A W and Upe A 2011 *Jurnal natur Indonesia* **13(3)** 269-275
[13] Ihesinachi KA, Augustus U B and John C U 2011 *Nature and science* **9 (3)**
[14] Todorova O, Vassileva P and Lakov L 1993 *Fresenius journal of analytical chemistry* **346** 943-946
[15] Lakov L, Vassileva P and Peshev O 1998 *Carbon* **37** 1655-57
[16] Daou T J, Pourroy G, Bgin-Colin S, Grenche J M, Ulhaq-Bouillet C, Legar P, Bernhardt P, Leuvrey C and Rogez G 2006 *Chemistry of Materials* **18** 4399-4404
[17] Narin I, Soylak M, Elci L and Dogan M 2000 *Talanta* **52** 1041-1046
[18] Kwon J H, Wilson LD and Sammynaiken R 2014 *Synthetic metals* **197** 8-17
[19] Waldron R D 1955 *Phys. rev.* **99** 1727-35
[20] Huheey J E, Keiter E A and Keiter R L 1993 *Inorganic chemistry principles of structure and reactivity.* America (US): HarperCollins
[21] Allen S J and Brown P A 1995 *J chem technol biotechnol.* **62(1)** 17-24

Outlier Detection on Hotspot Data in Riau Province using OPTICS Algorithm

N L Febriana[1] and I S Sitanggang[1]

[1] Department of Computer Science, Jalan Meranti Wing 20 Level 5 Kampus IPB, Bogor, West Java, Indonesia.

Email: nurlailaf76@gmail.com , imas.sitanggang@ipb.ac.id

Abstract. Hotspot is a point that shows the coordinate of an area with relatively higher temperature than another surrounding area. Hotspot is one of indicators of forest and land fires. An effort in forest and land fires prevention is detecting hotspot occurrences and its outlier. The purpose of this study is to detect outlier occurrences on hotspot data in Riau Province using the density based algorithm namely Ordering Point to Identify the Clustering Structure (OPTICS). The data used are hotspots in Riau Province for the period of 2001 to 2012. In order to find the best clustering results, OPTICS was executed on the parameter Eps of 0.01, 0.02, 0.03, 0.04, 0.05, 0.06, 0.1 and MinPts of 1 to 6. This study found that outliers are commonly occurred in hotspot data in 2007 at the parameter Eps of 0.01 and MinPts of 6. This study identifies 906 outliers in hotspot data in 2007 with SSE of clustering result of 0.0219. Outliers are mostly found in February spreading on several districts including Siak district in Riau Province.

1. Introduction

Detecting hotspot occurrence is an important activity in forest and land fire prevention. However, forest and land fire can be located far from the position of hotspot. In addition, some hotspot locations are difficult to be reached and it may also too far from the water resources. This condition causes the fire supression activities to be delayed, forest and land fires occur much longer, and fires propagate to the other regions. Therefore losses values due to fires are increased.

The study by [1] reports that in the period July to October 2015, trajectory CO and CO_2 pollutants from peat fires in Riau Province spread to the west, northwest, east, northeast, and north of the Riau Province to the province of West Sumatra, North Sumatra, Aceh and Riau Islands. On July 9, July 22, July 26, August 30 and October 21, 2015, high pollutants concentration spread in some districts in Riau province including Rokan Hilir, Bengkalis, Meranti island, Indragiri Hulu, and Siak. In the study by [1], haze trajectory patterns are obtained using the HYSPLIT model and the CO and CO_2 pollutants are analyzed using K-Means algorithm.

Because forest and land fires have negative impacts for the environment, monitoring hotspot distribution is required to minimize damages due to fires. In addition, identifying hotspot distribution especially those are located apart from other hotspots is important in forest and land fires prevention. Those hotspots which are occurred at isolated locations are considered as outliers. In addition to determine hotspots in dense clusters, outliers on hotspot data should be detected because those outliers are potentially to be fires. Several clustering algorithms including K-Means, DBSCAN and Kulldorff's

Scan Statistics Method have been applied to analyze hotspot distribution in the previous works [2], [3], [4], [5], [6]. Studies on outliers detection on hotspot datasets have been done by applied clustering algorithms. The work by [7] applied the DBSCAN algorithm to detect outliers on hotspot data in 2005 in Riau Province. This work find as many 1241 outliers appeared in the study area. In addition, the work of [8] applied k-means clustering algorithm to discover outliers on hotspot datasets. This study identifies 59 collective outliers and 30 global outliers on the hotspot dataset. These outliers on the hotspot data are mostly occurred in February, March, June, and August. Another study by [9] detected contextual outliers on hotspot data in Riau province for the period 2001 to 2009 based on climate context, i.e. rainfall. The result of this study showed that there were 54 objects detected as contextual outliers, many of them occurred in February, March, June, July, and August. The contextual outliers have an average of daily occurrences of 65.76 hotspots with rainfall in average of 37.15 mm. This research aims to detect outlier occurrences on hotspot dataset in Riau Province on the period of 2001 to 2012 using the OPTICS algorithm. The results are compared to the output of DBSCAN algorithm [7].

2. Data and Methods

2.1. Hotspot Data

Hotspot data used in this study were collected from the National Aeronautics and Space Administration (NASA) Fire Information for Resource Management (FIRMS) http://earthdata.nasa.gov/. Data preprocessing including data cleaning and data transformation was done in the previous work by [8]. The datasets for clustering has three attributes namely longitude, latitude, and date of hotspot occurrence (acq_date).

2.2. Outlier Detection using OPTICS Algorithm

Ordering Points to Identify the Clustering Structure (OPTICS) is one of the density-based cluster algorithm that was introduced by [10]. In density-based cluster algorithms, objects are in dense regions are assigned as members of clusters. Noises are objects in less dense regions which are separately located from members of clusters. OPTICS is an extension of Density-Based Spatial Clustering Algorithm with Noise (DBSCAN) algorithm. The key idea in DBSCAN is also applied in OPTICS in which for each object of a cluster the neighbourhood of a given radius (Eps) has to contain at least a minimum number of objects (MinPts) [10]. The algorithm has two parameters namely Epsilon (Eps) and *Minimum Points* (MinPts). Epsilon (*Eps*) is the distance of cluster centroid to its neighbourhoods and MinPts is minimum number of object in one cluster [11]. OPTICS does not create cluster explicitly. It creates ordering of all objects in the dataset based on the density-based clustering structure. There are two values are assigned to each object namely a core-distance and a reachability-distance which are defined as follows [10]:

- Given p as an object from database D; ε is a distance value; $N_\varepsilon(p)$ is ε-neighbourhood of p; MinPts is integer; MinPts_distance (p) is a distance between p to its MinPts' Neighbourhood. Core distance of the object p is a minimum distance (ε') between p and another object in its ε-neighbourhood. p *is* considered as a core object if this neighbour consists of $N_\varepsilon(p)$, and otherwise the core distance of the object p is considered as *UNDEFINED.*

$$core\_distance_{\varepsilon,MinPts}(p) = \begin{cases} UNDEFINED, if\ Card(N_\varepsilon(p)) < MinPts \\ MinPts_{distance(p)}, otherwise \end{cases} \quad (1)$$

- Given p and o are objects from the database D; $N_\varepsilon(o)$ is number of o's neighbours in ε-neighbourhood; MinPts is integer. The reachability-distance of p with respect to o is defined as follows

$$reachability\_distance_{\varepsilon,MinPts}(p,o) = \begin{cases} UNDEFINED, if\ |N_\varepsilon(p)| < MinPts \\ max(core\_distance(o), distance(o,p)), otherwise \end{cases} \quad (2)$$

Core distance of the object p is a minimum distance (ε') between p and another object in its ε-neighbourhood. p *is* considered as a core object if its neighbour consists of $N_\varepsilon(p)$, and otherwise the core distance of the object p is considered as *UNDEFINED.*

In the OPTICS algorithm, reachability-distance does not allowed to be smaller than the core distance. If this situation is happened then there will be no objects in o neighbourhood [10]. Euclidean distance id used to calculate the core-distance and reachability-distance.

Sum Square Error (SSE) was calculate to evaluate clusteing results in which the best clustering was selected based on minimum SSE. The formula of SSE is as follows [12]:

$$SSE = \sum_{i=1}^{K} \sum_{x \in c_i} d(p, m_i)^2 \quad (3)$$

where d: distance between the objects; m_i: a centroid of cluster i; $p \in C_i$: p is an object in cluster i.

3. Result and Discussion

OPTICS algorithm that is available in Weka was used to generate clusters on a hotspot dataset as well as to identify outliers on the dataset. For all objects in the dataset, OPTICS assigns the values of Core-Distance and Reachability-Distance. Table 1 provides some examples of objects as the results of OPTICS algorithm.

Table 1. Some objects as the result of OPTICS algorithm.

Id	Objects (longitude, latitude)	Core-Distance	Reachability-Distance
0	101.476, 1.673	UNDEFINED	UNDEFINED
1	102.807, 1.134	UNDEFINED	UNDEFINED
10	102.005, 0.796	0.00064	UNDEFINED
127	102.005, 0.795	0.00084	0.00064
173	102.004, 0.797	0.0007	0.00064

Core-distance in table 1 indicates the distance of an object within a certain radius that meets MinPts. Reachability-distance denotes distance between an object with another object. If number of objects within a certain radius does not meet MinPts, then the distance is considered as UNDEFINED. In this study, if core distance and reachability distance of an object is UNDEFINED then the object is considered as an outlier. An object is classified as an outlier or abnormal if it meets the following criteria (1) The object is not included in any cluster, (2) the object is far away from the nearest cluster, (3) the object is a member in a small cluster [12].

Several values of parameters Eps and MinPts are used to determine the best clustering results in which outliers are identified in the best cluster. The value of Eps are 0.01, 0.02, 0.03, 0.04, 0.05, 0.06 (in degree) whereas the MintPts starts from 1 to 6 as shown by the following table.

Table 2 represents clustering results on hotspot dataset in 2001 to 2012 at the Eps of 0.01 and MinPts of 6. Based on table 2 the outliers mostly appeared in 2007. Figure 1 shows number of cluster's member and outliers as the results of OPTICS algorithm in a histogram.

Table 2. Clustering results of OPTICS algorithm at the Eps of 0.01 and MinPts of 6.

Year	Number of object in cluster	Outlier	SSE
2001	1,264	410	0.0060
2002	5,181	773	0.0728
2003	6,041	833	0.0337
2004	7,498	890	0.0395
2005	22,369	671	0.0772
2006	10,245	879	0.0545
2007	**3,188**	**906**	**0.0219**
2008	4,843	807	0.0278
2009	10,167	728	0.0457
2010	3,414	686	0.0194
2011	5,997	843	0.0320
2012	7,034	786	0.0361

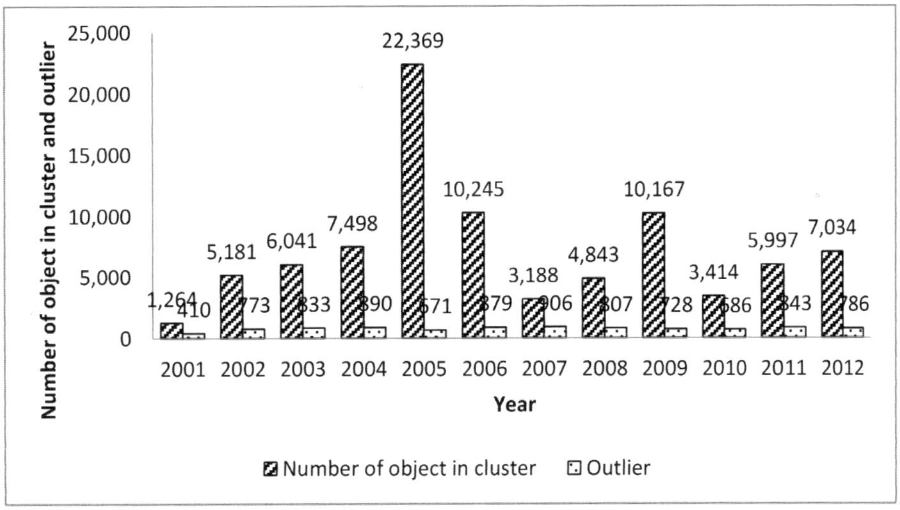

Figure 1. Number of objects in cluster and outliers.

From the figure 1 in average as many 767 outliers are detected on hotspot datasets each year in the period of 2001 to 2012. These outliers are hotspots that are located not in clusters meaning that distance between one a hotspot to another is greater to 0.01 or about 1.1132 km. The OPTICS algorithm results the highest number of outliers on hotspots in 2007, the outlier's distribution as shown in Figure 2.

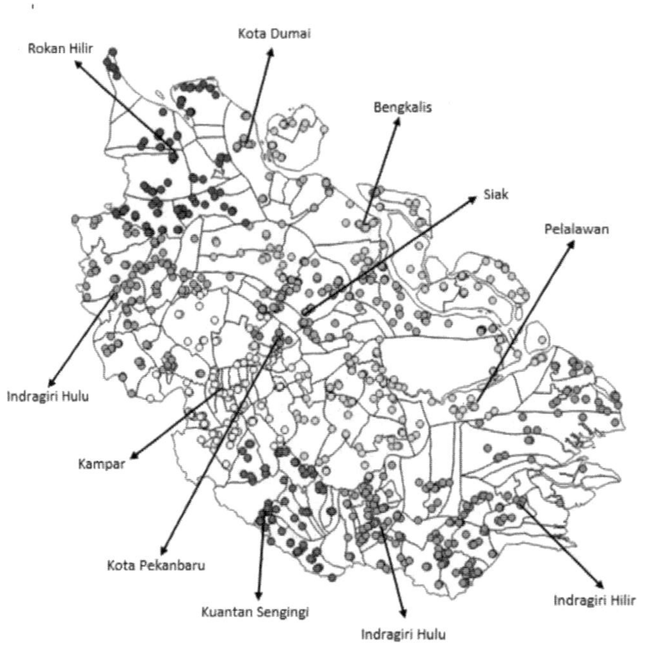

Figure 2. Outlier occurrences in district level, Riau Province in 2007.

Figure 2 shows a plot of 906 outliers on hotspot data in 2007 in which those outliers are mostly occurred in February. The outlier distribution in district level in 2007 is as follows: 100 outliers in Bengkalis District, 99 outliers in Indragiri Hilir District, 97 outliers in Indragiri Hulu District, 111 outliers in Kampar District. 26 outliers in Dumai City, 9 outliers in Pekanbaru City, 71 outliers in Kuantan Sengigi, 99 outliers in Pelalawan. 90 outliers in Rokan Hilir, 90 outliers in Rokan Hulu, and 112 outliers in Siak. Based on the distribution, in district level the ouliers mostly appeared in Siak and 71 of them appeared on February. The outliers in Siak plotted in sub districts level, as follows.

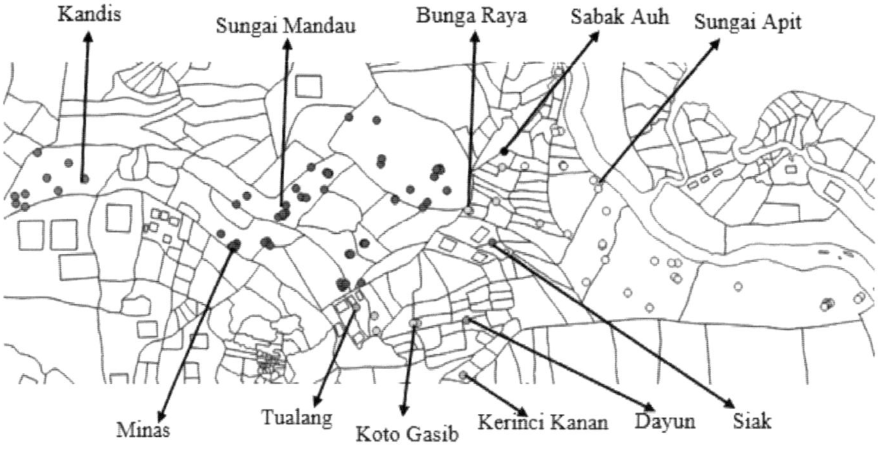

Figure 3. Outlier occurrences in sub districts level in Siak District, Riau Province in 2007.

Figure 3 shows 112 outliers distributed 6 in Bunga Raya, 1 in Dayun, 10 in Kandis, 1 in Kerinci Kanan, 5 in Koto Gasib, 15 in Minas, 2 in Sabak Auh, 1 in Siak, 31 in Sungai Apit, 38 in Sungai Mandau, and 2 in Tualang.

The previous research by [7] identified outliers on hotspot datasets in 2001 to 2012 using the DBSCAN algorithm. Table 3 provide the comparison of clustering results using DBSCAN and OPTICS algorithms at the Minpts of 2. The output of the DBSCAN algorithm obtained from the previous work [7].

Table 3. Clustering results of OPTICS and DBSCAN algorithms at Minpts of 2.

Year	Eps	OPTICS algorithm				DBSCAN algorithm			
		Number of object in cluster	Outlier	%	SSE	Number of cluster	Outlier	%	SSE
2001	0.02	1515	159	9.48%	0.024	175	188	11.21%	0.048
2002	0.02	5904	50	0.84%	0.068	457	769	12.91%	0.036
2003	0.02	6808	74	1.08%	0.1	497	375	5.45%	0.022
2004	0.1	8236	152	1.81%	0.027	25	14	0.16%	0.026
2005	0.01	22902	138	0.59%	0.041	**864**	**1241**	**5.38%**	**0.084**
2006	0.02	11054	70	0.63%	0.141	862	1229	11.04%	0.032
2007	0.02	4048	46	1.12%	0.070	446	409	9.99%	0.021
2008	0.02	5620	30	0.53%	0.069	489	366	6.48%	0.018
2009	0.02	10866	29	0.27%	0.083	768	990	9.09%	0.022
2010	0.02	4057	43	1.05%	0.055	389	311	7.58%	0.021
2011	0.02	6804	36	0.53%	0.071	500	349	5.10%	0.021
2012	0.02	7786	34	0.43%	0.074	742	946	12.09%	0.024

Table 3 shows the differences of clustering result using OPTICS and DBSCAN with the same parameters. OPTICS algorithm provide a number of object in cluster, while DBSCAN algorithm provide number of cluster. Both of the algorithm have the same parameters and similar process with a very difference result. OPTICS algorithm do not provide cluster explicitly.

4. Conclusion

The OPTICS algorithm has successfully applied on the hotspot datasets in the period 2001 to 2012 to identify outliers on the datasets. An outlier on hotspot datasets is a hotspot that is located far from the clusters of hotspots. The distance of those outliers to other nearest hotspots is about 1.11 km. Experimental results at the parameter Eps of 0.01 and MinPts of 6 shows that about 767 outliers are found on hotspot dataset per year in the period 2001 to 2012 in which the highest number of outliers are found in 2007. As many 906 outliers found in 2007 that mostly spread in the districts of Bengkalis, Indragiri Hilir, Indragiri Hulu, Kampar, Pelalawan, Rokan Hilir, Rokan Hulu, and Siak.

References

[1] Ni'am M, Sitanggang I S, Nuryanto D E 2016 *3rd Int. Symp. Of LISAT on Food Security and Environmental Monitoring* (October 25th-26th Bogor)
[2] Kirana A P, Sitanggang I S, Syaufina L 2015. *J. TELKOMNIKA* **13**(4) 1376-83
[3] Hermawati R, Sitanggang I S 2015 *Proc.2nd Int. Symp. On LAPAN-IPB Satellite for Food Security and Enviromental Monitoring* (Bogor) 317-23
[4] Usman M, Sitanggang I S, Syaufina L 2015 *Proc. 1st Int. Symp. On LAPAN-IPB Satellite for Food Security and Enviromental Monitoring* (Bogor) 132-40
[5] Fuad T, Sitanggang I S, Annisa 2010 *IEEE Int. Symp. On Information Technology* (June 15th-17th Kuala Lumpur) CFP1033E-CDR ISBN: 978-1-4244-6717-4

[6] Hayardisi G, Sitanggang I S, Syaufina L 2009 *IEEE 2nd Conf. On Data Mining & Optimization* (October 27th-28th) CFP0982H-PRT
[7] Sukmasetya P and Sitanggang I S 2016 *IOP. Conf. On Earth and Environmental Science* **31(1)** 1-4
[8] Sitanggang I S and Baehaki D A M 2015 *IEEE 2nd Proc. Int. Conf. On Spatial Data Mining and Geographical Knowledge Services* (July 8th-10th Fuzhou) p 66 CFP1584-ART ISBN 978-1-4799-7749-9
[9] Thah P H, Sitanggang I S 2015 *Proc.2nd Int. Symp. On LAPAN-IPB Satellite for Food Security and Enviromental Monitoring* (Bogor) 238-63
[10] Ankerst M, Breunig M M, Kriegel H P, Sander J 1999 *OPTICS: Ordering Points to Identify the Clustering Structure* (Philadelphia-Munich: Institute for Computer Science University of Munich) 3-4
[11] Ester M, Kriegel H P, Sander J, and Xu X *A Density-Based Algorithm for Discovering Clusters in Large Spatial Database with Noise* (Munich: Institute for Computer Scince University of Munich) 2-3
[12] Han J, Kamber M, Pei J 2012 *Data Mining Concepts and Technique* (San Diego: Morgan Kaufman Publisher) 3rd

IOP Conf. Series: Earth and Environmental Science **58** (2017) 012005 doi:10.1088/1755-1315/58/1/012005

Segmentation of Sharia Rural Banks based on Growth of the Business Performance

B Sartono[1*], S Rahardiantoro[1], C Suhaeni[1], E A Irianto[2] and U Maulidya[2]

[1]Department of Statistics, Bogor Agricultural University, Bogor, Indonesia
[2]Otoritas Jasa Keuangan

Email : bagusco@gmail.com , bagusco@ipb.ac.id

Abstract. The Otoritas Jasa Keuangan (OJK) is an institution that is responsible to supervise bank operational in Indonesia, including those which are BPRS's. As a part of their tasks, they do a monitoring procedure to recognize the performance progress of BPRS. A time series analytics could help in studying time-to-time change of certain performance indicators. In term of the performance indicators, not all BPRS has a good growth since they were established. Indeed many of them are growing positively, but some of them may have up-and-down pattern, even tend to be negatively diminishing. Recognizing the time series pattern of the performance indicators of BPRS is useful for OJK to identify working strategy. This recent paper focuses on segmentation process to identify groups of BPRS based on their growth pattern in multivariate time series cluster analysis cases. The result is there are 3 groups of 156 rural banks, 128 banks in the first group, 18 banks in the second, and 10 banks in the third. The general patterns of each group follow: cluster 1 is a group of banks whith positive growth in most of the variable; cluster 2 is a group that the banks that belong to this cluster have negative growth along the time, especially in the last two years; whereas Cluster 3 consist of rural banks with negative growth at the beginning of the observation period and then got rebound at the rest of the time.

1. Introduction

Micro and small enterprises are believed to be one of the most important entities that support the economy of Indonesia. In 2016, the Ministry of KUKM estimates that at least 58 million enterprises exist in Indonesia and absorb 110 million human resources. However, most of them are not accessible by financial institutions, especially banks, by the reason of poor quality on the record of their business activities. One of the breakthroughs for this problem is the establishment of rural bank which in Indonesia called as Bank Perkreditan Rakyat (BPR) since 1990s. This kind of bank need more relax requirement to the micro and small enterprise to have financing support. As the development of the economics, recently it also exist Sharia/Islamic Rural Bank (Bank Pembiayaan Rakyat Syariah, BPRS).

The Otoritas Jasa Keuangan (OJK) is an institution that is responsible to supervise bank operational in Indonesia, including those which are BPRS's. As a part of their tasks, they do a monitoring procedure to recognize the performance progress of BPRS. A time series analytics could help in studying time-to-time change of certain performance indicators. In term of the performance indicators, not all BPRS has a good growth since they were established. Indeed many of them are growing positively, but some of them may have up-and-down pattern, even tend to be negatively diminishing. Recognizing the time series pattern of the performance indicators of BPRS is useful for

OJK to identify working strategi. Gor example, it may help in determining which bank that should be supervised more and to be helped to win-back the business. It is also useful to learn more about the factor that influence the variation of performance among BPRS.

This recent paper focuses on segmentation process to identify groups of BPRS based on their growth pattern. The complication of the problem arises by the number of the series which is not one. So we would consider in handling a multivariate time series cluster analysis. Recall that what we would like to group is not the level of the indicators but the growth. Therefore the final segments would be uncorrelated with the business scale of the banks.

2. Time Series Cluster Analysis

In this section, would be described the main ideas of time series clustering which applied in this study. Based on Liao in 2005 [1], there are three time series clustering approaches, that are raw-data based, feature based, and the model based. The raw-based approach worked by calculating the distance among series in the raw-data, then the spicify clustering methods implemented using this distance. The second approach worked based on the feature cores of the raw-data. The feature extractions would be the first step in this approach. The model based approach implemented using the model based on raw data. The necessary thing from this model is usually the coefficients of model, or the residuals of models.

This study applied the most popular approach above to find the best suggestion of cluster, that is the raw-data based approach. In this case, we use clustering based on univariate correlation matrices in multivariate time series. This method developed based on the raw-data. The main idea of this approach is get the transformation of correlation matrices become distance matrices. That is the correlation matrices mention above are the correlation among BPRS in every variables. Because there are more than one variable (in context multivariate data) the correlation matrices built from the average of the correlation matrices in every variable.

3. Data

The data that are involved in the analysis consist of eight time series of the indicators. Those indicators are including:

- Return on Equity (ROE). It is a measure of profitability that calculates how many dollars of profit a company generates with each dollar of shareholders' equity. The formula for ROE is: ROE = Net Income/Shareholders' Equity. ROE is sometimes called "return on net worth."
- Return on Asset (ROA). It is obtained by dividing the BPRS's annual income by its total assets. This indicator measures how profitable a bank is relative to its total assets and how efficient management is at utilizing the assets to generate earnings.
- Third party funds (Dana Pihak Ketiga, DPK). It consists of demand deposit, time deposit, savings, certificates of deposit, and other short-term deposits at bank.
- Total Financing (Pembiayaan yang Dilakukan, PYD). It is the total amount of loan provide by the rural bank.
- Year-to-date profit and loss statement (Laba Rugi Berjalan). It is a document that shows the incoming and outgoing money from January 1 of the current year to the date the document is being prepared. This statement is a tool used by business owners to determine whether or not the business is lucrative.
- Total Assets. It refers to the total amount of assets owned by a person or entity. Assets are items of economic value, which are expended over time to yield a benefit for the owner. Total Assets is sum of all current and noncurrent assets and must equal the sum of total liabilities and stockholders' equity combined.
- Earning Assets (Aset produktif). Earning Assets are anything that directly generates income, such as stocks that pay dividends, bonds that pay interest, real estate properties that generate rents, copyright and patents that bring in licensing fees, and machinery that allows you to produce goods for sale at a profit.

- Core Capital (Modal Inti). It is the minimum amount of capital that a thrift bank, such as a savings bank or savings and loan company, must have on hand in order to comply with Federal Home Loan Bank regulations. Core capital consists of equity capital and declared reserves.

There are as many as 156 BPRS's that are of interest and the series of the data were on the period of Jan 2011 – Dec 2015.

4. Method

To be able to have a good segmentation of BPRS, we did the following steps:

1. General checking to the data quality which included logical violation to the values (mostly extremes) and sudden changes. A series of graphical exploration such as time series plots and histograms were generated to identify the anomaly within the dataset.
2. Constructing among-bank correlation matrix for each performance indicator, so that we had eight matrices. The correlation
3. Combining eight correlation matrices into a single matrix by averaging the correspondent elements
4. Calculating between-bank distance matrix. The between-bank distance is obtained by the following formula:

$$D(i,i') = \left(\frac{1-C(i,i')}{1+C(i,i')}\right)^{\beta};\ \beta > 0 \text{ (in this case we use } \beta = 0.8)$$

   where
   $C(i,i')$ is the correlation coeficients between i – th BPRS and i' – th BPRS
   $D(i,i')$ is the distance between i – th BPRS and i' – th BPRS
5. Implementing an agglomerative clustering algorithm using complete linkage strategy [2]
6. Evaluating the result of the clustering, using some measurement of cohesion and separation statistics, that is the numerical measurement to evaluate the cluster result [3]. This step determines the number of optimal clusters/groups. The formula of cohesion and separation by Tan et al 2005 [4] describe below

$$cohesion(Cluster_i) = \sum_{\substack{x \in Cluster_i \\ y \in Cluster_i}} proximity(x,y)$$

$$separation(Cluster_i, Cluster_j) = \sum_{\substack{x \in Cluster_i \\ y \in Cluster_j}} proximity(x,y)$$

7. Profiling the growth pattern of each group of banks

5. Results and Discussion

5.1. Data Distribution

Figure 1 presents histograms of univariate distribution of the indicators we are interested in. The histograms are summarized from 156 observations, one for each bank. The condition of each bank was summarized using the average value of the last six months.

We could see from the histograms that most of the indicators which are related to business scale or the size such as DPK, PYD, asset and capital are skewed to the right. It happens because there are a small number of banks whose very large scale of business compared to other banks. Whereas, the ROA and ROE tend to have a left-tailed distribution. A bank with a negative and large value of ROA and ROE is a bank which is almost bankrupt or has very bad business. From the graphs we could see that there are two-four banks of that bad.

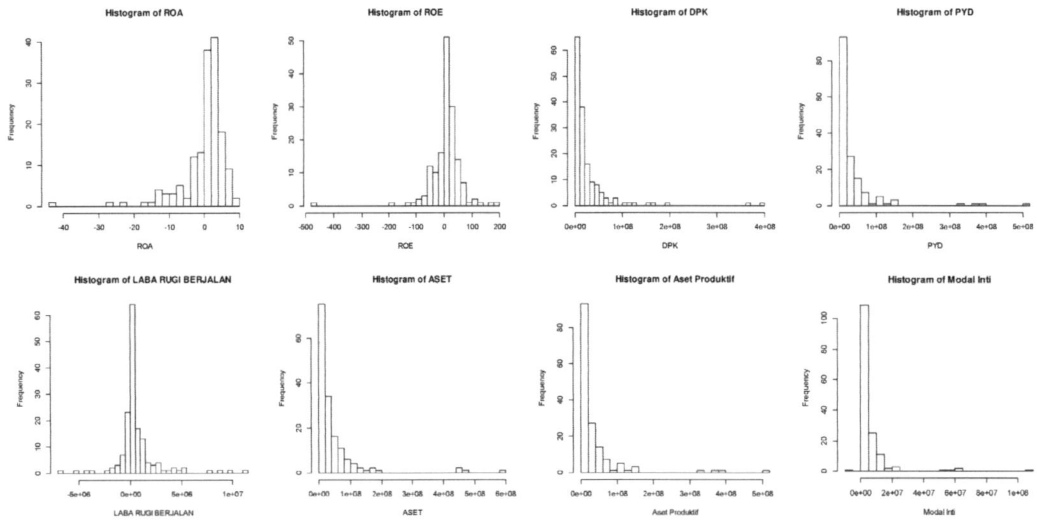

Figure 1. Histograms of the univariate distribution of the performance variables.

5.2. Clustering Results

The agglomerative clustering algorithm that was implemented to the distance matrix using complete linkage resulted a dendrogram that records the clustering process as shown in Figure 2. The dendrogram informs us which banks are grouped at the first iteration and what are at the next, until all banks are grouped together into a single cluster.

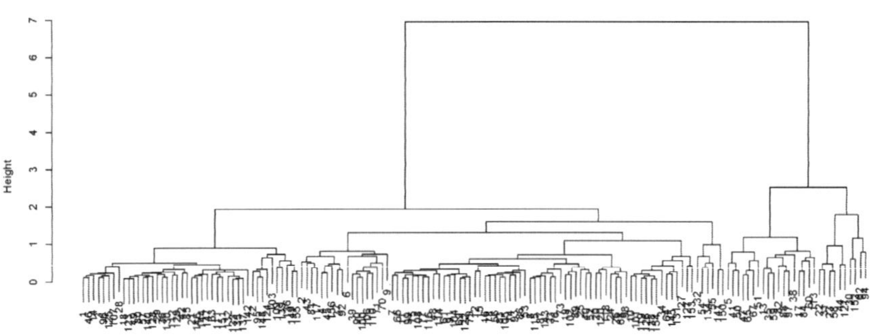

Figure 2. The dendrogram of the clustering process with complete linkage approach.

The chart of a dendrogram might be useful to determine the number of optimal clusters. Recall that a horizontal line in the dendrogram represents a merge step of two nearest clusters. Using the rule of largest jump between two horizontal lines in the dendogram, we can decide how many clusters should be produced. It is obvious that the jump between the most above line and the second above are the largest jump, so that it indicates that 2 (two) clusters would be good. However we prefer to use other assessment called cohesion and separation statistics to determine the number of clusters. To do that, we assessed using statistics if the number of clusters is 2, 3, 4 and 5. Then, we decide the cluster number with the number whose largest difference of cohesion and separation. Table 1 provides the statistics and 3 (three) clusters are the optimal using this criterion.

Table 1. The assessment of clustering in several numbers of clusters.

Number of Clusters	Cohesion	Separation	Difference
2	0.4421	-0.1195	0.5616
3	0.4528	-0.1164	0.5692
4	0.4907	0.1421	0.3486
5	0.4927	0.1420	0.3508

Based on the last evaluation, we decided to group the BPRS into three groups of banks. The numbers of banks in each group are as follow: 128 banks in the first group, 18 banks in the second, and 10 banks in the third. Table 2 provides us information about the distribution of banks in each group/cluster based on their location/zone and their age group. In the management of OJK, the zones of BPRS are distinguished into four zones: (1) Greater Jakarta, (2) Java and Bali Island, including Bogor, Depok, Tangerang, and Bekasi, (3) capital of Province in other island than Java and Bali, (4) the other locations. It can be seen form Table 2 that the members of each cluster distribute almost uniformly in every zone and every age group.

Table 2. The distribution of group membership based on location zones and age groups.

	Cluster			
Zone	**1**	**2**	**3**	**Total**
1	16	2	0	18
2	72	6	5	83
3	14	2	3	19
4	26	8	2	36
Total	**128**	**18**	**10**	**156**

	Cluster			
Age Group	**1**	**2**	**3**	**Total**
more than 15 years	49	11	8	68
8 – 15 year	38	6	1	45
less than 8 year	41	1	1	43
Total	**128**	**18**	**10**	**156**

5.3. Growth Profiling

It is now interesting to explore the growth pattern of each group resulted by the cluster analysis. Figure 3 depicts line charts, one for each variable, to examine the pattern of monthly growth of BPRS. To be able to have the general pattern, first we standardized the original values across the time for each bank, and then calculated an average within the group so that we have three series for every variable.

The charts on Figure 3 suggest the following general pattern. Cluster 1 is a group of banks whith positive growth in most of the variable. The banks in this cluster also tend to have low changes in term of ROA and ROE. By this profile the authors agree to say that it is a group of banks with good business.

A different pattern was found in Cluster 2. The banks that belong to this cluster have negative growth along the time, especially in the last two years. Whereas Cluster 3 consist of rural banks with negative growth at the beginning of the observation period and then got rebound at the rest of the time. The rural banks in this group tend to have higher monthly fluctuation in their businesses.

a. ROA

b. ROE

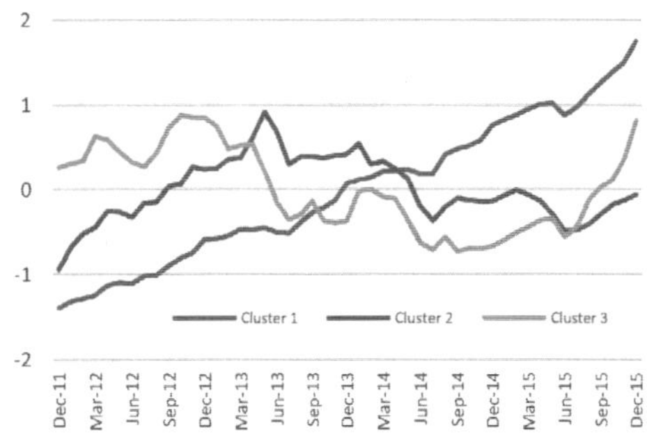

c. third party funding

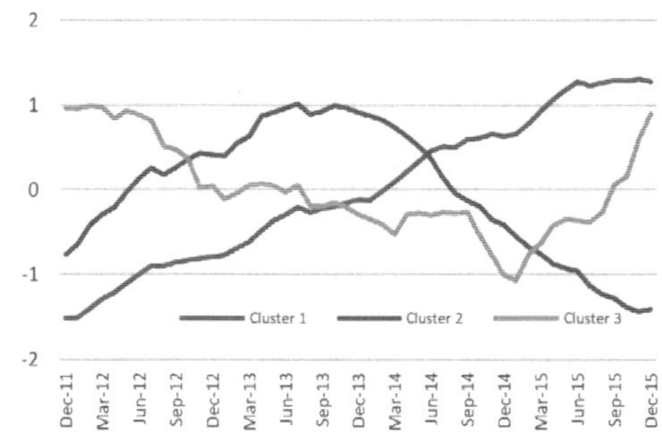

d. total financing

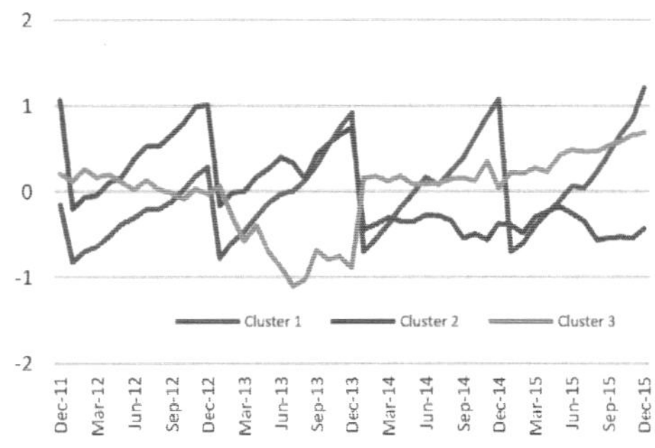

e. current profit/loss

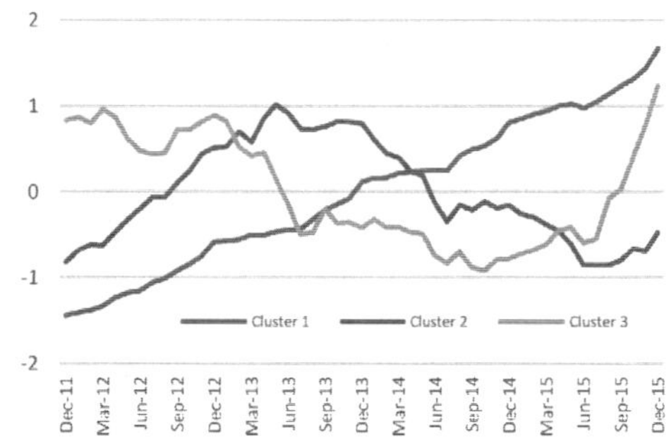

f. total assets

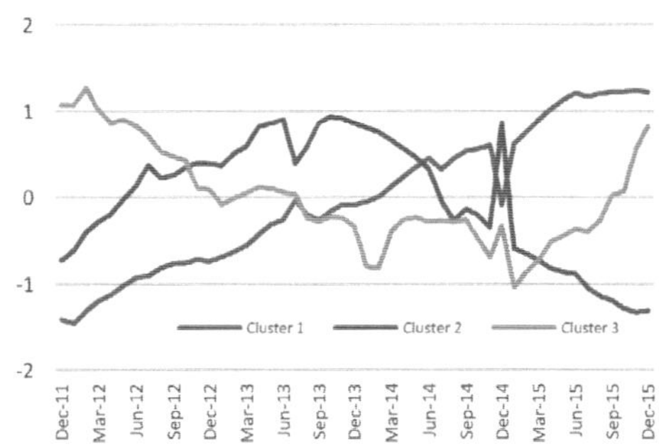

g. total productive assets

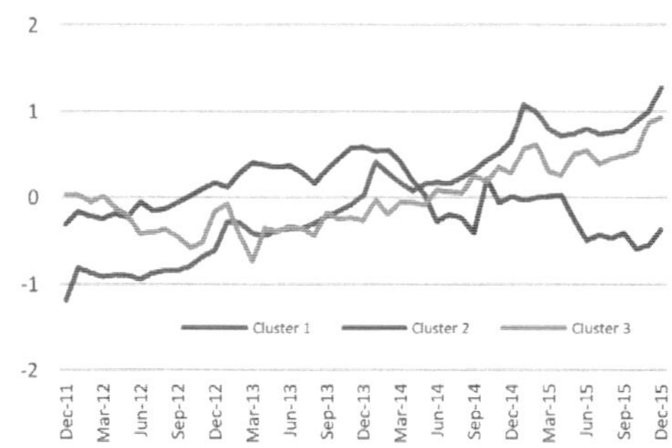

h. core capital

Figure 3. General growth pattern of groups of banks for eight variables under interest.

6. Conclusion

This paper discovered that there are three time series growth patterns of the Sharia Rural Banks in Indonesia. The result was generated by implementing a time series cluster analysis by utilizing an agglomerative algorithm based on correllational distance between banks. Profiling to the three groups shows that there are 128 banks with good growth, 18 banks with negative progress, and 10 banks with worsened at the beginning and tent to be recovered at the recent period.

References

[1] Liao TW 2005 *Pattern Recognition* **38(11)** 1857-1874
[2] Kaufman L and Rousseeuw PJ 1990 *Finding Groups in Data : An Introduction to Cluster Analysis* Wiley New York
[3] Johnson RA, Wichern DW 2007 *Applied Multivariate Statistical Analysis* 6th ed Pearson
[4] Tan PN, Steinbach M, Kumar V 2005 *Introduction to Data Mining* Addison-Wesley Longman Publishing Co. Inc Boston MA USA

Bayesian estimation of linear and nonlinear mixed models of fertilizer dosing with independent normally distributed random components

M Masjkur[1] and H Folmer[2, 3]

[1]Department of Statistics, Faculty of Mathematics and Natural Sciences, Bogor Agricultural University, Indonesia
[2]Faculty of Spatial Sciences, University of Groningen, The Netherlands
[3]Department of Agricultural Economics, College of Economics and Management, Northwest A&F University, Yangling, Shaanxi, China

Email : m_masjkur@yahoo.com

Abstract. Many linear and nonlinear mixed response models are proposed to predict the optimum dose of fertilizer. However, a major restriction of this class of models is the normality assumption of the random parameter component. The purpose of this paper is to analyze the performance of linear and nonlinear mixed models of fertilizer dosing with independent normally distributed random parameter components. We compare the Linear Plateau, Spillman-Mitscherlich, and Quadratic random parameter models with different random effects distribution assumption, i.e. the normal, Student-*t*, slash, and contaminated normal distributions and the random errors following their symmetric normal independent distributions. The method is applied to datasets of multi-location trials of potassium fertilization of soybeans. The results show that the Student-*t* Spillman-Mitscherlich Response Model is the best model for soybean yield prediction.

1. Introduction

Many linear and nonlinear response models are commonly used to predict the optimum dose of fertilizer. One modeling approach is to fit a general quadratic form to the data by means of least squares under the assumption of a fixed effects model with independent normally distributed random error term with constant variances ([1]-[2]). However, this approach is unrealistic because it neglects the variability that probably exists between sites and/or years.

An alternative model is the mixed effects approach ([3]-[6]). This approach allows the parameters to have a random effect component that represent between sites or years variability. The random parameter models have been found to outperform the fixed parameter models to model dose-response relationships ([5], [7]-[8]). Furthermore, the quadratic functional form commonly used is not always the best model. [7] and [9] showed that the stochastic linear plateau model and the Mitscherlich exponential type functions outperform the quadratic form. In a similar vein, [8] showed that the stochastic linear plateau function is more adequate than the stochastic quadratic plateau function for corn response to Nitrogen fertilizer.

The random parameter components and the errors are usually taken as normally distributed random variables ([5]-[8]). However, the normality and symmetry assumptions may be too restrictive because in practice departures from normality is common. Particularly, [10] and [11] concluded that the field crop yield distributions are in general non-normal or non-lognormal. The degree of skewness and kurtosis vary by crop and by the amount of nutrients uptake.

Rosa et al. [12] advocated the use of the Normal independent distribution for robust modeling of linear mixed models. Furthermore, [13] considered the Normal independent distribution for modeling of nonlinear mixed models. The Normal independent distribution is a class of symmetric, heavy-tailed distributions that includes the normal distribution, Student-*t*, slash, and contaminated distributions. The class of Normal independent distributions accomodate observations with heavy tails as well as the normal distribution.

Traditionally, fertilizer-dose response models are estimated by means of maximum likelihood estimation (ML) ([5]-[8]). However, for nonlinear models and small sample sizes ML is frequently biased ([14]). In addition, convergence can be a problem even with careful scaling and good starting values. Bayesian estimation is an alternative to ML. The advantages of Bayesian estimation are that the results are valid in small samples and that convergence in the case of nonlinear models is not an issue ([14]-[15]).

The purpose of this paper is Bayesian estimation of random parameter dose (fertilization)-response (yield) models for yield data that is Normally independently distributed.

2. Mixed effects model and normal independent distributions

2.1. The normal mixed effects model

In general, a Normal mixed effects model reads:

$$\boldsymbol{Y}_i = \eta(\boldsymbol{\phi}_i, \boldsymbol{X}_i) + \boldsymbol{\epsilon}_i, \qquad \boldsymbol{\phi}_i = \boldsymbol{A}_i\boldsymbol{\beta} + \boldsymbol{B}_i\boldsymbol{b}_i, \tag{1}$$

with

$$(\boldsymbol{b}_i, \boldsymbol{\epsilon}_i) \overset{ind}{\sim} N_{n_i+q}\left(\boldsymbol{0}, Diag(\boldsymbol{\Sigma}, \sigma_e^2 \boldsymbol{I}_{n_i})\right),$$

where the subscript i is the subject index, $i = 1, \ldots\ldots, n$; $\boldsymbol{Y}_i = (y_{i1}, \ldots, y_{in_i})^T$ is a $n_i \times 1$ vector of n_i observed continuous responses for subject i, $\eta_i(\boldsymbol{\phi}_i, \boldsymbol{X}_i) = \{\eta(\phi_i, X_{i1}), \ldots, \eta(\phi_i, X_{in_i})\}^T$ with $\eta(.)$ the nonlinear or linear function of random parameters $\boldsymbol{\phi}_i$, and covariate vector $\boldsymbol{X}_i$, $\boldsymbol{A}_i$ and $\boldsymbol{B}_i$ are known design matrices of dimensions $n_i \times p$ and $n_i \times q$, respectively, $\boldsymbol{\beta}$ is the $p \times 1$ vector of fixed effects, $\boldsymbol{b}_i$ is the $q \times 1$ vector of random effects, and $\boldsymbol{\epsilon}_i$ is the $n_i \times 1$ vector of random errors, and $\boldsymbol{I}_{n_i}$ denotes the identity matrix. The matrices $\boldsymbol{\Sigma} = \boldsymbol{\Sigma}(\boldsymbol{\alpha})$ with unknown parameter $\boldsymbol{\alpha}$ is the $q \times q$ unstructured dispersion matrix of $\boldsymbol{b}_i$, σ_e^2 the unknown variance of the error term. When $\eta(.)$ is a nonlinear parameter function, we have the Normal NonLinear Mixed Model (N-NLMM); if $\eta(.)$ is a linear parameter function, we have the N-Linear Mixed Model (N-LMM).

It follows that

$$\boldsymbol{b}_i \overset{ind}{\sim} \boldsymbol{N}_q(\boldsymbol{0}, \boldsymbol{\Sigma}) \text{ and } \boldsymbol{\epsilon}_i \overset{ind}{\sim} \boldsymbol{N}_{n_i}(\boldsymbol{0}, \sigma_\epsilon^2 \boldsymbol{I}_{n_i})$$

and they are uncorrelated, since $Cov(\boldsymbol{b}_i, \boldsymbol{\epsilon}_i) = \boldsymbol{0}$ ([13],[16]).

2.2. Normal independent (NI) distributions

A Normal independent distribution is defined as the *p*-dimensional random vector $\boldsymbol{y} = \boldsymbol{\mu} + U^{1/2}\boldsymbol{Z}$, where $\boldsymbol{\mu}$ is a location vector, $\boldsymbol{Z}$ is a multivariate normal random vector with location vector $\boldsymbol{0}$, scale matrix $\boldsymbol{\Sigma}$ and U is a positive random variable with cumulative distribution function (cdf) $H(u|v)$ and probability density function (pdf) $h(u|v)$, $\boldsymbol{v}$ is a scalar or vector of parameters indexing the distribution of the scale factor U ([12]-[13],[16]). Given U, $\boldsymbol{Y}$ follows a multivariate normal distribution with location vector $\mathbf{0}$, and scale matrix $u^{-1}\boldsymbol{\Sigma}$. Thus, the NI distributions are scale mixtures of the normal distributions denoted by $\boldsymbol{Y} \sim N_p(\boldsymbol{\mu}, \boldsymbol{\Sigma}, H)$. The marginal pdf of $\boldsymbol{Y}$ is

$$f(\boldsymbol{y}) = \int_0^\infty \phi_p(\boldsymbol{y}; \boldsymbol{\mu}, u^{-1}\boldsymbol{\Sigma}) dH(u|\boldsymbol{v}),$$

The class of normal independent distribution is a group of symmetric heavy-tailed distribution of robust alternative to the routinely used of normal distribution for mixed effects model.

2.3. The NI-mixed effects model

Using the general framework (1), the following general NI-Mixed Model (NI-MM) is defined as:

$$\boldsymbol{b}_i \overset{iid}{\sim} N_q(\mathbf{0}, Diag(\boldsymbol{\Sigma}), H) \text{ and } \boldsymbol{\epsilon}_i \overset{ind}{\sim} N_{n_i}(\mathbf{0}, \sigma_e^2 \boldsymbol{I}_{n_i}, H), \; i = 1, \dots\dots, n. \tag{2}$$

where the random effects are assumed to have a multivariate NI distributions and also the random errors.

3. Bayesian inference

3.1. Prior distributions and joint posterior density

Below, we apply a Bayesian framework based on the Markov Chain Monte Carlo (MCMC) algorithm to infer posterior parameter estimates. Following (1) and (2), the NI mixed model can be formulated in hierarchical for $i = 1, \dots\dots, n$, as:

$$\boldsymbol{Y}_i | \boldsymbol{b}_i, U_i = u_i \overset{ind.}{\sim} N_{n_i}\big(\eta(\boldsymbol{A}_i\boldsymbol{\beta} + \boldsymbol{B}_i\boldsymbol{b}_i, \boldsymbol{X}_i), u_i^{-1}\sigma_e^2 \boldsymbol{I}_{n_i}\big).$$

$$\boldsymbol{b}_i | U_i = u_i \overset{ind.}{\sim} N_q\big(\mathbf{0}, u_i^{-1}\boldsymbol{\Sigma}\big).$$

$$U_i \overset{iid.}{\sim} H(u_i|\boldsymbol{v}).$$

([13],[16]).

Let $\boldsymbol{Y} = (\boldsymbol{y}_1^T, \dots, \boldsymbol{y}_n^T)^T$, $\boldsymbol{b} = (\boldsymbol{b}_1^T, \dots, \boldsymbol{b}_n^T)^T$, $\boldsymbol{u} = (u_1, \dots, u_n)^T$. Then, the complete likelihood function associated with $(\boldsymbol{y}^T, \boldsymbol{b}^T, \boldsymbol{u}^T)^T$, is given by

$$L(\boldsymbol{\theta}|\boldsymbol{Y}, \boldsymbol{b}, \boldsymbol{u}) \propto \prod_{i=1}^n [\phi_{n_i}\big(\boldsymbol{y}_i; \eta(\boldsymbol{A}_i\boldsymbol{\beta} + \boldsymbol{B}_i\boldsymbol{b}_i, \boldsymbol{X}_i), u_i^{-1}\sigma_e^2 \boldsymbol{I}_{n_i}\big)\phi_q\big(b_i; \mathbf{0}, u_i^{-1}\boldsymbol{\Sigma}\big)\, h(u_i|\boldsymbol{v})].$$

To complete Bayesian specification, we need to consider prior distributions for all the unknown parameters $\boldsymbol{\theta} = (\boldsymbol{\beta}^T, \sigma_e^2, \boldsymbol{\alpha}^T, \boldsymbol{v}^T)^T$. We consider $\boldsymbol{\beta} \sim N_p(\boldsymbol{\beta}_0, \boldsymbol{S}_\beta)$, $\sigma_e^2 \sim IG(\tau/2, T/2)$, $\boldsymbol{\Sigma} \sim IW_q(\boldsymbol{M}, l)$ ([13],[16]). For v we take $v \sim Exp(\tau/2)\mathbb{I}_{(2,\infty)}$ for the Student-t (t) model, Gamma (a, b) for the slash (SL) model. Furthermore, $U(0,1)$ for v_1 and Beta (ρ_0, ρ_1) for v_2 for the contaminated normal (CN) model.

Assuming independency of the parameter vector, the joint prior distribution of all unknown parameters is

$$\pi(\boldsymbol{\theta}) = \pi(\boldsymbol{\beta})\pi(\sigma_e^2)\pi(\boldsymbol{\alpha})\pi(\boldsymbol{v})$$

Combining the likelihood function and the prior distribution, the joint posterior density of all unknown parameters is

$$\pi(\boldsymbol{\theta}, \mathbf{b}, \mathbf{u}|\mathbf{y}) \propto \prod_{i=1}^{n} [\phi_{n_i}(\boldsymbol{y}_i; \eta(\boldsymbol{A}_i\boldsymbol{\beta} + \boldsymbol{B}_i\boldsymbol{b}_i, \boldsymbol{X}_i), u_i^{-1}\sigma_e^2\boldsymbol{I}_{n_i})\phi_q(\boldsymbol{b}_i; \mathbf{0}, u_i^{-1}\boldsymbol{\Sigma})h(u_i|\boldsymbol{v})]\pi(\boldsymbol{\theta})$$

3.2. Model comparison criteria

The expected Akaike information criterion (EAIC) and the expected Bayesian information criterion (EBIC) are a deviance-based measure appropriate for Bayesian model selection ([17]-[18]).

Let $\boldsymbol{\theta}$ and $\boldsymbol{Y} = (y_1, \dots, y_n)^T$ be the entire model parameters and data, respectively. Define $D(\boldsymbol{\theta}) = -2\ln f(\boldsymbol{y}|\boldsymbol{\theta}) = -2\sum_{i=1}^{N} \ln f(\boldsymbol{y}_i|\boldsymbol{\theta})$, where $f(\boldsymbol{y}_i|\boldsymbol{\theta})$ is marginal distribution of y_i, then $E[D(\boldsymbol{\theta})]$ is a measure of fit and can be approximated by using the MCMC output in a Monte Carlo simulation. This index is given by $\overline{D} = \frac{1}{K}\sum_{k=1}^{K} D(\boldsymbol{\theta}^{(k)})$. Where $\boldsymbol{\theta}^{(k)}$ is the kth iteration of MCMC chain of the model and K is the number of iterations.

The expected Akaike information criterion (EAIC) and the expected Bayesian information criterion (EBIC) define as follows

$$\widehat{EAIC} = \overline{D} + 2p, \text{ and } \widehat{EBIC} = \overline{D} + p\,log(N)$$

where $\overline{D}$ is the posterior mean of the deviance, p is the number of parameters in the model, N is the total number of observations. These criteria penalizing models with more complexity. Smaller value of EAIC and EBIC indicate a better fit ([19]).

4. Case study

4.1. Data

The dataset is obtained from 19 multi-location trials of potassium fertilization of soybeans. The experiments were carried out between 2002 and 2014. The soil types are Ultisols, Inceptisols, Vertisols, and Oxisols with soil potassium contents varying from very low to very high. Common soybean varieties were used. Each experiments consisted of five levels of potassium fertilization. The doses applied were 0, 40, 80, 160 and 320 kg ha^{-1} of KCl. The plots were 6 by 5 m, or 4 by 5 m arranged in a randomized complete block design with three to nine replications. The response variable was soybean yield (t ha^{-1}). The yields reported are averages over replications ([20]-[22]).

4.2. Response functions

We consider three response functions: the Linear Plateau (LP), the Spillman-Mitscherlich (SM) and the Quadratic functions (Q).

The stochastic LP is defined as follows:

$$Y_i = \min(\alpha_1 + (\alpha_2 + b_{2i})X_i;\ \mu_p + b_{3i}) + b_{1i} + \varepsilon_i \quad (3)$$

where for location i, Y_i is the soybean yield; X_i the potassium fertilizer dose; α_1 the intercept parameter; α_2 the linear response coefficient; u_p the plateau yield; b_{1i}, b_{2i}, and b_{3i} are the random effects; and ε_i is the random error term. In term of (1), $\boldsymbol{\alpha} = (\alpha_1, \alpha_2, \alpha_3)^T$ $\boldsymbol{b}_i = (b_{1i}, b_{2i}, b_{3i})^T$; $\boldsymbol{b}_i \overset{iid}{\sim} N_q(\mathbf{0}, \boldsymbol{\Sigma}, H)$ and $\boldsymbol{\epsilon}_i \overset{ind}{\sim} N_{n_i}(\mathbf{0}, \sigma_e^2\boldsymbol{I}_{n_i}, H)$.

The stochastic SM reads:

$$Y_i = \beta_1 - (\beta_2 + b_{2i})\exp\left((-\beta_3 + b_{3i})X_i\right) + b_{1i} + \varepsilon_i \quad (4)$$

where β_1 is the maximum yield attainable by potassium fertilization; β_2 is the yield increase; β_3 is the ratio of consecutive increments of the yield; all other parameters, variables and distributions as in (3).

The stochastic Q is defined as:

$$Y_i = \gamma_1 + (\gamma_2 + b_{2i})X_i + (\gamma_3 + b_{3i})X_i^2 + b_{1i} + \varepsilon_i \quad (5)$$

where γ_1 is the intercept parameter whose position (value) can be shifted up or down by the random effect b_{1i}; γ_2 is the linear response coefficient with random effect parameter b_{2i}; γ_3 is the quadratic response coefficient whose position can be shifted up or down by the random effect b_{3i}; $\boldsymbol{\gamma} = (\gamma_1, \gamma_2, \gamma_3)^T$; all other variables and distributions as in (3) ([7]-[9]).

4.3. Statistical analysis

The datasets was used to identify the model with the best fit among the random parameter models of fertilizer dosing. Several statistical models with differing distribution in the random effects and random errors are compared. These models are :
Model 1: Normal distribution for the random effects and for the random errors (N-N)
Model 2: Student-*t* distribution for the random effects and for the random errors (*t-t*)
Model 3: Slash distribution for the random effects and for the random errors (SL-SL)
Model 4: Contaminated normal distribution for the random effects and for the random errors (CN-CN).

The following independent priors were considered to perform the Gibbs sampler, $\boldsymbol{\beta_k} \sim N(0,\ 10^3)$, $\sigma^2 \sim IG(0.1, 0.1)$, $\boldsymbol{\Sigma} \sim IG(0.1, 0.1)$, and $v \sim Exp(0.1)I(2,)$ for the *t-t* model, $v \sim Gamma(0.1, 0.01)$ for the slash model, $v_1 \sim Beta(1,1)$ and $v_2 \sim Beta\ (2,2)$ for the contaminated normal model, respectively.

For each of the models, we ran three parallel independent chains of the Gibbs sampler with size 50 000 iterations for each parameter with thinning of 5 and an initial burn in of 25 000. We monitored chain convergence using trace plots, autocorrelation plots and the Brooks-Gelman-Rubin scale reduction factor ($\hat{R}$) ([23]). To avoid non-convergence, we normalized the original doses (subtracted the mean and divided by the standard deviation) which gave: -1.06, -0.70, -0.35, 0.35, and 1.76, respectively ([24]). We fitted the models using the R2jags package available in R ([25]).

5. Results and discussion

5.1. Soybean yield data

Figure 1 shows the histogram and normal Q-Q plot of soybean yield data for 19 locations, while the boxplot is presented in figure 2. The figures indicates non-normality (heavy-tailed) pattern. The Q-Q plot does not show a straight line, while the boxplot shows asymmetry and an outlier. Thus, it seems appropriate to fit a heavy-tailed model to the data.

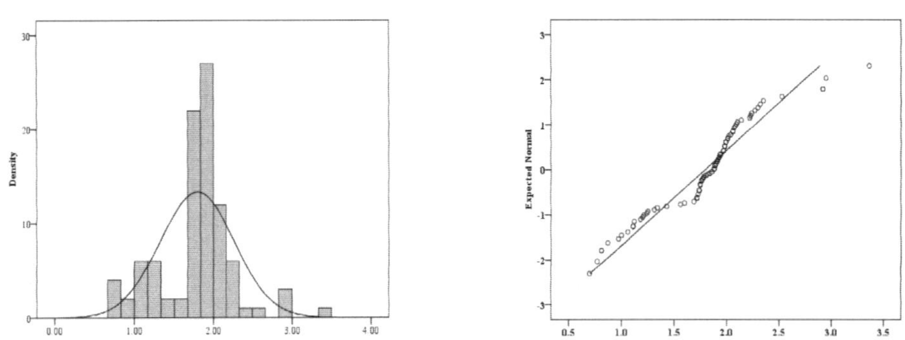

(a) (b)

Figure 1. Soybean yield data: (**a**) Histogram; (**b**) Normal Q-Q plot.

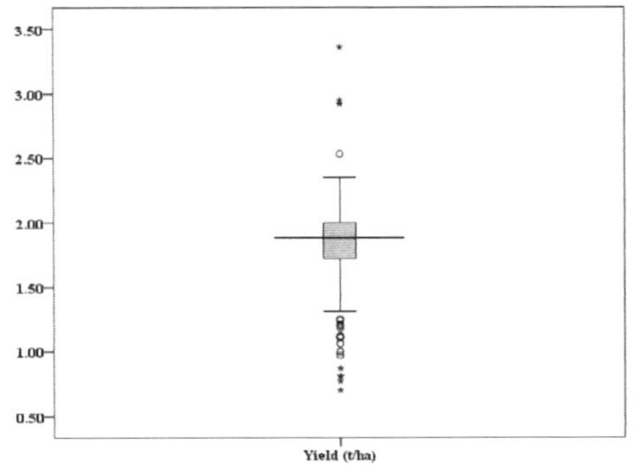

Figure 2. Boxplot of soybean yield data.

5.2. *Linear plateau response models*

Based on the EAIC and the EBIC in table 1, we find that among the NI models the Student-*t* (*t-t*) Model gives the best fit, followed by the contaminated normal (CN-CN), slash (SL-SL), and normal (N-N) Model. We furthermore find that the heavy-tailed distributions outperform the normal distributions. Thus, the *t-t* Model is the best Linear Plateau Response Model.

Table 1. The Linear plateau models

Parameter	N-N		*t-t*		SL-SL		CN-CN	
	Mean	SD	Mean	SD	Mean	SD	Mean	SD
α_1	1.473	0.114	1.555	0.109	1.482	0.110	1.521	0.125
α_2	39.968	20.482	29.274	19.480	30.780	19.899	29.061	19.030
μ_p	1.878	0.129	1.853	0.111	1.854	0.122	1.867	0.119
σ^2_ε	0.139	0.014	0.014	0.004	0.012	0.004	0.012	0.006
d_1	0.467	0.095	0.374	0.098	0.355	0.085	0.340	0.113
d_2	13.135	11.964	2.534	4.694	3.684	6.048	2.452	3.937
d_3	0.306	0.081	0.241	0.074	0.227	0.068	0.227	0.077
ν (ν_1)			5.043	3.039	2.697	1.620	0.515	0.253
ν_2							0.450	0.226
EAIC	-93.56		**-104.91**		-95.86		-97.21	
EBIC	-93.71		**-105.09**		-96.04		-97.41	

Table 1 furthermore shows that for the *t-t* Model, all the fixed effects, i.e., the intercept parameter (α_1), the linear response coefficient (α_2), the plateau yield u_p and the random effects (d_1, d_2, d_3) are significant.

5.3. *Spillman-Mitscherlich response models*

Based on the EAIC and EBIC in table 2 we find the following rankings of the NI models: *t-t* < N-N < SL-SL < CN-CN. Therefore, the *t-t* Model is the best Spillman-Mitscherlich Response Model.

Table 2. The Spillman-Mitscherlich models

Parameter	N-N		*t-t*		SL-SL		CN-CN	
	Mean	SD	Mean	SD	Mean	SD	Mean	SD
β_1	1.950	0.111	1.919	0.092	1.945	0.103	1.955	0.106
β_2	0.032	0.013	0.054	0.021	0.069	0.024	0.067	0.023
β_3	2.495	0.415	1.848	0.312	1.765	0.300	1.808	0.305
σ^2_ε	0.104	0.010	0.009	0.003	0.009	0.003	0.010	0.003
d_1	0.465	0.087	0.371	0.087	0.355	0.080	0.401	0.086
d_2	0.008	0.006	0.053	0.012	0.051	0.012	0.052	0.012
d_3	0.623	0.199	0.441	0.240	0.463	0.198	0.490	0.222
ν (ν_1)			4.754	2.731	3.139	1.735	0.321	0.235
ν_2							0.544	0.203
EAIC	-147.72		**-153.88**		-138.45		-135.19	
EBIC	-147.88		**-154.06**		-138.63		-135.39	

For the *t-t* Model, the fixed effects, i.e., the maximum yield coefficient (β_1), the increase in yield (β_2), the ratio of successive increment (β_3) and the random effects (d_1, d_2, d_3) are significant.

5.4. The Quadratic response models

Comparison of the EAIC and EBIC in table 3 leads to the following rankings: *t-t* < SL-SL < CN-CN < N-N. The results furthermore show that the heavy-tailed distributions outperform the normal distribution, and that overall the *t-t* Model is the best Quadratic Response Model.

Table 3. The Quadratic models

Parameter	N-N		*t-t*		SL-SL		CN-CN	
	Mean	SD	Mean	SD	Mean	SD	Mean	SD
γ_1	1.796	0.107	1.825	0.079	1.783	0.096	1.788	0.096
γ_2	0.510	0.072	0.330	0.056	0.398	0.072	0.391	0.076
γ_3	-0.386	0.072	-0.246	0.052	-0.297	0.067	-0.292	0.069
σ^2_ε	0.033	0.006	0.014	0.005	0.011	0.005	0.014	0.006
d_1	0.445	0.085	0.327	0.085	0.297	0.074	0.315	0.092
d_2	0.046	0.030	0.082	0.023	0.075	0.020	0.077	0.021
d_3	0.030	0.022	0.072	0.019	0.066	0.017	0.068	0.018
ν (ν_1)			2.516	1.329	1.617	0.868	0.366	0.160
ν_2							0.265	0.141
EAIC	-45.32		**-104.44**		-83.22		-80.59	
EBIC	-45.47		**-104.62**		-83.40		-80.79	

For the *t-t* Model, all the fixed effects, i.e., the intercept parameter (γ_1), the linear response coefficient (γ_2), the quadratic response coefficient (γ_3), and the variance component (d_1, d_2, d_3) are significant.

5.5. Comparing the Linear plateau, Spillman-Mitscherlich and Quadratic models

Comparing the Linear Plateau (LP), Spillman-Mitscherlich (SM) and Quadratic (Q) models under four distributional assumptions, we find that the *t-t* Spillman-Mitscherlich model has the smallest EAIC and EBIC values among the competing models indicating that this is the best fit model for the soybean yield data (table 4).

Table 4. Comparison of LP, SM and Q models

Distribution	LP		SM		Q	
	EAIC	EBIC	EAIC	EBIC	EAIC	EBIC
N	-93.56	-93.71	-147.72	-147.88	-45.32	-45.47
t	**-104.91**	**-105.09**	**-153.88**	**-154.06**	**-104.44**	**-104.62**
SL	-95.86	-96.04	-138.45	-138.63	-83.22	-83.40
CN	-97.21	-97.41	-135.19	-135.39	-80.59	-80.79

6. Conclusion

We investigated the performance of linear and nonlinear mixed response models with normal independent (NI) distributions of random effects. We applied the Bayesian estimation framework to datasets of multi-location trials of potassium fertilization of soybeans. We compared the Linear Plateau, Spillman-Mitscherlich, and Quadratic random parameter models with different distributions of the random parameter component, i.e. the normal, Student-*t*, slash, and contaminated normal distributions and the random errors following their symmetric normal independent distributions.

The overall results showed that for all three models of fertilizer dosing, the Student-*t* distributions outperform the normal distributions. The best model for soybean yield prediction is the Student-*t* Spillman-Mitscherlich Response Model.

References

[1] Sain G E and Jauregui M A 1993 *Agron. J.* **85** 934-37
[2] Shen J, Li R, Zhang F, Rengel Z and Tang C 2003 *Nutr. Cycl. Agroecosyst.* **65** 243–51
[3] Wallach D 1995 *Biometrics* **51** 338–46
[4] Makowski D, Wallach D and Meynard J M 2001 *Agron. J.* **93** 531–39
[5] Makowski D and Wallach D 2002 *Agronomie* **22** 179–89
[6] Makowski D and Lavielle M 2006 *J. Agric. Biol. Environ. Stat.* **11** 45-60
[7] Tumusiime E, Brorsen B W, Mosali J, Johnson J, Locke J, Biermacher J T 2011 *J. Agr. Appl. Econ.* **43** 541–52
[8] Boyer C N, Larson J A, Roberts R K, McClure A T, Tyler D D, Zhou V 2013 *J. Agric. Appl. Econ.* **45** 669-81
[9] Park S C, Brorsen B W, Stoecker A L and Hattey J A 2012 *.J. Agr. Appl. Econ.* **44** 593–606
[10] Day H R 1965 *J. Farm Econ.* **47** 713-41
[11] Ramirez O A, Misra S and Field J 2003 *Amer. J. Agr. Econ.* **85** 108-20
[12] Rosa G J M, Padovani C R and Gianola D 2003 *Biom. J.* **45** 573–90
[13] Lachos V H, Castro L M and Dey D K 2013 *Comput. Statist. Data Anal.* **64** 237–52
[14] Brorsen B W 2013 *Selected Paper prepared for presentation at the Southern Agricultural Economics Association (SAEA) Annual Meeting (Orlando, Florida)* 3-5 February
[15] Masjkur M and Folmer H 2016 *Proc. Basic Sci. 6* (Malang: Faculty of Mathematics & Sciences Brawijaya University) pp 372–76
[16] Lachos V H, Castro L M and Dey D K 2011 *Biometrics* **67** 1594-1604
[17] Arellano-Valle R B, Bolfarine H and Lachos V H 2007 *J. Appl. Stat.* **34** 663-82
[18] Baghfalaki T and Ganjali M 2015 *REVSTAT – Stat. J.* **13** 169–91
[19] Bandyopadhyay D, Lachos V H, Castro L M and Dey D 2012 *Biom. J.* **54** 405–25
[20] Nursyamsi D, Sutriadi M T and Kurnia U 2004 *J. Trop. Soils.* **10** 1-9
[21] Nursyamsi D 2006 Jurnal Ilmu Tanah dan Lingkungan **6** 71-81
[22] Masjkur M and Hartatik W 2014 Laporan Penelitian Institusi Institut Pertanian Bogor
[23] Gelman, A, Carlin J and Rubin D 2006 *Bayesian Data Analysis* (New York: Chapman & Hall/CRC)
[24] Kéry M 2010 *Introduction to WinBUGS for Ecologists: A Bayesian approach to regression, ANOVA, mixed models and related analyses* (Amsterdam: Academic Press Elsevier)
[25] Su Y S and Yajima M 2015 *R2jags: A package for running jags from R (R package ver.n 0.5-7)*

Comparison of Freshwater Mollusc Assemblages between Dry and Rainy Season in Situ Gede System, Bogor, Indonesia

W Priawandiputra[1*], D J Nasution[1] and T S Prawasti[1]

[1]Department of Biology, Faculty of Mathematics and Natural Sciences, Bogor Agricultural University, Indonesia.

Email: priawandiputra@apps.ipb.ac.id and priawandiputra_windra@yahoo.co.id

Abstract. Anthropogenic activities, which reduced and damaged natural *situ* (freshwater ponds), also reduced fauna diversity in its aquatic ecosystem. Freshwater molluscs in the *situ*, one of the largest numbers of animals group with documented extinction, may also be impacted. The aims of this study were to record and to compare the abundance and species composition of freshwater molluscs between dry and rainy season in three situ. The freshwater molluscs were determined by twelve sampling points in Situ Gede (SG), Situ Panjang (SP) and Situ Burung (SB). Samplings were conducted once during dry season (August 2015) and rainy season (February 2016). Total abundance of molluscs encountered was 4321 individuals, which was comprised of 76 bivalve individuals (1.75 %) and 4245 gastropods individuals (98.44%). The abundance of molluscs were generally higher in rainy season than in dry season in all situ, while species richness showed the contrary. The species composition was significantly different between dry and rainy season in SP and SB but no significant differences was found in SG. From eight dominant species, there were six dominant species such as *Filopaludina javanica*, *Melanoides tuberculata*, *Thiara scabra*, *Sermyla requeti*, *Pila scutata* (gastropods) and *Pilsbryoconcha exilis* (bivalve) which were found in high numbers during dry season while two gastropod species (*Pomacea canaliculata* and *Wattebledia crosseana*) was numbered higher in rainy season than dry season.

Keywords: Situ, Bogor, Gastropod, Bivalve, *Filopaludina javanica*.

1. Introduction

A natural lake, namely situ in Sundanese term, is not only important to protect biodiversity of aquatic organisms but is also important to prevent flooding [1]. Water trapped in situ are also commonly utilised for watering system, irrigation, fish cultivation and recreation. Unfortunately, the number of situ in Bogor area is slightly reduced and their conditioned damaged due to many factors such as drought, pollution and land-use change. Bogor has lost 29% of situ in the last 10 years [1,2]. This circumstance heavily impacted the diversity of aquatic fauna in its aquatic ecosystem. Anthropogenic activities such as agriculture, household waste and sand mining around the Cikini River decreased species richness and changed composition and abundance of benthic macroinvertebrates [3].

Benthic macroinvertebrates, dominated by freshwater molluscs, were also found not only in rivers but also in lakes. Extinction of freshwater molluscs, one of the largest number documented, may also be cause by similar conditions as found in some situ [4,5]. Freshwater molluscs play an important role in human welfare where many species provide food sources while other species are detrimental, for example by acting as the vector that can transmit parasites to human and animal [6]. However, the assessment of freshwater mollusc species are still lacking and less than 2% of mollusc species has been examined [4,5].

Investigations on abundance and species composition of freshwater molluscs in situ are barely studied in Indonesia, especially in Bogor. The recent research of molluscs around Bogor compared molluscs among 36 situ in the watersheds of Ciliwung and Cisadane rivers [2]. Unfortunately, collection time was not clearly differentiated between dry and rainy season [2]. Therefore, the aim of this study was to record and to compare the species composition of freshwater molluscs among four situ during dry and rainy season.

2. Methods

2.1. Study Areas

The study was conducted during dry season (August 2015) and rainy season (February 2016) in Situ Gede system, Bogor, Indonesia (figure 1). The situ gede system consists of three situ: Situ Gede (SG), Situ Panjang (SP) and Situ Burung (SB), whose water flows into Cisadane river. Those situ are approximately 3-5 km from Bogor Agricultural University's Dramaga Campus. SG is known as a recreation area and is surrounded mostly human habitats such as settlements, ponds, agricultural areas and some forest patches. SP is surrounded by agricultural areas especially paddy fields. In SB, some peoples cultivated fish in cage cultures.

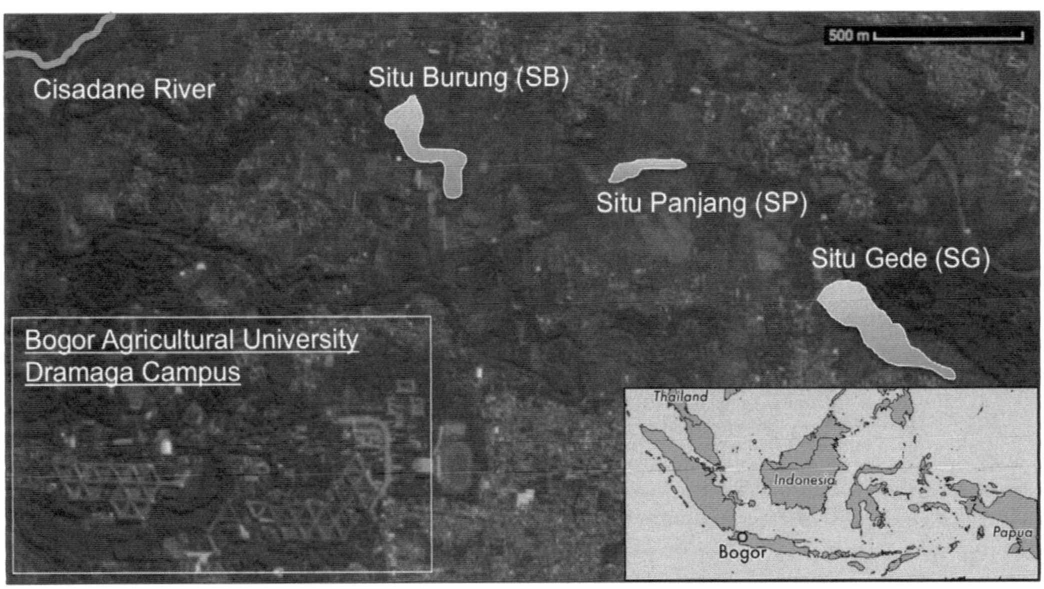

Figure 1. Map showing the position of Situ Gede (SG), Situ Panjang (SP) and Situ Burung (SB) in Bogor, Indonesia.

2.2. Sampling Techniques

Ten sampling stations were selected in each situ, representing all conditions of the situ. Abiotic factors such as temperature, pH, depth and Dissolved Oxygen (DO) were directly measured in field, while turbidity was recorded in BAU's Aquatic Productivity and Ecosystem (Proling) laboratory during study period. Freshwater molluscs were collected once a month using van veen grab sampler in littoral zone of each situ. All samples were kept in labelled plastic bags and were taken to laboratory for further studies. The samples were washed and sorted in laboratory. The mollusc samples were preserved with 70% ethyl alcohol. Preserved samples were identified using several mollusc identification books [7,8,9,10]. Identified molluscs were confirmed by Malacology Laboratory of the Indonesian Institute of Sciences (LIPI). Collected number of individuals represent the abundance of mollusc.

2.3. Analysis
Data analysis was computed using descriptive statistics and statistical analysis. The descriptive statistics calculated includes average and standard deviation, while statistical analysis includes non-parametric Kruskal-Wallis test with Mann-Whitney Pairwise Comparison, Analysis of Similarity (ANOSIM) and Non-Metric Dimensional Scaling (NMDS). Differences in species richness and abundance of total mollusc and each species among SG, SP and SB in dry and rainy season were examined using Kruskal-Wallis test with Mann-Whitney Pairwise Comparison. Analysis of Similarity (ANOSIM) and Non-Metric Dimensional Scaling (NMDS) using Bray-Curtis matrix with 1000 permutations were used to analyse the similarity of mollusc species between dry and rainy season in all situ. All analyses were calculated using PAST software version 1.95 [11].

3. Results

3.1 Abiotic Factors
The abiotic water variables water of the three situ were variously different during dry and rainy season (table 1). Precipitation was obviously higher in rainy season (610 mm) than in dry season (112 mm). It indirectly influenced other abiotic factors such as littoral depth, water temperature and turbidity. The littoral depth was higher in SG for both dry and rainy season than in the other situ. The littoral zone is generally deeper when during rainy season. Precipitation could decrease water surface temperature resulting in equal or lower temperature during the rainy season than during the dry season. Turbidity mostly decreased in SG and SP during rainy season. However, SB showed contrary results. The pH and DO was not clearly affected by precipitation where the data did not show any pattern.

Table 1. The differences of abiotic factors values between dry (MK) and rainy season (MH) in all situ. SG - Situ Gede, SP - Situ Panjang and SB - Situ Burung.

Parameters	SG		SP		SB	
	MK	MH	MK	MH	MK	MH
Littoral Depth (m)	0.60 ±0.08	1.04 ±0.13	0.31 ±0.13	0.29 ±0.12	0.40 ±0.14	0.42±0.12
Water temperature (℃)	32 ±1.15	32 ±0.74	33 ±0.13	29 ±1.14	29 ±1.53	29±2.35
pH	6 ±0.49	6 ±0.39	6 ±2.54	6 ±0.45	7 ±0.52	6 ±0.39
DO (mg/L)	6 ±0.57	5 ±0.46	6±2.50	6 ±0.60	5 ±0.76	6 ±0.54
Turbidity (NTU)	15 ±4.49	2 ±1.47	18 ±8.60	8 ±4.79	7 ±1.22	9 ±2.54
Precipitation (mm)	112	610	112	610	112	610

3.2 Abundance and Species Richness of Mollusc
A total 4321 individuals of mollusc from 18 species (14 species of gastropods and 4 species of bivalves) and 10 families were recorded in all situ (table 2). The highest number of individuals was found in SG (2289 individuals, 338.9 individuals/station) (Kruskal-Wallis Test, $P<0.05$), while the highest number of species was in SP (15 species). The number of individuals was higher in dry season than in rainy season at all sites, while the number of species was higher in rainy season than in dry season (figure 2).

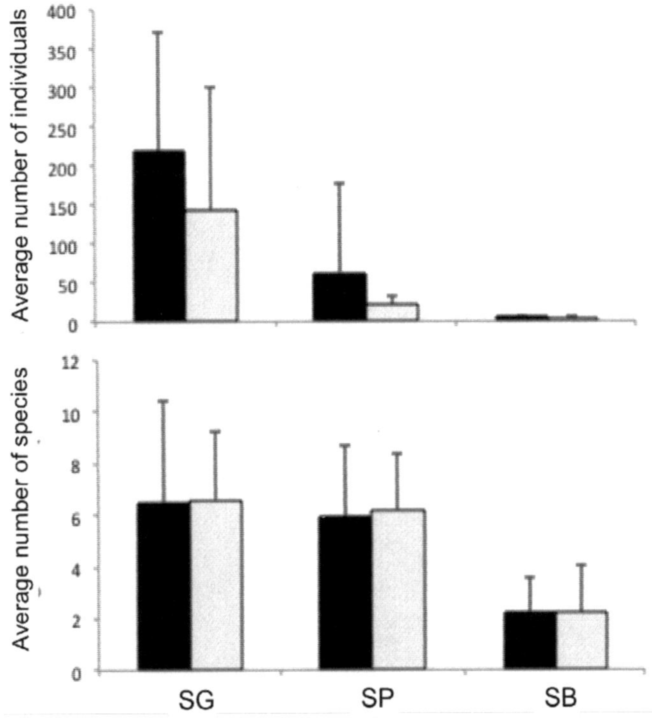

Figure 2. Comparison of average number of individuals (top) and species (below) between dry season (■) and rainy season (□) in Situ Gede (SG), Situ Panjang (SP) and Situ Burung (SB).

3.3 Species Composition of Mollusc

Figure 3 shows the NMDS ordination of the mollusc species composition between dry and rainy season in each situ. In each situ, the mollusc species composition variously overlapped between dry and rainy season. In SG, no significant difference was found in mollusc species composition between dry and rainy season (figure 3a, ANOSIM, P=0.1). Although mollusc species composition slightly overlapped between dry and rainy season in SP and SB, significant difference was found (figure 3b, ANOSIM, P=0.001 and figure 2c, ANOSIM, P=0.04).

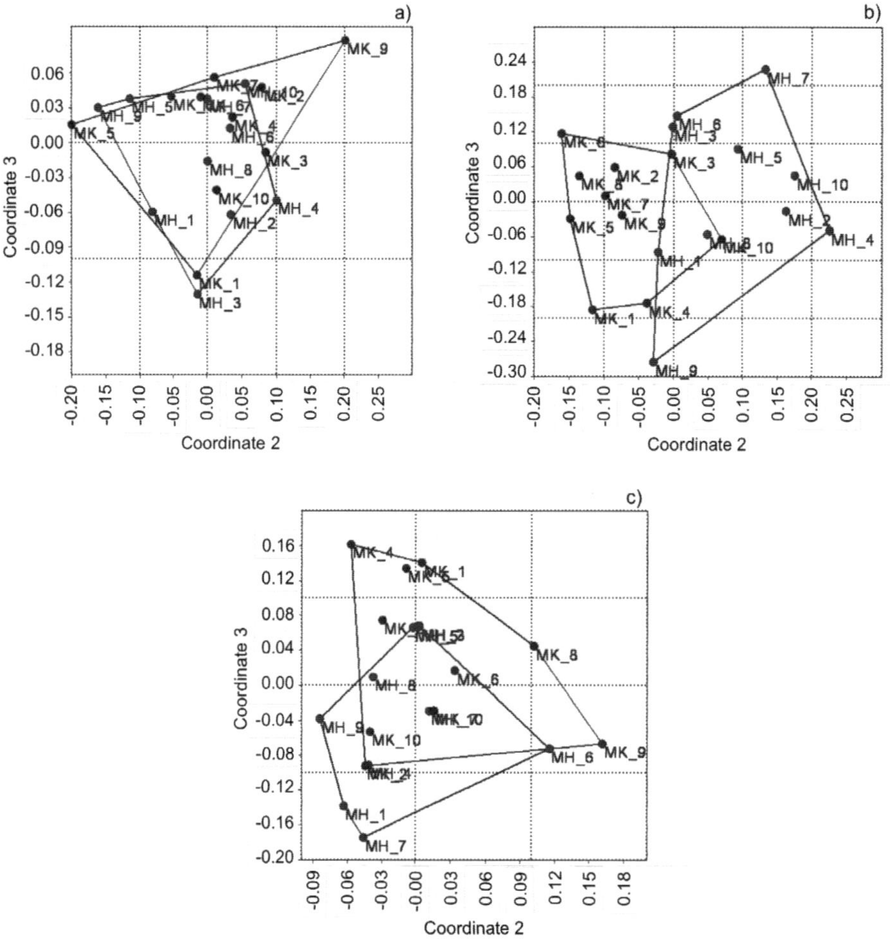

Figure 3. Species composition of molluscs between dry season (MK) and rainy season (MH) was illustrated by Non-Multi Dimensional Scaling (NMDS) in a) Situ Gede (SG), b) Situ Panjang (SP) and c) Situ Burung (SB). Stress of NMDS in SG, SP and SB were 0.04, 0.14 and 0.38, respectively.

Eight mollusc species (seven gastropod species and one bivalve species), recognized as dominant species (more than 50 individuals) were indicated to influence the NMDS pattern (figure 4). Those species, which were *Filopaludina javanica*, *Melanoides tuberculata*, *Thiara scabra*, *Sermyla requeti*, *Pila scutata*, *Pomacea canaliculata*, *Wattebledia crosseana* (gastropods) and *Pilsbryoconcha exilis* (bivalves), were significantly different in number of individuals among SG, SP and SB in dry and rainy season (Kruskal-Wallis Test, $P<0.05$). However, only two species (*F. javanica* and *M. tuberculata*) have relative abundance more than 20% (table 2 and figure 4). Five species (*F. javanica*, *M. tuberculata*, *T. scabra*, *P. canaliculata* and *P. exilis*) were found in dry and rainy season in all three situ. In all situ, all species except *P. canaliculata* and *W. crosseana* were easily collected during the dry season compared to rainy season. Molluscs of both species were found in higher number during dry season than in rainy season. Both of these species were also in greater numbers in SP than in SG. Nevertheless, mollusc species were mostly higher in SG than others.

Table 2. List of freshwater mollusc species in individuals.

Class	Family	Species	Sites			
			SG	SP	SB	Total
Bivalve	Corbiculaidea	*Corbicula javanicum* (Mousson 1849)	0	2	3	5
	Sphaeriidae	*Pisidium javanum* (Van Benthem Jutting 1931)	4	10	0	14
	Unionidae	*Anodonta woodiana* (Lea 1834)	0	0	2	2
		Pilsbryoconcha exilis (Lea 1838)	32	16	7	55
Gastropod	Ampullaridae	*Pila scutata* (Mousson 1848)	122	22	0	144
		Pomacea canaliculata (Lamarck 1819)	28	78	10	116
	Bihyniidae	*Wattebledia crosseana* (Wattebled 1884)	0	73	0	73
	Buccinidae	*Anentome helena* (Von Dem Bosch 1847)	5	14	1	20
	Lymnaidae	*Lymnaea rubiginosa* (Michelin 1831)	0	16	0	16
	Planorbidae	*Amerianna carinata* (H. Adams 1841)	14	0	0	14
		Gyraulus convexiusculus (Hutton 1849)	15	6	0	21
		Indoplanorbis exustus (Deshayes 1843)	12	3	2	17
		Physastra sp.	4	0	0	4
	Thiaridae	*Melanoides tuberculata* (Muller 1774)	940	199	34	1173
		Sermyla requeti (Grateloup 1840)	261	33	0	294
		Tarebria granifera (Lamarck 1816)	95	7	2	104
		Thiara scabra (Muller 1774)	621	103	9	733
	Viviparidae	*Filopaludina javanica* (Von Dem Busch 1844)	1236	257	23	1516
Total			**3389**	**839**	**93**	**4321**

4. Discussions

Abundance of freshwater molluscs in SG was significantly higher than SP and SG where SG is the main situ of Situ Gede System of Cisadane watershed. The highest abundance of molluscs was in upstream lake (SG) and gradually declined downstream (SB). Compared to the other situ, SG (6.2 ha) as the largest lake is surrounded by many habitats such as forests, paddy fields, small ponds, field and settlement, which contribute to organic and inorganic nutrient sources. Total Organic Matters (TOM) of each situ was measured in September and the results showed that SG, SP and SB contained 36.34±16.09, 29.07±6.92 and 16.11±1.21 mg $KmnO_4$/mL, respectively (Priawandiputra, unpublished data). TOM concentration follows a similar pattern with the abundance of molluscs from upstream lake to downstream lake, indicating that the organic matter available supported high abundance of some mollusc. Particulate organic matters are the main food of bivalves and gastropods, which could impact not only on the abundance but also on freshwater mollusc assemblage structure [12].

The data showed that total abundance of molluscs were clearly greater during dry season than in rainy season. Abundance of six dominant species such as *F. javanica*, *M. tuberculata*, *T. scabra*, *S. requeti*, *P. scutata* and *P. exilis* tend to occur in dry season. Our results are supported by research in Nigeria where the density of gastropods and *M. tuberculata* was higher in dry season (4.81 ± 3.07) than in rainy season (5.1 ± 2.63) [13]. Frequency of snail occurrences increased during dry season because their high rates of biological activities while rainy season is attributed to the flushing out of molluscs to different places following the dispersal of aquatic fauna [14]. The rates of precipitation affect population of freshwater molluscs by increasing habitat volume where the precipitation between dry season (112 mm) and rainy season (610 mm) were significantly different. In contrary, abundance of two dominant species (*P. canaliculata and W. crosseana*) was mostly higher in rainy season than dry season. This indicates that not all mollusc species prefer dry season and some mollusc have some exceptional preferences. Morphological and physiological comparison between dry season and rainy season mollusc should be examined for further study.

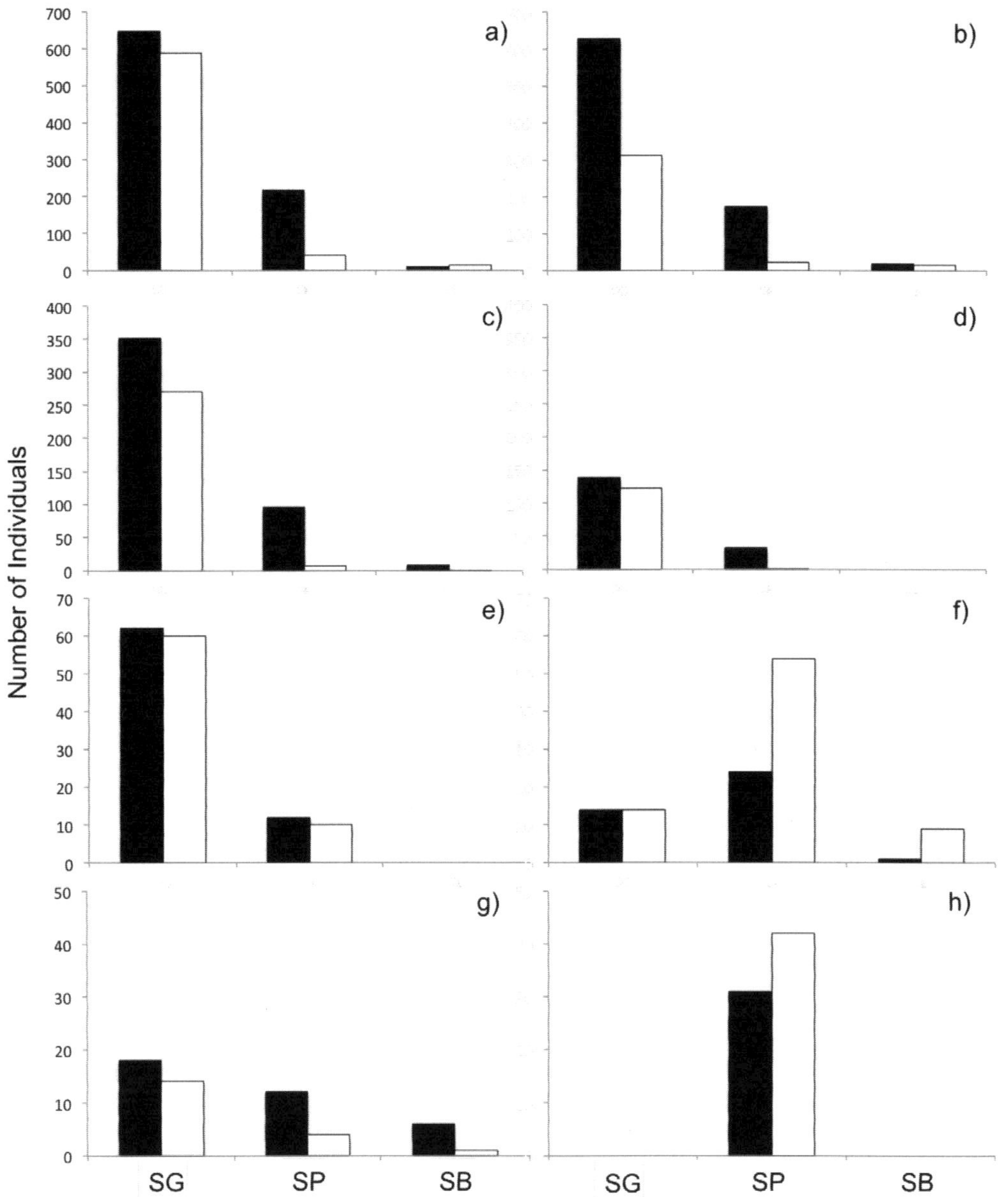

Figure 4. Comparison of mollusc abundance for each dominant species between dry season (■) and rainy season (□) in Situ Gede (SG), Situ Panjang (SP) and Situ Burung (SB). The dominant species with more than 50 individuals are a) *Filopaludina javanica*, b) *Melanoides tuberculata*, c) *Thiara scabra*, d) *Sermyla requeti*, e) *Pila scutata*, f) *Pomacea canaliculata*, g) *Pilsbryoconcha exilis* and h) *Wattebledia crosseana.*

From eight dominant species of molluscs in our study, only six species were recorded in situ of the Ciliwung and Cisadane watersheds, where *S. requeti* and *P. scutata* were not reported [2]. However, some mollusc species such as *Bellamya javanica* (*F. javanica*), *M. tuberculata P. scutata, P. canaliculata* and *L. rubiginosa* were known as common species, which can be found in many human habitation (lakes, ponds, paddy fields, ditches, garden, etc) [5,6,13]. *F. javanica* (35%) and *M. tuberculata* (27%) were the most dominant species of mollusc in our study. Those species are well distributed in most situ of West Java because they can tolerantly and adapt to varying environment and is potentially valuable economically as a food resource and for the ornament pet trade [2,5,15]. *F. javanica*, which is a native species of mollusc, is consumed as a delicacy by Sundanese. Meanwhile, *M. tuberculata* is the most common and most wide-ranging member of the family Thiaridae but is a vector of various parasites, which results in loss commercial fisheries and infects wild fish populations [5,13,15].

The notorious invasive mollusc species (*P. canaliculata*) was also found in our study [16], although the species has only 2% of relative abundance. This species originated from South America and is widely distributed in lakes, ponds, swamps [5]. The species is well a recognized threat to native species around the world and can adapt to habitat modification [16].

A total only 76 individuals (1.75 %) of bivalves was collected in our study. These molluscs barely migrate from one place to another and they remain under severe environmental stress [15]. They are also an important component of ecological food chain in aquatic ecosystems.

5. Conclusions

This study recorded 4 bivalve species (1.75% of relative abundance) and 14 freshwater gastropod species (98.44% of relative abundance) and determined precipitation during dry and rainy season which influenced their population abundance and composition in Situ Gede System. Abundance of mollusc was larger in SG than other situ where SG has high TOM and is the main situ of Situ Gede System. Generally, abundance of total mollusc and abundance of dominant mollusc species were higher during dry season than in rainy season. According to our result, it is recommended that further studies should be carried out to investigate season-driven monthly changes in mollusc composition.

Acknowledgements

This study is supported by Excellent Higher Education Institution Research categorized Excellent Mandate Division Research Bogor Agricultural University 2016 Grant No. 079/SP2H/LT/DRPM/II/2016 to TSP from Ministry of Research Technology and Higher Education the Republic of Indonesia.

References

[1] Priadie B 2011 *Jurnal Sumber Daya air,* **7**(1) 1-104 (*In Indonesian Language*)

[2] Marwoto R M and Isnaningsih N R 2014 *Berita Biologi* **13**(2) 181-189 (*In Indonesian Language*)

[3] Sudarsono Y, Yoga G P, Suryono T, Syawal M S and Yustiawati 2009 *Limnotek* **16**(2)153-166 (*In Indonesian Language*)

[4] Lydeard C, Cowie R H, Ponder W F, Bogan A E, Bouchet P, Clark S A, Cummings KS, Frest T J, Gargominy O, Herbert D G, Hershler R, Perez K E, Roth B, Seddon M, Strong E E and Thompson AG 2004 *BioScience* **54** 321–330.

[5] Ng TH, Tan SK, Wong WH, Meier R, Chan S-Y, Tan HH and Yeo CJ 2016 *PLoS ONE* **11**(8) e0161130 doi:10.1371/journal.pone.0161130

[6] Djajasasmita M 1989 *Bul. Penelit. Kesehat.* 17(2) 135-140

[7] Van Benthem Jutting W S S V 1953 *Treubia.* **22**(1)10-73

[8] Van Benthem Jutting W S S V 1956 *Treubia.* **23**(2) 259 – 477

[9] Pennak R W 1989 *Fresh-Water Invertebrates of the United States "Protozoa to Mollusca" 3rd Ed* (New York (US): John Wiley & Sons Inc)

[10] Marwoto R M, Isnaningsih N R, Mujiono N, Heryanto, Alfiah and Riena 2011 *Leaflet of Freshwater Gastropods in Java Island* (*Moluska, Gastropoda*) (Bogor (ID): LIPI) (*In Indonesian Language*)
[11] Hammer Ø, Harper D A T and Ryan P D 2001 *Palaeontologia Electronica* **4**(1) 9 Available: http://palaeo-electronica.org/2001_1/past/issue1_01.htm
[12] Colle A C and Callil C T 2012 *Braz. J. Biol.* 72(2) 249-256
[13] Ikpeze O O and Obikwelu M E 2016 *Int. J. Pure App. Biosci.* **4**(2) 91-102 doi: http://dx.doi.org/10.18782/2320-7051.2264
[14] Ndifon G T and Ukoli F M A 1989 *Hydrobiologia* **171** 231-253
[15] Sharma K K, Bangotra K and Saini M 2013 *International Journal of Biodiversity and Conservation* **5**(4) 240-249 doi:10.5897/IJBC12.127
[16] Global Invasive Species Database 2005 *Pomacea canaliculata* Available: http://www.issg.org/database/species/ecology.asp?si=135&fr=1&sts=sss&lang=EN

Amino acid and proximate composition of fish bone gelatin from different warm-water species: A comparative study

Y Atma

Department of Food Science and Technology, Universitas Trilogi. Jln. Taman Makam Pahlawan Kalibata No.1, South Jakarta 12760, Indonesia

Email: yoniatma@trilogi.ac.id

Abstract. Research on fish bone gelatin has been increased in the last decade. The quality of gelatin depends on its physicochemical properties. Fish bone gelatin from warm-water fishes has a superior amino acid composition than cold-water fishes. The composition of amino acid can determine the strength and stability of gelatin. Thus, it is important to analyze the composition of amino acid as well as proximate composition for potential gelatin material. The warm water fish species used in this study were Grass carp, Pangasius catfish, Catfish, Lizard fish, Tiger-toothed croaker, Pink perch, Red snapper, Brown spotted grouper, and King weakfish. There werre five dominant amino acid in fish bone gelatin including glycine (21.2-36.7%), proline (8.7-11.7%), hydroxyproline (5.3-9.6%), alanine (8.48-12.9%), and glutamic acid (7.23-10.15%). Different warm-water species has some differences in amino acid composition. The proximate composition showed that fishbone gelatin from Pangasius catfish has the highest protein content. The water composition of all fishbone gelatin was well suited to the standard. Meanwhile, based on ash content, only gelatin from gelatin Pangasius catfish met the standard for food industries.

1. Introduction

Gelatin has been widely used in food, pharmaceutical and cosmetic industries. Gelatin is mainly used to improve the gelation, water binding, foaming, emulsifying, elasticity and viscosity of food products. Furthermore, gelatin is applicable for encapsulation and edible film formation purposes, making it interesting for the biomaterial-based packaging, as well as photographic industries [1]. Nevertheless, commercial gelatin from the mammals such as pig skins, cattle hides, and cattle bones are unacceptable by some religions like as Judaism, Hinduism, and Islam. Therefore, the seeking of an alternative source of gelatin has been increasing in the last decade. The waste from the fish processing is potential source of gelatin. More than 30% fish weight comes from skin and bones [2]. The skins sometimes are still used in the further processing, but the bones are often discarded. Thus, utilization of fish bone as a source of gelatin can minimize waste as well as add value for this particular fishery products.

The indicator of fish bone gelatin production can be analysed both qualitatively and quantitatively. The quality of gelatin is mostly determined by physicochemical properties. It is expected to have the same quality as well as the commercial gelatin. The physicochemical characteristics of fish bone gelatin are gel strength, stability, visco-elastisity, amino acid and proximate composition. In the previous study, it is stated that fish bone gelatin from warm-water fishes has a superior amino acid than that cold-water fishes. It gives effect to the stability and gel strength of fishbone gelatin [3]. Moreover, Gomez-Guillen et al. stated that fish bone gelatin from warm-water fish has better thermo-stability and rheology than cold-water fish [4]. Wang and Regenstein stated that there are positive correlations between composition of amino acid hydroxyproline (Hyp) and gel strength [5]. High content of amino acids (Hyp+Proline) have been improve the rheological (viscoelasticity and gel strength) and stability of fish gelatin meanwhile glycine is important for gel strength [6]. The chemical quality of gelatin can also be indicated by proximate composition.

Protein is the most abundant compound in collagenous material, thus, it represents the maximum yield of gelatin extraction [7] In addition to protein content, ash and water in collagenous material contribute to the quality of gelatin, as the maximum level of ash and water for food application are 2.6% [8] and 15% [9], respectively. About 6-8% of water content leads to the hygroscopic gelatinous material. Meanwhile, 13% of water content leads to the reduction size of particle, since the glass transition temperature of gelatin is about 64 ^{o}C [10].

The objective of this study was to compare the amino acids and proximate compositions of fish bone gelatin from warm-water fish species. This study may give additional information regarding the correlation between amino acids and proximate composition toward physicochemistry properties of fishbone gelatin which can be used for further studies.

2. Research workflow and result

Three consecutive research steps were conducted for this study : (1) selection of related research publications, (2) clusterization uponwarm-water fish species and (3) comparison analysis of both amino acid and proximate compositions (figure 1). The clusterization was important to specify the warm water fish species used in this study, since numerous fish species has been used for the production of gelatin. Warm water fish species were choosen and clarified based on the data available on *fishbase.org*

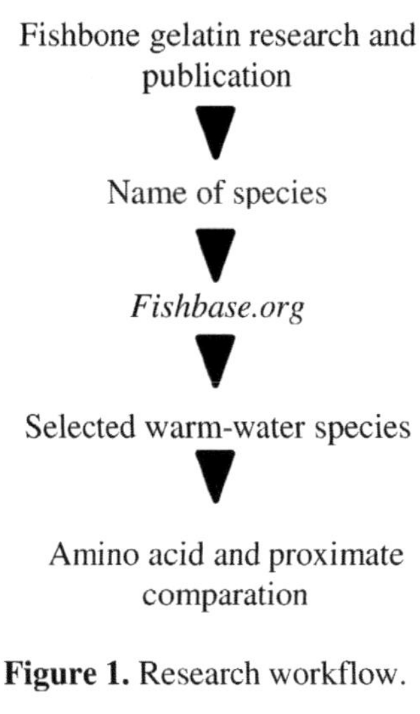

Figure 1. Research workflow.

2.1. *Warm-water fish*

The warm-water species were used in this study were Grass carp (*Ctenopharyngodon idella*), Pangasius catfish (*Pangasius sutchi*) [8], Catfish (*Clarias gariepinus*) [6], Lizard fish (*Saurida* spp.) [11], Tiger-toothed croaker (*Otolithes rubber*) and Pink perch (*Nemipterus japonicus*) , Red snapper (*Lutjanus campechanus*), Brown spotted grouper (*Epinephelus chlorostigma*), and King weakfish (*Macrodon ancylodon*) [10,12,13].

2.2. Amino acid composition

There are five dominant amino acids in fish bone gelatin including glycine (21.2-36.7%), proline (8.7-11.7%), hydroxyproline (5.3-9.6%), alanine (8.48-12.9%), and glutamic acid (7.23-10.15%) (table 1).

Table 1. Amino acid composition of fishbone gelatin form warm-water fish.

Amino acid (%) /Species	Code	Grass carp[5]	Pangasius catfish[1]	Catfish[2]	Lizard fish[7]	King Weakfish[6]	Comercial Gelatin[1]
Alanin	Ala	12.9	8.48	9.1	9.1	9.75	7.73
Arginin	Arg	5	8.11	6	5.36	7.25	7.01
Aspartic acid	Asp	4.7	4.62	6.5	5.36	0	4.97
Asparagin	Asn	0	0	0	0	5.35	0
Cysteine	Cys	0	0	0	0	0	0
Glutamic acid	Glu	7.7	8.16	9.1	7.23	10.15	6.19
Glycine	Gly	36.7	25.02	21.2	30.34	24.42	42.71
Histidine	His	0.5	0.8	-	0.9	0.92	1.02
Hydroxyproline	Hyp	7	5.97	5.3	8.13	9.6	6.49
Isoleucine	Ile	1	1.33	1.5	0.85	1.73	1.35
Leucine	Leu	2.1	2.23	2.5	1.9	2.54	2.42
Lysine	Lys	2.5	3.13	3.5	2.1	3.25	3.11
Methionine	Met	1.3	10.83	7.7	1.35	1.85	6.43
Phenylalanin	Phe	1.3	1.7	2.9	1.4	1.9	1.63
Proline	Pro	8.7	11.4	9	11.7	10.25	11.09
Serine	Ser	3.9	3.08	3.8	3.4	3.25	2.87
Threonine	Thr	2.5	2.45	3.9	1.53	2.82	1.43
Tyrosine	Tyr	0.4	0.53	0.6	0.3	0.75	0.4
Valine	Val	1.8	2.19	2.5	1.84	1.6	2.08
Hydroxylisin	Hyl	0	0	0	0	0.85	0
Total		100	100	100	100	100	100
Hydroxyproline Proline	Hyp+Pro	15.7	17.37	14.3	19.83	19.85	17.58

2.3. Proximate composition

Table 2. Proximate composition (%) of fishbone gelatin from warm-water fish and commercial gelatin.

Proximate properties (%) /species	Pangasius catfish[1]	Catfish[2]	Lizard fish[7]	Tiger-toothed croaker[3]	Pink perch[3]	Red snapper[4]	Brown spotted grouper[4]	King Weak-fish [6]	Commercial Gelatin[3,4,9]
Water	9.2	11.43	8.27	10.33	8.56	6.24	4.1	9.9	< 15
Protein	87.3	81.75	81.89	82.5	69.49	78.56	82.36	82.3	86.1*
Ash	2.6	5.6	11.17	2.7	2.8	10.32	6.58	3.8	< 2.6
Lipid	0.96	0.95	0.01	0.52	0.32	5.16	3.92	-	-

*) Bovine gelatin [14]

3. Discussion

From this study, it is likely that amino acid compositions varied in each warm-water species. Importantly, we noticed that Glycine (Gly) was the highest content of amino acid in most of the fishbone gelatin. Methionine was relatively high in fishbone gelatin of Pangasius cathfish and Catfish, but it was found in low concentration in Grass carp, Lizard fish and King Weakfish. Histidine was occurred in low level in most of the fish species and almost un-detectable in catfish. Interestingly, Hydroxyproline of Tiger toothed croaker was occurred wiithin 7.51% and Pink perch 7.41%.

Based on the proximate compositios, fishbone gelatin from Pangasius catfish has the highest protein content (table 2). The protein content, however, is not quite necessarry for new gelatin discoveries, since gelatin itself is considered as highly-abundant protein material. On the other hand, high protein would be important for an optimum extraction step [15]. The water composition in all fish bone gelatin from warm-water fish suited to the standard. Judging from water content alone, the gelatin from Red snapper and Brown spotted-grouper were considered as hygroscopic gelatin. Based on ash content, amongst the warm water fish tested, only gelatin from Pangasius catfish met the standard of that commercial gelatin for food application.

Based on our study, King weakfish and Lizard fish have the highest hydroxyproline and amino acid compositions among others, contrarily to their corresponding gel strength value (figure 2). Thus, previous reports regarding the correlation between the content of amino acids, particularly, hydroxyproline with gel can not be fully accepted. However, it has been reported the gel strength value can be influenced by the composition of several amino acids, extraction method and molecular weight distribution [6, 16, 17].

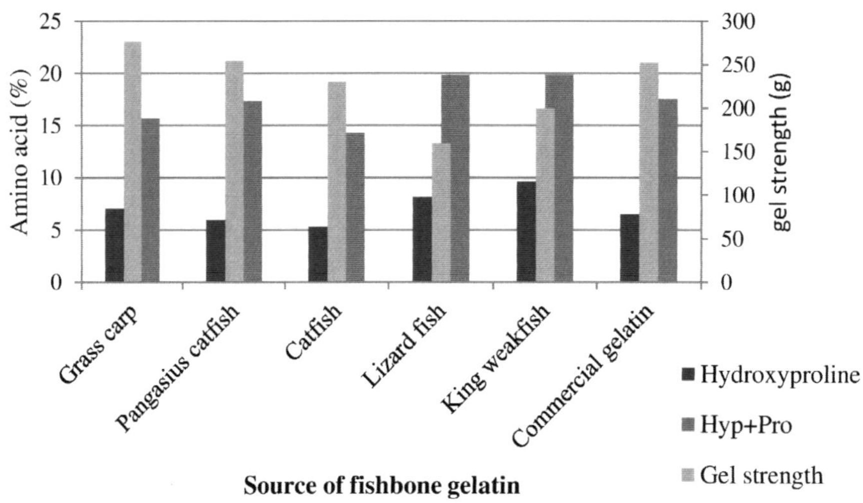

Figure 2. Hydroxyproline and amino acid composition (%) and gel strength (g) of fishbone gelatin from warm-water species.

References

[1] Karayannakidis P D and Zotos A 2014 *J Aquat Food Prod T* **25** 1
[2] Shyni K, Hema G S, Ninan G, Mathew S, Joshy C G, Lakshmanan P T 2014 *Food Hydrocolloid* **39** 68-76
[3] Zhang F, Xu S and Wang Z 2011 *Food Bioprod process* **89** 185–93
[4] Gomez-Guillen M C, Perez-Mateos M, Gomez-Estaca J, Lopez-Caballero E, Gimenez B and Montero P 2009 *Trends Food Sci Tech* **20** 3-16
[5] Wang Y and Regestein 2009 *Food Chem* **74** 426-31
[6] Sanaei A V, Mahmoodani F, See S F, Yusop S M and Babji A S 2013 *Int Food Res J.* **20** 1
[7] Muyoga J H, Cole C G B and Duodu K G 2004 *Food Chem* **85** 81-9
[8] Mahmoodani F, Sanaei A V, See SF, Yusop S M, and Babji A S 2014 *J Food Sci Technol* **51** 11
[9] GME 2005 Standard Methods for the Testing of Edible Gelatine, Gelatine Monograph, Gelatin Manufacturers of Europe

[10] Koli J M, Basua S, Nayaka B B, Patageb S B, Pagarkarb A U and Gudipatia V 2012 *Food Bioprod Process* **90** 555-62
[11] Taheri A, Kenari A A, Gildberg A and Behnam S 2009 *J. Food Sci* **74** 160-65
[12] Shakila R J, Jeevithan E, Vratharajakumar A, Jeyasekaran G and Sukumar D 2012 *Food Sci Technol* **48** 30-6
[13] Alfaro A T, Costa C S, Fonseca G G and Prentice C 2009 *Food Sci Tech Int.* **15** 6
[14] Chuaynukul K, Prodpran T and Benjakul S 2014 *Res J Chem Environ* **2** 4
[15] Sanei V A, Mahmoodani F, See S F, Yusop S M and Babji AS 2013 *Sains Malays* **42** 12
[16] Nurul A G and Sarbon N M 2015 *Int Food Res J* **22** 2
[17] Raja Mohd Hafidz R N, Yaakob C M, Amin I and Noorfaizan A 2011 *Int Food Res J* **18** 813-17

Enhancing Activity and Stability of Uricase from *Lactobacillus plantarum* by Zeolite immobilization

D Iswantini[1*], N Nurhidayat[2] and Sarah[1]

[1]Department of Chemistry, Faculty of Mathematics and Natural Sciences, Bogor Agricultural University, Bogor, 16680, Indonesia
[2]Division of Microbiology R&D for Biology, The Indonesian Institute of Sciences, Jalan Raya Jakarta-Bogor Km 46 Cibinong Bogor, 16911, Indonesia

Email: dyahprado@gmail.com

Abstract. *Lactobacillus plantarum* has been known be able to produce uricase for uric acid biosensor. Durability and stability of *L. plantarum* in generating uricase enzyme was low. Hence, we tried to enhance its durability and stability by immobilizing it onto activated 250 mg zeolite at room temperature using 100 µL *L.plantarum* suspension and 2.87 mM uric acid, while Michaelis–Menten constant (K_M) and V_{max} were obtained at 6.7431 mM and 0.9171 µA consecutively, and the linearity range was 0.1-3.3 mM (R^2 = 0.9667). Limit of detection (LOD) and limit of quantification (LOQ) value of the measurement were 0.4827 mM and 1.6092 mM respectively. Biosensor stability treatment was carried out in two different treatments, using the same electrode and using disposable electrode. The disposable electrode stability showed better result based on repeated measurements, but stability was still need improvement.

1. Introduction

Uric acid is the final metabolism product of purine and its derivatives from the body. Normally, uric concentration in blood is around 0.13-0.46 mM and 1.49-4.46 in the urine [1]. The increase of uric acid in blood above the normal level may lead to stimulate some disease, like gout, hyperurisemia, and Lesch-Nyhan syndrome. However, at low concentration it can lead to cause oxidative stress and multiple sclerosis [2]. Hence, a simple measurement method needs to be developed for monitoring the uric acid concentration in human body.

Some spectrophotometry methods have been developed for measuring uric acid concentration, like colorimetry [3], high performance liquid chromatography [4], chemiluminescence [5], and fluorescence [6]. This approach offers high precision and selectivity, however this method is highly influenced by light, needs trained operator, relatively high cost, complex sample preparation, and long time analysis [7]. Overcoming these difficulties, biosensor as an alternative method has been intruduced for further development [8].

Biosensor is an analytical device which applies bioreceptor as detector, to detect a signal that indicate biochemical response. Tranducer then transfers the signal produced by bioreceptor, transmitting it into analyzer. Frequently biosensor uses enzyme as bioreceptor, because it has specific and seletive reaction. For detecting uric acid, the applied enzyme which can be used is uricase, because it can catalyze uric acid, transforming it into allantonin facilitated by O_2 [9]. However, using pure enzyme is not economically applicable because of the difficulty in purification process, and in addition it only produces small portion of enzyme. The enzyme utilization can be substituted by a selective microbe, since the microbe can produce uricase enzyme and be regenerated easily so that the enzyme can be obtained easily [10]. Several former investigations have been conducted and one of the microbe candidate which easily produce uricase was *Lactobacillus plantarum* [11].

As analytical instrument for routine analysis, biosensor should have high stability and can be used for long time. To support this function, *L. plantarum* was immobilized with zeolite. In our previous research, we used base activated zeolite which was compatible with the enzyme range, but not with the optimum pH of *L. plantarum*. Furthermore, we used measurement at 40°C, hence it is unpractical for routine analysis. In addition, the avarage produced enzyme by microbe was unknown. Hence, in this research mainly we would like to enhance the biosensor stability. *L. plantaru* was still immobilized into zeolite. The measurement was conducted at room temperature condition. Therefore, applying activated natural zeolite we studied the optimum condition (pH, zeolite mass, uric acid) of this biosensor.

2. Research Methods

2.1. Tools and Materials

The tools used here were eDAQ potensiostat-galvanostat with computer installed *Echem* v.2.0.1 (3 electrodes system) and Minitab 16, incubator (Himawari), microwave (Sanyo), autoclave (Hirayama), pH meter (TOA DK HM-250), spectrophotometer BioSpec-1601 (Shimadzu), micropipette 1000µL, analytical scale, laminar airflow, and other glassware. And the used material were: *L. plantarum* isolate from LIPI, activated natural zeolite, uric acid (Nacalaic tesque, Japan), $K_3[Fe(CN)_6]$ (Merck, Germany), KCl (Merck, Germany), 2,3-dimethoxy-5-methyl-1,4-benzoquinone (Q_0, Merck, Germany), dialysis membrane, O-ring, calcium carbonate (Merck, Germany), NaAcO (Merck, Germany), NaCl (Merck, Germany), polysorbate 80, yeast extract, beef extract, agar, phosphate buffer (Merck, Germany), and deionized water.

2.2. Methods

2.2.1. L. plantarum preparation. Solid GYP media was made by mixing homogenously 5 g $CaCO_3$, 15 g agar, 10 g glucose, 5 g pepton, 1.4 sodium acetate, 5 mL NaCl solution, 10 mL Tween 80, 2 g beef extract, and 10 g yeast extract with 100 mL aquadest, autoclave at 121°C for 15 minutes. The liquid GYP media was made in the same procedure, but without adding agar and $CaCO_3$. The prepared GYP media then was poured into petri dishes and was let until became harden. Later, a loopful of *L. plantarum* was smeared onto its surface and incubated at 37°C for 1-3 days anaerobically. After the incubated microbe had been grown, it was relocated into liquid MRS media and reincubated aerobically at 37°C for 24 hours. Later, its liquid was measured for OD_{600} value until it had range value of 0.5-0.6, then separated by centrifugation (10 000 rpm, 10 minutes, 4 °C), to produce pellet. The pellet was rewashed twice then diluted into phosphate buffer, mixed with activated zeolite for immobilization process later.

2.2.2. Biosensor Preparation Measurement. Carbon paste electrode was prepared by mixing graphite, paraffin oil (2:1 m/m) and Q_0 as mediator (3% of graphite mass). Q_0 was diluted first into DMSO before mixing it with graphite. After Q_0 and graphite mixed homogenously (DMSO should has been evaporated), paraffin was added and mixed again, transforming it into carbon paste. Next, the modified carbon paste was put into the electrode compartment, packed densely and polished. After 5 days from the first preparation, the electrode characterized by cyclic voltammetry method using KCl

0.1 M and $K_3[Fe(CN)_6]$ 0.01 M. The parameters were: mode cyclic, initial 500 mV, final 500 mV, rate 250 mV/s, step W 20 ms, upper E 900 mV, lower E -600 mV, range 5 V, and arus 100 μA.

After the electrode was tested by voltammetry cyclic, the harvested microbe at the previous step was immobilized onto the electrode surface. Firstly, 10 mL phosphate buffer was prepared and mixed with several mg of zeolite. From the mixture, 10 μL mixture was mixed suspended *L. plantarum*, then 10 μL of the mixture was immobilized as bioreceptor onto the electrode's surface. After immobilization process was done, the electrode was stored into buffer phosphate solution (pH 7, 4 °C) until the usage. Biosensor measurement was conducted by setting the parameters in the computer as like in the electrode characterization process. In this step, we measured the oxidation peak of the current from every treatment.

For measurement analysis the optimum condition was set by combining 3 factors, zeolite, L. plantarum, and uric acid used for further test. The combinations of 3 variables are listed in table 1 by using Central Composite Design. Before optimizing 3 variables given, we searched the optimum the acidity of L. plantarum (pH 5, 6, 7). The measurement was conducted in room temperature, with blank sample (2 mL phosphate buffer). After adding some uric acid into the cell, the current was again measured.

Table 1. Combination of Optimation 3 Parameters.

No.	Zeolite (mg)	[Uric Acid] (mM)	*L. plantarum* (μL)
1	250	4	20
2	50	0.1	100
3	50	2.05	60
4	250	0.1	100
5	150	2.05	20
6	150	2.05	60
7	150	2.05	60
8	150	2.05	60
9	150	2.05	60
10	150	2.05	100
11	50	4	100
12	50	4	20
13	250	2.05	60
14	150	2.05	60
15	150	0.1	60
16	150	2.05	60
17	250	4	100
18	250	0.1	20
19	50	0.1	20
20	150	4	60

At the optimum point, we measure the kinetic parameters of uricase activity according to Michaelis-Menten equation, and transformed it into Lineweaver-Burk plot. Here is the formula:

$$I = \frac{I_m^{app}[\text{Uric acid}]}{K_M + [\text{Uric acid}]}$$

I_m^{app} : Response of maximum apparent current (A)
K_M : Michaelis-Menten constant
[Uric acid] : Uric acid concentration

There were 2 kinds of validation technique performed in this research: LOD (limit of detection) and LOQ (limit of quantization) value, and stability. LOD and LOQ value were counted according this equation:

$$Q = \frac{k.sb}{b}$$

Q : LOD or LOQ value
k : constant, LOD = 3, LOQ = 10
sb : standard deviation of linearity measurement
b : slope of linearity measurement

The parameter of stability was divided into 2 kinds of treatment. The first was by using the same electrode from the initial day, and the second was by using new-made electrode for each day. Both treatments were done for 20 days. From both treatments, current data were obtained and calculated using this formula:

$$\text{Stability (\%)} = (100\% - \frac{[I_0 - I_t]}{I_0} \times 100\%)$$

I_0 = oxidation current at day-0
I_t = oxidation current at day-t

3. Result and Discussion

In this research we used uricase from *L. plantarum*. Uricase activity was measured electrochemically. The enzymatic reaction oxidizes uric acid to allantoin facilitated by oxygene. It produces H_2O_2 as the product of O_2 reduction. This peroxide is very reactive and rapidly change to O_2 and transferring electron. This electron was transmitted to electrode and detected as signal. hereby the following reactions [12]:

$$\text{Uric acid} + O_2 + H_2O \xrightarrow{\text{Uricase}} \text{Alantoin} + CO_2 + H_2O_2$$
$$H_2O_2 \rightarrow O_2 + 2H^+ + 2\bar{e}$$

By using modified carbon paste electrode (2:1 graphyte/paraffine mass) we examined the characteristic of this biosensor. We used Q_0 (2,3-dimethoxy-5-methyl-1,4-benzoquinone) as the modifier for enhancing oxidation current detected by the potentiostat. Having the ready electrodes for further measurements, we tried to immobilized *L. plantarum* directly to the electrodes at various pH (5, 6, and 7) then we measured the produced current (using 0.5 mM uric acid). Obviously we obtained at pH 5.0 the electrode produced higher current compared the others. It means that *L. plantarum* was more stable at pH 5.0, like in the planktonic mode [13].

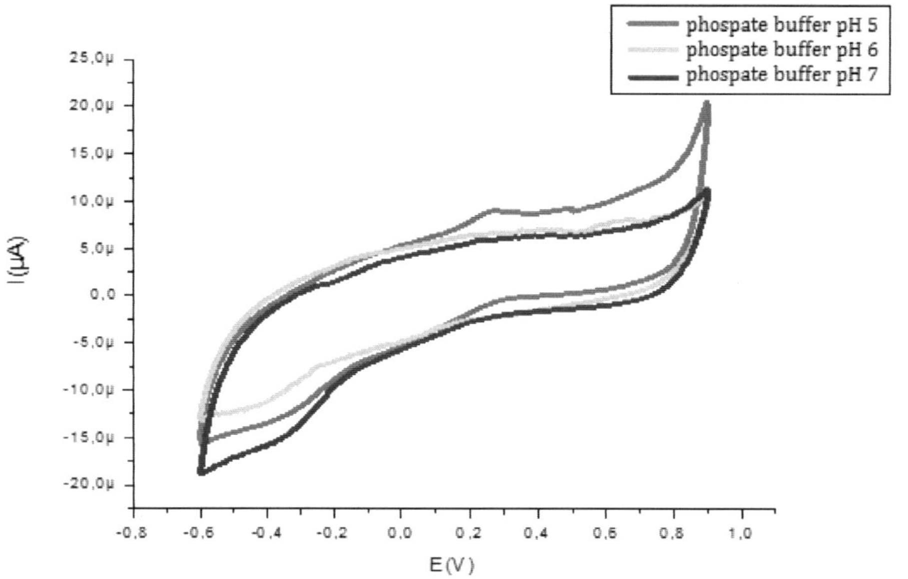

Figure 1. *L. plantarum* response to various pH (5, 6, and7).

Based on this pH, the amount of *L. plantarum* and zeolite applied onto the electrode surface was examined. After analyzing the result, we got the optimal current would be produced by mixing 250 mg zeolite with 100 µL *L. plantarum* at 2.87 mM of uric acid. Therefore, at this optimum point, we studied the uricase activity of *L. plantarum* after being immobilized onto the zeolite by observing the current after introducing various uric acid concentration (0.1-4.0 mM). At room temperature, it only produced 5.73 µA, the maximum current, at 3.3 mM uric acid (figure 2). The result was lower compared to the similar reseach which have been conducted before, because the pH affected the enzyme stability. And also, it used lower temperature compared to former study [14].

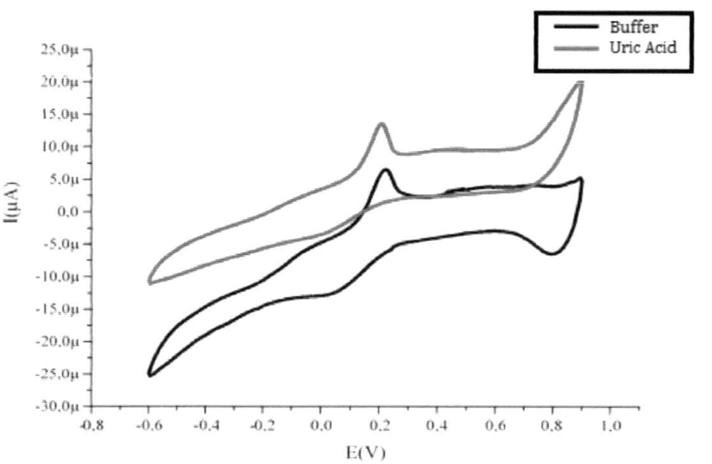

Figure 2. Voltammogram profile of immobilized *L. plantarum* onto zeolite.

Analytical properties of measurement method will describe how the proposed method works. In this research we examined LOD, LOQ, linearity, and range of measurement. Based on figure 3, this proposed method had best range between 0.1-3.3 mM uric acid. It is still inside the normal range of uric acid in blood[3]. This method also had good linearity and had strong correlation between signal and uric acid concentration, obviously seen from R^2 value (96.97%). It had low LOD value, 0.4827 mM which means it can be useful for detecting uric acid at small concentration. However, it has a higher value of LOQ, up to 1.6092 mM, indicating this method is surely good for detecting uric acid above stated LOQ value. We conducted also a test about the stability of this biosensor for 16 days, however it didn't produce good stability. Along that period. Some further enhancement are needed for developing this alternative method for measuring uric acid.

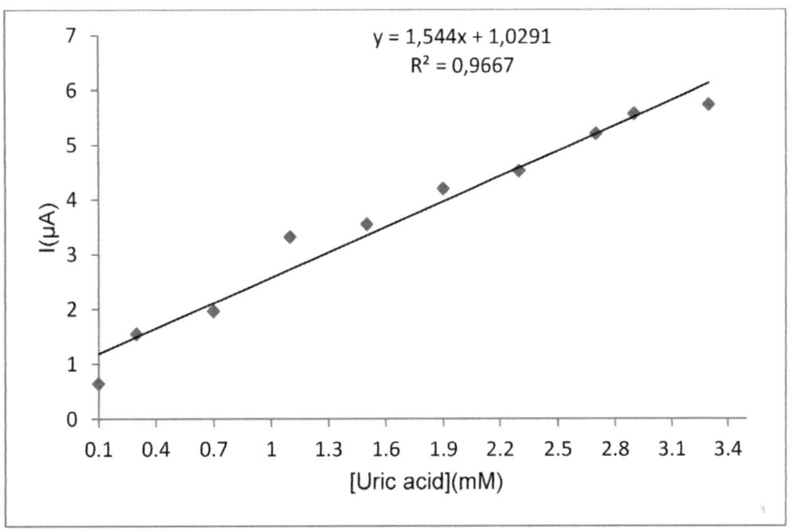

Figure 3. Linearity and measurement range of this uric acid biosensor.

References

[1] Lou YC, Do J S and Liu C C 2006 *Biosensors and Bioelectronics* **22** 482
[2] Ghosh T, Sarkar P and Turner A P F 2015 *Bioelechem.* **102** 1
[3] Barr, W G 1990 *Clinical Methods: The History, Physical, and Laboratory Examinations 3rd edition*, ed Walker HK, Hall WD, Hurst JW Butterworth (Boston (US): Butterworth Publisher)
[4] Zhou S, Zuo R, Zhu Z, Wu D, Vasa K, Deng Y and Zuo Y 2013 *Anal. Methods.* **5** 1307
[5] He D, Zhang Z, Huang Y, Hu Y, Zhou H and Chen D 2005 *Luminascence.* **20** 271
[6] Wang M, Niu W, Wu X, Li L, Yang J, Shuang S and Dong C 2014 *RCS Adv.* **4** 25183
[7] Arslan F 2008 *Sensor Articles* **8** 492
[8] Rawal R, Chawla S, Chauhan N, Dahiya T and Pundir C S 2012 *International Journal of Biological Macromolecule* **50** 112
[9] Erden PE, Kılıc E 2013 *Talanta* **107** 312
[10] Su L, Jia W, Hou C and Lei Y 2011 *Biosensors and Bioelectronics* **26** 1788
[11] Iswantini D, Nurhidayat N, Trivadila and Widyamoko O 2014 *Pakistan Journal of Biological Sciences* **17** 277
[12] Çete S, Yasar A and Arslan F 2006 *Artificial Cells, Blood Substitutes, and Biotechnology* **34** 367

[13] Vos P, Garrity G, Jones D, Krieg N R, Ludwig W, Rainey F A, Scleifer K H and Whitman W 2009 *Bergey's Manual of Systematic Bacteriology Volume 3: The Firmicutes* (New York (USA): Springer)
[14] Kamal F 2014 *Penentuan kestabilan dan linearitas pada biosensor asam urat menggunakan urikase dari Lactobacillus plantarum termodifikasi zeolit secara elektrokimia* (Bogor (ID): Fakultas Matematika dan Ilmu Pengetahuan Alam, Institut Pertanian Bogor)

Alcohol Dehydrogenase of *Bacillus* strain for Measuring Alcohol Electrochemically

D Iswantini[1*], N Nurhidayat[2] and H Ferit[1]

[1]Department of Chemistry, Faculty of Mathematics and Natural Sciences, Bogor Agricultural University, Bogor, 16680, Indonesia
[2]Division of Microbiology R&D for Biology, The Indonesian Institute of Sciences, Jalan Raya Jakarta-Bogor Km 46 Cibinong Bogor, 16911, Indonesia

Email: dyahprado@gmail.com

Abstract. Alcohol dehydrogenase (ADH) was applied to produce alcohol biosensor. The enzyme was collected from cultured *Bacillus sp.* in solid media. From 6 tested isolates, bacteria from fermented rice grain (TST.A) showed the highest oxidation current which was further applied as the bioreceptor. Various ethanol concentrations was measured based on the increase of maximum oxidation current value. However, a reduction value was happened when the ethanol concentration was higher than 5%. Comparing the result of spectrophotometry measurement, R^2 value obtained from the biosensor measurement method was higher. The new proposed method resulted a wider detection range, from 0.1-5% of ethanol concentration. The result showed that biosensor method has big potency to be used as alcohol detector in foods or bevearages.

1. Introduction

Alcohol is one of chemical substance which may be present in many kinds of foods and beverages. Introducing alcohol to biological system of human body exceeds the tolerable concentration can harm the biological function like nervous, transportation, and digestive system in human body [1], which can lead to many metabolism disfuctions and physiological deseases. Therefore, controlling the alcohol concentration in a processed food or beverage product becomes a major concerned. In addition, muslims consumers are also require assurance of halal products, since the Muslim people do not consume food or drinks that contain alcohol. Considering the people's need, Ulama's Council of Indonesia (MUI) released fatwa MUI No. 4/2003 stipulating that alcohol content inside foods or beverages should be below 1%. Therefore, there is necesity to have a new method that is able to measure alcohol content inside food materials sensitively and selectively [2].

For daily usage ethanol is frequently applied as part of food or beverage ingredient. Recently, several methods have been used in ethanol measurement, including gas chromatography [3], high performance of liquid chromatography [4], infrared spectroscopy [5] etc. These methods are very helpful for determining ethanol concentration based on its precision. However, it has some disadvantages because this method needs trained operator, has long time analysis, relatively high cost, and can be interferred with any substance which has same wavelength absorption. To solve this problem, we tried to develop other technique based on electrochemical reaction. One of the electrochemical measurement method is electrochemical biosensor. Biosensor has bioreceptor which can detect substance selectively since it works is based on a selective enzymatic reaction. Biosensor has an opportunity being developed as selective, applicative, and fast time analysis, as well as a portable kit analyser [6].

In this method, choosing the right bioreceptor is a challenging since it determines the further process of detecting the targetted analyt. In this research microbe was used as the main bioreceptor because it is readily and easily to be regenerated. Many reports have written that *Bacillus sp.*, a positive-Gram bacteria, has potency to be used for detecting alcohol, because this microbe produces alcohol dehydrogenase (ADH). Previous research has been conducted using *B. stearothermophillus* which was able to produce a stable ADH from heat of reversible oxidation reaction. In this reaction ADH required nicotinamide adenine dinucleotide (NAD^+) as cofactor. As the microbe becomes bioreceptor, during reaction, all biochemical reaction involving ADH would influence the charge of electrode. When proton is generated by turning NADH to NAD^+, it would be transfered into electrode directly or indirectly, by which the current changes then can detected [7]. Some experiments have been conducted in this basic concept and shown good result [8,9,10]. However, it still required more time for detection process and higher cost was needed for enzyme extraction. Therefore, using whole microbe cell as bioreceptor directly we expect that the technology may be able to reduce the fabrication cost of biosensor in the future.

This preliminary research aimed to develope an alcoholic biosensor using alcohol dehydrogenase (ADH) enzyme produced from *Bacillus sp.* for detecting ethanol content.

2. Research Methods

2.1. Tools and Materials

In this research we used a computer installed with software Echem v. 2.0.1 connected to E-Daq potentiostat E-Coder 410, spectrophotometer UV-Vis (Shimadzu 1700), centrifuge (Kokusan H-1500 F), autoclave (Hariyama), pH meter (TOA DK HM-250), micropipette (Gilson), oven, andglassware. The materials used in this research were *Bacillus* sp. isolate, solid agar media, graphite, ferrocene, liquid paraffin, simethyl sulfoxide (DMSO), buffer phosphate solution 0.05M pH 7.00, and ethanol (Merck).

2.2. Methods

2.2.1. Isolate preparation. There were 2 types of bacteria media: Heterotroph (HTR) and Yeast Mol (YM). HTR's preparation was carried out by pouring 1.5 g Bacto Agar, 1.5 g Peptone, 0.3 g Tripton, 0.5 g NaCl and 0.25 g K2HPO4 into 300 mL^3 erlenmeyer flask, dilluted with aquadest, while YM's preparation was prepared using 1.5 g Bacto Agar, 0.5 g Pepton, 0.3 g Yeast Extract, 1 g Glukosa, 0.05 g $MgSO_4.H_2O$, and 0.1 g K_2HPO_4 with the same aquadest volume. Both media were put into microwave and mixed homogenously. After mixing process the flasks were corked with aluminium foil. Later, it was put into autoclave 90 minutes and poured into several sterilized petri dishes in warm condition. There were 6 isolated bacteria of Bacillus sp., coded as follows: ET.OH, TST.A, BS 1, BS 2, BS 4, and BS 5. Those isolates were prepared by dethawing the isolate from freezer, taken a loopful and planted onto by smearing the solid agar media prepared before, and were stored into incubator for 24 hours (30°C). The first generation was replanted again with the same method, incubated for 24 hours. All the treatment was done inside steril laminar air flow. After 24 hours the second plantation, 10 loopfuls was put into eppendorf, diluted with 1 mL phosphate buffer 0.05 M pH 7.0, and washed by centrifugation. Later, the microbes was separated by 10 minutes centrifugation (10 000 rpm, 4°C). The pellet then was suspended into 10 minutes phosphate buffer 0.05 M pH 7.00.

2.2.2. Electrode preparation and microbe screening. Carbon paste electrode was made by mixing 3 mg ferrocene (diluted by 1 mL DMSO before) and 100 mg graphite, allowed to stand for 2 hours, and dried at 105 °C. The modified carbon paste then mixed with 30 μL liquid paraffin homogenously. The paste was put into cyllindrical glass compartment, desed, and polished. The electrode was ready for further used after 7-12 days. All of the electrodes shoudl be characterized by $K_3[Fe(CN)_6]$ (diluted in 0.1 M of KCl), observing redox peak by cyclic voltammetry.The well-tested electrode later being used for further measurement. Ag/AgCl was used as refference electrode and Pt as auxiliary electrode. These following parameters was input into computer installed with Echem v.2.0.1 and connected to eDAQ potensiostat:

- Mode : Cyclic
- $E_{Initial}$: 0 mV
- E_{Final} : 0 mV
- Rate : 250 mV/s
- Step W : 20 ms
- E_{Upper} : 1000 mV
- E_{Lower} : 0 mV

For measurement, 5 mL buffer phosphate solution 0.05 M pH 7 was put into a vial, and the blank voltammogram profile was obtained. Later, 10 µL bacteria isolate, 1 mL ethanol 1% were also poured into blank solution, then the current observed. Difference between the sample current and blank current was considered as net current. The applied isolate bacteria was taken from aforesaid bacteria (ET.OH, TST.A, BS 1, BS 2, BS 4, and BS 5). Those six bacteria samples were measured electrochemically and screened for obtaining the highest oxidation current produced. The selected bacteria later was used for forward experiments.

2.2.3. Electrochemical and UV-Vis Measurement of Alcohol. Selected bacteria was rejuvenated and reproduced massively for linearity test with the same procedure listed before. The electrochemical measurement of enzymatic reaction was performed as written before, but with various ethanol concentration (0.1%, 0.3%, 0.5%, 0.7%, 1%, 3%, and 5%). The net current was calculated after observing the blank and sample current. With the same ethanol concentration, we applied it to spectrophotometry measurement method, using UV-Vis spectrophotometer (2 beams). The first cuvette was filled with phosphate buffer 0.05 M pH 7, and the second cuvette filled by pure ethanol 98%. After searching maximum wavelength absorbance, the first cuvette was added by ethanol increasingly (0.1%, 0.3%, 0.5%, 0.7%, 1%, 3%, and 5%). The absorbance value then was recorded.

3. Result and Discussion

Bacteria were collected from 6 different sources, the two first were isolated from fermented alcohol (ET.OH) and fermented rice grain (TST.A), while the rest bacteria sources were taken from laboratory isolate (B1, B2, B3, and B4). Only ET.OH isolate was planted in the HTR media, while the rest was planted on YM. By smearing method, the bacteria were planted into each proper media, and the result was presented in figure 1. These bacteria were chosen electrochemically (using hardware from EDAQ potentiostat and E-Coder 410: cyclic voltammetry method, data analyzed with software *Echem* v. 2.0.1) by using mixture of 1% ethanol and buffer (pH 7.00) read as blank current. The oxidation peak current was measured after 10 µL of suspended bacteria poured into the blank solution. The measurement showed that TST.A produced the highest amount of targeted enzyme indicated the highest amount of oxidation current (figure 2). The current was produced as much as 2.34µA at 0.545V. Therefore, we used TST.A as bacterial source for detecting ethanol.

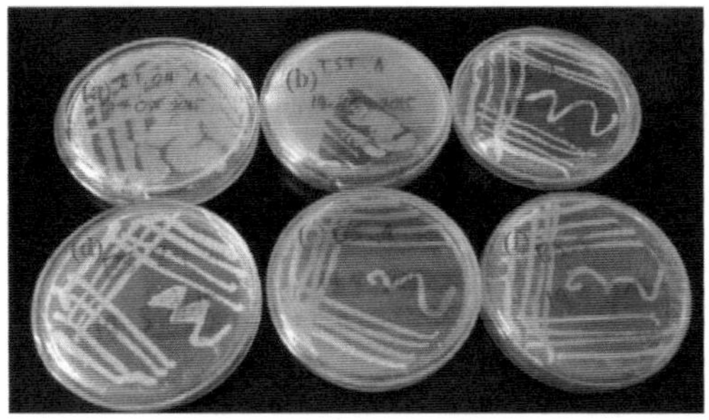

Figure 1 Isolated Bacteria: (a) ET.OH, (b) TST.A, (c) BS.1, (d) BS.2, (e) BS.3, (f) BS.4.

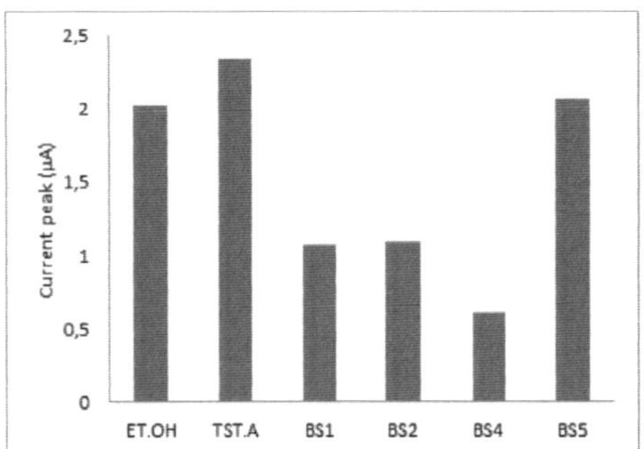

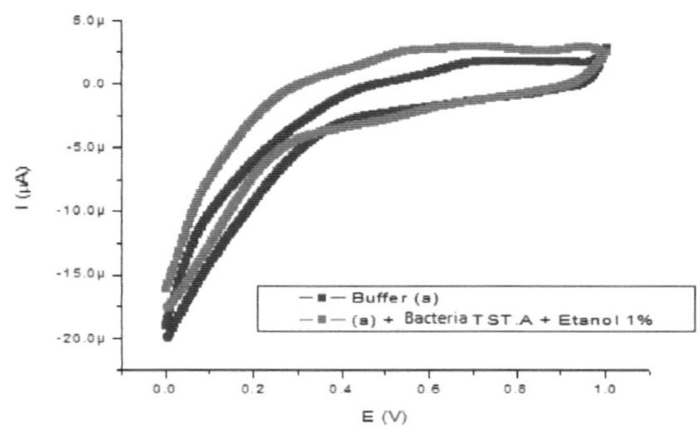

(a) (b)

Figure 2. (a) Oxidation peak current profiles of 6 bacteria; (b) voltammogram current profile of TST.A: blue line is blank solution, red line is after addition of TST.A isolate and ethanol.

For the next experiment, we measured the various concentration of ethanol (0.1%, 0.3%, 0.5%, 0.7%, 1.0%, 3.0%, and 5.0%) by using biosensor, so that we can measure the measurement linearity. For this measurement, we got the average of oxidation current for each concentration (figure 3.a). The voltammogram profile also exhibited an increasement of current after adding ethanol to the mixture of bacteria and blank solution (figure 3.b). The result showed that the more alcohol concentration presented into the solution, the more oxidation current produced. However, at the concentration of 7.0% or 10.0% ethanol the reaction did not give linear result. So, using the measurement range of 0.1-5.0%, the graph produced R^2 value up to 99.63%. The result showed that using more etanol into sample solution decreased the current that was produced. It probably because the higher ethanol concentration lead to several chemical and physical changes inside the microbe's cell. Thus, the metabolism become ruptured because alcohol has potency as bactericide [10].

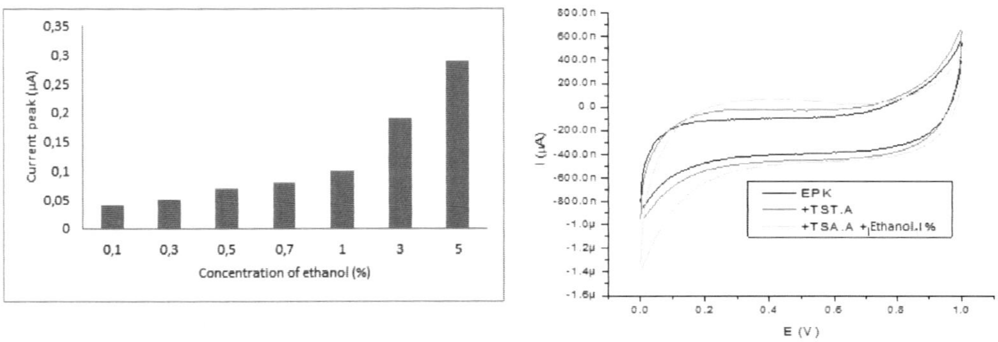

(a) (b)

Figure 3. (a) Average of thrice measurement on the same ethanol level concentration; (b) Voltammogram profile of ethanol measurement.

We also compared the linearity of the measurement with the electrochemical analysis using UV-VIS (Shimadzu 1700) at the same ethanol concentration. Than the R^2 value of both methods was evaluated. The analyzed results showed that both methods had good linearity range, however measurement using ethanol electrochemical method produced slightly higher value of R^2 (99.63%) compared to biosensor (98.57%, Figure 4). In addition, the measurement using biosensor only gave best result if the concentration of ethanol was below 1%. Above that concentration level, the graph was not linear anymore. Considering this, the proposed method had a good chance for being developed as alternative method for measuring ethanol even thogh some modification may need to be carried out.

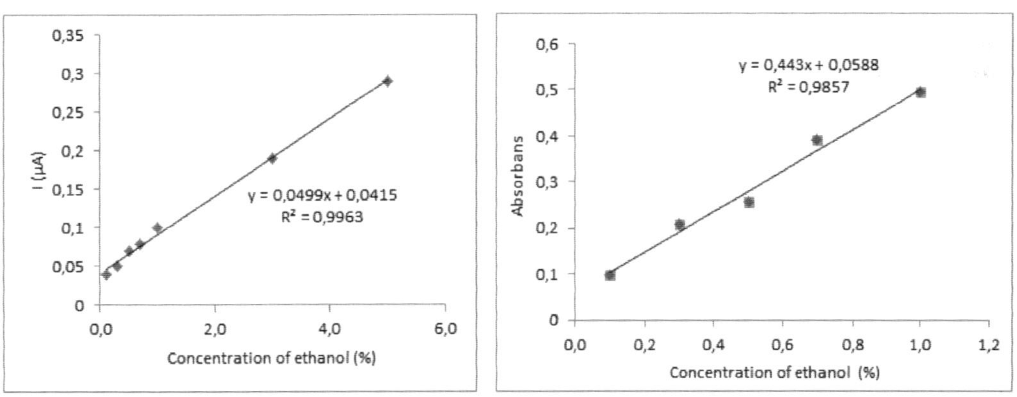

(a) (b)

Figure 4. Measuring ethanol: (a) biosensor; (b) UV-VIS.

4. Conclusion

Alcohol biosensor can be constructed using bacteria which produced alcohol dehydrogenase enzyme (ADH). From 6 tested *Bacillus* sp., the bacteria isolated from fermented rice grain (TST.A) produced the best result (2.34 μA at 0.545 V) as compared to other bacteria. At linearity test, there was current decreasement above 5% of ethanol. The used measurement range was 0.1-5.0%. Therefore, measuring alcohol using ADH producer microbe potentially can be developed more in the future.

References

[1] Rotariu L, Bala C and Magearu V 2004 *Analytica Chimica Acta* **513** 119
[2] Martinis R S, Ruzzene M A M and Martin C C S 2004 *J. Anal. Chem. Acta.* **522** 163
[3] Kitagawa Y, Kitabatake K, Suda M, Muramatsu H, Ataka T, Mori A, Tamiya E and Karube I 1991 *J. Ana.l Chem.* **63** 2391
[4] Liden H, Vijayakumar A R, Gorton L and Marko V G 1998 *J. Pharm. Biomed. Anal.* **17** 1111
[5] Gallignani M, Garrigues S and Delaguardia S 1993 *J. Of. Analyst.* **118** 1167
[6] Manurung R V, Kurniawan E D, Hidayat J, Aminuddin and Risdian C 2012 *J. Ilmiah Elite Elektro.* **3** 65
[7] Zhang L, Xu Z, Sun X and Dong S 2007 *J of Biosens. Bioelect.* **22** 1097
[8] Downey J M and Nieman T A 1992 *J Anal Chem.* **64** 261
[9] Jameison F, Sanchez R I, Dong L, Leland J K, Yost D and Martin M T 1996 *J Anal Chem.* **68** 1298
[10] Xu Z A, Guo Z H and Dong S J 2005 *Biosens. Bioelectron.* **21** 455
[11] Boyce J M 2000 *Inf. Conf. Host. Epidem.* **21** 438

Proposed Application of Fast Fourier Transform in Near Infra Red Based Non Invasive Blood Glucose Monitoring System

R P Jenie[1,2*], J Iskandar[2], A Kurniawan[2], E Rustami[2], H Syafutra[2], N M Nurdin[1], T Handoyo[4], J Prabowo[4], R Febryarto[4], M S K Rahayu[3], E Damayanthi[1], Rimbawan[1], D Sukandar[1], Y Suryana[4*], Irzaman[2], and H Alatas[2*]

[1]Department of Community Nutrition, Bogor Agricultural University, Indonesia
[2]Department of Physics, Bogor Agricultural University, Indonesia
[3]PT Tesena Inovindo, Indonesia
[4]Centre of Electronics Technology, Agency for The Assessment and Application of Technology, Indonesia

Email: qwerty.user1983@gmail.com , yaya.suryana@bppt.go.id, alatas@ipb.ac.id

Abstract. Worldwide emergence of glycaemic status related health disorders, such as diabetes and metabolic syndrome, is growing in alarming rate. The objective was to propose new methods for non invasive blood glucose level measurement system, based on implementation of Fast Fourier Transform methods. This was an initial-lab-scale-research. Data on non invasive blood glucose measurement are referred from Scopus, Medline, and Google Scholar, from 2011 until 2016, and was used as design references, combined with in house verification. System was developed in modular fashion, based on aforementioned compiled references. Several preliminary tests to understand relationship between LED and photo-diode responses have been done. Several references were used as non invasive blood glucose measurement tools design basis. Solution is developed in modular fashion. we have proven different sensor responses to water and glucose. Human test for non invasive blood glucose level measurement system is needed.

Keywords : fast Fourier transform, non invasive blood glucose monitoring system, proof of concept

1. Introduction

Worldwide emergence of glycaemic status related health disorders, such as diabetes and metabolic syndrome, is growing in alarming rate [1]. Worldwide current prevalence of aforementioned disorders is between 6.6 % [2] to as much as 30 % [3,4]. Indonesia reported 2.1 % prevalence of diabetes [5], but the real number possibly 5 times larger [1] due to different methods in diagnosing diabetes.

Blood glucose level measurement is integral part on glycaemic status management [6,7]. Current available methods include invasive and cause discomfort [8], need laboratory test [9], and expensive use of 1 time use biosensor [10,11]. Research for non invasive methods on blood glucose level measurement have been conducted [12,13], but the results were not yet satisfactory.

This research objective was to propose new methods for non invasive blood glucose level measurement system, based on implementation of Fast Fourier Transform methods, and to develop proof of concept to the solution.

2. Method

Data on non invasive blood glucose measurement are referred from Scopus, Medline, and Google Scholar, from 2011 until 2016, indexed using Docear [14] and Zotero [15] and was used as design references, combined with in house verification. System was developed in modular fashion, based on aforementioned compiled references. Several preliminary tests to understand relationship between LED and photo-diode responses has been done. All calculation needed done with Libre Office Calc [16] and R [17].

3. Result

3.1. Several references is used as non invasive blood glucose measurement tools design basis

The solution structure largely inspired by works of Farshad and Szabo [18,19]. The aforementioned tools are amperometric [20], micro controller based [21,22], and using piezo electric sensor [23]. They were both portable [24] and wearable [25]. It implemented machine learning algorithm [26,27]. Sensor works based on modulated spectrophotometry [28,29] in the Near Infra Red (NIR) range [30]. Tools were able to monitor blood glucose continuously [31,32].

NIR Light Emitting Diode (LED) and Photo-diode in 1000 nm – 1700 nm wavelength range are used, provided by Thorlabs [33,34]. Wavelength selected based on works by Smith and Goodarzi [35,36], which was confirmed by our own unpublished research. NIR LED and Photo-diode mounted on 3 dimensionally printed [37,38] probe, modulated by Raspberry Pi [39] through Inter-Integrated Circuit (I^2C) protocol [40]. Spectrum data are transformed using Fast Fourier Transform (FFT) [41] and will be inferenced using Cascade Correlation Artificial Neural Network [42]. Results are stored in SQLite database [43]. Whole system developed with Qt Software Development Kit (SDK) [44,45].

3.2. Solution is developed in modular fashion

System was developed in 5 independent modules (Figure 1). Sensor module / probe was printed from poly-lactic acid (PLA) [46,47], housing a pair of 1 or 2 NIR LED and a photo-diode. The LED is modulated in serial, in a period of certain seconds. The resultant spectrum data are pushed to repository module.

Analogue to Digital Converter (ADC) module implemented MCP 3424 [48] was used to enable 14 bit conversion at 60 measurement / s data rate. LED modulated using pulse width modulation (PWM) [49], and controlled with different intensity between each measurement period. It was attached to CPU module by data wire.

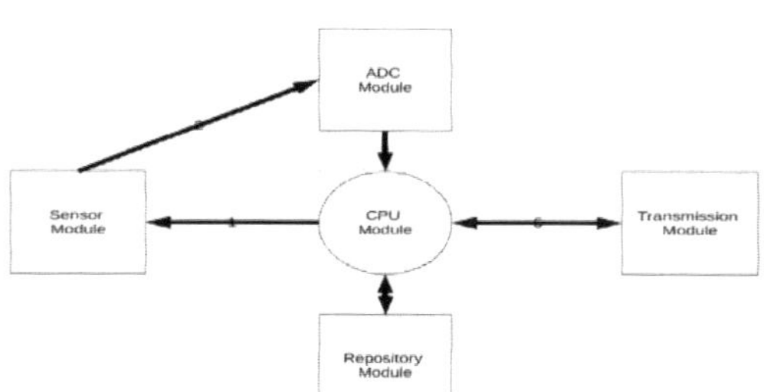

Figure 1. Non Invasive Blood Glucose Measurement System modular structure.

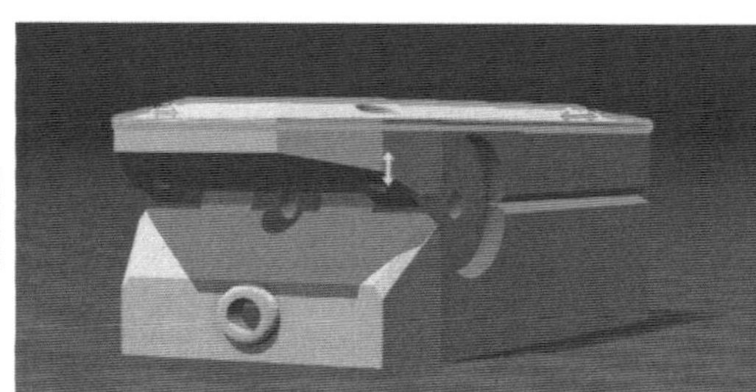

Figure 2. Non Invasive Blood Glucose Measurement probe.

Repository module was made in 2 parts. 1st part was implemented in each devices, as SQLite [43] database, which reside within SD Card. It stored both measurements spectrum and result of measurements, including measurement context such as user, measurement site, location, time, and device used.

Central Processing Unit (CPU) module houses the application built with Qt SDK [44,45]. The application sample the measurement in several seconds periods, and process the output of ADC Module using open source FFTW [41] class. The output would be fed to inferencing module implementing cascade correlation artificial neural network (CCANN) [42], to transform FFTW output into blood glucose level measurement values.

Transmission modules is a conglomerate of modules essential for serving human blood glucose measurement values, whether for monitor display for user view, or to be wireless transmitted over networking modules such as wireless Fidelity (Wi-Fi) or Blue-tooth modules.

Sensor and ADC module are housed together in probe module (Figure 2), and reminder modules are housed together in main module (Figure).

3.3. In preliminary test, we have proven different sensor responses to water and glucose.

We have done our preliminary testing against pure glucose and pure water (Figure 3). We have proven that the photo-diode gave different reading when given different LED intensity (Figure 4). Furthermore, the photo-diode gave different responses to different substances.

Research to find relationship between glucose meter reading, glucose water concentration, and photo-diode responses are still on going.

4. Discussion

4.1. This is the first research that propose Fast Fourier Transform as data transformation methods to measure blood glucose level

To our knowledge, this is the 1st time for Fast Fourier Transform based methods that is used for blood glucose level measurement spectrum transformation. The research combines strength of previous researches, mainly on the usage of modular spectrophotometry.

This research also confirm possibilities in non invasive blood glucose monitoring system, including wavelength used and one possible probe and case shape. We also proved that the system can be developed modularly, enabling different rate of development and reusing same core system to do different biomarker measurement.

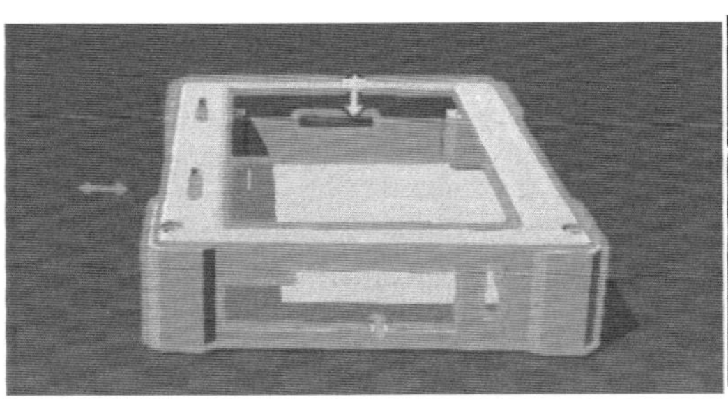

Figure 3. Non Invasive Blood Glucose Measurement outer case.

Figure 4. Non Invasive Blood Glucose Measurement Tools trial. Cases and Probe not shown. LED and photo-diode attached to dark tube for simulated response testing, which filled with substance cuvette.

4.2. While several methods exist, none of non invasive blood glucose level measurement methods are proven

Non invasive blood glucose measurement is a relatively new fields. Until the time when this manuscript written, only 13 trials indexed at ClinicalTrial.gov [50] running, most notable, trial by Lamberg and Nimri [51,52], and none of them have published their results. Furthermore, none of aforementioned non invasive glucose methods have been clinically trialled.

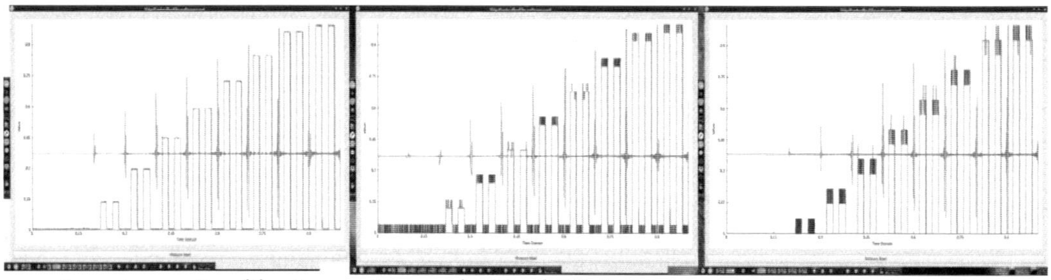

a) empty cuvette without substance **b)** pure double distilled water **c)** pure glucose

Figure 5. Different responses from photo-diode to different substances. In the photograph is the actual application view for testing. All chart is mapped to [0, 1] values. Real photo-diode reading in blue bar shaped lines, and Fast Fourier Transformed values in red fish bone shaped lines.

Even then, our internal testing show some promises. We have proven different photo-diode responses to pure water and pure glucose, which open possibilities of non invasive blood glucose measurement.

4.3. Best wavelength, measurement site, and probe shape still being researched

There has been a dispute for proper wavelength to detect and measure blood glucose level. Several values exist, from as low as 800 nm [35], to as high as 2500 nm [53]. We have confirmed several wavelength in our unpublished research, but only to glucose and water solution. Ethical clearance pending, so we not yet able to conduct human test for the solution. Furthermore, LED and Photo-diode upward of 1700 nm is too expensive at the time of this writing, rendering procurement and testing of those wavelength not yet possible.

While several body nodes have been used for non invasive blood glucose measurement in previous researches, commonly in fingers, none actually comparing between several possible nodes, like each finger or each ear. We cannot use nodes for finger prick based glucose meter sites as reference, because we did not face similar risks. This imply that we have yet to test optimal shape of measurement probe.

4.4. The proposed solution is a compromise between component capabilities

There are 5001 different integer values between 0.0 mg / dl to 500.0 mg /dl, assuming 1 value after decimal, which is the extreme range of human blood glucose level. That make usage of 13 bit [0 to 8191] ADC necessary. MCP 3424 [48] able to generate 14 bit data [0 to 16383] at 60 datum/s, but that too slow to flood simple serial port, making usage of Arduino [54,55] as intermediary is impossible, hence why direct I^2C control is used instead.

Another problem, Raspberry Pi [39] only have a pair of functional I^2C pins, hence made usage of Digital to Analogue Converter (DAC) module in tandem with ADC module impossible; they shall interference with each other. That means, the I^2C can only used for ADC, and the LED is controlled using PWM.

4.5. This proposed system can change the way glycaemic status treatment is conducted, but human tests is needed for confirmation.

Traditional glucose meter finger prick based blood glucose measurement can only measure as fast as 4 measurement/hour, mainly due to risk of inflammation and discomfort of patient. This system by concept can measure up to 60 measurements / hour, and wearable, enabling continuous telemetry.

We have yet to test whether different body composition on different body site, muscle, water, skeleton, blood composition, and others, and different medical condition, like hypo hydration, can change blood glucose level reading. Furthermore, We have yet to find proper measurement calibration for human blood glucose level in different body sites.

Acknowledgements

This is a join research between Department of Physics and Department of Human Nutrition, Bogor Agricultural University, PT Tesena Inovindo, and Centre of Electronics Technology, Agency for The Assessment and Application of Technology.

The research is funded by Program Insentif Riset Sistem Inovasi Nasional, as part of Research Consortium on Non Invasive HbA1c Measurement System.

RPJ, HA, I, ED, and R, design the whole study. RPJ design and implemented the software and analytical methods and running the whole research. RPJ, HS, HA, and R design the data transformation and inference module. JI, AK, ER, and I design the probe and electronics parts. RPJ and NMN handling system testing. TH, JP, RF, and YS handling electronics printing. MSKR provide information regarding proper medical testing equipment. DS observe the study. This manuscript prepared by RPJ and HA.

We declare non competing interests.

References

[1] Kanchi S, Sharma D, Bisetty K and Nuthalapati V N 2015 *J Env. Anal Chem* **2** e111

[2] Zhao X 2014 *J. Anesth. Clin. Res.* **5**

[3] Dutra E S, de Carvalho K M, Miyazaki É, Hamann E M and Ito M K 2012 *Diabetol Metab Syndr* **4** 20

[4] Vidigal F, Ribeiro A, Babio N, Salas-Salvadó J and Bressan J 2015 *Diabetol. Metab. Syndr.* **7** 6

[5] Kesehatan D and RI K K 2013 Riset Kesehatan Dasar *Jkt. Badan Penelit. Dan Pengemb. Kesehat. Dep. Kesehat. Repub. Indones.*

[6] Finfer S, Wernerman J, Preiser J-C, Cass T, Desaive T, Hovorka R, Joseph J I, Kosiborod M, Krinsley J, Mackenzie I and others 2013 *Crit Care* **17** 229

[7] Braga J R G, Ramos A C B and Queiroz A A A 2012 Using artificial neural nets to Hemo metabolites identification *e-Health Networking, Applications and Services (Healthcom), 2012 IEEE 14th International Conference on* (IEEE) pp 124–131

[8] Salacinski A J, Alford M, Drevets K, Hart S and Hunt B E 2014 *J. Diabetes Sci. Technol.* **8** 95–99

[9] Budak Y U, Huysal K and Polat M 2012 *Anesthesiol.* **12** 17

[10] Yeaw J, Lee W C, Aagren M and Christensen T 2012 *J Manag Care Pharm* **18** 21–32

[11] Turner A P F 2013 *Chem. Soc. Rev.* **42** 3184

[12] Govada A, Renumadhavi C and Ramesh K 2014 *Int. J. Adv. Res. Comput. Commun. Eng.* **3** 5122–5125

[13] Abdulhalim I S S 2015 *J. Biosens. Bioelectron.* **6**

[14] Beel J, Gipp B, Langer S and Genzmehr M 2011 Docear: An academic literature suite for searching, organizing and creating academic literature *Proceedings of the 11th annual international ACM/IEEE joint conference on Digital libraries* (ACM) pp 465–466

[15] Harding K 2014 Zotero *J. Can. Health Libr. Assoc. Assoc. Bibl. Santé Can.* **34** 41–43

[16] Smith J A, Weber J H and Fox M J 2013 Libre Office 4.0 The document foundation Calc Guide working with spreadsheets *Libre Off. 40 Doc. Found. Calc Guide Work. Spreadsheets*

[17] Dixit Y, Cama R, Sullivan C, Alvarez Jubete L and Ktenioudaki A 2014 *J NIR News* **25** 15
[18] Farshad K, Islam M, Leech J, Lin C and Chrostowski L 2006 Preventing Blood Glucose Level from Flying High
[19] Szabo I, Gyuk P and Vassanyi I 2014 Lifestyle Log Based Blood Glucose Level Prediction for Outpatient Care *Accepted as full paper for Int. Conf. e^{TEL}EMED, Barcelona, Spain*
[20] Cardosi M and Liu Z 2012 Amperometric Glucose Sensors for Whole Blood Measurement Based on Dehydrogenase Enzymes *Dehydrogenases* ed R A Canuto (InTech)
[21] Taneva L, Hristov V and Ibrahim F 2013 *Int J Open Probl. Compt Math* **6**
[22] Meshram N D and Dahikar P B 2015 *International Symposium on Ultrasonics* **22**
[23] Pattnaik P, Pradhan S K, Kamilla S K and Das D P 2012 Studies of Lead Free Piezo-Electric Materials Based Ultrasonic MEMS Model for Bio Sensor *Proceeding of COMSOL Conference*
[24] Athanasiou L V, Tsokana C N, Saridomichelakis M N, A$þeta$ανασ$\acuteιota$ου Λ, Τσοκαν$\acute\alpha$ K N and Σαρ$ιota$δομ$ιota$χελ$\acute\alpha$κης M N 2014 *J. Hell. Vet. Med. Soc.* **65** 273–288
[25] Chu M K, Iguchi S, Takahashi D, Arakawa T, Kudo H and Mitsubayashi K 2009 Wearable biosensor for monitoring tear glucose on rabbit eye as novel device of body sensor network *Intelligent Sensors, Sensor Networks and Information Processing (ISSNIP), 2009 5th International Conference on* (IEEE) pp 191–194
[26] Allam F, Nossair Z, Gomma H, Ibrahim I and Abdelsalam M 2012 *Int. J. Intell. Syst. Appl.* **4** 58–71
[27] Dicker L, Sun T, Zhang C-H, Keenan D B and Shepp L 2013 Continuous blood glucose monitoring: A Bayes-hidden Markov approach *Stat. Sin.*
[28] Chowdhury M K, Srivastava A, Sharma N and Sharma S 2014 *IOSR J. Electr. Electron. Eng.* **9** 34–41
[29] Srivastava A, Chowdhury M K, Sharma S and Sharma N 2014 *Adv. Bioresearch*
[30] Srivastava A, Chowdhury M K, Sharma S and Sharma N 2013 *Int. J. Adv. Eng. Technol. IJAET* **6** 1097–1108
[31] Cameron F, Buckingham B A, Wilson D M and Bequette B W 2014 Sensitivity Analysis of a Predictive Pump Suspension System to Treat People with Type 1 Diabetes *World Congress* vol 19pp 243–248
[32] Christiansen M, Bailey T, Watkins E, Liljenquist D, Price D, Nakamura K, Boock R and Peyser T 2013 *Diabetes Technol. Ther.* **15** 881–888
[33] Thorlabs, Inc. 2016 Unmounted LEDs
[34] Thorlabs, Inc. 2016 Unmounted Photodiodes
[35] Smith J L 2011 *Last Accessed May* **2**
[36] Goodarzi M, Sharma S, Ramon H and Saeys W 2015 *Trends Anal. Chem.* **67** 147–58
[37] Ciobanu O, Xu W and Ciobanu G 2013 *Fiability Durab.* **2013**
[38] Kulkarni A and Bhargava M R 2015 3D Printing
[39] Anwaar W and Shah M A 2015 *Energy* **4**
[40] Semiconductors P 2000 The I2C-bus specification *Philips Semicond.* **9397** 954
[41] fftw.org 2015 FFTW Home Page
[42] FANN 2015 Fast Artificial Neural Network Library (FANN)
[43] SQLite 2016 SQLite Home Page
[44] Akinkuolie B B, Lin C-F and Yuan S-M 2011 A cross-platform mobile learning system using QT SDK Framework *Genetic and Evolutionary Computing (ICGEC), 2011 Fifth International Conference on* (IEEE) pp 163–167
[45] Qt 2015 Qt - Download Open Source Step 3
[46] Mai F, Palatov K, Bilotti E and Peijs T 2012 Self-reinforced polylactic acid (PLA) biocomposites with degradation sensing *Proceedings of ECCM-15 Conference. Venice*
[47] Harikrishna M, Narasimhulu P, Ramu J, Shaikshavli P, Prasad V H, Sekhar K R, Kumar M A,

Venkateswarlu M, Sagar P and Venkataramana B 2016 Development of Poly-Lactic-Acid (PLA) Composite Films Filled with Aerosil/Wollastonite on Characterization
[48] Gupta A K 2006 *Design techniques for low noise and high speed A/D converters* (Texas A&M University)
[49] Baruah A D and Patel R M 2015 *Development* **2**
[50] Wong E and Williams R J 2012 *Regul. Focus*
[51] Lamberg H 2014 Measuring and Monitoring Blood Glucose Levels Utilizing Non-invasive Blood Glucose Monitoring Device - Full Text View - ClinicalTrials.gov
[52] Nimri R 2012 Assessing Continuous Non- Invasive Blood Glucose Sensor- Physical Logic AG - Full Text View - ClinicalTrials.gov
[53] Krushinitskaya O 2012 *Osmotic sensor for blood glucose monitoring applications* (Vestfold University College)
[54] El-Hamid A S A, Fetohi A, Amin R S and Hameed R A 2015 *Int. J. Softw. Hardw. Res. Eng.*
[55] Arduino 2015 Genuino Starter Kit [English] Code: GKX00007

Electrical Resistivity Tomography Using Wenner β-Schlumberger Configuration for Anomaly Detection in The Soil

Y Pebriyanto[1*], K Dahlan[1] and Y W Sari[1]

[1]Department of Physics, Bogor Agricultural University, Bogor, Indonesia
Jl. Meranti, kampus IPB Darmaga Bogor, Indonesia, 16680

Email: yunuspebriyanto@yahoo.com

Abstract. In the subsurface exploration investigations there are many methods used, one of them is Electrical Resistivity Tomography (ERT). ERT method is able to measure the electrical properties of the material below the earth surface based on the value of the resistivity of the material by injecting electric current and measure the potential at the surface. Based on the data obtained then will be inputted into RES2DINV software for final processing of 2D image. This research has been created by testing 2 configurations Wenner-Schlumberger and Wenner β-Schlumberger for detecting anomalies in homogeneous soil. A wooden box containing homogeneous soil is used for the test. Three anomalies (wood, stone, and wet soil) were placed in different positions and the variation of resistivity was detected. We found that the Wenner β-Schlumberger configuration results in a smaller resistivity value error than the Wenner-Schlumberger configurations.

Keyword: ERT, Wenner β-Schlumberger, anomaly detection

1. Introduction

In the field of imaging method, Electrical Resistivity Tomography (ERT) is a method that is widely used in exploration field especially subsurface exploration [1,2]. ERT can measure the electrical properties of the material below the surface that are based on the value of resistivity of the material by injecting electric current and measure the potential at the surface [3,4,5]. Improvement in their imaging methods leads to more popular use in the field of subsurface exploration. In addition, the popularity is also due to the easy-to-carry equipment, easy and cheap to operate, and the measurement of data accuracy is reliable [6,7,10].

In the application there are various configurations ERT methods used such as Wenner, Schlumberger, Dipole-Dipole, Pol-Dipole and incorporation of two types of configurations (Wenner-Schlumberger) [8,9]. This research has been created to develop the latest configurations. The purpose of this development is to get the image result for more efective than latest configurations.

Based from above, there is exist the incorporation of two configuration as Wenner and Schlumberger that become Wenner-Schlumberger. The inconporation of these configuration is to get maximal of the image result. But in this research, has been created the development from Wenner-Schlumberger configuration that was proven to produce images that are more effective and accurate image. This configuration is Wenner β-Schlumberger configuration. The development of this configuration is make a modification from the Wenner configuration becomes Wenner β. actually Wenner configuration has 3 types as Wenner α, Wenner β and Wenner γ. Wenner configuration used previously was Wenner α which has weaknesses that are less sensitive in detecting non-homogeneity of horizontal layers. By the modification of the wenner configuration and incorporated with Schlumberger (Wenner β-Schlumberger) was proven to produce images that are more effective and accurate image than Wenner-Schlumberger.

2. Method

2.1. Geoelectric Resisitivity of Method

Geoelectric method is an imaging method that can measure the electrical properties of the material below the surface that are based on the value of the resistivity of the material by detecting from the surface [11]. This method is performed by injecting electric current to a pairs electrodes and measuring the potential data to the other pairs electrodes. [12,13]. From the results of current and potential measurements then can be obtained resistance value each layer below the measuring point [14].

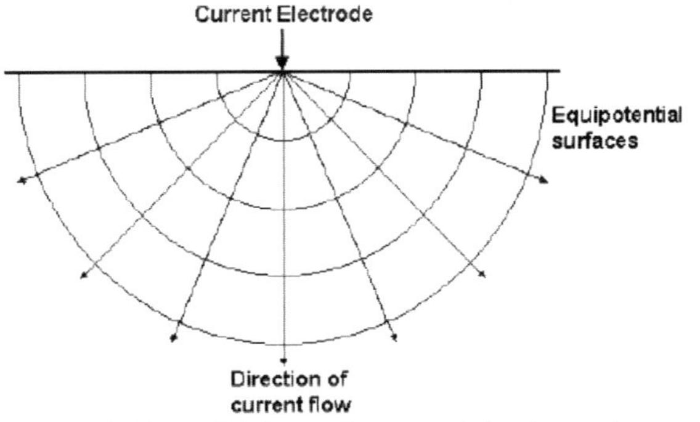

Figure 1. Potential Around current injecting point on Earth's surface. [8]

The distribution of current is spread out to hemispherical surface with an area $2\pi r^2$, so that:

$$J = \sigma E \tag{1}$$

$$J = \frac{I}{A};\ \sigma = \frac{1}{\rho};\ E = \frac{V}{r}$$

$$V_r = \frac{\rho I}{2\pi r} \tag{2}$$

ρ resistivity of the surface (Ωm), V is electric potential in distance r (Volt), I is electric current supply (A), J electric current density (A/m^2), σ is conductivity of the system (Ω^{-1}m^{-1}), and E is electric field (V/m).

2.2. Wenner-Schlumberger Configuration

Wenner-Schlumberger configuration is the configuration of the system which has constant spacing rules. For this configuration comparison distance between the current electrode C_1 and potential electrode P_1 (C_1-P_1) is placed at a distance na and so does the distance between C_2-P_1 electrodes. While the P_1-P_2 electrodes is placed at a distance a (can be seen in figure 2). Determining process of resistivity using four pieces of electrodes placed in a straight line [10].

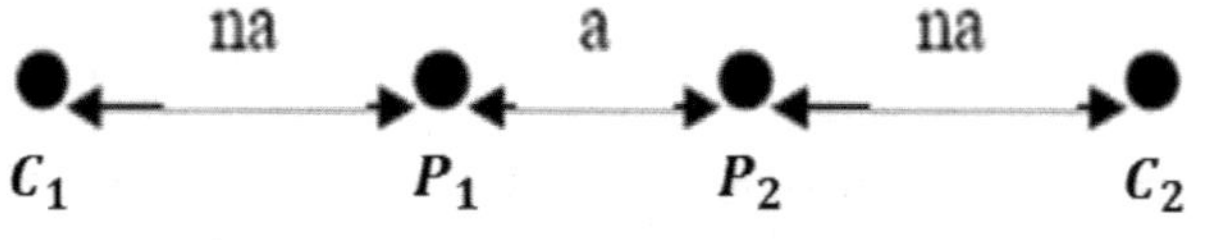

Figure 2. Wenner-Schlumberger configuration.

$$V=\frac{\rho I}{2\pi}\left(\frac{1}{R_1}-\frac{1}{R_2}\right) \quad (3)$$

By subtituting the value of electrodes distance, so that:

$$\Delta V=V_1-V_2=\frac{\rho I}{\pi}\left(\frac{1}{na(n+1)}\right)$$

$$\Delta V=\frac{\rho I}{\pi}\left(\frac{1}{na(n+1)}\right) \quad (4)$$

$$\Delta V=\frac{\rho I}{K}$$

$$K=\pi na(n+1) \quad (5)$$

ρ is resistivity (Ωm), ΔV is electric potential betweet electrodes P_1-P_2 (Volt), I electric current between electrodes C_1-C_2, dan K geometric factor (m).

2.3. Wenner β-Schlumberger Configuration

Wenner β-Schlumberger configuration is the configuration which has spacing pattern rules are almost same as the Wenner-Schlumberger configuration. but this configuration has the distinction on the laying of the current electrodes and potential electrodes. The electrodes C_2-C_1 (current electrodes) and electrode P_1-P_2 (potential electrodes) placed respectively at the same distance of na. While the electrode current C_1 and potential electrode P_1 (C1-P1) placed at distance a (as shown in figure 3).

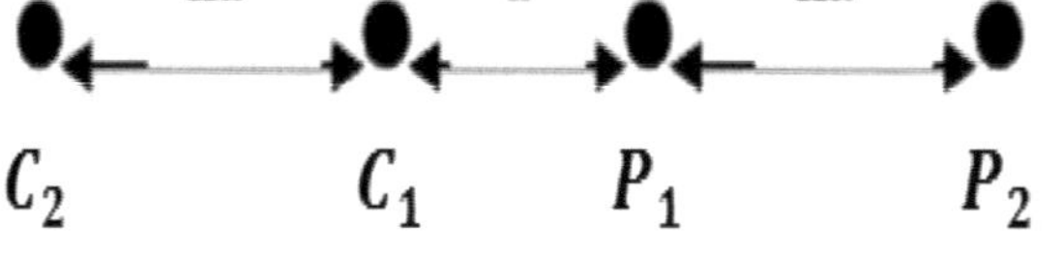

Figure 3. Wenner β-Schlumberger configuration.

$$V=\frac{\rho I}{2\pi}\left(\frac{1}{R_1}-\frac{1}{R_2}\right)$$

By subtituting the value of electrodes distance,

$$\Delta V=V_1-V_2$$

$$\Delta V=\frac{\rho I}{\pi}\left(\frac{n^2}{a(n+1)(2n+1)}\right) \quad (6)$$

$$\Delta V = \frac{\rho I}{K}$$
$$K\left(\frac{a(n+1)(2n+1)}{n^2}\right) \quad (7)$$

ρ is resistivity (Ωm), ΔV is electric potential betweet electrodes P_1-P_2 (Volt), I electric current between electrodes C_1-C_2, dan K geometric factor (m).

2.4. Procedure

This research is a laboratory scale that has been created in Laboratory of Electronic, Department of Physics Bogor Agricultural University. In the test, will be made a wooden box as the system (size 120 cm x 30 cm x 20 cm) that containing homogeneous soil. then, will be used two configurations (Wenner-Schlumberger dan Wenner β-Schlumberger) by measuring range is 100 cm of each configurations and the smallest spacing value is 4 cm (a = 4 cm). This measurement have been created by moving the potential electrode at the constant current electrodes, then the electrode displacement currents on the space n followed by the displacement of potential electrode along the system up on that last point of measurement in the system.

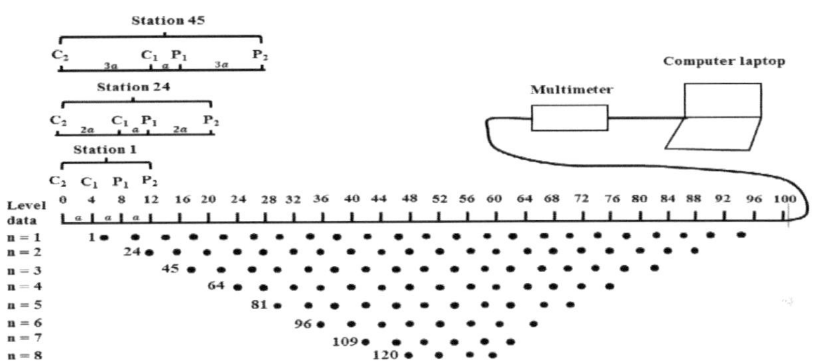

Figure 4. The procedure of measurement. [8]

3. Result and Discussion

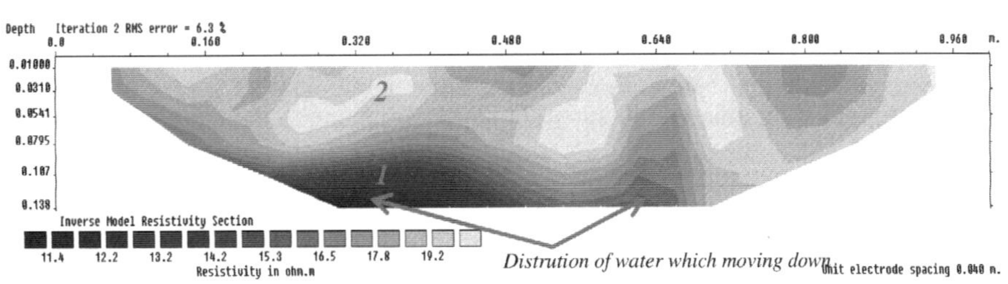

Figure 5. Result of the 2D image without anomalies by Wenner β-Schlumberger configuration.

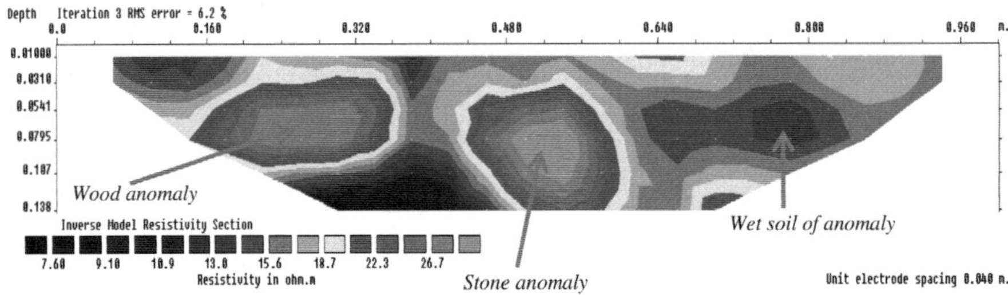

Figure 6. Result of the 2D image with anomalies by Wenner β-Schlumberger configuration.

Figure 5 describes the 2D image of system without anomalies with resistivity value measured in the range of 11,4 Ωm-19,2 Ωm. Based on the figure can be seen that the bottom of the system detected a small resistivity, it indicates the distribution of water seeping into the bottom of the system. It was proved by testing the water content in the soil by taking soil samples of 10 grams for each in points (point 1 and 2). The test has been done by heating the samples into oven at 105°C for 24 hours and then measure its mass. The result obtained that sample 1 has 28% of water content, while sample 2 is 21% water content. Based on these results can be seen that sample 1 has a high water distribution than the sample 2.

Figure 6 describes the 2D image of system with anomalies. The results obtained that the range of resistivity values measured between 7,60 Ωm-26,7 Ωm. This Changes was happen because the anomalies exist in the system. Clearly, the wood anomaly detected in position range 18 cm-34 cm with range of resistivity value between 18,7 Ωm-26,7 Ωm, the stone anomaly in the position range 44 cm-60 cm with range of resistivity values between 18,7 Ωm-31,1 Ωm, while the wet soil anomaly in the position range 72 cm-84 cm with range of resistivity values between 9,1 Ωm-15,6 Ωm. Based on these results can be seen that the differences of all anomalies resistivity value can be detected as wet soil anomaly that has the smallest resistivity values, followed by wood anomaly and stone anomaly that has the greatest resistivity values. Surely in this study also contained an error value in the detection of 6.3% for the system without anomalies and 6.2% for systems with anomalies. This error value will be a references for researchers in the stage of analyzing the data and accuracy of detection of the position and resistivity.

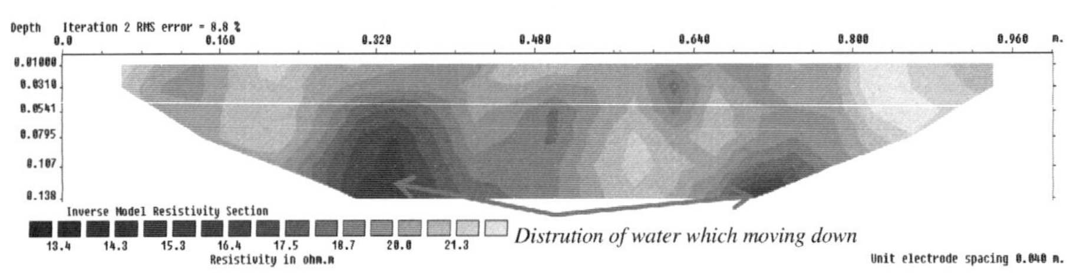

Figure 7. Result of the 2D image without anomalies by Wenner-Schlumberger configuration.

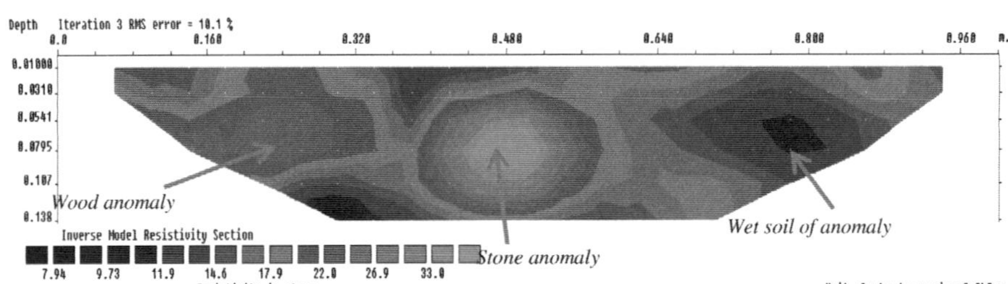

Figure 8. Result of the 2D image with anomalies by Wenner-Schlumberger configuration.

Figure 7 describes the 2D image of system without anomalies by Wenner-Schlumberger cinfiguration with resistivity value measured in the range of 14,3 Ωm-22,8 Ωm. Based on the figure can be seen that the bottom of the system also detected a small resistivity as well as the detection of Wenner β-Schlumberger configuration which indicates the distribution of water seeping into the bottom of the system. While in figure 8, describes that the results of the 2D image of the system with anomalies. The results obtained that the range of resistivity value measured between 7,94 Ωm-33,0 Ωm, the same as the previous explanation above that these changes was occured because of three anomalies in the system. Clearly, the wood anomaly detected in position range 12 cm-30 cm with the range of resistivity value 17,9 Ωm -26,9 Ωm, the stone anomaly in the position range 42 cm-54 cm 17,9 Ωm-33,0 Ωm, while the wet soil anomaly in the position range 70 cm-86 cm with the range of resistivity value 9.73 Ωm-14.6 Ωm. Surely in this study also contained an error value in the detection of 8,8% for the system without anomalies and 10,1% for systems with anomalies

4. Conclusion

Based on the results obtained that all of anomalies (wood, stone, and wet soil anomaly) can be detected by both configuration (Wenner-Schlumberger and Wenner β-Schlumberger), but have the error levels were different. Wenner β-Schlumberger configuration has an error value is relatively low at 6.2% than using Wenner-Schlumberger configuration that has error value 10.1%. Based on these differences, it can be concluded that the developeding of this configuration (Wenner β-Schlumberger), was proven to maximize the results of the image.

References

[1] Osazuwa I and Chii E C 2010 *International Journal of Physical Sciences* **5** 238-245
[2] Bery A A *et al* 2012 *Electronic Journal of Geotechnical Engineering* **17** 3223-3337
[3] Metwaly M and Alfauzan F 2013 *Geoscience Frontiers* **4** 469-476
[4] Bernard J and Valla P 1991 *Geoexploration, Elsevier* **1** 81-91
[5] Yadav G and Singh S 2007 *Journal of Applied Geophysics, Elsevier* **62** 301-312
[6] Loke M H and Dahlin T 2002 *Journal of Applied Geophysics, Elsevier* **49** 149-162
[7] Harikumar R *et al* 2013 *International Journal of Soft Computing and Engineering (IJSCE)*, ISSN **3** 2231-2307
[8] Loke M H 2002 Tutorial: 2D and 3D Electrical Imaging Survey. Malaysia: *Geotomo Software*
[9] Loke M H 2002 *RES2DINV* ver. 3.59 for Windows XP/Vista/7 Rapid 2-D Resistivity & IP inversion using the least-squares method. Malaysia: *Geotomo Software*
[10] Priambodo I C *et al* 2011 *Bulletin Vulkanologi dan Bencana Geologi* (Indonesia) vol 6 pp 1-10
[11] Ibrahim K O *et al* 2012 *Journal of Environment, ISSN* **1** 21-25
[12] Al-Garni M 2009 *Jordan Journal of Civil Engineering* **3** 118-136.
[13] Clark J and Page R 2011 *Journal of Water Resource and Protection* **3** 768-780.
[14] Talwani M and Kesingger W 2006 *Exploration Geophysics* (New York: *Rice University*)

A robustness study of student-*t* distributions in regression models with application to infant birth weight data in Indonesia

A Ubaidillah[1,3], K A Notodiputro[1]* , A Kurnia[1], A Fitrianto[1,4] and I W Mangku[2]

[1]Department of Statistics, Faculty of Mathematics and Natural Sciences, Bogor Agricultural University, Bogor, Indonesia
[2]Department of Mathematics, Faculty of Mathematics and Natural Sciences, Bogor Agricultural University, Bogor, Indonesia
[3]Institute of Statistics, Jakarta, Indonesia
[4]Department of Mathematics, Faculty of Science, Universiti Putra Malaysia, 43400 UPM Serdang, Selangor, Malaysia

Email: khairilnotodiputro@gmail.com

Abstract. In regression models, the use of least squares method may not appropriate in modelling the data containing outliers. Many robust statistical methods have been developed to handle such a problem. Lange *et al.* [1] developed robust models based on *t* distributions and using M-estimation approaches. In this recent article we evaluate the performance of M-estimation as well as investigated the robustness of *t* distribution models in linear regression by means of simulation. The models are then applied to infant birth-weight data in Indonesia. We show that the *t* distribution models with small degrees of freedoms have produced better estimates from perspectives of their performance and robustness when compared to other estimates.

1. Introduction

Classical methods in regression analysis, such as least squares regression, are widely used by researchers in many disciplines. In the usage of the least squares regression, the assumptions such as normality and homoscedasticity must be satisfied. Unfortunately, these assumptions are rarely met when analyzing real data. Micceri [2] examined 440 large datasets from the psychological and educational literature. None of the data were normally distributed. The use of least squares regression with violated assumptions can result in the inaccurate computation of p-values, effect sizes, and confidence intervals. This may lead to substantive errors in the interpretation of data.

Violation of the assumptions in the least squares regression can be due to the presence of outliers. The data values which are extremely different to the majority of the dataset are called outliers. Outliers may be correct observations, but they should always be checked for transcription errors. Since 1960, many robust and resistant methods have been developed to be less sensitive to outliers, i.e. robust with respect to outliers and stable with respect to small deviations from the assumed parametric model.

It is well-known that to screen the data, to remove the outliers and then to apply classical inferential procedures in many cases are not obvious. First, in multivariate or highly structured data, it can be difficult to identify outliers and influential observations. Second, it could be better to down-weight uncertain observations rather than to reject them, although we may wish to reject completely wrong observations. Moreover, rejecting outliers reduces the sample size and could affect the data distribution. The variances could also be underestimated from the cleaned data. Finally, empirical evidence shows that good robust procedures behave quite better than techniques based on the rejection of outliers [3].

Robust statistical procedures focus in estimation procedures, testing hypotheses and in regression models. It can proceed in two ways. First, analyst could design estimators so that a pre-selected behaviour of the influence function is achieved. There exist a great variety of approaches toward the robustness method. Among these, procedures based on M-estimators and high breakdown point estimators employ an important and complementary role. For the references, Huber [4], Hampel *et al.* [5] and Staudte and Sheather [6] are the main theoretical; see also Susanti *et al.* [7] for more practical one. Second, by replacing estimators that are optimal under the assumption of a normal distribution with estimators that are optimal for, or at least derived for, other distributions: for example using the t-distribution or with a mixture of two or more distributions. Lange *et al.* [1] proposed robust statistical modeling using the t ditsribution for both linear and nonlinear regression models.

This paper presents a comparative study of robustness in regression models. Section 2 reviews on robust method of estimation, and section 3 discusses robust regression using student-t distribution. Section 4 and 5 present the evaluation and application of robust methods. Simulation and implementation to real dataset are presented in these section. Section 6 discusses comparation with other study, advantage and interesting results, and section 7 states conclusions.

2. Review on robust method of estimation

Robust regression is an important tool for analyzing data affected by outliers so that the resulting models are stout against outliers. An anomalous observations may be dealt with by a preliminary of the data screening, but this is not possible with influential observations which can only be detected once the model has been fitted. In the light of the amount of data available nowadays and the automated procedures used to analyze them, robust techniques may be preferable as they automatically take possible deviations into account.

In regression problem , there are two possible sources of errors, the observations y_i, and the corresponding row vector of p regressors $\mathbf{x}_i$. Most robust methods in regression only consider the first. Consider a regression problem with n cases $(y_i, \mathbf{x}_i)$ from the model

$$y_i = \mathbf{x}_i^T \boldsymbol{\beta} + e_i,\ i = 1,\dots,n, \tag{1}$$

for a p-variate row vector $\mathbf{x}$.

2.1. M-estimation

M-estimation is one of the robust regression estimation methods. The letter M indicates an estimation of the maximum likelihood type. Let $\hat{\boldsymbol{\beta}} = \boldsymbol{\beta}_n(\mathbf{x}_1,\dots,\mathbf{x}_n)$ then

$$E\left(\boldsymbol{\beta}_n(\mathbf{x}_1,\dots,\mathbf{x}_n)\right) = \boldsymbol{\beta}. \tag{2}$$

Equation (2) shows that $\hat{\boldsymbol{\beta}}$ is unbiased and has minimum variance, so M-estimator has the smallest variance estimator.

$$\operatorname{var}\left(\hat{\hat{\beta}}_j\right) \geq \frac{\left[\hat{\beta}_j^{T}\right]^2}{nE\left(\dfrac{d}{d\beta}\ln f\left(x_{ij},\beta_j\right)\right)^2},\ j = 1,\dots,p, \tag{3}$$

where $\hat{\hat{\beta}}_j$ is other linear and unbiased estimator for β_j.

M-estimation is an extension of the maximum likelihood estimate method that is possible to eliminate some of the data. M-estimation principle is to minimize the function $\rho(.)$:

$$\hat{\beta}_j^M = \min_{\beta_j} \sum_{i=1}^{n} \rho\left(y_i - \sum_{j=0}^{p} x_{ij}\beta_j \right). \tag{4}$$

To obtain (4), we have to solve

$$\hat{\beta}_j^M = \min_{\beta_j} \sum_{i=1}^{n} \rho(u_i) = \min_{\beta_j} \sum_{i=1}^{n} \rho\left(\frac{e_i}{\sigma}\right) = \min_{\beta_j} \sum_{i=1}^{n} \rho\left(\frac{1}{\sigma}\left(y_i - \sum_{j=0}^{p} x_{ij}\beta_j \right)\right)$$

and we set estimator for σ :

$$\hat{\sigma} = \frac{MAD}{0.6745} = \frac{\text{median}\left|e_i - \text{median}\left(e_i\right)\right|}{0.6745}.$$

Furthermore, we get for the first partial derivative $\hat{\beta}_j^M$ to β_j so that

$$\sum_{i=1}^{n} x_{ij}\psi\left(\frac{1}{\hat{\sigma}}\left(y_i - \sum_{j=0}^{p} x_{ij}\beta_j \right)\right) = 0, \tag{5}$$

where $\psi = \rho'$, x_{ij} is i-th observation on the j-th independent variable and $x_{i0}=1$.

The Huber function finds the Huber M-estimator of a location parameter with the scale parameter estimated with the MAD. In this case

$$\psi_k(x) = \max\left(-k, \min(k, x)\right), \tag{6}$$

where k is a constant the user specifies. Its limit as $k \to 0$ is the median, and as $k \to \infty$ is the mean. The value $k = 1.345$ gives 95% efficiency at the Gaussian model.

Remember that huber's function has the drawback that large outliers are not down weighted to zero. This can be achieved with Tukey's *biweight* function which is a redescending estimator such that $\psi(x) \to 0$ for $x \to \infty$. In particular, it has

$$\psi_k(x) = x\left(k - x^2\right)^2, \tag{7}$$

for $-k \ge x \ge k$ and 0 otherwise, so that it gives extreme observations zero weights. The usual value of k is 4.685. In general, the standard error of these estimates are slightly smaller than in Huber fit.

For equation (5), Draper and Smith [8] give a solution by defining weighted function

$$w(e_i) = \frac{\psi\left(\frac{1}{\hat{\sigma}}\left(y_i - \sum_{j=0}^{p} x_{ij}\beta_j \right)\right)}{\frac{1}{\hat{\sigma}}\left(y_i - \sum_{j=0}^{p} x_{ij}\beta_j \right)},$$

so equation (5) becomes

$$\sum_{i=1}^{n} x_{ij}\psi\left(y_i - \sum_{j=0}^{p} x_{ij}\beta_j\right) = 0. \tag{8}$$

In matrix notation, equation (8) can be written as

$$\mathbf{X}^T \mathbf{W}_i \mathbf{X}\boldsymbol{\beta} = \mathbf{X}^T \mathbf{W}_i \mathbf{Y}, \tag{9}$$

where $\mathbf{W}_i$ is a $n \times n$ matrix with its diagonal elements are the weights. Equation (9), as known as weighted least squares (WLS) equation, can be solved by *iteratively reweighted least squares* (IRLS) method.

2.2. MM-estimation

M-estimators are not very resistant to leverage points, unless they have redescending ψ functions. The breakdown point of M-estimators is cannot exceed $1/p$ for other robust M-estimators (that is, it decreases with increasing dimension where there are more opportunities for outliers to occur). Several robust estimators of regression have been proposed in order to remedy this shortcoming, and are high breakdown point estimators of regression. Many of them are the *least median of squares* (LMS) regression, the *least trimmed squares* (LTS) regression and S-estimator. All of them have breakdown point of 50% (Rousseeuw [9]). The disadvantages of these high breakdown point estimators are the highly inefficiency of their parameter estimates and require heavy computational effort (Venables and Ripley [3]).

It is possible to combine the resistance of these high breakdown estimators with the efficiency of M-estimation. The MM-estimator proposed by Yohai, Stahel and Zamar [10] (see also Susanti *et al.* [7]) is an M-estimator starting at the coefficients and fixed scale given by the S-estimator. The MM-estimates have an asymptotic efficiency as close to one as desired, and simultaneously breakdown point 50%. Formally, the MM-estimate $\hat{\beta}_j^{MM}$ consists in many procedures: First, compute an S-estimate with breakdown point 50%, denoted by $\hat{\beta}_j^*$. Second, compute the residuals $e_i^* = y_i - \sum_{j=0}^{p} x_{ij}\beta_j^*$ and find $\hat{\sigma}^*$ as solution of $\sum_{i=1}^{n} \varphi\left(\frac{e_i}{\sigma}\right) = (n-p)k_2$. Third, find the minimum $\hat{\beta}_j^{MM}$ of $\sum_{i=1}^{n} \rho\left(\frac{1}{\hat{\sigma}^*}\left(y_i - \sum_{j=0}^{p} x_{ij}\beta_j\right)\right)$, where $\rho(.)$ is a Tukey's biweight function.

3. Robust regression using student-*t* distribution

Statistical inference based on the normal distribution is known to be vulnerable to outliers. The robustness procedures which discussed in Section 2 are mainly directed at detecting outliers. After editing outliers, subsequent analysis is often still restricted to least squares based on the normal linear model. A serious problem with this approach is that resulting inferences fail to reflect uncertainty in the exclusion process; in particular, standard errors tend to be too small. Lange *et al.* [1] proposed a method to robust inference on regression models using the *t* distribution. Its approach is to replace the normal distribution by the *t* distribution in statistical models.

Suppose that sample data $y_i\ (i = 1,\dots,n)$ are recorded for n units. Typically, one assumes that the y_i, are independent normal random vectors, then

$$y_i \underset{ind}{\sim} N(\mu_i(\theta), \Sigma_i(\varphi)). \tag{10}$$

The vector mean μ_i is a known form indexed by a set of unknown parameters θ, and the covariance matrix Σ_i is known form indexed by a set of unknown parameters φ. The method proposed by Lange *et al.* [1] is replacing (10) with model

$$y_i \underset{ind}{\sim} t\left(\mu_i(\theta), \Psi_i(\varphi), v\right), \tag{11}$$

where $t\left(\mu_i, \Psi_i, v\right)$ denotes the t distribution with location parameter μ_i, scale parameter Ψ_i, v degrees of freedom, and density

$$P\left(y_i \mid \mu_i, \Psi_i, v\right) = \frac{\left|\Psi_i\right|^{-1/2} \Gamma\left[(v+1)/2\right]}{\Gamma(1/2)\Gamma(v/2)v^{1/2}} \times \left(1 + \frac{1}{v}\sum_{i=1}^{n}\left(y_i - \mu_i(\theta)\right)^2 \Psi_i^{-1}(\varphi)\right)^{-(v+1)/2}. \tag{12}$$

Inferences about θ and φ in the univariate t distribution can be solved by likelihood methods. The log-likelihood for (11), ignoring constants, is $l(\theta, \varphi, v) = \sum_{i=1}^{n} l_i(\theta, \varphi, v)$, where

$$l_i(\theta, \varphi, v) = -\frac{1}{2}\ln\left|\Psi_i(\varphi)\right| - \frac{1}{2}(v+1)\ln\left(1 + \frac{\delta_i^2(\theta, \varphi)}{v}\right) - \frac{1}{2}\ln(v) + \ln\left[\Gamma\left(\frac{v+1}{2}\right)\right] - \ln\Gamma\left(\frac{v}{2}\right). \tag{13}$$

The letter Γ denotes the gamma function and

$$\delta_i^2(\theta, \varphi) = \sum_{i=1}^{n}\left(y_i - \mu_i(\theta)\right)^2 \Psi_i^{-1}(\varphi). \tag{14}$$

Suppose $g(s, v) = \frac{\Gamma\left[(v+1)/2\right]}{\Gamma(1/2)\Gamma(v/2)^{1/2}} \times \left(1 + \frac{s}{v}\right)^{-(v+1)/2}$, so (12) becomes

$$P\left(y_i \mid \mu_i, \Psi_i, v\right) = \left|\Psi_i\right|^{-1/2} g\left(\delta_i^2(\theta, \varphi), v\right). \tag{15}$$

The first and second partial derivatives of (15) with respect to θ, φ and v in order to obtain the score and expected information matrix can be found at Lange *et al.* [1].

Note that as $v \to \infty$, the univariate t distribution approaches the normal distribution. When $v < \infty$, maximum likelihood (ML) estimation of θ and certain functions of φ are robust in the sense that outlying cases with large Mahalanobis distances δ^2 are downweighted. In particular, ML estimates of θ (with q components, say) for the normal model (10) satisfy the likelihood equation $\frac{\partial l}{\partial \theta} = \sum_{i=1}^{n} A_i \Sigma_i^{-1}\left(y_i - \mu_i\right) = 0$, where l denotes the log-likelihood and A_i denotes q-1 matrix of partial derivatives of μ_i with respect to θ. ML estimate of θ under the t model (11) satisfy $\sum_{i=1}^{n} w_i A_i \Sigma_i^{-1}\left(y_i - \mu_i\right) = 0$, where

$$w_i = \frac{v+1}{v + \delta_i^2}. \tag{16}$$

is weight assigned to case i. Clearly, w_i decreases with increasing δ_i^2.

Although the t modeling is not a solution for all robustness problems, this model has many advantages. In particular, data with shorter-than-normal tails or asymmetric error distributions, varying degrees of long-tails among the variables, or extreme outliers are not well modeled by [11]. An advantage of the t modeling approach, however, is that a clear statement of assumptions that is incorporated in the model specification, and a critical assessment of them can yield modifications of the model that deal with some of its limitations by allowing different degrees of freedom parameters for different variables.

4. Simulation

This section presents two simulation studies. Simulation 1 is designed to analyze the behavior of parameter estimates from models using t distribution and M-estimation. Simulation 2 studies the robustness of parameter estimates from the models by using Sensitivity Curve.

4.1. Simulation 1

The target of simulation 1 is to investigate the comparativeness of the models using t distribution with M-estimation models. The procedures of a simulation study are denoted for the following steps.

1. Suppose the independently and identically distributed samples (x_i, y_i), $i = 1,...,n$ are sampled from the linear model.
2. Set X as a sequence of 30 number with minimum 1 and maximum 1000 ($x_1 = 1$, $x_2 = x_1 + I$, $x_3 = x_2 + I$, ... , $x_{29} = x_{28} + I$, $x_{30} = 1000$), where $I = 999/29$.
3. Generate the error $e_i \sim N(0,1)$, $i = 1,...,30$.
4. Consider the following eight cases for the contaminated error density of e_i:
   - *Case* 1: 1 contaminant from N(0,100) was added to replace 1 observation from point 3.
   - *Case* 2: 1 contaminant from N(50,100) was added to replace 1 observation from point 3.
   - *Case* 3: 3 contaminants from N(0,100) were added to replace 3 observations from point 3.
   - *Case* 4: 3 contaminants from N(50,100) were added to replace 3 observations from point 3.
   - *Case* 5: 6 contaminants from N(0,100) were added to replace 6 observations from point 3.
   - *Case* 6: 6 contaminants from N(50,100) were added to replace 6 observations from point 3.
   - *Case* 7: 30 observasions from $t_{v=3}$ to replace all observations from point 3.
   - *Case* 8: 30 observasions from $t_{v=1}$ to replace all observations from point 3.
5. In accordance with point 2 to 4, compute the value $y_i = a + bx_i + e_i$, where we set $a = 1$ and $b = 2$.
6. Estimate the parameter models with Huber M-estimation, Tukey's bisquare estimation, MM-estimation and t distribution with degrees of freedom v =1,3,5 dan $\hat{v}$ (estimated degrees of freedom).
7. Replicate the procedures 4 until 6 with the number of replicates is 1000.
8. Calculate and compare the performance of the models, which are the Mean value of Absolute Biases and Mean Residual Standard Error (RSE) as the following:

$$\text{mean Absolute Bias}\left(\beta_j\right) = \frac{1}{R}\sum_{r=1}^{R}\left|\hat{\beta}_j^{(r)} - \beta_j\right|,\ j = 1,2 \tag{17}$$

$$\text{mean RSE} = \frac{1}{R}\sum_{r=1}^{R}\left[\left(\frac{1}{n-p}\sum_{i=1}^{n}\left(y_i - \hat{y}_i\right)^2\right)^{1/2}\right], \tag{18}$$

where $r = 1,...,R$ (R =1000), $i = 1,...,n$ (n =30) and p =2.

Table 1 shows that the model with t distribution gives better results than the M-estimation. The mean value of absolute biases and mean residual standard error (RSE) of t distribution models (particularly with small degrees of freedom) is worth less than M-estimation. The interesting results can be obtained from the simulation. For the estimated degrees of freedom, $\hat{v}$, with the increasing number of outliers, the model t distribution will respond with smaller value of $\hat{v}$. It occurs both in the case of an error with the contaminants that the variance increase (Case 1, 3 and 5) as well as by shifting the mean and variance increase (Case 2, 4 and 6). As for the case of error with student-t distribution (Case 7 and 8), the model t distribution with $v = 1$ and $\hat{v} = 0.965$ still has the model performed better than the M-estimation.

Table 1. Mean of absolute biases and mean of residual standard error of models

Case	Criterion	Huber	Tukey	MM	$t_{v=1}$	$t_{v=3}$	$t_{v=5}$	$t_{v=\hat{v}}$
Case 1	Bias a	0.540	0.526	0.498	0.080	0.392	0.474	0.145
	Bias b	0.004	0.004	0.003	0.005	0.002	0.003	0.003
	RSE	1.056	1.072	0.897	0.506	0.760	0.853	0.590
	$\hat{v}$							1.267
Case 2	Bias a	0.520	0.527	0.498	0.081	0.393	0.475	0.139
	Bias b	0.004	0.004	0.003	0.005	0.002	0.003	0.003
	RSE	1.042	1.072	0.898	0.506	0.760	0.854	0.584
	$\hat{v}$							1.245
Case 3	Bias a	0.562	0.540	0.529	0.156	0.464	0.536	0.184
	Bias b	0.005	0.004	0.004	0.004	0.003	0.004	0.006
	RSE	1.094	1.114	1.034	0.591	0.955	1.258	0.485
	$\hat{v}$							0.708
Case 4	Bias a	0.501	0.538	0.528	0.150	0.458	0.528	0.184
	Bias b	0.005	0.004	0.004	0.004	0.003	0.004	0.006
	RSE	1.086	1.113	1.033	0.590	0.955	1.262	0.476
	$\hat{v}$							0.686
Case 5	Bias a	0.604	0.543	0.546	0.302	0.532	1.397	0.336
	Bias b	0.007	0.004	0.005	0.004	0.004	0.023	0.008
	RSE	1.245	1.209	1.263	0.750	1.714	8.946	0.453
	$\hat{v}$							0.486
Case 6	Bias a	0.468	0.541	0.545	0.297	0.520	1.301	0.342
	Bias b	0.007	0.004	0.005	0.004	0.004	0.023	0.008
	RSE	1.301	1.209	1.264	0.750	1.732	10.078	0.444
	$\hat{v}$							0.470
Case 7	Bias a	0.371	0.374	0.370	0.391	0.360	0.364	0.365
	Bias b	0.006	0.006	0.006	0.007	0.006	0.006	0.006
	RSE	1.109	1.103	1.166	0.654	0.959	1.084	0.988
	$\hat{v}$							3.539
Case 8	Bias a	0.585	0.526	0.528	0.429	0.506	0.581	0.437
	Bias b	0.010	0.009	0.009	0.008	0.008	0.010	0.008
	RSE	1.570	1.518	1.699	0.954	1.772	2.357	0.959
	$\hat{v}$							0.964

4.2. Simulation 2

The target of simulation 2 is to demonstrate the Sensitivity Curve to assess the stability and robustness of estimators from the models. The Sensitivity Curve $SC_n(y_m)$ is a translated and rescaled version of the empirical influence function. In many situations, $SC_n(y_m)$ will converge to the influence function when $n \to \infty$. The procedures for making Sensitivity Curve are:

1. Set X as simulation 1.
2. Generate the error $e_i \sim N(0,1)$, $i = 1,...,30$.
3. Compute the value $y_i = a + bx_i + e_i$, where we set $a = 1$, and $b = 2$.
4. Consider the arbitrary value of y_m, m=1,2,...,1000, where we set y_m as a sequence of 1000 numbers with minimum 100 and maximum 300.
5. Replace one observation of the sample by an arbitrary value y_m and count the value of $SC_n(y_m) = n[T_n(x_1,...,x_{n-1}, y_m) - T_{n-1}(x_1,...,x_{n-1}, y_m)]$, where $T_n(x)$ denotes the estimator of interest based on the sample x of size n.
6. Plot the value of $SC_n(y_m)$ as the y-axis dan y_m as the x-axis.

Figure 1 shows that the t distribution model with v = 1 looks more robust than the M-estimation. But with the increasingly large v (i.e v = 5), models of t distribution is no more robust than the M-estimate models.

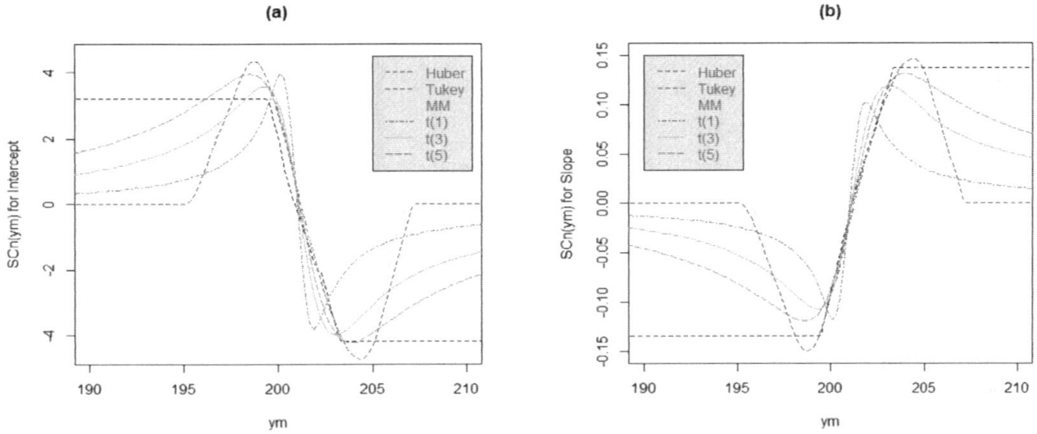

Figure 1. Sensitivity Curve for Intercept (a) and Slope (b) Parameter Estimates

5. Application to real data

In this paper we apply the models to the infant birth weight data in Indonesia at 2012. The infant birth weight data have been collected by the Indonesia Demographic and Health Survey (IDHS) which have been conducted regularly by Statistics Indonesia (BPS) [11] every five years. The sample size of data is 15124 infant which was born five years ago from the period of survey. The variables used in this study are the infant birth weight as dependent variable (Y) and the age of their mother as independent variable (X).

Table 2 shows the model t distribution with v = 1 has the smallest of standard error of estimates and residual standard error (RSE) value. The model t distribution with estimate of v seems not more efficient compared to M-estimates, but has smaller RSE value than M-estimates. For the model t distribution with v = 2, 3, 4, and 5 have smaller of both the standard error and RSE value than the M-estimation.

6. Discussion

In general, the results of this study indicate that the model using t distribution with degrees of freedom v =1 (as named Caucy distribution) has a performance and robustness are better than the models of M-estimation. Almost equal to the research conducted by Lange *et al.* [1] which analyzed Stack-Loss data which the model t distribution with $v = 1$ also showed better results than the M-estimation.

The interesting thing from this article is the pattern of robustness shown by Sensitivity Curve where the model t distribution with $v = 1$ looks more effective of robustness in estimate change of its fluctuations as compared to the model M-estimate. Moreover, with the decreasing of v, the performance and robustness of its models look better.

Tabel 2. Regression of infant birth weight data: estimated from 10 models

	β_0		β_1		
Model	Mean	Std. Error	Mean	Std. Error	RSE
LS	3018.154	21.192	5.320	0.750	567.938
Huber	3008.644	19.778	5.663	0.700	494.070
Tukey	3010.980	19.800	5.664	0.700	497.531
MM	3010.981	19.799	5.664	0.700	498.975
$t_{v=1}$	2981.949	16.286	6.016	0.576	308.620
$t_{v=2}$	3000.419	18.112	5.783	0.641	375.971
$t_{v=3}$	3005.356	18.845	5.712	0.667	412.357
$t_{v=4}$	3007.730	19.247	5.670	0.681	435.949
$t_{v=5}$	3009.202	19.507	5.639	0.690	452.734
$t_{\hat{v}=6.004}$	3010.226	19.917	5.614	0.706	465.411

7. Conclusion

This article shows the ability of models based on the t distribution to overcome outliers compared with M-estimates models. From the simulation and real data set application results we found that the best model fit to handle outliers is the models based on the t distribution with smallest degrees of freedom v. The t distribution model with small v also gives the good performance in robustness of its estimator.

Acknowledgments

The authors would like to express a sincere gratitude to Statistics Indonesia for providing funds for this research as a part of the PhD grant research program and for its human resource development projects in cooperation with Bogor Agricultural University, Indonesia.

References

[1] Lange K L, Little R J A and Taylor J M G 1989 *J. Am. Statist. Ass.* **84** 881-96
[2] Micceri T 1989 *Psyc. Bulletin* **105** 156-66
[3] Venables W N and Ripley B D 2002 *Modern applied statistics with S* 4th Ed. (New York: Springer) chapter 5 pp 119-26
[4] Huber P J 1981 *Robust statistics* (New York: John Wiley and Sons)
[5] Hampel F R, Ronchetti E M. Rousseeuw P J and Stahel W A 1986 *Robust statistics. The approach based on influence functions* (New York: John Wiley and Sons)
[6] Staudte R G and Sheather S J 1990 *Robust estimation and testing* (New York: John Wiley and Sons)
[7] Susanti Y, Pratiwi H, Sulistijowati S H and Liana T 2014 *Int. J. Pure and Appl. Math.* **91** 349-60
[8] Draper N R and Smith H 1998 *Applied Regression Analysis* 3th Ed (New York: Wiley–Interscience)

[9] Rousseeuw P J 1984 *J. Am. Statist. Ass.* **79** 871-81
[10] Yohai V, Stahel W A and Zamar R H 1991 A procedure for robust estimation and inference in linear regression *In Directions in Robust Statistics and Diagnostics* part 2. ed W A Stahel and S W Weisberg (New York: Springer-Verlag)
[11] BPS 2012 *Indonesia Demographic and Health Survey* (Jakarta: BPS)

Bias Reduction in Estimating Variance Components of Phytoplankton Existence at Na Thap River Based on Logistics Linear Mixed Models

R Arisanti[1], K A Notodiputro[2], K Sadik[2] and A Lim[3]

[1] Department of Statistics, Padjadjaran University, Indonesia
[2] Department of Statistics, Bogor Agricultural University, Indonesia
[3] Department of Mathematics and Computer Science, Faculty of Science and Technology, Prince of Songkla University, Thailand

Email : almiraza@gmail.com, khairilnotodiputro@gmail.com, kusmansadik@gmail.com, api_45@hotmail.com

Abstract. There are two approaches in estimating variance components, i.e. linearity and integral approaches. However the estimates of variance components produced by both methods are known to be biased. Firth (1993) has introduced parameter estimation for correcting the bias of the maximum likelihood estimates. This method is within the class of linear models, especially the Restricted Maximum Likelihood (REML) method, and the resulting estimator is known as the Firth estimator. In this paper we discuss the bias correction method applied to a logistic linear mixed model in analyzing the existence of Synedra phytoplankton along Na Thap river in Thailand. The Firth adjusted Maximum Likelihood Estimation (MLE) is similar to REML but it shows the characteristic of generalized linear mixed model. We evaluated the Firth adjustment method by means of simulations and the result showed that the unadjusted MLE produced 95% confidence intervals which were narrower when compare to the Firth method. However, the probability coverage of the interval for unadjusted MLE was lower than 95%, whereas for the Firth method the probability coverage is approximately 95%. These results were also consistent with the variance estimation of the Synedra phytoplankton existence. It was shown that the variance estimates of Firth adjusted MLE was lower than the unadjusted MLE.

1. Introduction

Breslow and Clayton [1] have introduced a combination of Linear Mixed Models (LMM) and Generalized Linear Models (GLM) in the form of Generalized Linear Mixed Models (GLMM). If X and Z are design matrices, b and b are the corresponding coefficients then the GLMM models can be described by $\boldsymbol{\eta} = \boldsymbol{X\beta} + \boldsymbol{Zb}$ with $\boldsymbol{b} \sim \boldsymbol{N}(\boldsymbol{0}, \boldsymbol{G}), \boldsymbol{\eta} = \boldsymbol{g}(\boldsymbol{\mu})$ and $\boldsymbol{y}|\boldsymbol{b} \sim \boldsymbol{G}(\boldsymbol{\mu}, \boldsymbol{R})$. $\boldsymbol{G}(\boldsymbol{\mu}, \boldsymbol{R})$ is a function that follows a specific distribution (not necessarily normal) with mean ($\boldsymbol{\mu}$) and the covariance matrix ($\boldsymbol{R}$). This model has three main components, namely fixed effects ($\boldsymbol{\beta}$), random effects (b), and components of variance ($\boldsymbol{\sigma}$). A model of GLM with binary responses can be approximated by logistic linear models [6]. Logistic regression models that involve random explanatory variables ~~is~~ are called Logistic Linear Mixed Models.

In mixed models, we are interested in estimation of variance components [5]. To estimate the variance components, it is common to employ linearity and integral approaches. The linearity method can result in REML (Restricted Maximum Likelihood Estimation) methods by substituting y in the linear mixed model with pseudo variables y*. The integral approach is more widely used since it uses the actual likelihood, hence the goodness of fit can be calculated. However, both methods have disadvantages in which the estimates of variance components are biased. The impact of this bias can be seen on the hypothesis testing and confidence interval produced by the model. If the variance estimation is too small, then the estimates will be inflated and tend to reject the null hypothesis. Moreover, the confidence intervals are also too narrow and the p-values are lower than the actual stated values.

Reference [4] introduced an estimation of parameters to adjust for the bias using maximum likelihood method. In this paper, we discussed estimation of variance components for logistic linear mixed models using the Firth correction methods. We then applied the models to the phytoplankton data, a data collected from Na Thap river in South Thailand.

2. Logistic Linear Mixed Model (LLMM)

In this study, we employed a logistic linear mixed model, especially a logit model with random intercepts to estimate the phytoplankton proportion parameter p. Then we choose r sites which consist of n samples for each site. The random intercept model is

$$\text{logit}\,(p_i) = \eta + b_i$$

where $\text{logit}(p_i) = \ln\left(\frac{p_i}{1-p_i}\right)$ is a canonical link, η is an intercept, and b_i is the effect of the ith subject and it is assumed distributed as $N(0, \sigma^2)$. Alternatively, we can use the following notation:

$$\text{logit}(p_i) = \eta + \sigma z_i$$

where z_i is the standardized random effects. Gauss-Hermite quadrature of estimation can be used when the r sites contain relatively small n [3].

Gauss-Hermite Quadrature is basically a numerical integral approximation based on normal distributions [7]. The conditional likelihood and log-likelihood equation for one subject can be written:

$$L(\boldsymbol{\theta}|\boldsymbol{y}, z) = \binom{n}{\boldsymbol{y}}\left(\frac{1}{1+e^{-(\eta+\sigma z)}}\right)^{y}\left(1-\frac{1}{1+e^{-(\eta+\sigma z)}}\right)^{n-y}\frac{1}{\sqrt{2\pi}}e^{\left(\frac{-\boldsymbol{z}}{2}\right)^2}$$

$$l(\boldsymbol{\theta}|\boldsymbol{y}, \boldsymbol{z}) = \ln\binom{n}{\boldsymbol{y}} + \boldsymbol{y}(\eta+\sigma\boldsymbol{z}) - n(\eta+\sigma\boldsymbol{z}) - n\ln\left(1+e^{-(\eta+\sigma\boldsymbol{z})}\right) - \ln\sqrt{2\pi} - \frac{\boldsymbol{z}^2}{2}$$

where $\boldsymbol{y}$ as response vector and $\boldsymbol{z}$ as random effect vector. Defined $h(\boldsymbol{z}, \boldsymbol{\theta})$ as the negative log-likelihood. To solve the fixed effect, we intergrate the random effects using Gauss-Hermite quadrature:

$$\int e^{-h(z,\boldsymbol{\theta})}d\boldsymbol{z}$$

The second order of Taylor expansion $h(z, \boldsymbol{\theta})$ is

$$q(z, \boldsymbol{\theta}) = h(\hat{z}(\boldsymbol{\theta}), \boldsymbol{\theta}) + \frac{1}{2}h''(\tilde{z}(\boldsymbol{\theta}), \boldsymbol{\theta})\left(z - \tilde{z}(\boldsymbol{\theta})\right)^2$$

$\tilde{z}(\boldsymbol{\theta})$ maximized the conditional likelihood $\boldsymbol{\theta}$, dan $h''(\tilde{z}(\boldsymbol{\theta}), \boldsymbol{\theta})$ is the second derivative $h(z, \boldsymbol{\theta})$ for the random effect which evaluated at $\tilde{z}(\boldsymbol{\theta}) = \hat{\sigma}\hat{z}$,

$$h''(z,\boldsymbol{\theta}) = \frac{1}{\sigma^2} + \frac{ne^{\eta+\sigma z}}{(1+e^{\eta+\sigma z})^2}$$

Then,

$$\int e^{-h(z,\boldsymbol{\theta})}dz = \int e^{-(h(z,\boldsymbol{\theta})-q(z,\boldsymbol{\theta}))}e^{-q(z,\boldsymbol{\theta})}dz$$

$$= e^{-h(\hat{z}(\boldsymbol{\theta}),\boldsymbol{\theta})}\int e^{-(h(z,\boldsymbol{\theta})-q(z,\boldsymbol{\theta}))}e^{\frac{h''(\hat{z}(\boldsymbol{\theta}),\boldsymbol{\theta})}{2}(z-\hat{z}(\boldsymbol{\theta}))^2}dz$$

$$= \frac{\sqrt{2\pi}e^{-h(\hat{z}(\boldsymbol{\theta}),\boldsymbol{\theta})}}{\sqrt{h''(\hat{z}(\boldsymbol{\theta}),\boldsymbol{\theta})}}\int e^{-(h(z,\boldsymbol{\theta})-q(z,\boldsymbol{\theta}))}\frac{e^{\frac{h''(\hat{z}(\boldsymbol{\theta}),\boldsymbol{\theta})}{2}(z-\hat{z}(\boldsymbol{\theta}))^2}}{\sqrt{2\pi}\left(\frac{1}{\sqrt{h''(\hat{z}(\boldsymbol{\theta}),\boldsymbol{\theta})}}\right)}dz$$

$$= \frac{\sqrt{2\pi}e^{-h(\hat{z}(\boldsymbol{\theta}),\boldsymbol{\theta})}}{\sqrt{h''(\hat{z}(\boldsymbol{\theta}),\boldsymbol{\theta})}}\int e^{-\left(h\left(\frac{z}{\sqrt{h''(\hat{z}(\boldsymbol{\theta}),\boldsymbol{\theta})}}+\hat{z}(\boldsymbol{\theta}),\boldsymbol{\theta}\right)-q\left(\frac{z}{\sqrt{h''(\hat{z}(\boldsymbol{\theta}),\boldsymbol{\theta})}}+\hat{z}(\boldsymbol{\theta}),\boldsymbol{\theta}\right)\right)}\frac{e^{-\frac{z^2}{2}}}{\sqrt{2\pi}}dz$$

Assume that (x_i, w_i) is the abscissa and Gauss-Hermite quadrature weight. Therefore,

$$\int e^{-h(z,\boldsymbol{\theta})}dz \cong \frac{\sqrt{2\pi}e^{-h(\hat{z}(\boldsymbol{\theta}),\boldsymbol{\theta})}}{\sqrt{h''(\hat{z}(\boldsymbol{\theta}),\boldsymbol{\theta})}}\sum_i e^{-\left(h\left(\frac{z}{\sqrt{h''(\hat{z}(\boldsymbol{\theta}),\boldsymbol{\theta})}}+\hat{z}(\boldsymbol{\theta}),\boldsymbol{\theta}\right)-q\left(\frac{z}{\sqrt{h''(\hat{z}(\boldsymbol{\theta}),\boldsymbol{\theta})}}+\hat{z}(\boldsymbol{\theta}),\boldsymbol{\theta}\right)\right)}w_i$$

$$= \frac{\sqrt{2\pi}e^{-h(\hat{z}(\boldsymbol{\theta}),\boldsymbol{\theta})}}{\sqrt{h''(\hat{z}(\boldsymbol{\theta}),\boldsymbol{\theta})}}\sum_i e^{-h\left(\frac{z}{\sqrt{h''(\hat{z}(\boldsymbol{\theta}),\boldsymbol{\theta})}}+\hat{z}(\boldsymbol{\theta}),\boldsymbol{\theta}\right)}e^{h(\hat{z}(\boldsymbol{\theta}),\boldsymbol{\theta})}e^{\frac{x_i^2}{2}}w_i$$

In the initialization step, $\hat{\eta} = \text{logit}\,(\hat{p})$ and $\hat{p} = \sum_i \frac{y_i}{n_i}$ can be used, and then followed by the gradien and Hessian by Newton-Raphson and Broyden method.

3. Firth Bias Reduction

Reference [4] proposed a technique to obtain estimates having less bias when compared to MLE. This technique combined the advantage of REML and inferences based on the true likelihood. Bias correction is resulted from the score equation, $S(\boldsymbol{\theta}) = \frac{\delta}{\delta\theta}L(\boldsymbol{\theta}|\boldsymbol{y})$. For parameter equals to θ_0, then the expected value of score is zero. Therefore, MLE is the invers score evaluated at zero:

$$E_{\theta_0}[S(\boldsymbol{\theta_0})|\boldsymbol{y}] = 0$$
$$\widehat{\boldsymbol{\theta}}_{MLE} = S^{-1}(0|\boldsymbol{y})$$

Firth proposed two adjustment. First, the adjustment uses the expected of Hessian, known as Fisher information matrix. The adjustment related to the jth parameter is

$$\boldsymbol{A}_{\theta j} = -\frac{1}{2}tr\left(\boldsymbol{F}^{-1}E\left[S_{\theta j}(\boldsymbol{H}-\boldsymbol{S}\boldsymbol{S}^{\boldsymbol{T}})\right]\right)$$

where $\boldsymbol{F}^{-1}$ is the invers of Fisher information matrix, **H** is Hessian matrix, **S** is score vector and $S_{\theta j}$ is the score vector element of parameter θj. The second adjustment is done by modifying the expected adjustment:

$$\boldsymbol{A}_{\theta j} = -\frac{1}{2}tr\left(\boldsymbol{H}^{-1}E\left[S_{\theta j}(\boldsymbol{H}-\boldsymbol{S}\boldsymbol{S}^{\boldsymbol{T}})\right]\right)$$

The double observed adjustment to reduce the expected adjustment is given by

$$A_{\theta j} = -\frac{1}{2} tr\left(\boldsymbol{H}^{-1} E\left[S_{\theta j}(\boldsymbol{H} - \boldsymbol{S}\boldsymbol{S}^T)\right]\right)$$

Let $\boldsymbol{S} = \sum_{i=1}^{r} \boldsymbol{S}(\boldsymbol{y}_i)$ and $\boldsymbol{H} = \sum_{i=1}^{r} \boldsymbol{H}(\boldsymbol{y}_i)$. The expected value of observed Firth adjustment is

$$E_y\left[\sum_{i=1}^{r} S_{\theta j}(\boldsymbol{y}_i)\left(\sum_{k=1}^{r} \boldsymbol{H}(y_k) - \sum_{k=1}^{r} \boldsymbol{S}(y_k)\sum_{l=1}^{r} \boldsymbol{S}^T(y_l)\right)\right]$$

Because $\boldsymbol{y}_i$ are independent and because $E_y\left[S_{\theta j}(y)\right] = 0$, then the expected value is

$$E_y\left[\sum_{i=1}^{r} S_{\theta j}(\boldsymbol{y}_i)\left(\boldsymbol{H}(y_i) - \boldsymbol{S}(y_i)\boldsymbol{S}^T(y_i)\right)\right]$$

The double observed adjustment uses the observed information matrix. Then, we use

$$A_{\theta j} = -\frac{1}{2} tr\left((\boldsymbol{H}^{-1})\sum_{i=1}^{r} S_{\theta j}(\boldsymbol{y}_i)\left(\boldsymbol{H}(y_i) - \boldsymbol{S}(y_i)\boldsymbol{S}^T(y_i)\right)\right)$$

To obtain the estimates iteratively, the Broyden method can be used.

4. Result and Discussion

In this study, we used survey data which was carried out by the Electricity Generation Authority of Thailand (EGAT) taken from June, 2004 until April, 2014. The data was collected every two months along the Na Thap river, Thailand. In this survey, EGAT divided the Na Thap river into ten sites from three zones as illustrated in Figure 1.

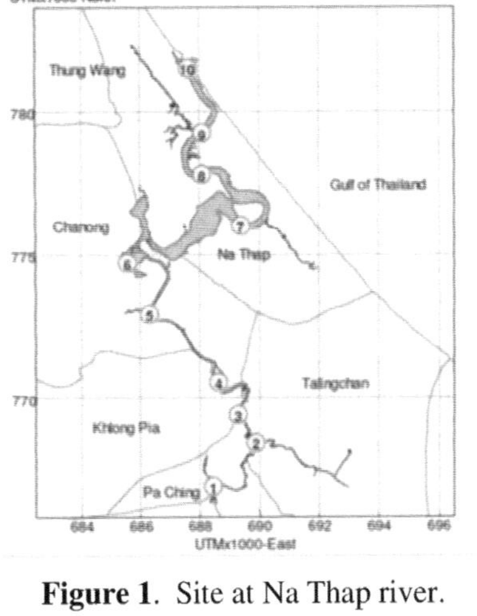

Figure 1. Site at Na Thap river.

Six divisions of phytoplankton have been found in Na Thap river. The number of genus was 53 genus [8]. The divisions are (1) 4 genus of Cyanophyta, (2) 14 genus of Chlorophyta, (3) 21 genus of Bacillariophyta, (4) 4 genus of Pyrrophyta, (5) 2 genus of Euglenophyta and (6) 1 genus of Cryptophyta. There are three dominant divisions, i.e Bacillariophyta, Chlorophyta and Pyrrophyta [2]. We used the Synedra phytoplankton data to evaluate the Firth method.

4.1. The Simulation

A simulation has been carried out to show the performance of Firth method in reducing the variance components. Table 1 showed the simulation results.

Table 1. Variance Components.

	Firth *adjusted* MLE	*Unadjusted* MLE
Standard deviation of sampling	0.4369	0.4352
Standard error mean	0.4361	0.3960
Estimated coverage probability	94.73%	91.52%

The table showed that the standard deviation of the unadjusted MLE (0.4352) was slightly lower than the standard deviation of the Firth adjusted MLE (0.4369). The table has provided evidence that there was bias in variance estimation in the logistic models. Moreover, the standard error of unadjusted MLE for the treatment effects (0.3960) was lower than the standard deviation of the sampling distribution (0.4352). This result produced narrow confidence intervals and low coverage probabilities (91.52%) for the unadjusted MLE. On the other hand, the standard error of the Firth adjusted MLE was 0.4361. This was much closer to the standard deviation of the sampling distribution (0.4369). The Standard error of Firth adjusted resulted in wider confidence intervals than those for the unadjusted MLE, but the coverage probabilities are close to the 95%.

4.2. The Existence of Synedra Phitoplankton at Na Thap River

Now, we will show the application of the Firth adjustment in estimating variance components of Synedra phytoplankton existence in Na Thap river. The application of this model to the phytoplankton data produced estimates of fixed effects and random effects for Synedra existence as shown in Table 2.

Table 2. Parameter estimation of fixed and random effects of Synedra Phytoplankton.

	Unadjusted MLE				Firth Adjusted			
	Estimates	Std. Error	t-stat	Pr(>|z|)	Estimates	Std. error	t_stat	Pr(>|z|)
Intercept	-0.5922	1.7165	-0.345	0.0071	-0.6860	1.4218	-0.4825	0.0063
DO	0.4243	0.3112	1.363	0.0173	0.4135	0.2330	1.7752	0.0007
BOD	-0.3716	0.2674	-1.389	0.0165	-0.3411	0.2316	-1.4728	0.0014
Salinity	0.0002	0.0007	0.304	0.0076	0.0002	0.2316	0.2816	0.0077
σ^2 residual	2.175				1.363			

The table showed that the variance of residuals using the Firth method was 1.363. This variance was smaller than residual variance of the unadjusted MLE (2.175). It means that the Firth method has reduced the bias of variance components of Synedra existence in the Na Thap river. The table also showed that three predictor variables have affected significantly to the Synedra existence either using unadjusted MLE and Firth adjusted method. The p-values of DO, BOD, and Salinity variables were less than 0.05.

5. Conclusion

We have showed that the mean estimates of the Firth adjusted MLE were closer to the true value than the unadjusted MLE. The Firth adjusted MLE was found to produced 95% probability coverage for the confidence interval whereas the unadjusted MLE produced less than 95% probability coverage. In general we showed that the Firth adjustment is effective to obtain lower variance component, and hence it ca be used to correct the underestimated variance components.

Acknowledgements

We thank Dr. Sarawuth Chesoh at Prince of Songkla University in Pattani for giving us permission to use the Na Thap river data as part of the international research collaboration with Bogor Agricultural University.

References

[1] Breslow NE and Clayton DG 1993 *Journal of the American Statistical Association* **88(421)** 925
[2] Chesoh S and A Lim 2015 Monitoring of Thermal Power Plant Operation on Aquatic Ecosystem in Tropical Estuarine River of Thailand, 2008-2013. IAHR World Congress. Netherlands
[3] Classen EA 2014 A Reduced Bias Method of Estimating Variance Component in Generalized Linear Mixed Model Lincoln: University of Nebraska
[4] Firth D 1993 *Biometrika* **80(1)** 27-38
[5] Henderson CR1953 *Biometrics* **9(2)** 226_252
[6] McCullagh P, Nelder JA 1989 Generalized Linear Models, 2nd. New York: Chapman and Hall
[7] Pinheiro JC, Chao EC 2006 *Journal of Computational and Graphical Statistics* **15(1)** 5881
[8] Saheem, N. 2015. Statistical Modelling of Aquatic Animal Abundance in the Na Thap River. A Thesis submitted in fulfillment of the requirements for the Degree of Doctor of Phylosophy in Research Methodology. Prince of Songkla University.

Robustness of location estimators under *t*-distributions: a literature review

C Sumarni[1,2], K Sadik[1], K A Notodiputro[1] and B Sartono[1]

[1] Department of Statistics, Bogor Agriculture University, INDONESIA.
[2] Statistics Indonesia, Bali Province.

Email: cucu_s@bps.go.id, kusmansadik@gmail.com, khairilnotodiputro@gmail.com, bagusco@gmail.com

Abstract. The assumption of normality is commonly used in estimation of parameters in statistical modelling, but this assumption is very sensitive to outliers. The *t*-distribution is more robust than the normal distribution since the *t*-distributions have longer tails. The robustness measures of location estimators under *t*-distributions are reviewed and discussed in this paper. For the purpose of illustration we use the onion yield data which includes outliers as a case study and showed that the *t* model produces better fit than the normal model.

1. Introduction

Estimation of parameters in a survey often assumes that the data is a random sample from a normally distributed population. Suppose that sample data y_i are recorded n units $(1 \leq i \leq n)$, and y_i are assumed as random samples from normal distribution,

$$y_i \sim \text{iid}\, N\left(\mu, \sigma^2\right) \tag{1}$$

In statistical modeling it is also common to assume normality of the error terms. The model can be written as (2) and called the normal model.

$$\begin{gathered} y_i = \boldsymbol{x}_i^T \boldsymbol{\beta} + \varepsilon_i, \text{ where } \varepsilon_i \sim N\left(0, \sigma^2\right) \text{or} \\ y_i \sim \text{iid}\, N\left(\mu(\boldsymbol{\beta}), \sigma^2\right) \end{gathered} \tag{2}$$

Assuming normal distribution implies that the estimates become inaccurate when there are outliers in the data. We may assume robust distribution of the error terms to solve this problem. The *t*-distribution is useful for statistical modelling when data contains outliers. Lange *et al.* [1] has successfully demonstrated that the estimation results were robust to outliers in linear models if the assumptions of normality has been replaced by assuming a *t*-distribution, $\varepsilon_i \sim t(0, \sigma^2, v)$. The *t*-model can be written as

$$y_i \sim \text{iid}\, t\left(\mu(\boldsymbol{\beta}), \sigma^2, v\right) \tag{3}$$

where $t(\mu(\boldsymbol{\beta}),\sigma^2,v)$ denotes the univariate t-distribution with location parameter $\mu(\boldsymbol{\beta})$ where $\boldsymbol{\beta}$ is a vector of coefficient's regression, scale parameter σ^2 and v degrees of freedom.

Lange *et al.* [1] has not shown explicit measures of the robustness of location estimators under t-distributions. In this paper, this robustness is reviewed by referring to the idea that the robustness of an estimator can be measured from the influence functions (IF), the asymptotic relative efficiency (ARE) and the breakdown point [2]. In this case, we focus on the influence functions and the asymptotic relative efficiency. We also study the influence of outliers toward modelling by illustrating the model of solid pesticides effects on the onion yield.

2. Location estimation under *t*-distributions

We assume in a robust estimation that y_i are random samples from t-distributed population which density function (pdf) is defined by (4).

$$f(y_i)=\frac{\Gamma((v+1)/2)}{\sqrt{\pi v\sigma^2}\,\Gamma(v/2)}\left(1+\frac{1}{v}\left(\frac{y_i-\mu}{\sigma}\right)^2\right)^{-(v+1)/2},-\infty\le y\le\infty \qquad (4)$$

Note that, If $\mu=0$ and $\sigma^2=1$, (4) is a density of univariate Student't distributions with v degrees of freedom (df), otherwise to be noncentral t-distributions with location parameter μ, scale parameter σ^2 and shape parameter v degrees of freedom [3]. Denote logarithm functions of (4) as $\ln f(y_i)=l_i$,

$$l_i=\ln\left(\Gamma((v+1)/2)\right)-\frac{1}{2}\ln(\pi v)-\frac{1}{2}\ln\left(\sigma^2\right)-\ln\left(\Gamma(v/2)\right)-\frac{(v+1)}{2}\ln\left(1+\frac{1}{v}\left(\frac{y_i-\mu}{\sigma}\right)^2\right) \qquad (5)$$

or it can be written as

$$l_i=\text{constant}-\frac{(v+1)}{2}\ln\left(1+\frac{1}{v}\left(\frac{y_i-\mu}{\sigma}\right)^2\right) \qquad (6)$$

The first derivative of l_i respect to μ is

$$\frac{\partial l_i}{\partial\mu}=\frac{v+1}{v+\left((y_i-\mu)/\sigma\right)^2}\left(\frac{y_i-\mu}{\sigma^2}\right) \qquad (7)$$

If σ^2 is assumed known and v is fixed, we would get the maximum likelihood estimator (MLE) of μ by finding the solution of this equation,

$$\sum_{i=1}^{n}\frac{\partial l_i}{\partial\mu}=0$$

$$\sum_{i=1}^{n}\frac{v+1}{v+\left((y_i-\mu)/\sigma\right)^2}\left(\frac{y_i-\mu}{\sigma^2}\right)=0$$

If we denote $w_i=\dfrac{v+1}{v+\left((y_i-\mu)/\sigma\right)^2}$, then $\sum_{i=1}^{n}w_i\left(\dfrac{y_i-\mu}{\sigma^2}\right)=0$

The MLE of μ under t-distributions is

$$\hat{\mu}_T=\left(\sum_{i=1}^{n}w_iy_i\right)\Bigg/\sum_{i=1}^{n}w_i \qquad (8)$$

This is a weighted mean with depending on the sample, so the location estimator under *t*-distributions is one of an M-estimator [2]. Since w_i is a function of parameters (μ), equation (8) has no closed form solution. The EM algorithm can be used to get the solution (for more detailed see Lange *et al.* [1]).

3. The measure of robustness under *t*-distributions

Robustness of an estimator can be measured quantitatively and qualitatively. The quantitative robustness can be measured by a breakdown point and the qualitative ones can be seen from efficiency and stability.

3.1. The Influence Function (IF)

The stability of robust estimators can be seen from the influence function. The influence function can be computed according to the ψ function. The influence function of an M-estimate (maximum likelihood type estimates) is defined as [2].

$$\text{IF}(.) = \frac{\psi(y;\mu)}{-E\left[(\partial/\partial\mu)\psi(y;\mu)\right]} \tag{9}$$

If the ψ function of an M-estimator is computed from the first derivative of logarithm function of pdf, $\psi(y;\mu) = \frac{\partial}{\partial\mu}(-\ln f(y))$, then the M-estimator is the usual MLE [4].

The location estimator under the *t*-distributions is a form of M-estimators with the ψ function defined as (10).

$$\psi(y;\mu) = \frac{v+1}{v+((y-\mu)/\sigma)^2}\left(\frac{y-\mu}{\sigma^2}\right) \tag{10}$$

The computing of the first derivative of the ψ function (10) can be seen in Appendix and the expectation is

$$E\left[(\partial/\partial\mu)\psi(y;\mu)\right] = E\left[\frac{\partial}{\partial\mu}\left[\frac{v+1}{v+((y-\mu)/\sigma)^2}\left(\frac{y-\mu}{\sigma^2}\right)\right]\right] = E\left[\frac{\frac{(v+1)}{\sigma^2}\left[\left(\frac{y-\mu}{\sigma}\right)^2 - v\right]}{\left(v+((y-\mu)/\sigma)^2\right)^2}\right]$$

$$= E\left[\frac{(v+1)}{v\sigma^2}\left[\frac{\frac{1}{v}\left(\left(\frac{y-\mu}{\sigma}\right)^2 - v\right)}{\left(\frac{v+((y-\mu)/\sigma)^2}{v}\right)^2}\right]\right]$$

Let $z = \frac{y-\mu}{\sigma}$, Lange *et al.* (1989) has computed that the integration yields,

$$E\left[\left(1+\frac{z^2}{v}\right)^{-m}\right] = \frac{\left(\frac{v}{2}+m-1\right)\text{L}\left(\frac{v}{2}\right)}{\left(\frac{v+1}{2}+m-1\right)\text{L}\left(\frac{v+1}{2}\right)} \text{ and}$$

$E\left[\frac{z^2}{v}\left(1+\frac{z^2}{v}\right)^{-2}\right]=E\left[\frac{\left(\frac{z^2}{v}+1\right)-1}{\left(1+\frac{z^2}{v}\right)^2}\right]=E\left[\left(1+\frac{z^2}{v}\right)^{-1}-\left(1+\frac{z^2}{v}\right)^{-2}\right]$. So we get

$$E\left[(\partial/\partial\mu)\psi(y;\mu)\right]=E\left[\frac{(v+1)}{v\sigma^2}\left[\frac{\frac{z^2}{v}-1}{\left(1+\frac{z^2}{v}\right)^2}\right]\right]=E\left[\frac{(v+1)}{v\sigma^2}\left(\frac{z^2}{v}\right)\left(1+\frac{z^2}{v}\right)^{-2}\right]-E\left[\frac{(v+1)}{v\sigma^2}\left(1+\frac{z^2}{v}\right)^{-2}\right]$$

$$=\frac{(v+1)}{v\sigma^2}\left[\frac{v}{(v+1)(v+3)}\right]-\frac{(v+1)}{v\sigma^2}\left[\frac{v(v+2)}{(v+1)(v+3)}\right]$$

$$=\frac{1}{\sigma^2(v+3)}-\frac{(v+2)}{\sigma^2(v+3)}$$

$$=-\frac{(v+1)}{\sigma^2(v+3)}$$

and

$$-E\left[(\partial/\partial\mu)\psi(y;\mu)\right]=\frac{(v+1)}{\sigma^2(v+3)} \tag{11}$$

Thus the influence function of an M-estimator under *t*-distributions is

$$\mathrm{IF}(\hat{\mu}_T)=\frac{v+3}{v+((y-\mu)/\sigma)^2}(y-\mu) \tag{12}$$

How to see that $\hat{\mu}_T$ (location estimator under *t*-distribution) is more robust than $\hat{\mu}_N$, the location estimator under normal distribution ($\hat{\mu}_N=\sum_{i=1}^{n} y_i/n$)? To answer this, we must know the ψ function and the influence function under normal distribution. The same way as before, we get under normal distribution the ψ function is

$$\psi^*(y;\mu)=\frac{y-\mu}{\sigma^2} \tag{13}$$

and the influence function $\mathrm{IF}(\hat{\mu}_N)$ is

$$\mathrm{IF}(\hat{\mu}_N)=y-\mu \tag{14}$$

The influence function of location estimators under normal distribution is a trend line and unbounded. The increasingly an observed value ($y_i\to\infty$) will produce a higher value of the influence function ($\mathrm{IF}(\hat{\mu}_N)\to\infty$) and vice versa. It means that the data's outlier will influence highly on the estimates. Meanwhile, the influence function of location estimators under *t*-distributions is a bounded function (see figure 1). That is since any outlier data is down weighted by

$$w_i=\frac{v+1}{v+((y_i-\mu)/\sigma)^2} \tag{15}$$

where v is a shape parameter, then influence of such data is reduced. An outlier will not affect the estimate significantly. That is why the location estimator under t-distributions is more robust than under the normal distribution.

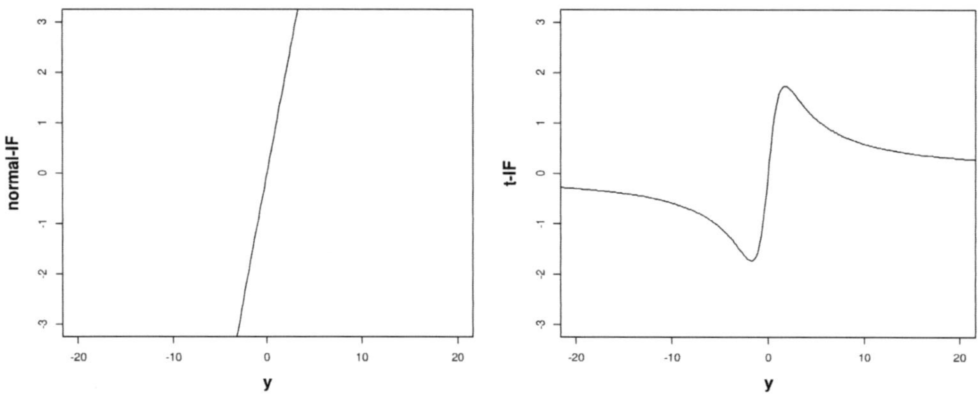

Figure 1. The influence function (IF) of location-estimators from normal distribution and t-distributions.

3.2. The Asymptotic distribution of location estimators under t*-distributions*
If the influence function (IF) of an M-estimates is proportional to the ψ function, then the asymptotic variance $(A(T))$ of $\sqrt{n}(T-\mu)$ satisfies: (Huber and Ronchetti 2009)

$$A(T) = E\left[\mathrm{IF}(.)^2\right] \geq \frac{1}{\mathrm{I}(\mu)} \quad (16)$$

where $\mathrm{I}(\mu)$ is the Fisher Information. In other words, $\sqrt{n}(T-\mu)$ is asymptotically normal with mean zero and variance,

$$A(T) = E\left[\mathrm{IF}(.)^2\right] \quad (17)$$

or we can write $\sqrt{n}\left(T-\mu\right) \rightarrow N\left(0, A(T)\right)$ (read: convergence in distribution), its mean,

$$T \rightarrow N\left(\mu, A(T)/n\right) \quad (18)$$

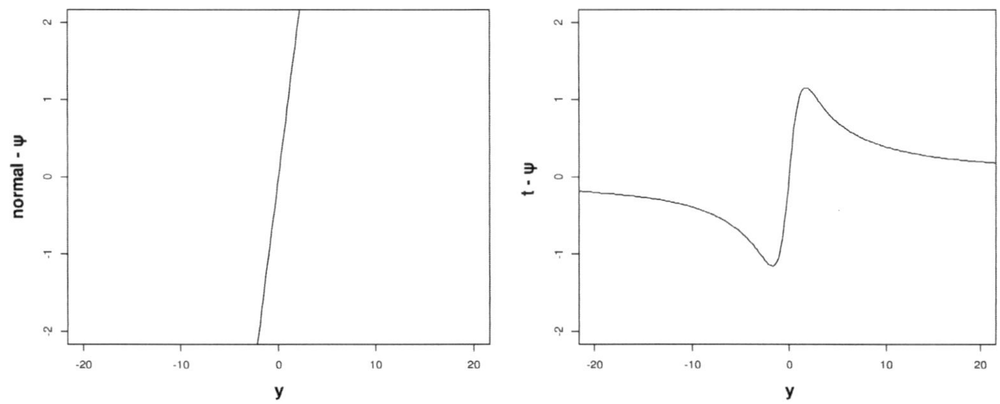

Figure 2. The ψ function of M-estimates from normal distribution and t-distributions.

Comparing figure 1 by figure 2 and according to (12), we can understand that $\mathrm{IF}(\hat{\mu}_T)$ is proportional to its ψ function. Now, we can find the asymptotic variance of $\hat{\mu}_T$ by (16) or (17).

$$A(\hat{\mu}_T) = E\left[IF(\hat{\mu}_T)^2\right] = \frac{E\left[\psi(y;\mu)\right]^2}{\left[-E\left[(\partial/\partial\mu)\psi(y;\mu)\right]\right]^2}$$

where $E\left[\psi(y;\mu)\right]^2 = E\left[\left(\frac{\partial}{\partial\mu}\left[\ln f(y)\right]\right)^2\right] = \mathrm{I}(\mu)$. Lange *et al.* (1989) showed that the expected information of location estimators under *t*-distributions was $\mathrm{I}(\mu) = \frac{v+1}{\sigma^2(v+3)}$. According to (11), $\left[-E\left[(\partial/\partial\mu)\psi(y;\mu)\right]\right]^2 = \left[\frac{v+1}{\sigma^2(v+3)}\right]^2 = \left[\mathrm{I}(\mu)\right]^2$. Thus, $A(\hat{\mu}_T) = \frac{\mathrm{I}(\mu)}{\left[\mathrm{I}(\mu)\right]^2} = \frac{1}{\mathrm{I}(\mu)} = \frac{(v+3)}{(v+1)}\sigma^2$ and the asymptotic variance of $\hat{\mu}_T$ is (19).

$$Var(\hat{\mu}_T) = \frac{(v+3)}{(v+1)}\frac{\sigma^2}{n} \tag{19}$$

The distribution of $\hat{\mu}_T$ converges to a normal distribution with mean μ and variance $Var(\hat{\mu}_T)$ as (19).

$$\hat{\mu}_T \rightarrow N\left(\mu, \frac{(v+3)}{(v+1)}\frac{\sigma^2}{n}\right) \tag{20}$$

Therefore the M-estimator under *t*-distributions is said to be asymptotically unbiased.

3.3. The Asymptotic relative efficiency

Definition (Casella and Berger 2002): if two estimator of μ, T and T' satisfy $\sqrt{n}(T-\mu) \rightarrow N(0, Var(T))$ and $\sqrt{n}(T'-\mu) \rightarrow N(0, Var(T'))$, the asymptotic relative efficiency (ARE) of T' with respect to T is defined as

$$ARE(T',T) = \frac{Var(T)}{Var(T')} \quad (21)$$

T is said to be asymptotically more efficient than T' if $ARE(T',T) \leq 1$ and T' is said to be asymptotically more efficient than T if $ARE(T',T) \geq 1$.

We know that the asymptotic variance of $\hat{\mu}_N$ is σ^2/n and according to (19) and (20), the asymptotic relative efficiency (ARE) of location estimators under t-distributions with v degrees of freedom ($\hat{\mu}_T$) respect to location estimators under normal distribution ($\hat{\mu}_N$) is

$$ARE(\hat{\mu}_T, \hat{\mu}_N) = \frac{Var(\hat{\mu}_N)}{Var(\hat{\mu}_T)} = \frac{\sigma^2/n}{(v+3)\sigma^2/(n(v+1))} \quad (22)$$

In normal data case (no outlier), an estimator under normal distribution of course will be better than t-distribution. The asymptotic relative efficiency of $\hat{\mu}_T$ with respect to $\hat{\mu}_N$ (22) will be

$$ARE(\hat{\mu}_T, \hat{\mu}_N, N(\mu, \sigma^2)) = \frac{v+1}{v+3} \quad (23)$$

We can see in table 1 that for normal data case, the efficiency of location estimators under t-distributions with 3 degrees of freedom ($v = 3$) is 0.67. If the degrees of freedom is raised to 30 ($v = 30$), the efficiency will increase to 0.94. Theoretically the ARE will tend to 1 as $v \rightarrow \infty$, and a location estimator under t-distributions as efficient as under the normal distribution.

Table 1. The Asymptotic Relative Efficiency of $\hat{\mu}_T$ respect to $\hat{\mu}_N$.

v	ARE
3	0.67
4	0.71
30	0.94
$\vdots$	
∞	1

In the case our data comes from t-distributions, the location estimator under t-distributions is more efficient than under the normal distribution ($ARE(\hat{\mu}_T, \hat{\mu}_N, F = t(\mu, \sigma^2, v)) > 1$). Back to table 1, if the data came from a t-distribution with 30 degrees of freedom then we would have got an $ARE = 1.06$, whereas if the degrees of freedom were equal to 3, we would have got $ARE = 1.5$.

It is interesting, if v is fixed, which degrees of freedom would be chosen? We know that an estimator under normal distribution is very sensitive to outlier which can be seen from an unbounded IF. We can see from figure 3 that the greater the degrees of freedom closer to normal IF. In other words, we can say that the smaller the degrees of freedom is more robust.

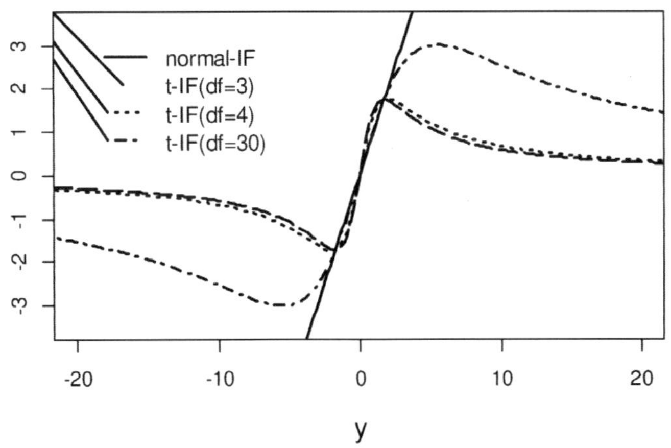

Figure 3. The Influence function of M-estimates under normal and *t*-distributions with some degrees of freedom (*df*).

4. Case study

One characteristics of Indonesian food is rich in flavors and a favorite seasoning is the onion. Brebes is the largest producer of onion in Indonesia with the best quality. The quality of onion is dependent on how treat the plants especially in dealing with pests. In this article, we studied how to fit the effects of solid pesticides on onion production when the data contains outliers. We used onion yield data presented in Listianawati [5] as a case study with little modification. The data is retrieved based on a survey of 67 onion farmers in the Kupu village, sub Regency Wanasari, Brebes Regency – the Middle Java Province at planting time May-August 2013. We used "heavy" package in R version 3.1.3 for the computing.

Figure 4(a) indicates that there is an outlier in the data. This happened on the 59th observation. He has produced 6.5 tons of onions when using solid pesticides as much as 2.6 kilograms, while the most other onion farmers harvested about 1 ton of onions and used about 1 kilogram of solid pesticides. We can also see in figure 4(a) that are likely to have a linear relationship between the use of solid pesticides and the onion productions. So we can use the linear models according to the normal model (2) or the t model (3). We set the degrees of freedom is equal to 4 ($v = 4$) for the t model.

Figure 4(b) shows the fitting model between the normal model and the t model. The regression line of the normal model is slightly above the regression line of the t model. We have just discussed in section before that the normal model is very sensitive to an outlier, and here we can see how an outlier influences the normal fit. If the outlier is removed, we will get two regression lines that overlap (figure 4(c)). It shows that the t model as good as the normal model when the data has no outlier. We can conclude from figure 4(b) and 4(c) that the t model was not influenced significantly by outliers or we can say that the t model is robust from outliers.

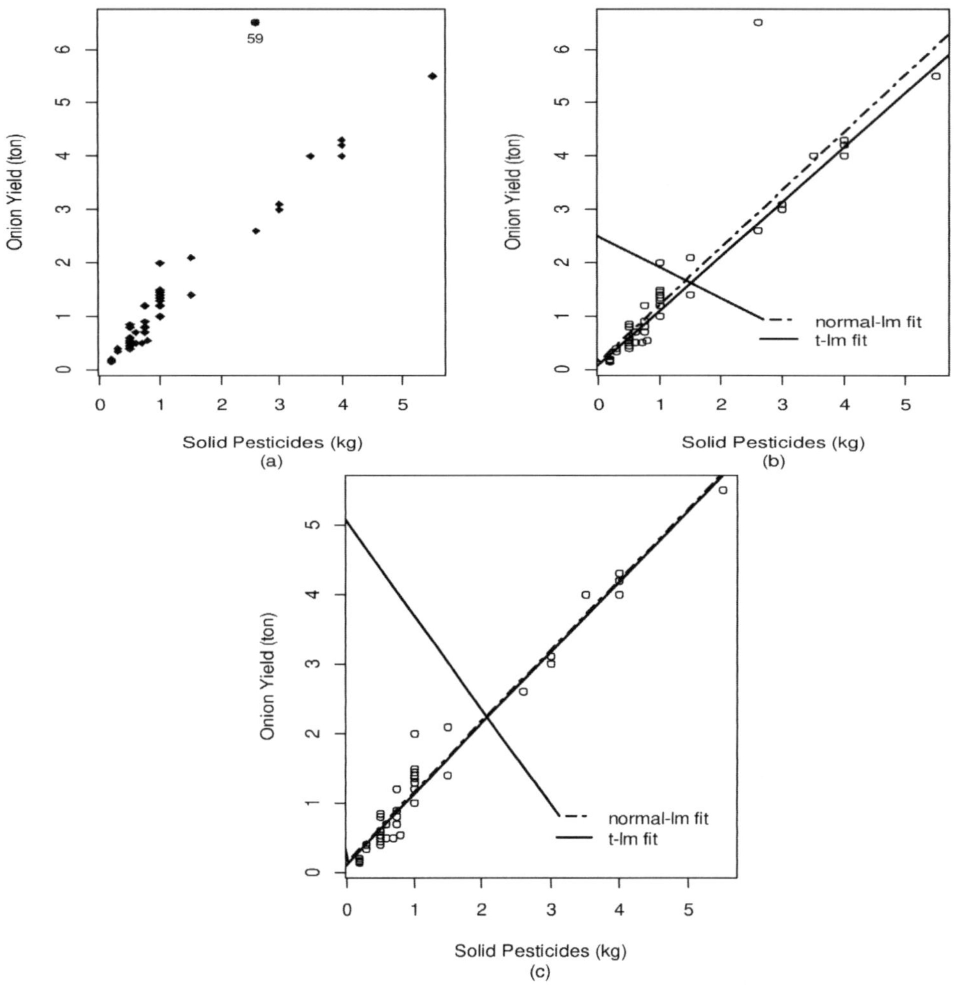

Figure 4. The plot of solid pesticides on the onion yield (a), the fitting model of normal and *t* models (b), the fitting model of normal and *t* models when the outlier is removed (c).

Table 2 presents that solid pesticides have significant effects against the onion yield both on the normal model and the *t* model. The addition of 1 kilogram of solid pesticides would increase production of the onion as much as 1.077 tons according to the normal model or 1.019 tons according to the *t* model. We see on a normal model would likely yield a higher estimation than the *t* model when there is an outlier to the right. The estimate of intercept in the normal model is also higher than the *t* model (0.131 for the normal model and 0.091 for the *t* model), but.it is not significant even at 90 percent of confidence level (*p-value*=0.146). While in the *t* model, it is significant at 95 percent of confidence level (*p-valu*e=0.018). This is because an outlier causing the estimated standard error of the regression coefficients in the normal model tend to be high so that it produces a high *p-value*.

The goodness of fit can be viewed from the value of log likelihood and/or the mean squared error (MSE). A good model would generate a maximum of log likelihood and/or minimum of MSE. Back to table 2, we see that a *t* model shows better fit than the normal model. The value of log-likelihood for the *t* model (-7.46) is greater than normal model (-49.34). The difference value is simply fantastic. It is about seven-times larger. It means that when the data includes an outlier, modelling which assumes *t*-

distributions will produce the log-likelihood maximum than normal model. The precision of the *t* model is also better than the normal model. The *t*-model produces smaller mean squared error (MSE=0.029) when compared to the normal model (MSE=0.255). The relative efficiency of the *t* model respect to the normal model is 8.8, or we may say the *t* model is about eight-times more efficient than the normal model in this cases. It is clear that the *t* model is more appropriate than the normal model.

Table 2. The estimate of the solid pesticides effect on the onion yield.

	normal-model			**t-model ($v = 4$)**		
	estimate	Std. error	*p-value*	estimate	Std. error	*p-value*
Parameters:						
Intercept	0.131	0.090	0.146	0.091	0.038	0.018
Solid Pesticides	1.077	0.054	0.000	1.019	0.023	0.000
Log Likelihood	-49.344			-7.460		
MSE	0.255			0.029		

5. Concluding Remarks

A location estimation which assumes *t*-distributions is one of robust estimations especially M-estimation because of the down weighted as discussed before. It has a bounded influence function and more efficient than under normal distribution when data contains outliers. Robustness of location estimators under *t*-distributions depends on its degrees of freedom. The smaller one is more robust. The case study on modeling of the onion yield has pointed out that the *t* model can overcome an outlier and preferably from normal models.

References

[1] Lange K L, Little R J A and Taylor J M G 1989 *Journal of the American Statistical Association* **408** 881-96
[2] Huber P J and Ronchetti E M 2009 *Robust Statistics Second Edition* (Hoboken, New Jersey: John Wiley & Son Inc.)
[3] Kotz S and Nadarajah S 2004 *Multivariate t Distribution and Their Applications* (Cambridge: Cambridge University Press)
[4] Casella G and Berger R L 2002 *Statistical Inference 2nd Edition* (Pacific Grove, USA: Duxbury)
[5] Listianawati N N 2014 The analysis of factors affecting the production of onion in the Kupu village sub Regency Wanasari Brebes Regency [under graduate Thesis] (Jakarta: Syarif Hidayatullah State Islamic University Jakarta)

Appendix

The derivative of the ψ function under *t*-distribution

$$\psi = \frac{\partial}{\partial \mu}\left[\ln f(y)\right] = \frac{v+1}{v+\left((y-\mu)/\sigma\right)^2}\left(\frac{y-\mu}{\sigma^2}\right)$$

$$\psi' = \frac{\partial \psi}{\partial \mu} = \frac{\partial}{\partial \mu}\left[\frac{v+1}{v+\left((y-\mu)/\sigma\right)^2}\left(\frac{y-\mu}{\sigma^2}\right)\right]$$

Note, $\mathrm{w}=\frac{\mathrm{u}}{\mathrm{v}}, \quad \mathrm{w}'=\frac{\mathrm{u'v-v'u}}{\mathrm{v}^2}$

$$\mathrm{u} = (v+1)\left(\frac{y-\mu}{\sigma^2}\right), \quad \mathrm{u}' = -\frac{v+1}{\sigma^2}, \quad \mathrm{v} = v + \left(\frac{y-\mu}{\sigma}\right)^2, \quad \mathrm{v}' = -2\left(\frac{y-\mu}{\sigma^2}\right)$$

$$\psi' = \frac{\left(-\frac{v+1}{\sigma^2}\right)\left(v+\left(\frac{y-\mu}{\sigma}\right)^2\right) - \left(-2\left(\frac{y-\mu}{\sigma^2}\right)\right)\left((v+1)\left(\frac{y-\mu}{\sigma^2}\right)\right)}{\left(v+\left(\frac{y-\mu}{\sigma}\right)^2\right)^2}$$

$$= \frac{\left(-\frac{(v+1)v}{\sigma^2}\right) - \left(\frac{(v+1)}{\sigma^2}\left(\frac{y-\mu}{\sigma}\right)^2\right) + \left(2\frac{(v+1)}{\sigma^2}\left(\frac{y-\mu}{\sigma}\right)^2\right)}{\left(v+\left(\frac{y-\mu}{\sigma}\right)^2\right)^2}$$

$$= \frac{\left(-\frac{(v+1)v}{\sigma^2}\right) + \left(\frac{(v+1)}{\sigma^2}\left(\frac{y-\mu}{\sigma}\right)^2\right)}{\left(v+\left(\frac{y-\mu}{\sigma}\right)^2\right)^2}$$

$$= \frac{\left(\frac{v+1}{\sigma^2}\right)\left(\left(\frac{y-\mu}{\sigma}\right)^2 - v\right)}{\left(v+\left(\frac{y-\mu}{\sigma}\right)^2\right)^2}$$

Characterization and Identification of Phosphate Solubilizing Bacteria Isolate GPC3.7 from Limestone Mining Region

D Fitriyanti[1], N R Mubarik[1] and A Tjahjoleksono[1]

[1] Department of Biology, Bogor Agricultural University, INDONESIA.

E-mail: nrachmania@ipb.ac.id

Abstract. Phosphate (P) are one of major macronutrients needed by plants. P in the soil are present in the organic and inorganic form. The amounts of P in marginal soil can be increased with plant growth promoting rhizobacteria (PGPR). The aim of this study was to characterize and identify P solubilizing bacteria (PSB) isolate GPC3.7 that characteristically could fix N from the soil around limestone mining area. There were 44 PSB isolates found from 15 soil samples around limestone mining area, Blindis mountain, Cirebon. The solubility index of all strain were measured about 0.125 to 2.375 on Pikovskaya media. There were 22 PSB isolates were grown on N-free bromothymol blue (NfB) medium and 19 isolates were grown on Congo Red Agar (CRA) medium. Only 10 isolates were indicated as symbiotic living microorganisms whereas 12 others were categorized as N-free fixing bacteria. Isolate GPC3.7 was chosen to be further observed, based on its P solubility index, N-fixing ability and growth stability. Phosphate quantitative estimation assay of isolate GPC3.7 was unmeasured. The P soluble concentration of GPC3.7 might be lower than 1 mg/L. The colony of GPC3.7 morphologically had round shape, entire margin, raised elevation and white color. Isolate GPC3.7 was Gram negative bacteria with coccus cell shape. Based on 16S rRNA gene, GPC3.7 was closely relative to *Acinetobacter baumannii*.

1. Introduction

Phosphate (P) and nitrogen (N) are macronutrient that is needed by plants, especially for various protein formation. Most of soil nitrogen is bound to organic materials, whereas the rest is used by plants in inorganic forms (NO_3^- and NH_4^+) [1]. Both organic and inorganic forms of soil phosphate are difficult to soluble, therefore the availability for organism especially plants are limited [2]. Chemical contents, including nitrogen and phosphate, in marginal or critical soil are lower than in fertile soil [3]. Statistics Institution of Indonesia [4] stated that critical land in Indonesia reached 27,296,000 hectares in 2012 whereas critical dry land that temporary unused in Indonesia on the same year was recorded as much as 14,252,383 hectares [5]. As marginal land, the ex-limestone mining soils physically, chemically and biologically has low fertility for agriculture activities [6].

Revegetation is needed for marginal soil such as ex-limestone mining soil. The use of NPK (nitrogen, phosphorus, and potassium) fertilizer was one of the alternatives most used. NPK fertilizer could increase nitrogen concentration in soil [7]. However, the excessive use could cause negative effects for plants such as root growth inhibition [8], and for environment, that is causing pollution and fast depletion of other nutrients (S, Ca, Mg, Zn, and Cu) [9]. Therefore, a specific and envinronmental friendly agent that can fix nitrogen and solubilize inorganic phosphate from soil are needed. One of biological agents that can fix nitrogen and solubilize phosphate is from plant growth promoting rhizobacteria (PGPR) group.

PGPR live around roots area and are able to produce plant growth promoting substances, such as hormone and enzyme. Some rhizobacteria group are able to fix nitrogen and live symbiotically with roots, forms nodul, such as *Bradyrhizoium japonicum* or free living as *Azotobacter* sp. [10]. Phosphate solubilizing bacteria can convert insoluble phosphate to soluble form that beneficial for plants. Phosphate solubilizing activity of *Arthrobacter ureafaciens* and *Delfia* sp. were associated with organic acid releasing and pH medium decreasing [11].

Nitrogen fixing and phosphate solubilizing bacteria can be isolated from mining area soil. Mursyida *et al.* [12] found phosphate solubilizing bacteria from limestone mining soil, Palimanan Quarry. Those are *Bulkholderia cepacia, Serratia marcescens* and *Pseudomonas putida.* Other bacteria from *Azotobacter* genus had been isolated from phosphate rock mining [13]. Natural phosphate deposit can be found around regions that contain abundant of lime [14]. One of limestone mining, PT Indocement is located in Cirebon, West Java. Revegetation is needed in the open ex-mining limestone areas there. PGPR potential bacteria as nitrogen fixer and phosphate solubilizer from PT Indocement limestone mining soil need to be further explored. One of the potential isolate coded as isolate GPC3.7. The aim of this study was to characterize and identify P solubilizing bacteria isolate GPC3.7 that characteristically could fix N from the soil round limestone mining area.

2. Material and Methods

2.1. Materials

Soil samples were obtained from Blindis mountain, around limestone mining area of PT Indocement, Palimanan, Cirebon.

2.2. Methods

2.2.1. Phosphate Solubilizing Bacteria Isolation. One gram of each soil sample was diluted in NaCl 0.85% solution until 10^{-4} dilution factor [15]. One mililiter of each dilution was spread on Pikovskaya media [16] and stored at 28 °C for 7 days incubation time. The colony that formed halo zone was purified and subcultured on Pikovskaya agar. Morphological characteristics of each colony were observed following Holt *et al.* [17].

2.2.2. Nitrogen Fixing Bacteria Selection. Each phosphate solubilizing isolates was grown on nitrogen free bromothymol blue (NfB) media [18], incubated at 28 °C for 7 days incubation time. Selected isolates were also grown on Congo Red Agar (CRA) [19]. Isolate that couldn't grow on CRA medium, but could grow on NfB medium was grown on Yeast Mannitol Agar (YMA), added with 0.0025% Congo red [20].

2.2.3. Qualitative Estimation on Phosphate Solubilization. Dot method was applied on qualitative estimation, using sterile tooth stick on Pikovskaya agar and was incubated in room temperature for 7 days. Solubility index (SI) was measured by subtracting the value of halo zone diameter toward the value of colony diameter. The resulted value divided by colony diameter [12].

2.2.4. Quantitative Estimation of Phosphate Solubilization. Isolate GPC3.7 were cultivated in 100 mL Pikovskaya broth, incubated in shaking incubator for 7 days at 37 °C. Every 24 h, 1.5 mL of each culture was centrifuged at 10600 ×*g* for 10 min. Each 1 mL were reacted with reagents consists of 2.5 mL sodium molybdate 2.5% and 1 mL hydrazine sulfate 0.3%. Formulation was made by following Lynn *et al.* [21].

2.2.5. Nitrogenase Activity Assay. Nitrogenase activity assay was measured by Acetylene Reduction Assay (ARA) method using gas chromatography instrument [22]. Isolate GPC3.7 was grown overnight in NfB medium at 30 °C. As much as 0.5 mL culture volume was inoculated to 5 mL NfB that was placed in 25 mL tube [23], incubated for five days at 30 °C [24]. The tube was covered by rubber cork and parafilm paper after 5 days of incubation. The ARA measurement were done by dropping the air in the tube out, used 1 mL sterile syringe, then the acetylene gas (C_2H_2) was injected into tube with the same volume as the dropped out air. After 2 hours of incubation, ethylene gas in the tube was measured, using gas chromatography instrument.

2.2.6. Hypersensitivity Assay. Isolate that was grown in nutrient broth (NB) (cell density $\pm 10^8$ cell/mL) was injected on the below surface of tobacco leaf by using syringe (without needle) [25]. *Pseudomonas syringae* was used as positive control whereas *Bacillus cereus* and sterile media were used as negative control. Hypersensitivity respond was observed after 18 hours of injection time [26].

2.2.7. Identification Based on 16S rRNA gene. The DNA of isolate GPC3.7 was isolated following the Presto™ Mini gDNA Bacteria Kit protocol. The 16S rRNA gene was amplified by Polymerase Chain Reaction (PCR) method, using specific primer for bacteria, 63f (5'-CAGGCCTAACACATGCAAGTC-3') and 1387r (5'-GGGCGGWGT-GTACAAGGC-3') [27]. The mix PCR was made up to 10 µL, consists of 5 µL GoTaq Green (Promega), 0.5 µL 63f primer, 0.5 µL 1387r primer, 1 µL DNA template and 3 µL ddH_2O. PCR was done following condition: predenaturation (94 °C, 5 min), denaturation (92 °C, 30 min), annealing (55 °C, 30 sec), elongation (72 °C, 1 min), and postelongation (72 °C, 5 min), made up to 30 cycles. The amplified DNA then was electroforated at 80 V for 30 min to see the pureness. The amplified DNA samples were sent to sequencing service provider. The DNA sequence was aligned by using Basic Local Alignment Search Tool-Nucleotide (BLASTN) from National Center for Biotechnology Information (NCBI). Phylogenetic tree were made by MEGA6 software [28].

2.2.8. Growth Curve Observation. One loop of bacteria isolate GPC3.7 was inoculated into 20 mL NB, incubated in shaker incubator until the cell density reached 10^8 CFU/mL. One mL of culture was inoculated to 100 mL NB, incubated in shaker incubator (120 rpm, 30 °C) from 0 – 48 h. Cell density was measured every 3 h by using spectrophotometer Genesys 20 at 620 nm wavelength. Cell amount was determined by comparing with log cell curve upon cell turbidity of isolate GPC3.7.

3. Result and Discussion

3.1. Result

There were 44 phosphate solubilizing bacteria (PSB) isolates found from 15 soil samples around limestone mining area, Blindis mountain, Cirebon. Preliminary screening was done by observing the clear zone on Pikovskaya media (Figure 1a). The solubility index of all isolate was measured about 0.125 to 2.375. There were 22 PSB isolates were grown on N-free bromothymol blue (NfB) medium and 19 isolates were grown on Congo Red Agar (CRA) medium. Colonies that grew on NfB medium showed that the bacteria could fix free nitrogen because media did not content nitrogen source. Nitrogen fixing bacteria could change the color of NfB medium from green to blue (Figure 1b). Colonies that could absorb Congo red were indicated as nitrogen free fixing bacteria. Only 10 isolates were indicated as symbiotic living microorganisms whereas 12 others were categorized as N-free fixing bacteria.

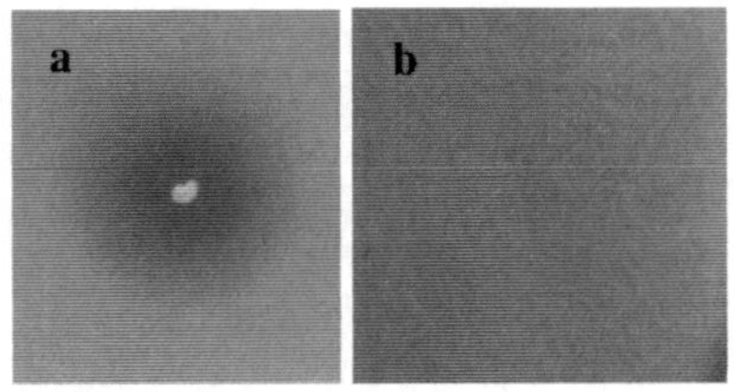

Figure 1 Isolate GPC3.7 colony on Pikovskaya agar after 7 days of incubation (a) and on NfB medium after 3 days of incubation (b).

There were 9 isolates that could not cause hypersensitivity symptoms (necrotic) on tobacco leaf after 48 h of incubation, compared to the hypersensitivity reaction caused by *P. syringae* as positive control, showed the forming of yellow-brownish spot on tobacco leaf. The description of 9 isolates are described in Table 1.

Table 1. Growth characteristics of nine isolates.

No	Isolate code	Gram	P Solubility Index	Growth on media			Hypersensitivity to tobacco leaves
				NfB	CRA	YMA	
1	GPA2.1	Negative	0.304	+++***	+++	0	Negative
2	GPA2.2	Negative	0.263	+*	++**	0	Negative
3	GPC1.3	Positive	0.444	+++	-	+++	Negative
4	GPC1.7	Positive	0.278	++**	+++	0	Negative
5	GPC3.7	Negative	2.375	++	+++	0	Negative
6	GPC4.2	Negative	0.714	+	+	0	Negative
7	GPC4.3	Positive	0.206	++	+++	0	Negative
8	GPC4.6	Negative	0.250	++	+++	0	Negative
9	GPC4.11	Negative	0.229	+	+++	0	Negative

Information: (*) small colony size, (**) moderate colony size, (***) big colony size, (-) not growing bacteria, (0) isolate was not grown on the media.

Isolate GPC3.7 was chosen to be further observed, based on its P solubility index, N-fixing ability and growth stability. Phosphate quantitative estimation assay of isolate GPC3.7 was unmeasured (Figure 2). The isolate GPC3.7 did not form blue color complex after reacted with reagent, but yellow that could not be well read on 830 nm wavelength. Nitrogenase activity of isolate GPC3.7 was determined by the amount of ethylene concentration produced. The concentration of ethylene produced by isolate GPC3.7 was 0.291 ppm/h.

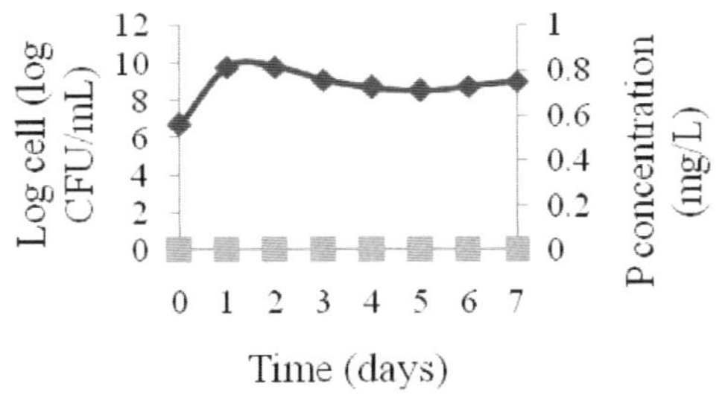

Figure 2. Amount of cell and phosphate solubility concentration of isolate GPC3.7 for 7 days observation. .▬■▬: P concentration (mg/L), ▬◆▬ : log cell (log CFU/mL).

The colony of GPC3.7 morphologically had round shape, entire margin, raised elevation and white color. Based on Gram staining, isolate GPC3.7 was Gram negative bacteria with coccus cell shape. The phylogenetic tree showed that isolate GPC3.7 was closely relative to *Acinetobacter baumannii* with similarty level 100% (Figure 3).

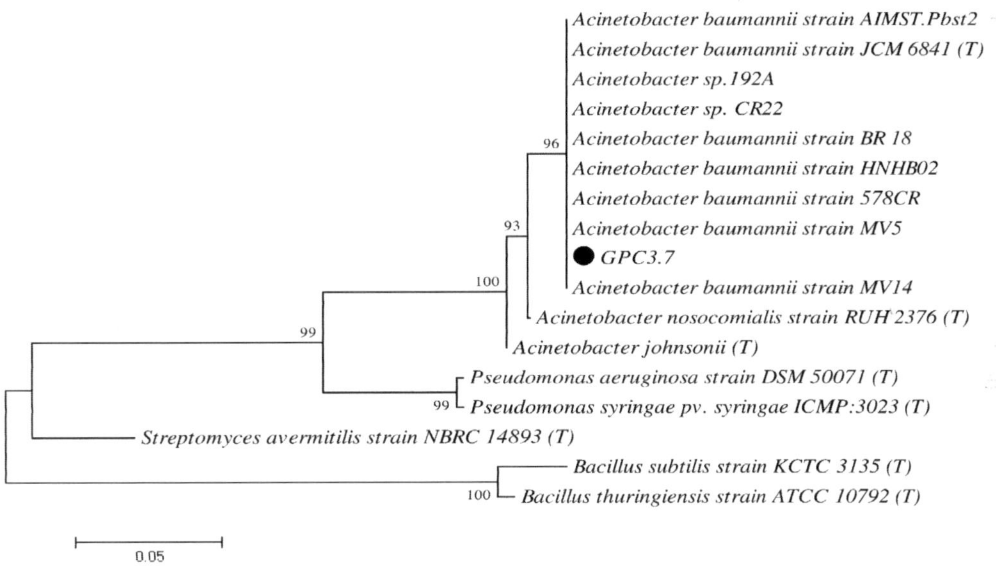

Figure 3. The phylogenetic tree of isolate GPC3.7

The growth curve of isolate GPC3.7 were observed every three hours. The lag phase was not observed. Logarithmic phase was observed at 30 hours (11.793 CFU/mL). Stationer phase was observed from 33 – 36 hours (Figure 4).

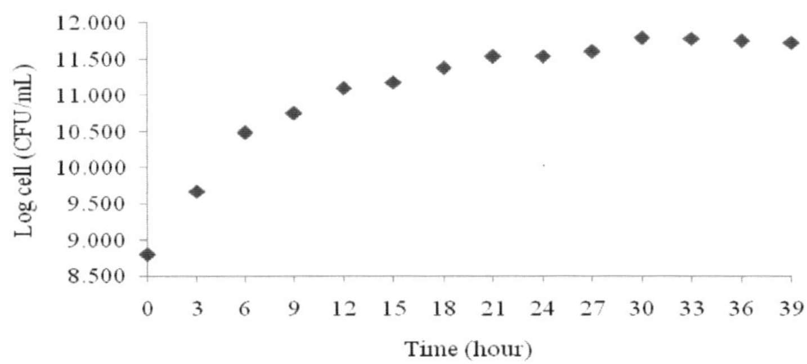

Figure 4. Growth curve of isolate GPC3.7

3.2. Discussion

Plant growth promoting rhizobacteria (PGPR) is a group of bacteria that live and colonize around plant root or rhizosphere. Some PGPR are able to solubilize phosphate and fix nitrogen that beneficial for plant. Phosphate solubilizing and nitrogen fixing bacteria can be isolated from different kind of soils and habitats. Chen *et al.* [11] isolated PSB from subtropical soil in Taiwan. Bacteria that could fix N and are able to solubilize phosphate could be isolated from some species of lichens, namely *Canoparmelia caroliniana*, *Canoparmelia crozalsiana*, *Canoparmelia texana*, *Parmotrema sancti-angeli* and *Parmotrema tinctorum* [29]. Siddike *et al.* [30] isolated PSB and nitrogen fixing halotolerant bacteria from soil around coastal area. Phosphate solubilizing psychrotolerant bacteria from extreme region such as Antarctic had also been reported [31].

Phosphate solubilizing bacteria could be isolated by using Pikovskaya media. Phosphate solubilizing bacteria isolate that grow on Pikovskaya media generally form clear zone around the colony. Clear zone forming indicates the phosphate solubilization activity and acid production by bacteria [32]. Tricalcium phosphate (Ca_3PO_4) in media could be absorbed by plant in ion form (Pi, HPO_4^{2-}, $H_2PO_4^-$) [33].

Nitrogen fixing bacteria was selected by three media, which were the NfB, CRA and YMA. NfB was a selected medium for the growth of nitrogen fixing bacteria. A number of researches used NfB as selected media for nitrogen free living bacteria [34, 35]. CRA was used to distinguish nitrogen fixing bacteria that live freely and symbiotically with plant root. Congo red in the media could distinguish rhizobia group with nitrogen free living bacteria *Azospirillum* [19]. Congo red stains amyloid and bacteria cellulose, also interact with β-D-glucan and polysaccharide on its rhizobia group capsule [36]. YMA plus Congo red medium has the same function as CRA medium. Generally, YMA medium is used for cultivating nitrogen fixing symbiotic rhizobia such as *Bradhyrhizobium japonicum* that could not absorb Congo red [37]. Selection process on NfB, CRA and YMA media could give us preliminary information about the ability of nitrogen fixation by bacteria. Isolate GPC3.7 could grow and change the color of NfB media from blue to green, as indication of pH changing. Isolate GPC3.7 also could not absorbs Congo Red that refer the isolate to root symbiotically bacteria.

Phosphate solubilization of isolate GPC3.7 was not quantitatively observed. Based on phosphate solubility curve (Figure 2), we could not define the correlation between cell growth (log cell) and the amount of soluble phosphate. The result implied that there was no correlation between qualitative and quantitative estimation, based on the observation of its growth on Pikovskaya agar and in Pikovskaya broth medium. Some researches had reported the same condition [12, 38]. The P soluble concentration of GPC3.7 might be lower than 1 mg/L.

Nitrogenase is the important key of N_2 transformation to ammonia (NH_3). Nitrogenase is encoded by a set of operon, including regulator genes (such as *nif*L dan *nif*A), structural genes (such as *nif*H, *nif*D and *nif*K) and other supporting genes. The *nif*H gene regulates the Fe protein whereas *nif*D and *nif*K genes regulates Mo-Fe protein [39]. From preliminary screening, we assumed that isolate GPC3.7 qualitatively could fix nitrogen and live symbiotically. Nitrogenase activity was measured by

Acetylene Reduction Assay (ARA). Acetylene (C_2H_2) in ARA method was used as alternative substrate. Nitrogenase complex reduces the three bond of N_2 to ammonia form. Nitrogenase activity is measured due to ethylene gas (C_2H_4) production [37]. The concentration of ethylene produced by isolate GPC3.7 was 0.291 ppm/h. Its activity was higher than bacteria that isolated from soil around Jambi forest, *Beijerinckia fluminensis* (0.094 ppm/h) [24].

The phylogenetic tree showed that isolate GPC3.7 was closely related to *A. baumannii* with similarity level 100% (Figure 3). *Acinetobacter* genus was known as strong phosphate solubilizer, but the name *A. baumannii* was not found as biofertilizer in any paper. Liba *et al.* [29] stated that chemoorganotropic bacteria that could fix N from genus *Acinetobacter* could be isolated from lichen. Isolate *Acinetobacter* strain CR 1.8 from rice rhizospheric soil in Northern Thailand had been reported was able to grow on up to 25% NaCl, between 25 and 55 °C, and pH 5–9. The isolate also had protease activity [40].

4. Conclusion

Phosphate solubilizing bacteria that could fix N could be found from soil around limestone mining area in Cirebon. Isolate GPC3.7 had the highest solubility index (2.375), however its quantitative estimation could not be determined because the concentration of soluble phosphate might lower than 1 mg/L. Based on nitrogense reduction assay, isolate GPC3.7 could convert acetylene to ethylene as much as 0.291 ppm/h. Isolate GPC3.7 was Gram positive bacteria. Identification based on 16S rRNA showed that isolate GPC3.7 was closely related to *Acinetobacter baumannii.*

References

[1] Murphy B W 2014 *Soil organic matter and soil function – Review of the literature and underlying data* (Canberra: Department of The Environment, Australia) p 73
[2] Saraswati R Husen E and Simanungkalit R D M 2007 *Soil Biology Analysis Method* (Bogor: BBPPSLP) pp 39-40
[3] Suharta N 2010 *J. Litbang Pertan.* **29** 139-146
[4] Statistics Indonesia 2014 *Statistical Yearbook of Indonesia 2014* (Jakarta: Statistics Indonesia) p 246
[5] Cakrabawa D N Hakim L and Gultom R 2003 *Statistics of Agricultral Land , 2008-2012* (Jakarta: Center for Agricultural Data and Information System Secretariate General-Ministry of Agriculture) p 190
[6] Prayudyaningsih R 2014 *J. Penel. Kehut. Wallacea* **3**13-23
[7] Mazzoncini M Sapkota T B Bárberi P Antichi D and Risaliti R 2011 *Soil Tillage Res.* **114** 165-174
[8] Herdiana N Lukman A H and Mulyadi K 2008 *J. Penel. Hutan Konserv. Alam* **5** 289-296
[9] Las I Subagyono K and Setiyanto A P 2006 *J. Litbang Pertan.* **25** 106-114
[10] Moat A G Foster J W and Spector M P 2002 *Microbial Physiology* (New York: Wiley-Liss Inc.) pp 475-480
[11] Chen Y P Rekha P D Arun A B Shen F T Lai W A and Young C C 2006 *Appl. Soil Ecol.* **34** 33-41
[12] Mursyida E Mubarik N R and Tjahjoleksono A 2015 *Res. J. Microbiol.* **10** 270-279
[13] Reyes I Valery A and Valduz Z 2006 *Plant and Soil* **287** 69-75
[14] Sastramihardja H Manalu F and Aprillani S E 2009 Natural Phosphate: Utilization of direct natural phosphate as P fertilizer source (Bogor: Balai Penelitian Tanah) pp 1-3
[15] Hadioetomo R S 1993 *Basic microbiology practical: Basic laboratory technique and procedure* (Jakarta: PT Gramedia Pustaka Utama) pp 76-81
[16] Nautiyal C S 1999 *Microbiol. Letters* **170** 265-270
[17] Holt J G Krig N R Sneath P Staley J and William S 1994 *Bergeys manual of determinative bacteriology* (Maryland: Lippincott Williams and Wilkins Company)
[18] Okon Y Albrecht S L and Burris R H 1977 *Appl. Environ. Microbiol.* **33** 85-88

[19] Cáceres E A R 1982 *Appl. Environ. Microbiol.* **44** 990-991
[20] Situmorang A R F Mubarik N R and Triadiati 2009 *HAYATI J. Biosci.* **16** 157-160
[21] Lynn T M Win H S Kyaw E P Latt Z K and Yu S S 2013 *Intern. J. Innovat. Appl. Studies* **3** 959-966
[22] Gibson A H and Turner G L 1980 Measurement of nitrogen fixation by indirect means. In: *Methods for Evaluating Biological Nitrogen Fixation*, ed F J Bergensen (New York: John Willey & Sons) pp 111-138
[23] Neilson A H and Sparell L 1976 *Appl. Environ. Microbiol.* **32** 197-205
[24] Harca N N 2015 Isolation and identification of nitrogen fixing and producing indole acetic acid bacteria from oil palm plantation in Jambi M Sc Thesis Bogor Agricultural Univ. Bogor, pp 9-15
[25] Zou L Wang X Xiang Y Zhang B Li Y Xiao Y Wang J Walmsley A R and Chen G 2006 *Appl. Environ. Microbiol.* **72** 6212 – 6224
[26] Wahyudi A T Meliah S and Nawangsih A A 2011 *Makara Sains* **15** 89-96
[27] Marchesi J R Sato T Weightman A J Martin T A Fry J C Hiom S J and Wade W G 1998 *Appl. Environ. Microbiol.* **64** 795-799
[28] Tamura K Stecher G Peterson D Filipski A and Kumar S 2013 *Molecul. Biol. Evol.* **30** 2725-2729
[29] Liba C M Ferrara F I S Manfio G P Fantinatti-Garboggini F Alburquerque R C Pavan C Ramos P L Moreira-Filho C A and Barbosa H R 2006 *J. of Appl. Microbiol.* **101** 1076-1086
[30] Siddikee M A Chauhan P S Anandham R Han G H and Sa T 2010 *J. Microbiol. Biotechnol.* **20** 1577-1584
[31] Selbmann L Zucconi L Ruisi S Grube M Cardinale M and Onofori S 2010 *Polar Biol.* **33** 71-83
[32] Tripti Kumar P and Anshumali 2012 *Internat. J. Engineer. Res. Develop.* **3** 1-6
[33] Rodriguez H and Fraga R 1999 *Biotechnol. Adv.* **17** 319-339
[34] Jolly S N Shanta N A and Khan Z U M 2010 *Internat. J. Botan.* **6** 117-121
[35] Kanimozhi K Panneerselvam A 2011 *Arch. Appl. Sci. Res.* **3** 525-536
[36] Kneen B E and Larue T A 1983 *Appl. Environ. Microbiol.* **45** 340-342
[37] Mubarik N R and Sunatmo T 2014 Symbiotic of nitrogen fixation between acid aluminum tolerant *Bradyrhizobium japonicum* and soybean. In: *Advances in biology and ecology of nitrogen fixation*, ed T Ohyama (Rijeka: InTech) pp 259-274
[38] Alikhani H A Saleh-Rastin N and Antoun H 2007 Phosphate solubilization activity of rhizobia native to Iranian soils. In: *First Meeting on Microbial Phosphate* Solubilization, ed E Vélazquez and C Rodríguez-Barrueco (Dodercht: Springer) pp 35-41
[39] Cheng Q 2008 *J. Integrat. Plant Biol.* **50** 784-796
[40] Chaiharn M and Lumyong S 2009 *World J Microbiol Biotechmol* **55** 305-314

A Comparative Study of Imputation Methods for Estimation of Missing Values of Per Capita Expenditure in Central Java

Y Susianto[1], K A Notodiputro[1], A Kurnia[1] and H Wijayanto[1]

[1]Department of Statistics, Faculty of Mathematics and Natural Sciences, Bogor Agricultural University

Email: susianto.sugiharto@gmail.com, khairilnotodiputro@gmail.com, akstk29@gmail.com, hari_ipb@yahoo.com

Abstract. Missing values in repeated measurements have attracted concerns from researchers in the last few years. For many years, the standard statistical methods for repeated measurements have been developed assuming that the data was complete. The standard statistical methods cannot produce good estimates if the data suffered substantially by missing values. To overcome this problem the imputation methods could be used. This paper discusses three imputation methods namely the Yates method, expectation-maximization (EM) algorithm, and Markov Chain Monte Carlo (MCMC) method. These methods were used to estimate the missing values of per-capita expenditure data at sub-districts level in Central Java. The performance of these imputation methods is evaluated by comparing the mean square error (MSE) and mean absolute error (MAE) of the resulting estimates using linear mixed models. It is showed that MSE and MAE produced by the Yates method are lower than the MSE and MAE resulted from both the EM algorithm and the MCMC method. Therefore, the Yates method is recommended to impute the missing values of per capita expenditure at sub-district level.

1. Introduction

Incomplete data problems as a result of missing values in repeated measurements have attracted concerns from researchers in the last few years. For many years, standard statistical methods for repeated measurements have been developed assuming the data was complete. In fact, if the data suffered substantially by missing values then the standard statistical methods cannot produce good estimates. To overcome this problem, proper statistical analysis methods are required.

This problem can be overcome by the complete case analysis with assuming the missing values are missing completely at random (MCAR). However, this approach can substantially reduce the sample size and consequently the estimates become inefficient. Furthermore, Schafer and Graham [1], Nakai and Weiming [2], and Buuren [3] have shown that the complete case analysis under the assumption of missing at random (MAR) produces poor estimates.

In practice, the missing values are usually missing at random (MAR). In this situation the available case analysis can be used. However, as mentioned by Donders *et al.* [4], the available case analysis produces biased estimates unless the missing values are imputed before analysis is carried out. Through the imputation technique, the missing values are replaced by the most likely numbers as if they were observed.

This paper discusses three imputation methods namely the Yates method, expectation-maximization (EM) algorithm, and Markov Chain Monte Carlo (MCMC) method. These three methods were used to estimate the missing values of per capita expenditure data at sub-districts level in Central Java. The performance of these imputation methods are evaluated by comparing the mean square error (MSE) and mean absolute error (MAE) of the resulting estimates using linear mixed models.

2. Methods

2.1. The Data

The national social economy survey (SUSENAS) has been designed by Statistics Indonesia (BPS) to produce social-economic indicators in Indonesia. In 2011 until 2014, SUSENAS was conducted quarterly with rolling samples. As a result of these rolling samples the aggregated data at sub-districts level became incomplete. For a particular quarter, some sub-districts had the observed data but in other sub-districts the data might not be available. In other words, at sub-districts level we had missing data problems especially missing data for repeated measurements.

In this paper, the per capita expenditure data at sub-district level from 2011 until 2014 were imputed. This paper provided a preliminary study of the data imputation applied to five districts in Central Java province, namely: Banyumas, Pati, Semarang, Brebes and kota Semarang. The imputation of the data was conducted separately for each district.

2.2. The Yates Method

This method is a classical approach developed by Yates in 1933 to estimate missing values in experimental design for the purpose of minimizing error variance. This approach consists of 3 stages, namely: estimation of missing values, replacement of the missing values with prediction values, and analysis of the complete data [5].

Since the problem structures were similar to the missing problem in a randomized complete block design then the formula for this design was used.

a). The formula for a single missing value is

$$y_{ij} = \frac{m\, y_{i.} - n\, y_{.j} - y_{..}}{(m-1)(n-1)} \tag{1}$$

Where y_{ij} is the prediction of missing value at the-i^{th} sub-district and the-j^{th} time, $y_{i.}$ is the total observed values at the-i^{th} sub-district containing a missing value, $y_{.j}$ is is the total observed values at the-j^{th} time containing missing values, $y_{..}$ is the total observed values for all observations, m is the number of subjects, and n is the number of times.

b). If number of missing values are more than one, we iterate formula (1) using a starting value

$$y_{\text{initial}(ij)} = \frac{\bar{y}_{i.} + \bar{y}_{.j}}{2} \tag{2}$$

In this formula, $\bar{y}_{i.}$ is the mean values of the-i^{th} subject, $\bar{y}_{.j}$ is the mean values of the-j^{th} time.

2.3. The EM Algorithm

This algorithm is a parametric method to impute missing values based on the maximum likelihood estimation. This algorithm is very popular in statistical literatures and has been discussed intensively by many researchers, such as : Dempster *et al.* [6], Little and Rubin [5,7], Schafer [8], and Watanabe and Yamaguchi [9].

This algorithm uses an iterative procedure to finding the maximum likelihood estimators of parameter vector through two step described in Dempster *et al.* [6], Schafer [8], and Little and Rubin [7] as follows:

a). The Expectation step (E-step)

The E step is the stage of determining the conditional expected value of the full data of log likelihood function $l(\theta|Y)$ given observed data. Suppose for any incomplete data, the distribution of the complete data Y can be factored as

$$f(Y|\theta) = f(Y_{mis}, Y_{obs}|\theta)$$
$$= f(Y_{obs}|\theta)\, f(Y_{mis}|Y_{obs}, \theta) \quad (3)$$

where $f(Y_{obs}|\theta)$ is the distribution of the data observed Y_{obs} and $f(Y_{mis}|Y_{obs}, \theta)$ is the distribution of missing data given data observed. Based on the equation (3), we obtained log likelihood function

$$l(\theta|Y) = l(\theta|Y_{obs}) + log\, f(Y_{mis}|Y_{obs}, \theta) \quad (4)$$

where $l(\theta|Y)$ is log likelihood function of complete data, $l(\theta|Y_{obs})$ is log likelihood function of observed data, and $f(Y_{mis}|Y_{obs}, \theta)$ is the predictive distribution of missing data given θ.

Objectively, to estimate θ is done by maximizing the log likelihood function (4). Because Y_{mis} not known, the right side of equation (4) can not be calculated. As a solution, $l(\theta|Y)$ is calculated based on the average value $log\, f(Y_{mis}|Y_{obs}, \theta)$ using predictive distribution $f(Y_{mis}|Y_{obs}, \theta^{(t)})$, where $\theta^{(t)}$ is temporary estimation of unknown parameters. In this context, an initial estimation $\theta^{(0)}$ be calculated using the complete case analysis. With this approach, the mean value of equation (4) can be expressed

$$Q(\theta|\theta^{(t)}) = l(\theta|Y_{obs}) + \int log\, f(Y_{mis}|Y_{obs}, \theta) f(Y_{mis}|Y_{obs}, \theta^{(t)}) \partial Y_{mis}$$
$$= \int [l(\theta|Y_{obs}) + \int log\, f(Y_{mis}|Y_{obs}, \theta)]\, f(Y_{mis}|Y_{obs}, \theta^{(t)}) \partial Y_{mis}$$
$$= \int l(\theta|Y) f(Y_{mis}|Y_{obs}, \theta^{(t)}) \partial Y_{mis} \quad (5)$$

The equation (5) basically a conditional expected value of log likelihood function for complete data $l(\theta|Y)$ given observed data and initial estimate of unknown parameter.

b). The maximization step (M-step)

The M step is to obtained the iteratively estimation $\theta^{(t+1)}$ with maximizes $Q(\theta|\theta^{(t)})$ as follow

$$Q(\theta^{(t+1)}|\theta^{(t)}) \geq Q(\theta|\theta^{(t)}) \quad (6)$$

Both E and M steps are iterated until convergent.

2.4. The MCMC Method

This method generates pseudo random draws from probability distributions via Markov chains. A Markov chain is a sequence of random variables in which the distribution of each element depends on the value of the previous one. MCMC is a multiple imputation methods be used to imputate the missing values of continuous data set. In application, MCMC assumes that data have a multivariate normal distribution, missing data is MCAR or MAR, and patten of missing data is monotone or arbitrary. Moreover, the inference of MCMC will be robust if the number of missing values are not too large [8].

MCMC consists of the two most popular methods namely Gibbs Sampling and Metropolis-Hastings algorithm. In Gibbs sampling, one draws from the conditional distribution of each component of a multivariate random variable given the other components. In Metropolis-Hastings, one draws from a probability distribution intended to approximate the distribution actually of interest, and then accepts or rejects the drawn value with a specified probability. In this paper, we use a Gibbs sampling algorithm to draw the missing values from the posterior predictive distribution.

In Bayesian inference perspective, information about unknown parameters is expressed in the form of a posterior probability distribution. It is a useful alternative approach to ML with to add a prior distribution for the parameters and compute the posterior distribution of the parameters of interest. Suppose that $\boldsymbol{Y_{mis}}$ and $\boldsymbol{Y_{obs}}$ are the missing and observed values, respectively. The observed data posterior can be express as

$$p(\theta|Y_{obs}) \propto p(\theta)\, p(Y_{obs}|\theta) \quad (7)$$

where $p(\theta)$ is the prior distribution and $p(Y_{obs}|\theta)$ is the observed likelihood function. The problem of incomplete data is the observed data posterior $p(\theta|Y_{obs})$ is intractable and cannot easily be summarized or simulated. To overcome this problem, Y_{obs} is augmented by an assumed value of the Y_{mis}. The resulting complete-data posterior $p(\theta|Y_{obs}, Y_{mis})$ becomes much easier to handle. The observed data posterior is related to the complete-data posterior distribution that would have been obtained if we had observed the missing data Y_{mis}, namely

$$p(\theta|Y_{obs}, Y_{mis}) \propto p(\theta)\, p(Y_{obs}, Y_{mis}|\theta) \tag{8}$$

From equation (7) and (8) can be obtained

$$\begin{aligned} p(\theta|Y_{obs}) &= \int p(\theta, Y_{mis}|Y_{obs})\, dY_{mis} \\ &= \int p(\theta|Y_{obs}, Y_{mis})\, p(Y_{mis}|Y_{obs})\, dY_{mis} \end{aligned} \tag{9}$$

In equation (9), the posterior predictive distribution $p(Y_{mis}|Y_{obs})$ cannot be simulated directly. However, it is possible by create random draws of Y_{mis} from $p(Y_{mis}|Y_{obs})$ using techniques of MCMC. In this regard, we use a Gibbs sampling algorithm to draw the missing values Y_{mis} from $p(Y_{mis}|Y_{obs})$. By assuming that data have a multivariate normal distribution, data augmentation is applied to Bayesian inference with missing data by repeating the following steps

a). The imputation I-step

Given a current guess $\boldsymbol{\theta}^{(t)}$ of the parameter, create random draws of the missing values $\boldsymbol{Y}_{mis}$ from the posterior predictive distribution $\boldsymbol{p(Y_{mis}|Y_{obs})}$

$$Y_{i(mis)}^{(t+1)} \sim p(Y_{i(mis)}|Y_{obs}, \theta^{(t)}) \tag{10}$$

b). The posterior P-step

Then, with conditional to $\boldsymbol{Y}_{i(mis)}^{(t+1)}$, draw a new value of θ from the complete data posterior

$$\theta^{(t+1)} \sim p(\theta|Y_{obs}, Y_{i(mis)}^{(t+1)}) \tag{11}$$

Given starting from a initial values $\boldsymbol{\theta}^{(0)}$ and $\boldsymbol{Y}_{mis}^{(0)}$, these two step define a Gibbs sampler. Repeating the Gibbs sampling algorithm with large enough number of iterations, it creates stochastic sequences $\{\boldsymbol{\theta}^{(t)}\}$ and $\{\boldsymbol{Y}_{mis}^{(t)}\}$ whose stationary distribution are $\boldsymbol{p(\theta|Y_{obs})}$ and $\boldsymbol{p(Y_{mis}|Y_{obs})}$, respectively. In regard to MCMC, we use the initial value of the EM algorithm for the posterior mode, and the resulting EM estimates are used to begin the MCMC method. Moreover, we also specify the prior parameter information using one of a noninformative or ridge prior. A noninformative prior is used when no strong prior information is available about $\boldsymbol{\theta}$, it is customary to apply Bayes's theorem with the improper prior which is the limiting form of the normal inverted-Wishart. Meanwhile, a ridge prior is used when the sample covariance matrix is singular or nearly so, either because the data are sparse or because such strong relationships exist among the variables that certain linear combinations of the columns of $\boldsymbol{Y}$exhibit little or no variability [8].

2.5. Evaluation of Missing Values Imputation Results

In this paper, we evaluated the performance of these imputation methods by comparing the mean square error (MSE) of the resulting estimates $\boldsymbol{\mu} = \mathbf{X}\boldsymbol{\beta} + \mathbf{Z}\boldsymbol{u}$ using linear mixed models (LMM) for each district. For each district and time, we also evaluated the performance by comparing the mean absolute error (MAE) of the resulting estimates $\boldsymbol{y_{ij}}$.

Let $\boldsymbol{y_{ij}}$ denotes the response of repeated mesurement at the tth time on the ith sub-district, $\boldsymbol{i = 1, 2, \ldots, m}$ and $\boldsymbol{t = 1, 2, \ldots, n}$, the LMM for response vector of $\boldsymbol{y_i}$ is

$$\begin{bmatrix} y_{i1} \\ y_{i2} \\ \text{M} \\ y_{in} \end{bmatrix} = \begin{bmatrix} 1 & 1 & 0 & \text{L} & 0 \\ 1 & 0 & 1 & \text{L} & 0 \\ \text{M} & \text{M} & \text{M} & \text{O} & \text{M} \\ 1 & 0 & 0 & \text{L} & 1 \end{bmatrix} \begin{bmatrix} \mu \\ \tau_1 \\ \text{M} \\ \tau_n \end{bmatrix} + \begin{bmatrix} 1 \\ 1 \\ \text{M} \\ 1 \end{bmatrix} \alpha_i + \begin{bmatrix} e_{i1} \\ e_{i2} \\ \text{M} \\ e_{in} \end{bmatrix} \quad (12)$$

In matrix notation, equation (12) can be expressed as

$$\boldsymbol{y_i} = \mathbf{X}_i\boldsymbol{\beta} + \mathbf{1}_n\alpha_i + \boldsymbol{e_i}, \quad i = 1,2,\dots,m \quad (13)$$

Where $\mathbf{X}_i = [\mathbf{1_n}\ \mathbf{I}_n]$ is the matrix of time factor fixed effect at the ith sub-districts, $\boldsymbol{\beta} = [\mu\ \tau_1 \dots \tau_n\ ]$ is the parameter vector of fixed effect, α_i is the random effect of sub-districts, $\mathbf{1_n}$ is an unity vector order n, and $\boldsymbol{e_i} = [e_{i1} \dots e_{in}\ ]$ is the vector of error model at i sub-districts. Assuming α_i and e_{it} are independenly distributed with mean 0 and varian σ_α and σ. The variance-covariance matrix of $\boldsymbol{y_i}$, is given by $\mathbf{V_i} = \mathbf{V}_i(\boldsymbol{\delta}) = \sigma_\alpha\mathbf{J}_n + \sigma\mathbf{I}_n$ where $\boldsymbol{\delta} = (\delta_1,\ \delta_2)' = (\sigma,\ \sigma_\alpha)'$, $\mathbf{J}_n$ is a unity square matrix order n and $\mathbf{I}_n$ is an identity matrix order n.
Another form of equation (13) is

$$\mathbf{y} = \mathbf{X}\boldsymbol{\beta} + \mathbf{Z}\boldsymbol{u} + \boldsymbol{e} \quad (14)$$

where $\mathbf{y} = (\mathbf{y}_1', \dots, \mathbf{y}_m')'$, $\boldsymbol{u} = (\alpha_1, \dots, \alpha_m)'$, $\mathbf{X} = \mathbf{1_n} \otimes [\mathbf{1_n}\ \mathbf{I}_n]$, $\mathbf{Z} = \mathbf{I}_n \otimes \mathbf{1_n}$, and $\boldsymbol{e} = (\boldsymbol{e}_1', \dots, \boldsymbol{e}_m')'$. Assuming $\boldsymbol{u}$ and $\boldsymbol{e}$ are independenly distributed with mean $\mathbf{0}$ and covariance $\mathbf{G}$ and $\mathbf{R}$. The variance-covariance matrix of $\mathbf{y}$ is $\mathbf{V} = \mathbf{V}(\boldsymbol{\delta}) = \mathbf{ZGZ}' + \mathbf{R}$.

Refering to Rao and Molina [10], we have found the empirical best linear unbiased prediction (EBLUP) of $\boldsymbol{\mu} = \mathbf{X}\boldsymbol{\beta} + \mathbf{Z}\boldsymbol{u}$ is

$$\widehat{\boldsymbol{\mu}} = \mathbf{X}\widehat{\boldsymbol{\beta}} + \mathbf{Z}\widehat{\boldsymbol{u}} \quad (15)$$

where

$$\widehat{\boldsymbol{\beta}} = \widehat{\boldsymbol{\beta}}(\widehat{\boldsymbol{\delta}}_{RE}) = (\mathbf{X}'\mathbf{V}^{-1}\mathbf{X})^{-1}\mathbf{X}'\mathbf{V}^{-1}\mathbf{y} \quad (16)$$

is the best unbiased estimator (BLUE) of $\boldsymbol{\beta}$,

$$\widehat{\boldsymbol{u}} = \widehat{\boldsymbol{u}}(\widehat{\boldsymbol{\delta}}_{RE}) = \mathbf{GZ}'\mathbf{V}^{-1}(\mathbf{y} - \mathbf{X}\widehat{\boldsymbol{\beta}}) \quad (17)$$

is the best unbiased prediction (BLUP) of $\boldsymbol{\alpha_i}$, and $\widehat{\boldsymbol{\delta}}_{\boldsymbol{RE}}$ the restricted maximum likelihood (REML) estimator of $\boldsymbol{\delta}$ is obatained iteratively using the Fisher-scoring algorithm, with updateing equation

$$\widehat{\boldsymbol{\delta}}_{RE}\widehat{\boldsymbol{\delta}}_{RE}^{(a+1)} = \widehat{\boldsymbol{\delta}}_{RE}^{(a)} + \left[\boldsymbol{I}\left(\widehat{\boldsymbol{\delta}}_{RE}^{(a)}\right)\right]^{-1} \boldsymbol{s}\left[\widehat{\boldsymbol{\delta}}_{RE}^{(a)}\right] \quad (18)$$

Note that $\boldsymbol{s}\left[\widehat{\boldsymbol{\delta}}_{\boldsymbol{RE}}^{(\boldsymbol{a})}\right]$ is the partial derivative of log-likelihood function

$$l(\boldsymbol{\delta}) = c - \tfrac{1}{2}\log|\mathbf{V}| - \tfrac{1}{2}\log|\mathbf{X}'\mathbf{V}^{-1}\mathbf{X}| - \tfrac{1}{2}\mathbf{y}'\mathbf{P}\mathbf{y} \quad (19)$$

with respect to $\boldsymbol{\delta}$, $\boldsymbol{I}\left(\widehat{\boldsymbol{\delta}}_{\boldsymbol{RE}}^{(a)}\right)$ is the matrix of expected second-order derivatives of $-\boldsymbol{l}(\boldsymbol{\delta})$ with respect to $\boldsymbol{\delta}$. In equation (19), $\mathbf{P} = \mathbf{V}^{-1} - \mathbf{V}^{-1}\mathbf{X}(\mathbf{X}'\mathbf{V}^{-1}\mathbf{X})^{-1}\mathbf{X}'\mathbf{V}^{-1}$, c denotes a generic constant, and $\mathbf{Py} = \mathbf{V}^{-1}(\mathbf{y} - \mathbf{X}\widehat{\boldsymbol{\beta}})$.

The MSE of $\widehat{\boldsymbol{\mu}}$ is given by

$$\begin{aligned} \text{MSE}(\widehat{\boldsymbol{\mu}}) &= E[\widehat{\boldsymbol{\mu}} - \boldsymbol{\mu}]^2 \\ &\approx g_1(\hat{\delta}) + g_2(\hat{\delta}) + g_3(\hat{\delta}) \end{aligned} \quad (20)$$

where

$$g_1(\hat{\delta}) = \mathbf{Z}'(\mathbf{G} - \mathbf{GZ}'\mathrm{V}^{-1}\mathbf{ZG})\mathbf{Z} \quad (21)$$

is MSE of $\mathbf{X}\boldsymbol{\beta} + \mathbf{B}(\mathbf{y} - \mathbf{X}\boldsymbol{\beta})$ with $\mathbf{B} = \mathbf{ZGZ}'\mathbf{V}^{-1}$,

$$g_2(\hat{\delta}) = \mathbf{D}(X'V^{-1}X)^{-1}\mathbf{D}' \tag{22}$$

is variance of $\mathbf{D}(\hat{\boldsymbol{\beta}} - \boldsymbol{\beta})$ with $\mathbf{D} = (\mathbf{I} - \mathbf{B})\mathbf{X}$, and

$$g_3(\hat{\delta}) = \text{tr}[(\partial B/\partial\delta)\, \mathbf{V}(\partial B/\partial\delta)'\, \overline{\mathbf{V}}(\hat{\delta})] \tag{23}$$

with $\overline{\mathbf{V}}(\hat{\boldsymbol{\delta}})$ is the asymptotic covariance matrix of $\hat{\boldsymbol{\delta}}$.

Then for each district, the MAE of $\boldsymbol{y}_{ij}$ in the jth time is formulated

$$\text{MAE}\,(y_{ij}) = \frac{1}{m}\sum_i^m |y_{ij} - \hat{y}_{ij}| \tag{24}$$

3. Results and Discussion

In this section, we showed the results of each imputation method to estimate the missing values of per capita expenditure data at sub-district level of five districts in Central Java province based on Susenas data from 2011 until 2014. After the data being completed then we estimated the mean of per capita expenditures for each district using the three methods. We also calculated the corresponding MSE. Moreover, for each district the mean of per capita expenditures by time as well as the mean absolute error of the resulting estimates of per capita expenditures were calculated.

Table 1 shows the results of each imputation method applied to the first ten sub-district of Banyumas district. The three methods produced different estimates of the missing values. Based on fifteen quarterly missing cells, the mean imputation results of the Yates method tend to be small, whereas the results of the MCMC method tend to be large. Nevertheless, we will evaluate these results by comparing the MSE after using these three methods.

Table 2 shows the estimates of means as well as the corresponding MSE for each districts based on the three different methods of imputation. The results showed that MSE produced by the Yates method was lower than the MSE resulted from both the EM and the MCMC methods.

In Figure 1 the estimates of imputed means of per capita expenditures for each quarter using the three methods of five districts were presented. Generally for these five districts, the mean imputation resulted from the Yates method tended to be small, whereas the MCMC method tended to be large.

Figure 2 showed the estimates of MAE for each quarter in five districts, based on the three different methods of imputation. The results showed that MAE produced by the Yates method was lower than the MAE resulted from both EM and MCMC methods. It is apparent that to impute the missing values of per capita expenditure data at sub-district level we should use the Yates method.

Table 1. Mean Imputation Results for the First Ten Sub-districts in Banyumas District Using the Yates, EM Algorithm, and MCMC Methods

Sub-districts	Yates	EM Algorithm	MCMC
1. Lumbir	379.04	293.31	296.68
2. Wangon	633.46	547.53	547.48
3. Jatilawang	555.85	804.55	804.67
4. Rawalo	338.18	621.57	571.42
5. Kebasen	473.88	395.49	425.25
6. Kemranjen	455.10	549.33	637.50
7. Sumpiuh	397.25	467.16	541.47
8. Tambak	422.88	624.95	581.48
9. Somagede	403.23	261.31	368.17
10. Kalibagor	443.20	575.97	655.74

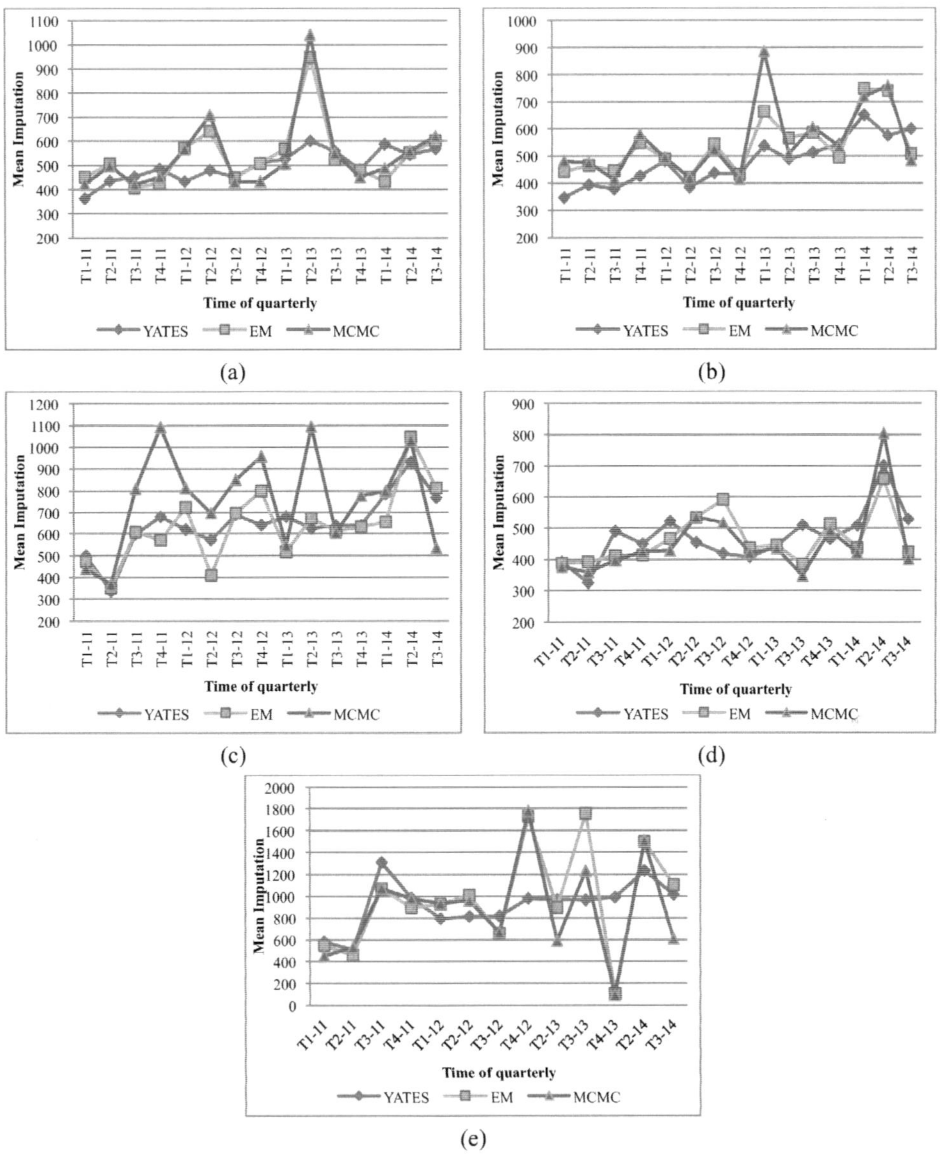

Figure 1. Comparison of Imputed Means among Yates, EM, and MCMC methods for five districts: (a) Banyumas, (b) Pati, (c) Semarang, (d) Brebes, and (e) Kota Semarang

IOP Conf. Series: Earth and Environmental Science **58** (2017) 012017 doi:10.1088/1755-1315/58/1/012017

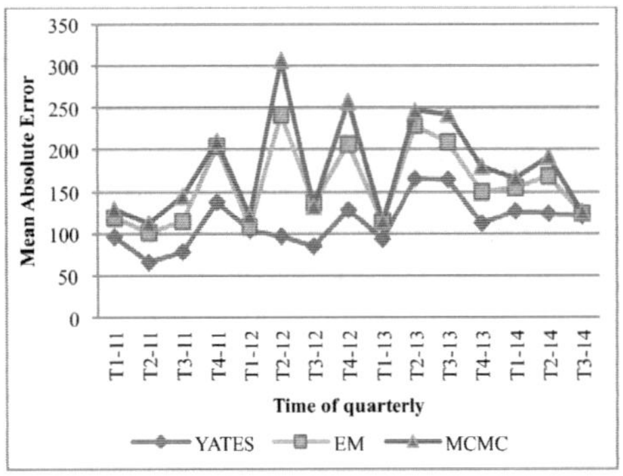

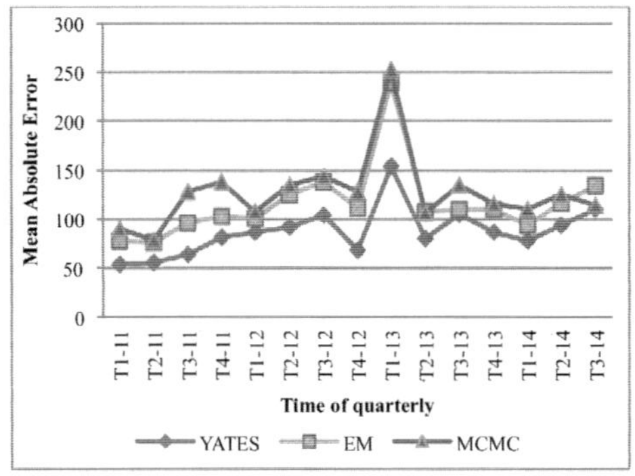

(a) (b)

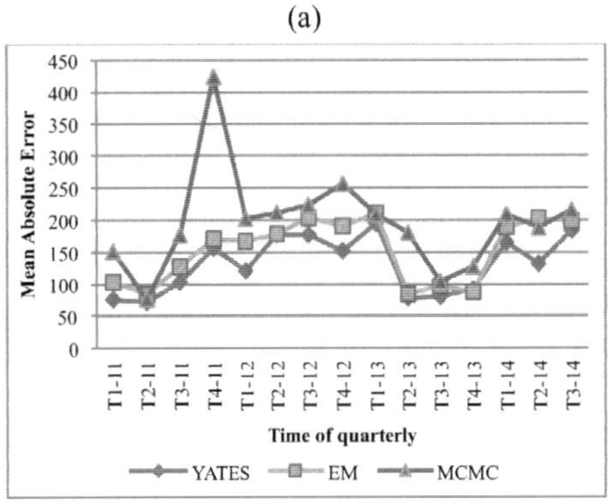

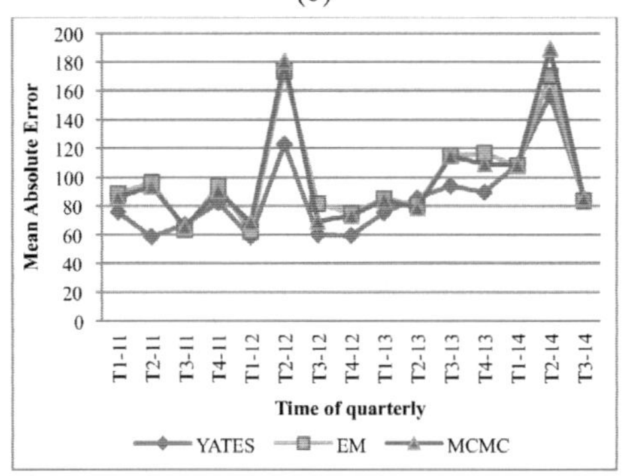

(c) (d)

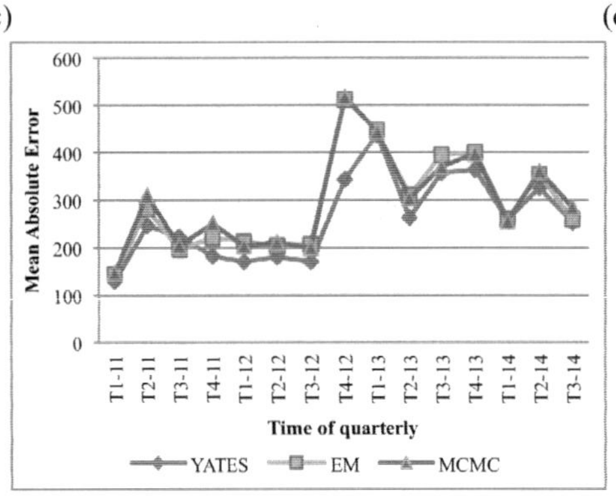

(e)

Figure 2. Comparison of Mean Absolute Error among Yates, EM, and MCMC methods for each district: (a) Banyumas, (b) Pati, (c) Semarang, (d) Brebes, and (e) Kota Semarang

Table 2. Comparison of Mean and MSE results Missing Values Imputation between Yates method, EM Algorithm, and MCMC per district

Districts	Yates		EM Algorithm		MCMC	
	Mean	MSE	Mean	MSE	Mean	MSE
1. Banyumas	490.03	3.5×10^4	527.71	5.6×10^4	531.80	6.6×10^4
2. Pati	460.96	2.7×10^4	520.30	3.4×10^4	533.48	3.8×10^4
3. Semarang	635.33	4.9×10^4	632.76	5.7×10^4	769.88	9.7×10^4
4. Brebes	456.01	1.8×10^4	460.80	2.1×10^4	450.62	2.2×10^4
5. Kota Semarang	885.61	15.5×10^4	959.71	18.0×10^4	915.84	18.2×10^4

Note : MSE were calculated based on residual sum of squares divided by their degree of freedom.

4. Conclusion

The results of this study revealed that the MSE produced by the Yates method was lower than the MSE resulted from both the EM algorithm and the MCMC method. These results were consistent with MAE of the Yates method which was also lower than the MAE resulted from the other two methods. Hence, based on those results, we concluded that to impute the missing values of per capita expenditure data at sub-district level we should use the Yates method.

Acknowledgments

We would like to thank Statistics Indonesia for giving us permission to use Susenas data for this paper.

References

[1] Schafer J L and Graham J W 2002 *Our View of the State of the Art* (*Psychological Methods* 7) **2** 147–177
[2] Nakai M and Weiming K 2011 *Int. Journal of Math. Analysis* **5** (1) 1-13
[3] Buuren S V 2012 *Flexible Imputation of Missing Data* (Boca Raton: Taylor & Francis Group, LLC)
[4] Donders A R T, van der Heijden G J M G, Stijnen T, and Moons K G M 2006 *Journal of Clinical Epidemiology* **59** 1087-1091
[5] Little R J A and Rubin D B 1987 *Statistical Analysis with Missing Data* (New York: John Wiley & Son Inc.)
[6] Dempster A P, Laird N M, and Rubin D B 1977 *Journal of the Royal Statistical Society* Series B **39** (1) 1-38
[7] Little R J A and Rubin D B 2002 *Statistical Analysis with Missing Data Second Edition* (Hoboken, New Jersey: John Wiley & Son Inc.)
[8] Schafer J L 1997 *Analysis of Incomplete Multivariate Data* (London: Chapman & Hall/CRC)
[9] Watanabe M and Yamaguchi K 2004 *The EM Algorithm and Related Statistical Models* (New York: Marcel Dekker, Inc.)
[10] Rao J N K and molina I 2015 *Small Area Estimation Second Edition* (Hoboken, New Jersey: John Wiley & Son Inc.)

Landscape Function of Post Tin-Mining Land After Reclamation in Bangka, Indonesia

H F Putra[1*], Sulistijorini[1] and N S Aryanti[1]

[1]Department of Biology, Faculty of Mathematics and Natural Sciences, Bogor Agriculture University, INDONESIA.

E-mail:hirmas_fp@yahoo.co.id

Abstract. Tin mining has a negative impact on the environment of Bangka, Indonesia. Efforts have being conducted to reclaim post tin-mining lands. However, the ecological function of the land have not been evaluated. The objective of this research was to assess the landscape function of post tin mining lands in Bangka. This researchwas conducted at three location that represented the tin mining conditions in Bangka.The landscape function analysis(LFA)was performed in each location for three types of land (A: unreclaimed tin-mining, B: reclaimed tin-mining, C: natural forest, as reference site). The results of LFA showed similar pattern for stability index among locations. The highest stability index (> 60%) occured at land C. On the other hand, the post tin mining area, land A and land B have low stability index (less than 38%). The infiltration index, high (>59%) in land C, low (33-36%) in land A and B).The nutrient cycling index also showed a value comparable to the stability index of land and infiltration index. It could be concluded that the post tin-mining landshas critical landscape function.However, by planting trees for reclaiming the area gradually improved the function of the post tin mining landscape.

Keywords:landscape function, post tin-mining land, Bangka.

1. Introduction

Tin mining in Bangka Belitung Province has lasted a long time. It started in 1711 at Bangka and started in 1852 at Belitung [1]. Over a period of more than 200 years these activities have a negative impact on environmental devastation, forests and farmlands were destroyed and turned into barren and cratered landscapes. The Forestry Service of Bangka Belitung Province reported that in 2009 the barren landscape covers 112,838,86 hectares or 6.93% of the land area of Bangka Belitung Province, shifting 99,146,97 ha of the forests in this province [2]. The impact of tin mining, more specifically, is degraded lands; the physical and chemical properties of soil reduced, the topography changed, the natural vegetation and wildlife habitat disappeared. The tin mined land is a marginal land since water holding capacity of the soils are low; consequently, the evaporation rate is high, the soil hardens and cracks, the organic material and nutrient is poor [3].

Efforts at the reclamation of tin mined landsby replanting the lands with various species of trees, such as acacia, pelawan, and sengon have been made by the government of Bangka Belitung Province and tin mining companies. Reclamation on 2,000 ha and 1,600 ha of tin mined landsin this province have been conducted respectively in 2007 and 2008 [2]. PT Timah, the largest tin mining company, has reclaimed its mined lands, covering 1379.55 ha at Bangka and 241.6 ha at Belitung in 2010[4].

The efforts at thereclamation by replanting the tin mined lands often fail as even the toughest trees, such as acacia, may struggle to grow on the poor soils left after mining. However, replanting the mined land with *Acasiamangium* which has lasting for six years by PT Timah is considered successful [5]. It indicates that this species is able to adapt with the acidic poor soils that characterized the mined lands. It is also a challenge for us to seek other species of plants that could survive well on the mined land. Hence, the diversity of plants on the mined land will be upgraded, the litters will multiply, the decomposed litters will improve nutrient cycle and physical properties of the soil.Reclamation success can be claimed based on analysis of the landscape function that reflects the quality of soil in three main indicators: land stability, water infiltration, and nutrient cycle. This study aimed to analyze the landscape function of reclaimed tin mining land, and to compare it with that of the non-reclaimed tin mining land and the natural land that is not mined.

2. Methods

This study was conducted at three sites in Central Bangka District, namely Air Pelempang, Air Jangkang, and Air Kabel. Three different land are, A) mined land that has not been reclaimed, B) mined land that has been reclaimed, and C) natural land that is considered representative of conditions prior to mining activities, were observed in each study site. Observations on general conditions, such as canopy density, presence or absence of understory plants, and litter conditions, were conducted in each landscape type. In addition, observation on the soil characteristicssuch as temperature, horizon, pH, moisture, porosity, organic and inorganic materials, and bulk density,were also performed. The microclimates data gathered in each landscape type was light intensity, temperature, and humidity.

Landscape Function Analysis (LFA) for each landscape type was performed by a method developed by Tongway and Hindley with modifications [6]. Data were collected by three line transects in each type of landscape. The line transect was laid out according to topography and its length was adapted to land conditions. Organizational landscapes characterization and soil surface assessment (SSA) was carried out along the line transects. The characterization of organizational landscape aimed to check the existence of patches and inter patch. The SSA was performed by observing and measuring the SSA's criteria, that is areas covered by pioneer vegetation (herbs, shrubs, trees) and litters, degree of decomposition, violence (severity) and type of erosion, material deposited, surface roughness and resistanceagainst tampering, land stability when moist, and soil texture [7]. The data gathered from the SSA are inserted into the worksheets of LFA and combined in different combinations to reflect three major soil habitat quality indices: stability or resistance to erosion, infiltration/runoff and nutrient cycling. The data are presented in a 0 to 100 scale.

3. Results

3.1 Microclimate And General Condition in The Landscape Studied

The conditions of diffrent types of land in the three study sites (Air Palempang, Air Jangkang, and Air Kabel) showed generally similar pattern. Land A (non reclimed mined land) has inclined flat and cratered topography, does not have a canopy cover, and there are only a few species of herb and creeping perennial plants at the ground. Land B (reclaimed mined land) has similar topography to the previous landscape, but it has a relatively opened canopy cover of the planted plant, litters were found under the plants. It also has relatively more numerous ground plants than land A. Land C (natural non-mined land) has a canopy cover more tightly than land B, the canopy derived from native plants of the landscape, understory plants (herbs and shrubs) were abundance, otherwise a lot of litterswerefound at the ground (table 1).

Different microclimates were found in the three types of landscape observed. Light intensity and air temperature in the land A and B is higher than these in the land C. It is followed by a lower humidity than land C (table 2).

IOP Conf. Series: Earth and Environmental Science **58** (2017) 012018 doi:10.1088/1755-1315/58/1/012018

Table1. The description of general condition at the three different lands, in the three study sitesin Bangka.

Study sites	Type of Land	Description of general condition
Air Pelempang	A : non reclaimed tin mining land	no canopy cover, bare soil substrates (sandy and partly pebbly bare soil), the ground is covered by a few perennial creeping herbs, there is a bit of litter on the soil surface, the topography is slightly rough
	B: reclaimed tin mining land	there is canopy cover of the ±2 m high of plants grown for reclamation, the litter and the ground plants is a bit more than the landscape A, the topography is similar to that in landscape A
	C : naturally vegetated land	The canopy is closed, the soil is covered by litters, there is a lot of trees species and understory plants (herbs, shrubs, small trees) that is spread and grow naturally, the topography is flat
Air Jangkang	A : non reclaimed tin mining land	no canopy cover, bare soil substrates (sandy and partly pebbly bare soil), the ground is covered by a few grasses and perennial creeping herbs, there is a bit of litter on the soil surface, the topography is flat and slightly slopping
	B: reclaimed tin mining land	there is canopy cover of the ±2 m high of plants grown for reclamation, the litter and the ground plants occur under the planted plants and is a bit more than the landscape A, the topography is flat and slightly sloppy similar to that in landscape A
	C : naturally vegetated land	The canopy is closed, there is a lot of trees species and understory plants (herbs, shrubs, small trees) that is spread and grow naturally, the soil is covered by litters,the substrate is sandy clay, the topography is flat
Air Kabel	A : non reclaimed tin mining land	no canopy cover, bare soil substrates (sandy and crusted bare soil), the ground is dominated the tussock of grasses, a small area of grass that is thicker or longer than the grass growing around it, the topography is slightly rough and pitted with several large pools of stagnant water called "kolong"
	B: reclaimed tin mining land	there is canopy cover of the *Acasiamangium*planted for reclamation, a lot of litter and the ground plants occur under the planted plants, the topography is flat
	C : naturally vegetated land	The vegetation has closed canopy withthe trees has larger diameter, the soil is covered by litters,the topography is flat

Table 2. The microclimates at the three different lands, in the three study sitesin Bangka.

Study sites	Type of Land	Parameters of microcimate		
		Light intensity (lux)	Air temperature (°F)	Air humidity (%)
Air Pelempang	A : non reclaimed tin mining land	45,97×10³	85,3 ± 1,0	89,3 ± 0,6
	B: reclaimed tin mining land	67,17×10³	87,8 ± 2,1	79,0 ± 0,5
	C : naturally vegetated land	1,87×10³	84,8 ± 0,3	89,3 ± 0,3
Air Jangkang	A : non reclaimed tin mining land	128,5×10³	89,5 ± 0,5	80,2 ± 1,8
	B: reclaimed tin mining land	117,7×10³	89,8 ± 0,8	81,7 ± 1,2
	C : naturally vegetated land	6,86×10³	84,2 ± 1,8	85,0 ± 3,7
Air Kabel	A : non reclaimed tin mining land	118 ×10³	89,5 ± 1,3	79,7 ± 4,2
	B: reclaimed tin mining land	130,6×10³	91 ± 3,5	77,0 ± 0,5
	C : naturally vegetated land	8,22×10³	83,7 ± 0,6	95,7 ± 2,1

3.2 Soil Characteristics of the Study Sites

There are differences on physical characteristics and chemical conditions of the soil inthreedifferent type of land in the study sites of Air Pelempang, Air Jangkang, and Air Kabel. The characteristics of soil in those study sites generally showed similar trends. Soil fertility in the land C is much better than that of the land A and B. In addition, the soil texture in the landA and B have over 90% of sand fractions, whereas the land C was dominated by clay fraction. This condition of land A and B was caused by tin mining activities that left expanse of tailings over 90% of sand [8]. Soil washing process on tin mining activities cause loss of nutrients in the soil; the level of C organic compounds in land A and B is lower (< 1%) than that in the land C (2-4%). The soil in the three study sites is commonly acidic (pH ≤5.5), however the value of cation exchange capacity (CEC) in the land C (4:25 to 11:29 cmol / Kg) is higher than that in the land A and B (0.20-0.99 cmol / Kg).

3.3 Landscape Function Analysis for Monitoring and Assessing the Reclaimed Tin Mined Lands

The result of landscape function analysis is demonstrated by the value of land stability index, the water infiltration index and nutrient cycling index. The value of land stability index isobtained from the indicators of the SSA, such as the level of crust destruction, surface resistance, slake test, the type and severity of erosion, deposition materials, cryptogams cover, ground cover from the rain, and litters cover.The lowest land stability index (respectively 37.2%, 37.7%, 35.3%) was found in the land A at the three study sites (Pelempang Water, Air Jangkang, Air Kabel). Land B, reclaimed tin mined land, have a stability index slightly higher than landA. These conditions are found in the three study sites, Pelempang Water, Air Jangkang, Air Kabel; stability index values respectively 39.6%, 37.7%, 46.1% (table 3).Land C, natural land that is not mined, has the highest stability index among the other lanscapes. Sabilitas index value of 61.7%, 75.3%, and 82% was obtained respectively in the study sites Air Pelempang, Air Jangkang, and Air Kabel.

Table 3. The indices of landscape functions analysis at three different lands, in the three study sites in Bangka.

Study sites	Type of Land	The indices of landscape function analysis		
		stability	infiltration	nutrient cycling
Air Pelempang	A : non reclaimed tin mining land	37.2±0.78	33.6±0,35	10.3±0,33
	B : reclaimed tin mining land	39.6±1.19	33.2±1.50	11.4±1,43
	C : naturally vegetated land	61.7±0	62.3±0	54.4±0
Air Jangkang	A : non reclaimed tin mining land	37.7±2.01	33.4±1.34	8.6±0.55
	B : reclaimed tin mining land	37.7±1.21	31.4±3.15	8.8±0.91
	C : naturally vegetated land	75.3±5.48	59.3±2.74	54.9±0.49
Air Kabel	A : non reclaimed tin mining land	33.6±0,35	30.3±6.62	8.6±0.31
	B : reclaimed tin mining land	33.2±1.50	35.2±2.75	15.5±7.69
	C : naturally vegetated land	62.3±0	60.2±2.92	55.1±3.25

The infiltration/runoff index is defined as how the soil divides rainfall into soil-water (water available for plants to use), and runoff water which is lost from the local system, or may also transport materials (soil, nutrients and seed) away [6]. Index infiltration in the land A and B is less than 35 and categorized as low (table 3). On the other hands, high infiltration index is found at land C, natural non-mined land, in the three study sites. The value of index infiltration in AirPelempang, Air Jangkang, and Air Kabel, was respectively 62.3%, 59.3%, 60.2%.

The highest score of index of nutrient cycling in the three study sites was found in the land C. This land has a score of the nutrient cycling of 54.4%, 54.9%, 55.1%; obtained in Air Pelempang, Air Jangkang, Air Kabel respectively (table 3).

4. Discussion

The low cover of vegetation on the land A and B causes high intensity of light irradiation. Conversely, thelight irradiation intensity in the landscape Cwas low due to dense canopy cover of the natural vegetation. Vegetation plays an important role to control the thermal environment (air temperature, air humidity and air movement) through the evapotranspiration mechanism to accelerate the cooling on the leaf surfacesand eventually lowered the air temperature.Vegetation also provides a cooling effect of the air by means of shadowing (canopy effect) and reducing the sunlight that reaches the ground [9].The opened area that covered only by a few ground plants and young trees on the landscape type A and B reduce the beneficial effects of such vegetation.

Tin post-mining land generally have lower pH values due to high concentrations of Al [10].It is thought to be happen because of the high air temperatures and high rainfall intensity resulting soil weathering processes take place quickly. Basic cations such as Ca, Mg, Kwere leached from the soil profile, leaving a more stable material that is rich in Al and Fe [11].The low content of the soil macronutrient and the high composition of the sand fraction in the land (A) and (B), which is a post-mining land of tin, shows the characteristics of critical land [12].

The low of stability index in land A indicates a low ability of tin mined lands that have not been reclaimed in terms with stand the soil erosion, and lack of ability to repair themselves after disruption [6]. Low stability occurred in land A was due to a lack of vegetation and the opened soil is more than 90%. Such conditions result in the least amount of litters on the ground and cover against rainfall on the soil is low. Land A, not reclaimed mined land, has a high potential to erode because it has a sandy soil that has cohesion among the particles that are not tight.The parameters contributed to the stability index in land B that is slightly higher than land A are the ground cover and cryptogam. The area covered by vegetation in land B is higher than land A. Cryptogamic plants, such as some species of mosses and ferns, were found in land B. However, the value of stability index in land B is still considered low and indicate a high potential for erosion. The existence of various species of trees that are naturally dispersed in land C, contributes most to the stability. These tree species have the potential to reclaim land mined tin, it is due to the local plant are naturally adapted to the local climate better than the introduced plant [14].

The low infiltration index in land A and land B is evidenced by the low level of perennial vegetation cover and the soil surface roughness that could cause in low retention of groundwater and the loss of water wasted out the ground through runoff water. High infiltration index in land C indicates that runoff water in this land is low. The runoff water will decrease with the increase of vegetation [15]. Land C has a high perennial vegetation cover. This lush vegetation has plenty of underground roots that help the soil retaining water, increasing infiltration and reducing runoff water. In addition, the soil surface in the land C has high degree of roughness. It also supports the water retentionin the process of infiltration. Sandy clay soil texture at this land also helps the process of infiltration, the soil has tight cohesion between the particles and high ability to retain water.

High nutrient cycling index in land C contributed by the lush perennial vegetation and a lot of decomposed litters. The lush vegetation supports the process of nutrient cycle by its contribution in providing a lot of litters. Decomposition of the litters eventually lead to the process of nutrient cycling. The decomposed litters released carbon into the atmosphere and nutrients into the soil, organic material of the litter are converted by microbes into inorganic compounds that are used by plants [15].

5. Conclusion

The results of landscape function analysis (LFA) showed that the reclamation on tin-mined lands have not been able to restore the natural function of the land, the biogeochemical processes in the tin-mined lands have not been going well. This can be affected to several factors, such as reclamation takes a long time to restore the function of the land. Additionally, the trees species planted can also determine the success of reclamation.

Acknowledgement

This study is supported byExcellent Higher Education Institution Research categorizedExcellent Mandate Division Research Bogor Agricultural University 2016 to Sulistijorini from Ministry of Research Technology and Higher Education the Republic of Indonesia.

References

[1] Anonim *Sekilas Sejarah PT Tambang Timah (Persero)* http://bumn.go.id
[2] Anonim 2011*Ribuan Hektar Lahan Belum Direklamasi* Bangka Pos, ed 25 April 2011
[3] Notohadiprawiro T 2006 *Pengelolaan Lahan dan Lingkungan Pasca Penambangan* http://www.soil.faperta.ugm.ac.id
[4] [PT. Timahtbk] 2010LaporanTahunan 2010*Meningkatkan Kualitas, Menggapai Kejayaan*
[5] Latifah S 2000 *Keragaman pertumbuhan Acacia mangium Willd. Pada lahan bekas tambang timah (studi kasus di areal kerja PT. Timah tbk)* (Bogor: Program Pascasarjana, InstitutPertanian Bogor)
[6] Tongway D J and Hindley N L 2005 *Landscape Function Analysis, Procedures for Monitoring and Assessing Landscap* (Canberra: CSIRO Sustainable Ecosystems)
[7] Lau I C, *et al* 2008 *Remote Mine Site Rehabilitation Monitoring Using Airborne Hyperspectral Imaging and Landscape Function Analysis (LFA)* Beijing: *The International Archives of the Photogrammetry, Remote Sensing, and Spatial Information Sciences* **XXXVII** B7
[8] Inonu, Ismed 2010 Pengelolaan Lahan Pascatambang Timah di Pulau Bangka: Sekarang dan yang akan Datang. http://www.ubb.ac.id
[9] Wonorahardjo S, *et al Studi Pengaruh Kualitas Vegetasi pada Lingkungan Termal Kawasan Kota di Bandung Menggunakan Data Citra Sateli* http://sappk.itb.ac.id/tb/templates/kk-tb/
[10] Hasnelly Z, *et al* 2006 Pemanfaatan Lahan Bekas Tambang Timah melalui Integrasi Tanaman dan Ternak. *Lokakarya Nasional Pengembangan Jejaring Litkaji Sistem Integrasi Tanaman Ternak* http://peternakan.litbang.deptan.go.id/fullteks/lokakarya/

[11] Hue N V and Ikawa H *Acid Soils in Hawaii: Problems and Management* *http://www2.hawaii.edu/~nvhue/acid.html*
[12] [Presiden RI] 2000 *Peraturan Pemerintah RI No.150 Tahun 2000*, tentang Pengendalian Kerusakan Tanah untuk Produksi Biomassa
[13] Putra H F 2013*Evaluasi fungsi ekologis dan tingkat revegetasi lahan pascatambang timah di Air Mungkui, kabupaten Belitung* (Bandung: Program Studi Magister Biologi, Institut Teknologi Bandung)
[14] Tordoff G M, Baker A J M and Willis A J 2000 *Chemosphere* **41** 219-228
[15] Chapin *et al 2002 Principles of Terrestrial Ecosystem Ecology* (New York: Springer-Verlag)

Model Averaging for Predicting the Exposure to Aflatoxin B_1 Using DNA Methylation in White Blood Cells of Infants

S Rahardiantoro[1], B Sartono[1] and A Kurnia[1]*

[1]Department of Statistics, Bogor Agricultural University

Email: anangk@apps.ipb.ac.id

Abstract. In recent years, DNA methylation has been the special issue to reveal the pattern of a lot of human diseases. Huge amount of data would be the inescapable phenomenon in this case. In addition, some researchers interesting to take some predictions based on these huge data, especially using regression analysis. The classical approach would be failed to take the task. Model averaging by Ando and Li [1] could be an alternative approach to face this problem. This research applied the model averaging to get the best prediction in high dimension of data. In the practice, the case study by Vargas *et al* [3], data of exposure to aflatoxin B_1 (AFB_1) and DNA methylation in white blood cells of infants in The Gambia, take the implementation of model averaging. The best ensemble model selected based on the minimum of MAPE, MAE, and MSE of predictions. The result is ensemble model by model averaging with number of predictors in model candidate is 15.

1. Background

In high dimensional data, which is popular in recent century, the classical regression analysis failed to bring the model of relationship two sets variables. This approach limited in dimension of data condition, number of observations has to higher than the number of predictors. Certainly, this condition is opposite with high dimensional data. Suppose the design matrix X with dimension $n \times p$, high dimension condition accurs when $p \gg n$. Therefore, when facing with high dimensional regression cases, classical approach would be a bad choice.

Ando and Li, in 2014, proposed the alternative approach in regression analysis cases using high dimensional data that called model averaging (MA) [1]. Moreover, the MA approaches also applied in experimental design for supersaturated topics [2]. The main purpose of MA is to get the predictions of response variable accurately. MA takes some candidate model by the splitting process of predictor variables in the order of each marginal correlation with the response variable. The final predictions are from the averaging of the predictions of each candidate models.

Generally, high dimensional data condition occurs in the biological research, furthermore in the research of genetics. The one of popular cases are in DNA methylation that use the microarray technologies. An array can contain up to 20000 spots, which is each spot has been implanted with target DNA sequence. Because of that, it can't be imagine how huge data used in this subject case.

This research focuses on the application of MA in the high dimensional data. The application taken based on the data of exposure to aflatoxin B_1 (AFB_1) and DNA methylation in white blood cells of infants in The Gambia by Vargas *et al* [3]. Aflatoxin is a contaminant of staple foods in sub-Saharan Africa, is a known human liver carcinogen and has been associated with stunting in infants. MA applied to get the best prediction of AFB_1 from DNA methylation based on MAPE, MAE, and MSE criteria. Number of predictors in the model candidates (m) has been set in some options, therefore the minimum of MAPE, MAE and MSE values indicates the predictions of MA with specify m to be the best choice. In practice, this research using R software in implementing all of process.

2. Methodology

In this section would be described about the data conditions, algorithm process of MA, and also the value of best-model criterias (MAPE, MAE, and MSE) [4].

2.1. Data

The data of this research is from the case based on Vargas *et al* [3] in the previous section. There were 124 pregnant mothers to be subject of research. Variable of AFB_1 took from those pregnant mothers in first trimester. Their 3rd – 6th month infants have been studied to take the DNA methylation in their white blood cell that contained 485577 profiles. Based of that, the response variable, AFB_1, has the number of observations $n = 124$, and the predictor variables, DNA methylation, have the number of profiles $p = 485577$, which is include in the high dimensional data $(p \gg n)$.

The first step before the analysis process is data cleaning. This process was taken to avoid the missing values in the analysis, that is guarantee the MA process well done. The dimension of data after the cleaning process is $n = 118$ and $p = 480050$. In the practices, these data divide into data training (for modeling process) and data testing (for prediction process). Therefore, MA approach applied in these data.

2.2. Model Averaging Methods

One of necessary point in the MA is determining the number of predictors in the candidate model (m). In this research take 8 numbers, which is $m = 5, 10, \dots, 35, 40$. The averaging process taken with the specify weight w_i, AIC weight which is based on the value of AIC in each model candidates. Suppose there are k model candidates, therefore the $i-$th AIC weight follows

$$w_i = \frac{exp\left(\frac{1}{2}a_i\right)}{\sum_{i=1}^{k} exp\left(\frac{1}{2}a_i\right)} \quad (1)$$

where a_i denotes the value of AIC in the $i-$th model candidates, and $w_i \geq 0$; $\sum_{i=1}^{k} w_i = 1$ [3].

The ensamble predictions of MA define by the weighted averaging of each model candidate predictions. Suppose $\hat{\boldsymbol{y}}_i$ is the prediction vector of $i-$th candidate model, therefore

$$\hat{\boldsymbol{y}} = \sum_{i=1}^{k} w_i \hat{\boldsymbol{y}}_i \quad (2)$$

become the ensamble prediction vector.

Furthermore, in the model candidate preparation, the selection of predictor variables is the essential process in MA. Based on the proposed MA by Ando and Li (2014), predictors selection in model candidate is done based on the rank of marginal correlation between all of predictors and the response variable [1].

2.3. Evaluation Criteria

The evaluation criteria in this research are based on the value of Mean Absolute Percentage of Error (MAPE), Mean Absolute of Error (MAE), and Mean Square Error (MSE). These criteria values selected because include in the most popular prediction quality measurements that useful in regression cases [5]. The minimum values of MAPE, MAE, and MSE indicate that the model selected give the best predictions. The formulas of MAPE, MAE, and MSE as follow

$$\text{MAPE}=\frac{\left(\sum_{t=1}^{n}\left|\frac{\hat{y}_t-y_t}{y_t}\right|\right)}{n}\times 100\% \tag{3}$$

$$\text{MAE}=\frac{\sum_{t=1}^{n}|\hat{y}_t-y_t|}{n} \tag{4}$$

$$\text{MSE}=\frac{\sum_{t=1}^{n}(\hat{y}_t-y_t)^2}{n} \tag{5}$$

where t indicates the index of number of observations, $t = 1,2,\dots,n$.

3. Result and Discussion

In $n = 118$ and $p = 480050$ data, and $m = 5,10,\dots,35,40$, table 1 shows the result of measurement accuracy of prediction.

Table 1. The result of measurement accuracy of AFB_1 prediction in several m

m	MAPE		MAE		MSE	
	Data training	Data testing	Data training	Data testing	Data training	Data testing
5	2.02	0.79	36.76	30.67	2289.87	2723.57
10	1.49	0.94	30.47	40.43	1560.06	4125.55
15	1.47	0.78	28.29	40.49	1166.02	8203.09
20	1.48	1.43	24.37	46.91	1036.51	3117.30
25	1.50	1.59	25.33	52.94	1023.13	4451.83
30	0.89	1.36	17.70	48.47	572.00	5570.61
35	1.09	1.72	18.02	72.43	549.17	17214.61
40	0.91	2.48	16.36	78.43	440.24	15721.91

*the yellow highlight indicate the lowest values

The best accuracy was determined based on the lowest MAPE, MAE and MSE in the data testing. Based on those results, m =15 has the lowest MAPE value, and $m = 5$ has the lowest MAE and MSE. Figure 1 shows the m =15 and $m = 5$ prediction with the actual of AFB_1 variable.

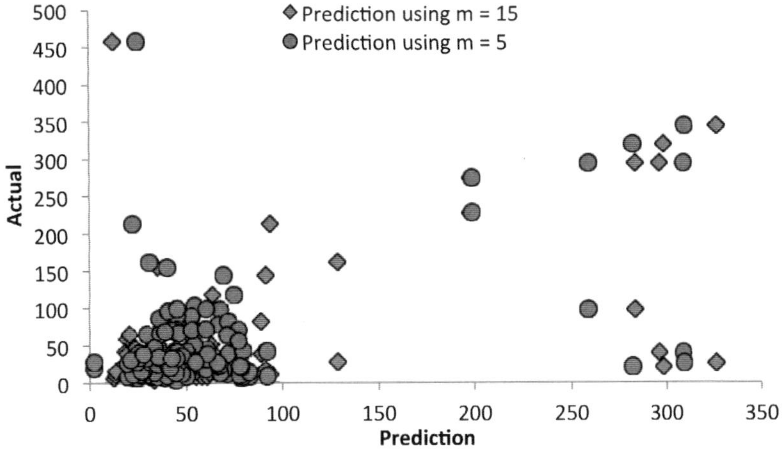

Figure 1. Scatter plot of the m =15 and $m = 5$ prediction with the actual of AFB_1 variable

The charts on Figure 1 suggest the following general pattern. Model with m =15 has better pattern than model with m = 5, it also supported by the correlations between the actual values and the prediction values 0.523. Based on this research, we suggest the model averaging as an alternative when the data has high dimension, in the prediction AFB_1 using DNA methylation of white blood cell, we can choose the m = 15 model averaging.

4. Conclusion

The MA can be an alternative approach in the case of high dimensional regression. The prediction of AFB_1 from pregnant mothers using the DNA methylation on white blood cells of their infants can be solved using MA. The best model based on the minimum value of MAPE, MAE, and MSE, also in addition the highest correlation between the prediction values and the actual values, is MA with m = 15.

References

[1] Ando T and Li K C 2014 *Journal of the American Statistical Association* **194** 254-265.

[2] Salaki D T, Kurnia A and Sartono B 2016 *IOP Conf. Series: Earth and Environmental Science.* **31** 1-7.

[3] Vargas H H, Castelino J, Silver M J, Salas P D, Cros M P, Durand G, Kelm F L C, Prentice A M, Wild C P, Moore S E *et al.* 2015 *International Journal of Epidemiology* 1-11.

[4] Myttenaere A D, Golden B, Grand B L, Rossi F 2015 Using the Mean Absolute Percentage Error for Regression Models. *Proceedings of the 23-th European Symposium on Articial Neural Networks, Computational Intelligence and Machine Learning (ESANN 2015).* hal-01162980.

[5] Claeskens G, Hjort N L 2008 *Model Selection and Model Averaging* (New York (US): Cambridge University Press)

Clustering Information of Non-Sampled Area in Small Area Estimation of Poverty Indicators

V Y Sundara[1], A Kurnia[1]* and K Sadik[1]

[1]Department of Statistics, Bogor Agricultural University, Indonesia

Email: anangk@apps.ipb.ac.id
Abstract. Empirical Bayes (EB) is one of indirect estimates methods which used to estimate parameters in small area. Molina and Rao has been used this method for estimates nonlinear small area parameter based on a nested error model. Problems occur when this method is used to estimate parameter of non-sampled area which is solely based on synthetic model which ignore the area effects. This paper proposed an approach to clustering area effects of auxiliary variable by assuming that there are similarities among particular area. A simulation study was presented to demonstrate the proposed approach. All estimations were evaluated based on the relative bias and relative root mean squares error. The result of simulation showed that proposed approach can improve the ability of model to estimate non-sampled area. The proposed model was applied to estimate poverty indicators at sub-districts level in regency and city of Bogor, West Java, Indonesia. The result of case study, relative root mean squares error prediction of empirical Bayes with information cluster is smaller than synthetic model.

1. Introduction

Survey is a method to collect data which has several advantages such as less time, less budget and other resources. Most of national surveys provide limited information and low precision in predicting the small area level [1]. Direct estimate of parameters in area with small sample size would have large variance and even cannott produce estimates when there is no sample unit for those area [2]. The science of small area estimation is developed to solve this problem. The indirect estimate or model-based in small area estimation was first introduced by Fay and Heriot [3] using anempirical Bayes (EB) method. The problem of the ordinary EB for non-sampled area using synthetic model is ignoring the area effects. Anisa et al (2014) studied the effect of adding the cluster information empirical unbiased best linear prediction on non-sampled area to produce better predictions [4]. The simulation study showed the use of factor analysis in clustering has increased the average percentage of accuracy particularly when the Ward method is implemented [5].

Poverty is one issue that became the focus of governments in many countries. Statistics Indonesia uses poverty indicator which is proposed by Foster, Greer and Thorbecke. The indicators are head count index - P_0 (the percentage of population below the poverty line), poverty gap index - P_1(the average size of each expenditure gap of the poor to the line poverty) and poverty severity index - P_2(provide a picture of the spread of expenditure among the poor).

In this paper, simulation study is carried out to evaluate the performance of the proposed models or cluster information with ordinary empirical Bayes or Molina and Rao method. This paper also presents application of the proposed model using data from Statistics Indonesia to estimate poverty indicators at sub-districts level in regency and city of Bogor.

2. Direct Estimation and Empirical Bayes Estimation of Poverty Indicators

Direct estimation is a method parameter estimation which is only based on sample [6]. Direct estimation has a large variance for small sample size. Direct estimation of FGT poverty measures can be defined as follows

$$P_{ai} = \frac{1}{N_i}\sum_{j=1}^{N_i} P_{\alpha ij}, \quad i = 1, \dots, m \tag{2.1}$$

with $P_{\alpha ij}$defined as

$$P_{\alpha ij} = \left(\frac{z - E_{ij}}{z}\right)^{\alpha} I(E_{ij} < z), \qquad j = 1, \dots, N_i, \qquad \alpha = 0,1,2 \tag{2.2}$$

where $I(E_{ij} < z) = 1$ if $E_{ij} < z$ (person under poverty) and $I(E_{ij} < z) = 0$ if $E_{ij} \geq z$ (person not under poverty). For α = 0 is the proportion of the population below the poverty line, called the head count index (P_0). For α = 1 represents the poverty gap index (P_1), and for α = 2 is the poverty severity index (P_2). Estimates of FGT poverty measure for each small area i is as follows

$$\hat{P}_{ai} = \frac{1}{n_i}\sum_{j=1}^{n_i} P_{\alpha ij}, \quad i = 1, \dots, m, \quad \alpha = 0,1,2 \tag{2.3}$$

Molina and Rao proposed EB method to estimate poverty measures in a small area [1]. This method assumes that welfare variable can be transformed to follows a normal distribution. Let $Y_{ij} = T(E_{ij})$ is a transformation on welfare, and $\boldsymbol{y} \sim N(\boldsymbol{\mu}, \boldsymbol{V})$. Thus FGT poverty measure for each small area i in equation (2.2) can be written as follows

$$P_{aij} = \left(\frac{z - T^{-1}(Y_{ij})}{z}\right)^{\alpha} I\left(T^{-1}(Y_{ij}) < z\right), \qquad j = 1, \dots, N_i, \qquad \alpha = 0,1,2 \tag{2.4}$$

Empirical Bayes estimation for the poverty measure P_{ai} formulated as follows [1]:

$$\hat{P}_{\alpha i}^{EB} = \frac{1}{N_i}\left[\sum_{j \in s_i} P_{\alpha ij} + \sum_{j \in r_i} \hat{P}_{\alpha ij}^{EB}\right] \tag{2.5}$$

where s_i is the sampled unit and r_i the non-sampled unit. $\hat{P}_{\alpha ij}^{EB}$ is an empirical Bayes estimators of $P_{\alpha ij} = h_\alpha(Y_{ij})$ are defined as follows:

$$\hat{P}_{\alpha ij}^{EB} = E_{\boldsymbol{y}_r}\left[h_\alpha(Y_{ij}) | \boldsymbol{y}_s\right] = \int_{IR} h_\alpha(y) f_{Y_{ij}}(y|\boldsymbol{y}_s)\, d_y, \ j \in r_i \tag{2.6}$$

where $f_{Y_{ij}}(y|\boldsymbol{y}_s)$ is a conditional probability density function of Y_{ij}. According to Molina and Rao [1], the complexity of the functions $h_\alpha(.)$, there is not explicit expression for the expectation in (2.6), but it can be approximated by Monte Carlo. For this, generate L replicates of $\boldsymbol{y}_r$ from the distribution of $\boldsymbol{y}_{ir}|\boldsymbol{y}_{is} \sim N(\boldsymbol{\mu}_{r|s}, \boldsymbol{V}_{r|s})$ or generate univariate value $Y_{ij}^{(l)}$ from the nested error linear regression model as follows:

$$Y_{ij}^{(l)} = \boldsymbol{x}_{ij}\hat{\beta} + \hat{u}_i + \varepsilon_{ij}, \quad u_i \sim N\left(0, \hat{\sigma}_u^2(1 - \hat{\gamma}_i)\right), \quad \varepsilon_{ij} \sim N(0, \hat{\sigma}_e^2) \tag{2.7}$$

where $\hat{\gamma}_i = \sigma_u^2(\sigma_u^2 + \sigma_e^2/n_i)^{-1}$ and n_i is the sample size. Then, an approximate to the best predictor of Y_{ij} is

$$\hat{P}_{\alpha ij}^{EB} \approx \frac{1}{L}\sum_{l=1}^{L} h_\alpha\left(\boldsymbol{y}_{ij}^{(l)}\right), j \in r_i \tag{2.8}$$

For $n_i = 0$ or domain i is not sampled, then $Y_{ij}^{(l)}$ for $j = 1, \dots, N_i$ are generated by bootstrap from $Y_{ij} = x_{ij}'\beta + u_i^* + e_{ij}^*$ where $u_i^* \sim iidN(0, \hat{\sigma}_u^2)$ and $e_{ij}^* \sim iidN(0, \hat{\sigma}_e^2)$ with assume u_i^* is independent of e_{ij}^*. Formula (2.6) is used to get the estimator $\hat{P}_{\alpha ij}^{EB}$ of $P_{\alpha ij}$ and the EB estimator P_{ai} is defined as

$$\hat{P}_{\alpha i}^{EB} = \frac{1}{N_i}\sum_{j=1}^{N_i} \hat{P}_{\alpha ij}^{EB} \qquad (2.9)$$

The estimator (2.9) is essentially a synthetic estimator since no sample observations are available from domain i if $n_i = 0$

3. Cluster Analysis

Cluster analysis is a multivariate technique to classify object based on similarities or dissimilarities. Characteristics between object can be measured by Euclidean distance, Mahalanobis distance, and others. There are two approaches in cluster method, hierarchical and non-hierarchical methods. Hierarchical method is used when the number of cluster is unknown. While non-hierarchical methods is used when the number of cluster is known [7]. Cluster analysis can be used when there is non-sampled area. The addition of cluster information to non-sampled areas shows that in general has a better prediction [4].

4. Proposed Model

The basic model which is used in this study is a nested error regression model. This model referred to Molina and Rao model. This study use unit level with i and j respectively denotes area and unit of sampled area, while i^* and j^* respectively denotes area and unit of non-sampled area. Response variable denoted by y_{ij}, auxiliary variable denoted by x_{ij}, random effect of area denoted by u_i, and sampling error for sampled area denoted by e_{ij}. Prediction of non-sampled area $(\hat{y}_{i^*j^*})$ is obtained from auxiliary variables of non-sampled area$(x_{i^*j^*})$. The Molina and Rao model for one auxiliary variable can be written as follows:

a) Model for population :

$$y_{ij} = \beta_0 + \beta_1 x_{ij} + u_i + e_{ij} \qquad (4.1)$$

b) Prediction model for sampled area:

$$\hat{y}_{ij} = \hat{\beta}_0 + \hat{\beta}_1 x_{ij} + \hat{u}_i \qquad (4.2)$$

c) Prediction model for non-sampled area :

$$\hat{y}_{i^*j^*} = \hat{\beta}_0 + \hat{\beta}_1 x_{i^*j^*} \qquad (4.3)$$

The cluster information model is modifying basic model by adding average the random effects of area on each clusters into prediction model for non-sampled area. The average of random effects of area is defined by $\bar{\hat{u}}_{(k)} = \frac{1}{m_k}\sum_{i=1}^{m_k} \hat{u}_i$ with m_k is the number of sample area on the kth cluster. The cluster information model can be written as follows:

a) Model for population :

$$y_{ijk} = \beta_0 + \beta_1 x_{ijk} + u_i + e_{ijk} \qquad (4.4)$$

b) Prediction model for sampled area :

$$\hat{y}_{ijk} = \hat{\beta}_0 + \hat{\beta}_1 x_{ijk} + \hat{u}_i \qquad (4.5)$$

c) Prediction model for non-sampled area :

$$\hat{y}_{i^*j^*k} = \hat{\beta}_0 + \hat{\beta}_1 x_{i^*j^*k} + \bar{\hat{u}}_{(k)} \qquad (4.6)$$

5. Simulation Study

A simulation study has been carried out to study the performance of the proposed model of non-sampled area. We simulate a population of size $N = 502$, composed of $M = 46$ areas with each area is 6-16 units. Response variable for a population generated by nested error regression. Auxiliary variables generated by the uniform distribution $U(0,10)$ with intercept $\beta_0 = 1.0$ and regression coefficient $\beta_1 = 0.5$. The population is divides into three clusters, with each cluster consisting of 18, 19 and 9 areas respectively. The intercept and regression coefficients for each cluster are shown in Table 1. Random effects of area (u_i) and sampling error (e_{ij}) respectively were generated from normal distribution $u_i \sim iid\ N(0\ ,0.15)$ and $e_{ijk} \sim iid\ N(0\ ,0.50)$. The response variable y_{ijk} were generate from nested error linear regression model.

Table 1. Intercept and Regression Coefficients on Each Cluster

Cluster	β_{0k}	β_{1k}
1	3.0	6.5
2	2.0	3.5
3	1.0	2.5

There are three cluster areas on population that has been generated with three non-sampled area. Population is shown in Figure 1.

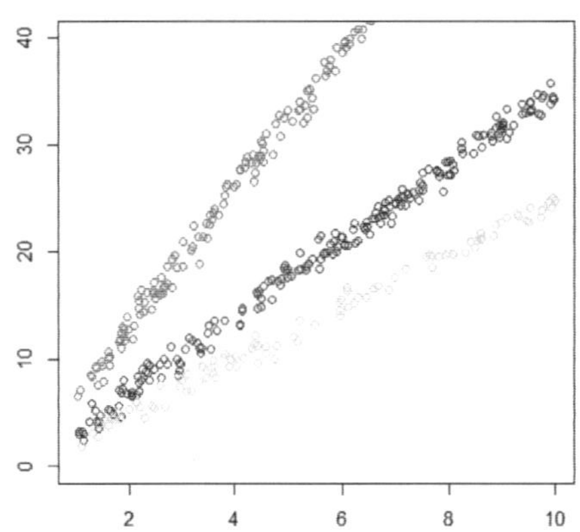

Figure 1. Population's Simulation Study.

Simulation study evaluated based on relative bias (RB) and relative root mean squares error (RRMSE). The formulas can be defined as follows:

$$RB_i = \frac{1}{ul}\sum_{i=1}^{ul}\left(\frac{\hat{\theta}_i - \theta_i}{\theta_i}\right)$$

$$RRMSE_i = \frac{1}{\theta_i}\sqrt{\frac{1}{ul}\sum_{i=1}^{ul}\left(\hat{\theta}_i - \theta_i\right)^2}$$

This process was repeated $ul = 1000$ times. Relative bias and relative root mean squares error of empirical Bayes of P_0, P_1, and P_2 for non-sampled area are shown in Table 2 and Table 3 respectively.

Table 2. Relative Bias of EB estimates of Poverty Indicators

Area	EB (Molina and Rao)			EB (Cluster Information)		
	P_0	P_1	P_2	P_0	P_1	P_2
13	0.895	1.614	2.452	-0.203	-0.411	-0.574
23	-0.227	-0.150	-0.251	-0.143	-0.066	-0.183
37	-0.321	-0.172	-0.154	-0.163	-0.030	-0.090
Mean	0.116	0.431	0.682	-0.170	-0.169	-0.282

Table 3. Relative Root Mean Squares Error of EB estimates of Poverty Indicators

Area	EB (Molina and Rao)			EB (Cluster Information)		
	P_0	P_1	P_2	P_0	P_1	P_2
13	0.899	1.623	2.468	0.214	0.419	0.579
23	0.230	0.158	0.257	0.160	0.090	0.195
37	0.322	0.181	0.171	0.168	0.085	0.136
Mean	0.484	0.654	0.965	0.181	0.198	0.303

In this simulation study, it can be seen empirical Bayes method with cluster information has a smaller relative bias and relative root mean squares error than empirical Bayes Molina and Rao (synthetic model) for non-sampled area.

6. Case Study

In this study we used data National Socio-Economic Survey 2013 and Village Potential 2014 in regency and city of Bogor. There are 40 sub-districts in regency of Bogor and 6 sub-district in city of Bogor. In regency of Bogor, there are 3 non-sampled areas($n_i = 0$): Megamendung, Tanjungsari and Parung Panjang sub-districts. The response variable is average expenditure per capita per month from National Socio-Economic Survey data.

Auxiliary variables obtained from Village Potential data. There are six selected auxiliary variables from Village Potential 2014. The auxiliary variables are proportion of the number of villages with the main source of income in agriculture, proportion of the number of villages with the main income source processing industry, proportion of villages with the main income source of the field of large trading / retail and restaurants, proportion of villages with the main source of income in services, proportion of store or grocery shop, and proportion of restaurant.

The problem is arises when estimates non-sampled sub districts. Molina and Rao method used a synthetic model that ignores the effect of area. In this study is added synthetic model with mean effect of area in each cluster sub-district. Auxiliary data is used to cluster sub-district and used Euclidean distance and Ward method. Sub-districts in Regency and city of Bogor are divided into three clusters and each cluster is shown in Table 4.

Characteristics sub-district in first cluster is a group of sub-districts with the main source of income in the large or retail trade and restaurants. Second cluster is group of sub districts with the main source of income in agriculture. The last cluster is group of sub-districts with education, economics, and health facilities are adequate

Poverty line 2013 that used in regency of Bogor is 271970 and in the city of Bogor is 360518. Direct estimate of non-sampled area ($n_i = 0$) can't be made. Summary statistics for empirical Bayes estimate both estimates and relative root mean squares error of P_0, P_1 and P_2 for non-sampled area are shown in Table 5 and Table 6 respectively.

Table 4. Cluster of Sub-districts in Regency and City of Bogor.

Cluster	Number of Sub-districts	Sub-districts
1	18	Nanggung, Leuwiliang, Leuwisadeng, Pamijahan, Ciampea, Dramaga, Ciomas, Tamansari, Caringin, Ciawi, Cisarua, Sukaraja, Tanjungsari, Jonggol, Kelapa Nunggal, Tajur Halang, Gunung Sindur, Tanah Sereal
2	19	Cibungbulang, Tenjolaya, Cijeruk, Cigombong, Megamendung, Babakan Madang, Sukamakmur, Cariu, Citeureup, Kemang, Ranca Bungur, Parung, Ciseeng, Rumpin, Cigudeg, Sukajaya, Jasinga, Tenjo, Parung Panjang
3	9	Cileungsi, Gunung Putri, Cibinong, Bojong Gede, Bogor Selatan, Bogor Timur, Bogor Utara, Bogor Tengah, Bogor Barat

Table 5. Empirical Bayes Estimates of Poverty Indicators in Regency and City of Bogor

Sub-districts	Sample Size	Direct	**EB (Molina and Rao)**			**EB (Cluster Information)**		
			P_0	P_1	P_2	P_0	P_1	P_2
Megamendung	0	-	0.113	0.026	0.009	0.140	0.034	0.012
Tanjungsari	0	-	0.150	0.034	0.012	0.151	0.032	0.011
Parung Panjang	0	-	0.140	0.033	0.011	0.180	0.042	0.015

Table 6. RRMSE predictions of Poverty Indicators in Regency and City of Bogor

Sub-districts	EB (Molina and Rao)			EB (Cluster Information)		
	P_0	P_1	P_2	P_0	P_1	P_2
Megamendung	0.333	0.945	1.059	0.288	0.719	0.786
Tanjungsari	0.393	0.982	1.337	0.424	1.111	1.534
Parung Panjang	0.460	0.356	0.484	0.354	0.287	0.359
Mean	0.396	0.761	0.960	0.356	0.705	0.893

It can be seen empirical Bayes method with cluster information has a smaller relative root mean squares error predictions than empirical Bayes Molina and Rao for non-sampled area.

7. Conclusion

Direct estimation cannott be implemented when the area is not selected as the sample units, while Empirical Bayes method is used to solve that problem. Both simulation study and case study show that empirical Bayes method with cluster information is better than empirical Bayes method proposed by Molina and Rao on non-sampled area in terms of relative root mean squares error. Cluster information can improve predictive ability of non-sampled area by modifying the synthetic model.

Acknowledgments

The authors gratefully acknowledge to Statistics Indonesia for their support and provide data in this research, to Penelitian Unggulan Sesuai Mandat Divisi (PUD) which has funded this research, and to Indonesia Endowment Fund for Education (LPDP) which has funded study of the first author.

References

[1] Molina I and Rao J N K 2010 *The Canadian J. of Statistics* **38** 369-385
[2] Sadik K 2009 *Forum Statistika dan Komputasi* **15** 8-13
[3] Fay R E and Herriot R A 1979 *J. of The American Statistical Association* **74** 269-277
[4] Anisa R, Kurnia A and Indahwati 2014 *IOSR J. of Mathematics* **10** 15-19
[5] Wahyudi, Notodiputro K A, Kurnia A and Anisa R 2016 *AIP Conf. Proc.* **1707** 080017-10 doi: 10.1063/1.4940874
[6] Rao J N K 2003 *Small Area Estimation* (New York: John Wiley and Sons)
[7] Johnson R A and Wichern D W 2007 *Applied Multivariate Statistical Analysis 6th Edition* (London: Prentice-Hall)

Infra Red Light Emitting Diode in 1200 nm Range have Moderate Performance in Detecting Glucose in Human Blood Glucose Model

S Robiah[1*], R P Jenie[2], M Dahrul[1], N M Nurdin[2], J Iskandar[1], A Kurniawan[1], E Rustami[1], H Syafutra[1], H Alatas[1*], Irzaman[1], E Damayanthi[2], Rimbawan[2], D Sukandar[2], Evriyanti[3] and S Budiarti[4]

[1]Department of Physics, Bogor Agricultural University, INDONESIA
[2]Department of Community Nutrition, Bogor Agricultural University, INDONESIA
[3]Department of Statistics, Bogor Agricultural University, INDONESIA
[4]Department of Biology, Bogor Agricultural University, INDONESIA

E-mail: obie.crocus55@gmail.com, s.h.alatas@gmail.com

Abstract. The diagnosis of diabetes in Indonesia is still invasive therefore hurting the patient in the examination process. Here we propose a new diabetes examination method using non-invasive optical technique. We apply for the detection a wavelength of 1200 nm a LED and photo-diode. The obatined data is then Fast Fourier Transformed (FFT) to get a better intepretation. Based on the data, we obtained a good correlation.

Keywords: 1200 nm, Infra-Red Light Emitting Diode (LED), Human Blood Glucose Model

1. Introduction

Diabetes Mellitus (DM) is characterized by elevated blood sugar levels as a result of the metabolic system disorders due to defects in insulin secretion, and causes hyperglycaemia with disturbances of carbohydrate, fat and protein metabolism [1]. DM can be classified into two main types: type 1 diabetes or Insulin Dependent Diabetes Mellitus (IDDM) and DM type 2 or Non-Insulin Dependent Diabetes Mellitus (NIDDM). DM type 1 is caused by loss of the insulin-producing beta cells in the islands of Langerhans of the pancreas [2]. Type 2 diabetes is due to the hormone insulin in the body that cannot function properly so that only a handful of glucose enter cells. As a result, cells deprived of glucose, resulting in glucose building up in the blood. Until recently, type 1 diabetes can only be treated by administration of insulin therapy performed continuously and sustainable so as to check their blood sugar levels every time [3].

DM disease have increased dramatically in recent decades, primarily due to changes in lifestyle. In Indonesia, the disease is increasing from year to year so that Indonesia is a country that ranks fourth in the number of diabetics in the world after India, China and the United States. This is because approximately 8.4% of the population in Indonesia suffer from diabetes in 2000 and would be expected to increase as many as 21.3 million people with diabetes in 2030. In addition, the current global projections estimate that the number of people with diabetes has increased by 50% year on year 2010, and will almost double by 2025 [4]. Word Health Organization (WHO) and the International Diabetes Federation (IDF) has predicted that the number of diabetics will increase significantly in 2030 to around 366 million, an increase of 214% compared to the percentage in 2006 [5].

Given the number of people with diabetes are growing each day, diabetes detection becomes regular namely by testing their blood sugar levels. A tool is needed that can detect blood sugar levels in the body that are practical, accurate, secure and cost-effective. Blood sugar measuring instrument that is circulating in the community is now a tool that works invasive glucometer [3]. Measurements of blood sugar by means of invasive is to put a needle into the skin of the patient's finger for a blood sample and measuring the glucometer can carry risks in addition to making pain from a needle prick can also cause infections of the skin, while other risks which may be a risk of causing the deployment infectious diseases such as hepatitis, HIV, etc. through contact with body fluids [6].

To keep blood glucose levels to avoid distractions such as Type 2 Diabetes, it would require an effective measurement tool. Tools biomarker measurements of blood glucose levels that have the potential to be developed for the future is a tool based Discrete Fourier transform and the Beer-Lambert Law. Additionally, the non-invasive measurement will be preferred by users as compared to invasive methods that had been used extensively.

One of the main problems in the measurement of blood glucose non-invasively is that there is no consensus on the ideal wavelength to detect glucose in human blood in the human body. A wavelength of 1200 nm is proposed [7, 8]. The purpose of this study was to assess the ability of wave of 1200 nm LED to detect glucose in human blood glucose models.

2. Method

Research has been conducted in the Laboratory of Materials Physics, Department of Physics, Bogor Agricultural University, from March to August 2016. The model of human blood glucose is made with pure glucose in the aqueous solution mimics human blood, at 50 mg / dl, 100 mg / dl, 150 mg / dl, 200 mg / dl, 250 mg / dl, 300 mg / dl, 350 mg / dl, 400 mg, dl, 450 mg / dl and 500 mg / dl. 1200 nm IR LED and photo-diode provided by Thorlabs [9, 10], is controlled using the Arduino [11], is used to measure the glucose concentration in the model in the cuvette. LED fired using twin table model, in Width Modulation (PWM) 10 Pulse different intensity, and improved value using Fast Fourier Transform [12]. The correlation between the concentration of glucose intended and photo-diode reading is calculated using Libre Office Calc [13].

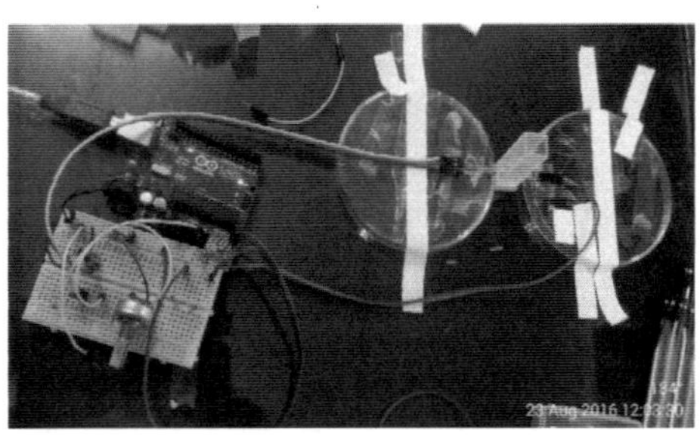

Figure 1. The circuit sensor blood glucose meter.

3. Results

3.1. Some are used as the reference blood glucose measurement tools, non-invasive design

Sensor works by spectrophotometry modulated [14, 15] near Infra-Red (NIR) [16]. Tool can continuously monitor blood glucose [17, 18]. NIR Light Emitting Diode (LED) and Photo-diode wavelength range used 1000 nm - 1700 nm, provided by Thorlabs [9, 10]. The wavelength selected by the work of Smith and Goodarzi [19, 20], which is confirmed by research that is not published on our own. Data modified spectrum using Fast Fourier Transform (FFT) [21] and will inference using correlation Cascade Artificial Neural Network [22]. The results are stored in a SQLite database [23]. The whole system developed with Qt Software Development Kit (SDK) [24, 25].

3.2. Sensor response is different for water and glucose.

We have tested our inception to pure glucose and pure water (Figure). We have proved that the photo-diodes give a different reading when given a different intensity LED (Figure). In addition, the photo-diodes respond differently to different substances. Research to find the relationship between glucose meter readings, the concentration of glucose water, and the photo-diode feedback.

4. Discussion

4.1 Research using Fast Fourier Transform as data transformation method for measuring blood glucose levels

Fast Fourier Transform used as a data transformation method for measuring blood glucose levels, this study confirm the possibility of monitoring wavelengths used in the system of non-invasion blood glucose.

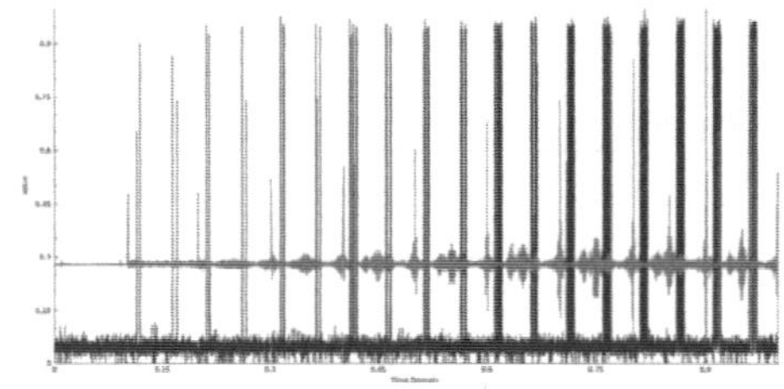

Figure 2. Correlation Results Fast Fourier Transform.

4.2 A Wavelength of 1200 nm LED

Has conducted testing of the glucose solution with various concentrations ranging from a concentration of 50 mg / dl, 100 mg / dl, 150 mg / dl, 200 mg / dl, 250 mg / dl, 300 mg / dl, 350 mg/ dl, 400 mg, dl, 450 mg / dl and 500 mg / dl using the 1200 nm LED and photo-diode provided by Thorlabs. The results obtained are as follows:

Table 1. Correlation between photo-diode readings at certain PWM intensity to blood glucose model concentration.

Correlations	Concentration (mg / dl)
Concentration (mg / dl)	1.000
Interval (cleaned) 01	-0.752
Interval (cleaned) 02	-0.341
Interval (cleaned) 03	0.239
Interval (cleaned) 04	-0.159
Interval (cleaned) 05	0.159
Interval (cleaned) 06	0.032
Interval (cleaned) 07	0.084
Interval (cleaned) 08	0.293
Interval (cleaned) 09	0.093
Interval (cleaned) 10	-0.003

4.3 Best Wavelength still being investigated

Reference wavelength is still in dispute to investigate and measure blood glucose levels. Ranging from 800 nm [19], to a high of 2500 nm [26]. We have confirmed some wavelengths in research that is not published, but only for glucose and water solution, so we're still doing tests with different wavelengths for a solution.

4.4 The solution proposed

Promoting testing glucose solution with varying concentrations using different LEDs ranging from 1050 nm, 1450 nm, 1550 nm, and 1600 nm, in order to find the best wavelength to produce optimal results. We did not use the LED above 1700 nm because the price is too expensive.

5. Conclusion

Based on research that has been done is made a conclusion that the measurement of glucose solution by using wavelength of 1200 nm LED that concentration value closer to 1 or -1 is said to be good.

Acknowledgements

The authors express their gratitude for the funding by the IPTEK development program under contact No. 079/SP2H/LT/DRPM/II/2016.

References

[1] Hovens FA, Van de Lar SC 2005 *Cochare Database of Systematic Reviews*
[2] Jensen EF 2007 Modeling of Simulation of Glukosa_insulin Metabilosm [tesis] Kongens Lyngby (DK): Technikal University of Denmark
[3] Ade K 2015 Penerapan Film $Ba_{0,5}Sr_{0,5}tio_3$ sebagai Sensor Cahaya pada Alat Ukur Gula Darah Berbasis Optik [Tesis] Bogor Agricultural University
[4] Zeev A and Shabat S 2008 *Archives of Gerontology and Geriatrics* **47(3)** 383-393
[5] Mubarak R, Marwan A, Alsuwyed 2010 *International Journal of Diabetes Mellitus*
[6] Yavad, Rani, Singh, Murari 2015 *Biomedical Signal Processing and Control*
[7] McEwen MP and Reynoldsk J 2014 *Opt. Appl.* **44**
[8] Smith JL 2011 *Last Accessed May* **2**
[9] Thorlabs, Inc. 2016 *Unmounted LEDs*
[10] Thorlabs, Inc. 2016 *Unmounted Photodiodes*
[11] Arduino 2015 Genuino Starter Kit [English] Code: GKX00007
[12] fftw.org 2015 FFTW Home Page
[13] Smith JA, Weber JH and Fox MJ 2013 Libre Office 4.0 The document foundation Calc Guide working with spreadsheets *Libre Off. 40 Doc. Found. Calc Guide Work. Spreadsheets*
[14] Chowdhury MK, Srivastava A, Sharma N and Sharma S 2014 *IOSR J. Electr. Electron. Eng. IOSR-JEEE* **9** 34–41
[15] Srivastava A, Chowdhury M K, Sharma S and Sharma N 2014 *Adv. Bioresearch*
[16] Srivastava A, Chowdhury M K, Sharma S and Sharma N 2013 *Int. J. Adv. Eng. Technol. IJAET* **6** 1097–1108
[17] Cameron F, Buckingham B A, Wilson D M and Bequette B W 2014 *World Congress* **19** 243–248
[18] Christiansen M, Bailey T, Watkins E, Liljenquist D, Price D, Nakamura K, Boock R and Peyser T 2013 *Diabetes Technol. Ther.* **15** 881–888
[19] Smith JL 2011 *Last Accessed May* **2**
[20] Goodarzi M, Sharma S, Ramon H and Saeys W 2015 *TrAC Trends Anal. Chem.* **67** 147–58
[21] fftw.org 2015 FFTW Home Page
[22] FANN 2015 Fast Artificial Neural Network Library (FANN)
[23] Akinkuolie B B, Lin C-F and Yuan S-M 2011 *Genetic and Evolutionary Computing* 163–167
[24] Qt 2015 Qt - Download Open Source Step 3
[25] Mai F, Palatov K, Bilotti E and Peijs T 2012 Self-reinforced polylactic acid (PLA) biocomposites with degradation sensing *Proceedings of ECCM-15 Conference Venice*
[26] Krushinitskaya O 2012 *Osmotic sensor for blood glucose monitoring applications* (Vestfold University Colleg.

IOP Conf. Series: Earth and Environmental Science **58** (2017) 012022 doi:10.1088/1755-1315/58/1/012022

Antioxidant and Cytotoxic Activity of *Phyllanthus acidus* Fruit Extracts

D Andrianto[1], W Widianti[1] and M Bintang[1]

[1]Biochemistry Department, Bogor Agricultural University, Bogor, Indonesia

Abstract. *Phyllanthus acidus* is an Indonesian plant belonging to the Euphorbiaccae family. Extraction of *P. acidus* was performed using the maceration method. Four-solvent extraction process by ethanol, 70% ethanol, 30% ethanol, and water was used. The antioxidant activity from this extract was determined using 2,2-diphenyl-1-pikrilhidrazil (DPPH) and cytotoxicity (biological potency from this extract) using Brine Shrimp Lethality Test (BSLT) method. We found that the extraction yields of samples ranging from 1.13% to 20.25%. The ethanol extract showed the highest yield, while the lowest yield was reported in the water extract of *P. acidus* fruit. Among the samples, water extract of *P. acidus* exhibited high antioxidant activity with IC_{50} 26.06 μg/mL. LC_{50} values for BSLT ranging from 473.26 to 908.98 μg/ mL, with the water extract having the lowest value and therefore the most potent, and the ethanol extract having the highest value

1. Introduction

Oxygen is a paradox in aerobic organism [1]. An organism needs oxygen for energy production by respiration. However, oxygen also undergoes reaction into superoxide anion radical, hydrogen peroxide, and hydroxyl radicals. Those molecules are free radicals and cause oxidative stress to cells. Oxidative stress causes biomolecular damage by attack of reactive oxygen species (ROS). In human, oxidative stress related to various degenerative process and diseases.

Phyllantus acidus (L.) Skeel species belongs to the Plantae kingdom, Spermatophyta division , Angiospermae sub division, Dicotyledonae class , Euphorbiales order ,Euphorbiaceae family , and *Phyllanthus* genus . It can be found in the Caribbean and Pacific Ocean part of Southeast Asia. Their local names are grosella (Puerto Rico), jimbilin (Jamaica), karamay (Philippines), and ciremai (Indonesia and Malaysia) [2]. *P. acidus* tree can grow up to 9 m in height. The tree has tough main branches and long branchlets. Leaves are densely arranged in the branch. The leaves size is 2 to 7.5 cm long with a light green color. The flowers are male, female, or hermaphrodite. Flower size is 5 to 12.5 cm long with pink color. Fruits are 2 cm in diameter and densely clustered. Fruit is edible and the color is white to yellow. Fruits contain high moisture and taste sweet and juicy [3].

In Indonesia, every part of the tree is believed to have different medicinal properties. The fruits are used as astringent, laxative, and to heal ulcer. The leaves are used to treat cancer, cough, asthma, scurvy, and reduce body weight. The bark is used for asthma and rash. Local people consume the fruit directly or as an ingredient in pickle. For medication, 25 g of plant material is mixed with 200 mL of water [3].

Cytotoxic, antibacterial, and antioxidant activity of *P. acidus* fruit petroleum ether extract was reported by Habib *et al.* [4]. *P. acidus* leaves activity as antiinflamatory, antinociceptive, and antioxidant have been reported by Chakraboty *et al* [5]. and antihyperlipidemic activity of *P. acidus* fruit acetone and methanol extract was reported by Andrianto *et al.* [2]. However, antioxidant and cytotoxic activity of *P. acidus* fruit extracts in ethanol and aqueous solvents have not yet been reported before. This research was carried out to evaluate antioxidant and cytotoxic activity of *P. acidus* fruit extracts.

2. Experimental

2.1 Plant materials
The plants were collected in Bekasi, Bogor, and Sumedang region of West Java Province, Indonesia. Plants were identified by Research Center for Biology, Indonesian Institute of Sciences, Bogor, Indonesia.

2.2 Materials
2-(N-morpholino)ethanesulfonic acid (MES) was from Dojindo Lab. (Tokyo, Japan). Trolox, 1,1-diphenyl-2-picrylhydrazyl (DPPH). Other chemicals used were analytical grade.

2.3 Extraction
Fresh fruits were washed and soaked in 5% NaCl to prevent browning. It was then blended into a fine juice and freeze-dried immediately. Dried samples were ground into powder and extracted using ethanol, 70% ethanol, 30% ethanol, or water (5 times volumes) at room temperature for 3 days. The macerates were filtered and the process was repeated one more time. Mixture was filtered using Whatman filter paper (No. 2), the solution was collected to become extract. Both acetone and methanol extract were concentrated under reduced pressure on rotary evaporator. The concentrated extracts were then used for the experiments.

2.4 DPPH radical scavenging activity
The antioxidant activities of the extracts were assayed for the effect of scavenging DPPH free radicals based on the method described by Blois [6]. Sample was added into a mixture (0.9 mL) of 0.4 mM DPPH solution, 20% methanol aqueous solution, and 0.2 M MES buffer solution to make final concentration 0, 30, 60, 90, 120, and 150 μg/mL. Absorbance was measured at 520 nm. Trolox was used as positive control. The inhibitory activity was calculated from the following equation:

$$\text{DPPH radical scavenging activity (\%)} = \left(\frac{1-A}{A_0}\right) \times 100\% \qquad (1)$$

where A_0 is the absorbance of the mixture without sample and A is the absorbance of the mixture with a sample.

2.5 Brine shrimp lethality test (Nurcholis et al.) [7]
Ten *Artemia salina* Leach were incubate in 100 μL sea water in a well plate. It was added with 100 μL sample in 0, 10, 100, 200, 500, and 1000 μg/mL concentration, respectively. Treatments were done with 3 replications. Mixture was gently shaken and incubates for 24 hours in room temperature. Dead *A. salina* from each well was counted and LC_{50} values were calculated based on probit analysis.

3. Result and Discussion

3.1 Extraction
Fresh *Phyllanthus acidus* fruit samples were used in the research. Water content in our samples were 85.55±3.00%. High water content supports the growth of the fruits while it attached on the trees. However, it will cause the fresh fruit easily broken after harvesting process.

Extraction processes were carried out by using ethanol, 70% ethanol, 30% ethanol, and water (figure 1). We obtained the yield of each extracts are 20.25%, 3.72%, 3.28%, and 1.13%, respectively.

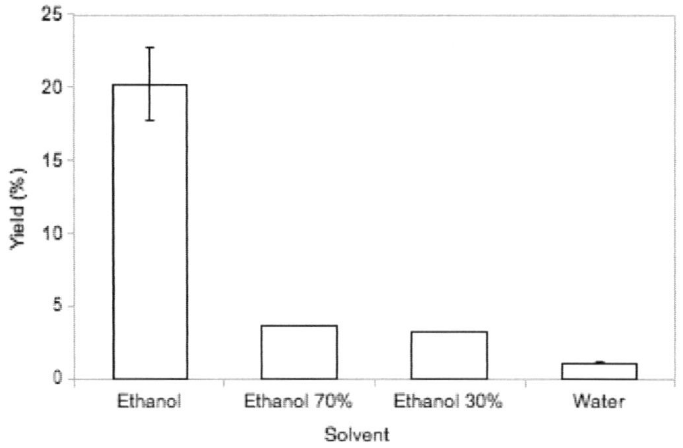

Figure 1. Yield percentage of *Phyllanthus acidus* fruit extract.

Results showed that the highest yield was given by the ethanol extract, while the lowest was given by the water extract. Extraction yields were different among the treatments because the bioactive components that are dissolved in those solvent also different. Wide ranges of chemical compounds easily dissolve in ethanol compared to water. Ethanol can dissolve less polar and polar compounds include alkaloids, steroids, saponines, flavonoids, quinones, and glicosides [8]. Natural products such as fruits contain those compounds. Water is commonly chosen due to its availability and price.

3.2 Phytochemical test

Secondary metabolites in *P. acidus* fruit were studied by qualitative phytochemical tests include flavonoids, alkaloids, tannins, saponines, steroids, terpenoids, phenolics, and glycosides (table 1). Results showed that all extracts contain flavonoids, alkaloids, phenolics, terpenoids, saponines, and glycosides. Although the components quantities were different in each solvent. Ethanolic extract contained more flavonoids, alkaloids, terpenoids, and saponines but less phenolics compare to the aqueous extract. All extract contained similar amount of glicosides and did not contain tannins and steroids.

Table 1. Phytochemical screening tests.

Phytochemical Tests	Extract			
	Ethanol	70% Ethanol	30% Ethanol	Water
Flavonoid	+++	+++	++	++
Tannin	-	-	-	-
Alkaloid	+++	+++	++	++
Phenolic	+	+	++	++
Terpenoid	+++	+++	++	++
Steroid	-	-	-	-
Saponine	+++	+++	+++	++
Glycoside	+	+	+	+

notes: (-) did not contain the components and (+) contains the components

Secondary metabolites are chemical components isolated from plant sources. These components act as protection for survival from predator and parasites. These compounds also can be acted as antioxidants [9]. Based on table 1, we suggest that flavonoid was the major constituents in our samples.

3.3 In vitro antioxidant activity

Among the samples, water extract of *P. acidus* exhibited high antioxidant activity with IC_{50} 26.06 µg/mL (figure 2). While ethanol extract did not give antioxidant activity. Lower value of IC_{50} means higher antioxidant activity, because we only need small amount of molecule to neutralize 50% of DPPH radicals in the solution.

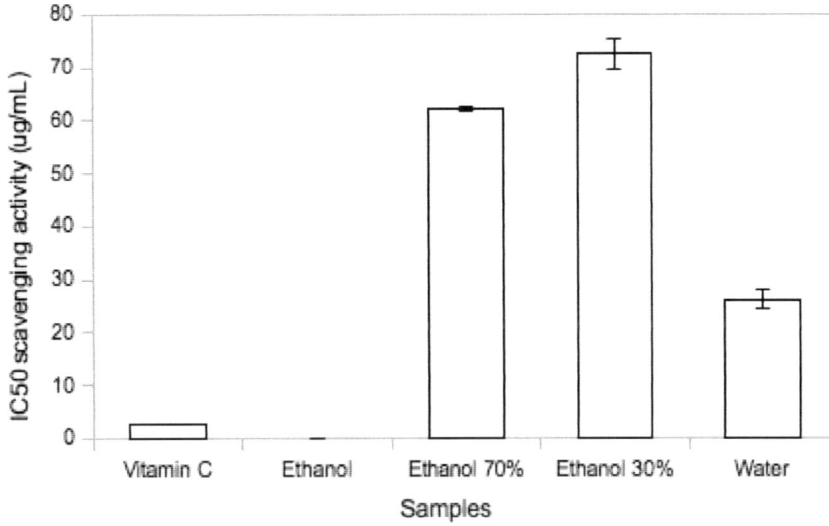

Figure 2. Antioxidant activity of *Phyllanthus acidus* fruit extract.

Natural products could be categorized as a strong antioxidant activity if is possess IC_{50} below 200 ⌠g/mL, while IC_{50} below 10 ⌠g/mL was considered as very strong antioxidant activity. Our results showed that ANOVA statistical test for all treatments different significantly ((<0.05). Hence, the solvents significantly gave effects in the number of bioactive chemical components extracted and furthermore give significant antioxidant activity differences.

3.4 Cytotoxicity test

LC_{50} values for BSLT ranged from 473.26 to 908.98 µg/ mL, with the water extract having the lowest value and therefore the most potent, and the ethanol extract having the highest value (figure 3). BSLT test is a simple bioassay test to study the acute toxicity of a chemical compound in *Artemia salina* Leach. The mechanism related with the ability of secondary metabolites in plant in poisoning the animal. 50% of lethal toxicity concentration (LC_{50}) was determined using probit analysis based on the number of mortality in *A. salina*. Juniarti [10] showed that potential bioactive crude extract posses LC_{50} value under 1000 ppm, while potential bioactive pure compounds have LC_{50} value below 30 ppm. *P.acidus* fruit extract has been reported to have antihyperlipidemic activity [2] and *P. acidus* leaves extract has activity as antiinflamatory, antinociceptive, and antioxidant have been reported by Chakraboty *et al.* [5].

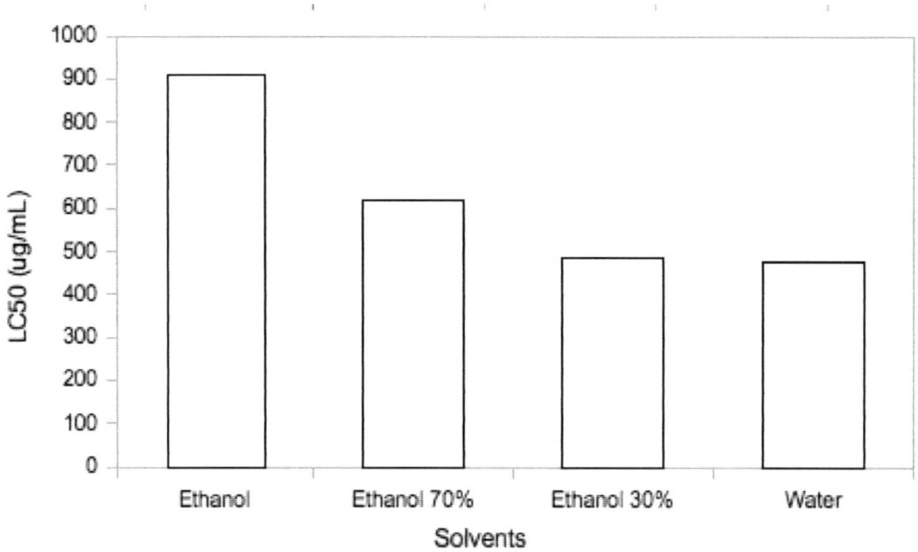

Figure 3. Cytotoxic activity of *Phyllanthus acidus* fruit extract.

4. Conclusion

This is the first report of antioxidant and cytotoxic activities of *Phyllanthus acidus* fruit ethanol and water extracts. We found that *P. acidus* water extract have the most potential as antioxidant and cytotoxic activities. Additional studies are needed to isolate and identify responsible compounds for those activities.

References

[1] Halliwell B 2007 *Biochemical Society Transaction* **35** 1147-1150
[2] Andrianto D, Katayama T and Suzuki T 2015 *J Forest Biomass Utilization Society* **10** 19-25
[3] Dalimartha S 2008 *Atlas Tumbuhan Obat Indonesia Jilid 1* (Jakarta: Trubus Agriwidya)
[4] Habib M R, Sayood M A, Rahman M, Hasan M and Saha A 2011 *Int. J. Pharmaceutical Sci. Res.* **2** 875-881
[5] Chakraboty R, De B, Devanna N and Sen S 2012 *Asian Pacific J. Trop. Med.* **2** S953-S961
[6] Blois M S 1958 *Nature* **181** 1099-1200
[7] Nurcholis W, Andrianto D, Falah S and Katayama T 2011 *International Symposium of Indonesian Wood Research Society* 87-95
[8] Harborne J B 1984 *Phytochemical Methods* (London: Chapman and Hall)
[9] Lu Y, Khoo T J and Wiart C 2014 *Pharmacology & Pharmacy* **5** 395-400
[10] Juniarti, Osmeli D and Yuhernita 2009 *Makara Sains* **13** 50-54

IOP Conf. Series: Earth and Environmental Science **58** (2017) 012023 doi:10.1088/1755-1315/58/1/012023

Influence of Explant Position on Growth of *Talinum paniculatum* Gaertn. Adventitious Root in Solid Medium and Enhance Production Biomass in Balloon Type Bubble Bioreactor

M H Solim[1], A N Kristanti[2] and Y S W Manuhara[1]*

[1]Department of Biology, Faculty of Science and Technology, Airlangga University, Surabaya, 60115, Indonesia
[2]Department of Chemistry, Faculty of Science and Technology, Airlangga University, Surabaya, 60115, Indonesia

Email: wulanmanuhara@gmail.com

Abstract. *Talinum paniculatum* Gaertn. is one of traditional medicinal plant in Indonesia as an aphrodisiac. This plant has various compounds which is accumulated in roots. In vitro culture of this plant can enhance production of adventitious roots. The aim of this research was to know the influence of explants position on growth of *T. paniculatum* Gaertn. adventitious root in MS solid medium and enhance the production of biomass in balloon type bubble bioreactor. Explants from leaf were cultured at abaxial and adaxial positions in solid MS medium supplemented with IBA 2 mgL^{-1}. Adventitious roots were cultured in bioreactor with various treatments (without IBA, supplemented with IBA 2 mgL^{-1} and supplemented with IBA 2 mgL^{-1} + buffer $NaHCO_3$). Result showed that the main growth of abaxial root was higher than adaxial, however, the total of adaxial root branch was higher than abaxial. The highest biomass production of adventitious root cultured was achieved by MS medium supplemented with IBA 2 mgL^{-1} + buffer $NaHCO_3$. This treatment has produced fresh biomass two fold of initial inoculum.

Keyword: Abaxial, adaxial, biomass, bioreactor, $NaHCO_3$, *Talinum paniculatum*

1. Introduction

Talinum paniculatum Gaertn. has grown in Indonesia and was used as traditional medicinal plant, but growth of this plant is low in natural habitat especially on root organ. It contains various bioactive compounds, including saponin. This compound has been revealed to bring benefit for vitality (aphrodisiac). The accumulation of saponin can be found in root organ, which can be increased with adventitious root culture. Many studies have used the adventitious root culture as one of the techniques to accelerate large-scale multiplication, to increase root biomass and secondary metabolites [1,2,3]. The adventitious root culture can be induced from parts of the plant. In the culture, explant position also can influence the root growth. Some authors reported that explant position can influence the growth of plant tissue culture [4,5]. Technology of root culture in bioreactor is a promising method to produce root biomass and secondary metabolites, because it can be controlled both physically and chemically. One of bioreactors for adventitious root culture is Balloon Type Bubble Bioreactor

(BTBB). Many research used bioreactor to culture various plants such as *Panax ginseng* [6], *Eurycoma longifolia* [7] and *Morinda citrifolia* [8].

One of the cultivation methods in bioreactor is batch cultivation which is characterized by constantly changing environmental conditions. Sajc *et al.* [9] explained it has potential to produce metabolites associated with any kinetic pattern. This cultivation can be used to determine the conditions for maximum biomass productivity. Plant growth hormone such as IBA has potent for induction of root formation and elongation [10,11].

Sodium bicarbonate ($NaHCO_3$) provides high buffer capacity in the cell culture medium thus keeping the pH-value in the physiological conditions during cultivation. The required CO_2 concentration is related to the desired pH-value and the $NaHCO_3$ concentration of the medium [12,13]. The aim of this research was to observe the influence of explants position on growth of *T. paniculatum* Gaertn. adventitious root in MS solid medium and enhance production biomass in BTBB with buffer and batch cultivation.

2. Material and Method

2.1. Plant material and optimization of the culture conditions

The study was carried out in the plant physiology laboratory at Departement of Biology, Faculty of Science and Technology, Airlangga University between December 2015 - July 2016. Adventitious roots were induced from leaves of *T. paniculatum* Gaertn. on MS [14] solid medium supplemented with 7 gL^{-1} agar, 30 gL^{-1} sucrose and 2 ppm indole-3-butyric acid (IBA). Samples of leaf (2 cm length) used were the second leaf from the tip. All of the explants were washed with detergent 10 % (v/v) and thoroughly under tap water for 15 minutes. Then surface of the explants were sterilized 5 % (v/v) $HgCl_2$ for 5 minutes in laminar airflow. After that, explants were washed 3-4 times with sterile distilled water. The cultures were planted in Petri dish containing 25 mL medium and pH was adjusted to 5.8 with 1.0 N HCl or 1.0 N NaOH before adding 7 gL^{-1} agar and autoclaving at 121 ^{0}C for 20 minutes. Different positions of leaf namely abaxial and adaxial were aapplied to verify the suitable culture method for accumulation of main root and root branch of adventitious root. After five weeks, the adventitious root from the first culture were transferred to liquid MS medium supplemented with 2 mgL^{-1} IBA in the balloon type bubble bioreactor (BTBB).

2.2. Balloon type bubble bioreactor (BTBB) Design

Bioreactor with a capacity of 1 L was filled with 350 mL of liquid MS medium supplemented with 2 ppm IBA and 30 mgL^{-1} sucrose and sterilized by autoclaving at 121°C for 15 min. The adventitious roots were cultured in bioreactor with various treatments (without IBA, supplemented with IBA 2 mgL^{-1} and supplemented with IBA 2 mgL^{-1} + buffer of $NaHCO_3$). Aeration volume was established at 0.2 vvm (volumes of gas per volume of liquid per minute) and the initial inoculum density 2 g/350 mL. Bioreactor was modified with net in middle to support growth of the adventitious root.

2.3. Determination of main root and root branch

After 5 weeks of culture, the main root and root branch were observed and calculated. The main root appeared from explant directly, meanwhile root branch was fixed of main root. The texture color of adventitious root was recorded. All the cultures were maintained for 35 days at 25 ± 2^{0}C, with a 16 hour light (40 μmol m^{-2} s^{-1})/ 8 hour dark photoperiod cycle provided by 40-W white fluorescent tubes.

2.4. Determination of fresh and dry biomass

The cultures were maintained for 35 days and sampling has done every seven days to determine the total sugar, electrical conductivity (EC) and hydrogen ion concentration (pH) of the medium. All the cultures were maintained at 25 ± 2^{0}C, with a 16 hour light (40 μmol m^{-2} s^{-1})/ 8 hour dark photoperiod cycle provided by 40-W white fluorescent tubes. After five weeks, the adventitious was harvested and separated from the medium through a stainless steel sieve. The fresh biomass was recorded after blotting away the surface water. The dry biomass was measured after drying root and shoot at 60^{0}C for 48 hour. The growth ratio (GR) was calculated as GR=[harvested DW(grams)–inoculated DW (grams)]/inoculated DW (grams).

2.5. Determination of total sugar, conductivity and hydrogen ion concentration in the medium
Measurement of total sugar, conductivity and hydrogen ion concentration has conducted to liquid MS medium supplemented 2 ppm IBA. The electrical conductivity (EC) was measured using a conductivity meter (Ezdo, Cond5021), total sugar was measured using a refractometer (Atago) and the hydrogen ion concentration (pH) of the culture medium was measured using a pH meter (Boeco, BT-600) at first week until fifth weeks.

2.6. Statistical analysis
For the first culture, the growth of leaves as explant was induced with two treatments (abaxial and adaxial positions) with six replicates. The data was analysed using SPSS software (v.17). Mean separation within column by Independent sample T-test at $P \leq 0.05$. For the second culture, the data of adventitious root was collected in BTBB and was analysed with descriptive analyses using the same software.

3. Result and Discussion

3.1. Determination of main root and root branch of T. paniculatum
In this study, the explant positions in the medium, both abaxial and adaxial of leaves could influence the production of adventitious root especially the main root and root branch. Based on statistic analysis, the main growth of abaxial root was higher than adaxial, but the number of adaxial root branches was higher than abaxial (table 1). The morphology of adventitious based on explant position exhibited similar texture and color (figure 1). Both the culture of abaxial and adaxial positions showed thick texture and white in color.

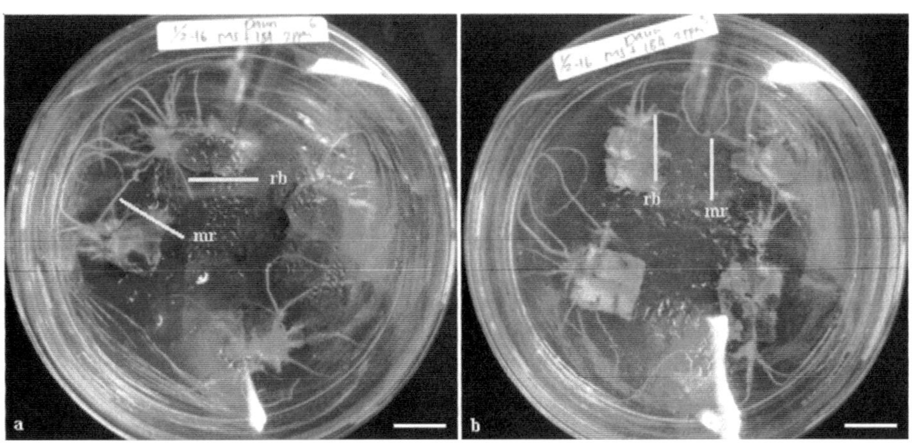

Figure 1. Adventitious root of *Talinum paniculatum* at different explant position treatments in MS medium supplemented 2 mgL^{-1} IBA: a) abaxial, b) adaxial. mr-main root; rb-root branch. (Bar = 1,5 cm).

The culture of adventitious root can be induced from leaf of *T. paniculatum*. In this study, the culture was developed from two different explant positions. It can influence growth of the adventitious root especially at the main root and branch root (table 1). Position of explants in the culture profound different between abaxial and adaxial (figure 1). Many studies also reported that explant position can improve the plant growth in plan tissue culture [4,5].

Table 1. Number of main roots and root branches from two position of explants (abaxial and adaxial) for five weeks in Petri dish containing MS solid medium with 2 mgL^{-1} IBA.

No.	Treatments	Main Roots (Mean ± SD)	Root Branches (Mean ± SD)
1	Abaxial	34 ± 11.78	18 ± 3.85
2	Adaxial	9.5 ± 2.35	37.2 ± 7.57

Data represents mean values (n=2) with six replicates. Mean separation within column by Independent sample T-test at P ≤ 0.05

3.2. Biomass of T. paniculatum adventitious root

The highest biomass production of *T. panicualtum* adventitious root was achieved by MS medium supplemented with 2 mgL^{-1} IBA + buffer of $NaHCO_3$. This treatment had produced fresh and dry biomass for 5.2 g and 0.26 g, respectively (figure 2), which was two fold of initial inoculums. Treatment of MS medium with 2 mgL^{-1} IBA without buffer resulted both fresh and dry biomass for 4.33 g and 0.21 g, respectively. The treatment without hormone and buffer produced both fresh and dry biomass lower than the others. Growth ratio in batch with IBA and buffer was 1.53, followed by batch with IBA for 1.13.

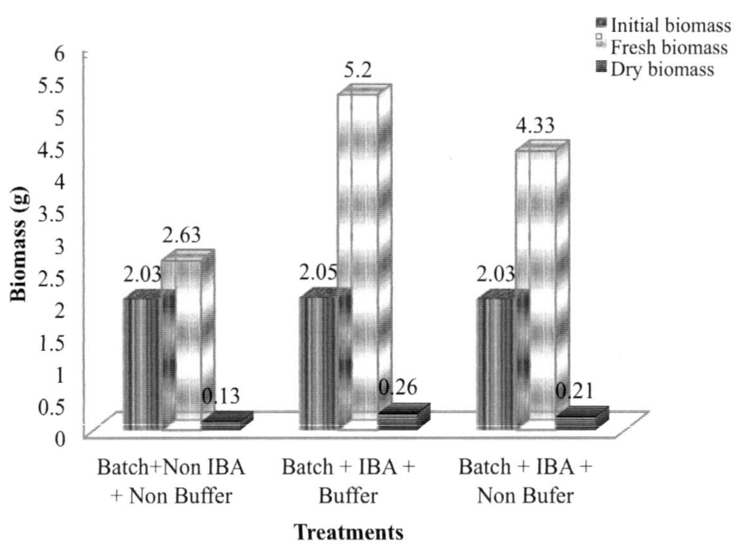

Figure 2. Biomass production of *T. paniculatum* adventitious roots in 1000 mL balloon type bubble bioreactor containing 350 mL MS liquid médium with three treatments for 35 days.

Studies revealed that the production of biomass in liquid culture is influenced by concentration of MS medium, culture method, inoculums, sucrose and plant growth homone [7,8,16,15,17]. Furthermore, Yan *et al.* [18] reported that the plant tissue culture may continuously be provided with better homogenization and hormones which improved the growth of the plantlets. Kuria *et al.* [19] also added that biomass accumulation in liquid medium is higher than that in solid medium. Similarly, Preece [20] reported that employment of liquid media could enhance the growth of root and shoot.

In this study, the tissue cultures were done in batch system. It showed a significant enhancement upon the growth of adventitious root in biorector. Bioreactor method for adventitious root cultures have been developed previously [21]. Sivakumar [22] explained that the bioreactor culture system is more advanced than the traditional plant tissue culture system because the culture conditions in a bioreactor can be optimized. The culture conditions that can be optimized including temperature, pH, concentrations of oxygen, carbon dioxide and nutrients in the medium. Moreover, nutrient uptake of roots can also be enhanced by continuous medium circulation. Furthermore, cell proliferation and regeneration rates can also be increased.

Figure 3. Adventitious root culture of *Talinum paniculatum* Gaertn.: a) balloon type bubble bioreactor ; b-c) fresh biomass in batch buffer and batch non-buffer ; d) culture process ; e-f) dry biomass in batch buffer and batch non-buffer. (Bar = 1,5 cm).

Other than physical parameters, the inoculum density can also influence the growth of adventitious roots. Inoculum density is an important factor affecting growth production in a number of plant cell culture systems. Baque *et al.* [15] reported that inoculum density significantly affect the accumulation of root biomass of *M. citrifolia* during 4 week of culture.

Plant growth hormone such as IBA has potentially induced root formation rather than IAA or synthetic auxins [10]. Yang and Davies [23] showed further evidence of IBA for basipetal transport. The authors stated that IBA can stimulate elongation of subtending nodes, suggesting IBA is transported basipetally in intact pea plants and it has been examined the radiolabeled distribution after application of a rooting solution to the explants base. Zolman *et al.* [11] also reported that auxin hormone such as IBA and IAA at root elongation and lateral root formation of *Arabidopsis* developmental processes profound sensitive. Moreover, Poupart and Waddell [24] explained that specificity of IBA transport tissue can support the possibility of endogenous auxin in growth and development of some *Arabidopsis* tissues.

Gertlowski and Petersen [25] explained that addition of sucrose can also affect cells metabolisms and the production of metabolites. Lee *et al.* [26] adescribed that sucrose can be hydrolyzed into two monosaccharides, glucose and fructose, by invertase which bounds extracellular and/or cell wall during the initial culture period. In addition, sodium bicarbonate ($NaHCO_3$) provides high buffer capacity in the cell culture medium thus maintaining the pH-value in the physiological conditions during cultivation. The required CO_2 concentration is related to the desired pH-value and the $NaHCO_3$ concentration of the medium. The bicarbonate and carbon dioxide buffering system is the most commonly used to maintain physiological pH. The bicarbonate is a weak buffer with a pKa of 6.1 making a pH range of 7.2-7.6 more difficult to prevent rapid pH changes. This buffer, however, is non-toxic and has nutritional value [12,13].

Adelberg *et al.* [27] had shown that with gelling agents the nutrients dissolved in the liquid phase are not completely available as water is bound to the gel by a matrix force. That study compared the spent gelled medium and spent the liquid medium and revealed that nutrients remained in agar, and therefore nutrients were somehow unavailable to the plant. It also has been shown that in presence of agar, the resistance of sucrose transfer from medium to the plant is 300× greater than that of sucrose transfer from a shaken liquid to the plant per unit of surface area [28]. Thus, the sucrose availability in liquid medium is higher than in gelled medium [27].

The liquid culture has also beeb reported to be more susceptible with shear stress. Sajc *et al* [9], in their review stated that shear levels which were below levels of causing cell damage while still high enough for efficient mixing in the liquid phase and dispersion of air are considered as important criteria of bioreactor design. Vogelmann *et al.* [29] reported that the critical shear stress is between 80 and 200 N/m2. During the process scale-up, the main goal is to reproduce on a larger scale the conditions found to be optimal on a small scale. The perceived sensitivity of plant cells to hydrodynamic stress associated with agitation and aeration. In addition to damaging and breaking the cells through the hydrodynamic stress generated by aeration, agitation, shaking, pumping, and other operations. The immediate consequence of the shear effect on plant cells is cell damage, which has been quantitatively measured by using a number of system responses such as reduction in cell viability [30], changes in morphology and/or aggregate patterns [31], release of intracellular compounds [32] and changes in metabolism. Liquid culture is often associated with hyperhydricity [33]. In contrast, our results showed that the adventitious roots in the liquid medium did not show any hyperhydricity symptoms. This phenomenon was likely caused by the application of the net in middle of bioreactor. Furthermore, the roots were floated in filter sieve and were not sunk in liquid medium. Therefore, the shear levels in the bioreactor was suitable for a good bioreactor design.

3.3. Total sugar, conductivity and hydrogen ion concentration in the medium

The determination of hydrogen ion concentration (pH), total sugar and electrical conductivity (EC) of the médium was conducted every seven days. Total sugar, pH and EC were decreased, but pH and EC were increased by using buffer for every two weeks (figure 4). Several studies described that liquid medium was examined by electrical conductivity, pH and total sugar. The high conductivity of medium showed that the adventitious root was unavailable to absorb inorganic compounds. Meanwhile, Thanh *et al.* [34] explained that electrical conductivity was used as indirect method to estimate biomass in the cell culture, because it represents the uptake of inorganic compounds by the cell during cultivation. Decreasing amount of organic compounds indicated that cells work and increase biomass of cell, instead. Furthermore, this conditions were supported by the decreasing pH of medium at the end of cultivation. Decreasing pH can decrease the ability of the cells to absorb inorganic compound from the medium [35]. It was likely that the decreasing pH during adventitious root culture was caused by ammonium in MS medium. Ammonium in MS medium is derived from

 doi:10.1088/1755-1315/58/1/012023

ammonium nitrate as source of nitrogen whichi is also important as a buffer. When cells need nitrogen, cells will release H^+ to uptake nitrogen source, amonium. Thus, continuous release of H^+ into medium will result acid conditions.

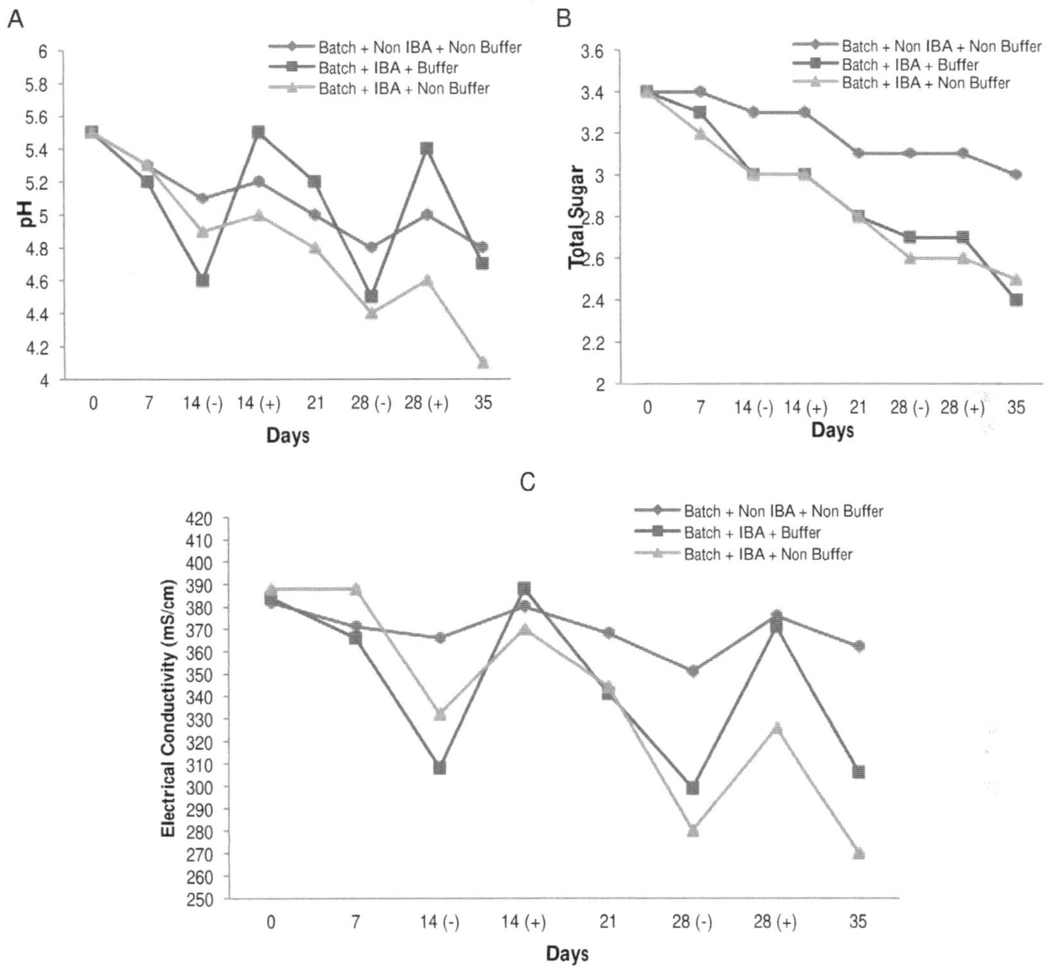

Figure 4. Adventitious root culture condition for 35 days. Culture condition change in (A) hydrogen ion (pH); (B) Total sugar; (C) Electrical conductivity (EC).

The source of carbon in liquid MS medium is sucrose. Sucrose will be hydrolyzed to glucose and fructose in the beginning of tissue culture. During cultivation, cell will consume monosacharides, therefore the measurement of total sugar can be indicated by cells growth [36]. Furthermore, oxygen for aeration in bioreactor should be supplied throughout the cultivation. Indeed, productivity of high-density and high-viscosity cell suspensions is proportional to cell concentration generally, as well as aeration is a major concern in bioreactor design and scale up for plant cell culture [9].

4. Conclusion

In this study, the growth of abaxial main root was higher than adaxial, contrarily to that total number of root branch. Meanwhile, the highest biomass production of adventitious root cultured was achieved by MS medium supplemented with IBA 2 mgL^{-1} + buffer $NaHCO_3$. This treatment has produced fresh biomass in two fold of the initial inoculum.

Acknowledgements
Thanks to Indonesia Endowment Fund for Education (LPDP), Ministry of Finance, Republic of Indonesia and Plant Physiology Laboratory at Departement of Biology, Faculty of Science and Technology, Airlangga University.

References

[1] Choi S M, Son S H, Yun S R and Paek K Y 2000 *Plant Cell Tiss. Org. Cult.* **62** 187–193
[2] Yu K W, Gao W Y, Hahn E J and Paek K Y 2001 *J. Plant Biol.* **44** 179–184
[3] Cui X H, Chakrabarty D, Lee E J and Paek K Y 2010 *Bioresource Technology* **101** 4708–4716
[4] Niederwieser J G and Vcelar B M 1990 *HortScience* **25**(6) 684-687
[5] Chen J-T and Chang W-C 2002 *Plant cell, tissue and organ culture* **69**(1) 41-44
[6] Thanh N T, Murthy H N, Paek K Y 2014 *Ind Crop and Prod* **60** 343-348
[7] Lulu T, Park S Y, Ibrahim R, Paek K Y 2015 *J Biosci and Bioeng* **119**(6) 712-717
[8] Baque M A, Shiragi M H, Moh S H, Lee E J and Paek K Y 2014 *In Vitro Cell. Dev. Biol. Plant* 49 p 737-749
[9] Sajc L, Grubisic D and Novakovic G V 2000 *Biochemical Engineering Journal* **4** 89–99
[10] Ludwig-Muller J 2000 *J. Plant Growth Regul.* **32** 219–230
[11] Zolman B K, Yoder A and Bartel B 2000 *Genetics* **156** 1323–1337
[12] Freshney R I 1992 *ed. Animal Cell Culture: A Practical Approach* 2nd ed (New York: Oxford University Press) p 54
[13] Davis J M, 1994 ed. *Basic Cell Culture: A Practical Approach* (New York: Oxford University Press) p 59-60
[14] Murashige T and Skoog F 1962 *Physiologia Plantarum* **15** 473-497
[15] Baque M A, Shiragi M H K, Moh S H, Lee E J and Paek K Y 2013 *In Vitro Cell.Dev.Biol.* **49** p737–749
[16] Cui H Y, Murthy H N, Moh S H , Cui Y Y, Lee E J and Paek K Y 2014 *Ind. Crops and Products* **53** 28-33
[17] Mariateresa C, Maria C S C and Giuseppe C 2014 *Industrial Crops and Products* **55** 194-201
[18] Yan H, Lian C and Li Y 2010 *Plant Cell Tissue Org. Cult.* **103** 131–135
[19] Kuria P, Demo P, Nyende A and Kahangi E 2008 *African Journal of Biotechnology* **7**(3) 301-307
[20] Preece J E 2011 *Propagation of Ornamental Plants* **10**(4) 183-187
[21] Paek K Y, Chakrabarty D and Hahn E J 2005 *Plant Cell Tissue Organ Cult.* **81** 287–300
[22] Sivakumar G, Yu K W, Paek K Y 2005 *Eng. Life Sci.* **5** 527–533
[23] Yang T and Davies P 1999 *J Plant Growth Regul.* **27** 157–160
[24] Poupart J and Waddell C S 2000 *Plant Physiol* **124** 1739–1751
[25] Gertlowski C and Petersen M 1993 *Plant CellTissue Org. Cult.* **34**(2) 183–190
[26] Lee E J, Mobin M, Hahn E J and Paek K Y 2006 *J. Plant Biol.* **49** 427-431
[27] Adelberg J, Delgado M P and Tomkins J P 2010 *In Vitro Cellular and Developmental Biology Plant* **46** 95-107
[28] Adelberg J and Fári M G 2010 *Propag. Ornam. Plants* **10**(4) 205–219
[29] Vogelmann H, Bischof A, Pape D and Wagner F 1978 *Some aspects of mass cultivation*, in: A.W. Alfermann, E. Reinhard (Eds.), Production of Natural Compounds by Cell Culture Methods, GSF, Munich p 130–146
[30] Scragg A H, Allan E J and Leckie F 1988 *Enzyme Microb. Technol.* **10** 361-367
[31] Kieran P M, O'Donnell H J, Malone D M and MacLoughlin P F 1995 *Biotechnol. Bioeng.* **45** 415- 425
[32] Meijer J J, ten Hoopen H J G, Luyben K C A M and Libbenga K R 1993 *Enzyme Microb.*

Technol. **15** 234-238
[33] Shaik S, Dewir Y H, Singh N and Nicholas A 2010 *Afr. J. Bot.* **76** 180–186
[34] Thanh N T, Murthy H N, Yu K, Jeong C S, Hahn E and Paek KY 2006 *J. Plant Physiol.* **163** 1337-1341
[35] Manuhara Y S W, Kristanti A N and Utami E S W 2015 *Asian Journal of Biological Sciences* **8**(2) 83-92.
[36] Gorret N, Rosli S K B, Oppenheim S F, Wilis L B, Lessard P A, Rha C K and Sinskey A J 2004 *J. Biotechnol.* **108** 253-263

Regional Analysis of Prone Drought Areas under Future Climate Change Scenarios: Case Study Agropolitan of Malang District

Perdinan[1,2,3*], R F Adi[2,3] and E Y Arini[2,3]

[1] Department of Geophysics and Meteorology, Bogor Agricultural University
[2] Prima Intelektual Area (PI AREA)
[3] Generasi Hijau Indonesia

Abstract. Global climate change challenges livelihoods across the globe concerning its potential impacts. Consequently, climate change adaptation (CCA) strategies should be devised to minimize its negative and maximize its positive impacts. In response to the needs for developing CCA strategies, climate risk assessment should be conducted as an approach to identify contributed factors to the climate risks in a region with which the adaptation strategies can be devised. This study focused on modeling prone areas to drought over a specialized area called Agropolitan located in Malang district. The model was developed based on the concept of drought in combination with spatial analysis. The model includes climate component (i.e., rainfall and evapotranspiration) and biophysical component (i.e., land cover/use, green open space, slope/elevation). The regional climate data for the baseline and future condition were processed using the gridded datasets of WorldClim, which also included the outputs of six global climate models, namely: BCC, CCCMA, CSIRO, GISS, MIROC and NCAR, for projecting future climate of the study region. The analysis shows that prone areas to drought in Malang district was projected to be larger under future climate scenarios. The expanded prone areas were Kalipare sub-district and Pagak sub-district, which prone areas were expanded towards to the south. For the sub-districts of Donomulyo, Bantur, Singosari, and Gedangan, the prone areas were expanded towards to the north. The prone areas of drought in Poncokusumo sub-district remained constant. The prone areas showed within the Agropolitan areas of Poncokusumo were in the villages of Ngadas, Gubukklakah, Pajaran, Argosuko, Ngebruk and Jambesari, and were projected to expand under future climate scenarios. This assessment was an essential step for further study focusing on defining proper CCA strategies for the study region.

Keywords: climate change, extreme, drought, agropolitan, malang

1. Introduction

Global climate change has been estimated to affect regional climate of many areas around the globe. The recent report of IPCC published in 2014 estimated that rainfall pattern and magnitude for Asia, including Indonesia, was changing [1]. This new climate regime may pose a challenge particularly on water supply throughout the country given the possibility of lower rainfall over a year. As an approach to address the challenge, climate risk assessment can be conducted to identify contributed factors to the climate risks in a region with which the adaptation strategies can be devised. The strategies were directed to minimize the potential negative impacts and maximize the potential positive impacts of climate change [1].

This research selected the district of Malang, particularly the Agropolitan area, as a case study to observe the impacts of climate change on prone areas to drought in the district. The impacts of drought were diverse and can be broadly classified as economic, environmental and social [8] The Agropolitan area is located in one of sub-districts of Malang, named Kecamatan Poncokusumo. This special area characterized by the complexity of topographical terrain and economic activities in the region (e.g., agriculture, livestock, and tourism). Based on the records of Data and Information on Disasters in Indonesia named Data Indeks Bencana Indonesia (DIBI) in 2011, it was reported that potential climate related hazards that threaten the Malang region were flooding, droughts, landslides, and wind speed. Specifically for drought occurred in 2015, the disaster agency of Malang named *Badan Penanggulangan Bencana Daerah* (BPBD) reported that the drought in 2015 caused the water shortage of about 1.5 million liters. As for floods, in July 2016, there was a flood disaster caused 73 houses were damage. The incidences of climate related hazards are increasing which are in line with increased climate variability induced by climate change causing extreme weather and/or climate extreme events.

The incidents of climate related hazards have been alarming because the extreme events may threat human livelihoods (e.g, water shortage and crop production failure). This situation urges the needs to understand factors contribute to the formation of a climate related hazard, such as drought and flood. This study focuses on the development model of prone areas to drought, which is an essential step for climate risk assessments. The model was constructed based on climate and biophysical factors. The climate component composed of two essential climate variables, i.e. rainfall and air temperature; while, the biophysical component included information on regional characteristics, i.e, land use, slope, and topography. The variables for each component were also chosen considering the availability of data required for each component. The model was expected to provide information on prone areas to drought and the most contributed factors causing drought in that area. Consequently, anticipation strategies can be devised properly by considering factors contributed to drought in a region.

2. Related Research

Drought is one of climate extreme events occurred as a result of water deficit, i.e, the amount of evapotranspiration is higher than the amount of rainfall. The development of regional drought models using a variety of techniques has been done in Indonesia. Mujtahiddin et al. [2] assessed drought incidences in the district of Indramayu using the ThornthWaite-Matter that were applied to monthly rainfall data of 13-rainfall stations over a 30-year period (1980-2009). The other study assessed drought occurred in 15 water catchment areas in eastern Java using the threshold level method, which assumed the value of discharge as a drought index to indicate drought incidences [3]. Yusron [4] evaluated drought area based on interpolation of standardized precipitation index (SPI). Another study identified various estimation methods developed for drought analysis, namely: SPI and SPEI applied to monthly climate data [5]. Rosanne *et al* [9] reconstruction of PDSI for Java,Indonesia, that reflects past monsoon drought variability and the influence of ENSO in this region. A considerably skilful simple empirical forecast model for predicting drought over Java Island has been developed using the DMI in combination with Ni no-3.4 [10]. Recent study [6] has developed a method for predicting prone areas to drought based on the concept of net balance between rainfall and evapotranspiration using climate data (i.e., temperature and rainfall) and biophysical data (i.e., topography, slope, and land use).

3. Methods

Malang District has an area of 3347 km^2 with formation of lowlands and highlands. Climatologically, Malang has a monsoonal precipitation type with a peak of rainy seasons from December to February (during the winter Asia) and the dry season from June to August (during the Australian winter). The prone areas to drought for the study region was modeled based on the concept of net balance between rainfall and evapotranspiration, and spatially analyzed using geographical information system. The model of simulated prone áreas was calculated based on regional climate information on rainfall and temperature, and biophysical characteristics of the region. The climate observations were only available from two climate stations, namely: Karangploso and Karangkates. These limited data available from these stations were employed to analyze climate averages such as rainfall pattern over a year and trends as presented in figure 1.

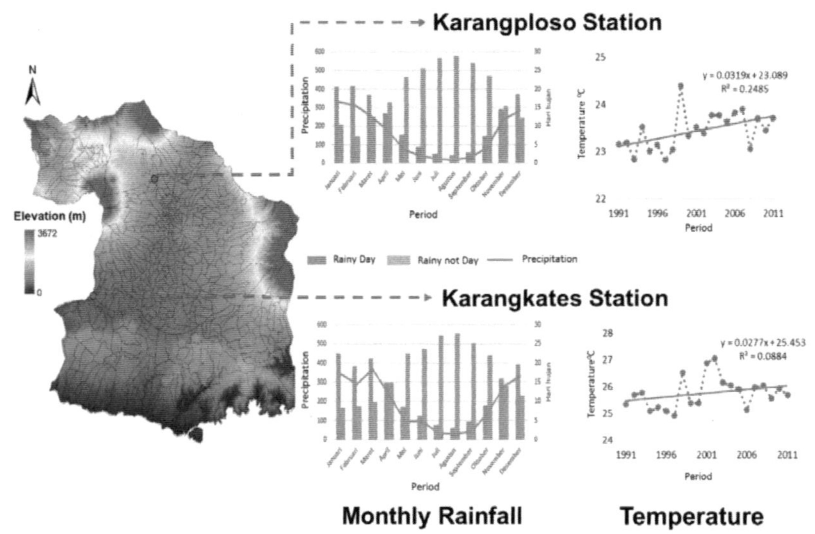

Figure 1. The district of Malang (left), rainfall averages for the period of 1991-2011 at Karangploso station (middle-above) and at Karangkates station (middle-below), and the trend of air temperature for the same period for the two stations (right).

The limited number of climate stations limit the ability to capture regional characteristics of climate over the studied regiön. This situation offered an opportunity to utilize the outputs of available gridded climate data. This research employed gridded data named WorldClim developed by Hijmans et al. [7] (figure 2). The WorldClim produced the gridded climate data using spatial interpolation techniques at a spatial resolution of 1 km. Climate variables include rainfall and air temperature (i.e., minimum, average, and maximum temperature), and solar radiation for the baseline period (1971-2000) and the future period (2021-2050) under emission scenario of RCP 4.5 projected using six global climate models included in the WorldClim. The included climate models are BCC, CCCMA, CSIRO, GISS, NCAR, and MIROC.

The rainfall data of WorldClim showed that the high rainfall ranging from 3300-3600 mm/year were in the plateau territory with elevation of 1700-2500 meters. The regions include the sub-districts of Poncokusumo, Jabung, Karangploso, Dau, Pujon and Ngantang. The value of the lowest rainfall occurred in the coastal area or the south of Malang, which ranged from 1500-1700 mm/year. These regions include the southern part of the sub-district of Donomulyo, the sub-districts of Bantur and Gedangan, and the northern part of the sub-district Kalipare Pagak. The annual temperature averages of Malang ranged from 25-27°C. As the topography rises above 1500 meters, air temperature decreases significantly (figure 2).

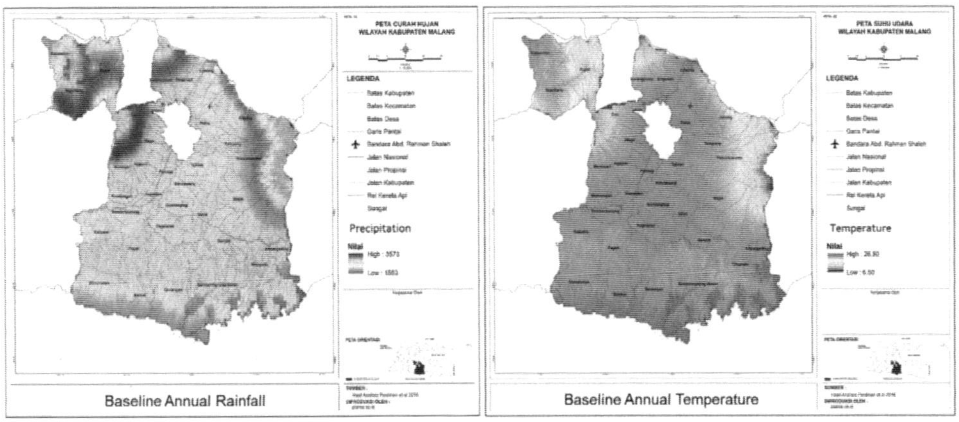

Figure 2. Rainfall (left) and air temperature (right) for the periode of 1971-2000 produced based on the Worldclim gridded data at 1 km resolution. Source of WorldClim: Hijmans et al. [7]

The biophysical components including land cover/use, green open space, and slope/elevation were analyzed. Land cover were analyzed using the outputs of Landsat 8 trained using supervised classification method with the Earth Indonesia map and Google Earth. The Landsat 8 was also employed to produce the green open space map using the Normalized Difference Vegetation Index. Slope was determined utilizing the SRTM data. Threshold values were determined to categorize values for each variable employed in determining the climate and biophysical components as described by Perdinan et al. [6]. The threshold values were parameterized uniquely for the district of Malang on the basis of field surveys and literature review.

4. Results and Discussion

The developed drought model for Malang district produced a map of prone areas to drought in the range of zero to one (0-1), indicating the potential areas to drought was increasing as the index value increased. Categorization was then performed to understand which areas within Malang district and the sub-district of Poncokusumo were prone to drought. The categorizations were 0-0.2 (Very Low), 0.21-0.40 (Low), 0.41-0.60 (Medium), 0.61-0.80 (High) and 0.81-1.00 (Very High). The analysis showed that Malang has the potential Low to Very High prone areas to drought. The prone areas to drought categorized into Very High were in sub-district of Poncokusumo (i.e., the village of Ngadas). The regions categorized into High were sub-districts of Kalipare (i.e., the villages of Kalirejo, Arjosari, Sukosari, and Argowinagun), Pagak (i.e., the villages of Tlogorejo, Gampingan, and Sumberejo), Bantur (i.e., the villages of Karangsari and Rejoyoso), Donomulyo (i.e., the villages of Sumberoto, Banjarejo, and Tulungrejo), Singosari (i.e., the villages of Randuagung, Candirenggo, Dengkol, and Toyomarto), Lawang (i.e., the villages of Lawang, Sumberporong, and Mulyoarjo) and Pakis (i.e., the villages of Buluwetan and Mangliawan). The rest areas were in the category of Medium and Low (figure 3-left). Specifically for the district of Poncokusumo (i.e, the Agropolitan region), the category of Very High was given for the village of Ngadas. The other regions with High category were the villages of Gubukklakah, Pajaran, Argosuko, Ngebruk, and Jambesari. While the rest villages were in the category of Medium to Low (figure 3-right).

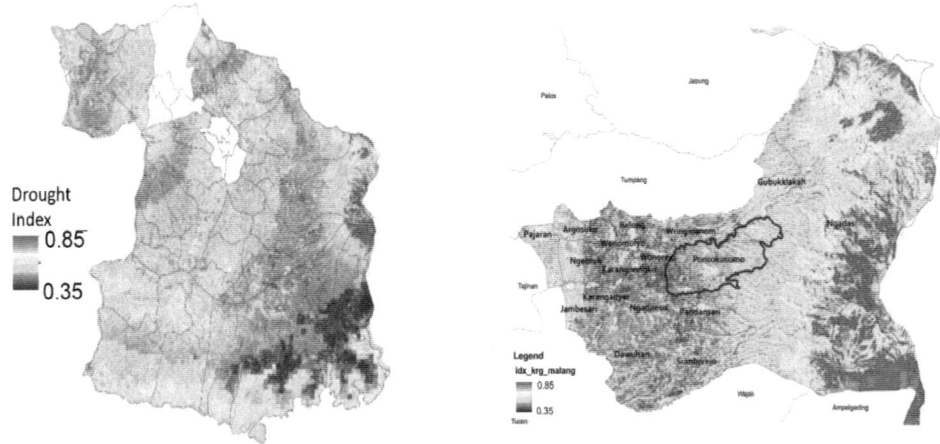

Figure 3. Potential prone areas to drought in the district of Malang (left) and the sub-district of Poncokusumo, i.e., the Agropolitan region (right).

The future climate was projected using the outputs of six (6) available global climate models (GCMs) within the WorldClim database, i.e., BCC, CCCMA, CSIRO, GISS, NCAR, and MIROC, under emission scenario of RCP4.5. The projected climatic condition was performed on rainfall and air temperature for the period of 2021-2050. Annually, the amount of rainfall may change in the range of -15 to 27%. The projected outputs of CCCMA, GISS, NCAR, BCC show a projected decrease in the rainfall amount ranging from -15 to -0.3% for the district of Malang. The CCCMA, BCC and NCAR indicated that the amount of rainfall was projected to decline for areas within the sub-districts of Dampit, Tirtoyudo, Ampelgading, and Sumbermanjing Wetan. The GISS model shows a decline of projected rainfall for areas within the sub-districts of Kalipare, Pagak, Sumberpucung, Kromengan, and Kepanjen. The CSIRO wss projected decreasing rainfall for areas within the sub-district of Poncokusumo and Singosari. As for GFDL, the rainfall amount was projected to increase for the district of Malang, especially for areas in the coastal and mountainous region (figure 4).

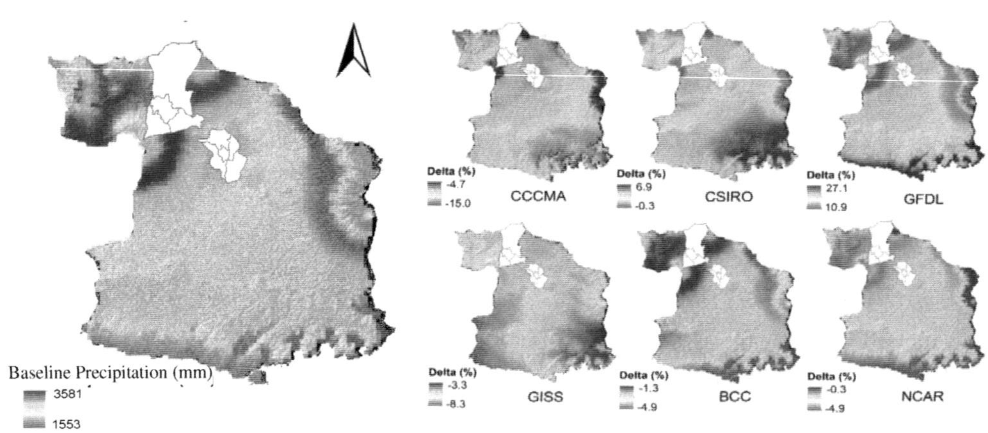

Figure 4. Projected of rainfall resulted from six global climate models simulated under RCP 4.5 archieved in WorldClim and extracted for the Malang district for the period of 2021-2050.

As rainfall data, the projection of air temperature was performed using the same models and period archived in WorldClim. The results show that changes in air temperature ranged from 0.5 to 1.4◦C. The CCCMA model exhibited the highest increase, whereas, the BCC model displayed the smallest increase. The projections of increasing temperature for the CCCMA, GFDL and GISS were distributed evenly throughout the district of Malang. The CSIRO and BCC display the highest increased temperature were in the territory of North Malang; while, the NCAR showed the highest increase for the southern areas of the study region (figure 5).

Employing the projections of rainfall and temperature, the prone areas to drought in Malang were simulated. The results suggested that there were potential extended drought areas. For example, the prone areas within the sub-district of Kalipare and Pagak showed the widespread moved towards the south. For the sub-districts of Donomulyo, Bantur, and Gedangan, the prone areas to drought may expand towards the north. For the sub-district of Singosari, the simulated drought index was increasing, but the index range will be narrowed as with those for the areas of the sub-district Poncokusumo. Specifically for the Agropolitan- Poncokusumo, the potential prone areas to drought include areas categorized into the Medium (M) to Very High (ST). Generally, the villages of Ngadas, Gubukklakah, Pajaran, Argosuko, Ngebruk and Jambesari were prone to drought for the baseline and future period, indicating future climate may expand prone areas to drought (figure 6).

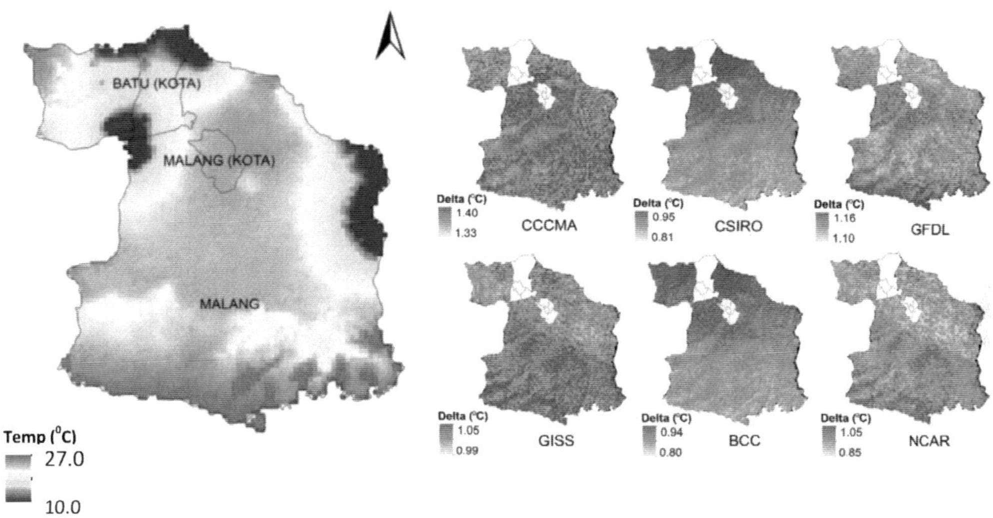

Figure 5. Projected of air temperature resulted from six global climate models simulated under RCP 4.5 archieved in WorldClim and extracted for the Malang district for the period of 2021-2050.

The analysis of prone areas to drought (figure 6) indicates that the district of Malang, particularly the sub-district of Agropolitan-Poncokusumo, should consider the prone areas when designing the development plan. Further exploration through field visits to the prone areas was also suggested to identify the exact factors contributed to drought in the prone areas, so proper anticipation strategies can be devised. The strategy should also be mainstreamed into development plan in order to support the achievement of the development targets.

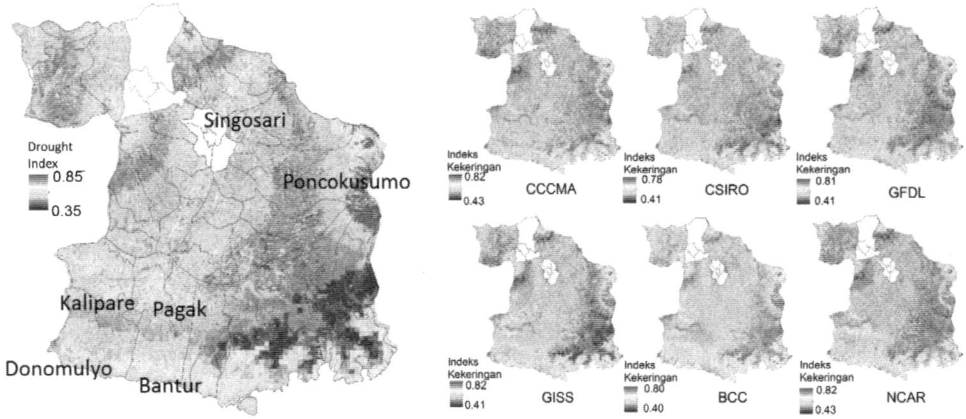

Figure 6. Analysis of prone areas to drought in the district of Malang for the baseline period of 1971-2000 and the future period of 2021-2050.

5. Conclusion

Modeling of prone areas to climate-related hazards such as drought essentially required climate information, e.g., rainfall and air temperature, as inputs. The modeling approach were developed by considering the drought concept in combination with geographic information system. The model simulateed the interactions between climate (i.e., rainfall and evapotranspiration) and biophysical factors (e.g., land use/cover, topography). Climate data for the baseline and future condition were obtained from gridded datasets available worldwide called WorldClim, and the future climate were projected based on the outputs of six global climate models, namely: BCC, CCCMA, CSIRO, GISS, MIROC and NCAR, archieved in WorldClim.

The analysis showed that prone areas to drought in Malang district were projected to be expanded under future climate scenarios. The expanded prone areas were for the sub-districts of Kalipare and Pagak, in which the prone areas were expanded towards the south. For the sub-districts of Donomulyo, Bantur, Singosari, and Gedangan, the prone areas were expanded towards the north. The prone areas to drought in Poncokusumo sub-district were not change. The employment of various global climate models exhibited diffferent expanded prone areas. This difference was commonly evaluated to show uncertain future climate condition, which may affect the hazard analysis. For example: CCCMA, GISS and NCAR show the largest increase, while the lowest is showed by CSIRO. Furthermore, specifically for the Agropolitan-Poncokusumo, the villages of Ngadas, Gubukklakah, Pajaran, Argosuko, Ngebruk and Jambesari were considered prone to drought for the baseline and future period, indicating future climate may expand prone areas to drought. The results provide an early warning information on which areas should be prioritized in addressing the potential impacts of drought which may pose adverse impacts on livelihoods in the district of Malang and specifically the Agropolitan areas of Poncokusumo. This urges proper anticipation strategies should be devised and mainstreamed into the development plan.

Acknowledgment

This study is supported by the Third National Communication (TNC) Project No. 00087506. The authors highly appreciate the TNC national project team and the officers of Malang district government involved in this study for their valuable contribution.

References

[1] [IPCC] 2014 Climate change impact, adaptation, and vulnerability. Part A: Global and Sectoral Aspects Contribution of Working Group II to the Fifth Assessment Report of the Intergovernmental Panel on Climate Change IPCC, Geneva, Switzerland

[2] Mujtahiddin M I 2014 *Jurnal Meteorologi dan Geofisika* 99-107

[3] Yusron A 2013 *Analisis indeks kekeringan pada 15 DAS di wilayah Jawa Timur menggunakan metode ambang batas* (Universitas Jember: Jember)

[4] Nurraahman, F I and A Pamungkas *Jurnal Teknik Pomits* 22-27

[5] Nugraha and Reza P 2015 *Aplikasi SPEI dan SPI sebagai indeks kekeringan meteorologis. Geofisika dan Meteorologi* (IPB: Bogor)

[6] Perdinan, Yon S, Tri A, Ujang S, Ryco F A and Enggar Y A 2015 *Climate Change Child Centered Risk Assessment in Surabaya* (FMIPA IPB: UNICEF)

[7] Hijmans R et.al. 2005 Very high interpolated climate surfaces for global land area *International Journal of Climatologi* **25** 1965-1978

[8] Wilhite D A and Pulwarty R S 2005 *Drought and water crisis: Lessons learned and the road ahead. In Drought and Water Crisis: Science, Technology, and Management Issues* (New York, NY, USA: Wilhite, D.A., Ed.; Taylor & Francis Group, CRC Press)

[9] Rosanne et. al. *Monsoon drought over Java, Indonesia during the two past centuries Geophysical Research Letters* **33** L04709 American Geophysical Union: New York

[10] Darrigo R and Wilson R 2008 *International Jurnal Climatology* 1654

[11] [BAPPENAS] 2014 National Action Plan for Climate Change Adaptation (RAN-API) Jakarta: Ministry of National Development Planning/National Development Planning Agency

Two-strain Tuberculosis Transmission Model under Three Control Strategies

S N Rayhan[1], T Bakhtiar[1] and Jaharuddin[1]

[1]Department of Mathematics, Faculty of Mathematics and Natural Sciences, Bogor Agricultural University, Bogor 16680, Indonesia

Email: syadza.nr@outlook.com, tbakhtiar@ipb.ac.id, jaharipb@yahoo.com

Abstract. In 1997, Castillo-Chavez and Feng developed a two-strain tuberculosis (TB) model, which is typical TB and resistant TB. Castillo-Chavez and Feng's model was then subsequently developed by Jung et al. (2002) by adding two control variables. In this work, Jung et al.'s model was modified by introducing a new control variable so that there are three controls, namely chemoprophylaxis and two treatment strategies, with the application of three different scenarios related to the objective functional form and control application. Pontryagin maximum principle was applied to derive the differential equations system as a condition that must be satisfied by the optimal control variables. Furthermore, the fourth-order Runge-Kutta method was exploited to determine the numerical solution of the optimal control problem. In this numerical solution, it is shown that the controls treated on TB transmission model provide a good effect because latent and infected individuals are decreasing, and the number of individuals that is treated effectively is increasing.

1. Introduction

Since 1993, WHO said that tuberculosis (TB) was one of global emergency issues. Since 2003 until 2011, it was estimated that there are still approximately 9.5 million new cases of TB at present, and almost 0.5 million people died because of TB around the world[3]. Meanwhile, controlling TB offers new challenges such as TB/HIV co-infection, drug-resistant TB, and other challenges with a higher complexity for humanity.

In human beings, TB is caused by *Mycobacterium tuberculosis* bacteria (Mtb) and it is an airborne transmitted disease. Mtb droplets are released into the air by sneezing or coughing infectious individuals, and so TB can be transmitted by it. TB is not highly infectious and so occasional contacts with an infectious individual that rarely leads to infection. TB is described as a slow disease because of its long and variable latency period distribution and its short and relatively narrow infectious period distribution. Latent stage of TB is a phase where Mtb lives inside a person's body but don't make them feel any pain, don't show TB symptoms, and can't spread the disease. However, if the bacteria inside their body became active and developed, they can be infected with active TB[2]. Individuals who are latently infected are neither clinically ill nor capable of transmitting TB[1]. Most latently infected individuals do not become infectious (active TB).

Treatment on latent TB plays an important role in reducing the risk of TB become active. Every effort must be done to initiate an appropriate treatment and to make sure the treatment is done completely to individuals with latent TB. It is really important for infected individuals that are being treated to consume their medicine as prescribed. If an individual stop consuming their medicine before the preferred time, the disease will relapse, and if they don't consume it regularly, the bacteria that still living inside their body will be resistant to the drugs. This condition can lead to a new disease called resistant TB, and this will be harder and require much more expenses to be treated.

TB transmission can be controlled by reducing the number of patients and doing actions that can prevent susceptible individuals become infected by infectious individuals. One of the prevention that can be done is applying chemoprophylaxis, that is giving drugs and therapy routinely to prevent the infection on latently infected individuals, and also avoid the disease transmission on infected individuals. Incomplete TB treatment can cause the disease to relapse, even on recovered individuals.

In 1997, Castillo-Chavez and Feng developed a two-strain TB transmission model, that is typical TB and resistant TB. Typical TB is a TB disease in general, while the resistant TB is caused by infectious individuals who didn't obey and complete their treatment so that the disease inside is resisting the drugs that were given. Afterwards, this model is developed by Jung *et. al* (2002) by adding two control variables.

In this paper, the authors integrated the models developed [4] and [1]. The model is taken from [4], and then added one control variable from [1] to the model in [4], then a model with three control variables, that are chemoprophylaxis and two treatment strategies, is obtained. This model represents the rate of typical TB and resistant TB transmission, and with these controls will be studied what are the impacts to the transmission model.

2. Mathematical model

2.1. Tuberculosis transmission model

The tuberculosis model from Jung *et. al* (2002) divides the total human population into the following sub-groups that are susceptible individuals S, those latently infected with typical TB and resistant TB respectively L_1 and L_2, those infected with typical TB and resistant TB respectively I_1 and I_2, and those who are treated (effectively) T. The total population size at time t is given by $N(t) = S(t) + L_1(t) + I_1(t) + L_2(t) + I_2(t) + T(t)$.

The state system is the following system of six differential equations in [4] and the relapse parameter q^* is added to it:

$$
\begin{aligned}
\frac{dS}{dt} &= \Lambda - \beta_1 S\frac{I_1}{N} - \beta^* S\frac{I_2}{N} - \mu S,\\
\frac{dL_1}{dt} &= \beta_1 S\frac{I_1}{N} - (\mu + k_1)L_1 + pr_2 I_1 + \beta_2 T\frac{I_2}{N} - \beta^* L_1\frac{I_2}{N},\\
\frac{dI_1}{dt} &= k_1 L_1 - (\mu + d_1)I_1 - r_2 I_1 + q^* T,\\
\frac{dL_2}{dt} &= qr_2 I_1 - (\mu + k_2)L_2 + \beta^*(S + L_1 + T)\frac{I_2}{N},\\
\frac{dI_2}{dt} &= k_2 L_2 - (\mu + d_2)I_2,\\
\frac{dT}{dt} &= \big(1 - (p + q)\big)r_2 I_1 - (q^* + \mu)T - \beta_2 T\frac{I_1}{N} - \beta^* T\frac{I_2}{N},
\end{aligned}
\tag{1}
$$

with initial conditions:

$$S(0) = S_0, L_1(0) = L_{10}, I_1(0) = I_{10}, L_2(0) = L_{20}, I_2(0) = I_{20}, T(0) = T_0, \tag{2}$$

and $S(t_f), L_1(t_f), I_1(t_f), L_2(t_f), I_2(t_f), T(t_f)$ are undetermined.

We assume that an individual may be infected only through contacts with infectious individuals. It is also assumed that susceptible humans are recruited into the population at per capita rate Λ. β_1 and

β_2 are the rates at which susceptible and treated individuals become infected by an infectious individual with typical TB, respectively. β^* is the rate at which an uninfected individual becomes infected by one resistant-TB infectious individual. The per capita natural death rate is μ while the per capita disease induced death rates are d_1 and d_2 for the typical TB and resistant TB, respectively. The rates at which an individual leaves the two latent classes by becoming infectious are k_1 and k_2. Some treated individuals relapse back into the infective state at rate q^*. r_1 and r_2 are the treatment rates of individuals with latent and infectious typical TB, respectively, and $p + q$ is the proportion of those treated infectious individuals who did not complete their treatment $(p + q \leq 1)$. Some treated individuals relapse back into the infective state at rate q^*.

There are three control strategies used to control the spread of the two-strain TB; applying chemoprophylaxis to latently infected individuals of typical TB, treatments for infected individuals of typical TB, and treatments for treated individuals to reduce the probability of disease relapse. With these controls, model (1) becomes:

$$
\begin{aligned}
\frac{dS}{dt} &= \Lambda - \beta_1 S\frac{I_1}{N} - \beta^* S\frac{I_2}{N} - \mu S,\\
\frac{dL_1}{dt} &= \beta_1 S\frac{I_1}{N} - (\mu + k_1)L_1 - u_1 r_1 L_1 + (1 - u_2)pr_2 I_1 + \beta_2 T\frac{I_1}{N} - \beta^* L_1\frac{I_2}{N},\\
\frac{dI_1}{dt} &= k_1 L_1 - (\mu + d_1)I_1 - r_2 I_1 + (1 - u_3)q^* T,\\
\frac{dL_2}{dt} &= (1 - u_2)qr_2 I_1 - (\mu + k_2)L_2 + \beta^*(S + L_1 + T)\frac{I_2}{N},\\
\frac{dI_2}{dt} &= k_2 L_2 - (\mu + d_2)I_2,\\
\frac{dT}{dt} &= u_1 r_1 L_1 + \big(1 - (1 - u_2)(p + q)\big)r_2 I_1 - \big((1 - u_3)q^* + \mu\big)T - \beta_2 T\frac{I_1}{N}\\
&\quad - \beta^* T\frac{I_2}{N}.
\end{aligned}
\tag{3}
$$

The coefficient u_1 represents a routine therapy for individuals latently infected with typical TB who are carrying on treatments, $1 - u_2$ represents the effort that prevents the failure of the treatment in typical TB infectious individuals, such as supervising individuals who are undergoing treatments, and $1 - u_3$ represents the effort that prevents the disease relapse of treated individuals. Basically, model (1) can be obtained by assigning $u_1 = u_2 = u_3 = 0$.

2.2. *Analysis of optimal control*

The optimal control problem faced is to determine the control functions u_1, u_2, and u_3 that are carrying the system from initial state $(S_0, L_{10}, I_{10}, L_{20}, I_{20}, T_0)$ to an undetermined terminal state $\left(S_{t_f}, L_{1t_f}, I_{1t_f}, L_{2t_f}, I_{2t_f}, T_{t_f}\right)$. Optimal control u_1^*, u_2^*, and u_3^* are seek to be found so that it minimize the objective function:

$$J(u_1, u_2, u_3) = \int_0^{t_f} [aL_1 + bI_1 + cL_2 + dI_2 + C_1 u_1^2 + C_2 u_2^2 + C_3 u_3^2]\, dt, \tag{4}$$

with coefficients a, b, c, and d are non-negative, and C_1, C_2, and C_3 are positive constants which are weight parameters on controls. The objective function is to minimize the latent and infectious individuals with typical TB and resistant TB. Therefore, the optimal control problem can be expressed as:

$$\min J,$$

with dynamical system constraints (3), the state variables initial values are known and the terminal values are independent from the following control variables:

$$0 \leq a_1 \leq u_1 \leq b_1 \quad (5)$$
$$0 \leq a_2 \leq u_2 \leq b_2 \quad (6)$$
$$0 \leq a_3 \leq u_3 \leq b_3 \quad (7)$$

The necessary conditions that an optimal control must satisfy come from the Pontryagin's maximum principle. In general, Hamiltonian function H is defined as follows:

$$H = aL_1 + bI_1 + cL_2 + dI_2 + C_1u_1^2 + C_2u_2^2 + C_3u_3^2 + p_1\left(\Lambda - \beta_1 S\frac{I_1}{N} - \beta^* S\frac{I_2}{N} - \mu S\right)$$
$$+ p_2\left(\beta_1 S\frac{I_1}{N} - (\mu + k_1)L_1 - u_1r_1L_1 + (1 - u_2)pr_2I_1 + \beta_2 T\frac{I_1}{N} - \beta^* L_1\frac{I_2}{N}\right)$$
$$+ p_3(k_1L_1 - (\mu + d_1)I_1 - r_2I_1 + (1 - u_3)q^*T)$$
$$+ p_4\left((1 - u_2)qr_2I_1 - (\mu + k_2)L_2 + \beta^*(S + L_1 + T)\frac{I_2}{N}\right) + p_5(k_2L_2 - (\mu + d_2)I_2)$$
$$+ p_6\left(u_1r_1L_1 + \big(1 - (1 - u_2)(p + q)\big)r_2I_1 - ((1 - u_3)q^* + \mu)T - \beta_2 T\frac{I_1}{N}\right.$$
$$\left. - \beta^* T\frac{I_2}{N}\right),$$

with $p_1(t), p_2(t), p_3(t), p_4(t), p_5(t)$, and $p_6(t)$ are adjoin functions to be determined.

The optimality conditions from Pontryagin's maximum principle is used to obtain the controls u_1^*, u_2^*, and u_3^*, so it gives:

$$\frac{\partial H}{\partial u_1} = 0 \Leftrightarrow 2C_1u_1 - p_2r_1L_1 + p_6r_1L_1 = 0,$$
$$\frac{\partial H}{\partial u_2} = 0 \Leftrightarrow 2C_2u_2 - p_2pr_2I_1 - p_4qr_2I_1 + p_6(p + q)r_2I_1 = 0,$$
$$\frac{\partial H}{\partial u_3} = 0 \Leftrightarrow 2C_3u_3 - p_3q^*T + p_6q^*T = 0,$$

hence, the controls obtained are:

$$u_1^*(t) = \frac{(p_2 - p_6)r_1L_1(t)}{2C_1},$$
$$u_2^*(t) = \frac{p_2p + p_4q - p_6(p + q)r_2I_1(t)}{2C_2},$$
$$u_3^*(t) = \frac{(p_3 - p_6)q^*T(t)}{2C_3}.$$

Since the optimal control problems above are using bounds on the controls, we conclude for control u_1:

$$u_1^* = \begin{cases} a_1 & ; \quad \dfrac{(p_2 - p_6)r_1L_1}{2C_1} \leq a_1 \\ \dfrac{(p_2 - p_6)r_1L_1}{2C_1} & ; \quad a_1 < \dfrac{(p_2 - p_6)r_1L_1}{2C_1} < b_1 \\ b_1 & ; \quad \dfrac{(p_2 - p_6)r_1L_1}{2C_1} \geq b_1 \end{cases}$$

In compact form

$$u_1^* = \min\left\{\max\left\{a_1, \frac{(p_2 - p_6)r_1 L_1}{2C_1}\right\}, b_1\right\}.$$

Similarly, for control u_2 and u_3 we have

$$u_2^* = \min\left\{\max\left\{a_2, \frac{p_2 p + p_4 q - p_6(p+q)r_2 I_1}{2C_2}\right\}, b_2\right\}, \tag{8}$$

and

$$u_3^* = \min\left\{\max\left\{a_3, \frac{(p_3 - p_6)q^* T}{2C_3}\right\}, b_3\right\}.$$

Then the adjoin system based on (4) can be written as:

$$\begin{aligned}
\dot{p_1} &= p_1\left(\beta_1\frac{I_1}{N} + \beta^*\frac{I_2}{N}\right) - p_2\beta_1\frac{I_1}{N} - p_4\beta^*\frac{I_2}{N},\\
\dot{p_2} &= -1 + p_2\left(\mu + k_1 + u_1 r_1 + \beta^*\frac{I_2}{N}\right) - p_3 k_1 - p_4\beta^*\frac{I_2}{N} - p_6 u_1 r_1,\\
\dot{p_3} &= -1 + p_1\beta_1\frac{S}{N} - p_2\left(\beta_1\frac{S}{N} + (1-u_2)pr_2 + \beta_2\frac{T}{N}\right) + p_3(\mu + d_1 + r_2)\\
&\qquad - p_4(1-u_2)qr_2 - p_6\left(1 - (1-u_2)(p+q))r_2 + \beta_2\frac{T}{N}\right),\\
\dot{p_4} &= -1 + p_4(\mu + k_2) - p_5 k_2,\\
\dot{p_5} &= -1 - p_1\beta^*\frac{S}{N} + p_2\beta^*\frac{L_1}{N} - p_4\beta^*\frac{S + L_1 + T}{N} + p_5(\mu + d_2) + p_6\beta^*\frac{T}{N},\\
\dot{p_6} &= -p_2\beta_2\frac{I_1}{N} - p_3 q^* - p_4\beta^*\frac{I_2}{N} + p_6\left((1-u_3)q^* + \mu + \beta_2\frac{I_1}{N} + \beta^*\frac{I_2}{N}\right).
\end{aligned} \tag{9}$$

with transversality conditions:

$$p_1(t_f) = 0, p_2(t_f) = 0, p_3(t_f) = 0, p_4(t_f) = 0, p_5(t_f) = 0, p_6(t_f) = 0. \tag{10}$$

3. Numerical results and discussion

In this section, the optimal treatment strategies for the two-strain TB model will be studied numerically. For solving the optimality system we use an iterative method, starting from solving the state equation using a forward fourth order Runge-Kutta scheme. Because of the transversality conditions (10), the adjoin system is solved by a backward fourth order Runge-Kutta scheme using the current iteration solution of the state equations. All numerical solutions are carried out using Scilab.

There are three scenarios examined in this problem associated with the objective functions formulation and controls application, as described in table 1. These scenarios aim to see the impacts of the objective functions formulation to the controls effectiveness and population dynamics in some compartments.

Table 1. Scenarios of the objective function formulation and controls application.

Scenarios	Coefficients				Controls		
	a	b	c	d	u_1	u_2	u_3
I	0	0	1	1	on	on	off
II	1	1	0	0	on	on	on
III	1	1	1	1	on	on	on

With these different scenarios, the effectiveness of the controls will be demonstrated. Time observed is 5 years, and the numerical parameters used in the system are adopted from [4] and [1] with $N = 12000$, given in table 2.

Table 2(a), (b) and (c). Parameters and their values.

Parameters	Values
Λ	171.6
β_1	13
β_2	13
β^*	0.0131
μ	0.0143
d_1	0
d_2	0

Parameters	Values
k_1	0.5
k_2	1
r_1	2
r_2	1
p	0.4
q	0.1
q^*	0.005

Parameters	Values
$S(0)$	7,600
$L_1(0)$	3,600
$I_1(0)$	400
$L_2(0)$	200
$I_2(0)$	100
$T(0)$	100

Coefficients used in all scenarios are given in table 3 below.

Table 3. Coefficients and their values.

Coefficients	Values
a_1	0.05
b_1	0.95
a_2	0.05
b_2	0.95
C_1	50
C_2	500
C_3	55

For the first scenario, the objective function is to minimize the number of the latent and infectious individuals of resistant TB, and control u_3 is deactivated. The optimal control problem formed is:

$$\min J(u_1, u_2) = \int_0^{t_f} [L_2 + I_2 + C_1 u_1^2 + C_2 u_2^2]\, dt,$$

subject to system (3). Meanwhile the optimal control problem for the second scenario can be written as:

$$\min J(u_1, u_2, u_3) = \int_0^{t_f} [L_1 + I_1 + C_1 u_1^2 + C_2 u_2^2 + C_3 u_3^2]\, dt,$$

subject to system (3). The above equation is expressing the population to be minimized are the latent and actively infectious individuals from typical TB, and all controls are activated. And in the last scenario, all latent and infectious individuals from both typical TB and resistant TB is minimized, and all controls are activated with each lower and upper bounds. The optimal control problem for this scenario is given by:

$$\min J(u_1, u_2, u_3) = \int_0^{t_f} [L_1 + I_1 + L_2 + I_2 + C_1 u_1^2 + C_2 u_2^2 + C_3 u_3^2]\, dt,$$

subject to system (3).

Figure 1 to 3 shows the impacts of the scenarios applied to the system. From all scenarios, it can be concluded that there are considerable differences between the presence and the absence of controls. With the first scenario, the total number of individuals $L_2 + I_2$ at the final time $t_f = 5$ years is 453 in the case with control and 1,498 without control. Meanwhile, with the second scenario, the total number of individuals $L_1 + I_1$ at final time t_f is 8,432 with control and 11,412 without control. With the last scenario, the total number of individuals $L_1 + I_1 + L_2 + I_2$ after 5 years is 8,966 with control and 13,304 without control. So the total cases prevented at the end of control program are 1,045 for the first scenario, 2,980 for the second scenario, and 4,388 for the third scenario. From these results, it can be concluded that the third scenario is the best strategy for our problem.

Figure 4 illustrates the optimal control strategies for the third scenario. The control u_1 is applied over the most for 5 years, while the control u_2 and u_3 are applied at the upper bound during almost 3.6 years and then it decreases to the lower bound until the fifth year.

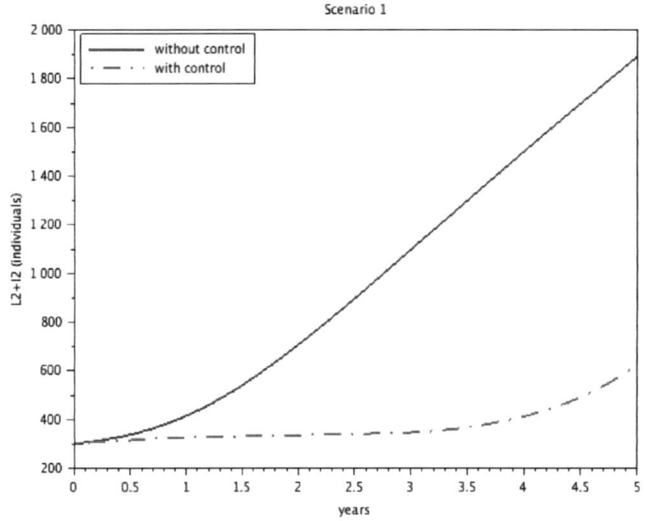

Figure 1. The population dynamics with and without control in scenario 1.

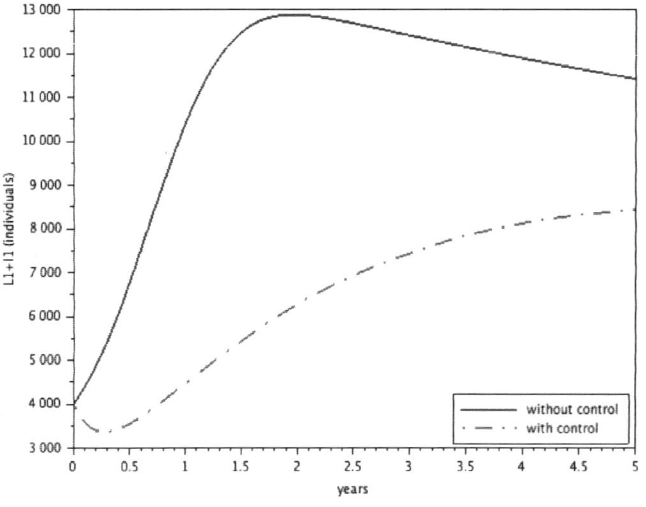

Figure 2. The population dynamics with and without control in scenario 2.

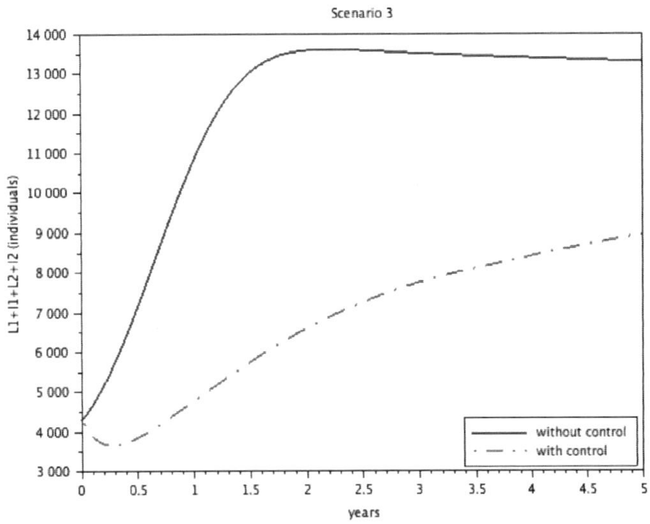

Figure 3. The population dynamics with and without control in scenario 3.

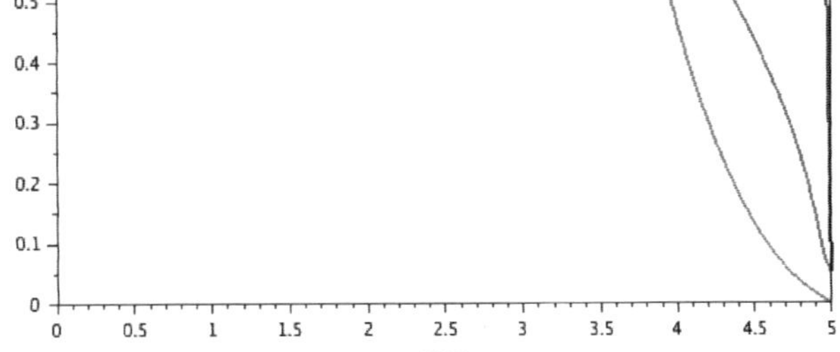

Figure 4. The optimal control strategy for scenario 3.

4. Conclusion

In conclusion, the application of controls on two-strain TB transmission model gives positive influences for it reduces the number of latent and infected individuals from both typical TB and resistant TB. From three scenarios studied, the best scenario is the third one, so we can say that minimizing the population with susceptible, latently infected and actively infected with typical TB, and treated individuals plays an important role in giving those positive influences.

References

[1] Agusto F B 2009 *World Journal of Modelling and Simulation* **5(3)** 163-173
[2] CDC 2012 *Latent TB Infection and TB Disease* available at: http://www.cdc.gov/tb/topic/basics/default.htm
[3] Jung E, Lenhart S and Feng Z 2002 *Discrete and Continous Dynamical Systems – Series B* **2**(4) 473-482
[4] [RI Ministry of Health] Kementerian Kesehatan Republik Indonesia Direktoral Jenderal Pengendalian Penyakit dan Penyehatan Lingkungan 2011 *Strategi pengendalian TBC di Indonesia 2010-2014* (Jakarta: Kemenkesnas) p 1

Effect of Application Propolis as Biocoating on the Physical and Chemical Properties of Tomatoes Stored at Room Temperature

R E Putra[1], S Khairannisa[1] and I Kinasih[2]

[1] School of Life Sciences and Technology, Insitut Teknologi Bandung, Jalan Ganesha No. 10 Bandung
2 Departement of Biology, Faculty of Science and Technology, Universitas Islam Negeri Sunan Gunung Djati Bandung, Jalan A.H. Nasution No. 105 Bandung

Abstract. Tomatoes is considered as one of important horticulture commodities which highly consume by Indonesia consumer. However, this horticulture product is perishable with high rate which reduce quantity and quality of marketable products. One of the method could be used to prevent this problem by application of edible biocoating. In this study, various concentration of ethanolic propolis extract was applied to tomatoes in order to find the effect of propolis coating in tomato preservation and best concentration for application. Tomatoes were grouped into 5 group, namely control group (no coating application), ethanol group (tomato wash with ethanol), and application group (coated with 5%, 10%, and 15% propolis). Variables observed during study were weight change, fruit firmness, total soluble solute, vitamin C, and lycopene. All tomatoes were kept in room temperature for 14 days and observation conducted every 2 days. Result showed that application of 10% propolis as biocoating reduced rate of weight loss and maintained fruit firmness while other variables relatively unaffected by propolis coating.

1. Introduction

Tomato is one of major horticulture commodity in West Java with significant production level and economic value [1]. However, due to their soft skin and their characteristic as climateric fruit, quality of tomatoes rapidly deteriorated which produces weight depreciation, mold development, and loss of attractiveness hence decreasing the economic value [2]. This loss could be prevented through low temperature storage. Unfortunately, not all area in Indonesia has continuous energy supply to enpowered provide power to low temperature storage facilities. Thus it is necessary to develop another preservative strategy which free from continuous need of energy, such as biocoating.

Biocoating is a thin layer that can be consumed and is serving as protection against humidity, oxygen, and solute transition in fruits [3]. Since biocoating mean to be consumed along with fruit, it is necessary to apply natural materials which safe, such as propolis. Propolis is a sticky gummy resinous substance collected by *Trigona* sp., in tropical regions, from the young shoots and buds of certain trees and shrubs. This substance known for having strong anti-bacterial, anti-fungal and anti-viral properties and has been used on various agricultural product for post-harvest protection. Even though it is proven that it can protect some horticulture products, the application of as post-harvest protection in Indonesia is rarely found.

The purpose of this reasearch is to observe effect of application of biocoating made of propolis with concentrations 5%, 10%, and 15% on the physical and chemical properties of tomato fruits in room temperature condition. Changes on physical parameter (weight depreciation and fruit firmness) and chemical (total soluble solid, vitamin C, and lycopene) were recorded every two days for 14 days.

2. Methods

2.1 Tomatoes
Cultivar of tomato used in this study was Warani. All tomatoes were supply by local farm in Ciwidey, West Java which harvested in same day and have same standard based on supplier standard for one day old tomatoes.

2.2 Propolis Extraction
Propolis was originated from local *Trigona* sp. farm in North Bandung. Raw propolis was extracted by ethanolic extraction to obtain highest amount of phenolic compound [4]. Propolis extract then diluted with propylen glycol to obtain 5%, 10%, and 15% solution. Unlike ethanol, propylen glycol disolved beeswax inside the propolis and produced gummy oily solution [5] which easier to be apply as coating material.

2.3 Coating Mechanism
In this study, 90 tomatoes were used as sample. Tomatoes were divided into 5 groups of application. One group was dedicated as control which did not receive any treatment, one group as control positive in which tomatoes was dipped into 70% ethanol, others were treatment group in which tomatoes were dipped into 5%, 10%, and 15% solution respectively.
All tomatoes were kept for 15 days in room temperature (25-27°C) with relative humidity between 75-85%.

2.4 Measurement of Physical and Chemical Properties
Physical and chemical properties were measured every 2 days and variabels measured were weight, fruit firmness, total soluble solid, glucose content, vitamin C, and lycopene.

2.4.1 Physical Properties. Weight of tomatoes measured by digital weight balance while fruit firmness measured by fruit firmness tester.

2.4.2 Chemical Properties
2.4.2.1 Total Soluble Solid and Sucrose Content. Total soluble solid was measured by refraktrometer (Atago) based on Abbe principle and determined by degree brix (^{o}B) [6].

2.4.2.2 Vitamin C Content. Vitamin C content was measured by iodometric titration method [7]. In this method, 10 g sample were collected from each tomato. Sample was grinded and mixed with 30 mL aquadest then filtered with Whatmann paper No. 1. 2 mL of 1% amylum add to solution. Resulted solution then titrated with iodine 0.01 N until the colour of solution change into dark colour, amount of iodine required for this process was measured to calculate Vitamin C content by formulae

$$mg\ Vit.C/_{100\ gram} = \frac{vol.iodin\ (ml) x\ A\ x\ fp}{massa\ \ of\ sampel\ (g)}$$

Whereas
- A is equivalency of 1 ml iodine 0.01 N equal to 0.88 mg ascorbic acid (Vitamin C)
- Fp is dillution factor

2.4.2.3 Lycopene Content. Lycopene content was measured by method of Sharma and Le Maguer [8].

3. Results and Discussion

3.1 Physical Properties

This study showed application of propolis as biocoating able to prevent significant weight loss of tomato and slightly maintained fruit firmness, two component used to determined fruit quality (Fig. 1).

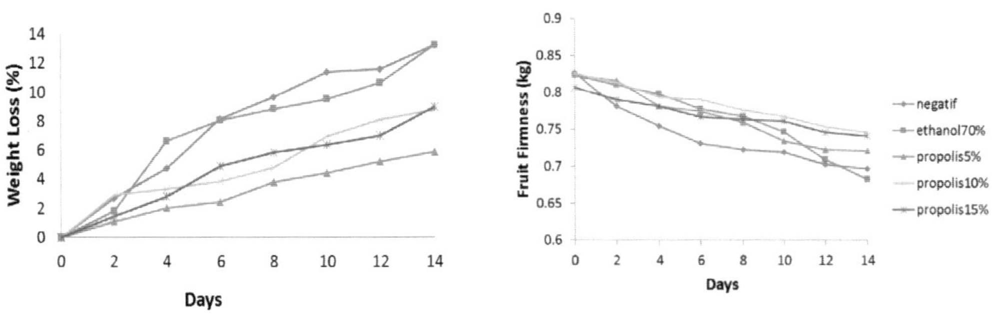

Figure 1. Changes of physical properties of tomato coated with propolis extract.

In general, weight loss of tomatoes coated with propolis were lower than uncoated tomatoes (both control and ethanol application). The lowest weight loss was recorded on tomatoes coated with 15% propolis solution.

Weight loss is main factor determine quality of horticulture products [9]. Most of weight loss caused by water loss from internal tissues. Based on the result, it could be concluded that propolis coat able to prevent water loss through transpiration and respiration [10]. Hydrophobic property of propolis may responsible for prevention of water loss from tomato tissues through transpiration and respiration. Another possibility is propolis coating produced several selective barriers for O_2 and CO_2 which also prevent water loss and maintained fruit weight [11]. This barrier may also reduce pectinase activity through imbalance amount of oxygen and carbondioxide in the fruit tissues [12]. Pectinase is responsible for hydrolisis of pectin and starch, a process that reduce internal cell structure, cell wall composition, and intercelular material which in the end reduce fruit firmness [13][14]. This condition could be found on the tomatoes coated with propolis as the fruit firmness of this group relatively higher than control group (Fig. 1).

3.2 Chemical Properties

This study showed application of propolis as biocoating produced significant higher total soluble solute but not vitamin C and lycopene (Fig. 2).

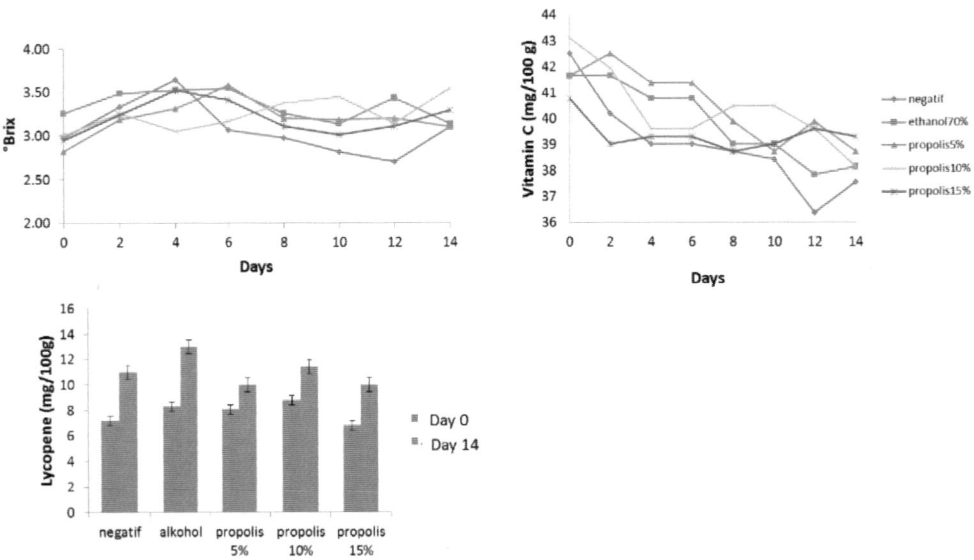

Figure 2. Changes of chemical properties of tomato coated with propolis extract.

Result indicated that all tomatoes coated with propolis had higher total soluble solute than control group (Fig. 2). It seem application of biocoating in tomato reduce respiration rate of fruit which responsible for increasing total soluble solute inside fruit [15].

Interestingly, application of propolis coating did not effected amount of Vitamin C (ascorbic acid) which usually related with TSS [16]. Total vitamin C content of tomatoes of both control and application groups decreased with longer storage period. Eventhough result did not indicated significant different among tomato groups, control group had the lowest vitamin C with highest rate of reduction. Higher amount water and possible imbalance amount oxygen and carbondioxide due to alternation of respiration process may cause fluctuation of Vitamin C content on tomato. Vitamic C is water soluble and could easily oxidate into L-dehydroascorbat [17]. It seem propolis coating reduce respiration rate of tomato thus reduce oxidation reaction of ascorbic acid. Rate of vitamic C reduction also depend on the variety of tomato [18].

On the other hand, lycopene which produces as response to ripening seems not effected by coating at the end of observation. However, differences between early and final lycopene measurement was higher in control and ethanol group (Fig. 2). During ripening process chlorophyll content decreases followed by rapid synthesis of the red pigment lycopene [19]. Different studies have shown that applying various types of coating reduces tomato metabolism, thus increasing shelf life [20][21]. Therefore, application of propolis might slightly reduce metabolism of tomatoes and delaying ripening process.

As conclusion, propolis coating might be used as preservative techniques for tomatoes. However, further studies required to improve effectiveness of propolis as biocoating. In general, 10% propolis is best concentration for application as biocoating.

Acknowledgments
This study is supported by the Riset Inovasi ITB 2015 granted to first author.

References

[1] Badan Pusat Statistik. 2014. Produksi Tomat 2010-2014 di Indonesia. Diakses dari http://www.pertanian.go.id/ap_pages/mod/datahorti tanggal 11 Desember 2015

[2] Arah IK, Amaglo H, Kumah EK and Ofori H 2015 *Int. J. Agron.* **2015** Article ID 478041, 6 pages
[3] Tharanathan RN 2003 *Trends Food Sci. Technol.* **14** 71-8.
[4] Ramanauskiene K, Inkeniene AM, Petrikaite V and Briedis V 2013 *Evid. Based Complement. Alternat. Med.* **2013**, 5 pages.
[5] Pujirahayu N, Ritonga H and Uslinawaty Z 2014 *Int J Pharm Pharm Sci* **6(6)** 338-40.
[6] Genanew T 2013 *World J Agric. Sci.* **9(1)** 29-37.
[7] Bekele DA and Geleta GS 2015 *Res. J. Chem. Sci.* **5(1)** 60-3.
[8] Sharma SK and Le Maguer M 1996 *Ital. J. Food Sci.* **8(2)** 107–13.
[9] Ali A, Maqbool M, RamachandranS and Alderson PG 2011 *Postharvest Biol. Technol.* 58 42–7
[10] Krochta JM, Baldwin EA and Nisperos-Carriedo MO 1994 *Edible Coating and Film to Improve Food Quality* (Technomic Publishing Company, New York).
[11] Krochta JM 1992 Control of Massa Transfer in Foods with Edible Coatings and Films (CRP Press: Boca Raton, FL), pp 517-38.
[12] Hernandez SM 2007 *Eur. Food Res. Technol.* **226** 423-35.
[13] Salunkhe DK, Boun HR and Reddy NR 1991. *Storage Processing and Nutritional Quality of Fruits and Vegetables* (CRC Press Inc., Boston, MA) pp. 156–61.
[14] Seymour GB, Taylor JE and Tucker GA 1993. *Biochemistry of fruit ripening.* (Chapman and Hall Publishers, London) 454 pp
[15] Hernández-Munoz P, Almenar E, Valle VD, Velez D and Gavara R 2008 *Food Chem.* **110** 428-35.
[16] Kays S. 1991 *Postharvest physiology of perishable plant product.* (New York.AVI Book).
[17] Winarno FG 2002 *Fisiologi Lepas Panen Produk Hortikultura.* (M-Brio Press, Bogor).
[18] Tigist M, Workneh T and Woldetsadik K 2011 *J. Food Sci. Technol.* doi:10.1007/s13197-011-0378-0.
[19] Frusciante L, Carli P, Ercolano MR, Pernice R, Di Matteo A, Fogliano V and Pellegrini N 2007 *Mol. Nutr. Food. Res.* **51** 609–617.
[20] KA Athmaselvi KA, P Sumitha, B Revathy 2013 *Int. Agrophys.* **27** 369-75.
[21] Dávila-Aviña JE, Villa-Rodríguez JA, Mónica A, Villegas-Ochoa MA, Tortoledo-Ortiz O, Olivas GI, Ayala-Zavala JF and González-Aguilar GA 2014 *J. Food. Sci. Technol.* **51(10)** 2706-12.

Optimal Control of Malaria Transmission using Insecticide Treated Nets and Spraying

D Athina[1], T Bakhtiar[1] and Jaharuddin[1]

[1] Department of Mathematics, Faculty of Mathematics and Natural Science, Bogor Agricultural University, Indonesia

E-mail: dwilaras@outlook.com, tonibakhtiar@yahoo.com, jaharmath@gmail.com

Abstract. In this paper, we consider a model of the transmission of malaria which was developed by Silva and Torres equipped with two control variables, namely the use of insecticide treated nets (ITN) to reduce the number of human beings infected and spraying to reduce the number of mosquitoes. Pontryagin maximum principle was applied to derive the differential equation system as optimality conditions which must be satisfied by optimal control variables. The Mangasarian sufficiency theorem shows that Pontryagin maximum principle is necessary as well as sufficient conditions for optimization problem. The 4th-order Runge Kutta method was then performed to solve the differential equations system. The numerical results show that both controls given at once can reduce the number of infected individuals as well as the number of mosquitoes which reduce the impact of malaria transmission.

1. Introduction

Epidemiology of malaria is one of quite complicated biological system. The parasites and its vertebrate and the arthropods hosts interact in close dependence on environmental conditions. Malaria is caused by one of the four species of protozoa in the genus of Plasmodium, namely *P. falcifarum*, *P. vivax*, *P. malariae*, and *P. ovale* [1]. The parasites are transmitted to humans by female mosquitoes belonging to certain species of the genus Anopheles each time the infected insect takes human blood. Conversely, the female mosquito can pick up the infection when they bite infected humans. Mosquito density largely depends on the longevity of the adult mosquitoes and the availability of suitable breeding places.

Environmental condition in the tropics is the prime factor for it being endemic. The moderate-to-warm temperatures, high humidity and water bodies allow mosquito and parasites to reproduce. The epidemiological pattern of malaria usually vary with season because of the dependency of transmission on mosquitoes. Each year 1-3 million deaths are attributed worldwide to malaria, out of which one-third are children. However, malaria can be prevented and controlled if it is detected early. Therefore, everyone needs to understand the main parameters in the transmission of the disease and to develop strategies for effective solutions to the prevention and control [3].

Malaria control is carried out through the following recommended malaria treatment and prevention interventions. The choice of interventions depends on the malaria transmission level in the area. In most malaria-endemic countries, four interventions—case management (diagnosis and treatment), ITNs (insecticide-treated nets), IPTp (intermittent preventive treatment of malaria in pregnant women), and IRS (indoor residual spraying) make up the essential package of malaria interventions. Occasionally, other interventions are used: larval control and other vector control interventions and mass drug administration and mass fever treatment [2].

Sir Ronald Ross, while working in the Indian Medical Service in 1890's, demonstrated the-life cycle of the malaria parasite in mosquito. He developed a simple model, now known as 'Ross Model', which explains the relationship between the number of mosquitoes and malaria in humans. Ross introduced a deterministic model of malaria by the differential equations by dividing the susceptible human population (S_h) and infected (I_h), with classes being reinfected again to the vulnerable class and leads to the structure of SIS. The mosquito population also has only two compartments (S_m, I_m), but they do not recover from the infection because of their short lifespan, and leads to the structure of SI [4].

2. Controlled model

Silva and Torres [5] modify the Ross model with an additional possibility of reducing cases of malaria caused by the education. In this study, we consider a model of the transmission of malaria which is developed by Silva and Torres equipped with two control variables, namely the use of insecticide treated nets (ITN) to reduce the number of human beings infected and spraying to reduce the number of mosquitoes.

We consider a malaria transmission model which consists of human population as the host and mosquito population as the vector, each is divided into two classes namely the susceptible and infected compartments. We denote by S_h and S_v the classes of susceptible humans and mosquitos, respectively, by I_h and I_v the classes of infected humans and mosquitos, respectively. So that total population for human (N_h) is given by $N_h = S_h + I_h$, and total population for mosquito (N_v) is given by $N_v = S_v + I_v$. The model is constructed under the following assumptions: all newborns individuals are assumed to be susceptible and no infected individuals are assumed to come from outside the community. The human and mosquito recruitment rates are denoted by Λ_h and Λ_v, respectively. The disease is fast progressing, thus the exposed stage is minimal and is not considered. Infectious individuals can die from the disease or become susceptible after recovery while the mosquito population does not recover from infection. ITN contributes for the mortality of mosquitoes. The average number of bites per mosquito, per unit of time (mosquito-human contact rate), is given by

$$\beta = \beta_{\max}(1 - b)$$

where $\beta_{\max}$ denotes the maximum transmission rate and b the proportion of ITN usage. It is assumed that the minimum transmission rate is zero. The value of β is the same for human and mosquito population, so the average number of bites per human per unit of time is $\beta N_v/N_h$. Thus, the force of infection for susceptible humans (λ_h) and susceptible vectors (λ_v) are given by

$$\lambda_h = \frac{m_1 \beta I_v}{S_h + I_h} \qquad \text{And} \qquad \lambda_v = \frac{m_2 \beta I_h}{S_h + I_h},$$

where m_1 and m_2 are the transmission probability per bite from infectious mosquitoes to humans, and from infectious humans to mosquitoes, respectively. The death rate of the mosquitoes is modeled by $\mu_{vb} = \mu_v + \mu_{max}$, where μ_v is the natural death rate and μ_{max} is the death rate due to pesticide on ITNs.

The coefficient $(1 - u(t))$ on Silva and Torres, represents the effort of susceptible humans that become infected. The proportion of ITN usage (b) is static. In this study, u_1 as control of proportion of nets usage that is dependent on t or dynamic. The new control variable u_2 added to the model represents spraying to reduce the number of mosquitoes. If $u_1 = 1$ then the proportion of nets usage is maximum. The equation of motion of the disease transmission control is provided by the following first order nonlinear differential equation. Table 1 provides the symbols and descriptions for the differential equation system.

$$\begin{aligned}
\frac{dS_h}{dt} &= \Lambda_h - (1-u_1)\frac{m_1\beta_{\max}I_v}{S_h+I_h}S_h + \gamma_h I_h - \mu_h S_h,\\
\frac{dI_h}{dt} &= (1-u_1)\frac{m_1\beta_{\max}I_v}{S_h+I_h}S_h - (\mu_h+\gamma_h+\delta_h)I_h,\\
\frac{dS_v}{dt} &= \Lambda_v - (1-u_1)\frac{m_2\beta_{\max}I_h}{S_h+I_h}S_v - \mu_{vb}S_v - u_2 S_v,\\
\frac{dI_v}{dt} &= m_2(1-u_1)\frac{m_2\beta_{\max}I_h}{S_h+I_h}S_v - \mu_{vb}I_v - u_2 I_v.
\end{aligned} \tag{1}$$

Table 1. Symbol and description for differential equation system.

Symbol	Description	Value
Λ_h	Recruitment rate in humans	$10^3/(70\times 365)$
Λ_v	Recruitment rate in mosquitoes	$10^4/21$
μ_h	Natural mortality rate in humans	$1/(70\times 365)$
δ_h	Disease induced mortality rate in humans	10^{-3}
γ_h	Recovery rate of infectious	$1/4$
μ_v	Natural mortality rate of mosquitoes	$1/21$
$\mu_{\max}$	Mortality rate of mosquitoes due to treated net	$1/21$
$\beta_{\max}$	Maximum mosquito-human contact rate	0.1
m_1	Probability of disease transmission from mosquito to human	1
m_2	Probability of disease transmission from human to mosquito	1
$S_h(0)$	Susceptible individuals initial value	800
$I_h(0)$	Infectious individuals initial value	200
$S_v(0)$	Susceptible vectors initial value	4000
$I_v(0)$	Infectious vectors initial value	900

3. Optimal control problem

Our control effort aims to seek the optimal controls u_1 and u_2, that are carrying the system from initial state $(S_{h0}, I_{h0}, S_{v0}, I_{v0})$ to undetermined state $(S_h(t_f), I_h(t_f), S_v(t_f), I_v(t_f))$, such that the controls minimize the following control performance

$$J(u_1,u_2) = \int_0^{t_f} [A_1 I_h + A_2 S_v + A_3 I_v + (C_1 u_1^2 + C_2 u_2^2)]dt, \tag{2}$$

where A_1 , A_2, dan A_3 are weight constant on infectious human, suspected mosquitoes, and infectious mosquitoes, respectively. Weight coefficient, $C_1$and C_2 are measure of the relative cost of the interventions associated to the control u_1 and u_2. Minimizing the objective function is to minimize the number of I_h, S_v and I_v. Optimum control problem can be written as follows:

$$\min J,$$

with constraints:

$$\begin{aligned}
\frac{dS_h}{dt} &= \Lambda_h - (1-u_1)\frac{m_1\beta_{\max}I_v}{S_h+I_h}S_h + \gamma_h I_h - \mu_h S_h,\\
\frac{dI_h}{dt} &= (1-u_1)\frac{m_1\beta_{\max}I_v}{S_h+I_h}S_h - (\mu_h+\gamma_h+\delta_h)I_h,\\
\frac{dS_v}{dt} &= \Lambda_v - (1-u_1)\frac{m_2\beta_{\max}I_h}{S_h+I_h}S_v - \mu_{vb}S_v - u_2 S_v,\\
\frac{dI_v}{dt} &= m_2(1-u_1)\frac{m_2\beta_{\max}I_h}{S_h+I_h}S_v - \mu_{vb}I_v - u_2 I_v,\\
&S_h(0) = S_{h0}, I_h(0) = I_{h0}, S_v(0) = S_{v0}, I_v(0) = I_{v0},\\
&S_h(t_f), I_h(t_f), S_v(t_f), I_v(t_f) \text{ free and } 0 \le u_i \le u_{i_{\max}}, i = 1,2.
\end{aligned} \tag{3}$$

Pontryagin maximum principle is used to derive the differential equation system as optimality conditions which must be satisfied by optimal control variables. In general Hamilton function is defined by equation (3) and objective function (2) as follows:

$$\begin{aligned} H = {} & A_1 I_h + A_2 S_v + A_3 I_v + C_1 {u_1}^2 + C_2 {u_2}^2 \\ & + p_1 \left(\Lambda_h - (1-u_1) \frac{m_1 \beta_{\max} I_v}{S_h + I_h} S_h + \gamma_h I_h - \mu_h S_h \right) \\ & + p_2 \left((1-u_1) \frac{m_1 \beta_{\max} I_v}{S_h + I_h} S_h - (\mu_h + \gamma_h + \delta_h) I_h \right) \\ & + p_3 \left(\Lambda_v - (1-u_1) \frac{m_2 \beta_{\max} I_h}{S_h + I_h} S_v - \mu_{vb} S_v - u_2 S_v \right) \\ & + p_4 \left(m_2 (1-u_1) \frac{m_2 \beta_{\max} I_h}{S_h + I_h} S_v - \mu_{vb} I_v - u_2 I_v \right), \end{aligned}$$

with $p_1(t), p_2(t), p_3(t), p_4(t)$ are adjoin functions to be determined.

Control function u_1^* and u_2^* are determined based on the condition given by the Pontryagin maximum principle are:

$$u_1^* = \frac{(p_2 - p_1)\beta_{\max} m_1 S_h I_v + (p_4 m_2 - p_3)\, \beta_{\max} m_2 S_v I_h}{2C_1 (I_h + S_h)},$$

$$u_2^* = \frac{p_3 S_v + p_4 I_v}{2C_2}.$$

Use limited control function, we consider the following cases:

Case 1. $0 \le u_1 \le u_{1_{max}}$

$$u_1^* = \begin{cases} 0 & ; \quad k < 0 \\ k & ; \quad 0 \le k \le u_{1_{max}} \\ u_{1_{max}} & ; \quad k > u_{1_{max}} \end{cases}.$$

with

$$k = \frac{(p_2 - p_1)\beta_{\max} m_1 S_h I_v + (p_4 m_2 - p_3)\, \beta_{\max} m_2 S_v I_h}{2C_1 (I_h + S_h)},$$

or

$$u_1^*(t) = \min\left\{u_{1_{max}}, \max\left\{0, \frac{(p_2 - p_1)\beta_{\max} m_1 S_h I_v + (p_4 m_2 - p_3)\, \beta_{\max} m_2 S_v I_h}{2C_1 (I_h + S_h)}\right\}\right\}. \quad (4)$$

Case 2. $0 \le u_2 \le u_{2_{max}}$

$$u_2^* = \begin{cases} 0 & ; \quad l < 0 \\ l & ; \quad 0 \le l \le u_{2_{max}} \\ u_{2_{max}} & ; \quad l > u_{2_{max}} \end{cases}.$$

with

$$l = \frac{p_3 S_v + p_4 I_v}{2C_2},$$

or

$$u_2^*(t) = \min\left\{u_{2_{max}}, \max\left\{0, \frac{p_3 S_v + p_4 I_v}{2C_2}\right\}\right\}. \quad (5)$$

Adjoin functions are determined from $\dot{p}^*(t) = -\frac{\partial H}{\partial x}$ with $x \in \{S_h, I_h, S_v, I_v\}$, then we will get differential equations for $p_i(t)$:

$$
\begin{aligned}
\dot{p_1} &= (p_4 m_2 - p_3)\left(\frac{m_2(1-u_1)\beta_{\max} S_v I_h}{(I_h+S_h)^2}\right) \\
&\quad + (p_1 - p_2)\left(\frac{m_1(1-u_1)\beta_{\max} S_h I_v}{I_h+S_h}\right)\left(1 - \frac{S_h}{I_h+S_h}\right) + p_1\mu_h, \\
\dot{p_2} &= -A_1 + (p_4 m_2 - p_3)\left(\frac{m_2(1-u_1)\beta_{\max} S_v I_h}{I_h+S_h}\right)\left(\frac{I_h}{I_h+S_h} - 1\right) \\
&\quad + (p_2 - p_1)\left(\frac{m_1(1-u_1)\beta_{\max} S_h I_v}{(I_h+S_h)^2}\right) - p_1\gamma_h \\
&\quad + p_2(\gamma_h + \delta_h + \mu_h), \\
\dot{p_3} &= -A_2 + (p_3 - p_4 m_2)\left(\frac{m_2(1-u_1)\beta_{\max} I_h}{I_h+S_h}\right) + p_3(u_2 + \mu_{\text{vb}}), \\
\dot{p_4} &= -A_3 + (p_1 - p_2)\left(\frac{m_1(1-u_1)\beta_{\max} S_h}{I_h+S_h}\right) + p_4(u_2 + \mu_{\text{vb}}).
\end{aligned} \tag{6}
$$

By assuming $S_h(t_f), I_h(t_f), S_v(t_f)$ and $I_v(t_f)$ are free, then the following transversality requirement must hold:

$$p_1(t_f) = 0, p_2(t_f) = 0, p_3(t_f) = 0 \text{ and } p_4(t_f) = 0.$$

Finally, because f_0 and $f_i, \forall i$ are nonlinear and convex functions on (x, u) and $p_i(t)$ are assumed nonnegative $\forall t$, show that Pontryagin maximum principle is necessary as well as sufficient conditions for minimize $J(x)$.

4. Numerical result and discussion

This section will show the effectiveness of control. Scenario made is

- Without control ($u_1 = 0, u_2 = 0$),
- With control usage insecticed treated net ($u_1 \neq 0, u_2 = 0$),
- With control spraying ($u_1 = 0, u_2 \neq 0$), and
- With control usage insecticed treated net and spraying ($u_1 \neq 0, u_2 \neq 0$).

Parameter values used are given in table 2 as follows:

Table 2. Parameter values.

Symbol	Value
A_1	25
A_2	0
A_3	25
t_f	100
u_{1max}	1
u_{2max}	1
C_1	25
C_2	40

In figure 1, we can see number of susceptible individuals without control, with u_1 control (nets usage), with u_2 control (spraying), and both. Using the mosquito nets S_h increased almost the same as using nets and spraying at once. Spraying also can raise S_h, but the results are not optimal. Giving only u_1 control will result almost the same by giving control u_1 and u_2 at once. Giving spraying control can also reduce the number of susceptible individuals, but the results are not optimal.

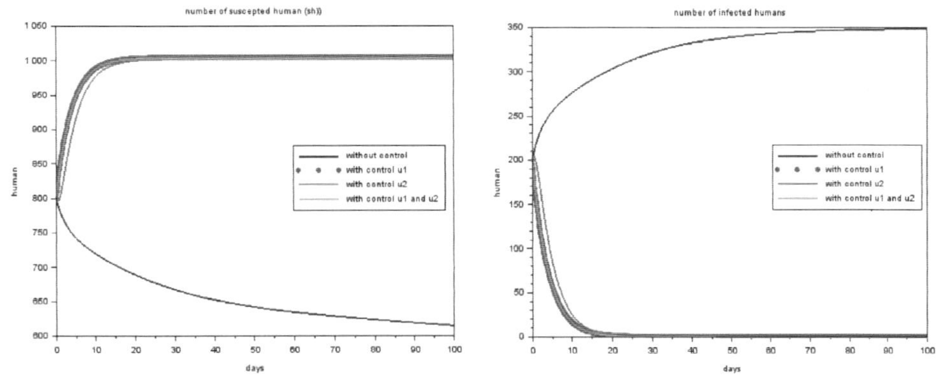

Figure 1. The number of susceptible and infected humans, with and without control.

From figure 2, it can be seen that the number of susceptible mosquitoes is most optimal with u_2 control than others. Conversely, giving u_1 control will increase the number of mosquitoes. The decline in the number of susceptible mosquitoes is most optimal when giving u_2 control at $t \in [0,10]$ because maximum spraying occurs in that time. The nets usage and spraying can reduce the number of infected mosquitoes (figure 2). The result will be more optimal when using nets and spraying at once rather than just with one of the two controls.

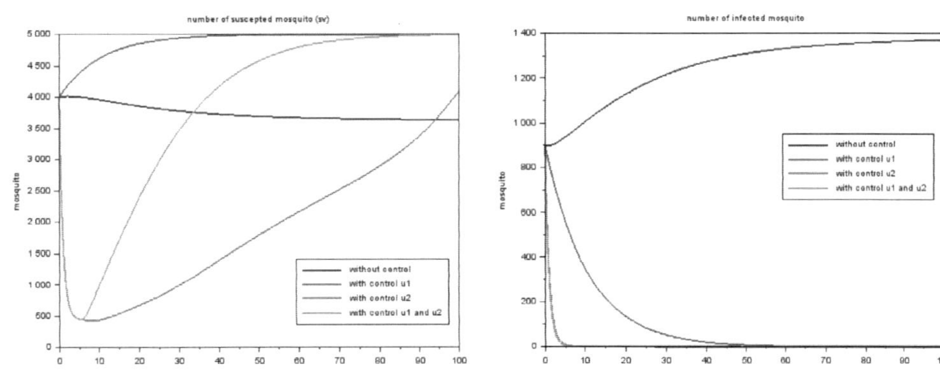

Figure 2. The number of susceptible and infected mosquitoes, with and without control.

Giving u_1 and u_2 control is illustrated in figure 3. For $t \in [0,100]$, $u_1 = 1$, means that nets are used every day. Maximum spraying is only at interval [0,7]. Furthermore, spraying can be reduced every day.

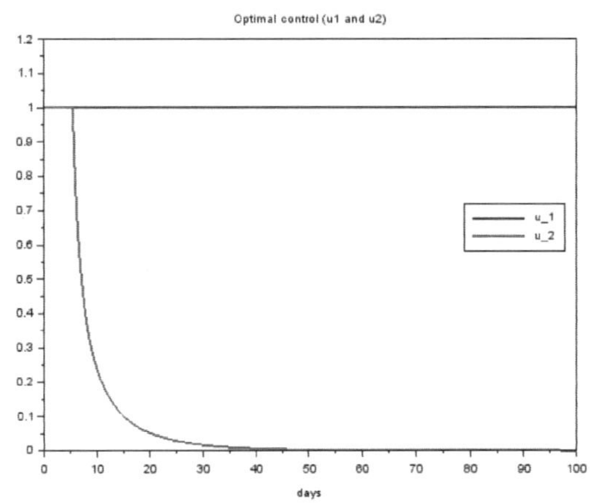

Figure 3. Control variables.

5. Conclusion

Giving u_1 control (nets usage) is very important in order to reduce the number of human beings infected. Giving u_2 control (spraying) to reduce the number of mosquitoes is also important. When giving the two controls at once, after 20 days, nobody is infected and the same will occur when only giving control 1 after 23 days. This means that only by giving control 1 is sufficient to reduce the number of infected individuals. Meanwhile the population of infected mosquitoes after being given controls at once is close to 0 after 8 days, if only giving control 2 it will take after 14 days, but if only giving control 1 it will take 74 days. It means giving control 2 is more optimal than control 1 to reduce the number of infected mosquitoes. In this case, using nets every day and spraying at first time is most optimal to reduce malaria cases.

References

[1] Arsin A A 2012 *Malaria di Indonesia – Tinjauan Aspek Epidemiologi* (Makasar: Masagena Press) p 37
[2] Centers for Disease Control and Prevention 2012 *How Can Malaria Cases and Deaths Be Reduced?*
[3] Hazarika G C and Bhattacharjee A 2011 *Proc. Indian Acad. Sci* **121(1)** 93–109
[4] Mandal S, Sarkar R R, and Sinha S 2011 *Malaria Journal* **10(202)** 1-19
[5] Silva C J and Torres D F M 2013 *Conference Papers in Mathematics* **2013** 1-8

Multivariable Semiparametric Regression Model with Combined Estimator of Fourier Series and Kernel

Khaerun Nisa[1]*, I Nyoman Budiantara[1] and Agnes Tuti Rumiati[1]

[1]Department of Statistics, Faculty of Mathematics and Natural Sciences, Institut Teknologi Sepuluh Nopember (ITS). Arif Rahman Hakim Street, Surabaya 60111 Indonesia

Email: khaerunnisaramli@yahoo.co.id

Abstract. We purpose a combined estimator of Fourier series and Kernel on semiparametric regression. This method is used to resolve the problem of regression modeling, when the relationship between the response variable and the predictor variables most follow a certain pattern, partly have a repetitive pattern, and some others not follow a specific pattern. Moreover, this method depends on oscillation, smoothing parameter and bandwidth. The purpose of this research are to obtained the estimator of semiparametric regression model with combined estimator of Fourier series and Kernel using Penalized Least Square method (PLS). The result show that the PLS estimation produces the estimator of parametrik linier regression, the estimator of Fourier series, the estimator of Kernel, and also the combined estimator of Fourier series and Kernel in semiparametric regression model.

1. Introduction

Semiparametric regression is a statistical methods used to estimate the relationship pattern of the predictor variable and response variable, when a case in regression analysis contains two components; parametric component and nonparametric component. Based on several previous research that has been conducted by researchers, mostly they use the same estimator approach for all or some of the predictor variables. Meanwhile, in many cases, the data pattern of each predictor variables are not always identical. Therefore, to solve these problems, we need a more proper estimator that can be used to approximate the data pattern. In semiparametric regression, there are many functions that can be used to approximate the data pattern, especially Fouries series and Kernel.

According to [1], estimation of Fourier series is capable of handling data that is smooth character and follow the pattern repeated at certain interval, previous research have been investigate by researchers as Amato [2], Asrini [3], Bilodeau [1], Pane [4], and Sudiarsa [5]. The kernel is one of the frequenlty used estimators in semiparametric and nonparametric regression. This estimator has more flexible shape, simple mathematics calculation, and achieves convergence more rapidly. Several researches on Kernel estimator have been conducted by some researchers such as Nadaraya [6], Budiantara et al [7], and Speackman [8].

Therefore, this research focus on semiparametric regression model with combined estimator between Fourier series and Kernel obtained through PLS method.

2. Semiparametric Regression Model

Given the data $\left(x_{1i}, x_{2i}, \ldots x_{pi}, t_i, z_{1i}, z_{2i}, \ldots, z_{ri}, y_i\right)$ with predictor $\left(x_{ji}, t_i, z_{mi}\right)$ and response variables y_i it can be approximated by a semiparametric regression model as follows :

$$y_i = \sum_{j=1}^{p} f_j(x_{ji}) + g(t_i) + \sum_{m=1}^{r} h_m(z_{mi}) + \varepsilon_i,\ i = 1,2,\ldots,n \tag{1}$$

The $f_j(x_{ji})$ curve is assummed referring to linear pattern. Meanwhile regression curve $g(t_i)$ and $h_m(z_{mi})$ are assummed unknown and smooth that can be approached by Fourier series function and Kernel function, then ε_i is random error that follows normally and independently distributed with mean zero and variance σ^2. The linear parametric components, Fourier series component and kernel component are defined as follows :

Linear function with *p* predictor variable can be written as :

$$f_j\left(x_{1i}, x_{2i}, \ldots x_{pi}\right) = \beta_0 + \beta_1 x_{i1} + \beta_2 x_{i2} + \cdots + \beta_p x_{ip} \tag{2}$$

where :

$\beta_0, \beta_1, \beta_2, \cdots, \beta_p$ are parameters of the linier function

Fourier series function with one predictor variable can be written as :

$$g(t_i) = bz_{ki} + \frac{1}{2}a_0 + \sum_{k=1}^{K} a_k \cos kz_i \tag{3}$$

where :

$b,\ a_0, a_k,\ k = 1,2,\ldots,K$ are parameters of the Fourier series function

and then Kernel function with *r* predictor variable can be written as :

$$\hat{h}_m(z_{1i}, z_{2i}, \ldots, z_{ri}) = n^{-1}\sum_{j=1}^{n} W_{\alpha i}(z) y_i \tag{4}$$

where :

$$W_{\alpha i}(z) = \frac{\frac{1}{\alpha}K\left(\frac{z-z_i}{\alpha}\right)}{n^{-1}\sum_{j=1}^{n}\frac{1}{\alpha}K\left(\frac{z-z_i}{\alpha}\right)},$$

Kernel estimator depends on the Kernel function K and bandwidth parameter α. There are several types of Kernel functions included a Gaussian Kernel of the form, as follows :

$$K(x) = \frac{1}{2}e^{\frac{x^2}{2}}, -\infty < x < \infty$$

Hence, a semiparametric regression model in equation (1) can be written as follows :

$$\tilde{y} = X\vec{\beta} + G\vec{a} + D\vec{y} + \vec{\varepsilon} \tag{5}$$

where :

$$X=\begin{bmatrix}1 & x_{11} & x_{12} & \cdots & x_{1p}\\ 1 & x_{21} & x_{22} & \cdots & x_{2p}\\ 1 & x_{31} & x_{23} & \cdots & x_{3p}\\ \vdots & & & \ddots & \\ 1 & x_{n1} & x_{n2} & \cdots & x_{np}\end{bmatrix}\ \vec{\beta}=\begin{bmatrix}\beta_0\\ \beta_1\\ \beta_2\\ \vdots\\ \beta_p\end{bmatrix}$$

$$G=\begin{bmatrix}t_{11} & \frac{1}{2} & \cos t_{11} & \cdots & \cos Kt_{11}\\ t_{12} & \frac{1}{2} & \cos t_{12} & \cdots & \cos Kt_{12}\\ \vdots & & \vdots & \ddots & \vdots\\ t_{1n} & \frac{1}{2} & \cos t_{1n} & \cdots & \cos Kt_{1n}\end{bmatrix}$$

$$\vec{a}=\begin{bmatrix}b & a_0 & a_{11} & \cdots & a_K\end{bmatrix}^T$$

$$D=\begin{bmatrix}n^{-1}W_{\alpha 1}(z_1) & n^{-1}W_{\alpha 2}(z_1) & \cdots & n^{-1}W_{\alpha n}(z_1)\\ n^{-1}W_{\alpha 1}(z_2) & n^{-1}W_{\alpha 2}(z_2) & \cdots & n^{-1}W_{\alpha n}(z_2)\\ \vdots & \vdots & \ddots & \vdots\\ n^{-1}W_{\alpha 1}(z_n) & n^{-1}W_{\alpha 2}(z_n) & \cdots & n^{-1}W_{\alpha n}(z_n)\end{bmatrix},\ \vec{y}=\begin{bmatrix}y_1\\ y_2\\ \vdots\\ y_n\end{bmatrix}.$$

3. Multivariable Semiparametric Regression with Combined Estimator between Fourier Series and Kernel

The combination estimator of Fourier series and Kernel in semiparametric regression, $\hat{\mu}\left(x_{1i},x_{2i},\ldots x_{pi},t_i,z_{1i},z_{2i},\ldots,z_{ri}\right)$ is obtained by employing PLS optimization, as follows :

$$\underset{f,g,h}{Min}\left\{R(f,g,h)+\lambda J(g)\right\}=\underset{f,g,h}{Min}\left\{n^{-1}\sum_{i=1}^{n}\left(y_i-\sum_{j=1}^{p}f_j(x_{ji})-g(t)-\sum_{m=1}^{r}h_m(z_{mi})\right)^2+\lambda\int_0^{\pi}\frac{2}{\pi}\left(g''(t)\right)^2dt\right\} \quad (6)$$

Several lemmas are needed to complete this PLS optimization.

Lemma 1

If the function of Fourier series is $g(t)$ as in equation (3), then the penalty is

$$J(g)=\int_0^{\pi}\frac{2}{\pi}\left(g^{(2)}(t)\right)^2dt=\sum_{k=1}^{K}k^4a_k^2, \quad (7)$$

this lemma has been proven by [3] and [8].

Lemma 2

If $f_j(x_{ji})$ is approach by multiple linier function as in equation (2), $g(t)$ is approach by Fourier series function as in equation (3) and $h_m(z_{mi})$ is approach by Kernel as in equation (4), then the goodness of fit $R(f,g,h)$ is obtained as follows :

$$R(f,g,h)=n^{-1}\left(\vec{y}-X\vec{\beta}-G\vec{a}-D\vec{y}\right)^T\left(\vec{y}-X\vec{\beta}-G\vec{a}-D\vec{y}\right)$$

Proof

In general, goodness of fit $R(f,g,h)$ is defined as follows :

$$R(f,g,h)=n^{-1}\sum_{i=1}^{n}\left(y_i-\sum_{j=1}^{p}f_j(x_{ji})-g(t)-\sum_{m=1}^{r}h_m(z_{mi})\right)^2$$

It could be shown

$$R(f,g,h)=n^{-1}\sum_{i=1}^{n}\left(\begin{array}{l} y_i-\beta_0-\beta_1x_{i1}-\beta_2x_{i2}-\ldots-\beta_px_{ip}-bt_{1i}-\frac{1}{2}a_0-\sum_{k=1}^{K}a_k\cos kt_{1i} \\ -\ldots-bt_{si}-\frac{1}{2}a_0-\sum_{k=1}^{K}a_k\cos kt_{si}-\sum_{i=1}^{n}\frac{\frac{1}{\alpha}K\left(\frac{z-z_i}{\alpha}\right)}{\sum_{i=1}^{n}\frac{1}{\alpha}K\left(\frac{z-z_i}{\alpha}\right)}Y_i \end{array}\right)^2$$

Hence, goodness of fit $R(f,g,h)$ can be written as follows :

$$R(f,g,h)=n^{-1}\left(\vec{y}-X\vec{\beta}-G\vec{a}-D\vec{y}\right)^T\left(\vec{y}-X\vec{\beta}-G\vec{a}-D\vec{y}\right) \tag{8}$$

Optimization using PLS can be solved by combining goodness of fit $R(f,g,h)$ and penalty $J(g)$ as follows :

$$\begin{aligned}\underset{f,g,h}{Min}\left\{R(f,g,h)+\lambda J(g)\right\}&=\underset{\vec{\beta},\vec{a}}{Min}\left\{n^{-1}\left(\vec{y}-X\vec{\beta}-G\vec{a}-D\vec{y}\right)^T\left(\vec{y}-X\vec{\beta}-G\vec{a}-D\vec{y}\right)\right\}\\&=\underset{\vec{\beta},\vec{a}}{Min}\left\{Q(\vec{\beta},\vec{a})\right\}\end{aligned}$$

The estimation of $\hat{\beta}$ and $\hat{a}$ are obtaining by using partial derivative of $Q(\vec{\beta},\vec{a})$ to $\vec{\beta}$ and $\vec{a}$. First, consider $Q(\vec{\beta},\vec{a})$ the function as follows :

$$\begin{aligned}Q(\vec{\beta},\vec{a})&=n^{-1}\left(\vec{Y}-X\vec{\beta}-G\vec{a}-D\vec{y}\right)^T\left(\vec{Y}-X\vec{\beta}-G\vec{a}-D\vec{y}\right)+\lambda\vec{a}^TU\vec{a}\\
&=n^{-1}\left[(I-D)\vec{y}-X\vec{\beta}-G\vec{a}\right]^T\left[(I-D)\vec{y}-X\vec{\beta}-G\vec{a}\right]+\lambda\vec{a}^TU\vec{a}\\
&=n^{-1}\left[\vec{y}^T(I-D)^T-\vec{\beta}^TX^T-\vec{a}^TG^T\right]\left[(I-D)\vec{y}-X\vec{\beta}-G\vec{a}\right]+\lambda\vec{a}^TU\vec{a}\\
&=n^{-1}\vec{y}^T(I-D)^T(I-D)\vec{y}-n^{-1}\vec{y}^T(I-D)^TX\vec{\beta}-n^{-1}\vec{y}^T(I-D)^TG\vec{a}\\
&\quad-n^{-1}\vec{\beta}^TX^T(I-D)\vec{y}+n^{-1}\vec{\beta}^TX^TX\vec{\beta}+n^{-1}\vec{\beta}^TX^TG\vec{a}-n^{-1}\vec{a}^TG^T(I-D)\vec{y}\\
&\quad+n^{-1}\vec{a}^TG^TX\vec{\beta}+n^{-1}\vec{a}^TG^TG\vec{a}+\lambda\vec{a}^TU\vec{a}\\
&=n^{-1}\vec{y}^T(I-D)^T(I-D)\vec{y}-2n^{-1}\vec{\beta}^TX^T(I-D)\vec{y}-2n^{-1}\vec{a}^TG^T(I-D)\vec{y}\\
&\quad+n^{-1}\vec{\beta}^TX^TX\vec{\beta}+n^{-1}\vec{\beta}^TX^TG\vec{a}+n^{-1}\vec{a}^TG^TG\vec{a}+\lambda\vec{a}^TU\vec{a}\end{aligned}$$

The derivation of $Q(\vec{\beta},\vec{a})$ to $\vec{\beta}$ and $\vec{a}$ as follows :

$$\frac{\partial Q(\vec{\beta},\vec{a})}{\partial\vec{\beta}}=-2n^{-1}X^T(I-D)\vec{y}+2n^{-1}X^TX\vec{\beta}+2n^{-1}X^TG\vec{a}$$

$$\frac{\partial Q(\vec{\beta},\vec{a})}{\partial\vec{a}}=-2n^{-1}G^T(I-D)\vec{y}+2n^{-1}G^TX\vec{\beta}+2n^{-1}G^TG\vec{a}+\lambda U\vec{a}$$

Equalizing the first derivatif of sum square error with zero, for $\vec{\beta}$ and $\vec{a}$ are obtained :

$$\hat{\vec{\beta}} = \left(X^T X\right)^{-1} X^T (I-D)\vec{y} - \left(X^T X\right)^{-1} X^T G\vec{a}$$

$$\hat{\vec{a}} = \left(X^T X\right)^{-1} X^T (I-D)\vec{y} - \left(X^T X\right)^{-1} X^T G\left[\left(G^T G + n\lambda U\right)^{-1} G^T (I-D)\vec{y} - \left(G^T G + n\lambda U\right)^{-1} G^T X\vec{\beta}\right]$$

Substitution $\vec{a}$ to $\vec{\beta}$, so :

$$\vec{\beta} = A\vec{y} \tag{9}$$

where :

$$A = P_1 M_1$$

$$P_1 = \left[I - \left(X'X\right)^{-1} X'G\left(G^T G + n\lambda U\right)^{-1} G^T X\right]^{-1}$$

$$M_1 = \left[\left(X^T X\right)^{-1} X^T (I-D) - \left(X^T X\right)^{-1} X^T G\left(G^T G + n\lambda U\right)^{-1} G^T (I-D)\right]$$

Then substitution $\hat{\vec{\beta}}$ into $\vec{a}$, so :

$$\vec{a} = B\vec{y} \tag{10}$$

where :

$$B = P_2 M_2$$

$$P_2 = \left[I - \left(G^T G + n\lambda U\right)^{-1} G^T X\left(X^T X\right)^{-1} X^T G\right]^{-1}$$

$$M_2 = \left[\left(G^T G + n\lambda U\right)^{-1} G^T (I-D) - \left(G^T G + n\lambda U\right)^{-1} G^T X\left(X^T X\right)^{-1} X^T (I-D)\right]$$

So the estimator of linear parametric curve is :

$$\vec{f}(x) = X\vec{\beta} \tag{11}$$

$$= C\vec{y},$$

where :

$$C = XP_1 M_1$$

the estimator of Fourier series curve is :

$$\vec{g}(t) = G\vec{a} \tag{12}$$

$$= K\vec{y},$$

where :

$$K = DP_2 M_2$$

then, the estimator of Kernel curve is :

$$\vec{h}(z) = D\vec{y} \tag{13}$$

The estimator of semiparametric with combined Fourier series and Kernel, obtained :

$$\hat{\mu}(x_{1i}, x_{2i}, \ldots, x_{pi}, t_i, z_{1i}, z_{2i}, \ldots, z_{ri}) = \hat{f}(x) + \hat{g}(t) + \hat{h}(z) \qquad (14)$$

$$= A\tilde{y} + K\tilde{y} + D\tilde{y}$$

$$= (A + K + D)\tilde{y}$$

$$= N\tilde{y}$$

where :

$$N = A + K + D$$

4. Conclusion

The results shows that the estimator of combination between Fourier series and Kernel function in semiparametric regression was obtined through the PLS optimization, as follows :

$$\underset{f,g,h}{Min}\left\{R(f,g,h) + \lambda J(g)\right\} = \underset{f,g,h}{Min}\left\{n^{-1}\sum_{i=1}^{n}\left(y_i - \sum_{j=1}^{p} f_j(x_{ji}) + g(t) + \sum_{m=1}^{r} h_m(z_{mi})\right)^2 + \lambda\int_0^{\pi}\frac{2}{\pi}\left(g''(t)\right)^2 dt\right\}$$

The estimator of combination between Fourier series and Kernel function is

$$\hat{\mu}(x_{1i}, x_{2i}, \ldots, x_{pi}, t_i, z_{1i}, z_{2i}, \ldots, z_{ri}) = \hat{f}(x) + \hat{g}(t) + \hat{h}(z)$$

where

$$\hat{f}(x) = \hat{\beta}_0 + \hat{\beta}_1 x_{i1} + \hat{\beta}_2 x_{i2} + \cdots + \hat{\beta}_p x_{ip}$$

$$\hat{g}(t) = \hat{b}z_i + \frac{1}{2}\hat{a}_{k0} + \sum_{k=1}^{K}\hat{a}_k \cos kz_i .$$

References

[1] Bilodeau M 1992 *The Canadian Journal of Statistics* **3** 257-269

[2] Amato U, Antoniadis A, and De-Feis I 2002 *Journal of Computational and Applied Mathematics* **146** 459-479

[3] Asrini L J, Budiantara I N 2014 *ARPN Journal of Engineering and Applied Sciences* **9** 9 1501-1506

[4] Pane R, Budiantara I N, Zain I, and Otok B J 2014 *Apllied Mathematical Sciences* **8** 102 5053-5064

[5] Sudiarsa I W, Budiantara I N, Suhartono, Purnami S W 2015 *Applied Mathematical Science* **9** 100 4997-5010

[6] Nadayara EA 1964 *Theory of Probability and its Applications* 141-142

[7] Budiantara I N, Ratnasari, Ratna and Zain 2015 *Applied Mathematical Sciences* 9 **122** 6038-6094

[8] Speackman P 1988 *Journal Royal Statistical Society* B **50** 413-436

Ensemble ROCK Methods and Ensemble SWFM Methods for Clustering of Cross Citrus Accessions Based on Mixed Numerical and Categorical Dataset

Alvionita[1]*, Sutikno[1], and A Suharsono[1]

[1]Department of Statistics, Sepuluh Nopember Institute of Technology, Indonesia.

Email: alvionita110293@gmail.com

Abstract. Cluster analysis is a technique in multivariate analysis methods that reduces (classifying) data. This analysis has the main purpose to classify the objects of observation into groups based on characteristics. In the process, a cluster analysis is not only used for numerical data or categorical data but also developed for mixed data. There are several methods in analyzing the mixed data as ensemble methods and methods Similarity Weight and Filter Methods (SWFM). There is a lot of research on these methods, but the study did not compare the performance given by both of these methods. Therefore, this paper will be compared the performance between the clustering ensemble ROCK methods and ensemble SWFM methods. These methods will be used in clustering cross citrus accessions based on the characteristics of fruit and leaves that involve variables that are a mixture of numerical and categorical. Clustering methods with the best performance determined by looking at the ratio of standard deviation values within groups (S_W) with a standard deviation between groups (S_B). Methods with the best performance has the smallest ratio. From the result, we get that the performance of ensemble ROCK methods is better than ensemble SWFM methods.

1. Introduction

Cluster analysis is a technique in multivariate analysis that reduces (classifying) data. This analysis has main purpose to classify objects into groups based on the characteristics possessed by the observatios. The analysis will classify objects so that each object with high degree of similarity to other objects are in the same group [1]. Problems encountered in many case is when the dataset contains mixed scale variables, i.e. it contains some variable with numerical scale while the other are categorical. In its development, analytical methods which are often performed for scale data clustering mixture, i.e. to transform categorical data into numeric data and vice versa. However, this analysis has weaknesses that led to the difficulty in determining the exact transformation so as not to lose a lot of information from the original data.

Apart from the clustering with the transformation method, it developed a ensemble method for clustering mixed data [2]. The ensemble clustering (clustering ensemble) is a clustering technique to combine the results of clustering several clustering algorithms to get a better group [2], One approach that can be used is a ROCK (Robust Clustering using links) method. ROCK method is developed by [3]. ROCK method uses the concept of distance to measure the similarity / proximity between a pair of data points [3].

To overcome the clustering on mixed data can also be used clustering method based on the similarity ensemble of weight and filter method (SWFM). SWFM method was first developed by [4]. This method is an outgrowth of clustering ensemble has a difference in the final stages of cluster formation. There are two main concepts in this clustering is similarity weight and filter method [4].

In this study, the analysis was done comparing performance of clustering ensembles ROCK and ensemble SWFM method. The analysis in this study begins by separating the two types of data are then performed clustering independently. Clustering numerical data was calculated using agglomerative hierarchical cluster, while categorical data are grouped using methods ROCK (Robust Clustering using links).. The criterion of better performance for the clustering is measure by the ratio of standard deviation values within groups (S_W) with a standard deviation between groups (S_B). The better the method, the smaller the ratio [5]. The purpose of this study is to compare performance of the ensemble ROCK methods with ensemble SWFM methods using a case study on citrus accessions clustering protoplasm fusion product.

2. Literature of Cluster Analysis for Mixed Numerical and Categorical Dataset

This section will explain the theories used in clustering to complete a case with a mixed numerical and categorical dataset. Cluster analysis for mixed data can't be done directly, but in its analysis requires a stage ensemble. Ensemble method can be done in several approaches, including ensemble Robust Clustering using Links (ROCK) method and Similarity Weight and Filter Methods (SWFM) method. Some of the benefits of cluster analysis is a double variable data exploration, data reduction, and predicted state of the object.

2.1. Clustering Methods

The technique used for clustering includes methods of hierarchical and non-hierarchical methods. Clustering hierarchy starts with two or more objects that have the closest similarity, then the process is passed to other objects that have the proximity of the two. The analysis was performed by the group will be forming a "tree", where there is a hierarchy (levels) which clearly between objects, from the most similar to least similar [6]. There are two techniques in the analysis group clustering hierarchy, namely division techniques (divisive) and merging techniques (agglomerative). The advantages of using hierarchical method in the analysis group is speeding up processing and save time as input data will form a hierarchy or precipitated separately, making it easier to interpretation, but the weakness of this method is often contained errors in the data outliers, the difference measure of distance used, and the presence of variables that are not relevant.

2.2. Cluster Analysis for Mixed Numerical and Categorical Dataset

Cluster analysis for mixed data is done by ensemble clustering. In the ensemble method, the final formation of the cluster is done using clustering algorithm that similar with clustering method for categorical data. In addition, other method is developed for clustering mixed data, namely ensemble SWFM.

2.2.1 Ensemble Method

Ensemble Clustering is a method that combines several different algorithms to obtain a general partition of the data, which is aimed at the consolidation of the portfolios of individual clustering. Interest ensemble clustering is to combine the results of several clustering algorithms to get the better clustering and robust. Ensemble clustering consists of two stages of the algorithm. The first stage is the clustering with some algorithm and storing the results of these clustering. Second, use the function to determine the final cluster of groups of the first stage results [3].

Steps in the analysis of mixed data using a method called ensemble clustering algorithm CEBMDC has the following stages, [3]

a. Gathering data consisting of numerical and categorical variables, divided into two sub-data, namely pure numerical and pure categorical. Suppose that there are m variables, with $m_{numerical}$

represent the number of numerical variables, and $m_{categorical}$ represent the number of categorical variables, in order to obtain $m = m_{numerical} + m_{categorical}$. Furthermore, clustering is done by clustering algorithm according to the type of data separately.
b. Clustering objects with numeric variables with numeric data clustering algorithm, and clustering objects with categorical variables with categorical data clustering algorithm.
c. Combining (combining) the results of the clustering of numerical and categorical variables, a process called ensemble.
d. Clustering ensemble using categorical data clustering algorithm to obtain the final group (final cluster).

2.2.2 Similarity Weight and Filter Method (SWFM)

SWFM clustering has a concept similar analysis by clustering ensemble in general. This method is an outgrowth of the clustering ensemble that has a difference algorithm to get the final stages of cluster formation. In the ensemble method, the final formation of the cluster is done using clustering algorithm similar to clustering on categorical data, while the ensemble SWFM using an algorithm based on the similarity weight and filter method. At the stage of weight similarity method used similarity measure that would include a weighting factor in the similarity formula.

The steps in clustering by SWFM are,
a. Dividing (splitting) that the original data,
b. Clustering accession by numerical variables using numeric data clustering algorithm,
c. Clustering accession by categorical variables using categorical data clustering algorithm,
d. Merging clustering results into the final cluster (stage ensembles). After getting the optimum group of rock and agglomerative method results, the next step is merging the group. At this stage, clustering is done by using the distance in equation (1), [4]

$$sim(X_i, X_j) = \sum_{i \le n_i, j \le n_i} \frac{S_{ij}}{\max(n_i, n_j)}, \quad i \ne j \tag{1}$$

where $S_{ij} = \dfrac{|X_i \cap X_j|}{|X_i \cup X_j|}$.

The input for this stage are group of agglomerative method results (output 1) and group of ROCK methods results (output 2). Output 1 and output 2 is considered as categorical variables are used to compile the final cluster using filter algorithm in equation (2). [4]

$$F(x_i, x_j) = \sum_{i=1}^{n_i} \sum_{j=1}^{n_j} w_{ij} d(x_i, x_j) \tag{2}$$

with w_{ij} is the weight between the groups to i and j.

2.3. Cluster Analysis for Numerical Dataset

Clustering for numerical data is done with the assistance of a distance, a distance that is often used is the euclidean distance. The concept of the euclidean distance measuring distances between observation $\mathbf{x}_i$ and $\mathbf{x}_j$, expressed in equation (3) as follows,[1]

$$d_{ij} = \sqrt{(\mathbf{x}_i - \mathbf{x}_j)^T (\mathbf{x}_i - \mathbf{x}_j)} \tag{3}$$

If the value of dissimilarity between two objects is greater, then the difference between the two objects will be even greater, so the two objects are increasingly not exist in the same group. Some techniques clustering between the groups stated as follows,[1]
a. Single linkage, this procedure is based on the smallest distance.

$$d_{w(u,v)} = \min\left(d_{wu}, d_{wv}\right) \tag{4}$$

b. Complete linkage, this procedure is based on the greatest distance.

$$d_{w(u,v)} = \max\left(d_{wu}, d_{wv}\right) \quad (5)$$

c. Average linkage, this procedure is based on the average distance

$$d_{w(u,v)} = \frac{n_u}{n_u + n_v} d_{wu} + \frac{n_v}{n_u + n_v} d_{wv} \quad (6)$$

2.4. Cluster Analysis for Categorical Dataset

One algorithm for clustering categorical data is ROCK Method. ROCK methods developed from agglomerative hierarchical clustering method used for categorical data with a new concept that is links to measure similarity/proximity between a pair of data points. Observations that have great links will be combined into one group, while the other has a small links will be separated from the group. The steps undertaken in ROCK method is as follows, [3]

a. Initializing the object as a group with a single member,
b. Forming similarity between objects to the criteria using equation (7),

$$sim\left(X_i, X_j\right) = \frac{\left|X_i \cap X_j\right|}{\left|X_i \cup X_j\right|}, \quad i \neq j \quad (7)$$

with $\left|X_i \cap X_j\right|$ is the number of intersection between X_i and X_j, and $\left|X_i \cup X_j\right|$ is the number of union between X_i and X_j.

c. Determining the threshold (θ) in the adjacency matrix, the threshold value (θ) is a parameter defined by the user that can be used to control how close the relationship between objects. The value of the input θ is $0 < \theta < 1$,
d. Calculating the value of a link between the observations,
e. Calculating the local heap goodness measure value using equation (8),

$$g(C_i, C_j) = \frac{link\left[C_i, C_j\right]}{\left(n_i + n_j\right)^{1+2f(\theta)} - n_i^{1+2f(\theta)} - n_j^{1+2f(\theta)}} \quad (8)$$

where $link\left[C_i, C_j\right] = \sum_{x_i \notin C_i, x_j \notin C_j} link(X_i, X_j)$ stating the number of links from all possible pairs of objects that exist in C_i and C_j, n_i and n_j each state the number of members in the group to i and j, while $f\left(\theta\right) = \frac{1-\theta}{1+\theta}$.

f. Determining the heap global value is the maximum goodness measure between columns in the ith row,
g. Repeat rare (e) and (f) in order to obtain the maximum value in the global heap and local heap,
h. During the data size> k, where k is the number of classes specified then did the merger group has a the largest value of local heap with the largest global heap into one group, and then add the link between the two groups are combined, then remove the group that emerged from the local heap and update values global heap with the result of merging,
i. Perform step (h) in order to obtain the expected number of groups or there are not any link between groups,
j. Repeating steps (a) through (i) the value of θ is different,
k. Calculate the ratio between S_W and S_B for each value of θ.
l. Comparing the results of step (k) for each value of θ and determine the optimum number of groups with a criteria the smallest ratio between S_W and S_B.

2.5. Validation Clustering
Determining the optimum number of groups is an important step after the clustering process. This stage is referred to as validation clustering [7]. In addition determining the optimum number of clusters, cluster validation phase can also measure the performance results of the grouping. Good group is to have a high homogeneity between members in the group and the high heterogeneity between groups [6]. Performance results for the variable clustering with scale numerical data can be determined from the ratio of the value of the standard deviation in groups or within (S_W) and the standard deviation between groups or between (S_B). The performance of a method of clustering would be better if the smaller the ratio between SW and SB, which means that there is a maximum homogeneity in the group and maximum heterogeneity between groups [5].

By using the average value of the variable, S_W and S_B can be formulated as in equation (9) and (10) [5].

$$S_W = \frac{1}{C}\sum_{c=1}^{C} S_c \tag{9}$$

with S_c is a standard deviation of all groups C and C is the number of groups formed.

$$S_B = \left[\frac{1}{C-1}\sum_{c=1}^{C}\left(\overline{x}_c - \overline{x}\right)^2\right]^{1/2} \tag{10}$$

with $\overline{x}_c$ is the average of all groups C and was average overall group. That performance measurement ratios S_W and S_B are only used for numeric data, while for categorical data is to use contingency tables which are equivalent to doing ANOVA (Analysis of Variance). Size diversity for categorical data one developed by [8]. So that the standard deviation within the group (S_W) and the standard deviation between groups (S_B) for categorical data can be written as in equation (11) and (12). [8]

$$S_W = \left[\frac{WSS}{(n-c)}\right]^{1/2} \tag{11}$$

$$S_B = \left[\frac{BSS}{(c-1)}\right]^{1/2} \tag{12}$$

with $WSS = \frac{n}{2} - \frac{1}{2}\sum_{c=1}^{C}\frac{1}{n_{.c}}\sum_{k=1}^{K} n_{kc}^2$ and $BSS = \frac{1}{2}\left(\sum_{c=1}^{C}\frac{1}{n_{.c}}\sum_{k=1}^{K} n_{kc}^2\right) - \frac{1}{2n}\sum_{k=1}^{K} n_{k.}^2$.

3. Case Study
Citrus plant breeding with protoplasm fusion between two kinds of oranges has been started since 2006 in Indonesian Citrus and Subtropical Fruits Research Institute (ICSFRI) in cooperation with the Indonesian Center for Agricultural Biotechnology and Genetic Resources Research and Development (ICABIOGRAD). Expected crop yield is superior seedless orange with a sweet taste, skin easy to peel and has an attractive color [9]. Protoplasm fusion is one of the techniques that use-right in the activities of crossbred plants by combining two different plant cells, so as to obtain hybrid plants. Plant cell that have incorporated parts of the cell wall is removed so that the genetic material contained in it is only protected by a plasma membrane [9]. Thus, it is possible the merger of the two parent protoplasm, so it cans produce new individuals are more diverse than the result of a combination of characteristics of both parents.

Results of protoplasm fusion is a new of plants. The new plants thus obtained are then observed by citrus fruit produced. However, due to the incorporation of protoplasts can occur involving two or more protoplasts, then the newly generated individual diversity is very high. The results of breeding is 120 plants hereinafter referred accession, but there are only 25 of 120 accessions which can be observed due to the accession of 95 more unfruitful mature or crop failure. The number of accessions affect the selection process more difficult due to an accession have different characteristics, although there is also an accession that have the same characteristics.

Citrus fruit morphological characteristics have two types of variables, the scale variables categorical and numerical scale variables. Citrus fruit morphology numeric variables include variables that can be measured numerically like fruit diameter, fruit height, weight of the fruit, and others. While categorical variables include fruit morphological variables that can't be measured numerically and can only be given a score such as skin color, fruit shape, color pulp and others.

The observed characteristics can't be done partially analysis between numerical and categorical variables. It is because of these characteristics affect each other and come from the same research object, so the analysis should be performed multivariate and involving all good numerical and categorical variables simultaneously. If these characteristics are analyzed partially it will give maximum results. The results are obtained if the analysis is partially inconsistencies existing group members in the group that is formed, so that the occurrence of multiple groups of accession the same, so it can't be done definite clustering of the accessions were observed. One method of multivariate analysis can be used for clustering is a cluster analysis. The data used in this research is secondary data obtained from the observed data at Indonesian Citrus and Subtropical Fruits Research Institute (ICSFRI). Here are the variable observations for analysis data are presented in Table 1.

Table 1. The variable observations.

Numerical Variable		Categorical Variable	
Fruit Diameter	(Diameter)	Fruit Shape	(Shape)
Insensitivity	(Insensitivity)	Skin Color	(Color)
Total Segment	(Segment)	The Surface Of The Skin	(Surface)
Thick Skin Meat	(Thick)	Fruit Axis	(Axis)
Diameter Axis	(D. Axis)	Flesh Color Of Skin	(Flesh C.)
Number Of Seeds	(Seed)	Looked Axis	(Looked A.)
Weight Fruit	(Weight)	Texture Of The Pulp	(Pulp)
Brix	(Brix)		

4. Results and Discussion

The first step in this analysis is to separate the data into two groups. The first group is a group of numeric data and the second group is a group of categorical data. For numeric data, we get a breakdown into five groups. From some consideration for the number of clusters, we chose the clustering results that have outcomes that are grouped into five groups that have R-square value around 90%. This value indicates that the clustering is good enough and able to represent the whole observation. By looking at the ratio between the standard deviation value within a group with the standard deviation between groups, for clustering numeric data used single linkage methods that has a smaller ratio than average linkage and complete linkage. Ratio value that is formed can be seen in Table 2.

Based on the explanation above, it can be concluded that the analysis of numerical data is done by agglomerative hierarchical clustering methods with single linkage and the number of clusters formed as 5. Members of clusters formed are presented in Table 3.For categorical variable data is grouped using ROCK methods, with theta trial had values between 0.05 to 0.30. From the theta trial, we get largest number of the clusters is 3 clusters. The value of θ that produces the number of clusters as much as 3 are 0.15, 0.17, and 0.20, so that these groupings of 3 results calculated ratio value between S_W and S_B. Ratio value that is formed can be seen in Table 4. Based on the explanation above, it can be concluded that the analysis for categorical data was conducted by ROCK with a value of θ at 0.15 and the number of clusters formed as 3. Members of the clusters that are formed can be seen in Table 5.

Table 2. Summary ratio value of single linkage method, complete linkage and average linkage with the number of clusters as much as 5 clusters.

Method	S_W	S_B	Ratio
Single linkage	0.681356	1.764561	0.386134
Complete linkage	0.955187	1.427995	0.668901
Average linkage	1.029224	1.361048	0.756200

Table 3. Cluster membership of the clustering on numerical data using hierarchical method agglomerative using single linkage with 5 clusters.

Cluster					
1		2	3	4	5
FS01	FS70	FS07	FS14	FS16	FS64
FS10	FS73	FS22	FS66	FS18	
FS15	FS89	FS29		FS32	
FS20	FS103	FS56			
FS57	P III i	FS84			
FS68		FS107			
		TL 3			
		TL 5			

Note : FS01, FS10, and the others are name code for the new citrus fruit.

Table 4. The ratio value of ROCK method for clustering categorical data with the number of clusters as much as 3 clusters.

	WSS	BSS	S_W	S_B	Ratio
0.15	12.5263	16.5137	0.754571	2.873474	0.262599
0.17	15.4673	2.2927	0.838486	1.070677	0.783136
0.20	20.8891	6.8709	0.974425	1.853497	0.525723

After obtaining optimum group of ROCK and agglomerative methods results, the next step is to merge the group. In the ensemble ROCK methods, this step just like clustering of numeric data using ROCK methods, wherein the input at this stage is the result of agglomerative hierarchical clustering methods as Output 1 and results clustering ROCK methods as Output 2. Output 1 and output 2 were expressed as categorical variables that will be used to draw up the final cluster using ROCK method. The optimum number of clusters is the number of groups that have smallest ratio between S_W and S_B. Clustering is done with theta trial had values between 0.05 to 0.30. Ratio value that is formed can be seen in Table 6. Based on the ratio values in Table 6, it is known that optimum clustering done with a value of θ at 0.12.

Table 5. Cluster membership of the clustering on categorical data using ROCK method using θ = 0.15 with 3 clusters.

Cluster				
1		2		3
FS07	FS29	FS01	FS70	FS66
FS14	FS64	FS10	FS84	FS73
FS15	FS68	FS16	FS89	
FS18	FS107	FS22	FS103	
FS20	TL 5	FS32	P III i	
		FS56	TL 3	
		FS57		

Table 6. The ratio value of ensemble ROCK method for clustering categorical data with the number of clusters as much as 4 clusters.

	WSS	BSS	S_W	S_B	Ratio
0.05	7.5090	13.8510	0.5842	2.6316	0.2220
0.07	4.3144	13.6856	0.4428	2.6159	0.1693
0.10	9.1380	26.6220	0.6445	3.6484	0.1766
0.12	4.4870	21.5130	0.4516	3.2797	0.1377
0.15	10.3820	25.6180	0.6870	3.5790	0.1919
0.17	10.3820	25.6180	0.6870	3.5790	0.1919
0.20	10.3820	25.6180	0.6870	3.5790	0.1919
0.22	11.2690	26.1710	0.7157	3.6174	0.1979
0.25	11.2690	26.1710	0.7157	3.6174	0.1979
0.27	11.2690	26.1710	0.7157	3.6174	0.1979
0.30	11.2690	26.1710	0.7157	3.6174	0.1979

For the ensemble SWFM methods, input to this step as the ensemble ROCK methods but these steps for clustering is used is the distance in equation (1) and filter algorithm in equation (2). Clustering is done by forming groups of 3 to 6, and then to determine the optimum number of clusters used ratio value between S_W and S_B. Ratio value that is formed can be seen in Table 7.

Table 7. Summary ratio value of ensemble SWFM method.

	WSS	BSS	S_W	S_B	Ratio
k=3	14.7046	1.1354	0.817552	0.753459	1.085065
k=4	16.9003	6.1397	0.876468	1.752099	0.500239
k=5	17.2272	24.5328	0.884904	3.502342	0.252661
k=6	34.9443	37.6157	1.260308	4.336802	0.290608

Results obtained from the final clustering of the ensemble ROCK methods is clustering the 25 observations into four groups with the ratio is 0.1377, while the results obtained from the clustering ensemble SWFM methods is clustering the 25 observations into five groups with the ratio is 0.2526. Members of the clusters that are formed from these methods can be seen in Table 8 and Table 9. However, the results revealed from ensemble ROCK method that there is a grouping formed one's own observations into clusters so must do more observations on those observations.

Table 8. Cluster membership of the clustering on mixed data using ensemble ROCK method with 4 clusters.

Cluster					
1		2	3		4
FS14	FS84	FS07	FS16	FS70	FS01
FS22	FS107		FS18	FS103	FS10
FS29	TL 5		FS20	P III i	FS15
FS32	FS68		FS57	TL 3	FS66
FS56			FS64		FS89
FS73					

Table 9. Cluster membership of the clustering on mixed data using ensemble SWFM method with 5 clusters.

Cluster				
1	2	3	4	5
FS01	FS07	FS16	FS56	FS57
FS10	FS14	FS18	FS84	FS70
FS20	FS15	FS22	FS107	FS89
FS68	FS32	FS29		
FS103	FS66	FS64		
P III i	FS73	TL 3		
		TL 5		

5. Conclusion

The value of the ratio from ensemble ROCK method is smaller than ensemble SWFM method, which shows that the performance ensemble ROCK method is better than ensemble SWFM method.. From the research conducted, it can be stated that ensemble ROCK method is more suitable for resolving existing case studies.

Acknowledgments

On this occasion, the author would like to thank to Indonesian Citrus and Subtropical Fruits Research Institute (ICSFRI) which has been granted permission to use their data as a case study. In addition, the authors also thank to other parties who contributed to the completion of this analysis.

References

[1] Johnson R A and Winchern D W 2007 *Applied Multivariate Statistical Analysis* (sixth ed.) (New Jersey: Pearson Education Inc)

[2] He Z, Xu X, and Deng S 2005 *CoRR* abs/cs/0509011
[3] Guha S, Rastogi R, and Shim K 2000 ROCK: A Robust Clustering Algorithm for Categorical Attributes *Proceedings of the 15th International Conference on Data Engineering*
[4] Reddy M V J and Kavitha B 2012 Clustering the Mixed Numerical and Categorical Dataset using Similarity Weight and Filter Method *International Journal of Database Theory and Application* **5** 1
[5] Bunkers M J and James R M 1996 *J.Climate* 130-146
[6] Hair J F, Black W C, Babin J B, and Anderson E R 2009 *Multivariate Data Analysis* (seventh ed.) (New Jersey: Prentice Hall Inc)
[7] Halkidi M, Batistakis Y, and Vazirgiannis M 2001 *Journal of Intelligent Information Systems* **17** 2/3 107-145
[8] Kader D G and Perry M 2007 *Journal of Statistics Education* **15** 2
[9] Martasari C 2014 *Kajian Genetik dan Percepatan Pembungaan Tanaman Hasil Fusi Protoplasma Jeruk Siam Madu dan Satsuma Mandarin* [Dissertation] (Malang: Universitas Brawijaya Malang)

Climate Changes Projection for Land and Forest Fire Risk Assessment in West Kalimantan

S D Jadmiko[1]*, D Murdiyarso[2, 3] and A Faqih[1, 3]

[1] Center for Climate Risk and Opportunity Management in Southeast Asia and Pacific (CCROM-SEAP), Bogor Agricultural University, Bogor-Indonesia.
[2] Center for International Forestry Research (CIFOR), Bogor-Indonesia.
[3] Department of Geophysics and Meteorology, Bogor Agricultural University, Bogor-Indonesia.

E-mail: djsyamsoe@gmail.com

Abstract. Risk analysis is a method used to determine the probability of disaster in the current and future. This research analyse of fire risk in West Kalimantan by using extreme climate and vulnerability analysis. Extreme climate was calculated based on the extreme dry rainfall from regional climate model RegCM4.4 outputs. Vulnerability analysis was conducted by using a composite mapping analysis used hotspot data and eleven indicators of vulnerability. We found that very high level of extreme dry rainfall located in the southern region and the western coast area of West Kalimantan. This condition was influenced by environment factors such as topography and land use. Extreme dry rainfall also associated with the pattern of annual rainfall in West Kalimantan which ranges between 1753-4861 mm. Modelling of the vulnerability of land and forest fires in West Kalimantan showed that the land use has impact 24% on the vulnerability model. The results of the vulnerability model analysis shows that the plantations areas and secondary swamp forests are highly vulnerable, particularly on the peat area with depth about 50-200 cm. The analysis of land and forest fires risk found that the vulnerable areas have high risk which is largely unmanaged plantation areas and peatlands.

1. Introduction

Forest and land fires have become the annual disaster phenomena in Indonesia, especially in Sumatra and Kalimantan. In the last two decades, the massive event of land and forest fires occurred during the El-Niño year at 1982/83, 1987, 1991, 1994 and 1997/98 [1]. El-Niño affects the drought and this condition cause the flammable areas [2-3]. The fires of 1997/1998 had become the largest land and forest fire event in Indonesia and caused large losses [4]. The main impact of forest and land fires is the haze, which is not only impacted of human health, but also socio-economic activities. Forest and land fire also caused land degradation and changes of forest function [5]

Forest and land fire is influenced by climates, fuels, ignition agents and human activities [6]. However, climates and human activities have become the dominant factor in Indonesia [5]. Climate (especially rainfall) will be affected the land drought. Strongly drought relates to fires event in Indonesia. For example, forest and land fire in 1997/1998 in Indonesia caused drought due to El-Niño events [7].

Climate change has impacts on rainfall changes which is can be changes the pattern of future droughts and the fires risk. Risk is a tools to assess the probability impact of disaster likes fires [8]. The level of risk can be determined by a probability of disaster (e.g. extremes climate) and that impact.

The impact caused by a disaster is determined by the level of vulnerability. Therefore, the risk can be calculated based on the relationship between probabilities of the hazard event and the level of vulnerability (risk = hazard x vulnerability) [9]. The concept of risk is widely used to assess in the current condition and the future due to climate changes. Understanding of fires risk would be minimize the potential impact of land and forest fires.

The objective of this research is to analyse the forest and land fires risk in West Kalimantan Province by using extreme climate and vulnerability analysis. Extreme climate was calculated based on the extreme dry rainfall with threshold value using 5th and 10th percentile. The extreme climate analysis, historical climate data information and projections were obtained through dynamical downscaling by using regional climate model RegCM4.4. Vulnerability analysis was conducted by using a composite mapping analysis used hotspot data and eleven indicators of vulnerability.

2. Fire Risk Analysis

Assessment of fire risk based on a matrix of risk that combines vulnerability analysis and the extreme dry rainfall (Table 1).

Table 1. Matrix of Fire Risk with five level (Very Low; VL, Low; L, Medium; M, High; H, Very High; VH). Category of monthly extreme dry rainfall: (1) 0-40 mm, (2) 41-80 mm, (3) 81-120 mm, (4) 121-160 mm dan (5) 161-200 mm.

Vulnerability Level	Category of Extreme Dry Rainfall				
	5	4	3	2	1
VL	VL	VL	L	L	M
L	VL	L	L	M	H
M	L	L	M	H	H
H	L	M	H	H	VH
VH	M	H	H	VH	VH

The vulnerability level could be calculated based on statistical approach of scores and weights of vulnerability indicators. We used the composite mapping analysis (CMA) to calculate of scores and weight of the level of vulnerability. This method was widely used in spatial analysis, not only in determining fires vulnerability but also can be used in other spatial vulnerability [10-11]. Scores for each indicator was calculated by using formula:

$$X_i = \left[\frac{O_i}{E_i}\right] \times \frac{100}{\sum_{i=1}^{n} \frac{O_i}{E_i}} \quad (1)$$

$$E_i = \left[\frac{T \times F}{100}\right] \quad (2)$$

With the Xi as Actual score for each indicator, Oi as number of hotspots at each classes of indicator (observed hotspot), Ei as expected hotspot for each indicator, T as total of number of hotspot for each indicator, F as percentage of areas (Ha) for each classes of indicator and i as classes of indicator. The rescaling scores was created with scale about 10-100 following the formula:

$$Score\ R_{out} = \left[\left(\frac{ScoreE_{input} - ScoreE_{min}}{ScoreE_{max} - ScoreE_{min}}\right) \times (ScoreR_{max} - ScoreR_{min})\right] + ScoreR_{min} \quad (3)$$

With the Score R_{out} as rescaling score, Score E_{input} as estimate score input, ScoreE_{min} as minimum estimated scores, ScoreE_{max} as maximum estimated scores, ScoreR_{max} as maximum of rescaling scores (100) and ScoreR_{min} as minimum of rescaling scores (10). The level of vulnerability used in this analysis included 5 levels (very low, low, medium, high and very high).

Analysis of extreme climate is more emphasized on rainfall due to its significant impacts on drought condition. There are several ways to define the climate extremes. Chu et al. [12] suggested three common approaches in defining extreme based on, i.e. i) actual rainfall values, ii) data probability distributions exceeding 95th or 99th percentile and less than 10th or 5th percentile, and iii) return periods of maximum annual rainfall. In this study, we used the probability analysis for analyzing the extreme dry rainfall with threshold 10th and 5th percentile. Extreme dry rainfall calculated for dry season at June-July-August (JJA).

The climate data for calculated the extreme dry rainfall generated by the dynamical downscaling analysis using model RegCM4.4. Simulations of RegCM4.4 conducted to obtain the climate information with high resolution. The outputs of simulations validated by using observations data in West Kalimantan. Simulation model RegCM4.4 intended to projection of climate in the future based on RCP-4.5 climate scenarios [13]. Projections of climate made for the 2016-2040 with the baseline year at the 1981-2005.

3. Result

West Kalimantan has equatorial pattern of rainfall with two peaks of rainfall in April and October. The high rainfall in occurred in the eastern part e.g. at Lake Sentarum National Park, while in the western part generally have lower rainfall. The pattern of projected rainfall distribution generally does not significantly changes compared to the baseline period (Figure 1).

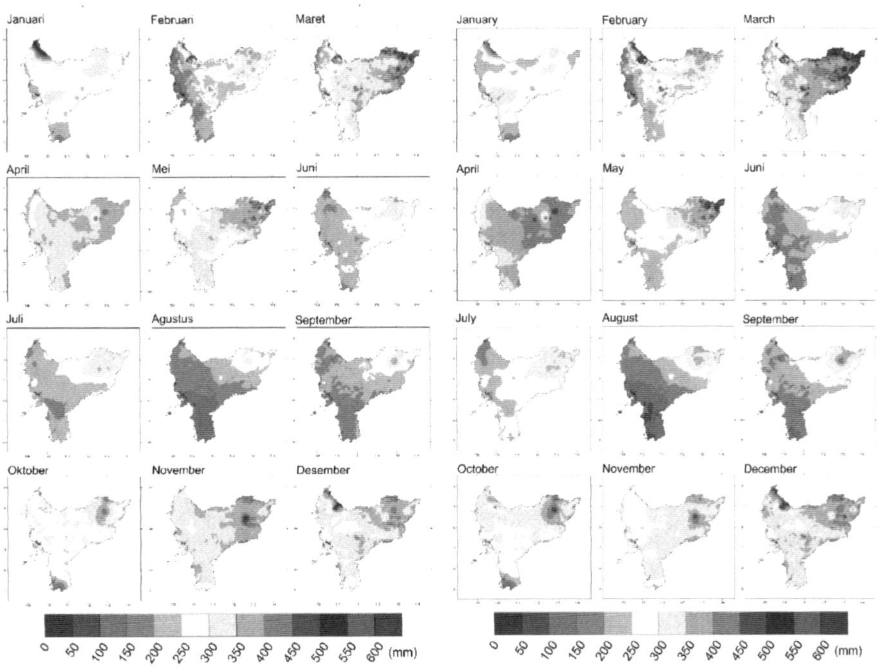

Figure 1. Comparison of spatial pattern of monthly rainfall for baseline (left) and projection (right).

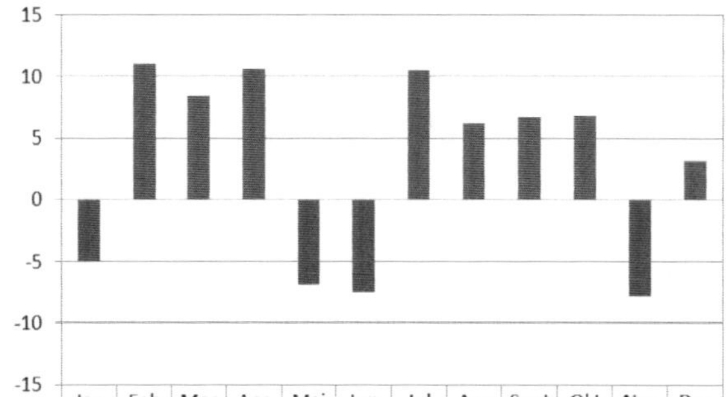

	Jan	Feb	Mar	Apr	Mei	Jun	Jul	Ags	Sept	Okt	Nov	Des
Δ (%)	-4.9	11.1	8.4	10.6	-6.9	-7.5	10.5	6.1	6.7	6.8	-7.8	3.2

Figure 2. Projected monthly rainfall changes (%) at 2016-2040 relative to 981-2005 for West Kalimantan.

The results of the rainfall changes analysis indicate decreased in January, May, June and November and increased in other months. The highest decrease rainfall occurs in November, which reached 7.8% compared with the baseline period, while the highest increase occurred in February, which reached 11.1% (Figure 2). Based on the pattern of these changes, the changes of rainfall (both increases and decreases) can occur in wet or dry season. The condition is certainly interesting to see how the impact of this change on extreme climate in the future.

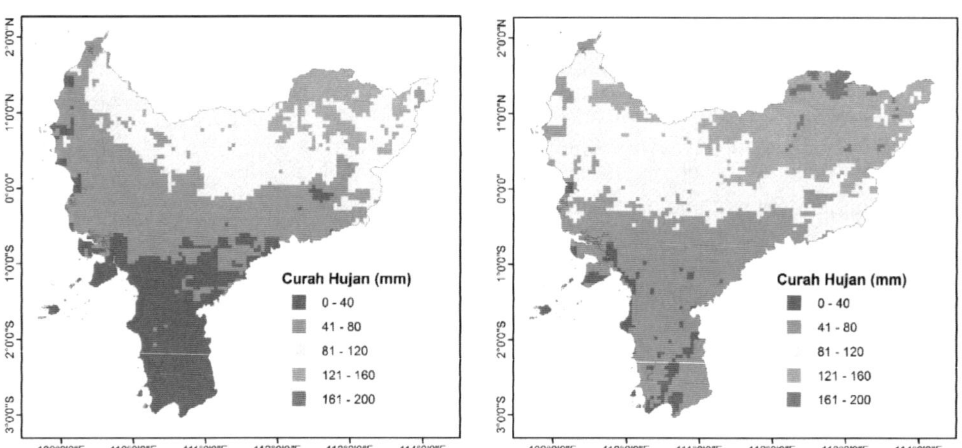

Figure 3. Pattern of baseline extreme dry rainfall at June-July-August (JJA) based on threshold by 5th percentile (left) and 10th percentile (right). The very extreme dry conditions generally occur in the southern part of West Kalimantan.

Analysis of extreme dry rainfall shows that areas of dry conditions due to low rainfall in the baseline period occurred in southern and western parts. Northern area has a lower levels of dry extreme rainfall than in other regions. The dry season (JJA) rainfall at the northern part generally above 100 mm. The pattern of extreme dry condition above happen to the threshold with 5th and 10th percentile (Figure 3).

Projected of extreme dry rainfall projection shows that an increase in the area of extreme drought. With the 5th percentile, the very extreme dry rainfall areas becomes more widespread until the middle part of West Kalimantan. Similarly, in the 10th percentile value where most of the southern part of

West Kalimantan become too extreme dry in June-July-August (JJA). The increase in the global area dry extreme potential to cause an increased risk of forest fires and land in the future. The pattern of projected extreme dry rainfall shows changes of future rainfall in dry season period at June-July-August (JJA).

The pattern of extreme dry rainfall as above when linked with land-use, areas of natural forest has lower level extreme dry rainfall than other region with agricultural area, plantation areas and peat areas which is located along the western coastline from southern part to middle areas of West Kalimantan.

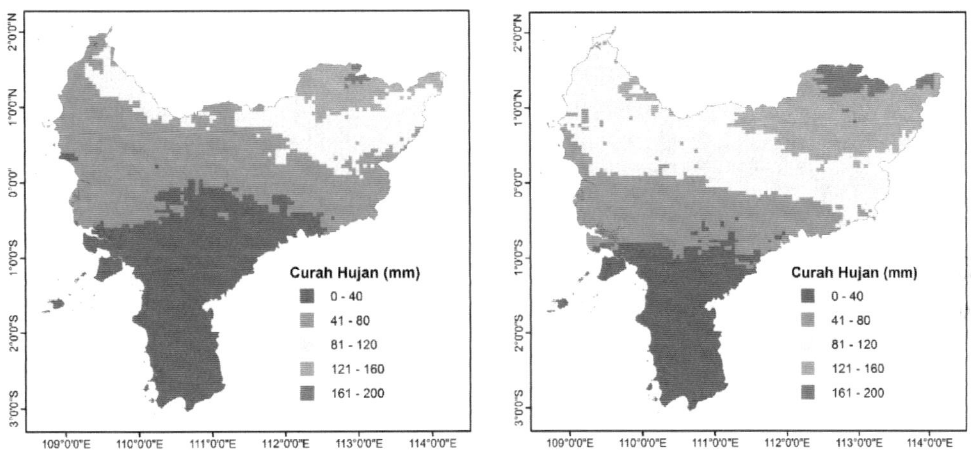

Figure 4. Pattern of projected extreme dry rainfall at June-July-August (JJA) based on threshold by 5th percentile (left) and 10th percentile (right). The very extreme dry conditions generally occur in the southern part of West Kalimantan.

Modelling of the vulnerability shows the land use has impact 24% on the vulnerability model. Plantation areas and scrub swamp areas are the region of frequent forest and land fires. However, based on eleven indicators, biophysical factors (as deep of the peat, land cover and land systems indicator) provides 49.85% compared to the 16% of the effect of human activity factor (the distance from the road, distance from the river and the distance from the center of the village / town) and a socio-economic factor (population density, the GDP, HTI, HGU and HPH) amounted to 34 %. The results of the vulnerability model analysis shows that plantations areas and secondary swamp forests are highly vulnerable, particularly on the peat area with depth about 50-200 cm. The equation for vulnerability model of forest and land fire shows at Equation 4.

$$VM = 0.0758\ PEA + 0.2409\ LUS + 0.0193\ STR + 0.0116\ RVR + 0.1291\ VLG + 0.1818\ LSY + 0.0481\ HTI + 0.1291\ HGU + 0.0873\ HPH + 0.0297\ POP + 0.0474\ PDRB \quad (4)$$

VM = Vulnerability model for land and forest fire in West Kalimantan
PEA = Peat depth
LUS = Land-use
STR = Distance from street
RVR = Distance from river
VLG = Distance from village
LSY = Land system
HTI = Percentage of area of HTI (Hutan Tanaman Industri/Industrial forest)
HGU = Percentage of area of HGU (Hak Guna Usaha)
HPH = Percentage of area of HPH (Hak Pengusahaan Hutan)

POP = Population density
PDRB = Produk Domestik Regional Bruto (Gross Regional Domestic Bruto)

Spatial planning needs to consider the condition of land use in view of the condition that affects the vulnerability of land and forest fires. Changes in land-use will affect the vulnerability of forest and land fires. If the type of land-use changes with the area which has a strong correlation with the presence of hotspots, then most likely will increase the level of vulnerability. Conversely, when the land-use be returned in their original function, the vulnerability of land and forest fires will likely decrease.

Figure 5 shows the pattern of vulnerability map of forest and land fires for West Kalimantan. Most areas of West Kalimantan has a low level of vulnerable which is about 43% of the total area, while the regions with high and very high level of vulnerable approximately 13% of the area of West Kalimantan. Areas with high level of vulnerability occurs in the western region as Kubu Raya and Mempawah district as well as the southern part at the Ketapang district.

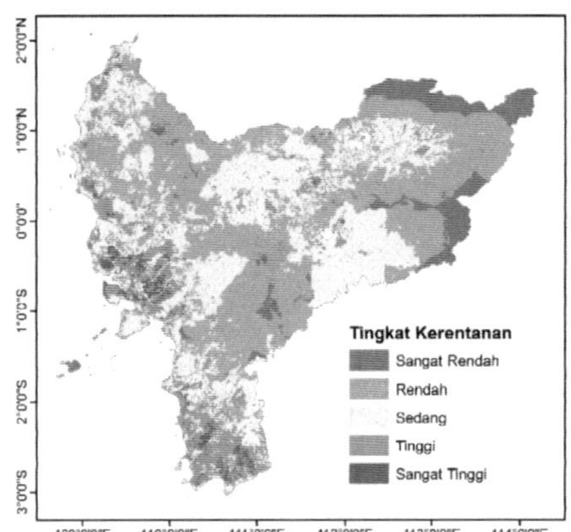

Figure 5. Vulnerability map for fire in West Kalimantan.

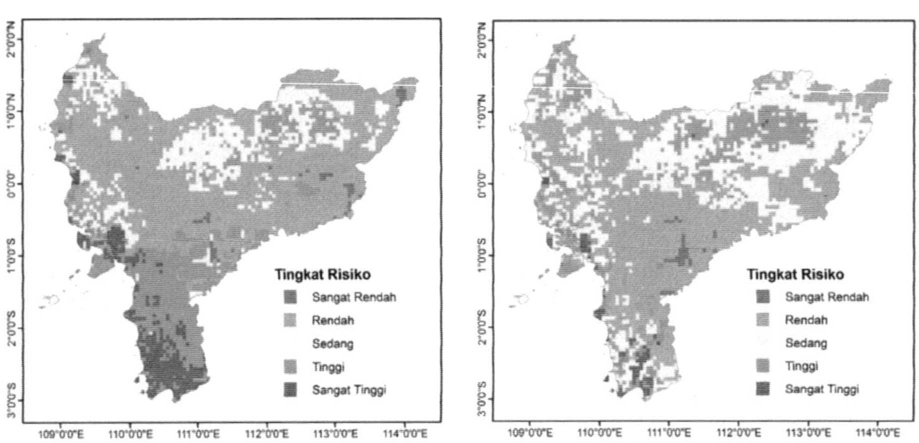

Figure 6. Pattern of baseline fire risk based on threshold by 5th percentile (left) and 10th percentile (right). The areas with very high level of risk generally occur in the southern part of West Kalimantan.

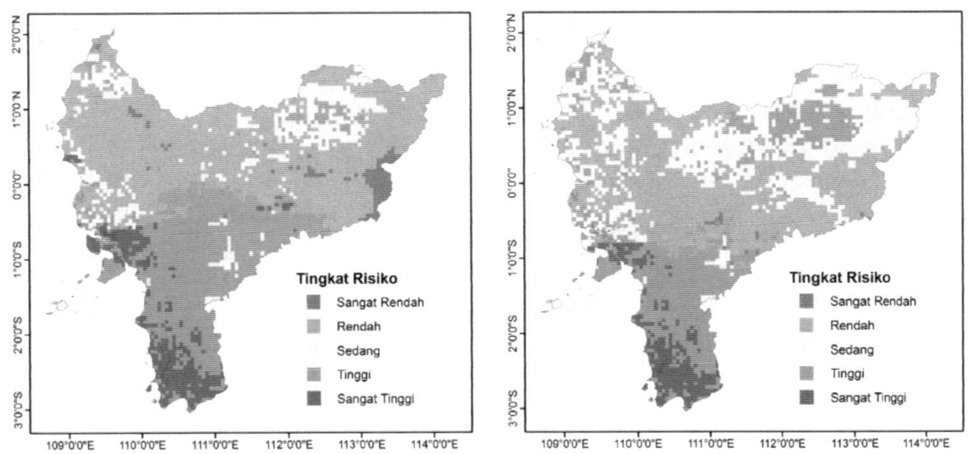

Figure 7. Pattern of projected of fire risk based on threshold by 5th percentile (left) and 10th percentile (right). The areas with very high level of risk generally occur in the southern part of West Kalimantan.

The very high level of fires risk occur in the southern part and the western part of West Kalimantan with land-use as peat areas and shrub swamps. In an area of natural forest, the level of fires risk has low level. The extreme dry rainfall value by 5th percentile indicates a higher risk than the 10th percentile. Projected of fires risk showed an increased risk in the southern part of West Kalimantan and decreased of fires risk in the northern part and eastern part. With 5th percentile a decrease in the fires risk in the northern region, while with 10th percentile can be clearly seen that the southern part has increased of fires risk in the future (Figure 7).

The very high level of fires risk generally located at secondary swamp forest land and shrub swamp in the southern part of West Kalimantan. These region also the location of peatlands, which has a depth about 50-200 cm. Areas with high level of vulnerability generally located on plantation areas and peat swamp forests rounded the Danau Sentarum National Park (DSNP). Normally, the DSNP region is a protected national park. However, there is a tendency in the communities around these location will conduct fishing activities during the dry season. These activities allegedly affect the fires risk where they will do the expansion of fishing regions in a way that could be caused forest and land fires [14].

4. Conclusion

The rainfall pattern in West Kalimantan is equatorial pattern with high rainfall locate in the northern and eastern part, while low rainfall in the southern and western part. Projected rainfall in 2016-2040 relative to the 1981-2005 indicate that an increase rainfall in the rainy season and decrease during the dry season. Projected of the extreme dry rainfall will increased and over most of West Kalimantan areas.

Vulnerability models of fire showed that some indicators such as land cover, distance from the center of government / village and land systems have a strong influence at the number of hotspot. Areas with the peat and agricultural areas has a propensity of a higher level of vulnerability compared to other land-use. Areas with high and very high level of vulnerability generally located on the southern and western parts of West Kalimantan province, while the level of vulnerability of low and very low generally in northern and eastern part.

The risk of forest and land fires for baseline condition have the same pattern with the patterns of vulnerability. In the projection period, the risk of forest and land fires increased with the wider region. The trend of decreased rainfall during the dry season in the future will contributed to the increased of risk level of forest and land fires.

References

[1] Bowen M R, Bompard J M, Anderson I P, Guizol P, and Gouyon A 2001 *Nova Science* 41–66
[2] Wooster M J, Perry G L W and Zoumas A 2012 *Biogeosciences* **9** 317-340
[3] Fuller D O and Murphy K 2006 *Climatic Change* **74** 435-455
[4] Taconi L 2003 *Kebakaran Hutan di Indonesia: Penyebab, Biaya dan Implikasi Kebijakan* CIFOR Occasional Paper No 38i Bogor: CIFOR
[5] Varma A 2003 *Ecological Economics* **46**159-171
[6] Swetnam T W 1993 *Science* **262** 885–889
[7] Murdiyarso D and Adiningsih E S 2005 *Mitigations and Adaptation Strategies for Global Change* **12** 101-112
[8] Brooks N 2003 Vulnerability, risk and adaptation: a conceptual framework *Working Paper 38* Tyndall Centre for Climate Change Research University of East Anglia Norwich www.tyndall.ac.uk
[9] Jones R and Boer R 2003 Assessing current climate risks. Adaptation Policy Framework: a Guide for Policies to Facilitate Adaptation to Climate Change UNDP http://www.undp.org/cc/apf.htm.
[10] Boonyanuphap J 2005 Spastial Model for Determining Risk Area of Deforestation. *Suranaree J. Sci. Technol* **12** 145-159
[11] Jaya I N S, Purnama E S, Arianti I, and Boonyanuphap J 2007 Forest fire risk assessment model and post-fire evaluation using remote sensing and GIS: A case study in Riau, West Kalimantan and East Kalimantan Provinces, Indonesia Paper presented at The Forest Restoration and Rehabilitation Training Course and Workshop in the Viiki Tropical Resources Institute (VITRI) of the University of Helsinki Finland 13-19 May 2007, Helsinki (FIN)
[12] Chu P S, Chen Y R and Schroeder T A 2010 *Journal of Climate* **23** 4881-4990
[13] Moss R H, Edmonds J A, Hibbard K A, Manning M R, Rose S K, van Vuuren D P, Carter T R, Emori S, Kainuma M, Kram T, Meehl G A, Mitchel J F B, Nakicennovic N, Riahi K, Smith S J, Stouffer R J, Thomson A M, Weyant J P and Wilbanks T J 2010 *Nature* **463** 747-756
[14] Onrizal C, Kusmana, Saharjo B H, Handayani I P and Kato T 2005 *Biodiversitas* **6** 220-223

Synthetic Over Sampling Methods for Handling Class Imbalanced Problems : A Review

B Santoso[1], H Wijayanto[1], K A Notodiputro[1] and B Sartono[1]

[1] Department of Statistics, Faculty of Mathematics and Natural Sciences, Bogor Agricultural University, Indonesia.

Email : budsant.bs@gmail.com, hari_ipb@yahoo.com, khairilnotodiputro@gmail.com, bagusco@gmail.com

Abstract. Class imbalanced commonly found in any real cases. Class imbalanced occur if one of the classes has smaller amount, called minority class, than other class (majority class). The problem of imbalanced data is usually associated with misclassification problem where the minority class tends to be misclassified as compared to the majority class. There are two approaches should be performed to solve imbalanced data problems, those are solution at data level and solution at algorithm level. Over sampling approach is used more frequently than the other data level solution methods. This study gives review of synthethic over sampling methods for handling imbalance data problem. The implementation of different methods will produce different characteristics of the generated synthetic data and the implementation of appropriate methods must be adapted to the problems faced such as the level and pattern of imbalanced data of data available. Results of the review show that there is no absolute methods that are more efficient in dealing with the class imbalance. However, the class imbalance problem depends on complexity of the data, level of class imbalance, size of data and classifier involved. Determination of over sampling strategy will affect the outcome of the over sampling. So it is still open better development oversampling methods for handling the class imbalance. The selection classifier and evaluation measures are important to get the best results. Statistical test approach is needed to assess the theoritical propertis of synthetic data and evaluate missclassification in addition to the evaluation methods that have been used.

1. Introduction

Class imbalanced commonly found in any real cases. Class imbalanced occur if one of the classes has smaller amount, called minority class, than other class (majority class). The problem of imbalanced data is usually associated with misclassification problem where the minority class tends to be misclassified as compared to the majority class. Problem rises when data of minority class contain important information and become the focus of attention of research so that errors in classification will lead to errors in decision-making especially in prediction accuracy of minority class.

There are two approaches should be performed to solve imbalanced data problems, those are solution at data level and solution at algorithm level [1]. Solution at data level is applied by balancing the distribution of the majority and minority class through methods of under sampling, over sampling or combination of both methods. Solution at algorithm level is applied by modification in classifier methods or optimize the performance of learning algorithm. The advantage of the data level is can use independent of the classifier selected.

In under sampling process, data on majority class are reduced or eliminated some data to balance the class distribution, while over sampling process is done by adding the data on minority class. The process of balancing the distribution of data can be achieved by performing under sampling and over sampling simultaneously, termed combination or hybrid. Over sampling approach is used more frequently than under sampling since under sampling method will eliminate data in the majority class thus causing lost of important information of the data. According to Batista *et al.* [2] in general, over sampling method gives better results than under sampling method. The famous over sampling method is the Synthetic Minority Over Sampling Technique (SMOTE). However, the SMOTE method still has some weaknesses hence it motivates the research development to overcome the problem of SMOTE.

This paper will review several over sampling methods which used to solve the imbalanced data problems, those were SMOTE and development methods of SMOTE such as SMOTE-TL-ENN, SMOTE RSB, Borderline-SMOTE and Safe Level SMOTE.

2. Evaluation measures

Most of evaluation measures for two class problem are built by confusion matrix as illustrated in Tabel 1. The class label of the minority class is positive and the class label of the majority class is negative. The first row of the table is the actual class label and the first column present their predicted class label. TP and TN denote the number of positive and negative examples that are classified correctly, while FN and FP denote the number of misclassified positive and negative examples respectively.

Table 1. Confusion matrix.

		Predicted class	
		Yes	No
Actual Class	Yes	TP: True positive	FN: False negative
	No	FP: False positive	TN: True negative

To evaluate class imbalanced, classification accuracy is not sufficient as a standard performance measure. Classification accuracy can not be used because it does not specifically describe the classification of a particular class. Receiver Operator Characteristics (ROC) analysis, Area Under the ROC Curve (AUC) and metrics such as precision, recall and F-value more better to understand the performance on the minority class.

Precision, recall and F-value can be expressed as:

$$Accuracy = \frac{TP + TN}{(TP + FP + TN + FN)}$$

$$Precision = \frac{TP}{(TP + FP)}$$

$$Recall = \frac{TP}{(TP + FN)}$$

$$F - value = \frac{(1 + \beta^2)\, Recall\ x\ Precision}{\beta^2\, Recall + Precision}$$

Where β corresponds to relative importance of precision vs. recall and it is usually set to 1.

The goal of evaluation in class imbalance is to improve the recall, without sacrificing the precision. However, the recall and precision goals are often conflicting and attacking them simultaneously may not work well, especially when one class is rare. The F-value incorporates both precision and recall, and the "goodness" of a learning algorithm for the minority class can be measured by the F-value. While ROC curves represent the trade-off between values of TP and FP, the F-value basically incorporates the relative effects/costs of recall and precision into a single number [3]. ROC curve value under ideal conditions (0.100) indicates that all the positive samples were classified correctly and no negative sample were classified as positive sample.

3. Current approaches for handling class imbalance

3.1. Solution at data level

Solution at data level is applied by balancing the distribution of the majority and minority class through methods of under sampling, over sampling or combination of both methods.

- Under sampling
  In under sampling, data on majority class are reduced or eliminated some data to balance the class distribution. The simple approach of under sampling method namely RUS (Random Under Sampling) where data on the majority of reduced class randomly. Furthermore, various developing methods under sampling conducted by Tomek [4] is by way of modification of Condensed Nearest Neighbor (CNN) and Tomek-Link. Furthermore Kubat and Matwin [5] using the method of One Side Selection (OSS). Laurikkala [6] proposed a new method, namely Neighborhood Cleaning Rule (NCR) which proved to be better than simple random and OSS. Cluster-based approach proposed by Yoon [7] and followed by the Yen and Lee [8].
- Over sampling
  Procedure in over sampling by adding the data on minority class to balance class distribution. Over sampling approach is used more frequently than under sampling since under sampling method will eliminate data in the majority class thus causing lost of important information of the data. The simple over sampling method is the Random Over Sampling (ROS), that is carried out by balancing the distribution of data through the application of minority data duplication at random. However, the method has the disadvantage that ROS cause problems overfitting [9]. To solve this problem, many research on methods over sampling conducted among others Chawla *et al.* [3], Batista *et al.* [2], He *et al.* [10], Han *et al.* [11], Bunkhumpornpat *et al.* [12], Tang and Chen [13], Stefanowski and Wilk [14], Cohen *et al.* [15], Napierala *el at* [16] and Ramentol *et al.* [1].
- Hybrid
  Hybrid methods combine the under sampling and over sampling methods by eliminating the data from majority class and adding the data from minority class to balance class distribution. Research on hybrid method conducted by Barandela *et al.* [17].

3.2. Solution at algorithm level

Solution at algorithm level is applied by modification in classifier methods or optimize the performance of learning algorithm on unseen data.

- Cost sensitive learning
  Another way to improve the performance of classifier is by applying cost for minority class distribution. In solution at data level, classifier try to minimize the number or errors with new data when the cost of different errors are equal. In many real-world applications the cost of different errors are often unequal. For example in medical diagnosis, the cost of erroneously diagnosing a patient to be healthy may be much bigger than that of mistakenly diagnosing a healthy person as being sick. Research on cost sensitive learning carried out by Veropoulos [18], Ting [19] and Zhou and Liu [20].
- Ensemble
  Ensemble is a combination of multiple classifiers to improve and increase the prediction accuracy. In past few yers, ensembles have emerged as a promising technique with the ability

to improve the performance of weak classification algorithms [21]. In the field of class imbalance, ensembles have mainly been used to combine the result of several classifiers. Research using ensemble approach undertaken by Sun *et al.* [22], Freund and Schapire [21], Schapire and Singer [23], Guo and Viktor [24], Liu *et al.* [25] and Chawla *et al.* [3].

4. Selected synthetic over sampling approaches

Over sampling approach is a widely used method to solve the imbalanced data. According to Batista *et al.* [2] in general, over sampling method gives better results than under sampling method. When data is highly imbalance, significant differences between majority dan minority class can be handled by over sampling methods.

The simple over sampling method is the random over sampling, that is carried out by balancing the distribution of data through the application of minority data duplication at random. Chawla *et al.* [26] proposed a new technique, named SMOTE, to generate synthetic data based on the distance between the minority data and the closest minority data therefore the new synthetic data will be formed between the two minority data.

Although SMOTE is quite effective in improving the classification accuracy of the minority data, but there are still problems, among others, that is the occurrence of overgeneralization. Data synthetic result of SMOTE is still possible to spread on both minority and majority data, hence it will reduce the performance of classification.

Formula to generate synthetic data by SMOTE can be expressed as

$$D_{new} = D_i + \left(\widehat{D}_\iota - D_i\right) \times \delta$$

Where D_{new} = synthetic data, D_i = examples from minority, $\widehat{D}_\iota$ = one of k-nearest neighbor from D_i , δ = random number between 0 and 1

To solve the SMOTE problems, some researches are conducted to modify SMOTE in order to create more effective technique in improving the classification performance. The development of SMOTE is mostly done in two ways, the first is to improve the synthetic data results of SMOTE [2,1]. The second way is to first determine which location will be choose for data generating as it is expected that the result of data generating will exactly found in the data minority area [11,12].

Batista *et al* [2] succeeded in correcting the SMOTE data by deleting the synthetic data located in the area of minority data through the Tomek Links (SMOTE-Tomeks links). In addition to the method of Tomek Links [4], Batista *et al.* [2] also improved the synthetic data on both data (majority and minority) by using the Edited Nearest Neighbor (SMOTE-ENN). Meanwhile Ramentol *et al.* [1] used the Rough Set theory to improve synthetic data generated by SMOTE, that is the SMOTE-RSB.

Han *et al.* [11] divided three location based on the amount of majority data in the nearest neighbors of minority data. If the majority data are all around the nearest neighbors, then the area is called noisy. If the majority data around the nearest neighbors are found higher than or equal to the minority, the area is called borderline. Meanwhile if most of/all data around the nearest neighbors are the minority data, then the area is called safe. Han *et al.* [11] focused on the borderline area located on the boundary between the minority and the majority of data. Data generating through SMOTE is performed in the borderline areas, called the Borderline SMOTE.

Moreover, Bunkhumpornpat *et al.* [12] paid attention to the safe area to perform over sampling based on the ratio between the number of minority data and the nearest neighbors. This method wants to make sure that the synthetic data that will be generated through the SMOTE are in areas that are completely safe. Bunkhumpornpat *et al.* [12] defines five criteria based on the ratio of safe level. Therefore the method is called the Safe Level SMOTE.

5. Discussion

We can compare several over sampling methods by located, level of class imbalance, amount of over sampling, number of k-nearest neighbour and choosen classifiers. Based location of examples, we can divided into 3 categories: safe, borderline and noisy examples [5,11,16,1].

- Borderline examples are located in the area surrounding class boundaries, where the minority and majority classes overlap.
- Safe examples are placed in relatively homogeneous areas with minority class.
- Noisy examples are individuals from one class occurring in areas of the other class.

SMOTE, SMOTE-TL-ENN and SMOTE RSB methods using all of minority examples in whole location to create to generate synthetic data. Batista *et al.* [2] use application of Tomek Links and ENN as cleaning method over the dataset obtained by the application of SMOTE. Ramentol [1] only selects the minority synthetic data that belong to the lower approximation using Rough Set Theory until the dataset is balanced.

Borderline-SMOTE and Safe Level SMOTE only use specific location to generate synthetic data. After determining an example of a minority in a specific location and then the process of SMOTE done to generate synthetic data.

Table 2. Comparative review.

	Location	Cleaning data	Number of K-NN	Amount OS	Random number
SMOTE	Borderline, noisy and safe	No	5	various	0 to 1
SMOTE-TL-ENN	Borderline, noisy and safe	Yes	3	various	0 to 1
SMOTE RSB	Borderline, noisy and safe	Yes	5	balance	0 to 1
Borderline-SMOTE	Borderline	No	5	various	0 to 0.5
Safe Level-SMOTE	Safe	No	5	various	various

Table 2 indicate that various methods use different strategy to handle class imbalance. Several methods using number of k-nearest neighbour set to 5 except SMOTE-ENN. Nearest neighbor indicating the similarity of the data. The problem is that it could be the nearest neighbor has a considerable distance. It can lead to overgeneralization. Determination of nearest neighbors can be evaluated further by using alternatives such as radius size or percentage.

In over sampling methods, we can add data on the minority so close to the amount of data on the class of the majority or equal as the majority class. In SMOTE, Borderline-SMOTE and Safe Level methods, the minority class was oversampled at 100%, 200%, 300%, 400%, 500%. The amount of data in oversampling depend on level of imbalance. Variations in the amount of oversampling cause changes in accuracy and other measures. In SMOTE RSB method, minority data was added so that the amount of data equal to a majority data.

Based on the formula in SMOTE, the distance between the data minority and its nearest neighbors multiply by random number between 0 and 1. If the random number is close to 0 then the synthetic data will be similar to the origin minority data. Conversely, if the random number is close to 1 then the synthetic data will be similar to nearest neighbors. Problems that can occur is if the random number is about 0.5 it is possible that the synthetic data similar to the data of the majority. This is the cause of overgeneralization.

In the Borderline-SMOTE, the random number is determined between 0 to 0.5 in hopes of synthesis data will be adjacent to the minority data. Different strategies used in other oversampling methods. Safe level SMOTE use RSL (Ratio of Safe Level) to determine random number. SMOTE-TL-ENN and SMOTE-RSB use cleaning data to improve synthetic data generated by SMOTE.

Table 3. Winner of AUC Result by percentage of minority

	Winner of AUC Result*	
	≤ 5 %	>5 %
SMOTE	-	2
SMOTE-TL-ENN	3	1
SMOTE RSB	10	5
Borderline-SMOTE	2	3
Safe Level-SMOTE	5	2

*) Ramentol *et at.* [1] using 44 datasets

Table 3 shows the effectiveness of information over sampling method using several datasets based Area Under the ROC Curve (AUC). The table is divided into two levels of class imbalance, namely class imbalance below 5 percent and above 5 percent.

Based on Table 3, we can't find best method in handling the class imbalance. Ramentol *et al.* [1] using 44 datasets with several over sampling methods. As a result, there is no better method on all datasets. SMOTE only effective on 2 dataset, SMOTE-TL-ENN effective in four datasets, Borderline-SMOTE effective at 5 dataset. An effective methods in many datasets are Safe Levels-SMOTE (7 dataset) and SMOTE RSB with 15 datasets. SMOTE only effective in percentage of minority up to 5 percent, while others methods effective in any imbalance level. The same result is done by Batista *et al.* [2] which by using 16 datasets, there are not the most efficient method. SMOTE efficient at 2 dataset, SMOTE-TL efficient in four data sets and SMOTE-ENN efficient at 9 datasets.

The results of the research by Han *et al.* [11] indicate that Borderline-SMOTE more efficient than SMOTE and ROS using the F-value and Recall. However there is one dataset that showed the opposite results. The experiments by Bunkhumpornpat *et al.* [12] show that the performance of Safe-Level-SMOTE evaluated by Precision and F-value are better than that of SMOTE and Borderline-SMOTE when decision trees C4.5 are applied as classifiers.

The use of classifier method affects effectiveness. In SMOTE method, the effectiveness of different classifier methods based on the dataset used and evaluation measures. The use of C.45 method, Naive Bayes and Ripper show different efficiencies according to the different datasets and different measures matrices. The same situation also happened on research conducted by Han *et al.* [11] and Batista *et al.* [2].

Further research is needed to evaluate the classification method is more appropriate in dealing with class imbalance. Exploration statistically in the data results over sampling is required to determine the characteristics of the synthetic data is generated. Based on synthetic data characteristics can be determined more precise method used to overcome the problem of class imbalance.

6. Conclusion

Results of the review show that there is no absolute methods that are more efficient in dealing with the class imbalance. However, the class imbalance problem depends on complexity of the data (located of minority data), level of class imbalance, size of data and classifier involved. Determination of the distance and the number k-nearest neighbors will affect the outcome of the over sampling. So it is still open better development oversampling methods for handling the class imbalance. The selection classifier is important to get the best results.

In general, the evaluation measure is still much to do with the measures on confusion matrices such as accuracy, precision, recall, F-value and ROC. Statistical test approach is needed to assess the theoritical propertis of synthetic data and evaluate missclassification.

References

[1] Ramentol E, Caballero Y, Bello B and Herrera F 2011 SMOTE-RSB: A Hybrid Preprocessing Approach based on Oversampling and Undersampling for High Imbalanced Data-Sets using SMOTE and Rough Sets Theory, *Knowledge and Information Systems* (London: Springer) 245
[2] Batista GEAPA, Prati RC and Monard MC 2004 *SIGKDD Exploration* **6(1)** 20
[3] Chawla NV, Bowyer KW, Hall LO and Kegelmeyer WP 2002 *Journal of Artificial Intelligence Research* **16** 321
[4] Tomek I 1976 *IEEE Transactions on Systems, Man and Cybernetics* **6** 769
[5] Kubat M and Matwin S 1997 Addressing the Curse of Imbalanced Training Sets: One-Sided Selection, *14th International Conference on Machine Learning (ICML97)* (USA: Tennessee) 179
[6] Laurikkala J 2001 Improving Identification of Difficult Small Classes by Balancing Class Distribution, 8th Conference on AI in Medicine in Europe AIME01 (Portugal: Cascais) 63
[7] Yoon K, Kwek S 2005 An unsupervised learning approach to resolving the data imbalanced issue in supervised learning problems in functional genomics, *5th International Conference on Hybrid Intelligent Systems HIS05* (Brazil: Rio de Janeiro) 303
[8] Yen S, Lee Y 2006 Under-sampling approaches for improving prediction of the minority class in an imbalanced dataset, *International Conference on Intelligent Computing ICIC06* (China: Kunming) 731
[9] Tetko I, Livingstone D and Luik A 1995 *Chemical Information & Computer Sciences* 826
[10] He H, Bai Y, Garcia EA and Li S 2008 *IJCNN08* 1322
[11] Han H, Wang WY and Mao BH 2005 *ICIC05* 878
[12] Bunkhumpornpat C, Sinapiromsaran K and Lursinsap C 2009 *PAKDD09* **5476** 475
[13] Tang S and Chen S 2008 The Generation Mechanism of Synthetic Minority Class Examples, *5th Int. Conference on Information Technology and Applications in Biomedicine (ITAB)* (China: Shenzhen) 444-447
[14] Stefanowski J and Wilk S 2008 Selective pre-processing of imbalanced data for improving classification performance, 10th International Conference in Data Warehousing and Knowledge Discovery (Springer)
[15] Cohen G, Hilario M, Sax H, Hugonnet S, Geissbuhler A 2006 *Artificial Intelligence in Medicine* **37** 7
[16] Napierala K, Stefanowski J and Wilk S 2010 Learning from Imbalanced Data in Presence of Noisy and Borderline Examples, *7th International Conference on Rough Sets and Current Trends in Computing (RSCTC2010)* (Poland: Warsaw) 158
[17] Barandela R, Valdovinos RM, Sánchez JS and Ferri FJ 2004 *LNCS* **3138** 806
[18] Veropoulos K, Cristianini N and Campbell C 1999 Controlling the sensitivity of support vector machines, *16th International Joint Conferences on Artificial Intelligence IJCAI99* (Sweden: Stockholm) 281
[19] Ting K M 2002 *IEEE Transactions on Knowledge and Data Engineering* **14(3)** 659
[20] Zhou Z, Liu X Y 2006 *IEEE Transactions on Knowledge and Data Engineering* **18(1)** 63
[21] Freund Y, and Schapire RE 1997 *Journal of Computer and System Sciences* **55(1)** 119
[22] Sun Y, Kamel M, Wong A, Wang Y 2007 *Pattern Recognition* **40** 3358
[23] Schapire RE and Singer Y 1999 *Machine Learning* **37** 297
[24] Guo H and Viktor HL 2004 *SIGKDD Explorations* **6(1)** 30
[25] Chawla NV, Lazarevic A, Hall LO and Bowyer KW 2003 SMOTEBoost : Improving Liu X Y, Wu J, Zhou Z 2009 *IEEE Transactions on Systems, Man, and Cybernetics, Part B* **39(2)** 539

[26] Prediction of Minority Class in Boosting, 7th *European Confrence on Principles and Practice Of Knowledge Discovery in Database* (Croatia: Dubrovnik) 107

Clustering box office movie with Partition Around Medoids (PAM) Algorithm based on Text Mining of Indonesian subtitle

A D Alfarizy[1], Indahwati[1] and B Sartono[1]

[1]Department of Statistics, Faculty of Mathematics and Natural Sciences, Bogor Agricultural University, Indonesia

Email: athikah.dalfarizy@gmail.com, indah.stk@gmail.com, bagusco@apps.ipb.ac.id

Abstract. Indonesia is the largest Hollywood movie industry target market in Southeast Asia in 2015. Hollywood movies distributed in Indonesia targeted people in all range of ages including children. Low awareness of guiding children while watching movies make them could watch any rated films even the unsuitable ones for their ages. Even after being translated into Bahasa and passed the censorship phase, words that uncomfortable for children to watch still exist. The purpose of this research is to cluster box office Hollywood movies based on Indonesian subtitle, revenue, IMDb user rating and genres as one of the reference for adults to choose right movies for their children to watch. Text mining is used to extract words from the subtitles and count the frequency for three group of words (bad words, sexual words and terror words), while Partition Around Medoids (PAM) Algorithm with Gower similarity coefficient as proximity matrix is used as clustering method. We clustered 624 movies from 2006 until first half of 2016 from IMDb. Cluster with highest silhouette coefficient value (0.36) is the one with 5 clusters. Animation, Adventure and Comedy movies with high revenue like in cluster 5 is recommended for children to watch, while Comedy movies with high revenue like in cluster 4 should be avoided to watch.

Keywords: gower similarity, partition around medoids, text mining, silhouette coefficient

1. Introduction

Hollywood is the biggest movie industry in the world with the fastest growth and the highest income from distribution in all countries. Motion Pictures Association of America [1] recorded 2015's global income from all Hollywood box office movies released in each country over the world combined, reached the number of US$38 billion. The highest portion is contributed by Asia Pacific region, including Indonesia which reached US$ 250 million at 2015, topping all South-East Asia countries. It is supported by the fact that Indonesia has huge number of population spread in wide areas which almost all of the areas has cinemas.

Movies distribution in Indonesia targeted all people from any range of age and gender, including children. They can be accessed through many devices and any places like in cinemas, local television channel, DVD, and even smartphones. All these devices are on daily life and can be accessed easily by children. The movies has information that can reach out and influence the people who watchs them, so it's necessary for children to have adults as a company while watching movies in order that they would not have wrong perception of the information from the movies.

However, adults' awareness to the importance of guiding children while watching movies is yet very low though it is very crucial. As a matter of fact, children who are roaming alone or with group of same age friends can be spotted easily in cinemas. Even parents who should guide their children, bring them along to watch unsuitable movies for their children's age. This resulted in children who can watch any kind of movies which are not appropriate for them. We know that movies that are distributed in Indonesia should have passed the censorship phase and translated to Indonesian. However there are still words or pictures that troubling for children to watch. So, adults need references to choose suitable movies for the children to watch.

Nowadays, most of the natural information retrieval is texts, like social media, movie subtitles, messages and any other. Texts are unstructured data that can't be processed by machine, because they can only be read by human (human learning). Text need to be treated so that it can change into structured one and read by the machine to find a lot of information from it. This method can be done by a method called text mining, which can be broadly defined as a knowledge-intensive process in which a user interacts with a document collection over time by using a suite of analysis tools to extract useful information from data sources through the identification and exploration of interesting patterns [2].

There are already many researches in any areas applying text mining method, [3] used text mining to movie rating classifications depended on the keywords in English, that are used in the movies and other attributes such as genres. They focused on the movies which have rating PG, PG-13 and R. They used text mining as preprocessing step to create prototype model based on decision tree (J48).

In this research, text mining will be applied to extract information from Indonesian language subtitle of 624 box office movies from 2006 until first half of 2016, to cluster the movies. Some words from the outcome of the method, will be grouped into three groups of words which are bad words, sexual words and terror words. These attributes and other attributes like IMDb user rating, genres and revenue, will be used to cluster movies using Partition Around Medoids (PAM) or *K*-Medoids algorithm. This algorithm is used since it is robust with outlier [4], while assumed that data consists of many movies with huge variances. The proxy matrix is necessary to do clustering, since data consist of mixed type data, [5] gower similarity is suitable to use in this research. Then after some numbers of cluster is created, the best number of clustered will be chosen based on silhouette coefficient value [6].

2. Backgrounds

In this section, we present the background of the works as following

2.1. Text Mining

Text mining is a method that can be defined as a process to extract useful information from groups of documents which consist of unstructured data, which is text. This method processes those unstructured data which can only be read by human into structured data, through some identification and exploration of some patterns.

According to [2] the most important process from text mining is the pre-process, which focused on identification and extraction representative feature for natural language documents to make the structured data. The text mining process usually needs input text making based on grammar of the language used in the documents [4]. This would be then followed by extraction of already structured data, evaluation and interpretation. The text mining usually used are text categorization, text clustering, text classification and any other analysis process.

2.2. Gower Similarity

Almost all of matrix proximity measurement can only be used for one type of data, but condition of data used in many researches is mostly consisting of mixed type data. Gower similarity is proximity matrix measurement that suitable for mixed type data [5].

Gower similarity coefficient s_{ij} compares two observations i and j, and is defined as follows:

$$s_{ij} = \sum_{k=1}^{n} s_{ijk}\omega_{ijk} \Big/ \sum_{k=1}^{n} \omega_{ijk} \qquad (1)$$

where s_{ijk} is the contribution provided by k-th attribute and ω_{ijk} is the weight for k-th attribute between two observation i and j. If s_{ijk} can be defined or there's no missing data, we can use $w_{ijk} = w_k$, where one value of weight is for one attribute. Usual combination of the weights that is used is 1, for all attribute. If there's missing data on dataset, then $w_k = 0$ for k-th attribute. Weights can be specified beforehand and the value can be assigned according to user.

Gower similarity concept is for each attribute with each type of data will use suitable measurement depends on the data type of attribute. Then linearly combined using weight for each attribute to calculate final distance. If attribute's type of data is continuous, following equation is used :

$$s_{ijk} = 1 - \left|x_{ik} - x_{jk}\right| \Big/ R_k \qquad (2)$$

where x_{ik} is i-th observation of k-th attribute, x_{jk} is j-th observation of k-th attribute and R_k is range of k-th attribute.

If attribute's type of data is nominal, the value of s_{ijk} is 1 if $x_{ik} = x_{jk}$ and valued 0 if $x_{ik} \neq x_{jk}$. Or in other words, $s_{ijk} = 1$ if i and j have same state for k-th attribute and , $s_{ijk} = 0$ if i and j have different state. If attribute's type of data is binary, Gower defines as the table below :

Table 1. Component of similarity and weight

	Value in k-th attribute			
i-th observation	+	+	-	-
j-th observation	+	-	+	-
s_{ijk}	1	0	0	0
ω_{ijk}	1	1	1	0

where "+" denotes that present and "-" denotes that absent, for the observation in k-th attribute.

2.3. Partition Around Medoids (PAM) Algorithm

Clustering is the process of dividing a set of data into some subsets, which called cluster, so that objects in a cluster are similar to each other and dissimilar with other objects in other clusters [4]. One of the major fundamental categories is partitional methods, which creates an initial partitioning then uses an iterative relocation technique that attempts to improve the partitioning by moving objects from one group to another. Partition Around Medoids (PAM) or usually called as K-Medoids is one of the partitional clustering methods that choose one of the object (medoids) as initial. Medoids has similar concept as mean or centroid, but Medoids is always an object from dataset. This method is more robust to outliers not like K-Means, because it minimize the absolute error.

Below is the algorithm of Partition Around Medoids (PAM) [4]:

(1) Arbitrarily choose k objects as initial representative objects of each cluster
(2) Assign remaining objects to the cluster with nearest representative objects
(3) Randomly select nonrepresentative objects as medoids candidate, O_{random}
(4) Calculate total cost, S_{random}, of the distance between remaining objects with initial object, O_j, and total cost, S_j, of the distance between remaining objects with O_{random}
(5) If $S_{random} < S_j$, then swap O_j with O_{random} as new medoids of k representative object
(6) Repeat step 2-5 until there is no change

2.4. Silhouette Coefficients

Silhouette coefficient is a method of interpretation and validation of consistency within clusters of data. The technique provides a succinct graphical representation of how well each object lies within its cluster [6]. Silhouette value used to measure how similar an object to its own cluster compared to other cluster. The value range from -1 to 1, where high value indicates that the object is well matched to its own cluster and poorly matched to other cluster.

After the data have been clustered by any methods into k cluster, we can evaluate with silhouette coefficient defined by the following equation :

$$s(i) = \frac{b(i) - a(i)}{\max\{a(i), b(i)\}} \tag{3}$$

where $a(i)$ is average of dissimilarity between i-th object with other object in its own cluster, $b(i)$ is the lowest average of dissimilarity between i-th object with other object in its other clusters. Cluster with the lowest average of dissimilarity is the neighbor of cluster where i-th object is and $s(i)$ is silhouette coefficient for i-th object. Coefficient for each object will be calculated and the average of silhouette value of entire dataset will be used to compare, to choose the best number of cluster.

3. Methodology

The most important and the key to do text mining is the data preprocessing. Indonesian language subtitle is used as it contents the words used in the movie. Since even after get through censorship process, there are still words that are uncomfortable for children to watch.

First, we collect the box office movies data from IMDb and subtitles data for each movie from various places like Subscene, Open Subtitle and other site that allow the user to free download the content. There are many records in the IMDb and we choose some of the records to help us clustering the movie. They are IMDb user rating, which the value given by the user to score movie after they watched it, Revenue, which is the income of a movie and genres of movies. Each movie can have more than one genre.

We manually extract these records of the movies from the website and inspect them. We strain the movies that released from 2006 until first half of 2016. The movies that later be used are box office movies with number of revenue more than US$ 50 million. Then, because each movie has possibly more than one genre, we divided each kind of the genre into each attribute consist of binary data.

Second is the preprocessing data of the Indonesian subtitles using text mining method, to prepare text data into structured data. The step of preprocessing, can be vary according to the needs of the user. In this research, follow the preprocessing step from [7]. Figure 1 shows the step of preprocessing.

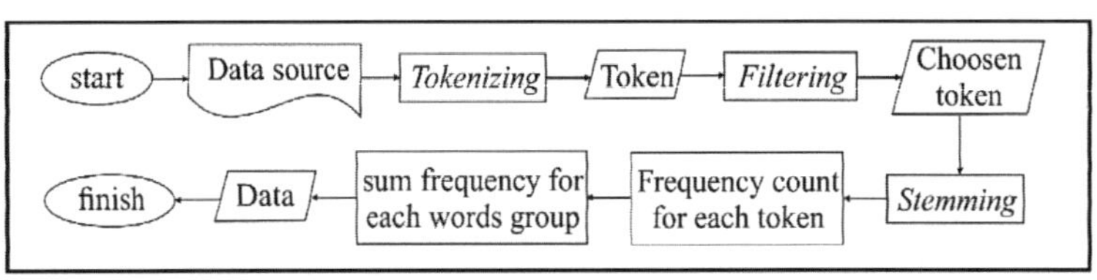

Figure 1. Step of preprocessing Indonesian subtitle texts

From group of documents sources, tokenizing step is to remove any objects beside letter, like number and punctuation, and transform all the letter into lowercase. For each sequence of letters will be called as token. Then we do the filtering, to remove any tokens that are included in stopwords. Stopwords are the words usually refered to the most common words in a language. Any group of words can be chosen as stopwords for a given purpose. In this research, we are using group of stopwords listed by [8].

The next step is stemming, to transform any tokens into their root words. In Indonesian language there are many conditions to transform a word into its root word according to the dictionary, like remove suffix or prefix. So we are using algorithm of stemming Indonesian language developed by

[9]. After all tokens in its root word formed, we count the the frequency of each words. Then it will be divided into three attribute which each attribute consist of group of chosen words. The three groups are group of words related to bad words, sexual words and terror words, the list of words is given in the *Appendices*. Then these attributes will be integrated with previous data containing information of the movies and can be analysed.

4. Results and Discussion

Based on the data obtained from previous step, a set containing 624 movies data is analysed. They included nine genres of popular movies which are Sci-Fi, Family, Fantasy, Adventure, Animation, Comedy, Drama, Horror and Action. The total frequency of appearance of three groups of words are calculated. Some most frequent words are shown in Figure 2.

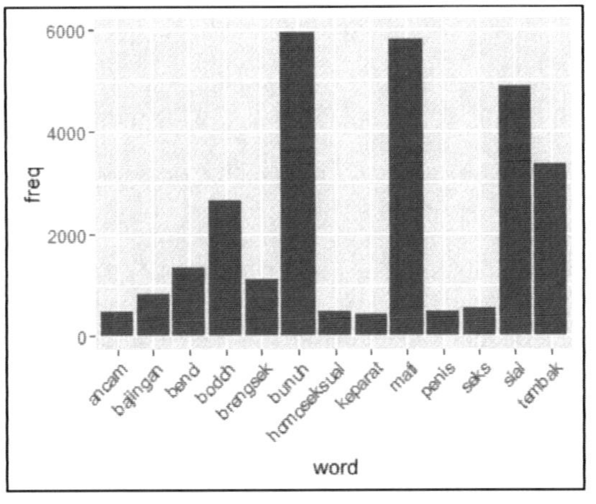

Figure 2. Most frequent words from group of words

Figure 2 shows the frequency of some words from the group words that appear at least 300 times in total from all movies. Two words of the highest frequency are words from terror words group, which are "bunuh" and "mati", that appear almost 6000 times. The third most used word is "sial' from bad words group that appear almost 5000 times. It can be seen that beside the censorship process, there are still many uncomfortable words making appearance in movies that can influence the filmgoers, especially children, if they watch them without any guidance from adults.

We need to divide these dataset into some clusters that has different characteristics, Partition Around Medoids (PAM) algorithm is used. This method is used to cluster the dataset into some variation number of cluster. Variations of number of clusters obtained from 2 clusters untiil 10 clusters. Then, silhouette coefficient is used to compare the effectiveness of the method in each variations. It can be seen in figure 3, that number of cluster which has the highest score of silhouette coefficient is the data with 5 clusters with the value of 0.29. So, this 5 clustered data is the one that will be used for the next analysis.

Since the dataset are high dimensional data, to help visualize the distribution of the clusters into two dimensional data using t-distributed stochastic neighbor embedding (t-SNE) [10]. It can help reduce the high dimensional data into two or three dimensions, which can then be visualized in a scatter plot. Specifically, it models each high-dimensional object by a two- or three-dimensional point in such a way that similar objects are modeled by nearby points and dissimilar objects are modeled by distant points.

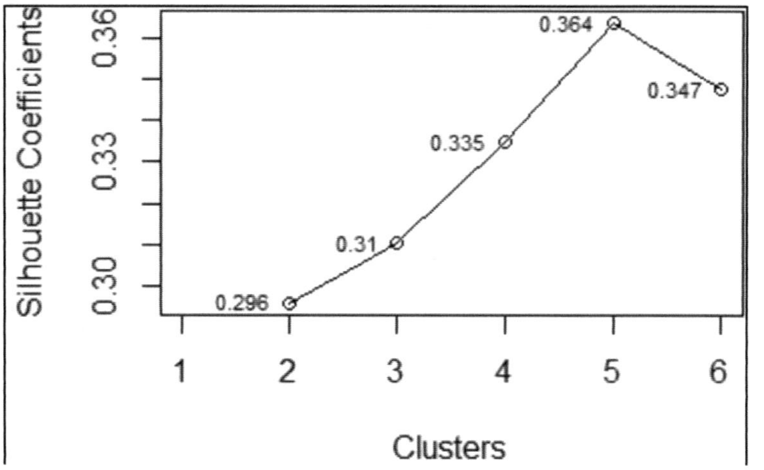

Figure 3. Comparison visualization between variations number of clusters

From figure 4, it can be seen that objects is grouped into 5 different clusters which represented in different colors. Each cluster has its own area consist of data objects included in their cluster. Though there are some objects that positioned quite far from the area of its cluster. It is reflected in silhouette value as evaluation for the making of cluster with the algorithm, 0.36, that is still quite far from value of 1. It can be happened because movies data have huge variation of information that need more advanced method to cluster, such as two-steps clustering or any other, or the number of clusters is too few. But just by seeing the plot, the clusters are quite distanced from another cluster and quite similar with other objects in its own cluster.

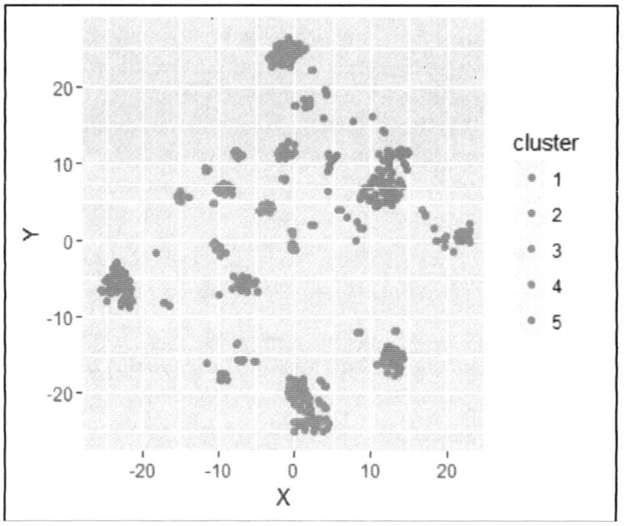

Figure 4. Scatter plot of objects data divided into 4 clusters

Here are the list of characteristics for each clusters shown in table 2.

Table 2. Characteristics for each clusters

Characteristics	Cluster				
	1	2	3	4	5
Number of movie	162	40	193	143	86
Average Revenue (US$ Million)	113.617	70.019	172.476	93.719	153.743
IMDb Rating	7.03	6.09	6.71	6.23	6.61
Movie with genre Sci-Fi	10.49%	15.00%	30.57%	0.70%	0.00%
Movie with genre Family	10.49%	0.00%	11.40%	13.29%	30.23%
Movie with genre Fantasy	9.26%	25.00%	18.65%	8.39%	2.33%
Movie with genre Adventure	14.20%	2.50%	81.87%	2.10%	76.74%
Movie with genre Animation	0.62%	0.00%	8.29%	0.70%	89.53%
Movie with genre Comedy	11.73%	7.50%	7.25%	100.00%	93.02%
Movie with genre Action	97.53%	2.50%	6.74%	18.18%	2.33%
Movie with genre Horror	12.35%	12.50%	89.64%	16.78%	3.49%
Movie with genre Drama	3.09%	95.00%	1.55%	0.00%	0.00%
Average Bad words	14.57	19.88	12.37	33.69	7.47
Average Sexual words	1.86	0.83	0.65	17.50	0.27
Average Terror words	30.04	30.60	36.56	26.81	16.19

From the characteristics above, shown that all cluster has similar average of IMDb user rating, so this attribute is not too significant to characterize this dataset. Cluster 4 has the highest average of bad words and sexual words appearance. It consist of movies with average revenue US$93 million and 100% of movies with Comedy genre. Movies with comedy genre itself actually vary from adult comedy until global comedy. Since the appearance of sexual words are the highest from other cluster combined, it can be assumed that the comedy movie in cluster 4 is mostly adult comedy.

In cluster 2, the characteristics of mostly the movie in this cluster, about 95%, have genre Drama and have income average about US$70 million or not quite popular. There are still many appearances of bad words and terror words in the movies in these cluster. So the movies in these cluster are still quite unsuitable for children.

In cluster 3, the characteristics of mostly the movie in this cluster, about 89%, have Horror genre and 82% have Adventure genre and have income average about US$172 million or popular movies. There are still many appearances of bad words and terror words in the movies in these cluster. So the movies in these cluster are still quite unsuitable for children.

Cluster 1 mostly consisted by movies with huge income with average US$113 million, it means that many people including children is interested in watching these movies. It mostly consisited by movies with Action genre, about 97%, which mostly require many violence scene followed with many terror words. So it's still best for adults to accompany their children while watching these kind of movies.

Cluster 5 has the least appearance of those three groups of words. Since this cluster mostly consist of high income movie which popular among children with genre Adventure with 77%, Animation with 89% and Comedy 93%. Animation movies are mostly targeted children, though there are still these bad words appearances, these movie are quite suitable for children. One of the most appeared genre movies are ones with Comedy movies, while quite similar with cluster 4 that have the most bad words and sexual words appearance, in this cluster can be assumed that the comedy contained is kind of comedy that appeal to children.

5. Conclusion

There are 5 cluster formed with PAM in this dataset, in which cluster 4 has the most appearances of bad words and sexual words, that movies in this cluster should not be watched by children without guidance or better avoiding them. The cluster with the least appearances of the words is cluster 5, which consisted of popular movies with children targeted genre like animation. So it is best to choose movie from cluster 5 for the children to watch. Even though movies in cluster 5 are quite comfortable to watch, there are still bad words that appear. The best way is to guide children to watch every movies they watch, even though the movie look harmless for them. So they can process the information correctly and will not influenced by bad things shown in movies.

References

[1] Motion Picture Assosiation of America (MPAA) 2015 *Theatrical Market Statistics*. http://www.mpaa.org/wp-content/uploads/2016/04/MPAA-Theatrical-Market-Statistics-2015_Final.pdf
[2] Feldman R and Sanger J 2007 *The text mining handbook: advanced approaches in analyzing unstructured data* (New York: Cambridge University Press)
[3] Kabinsingha S, Chindasorn S and Chantrapornchai C 2012 *International Journal of Engineering and Innovative Technology (IJEIT)* **2** *1*
[4] Han J, Kamber M and Pei J 2012 *Data mining concepts and techniques third edition*. (Massachusetts: Morgan Kaufmann Publishers.)
[5] Gower J C 1971 *Biometrics* **27** 4 857-871
[6] Rousseeuw P J 1987 *Computational and Applied Mathematics* **20** 53–65.
[7] Palupiningsih P 2013 *Pemodelan klasifikasi SMS berindikasi tindak penipuan menggunakan text mining* [Thesis] (Bogor: Bogor Agicultural University)
[8] Tala F Z 2003 A study of stemming effects on information retrieval in Bahasa Indonesia M.S. M.Sc. [Thesis] (Amsterdam: Universiteti van Amsterdam The Netherlands)
[9] Adriani M Asian J Nazief B Tahaghoghi SM. And Wiliams HE 2007 Stemming Indonesian: a confix-stripping approach. *ACM Transactions on Asian Language Information Processing*. **6** 4 1-33
[10] Van der Maaten L J P and Hinton G E 2008 *Journal of Machine Learning Research* **9** 2579–2605

Appendices

Table A1. List of bad words

bajingan	banci	bangsat	bedebah	bencong	bego	berandal
bodoh	brengsek	cebol	cecunguk	cundang	goblok	idiot
jalang	keparat	lacur	persetan	sial	sialan	tolol

Table A2. List of sexual words

anal	aseksual	asusila	berahi	biseksual	cabul	cumbu
ejakulasi	ereksi	erotika	erotis	erotisme	heteroseksual	homoseksual
homoseksualitas	kondom	lesbian	masturbasi	payudara	pemerkosa	penis
prostitusi	seks	sekstan	seksual	seksualitas	selingkuh	sensual
sensualitas	sodom	testis	vagina	zina		

Table A3. List of terror words

ancam	bantai	begal	benci	bengal	bengis	brutal	teror
bunuh	cambuk	cekik	culik	eksekusi	jagal	korupsi	teroris
koruptor	kriminal	mati	mutilasi	narapidana	overdosis	penggal	terorisme
psikopat	pukul	mesum	siksa	takut	tampar	tembak	

Application of hourly radar-gauge merging method for quantitative precipitation estimates

A Wardhana[1], H Pawitan[2] and B D Dasanto[2]

[1]Postgraduate student at Bogor Agricultural University, Indonesia
[2]Department of Geophysics and Meteorology, Bogor Agricultural University, Indonesia

Email : ali.wardhana@bmkg.go.id, vadya.wardhana@gmail.com, hpawitan@gmail.com, bambangdwi@ipb.ac.id

Abstract. Gridded quantitative precipitation estimation (QPE) of high spatial and temporal resolution is now being increasingly used for instance for hydrological modelling. The combination between observations from radar data and in-situ rain-gauge measurement is a popular mean to produce QPEs. Since the QPE generated from radar data only is considered to be not satisfactory, some corrections should be introduced in order to achieve better accuracy of the instant precipitation estimation. The correction method that were used were Mean Field Bias Correction (MFB) and static adjustment. MFB was applied by the assumption that the radar estimation was affected by uniform multiplicative error. The static adjustment was applied to correct radar rainfall data through plotting between radar rainfall and rain gauge so that the regression line between radar and rain gauge was approaching one. The purpose of radar QPE corrections was for hydrological modelling applications.

Keywords: QPE; weather radar; Mean Field Bias (MFB) correction; static adjustment; rain gauge measurement; hydrological modelling.

1. Introduction

Rainfall is one important variable in the success of applications such as hydrological modeling and flood forecasting for short-term studies. Resolution requirement of space-time rainfall estimation varies greatly depending on the application, the catchment area, types of rain events and the type of model used. Rainfall-runoff "*lumped*" models generally requiring area precipitation estimation while distributed model needs rainfall estimation based on grid. The shape of the catchment area and characteristics of rain influence temporal resolution of rainfall that will be used and might be varied starting from interval of 5 minutes to daily interval or even more [1].

It is not easy to estimate the value of precipitation if our calculation is only rely on one instrument. A rain gauge measurement although has a high accuracy in absolute values, the recorded data were often not distributed equally so that the representation level was still limited [2, 3]. One of the complementary instruments to estimate the spatial distribution of precipitation accurately is utilization of weather radar. Rainfall intensity can be estimated by weather radar through equation of reflectivity and rain intensity (rain rate).

There is increasing demand for accurate gridded quantitative precipitation estimation (QPE) with high spatial and temporal resolution for many applications. The scales of interest often very challenging because sometimes at scale of a single kilometer and only for some hours. Gridded QPE usually used in meteorology and climatology. On the other side, spatial precipitation analyses are increasingly used by other user groups as input data for models of natural system for instance for hydrological modelling that requires high-resolution QPE grids [2].

Evaluation and correction of the radar data is generally performed using a rain gauge measurement. Several methods have been proposed for correcting and validating radar precipitation products, from simple methods such as simple static adjustment, a multiplicative calibration, multi quadratic surface fitting to the more complex method e.g. geo statistics, co-Kriging or dynamic adjustment [1, 2, 4, 5].

The aim of this article was to perform a long-term verification with Mean Field Bias (MFB) correction method and static bias adjustment. Verification of these methods faced the problem that the real precipitation field was unknown. A traditional approach is to compare precipitation estimated with rain gauges measurement. Merging method was carried out on hourly basis accumulations. Several statistical analysis were performed followed by computational evaluation and comparing these two methods.

2. Radar and gauge observation

The output of the weather radar was reflectivity, Z, which depended on sum of sixth power of drop diameter. The real drop size distribution was highly variable depending on the type of precipitation, but because it is usually unknown, a default drop size distributions was used [6]. Various studies has been done to connect reflectivity radar data with rainfall intensity with the general equation:

$$Z = N(D)\,D6 = aRb \qquad (1)$$

where Z is the radar reflectivity (mm^6 m^{-3}); D is the diameter of raindrops (mm); $N(D)$ is the number of droplets in a diameter per cubic meter; R is the rainfall at the surface (mm h^{-1}); and a, b are coefficients, depending on the latitude, longitude, and kind of rainy season (convective, stratiform). Coefficient values (a, b) calculated using the distribution $N(D)$ or taken from the literature. Some Z-R relationship equation: $Z = 200R^{1.6}$ is a Marshall and Palmer equation for stratiform rain; $Z = 31R^{1.71}$ is Blanchard equation for orographic rain; $Z = 500R^{1.5}$ for the thunderstorm; and $Z = 350R^{1.4}$ for the convective rain.

Indonesia Meteorological, Geophysical and Climatological Agency or BMKG operates the C-Band weather radar at Tangerang District. The radar observation routinely used for operational short-term precipitation forecasting and for detecting the severe thunderstorm. BMKG also operates five rain gauges with hourly basis data accumulation surrounding Tangerang weather radar site (figure1).

Figure 1. Location of radar site (yellow pin) and five rain gauges (red balloon).

3. Description of the methods

Comparing between radar rainfall data and rain gauges measurement become a challenge due to the different sampling size of the instruments. Radar measurement volume can be several kilometers wide and thick, while the measurement area of a gauge is 400 cm^2 (weighting gauges) or 100 cm^3 (optical instruments). During the case study periods (June 2013 – June 2016), radar data were available at 10 minutes time interval and total of five rain gauges that provide detailed point information, which was used to correct the radar rainfall field. The spatial sampling issue is a crucial importance when the radar areal estimation is compared with rain gauge point measurement. A systematic difference or "bias" between radar rainfall and rain gauge measurement can be progressively removed using information provided by rain gauges. This is performed through an adjustment factor that is estimated as the ratio of the accumulated rain gauge rainfall, G, and the accumulated radar rainfall, R. The simplest method of bias correction is to multiply the R values by G/R. Note that the expected value of G/R is 1.0 [7].

3.1. Mean field bias correction (MFB)

A mean field bias correction algorithm aims to reduce the gross error in the truncated 1-h precipitation accumulation as observed against the automatic rain gauge observation. The assumption here is that the radar estimation affected by uniform multiplicative error. The bias-adjusted precipitation estimation calculated on hourly basis from the uncorrected radar data:

$$C_{MFB} = \frac{\sum_{i=1}^{n} G_i}{\sum_{i=1}^{n} R_i} \tag{2}$$

$$Ri_{adj} = C_{MFB} . Ri \tag{3}$$

where C_{MFB} is the correction factor, n is the number of valid radar-gauge pairs, G_i is the measured rainfall at gauge *i* and R_i is radar measured rainfall values coincident with *i* gauge location. Ri_{adj} is radar bias adjusted at *i* gauge location and R_i is uncorrected (raw) radar data before correction. The radar rainfall measured at the gauge taken as the spatial integration of rainfall for the radar grid above the rain gauge location. The correction factors are obtained at a set time step (e.g., hourly, daily, etc.). The correction factor then applied to the entire spatial domain of the radar, as it is multiplied with the radar value at each grid location in order to develop the adjusted radar image. Only the pairs for which both *G* and *R* exceed 1 mm h^{-1} are considered as valid. This minimizes discretization errors and the influence of anomalous propagation. MFB correction has become a widely recognized and applied technique for adjusting radar rainfall grids due to its simplicity and ease in implementation in near-real time.

3.2. Static gauge-adjustment

This method attempts to improve the radar data by identifying the long-term bias of a radar dataset then correcting for it. For the *i*'th rain gauge, the ratio bias, *Bi*, defined as the long-term arithmetic mean ratio of rain gauge and radar rainfall estimates over n 60 minutes time interval:

$$B_i = \frac{1}{n} \frac{\sum R_g^i}{\sum R_r^i} \tag{4}$$

where R_g^i is the raw radar estimation for the grid square coincident with the *i*'th rain gauge. The ratio is only calculated if both $R_g^i$and R_r^i are exceeds 1 mm h^{-1}. Averaging this over the *N* rain gauges gives the ratio bias, *B*, of the radar as:

$$B = \frac{1}{N} \sum B_i \tag{5}$$

Applying this factor to the entire radar dataset gives the static gauge adjusted radar rainfall estimates. Figure 2 showing the whole procedure sequence.

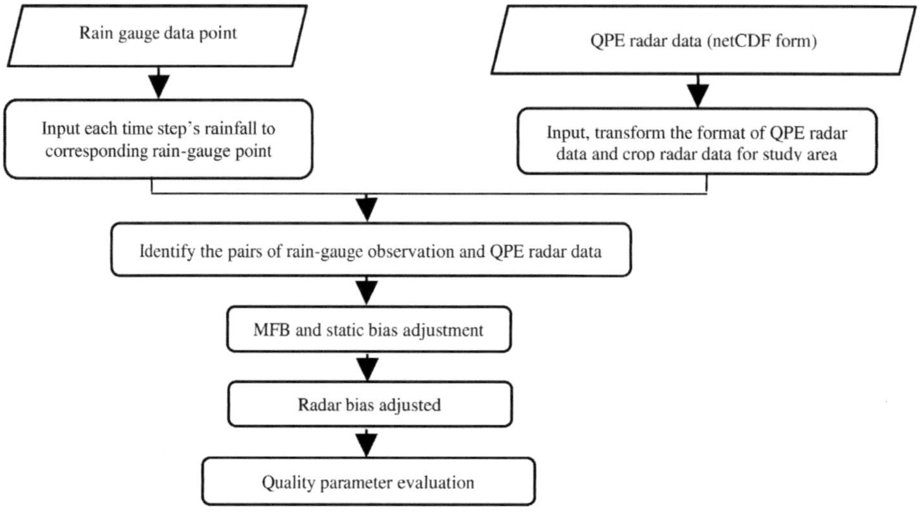

Figure 2. Work flowchart of the radar bias adjustment.

4. Result and Discussion

4.1. Mean field bias correction (MFLong-term verification methodology

The performance of the different methods verified against to surface rain gauge observation of rainfall accumulation data. The testing period extends from June 2013 – May 2016. The statistical quantification commonly based on the Root Mean Square Error (RMSE) and Mean Absolute Error (MAE), calculated with these data sets:

$$RMSE = \sqrt{\frac{\sum_{i=1}^{n}(R_i - G_i)^2}{N}} \quad (6)$$

$$MAE = \frac{\sum_{i=1}^{n}(|R_i - G_i|)}{N} \quad (7)$$

4.2. Results

The correction methodology with MFB was applied for three years data from five locations. As shown in table 1, the bias, RMSE and MAE of the original radar data decreased significantly after MFB correction. RMSE reduced about 20 – 25% for three locations (Serang, Darmaga and Pondok Betung) and about 45% for Kemayoran. In the contrary, for Citeko the RMSE increased about 25%. For MAE, all of five locations showed that MAE decreased about 28 % - 50%. As shown in figure 3, the MFB correction method provided the greatest reduction in error as compared to the raw radar data before MFB correction. Kemayoran had a significant increasing with R^2 (from 0.12 to became 0.60) while Citeko had a smallest R^2 (from 0.0 to became 0.17). A visualization for MFB correction method can be seen at figure 3.

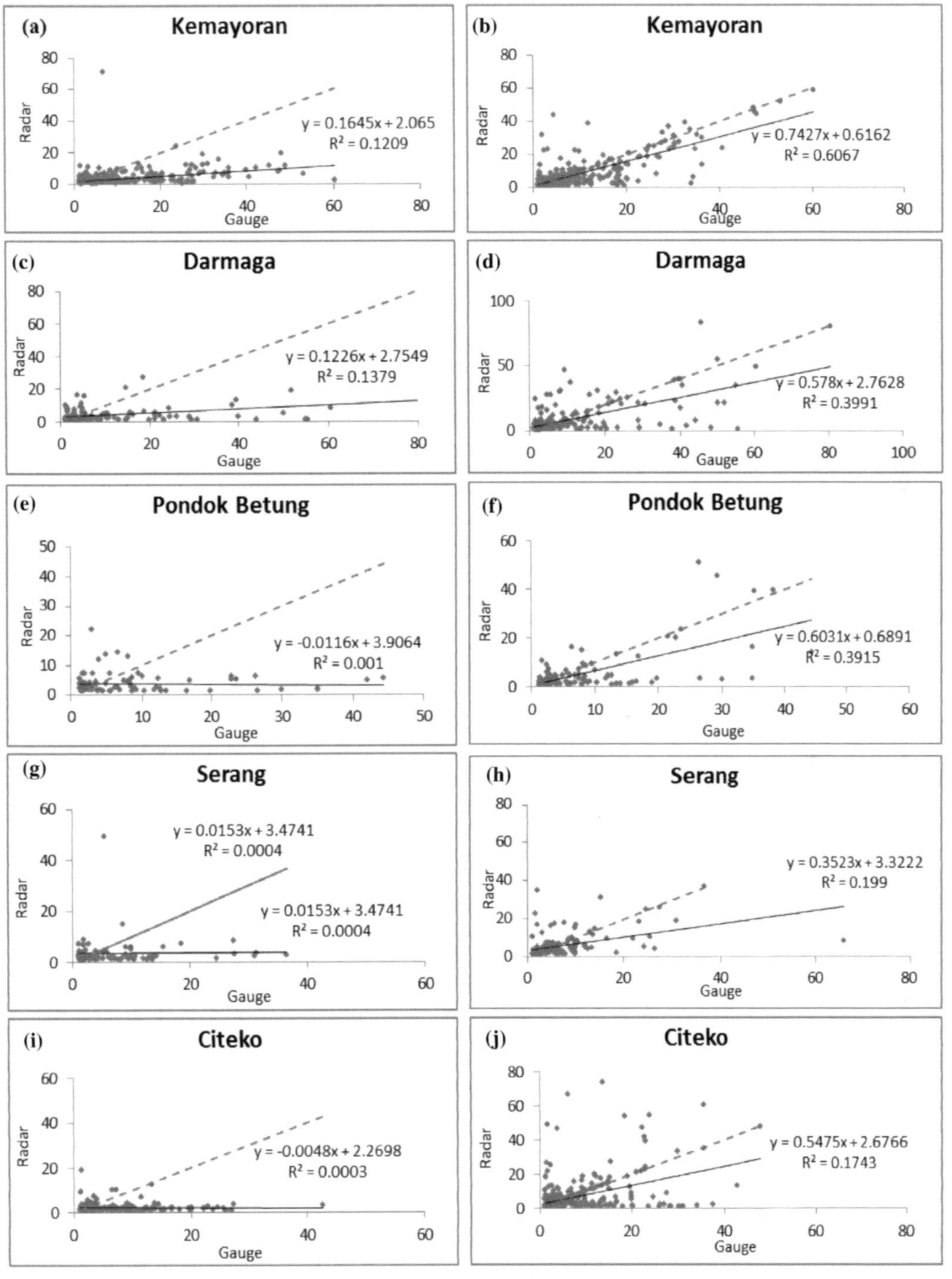

Figure 3. Plots of QPE radar data (x-axis) against observed rain-gauge values (y-axis) for 5 locations. The continuous black line is a linear fit to the data set and the dashed red line represents the perfect 1:1 fit in the plots. Left graph (a, c, e, g and i) showing data before MFB corrections while right graph (b, d, f, h and j) after MFB correction.

Table 1. Error statistics for non-MFB and MFB correction between five rain gauges and radar data.

	Serang		Kemayoran		Darmaga		Pondok Betung		Citeko	
	Non MFB	With MFB	Non MFB	With MFB	Non MFB	With MFB	Non MFB	With MFB	Non MFB	With MFB
Bias	3.4	1.8	3.6	1.8	4.5	2.4	4.0	2.6	3.7	2.3
RMSE	10.4	8.3	12.4	6.6	15.7	11.8	11.7	8.6	8.0	10.0
MAE	6.5	4.1	7.9	3.9	9.2	6.6	7.3	5.2	4.9	4.7

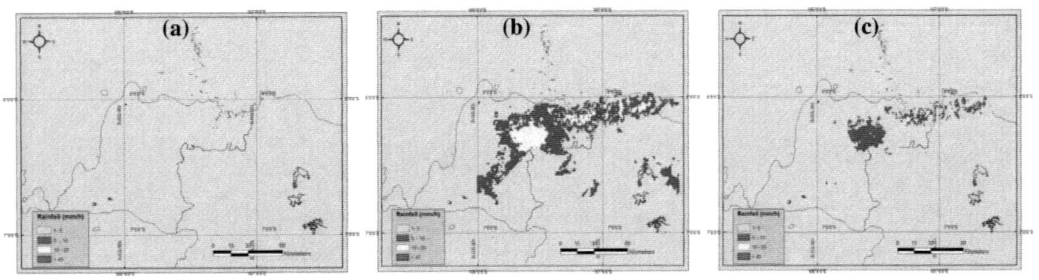

Figure 4. One hour raw radar rainfall accumulations data (a), rainfall accumulations derived using MFB correction (b) and rainfall accumulation derived using static gauge adjustment at 21 UTC January 1st, 2014 (c).

Over the 3 years case study, the ratio bias of the QPE radar data was calculated to be approximately 3.4. The monthly ratio biases calculated using individual rain-gauges, and as a mean over all rain-gauges are presented in figure 5 along with the ratio biases for the whole period (also listed in table 2). There does not appear to be a seasonal variation but there is some positive association with range and elevation with the higher rain-gauges having larger ratio biases.

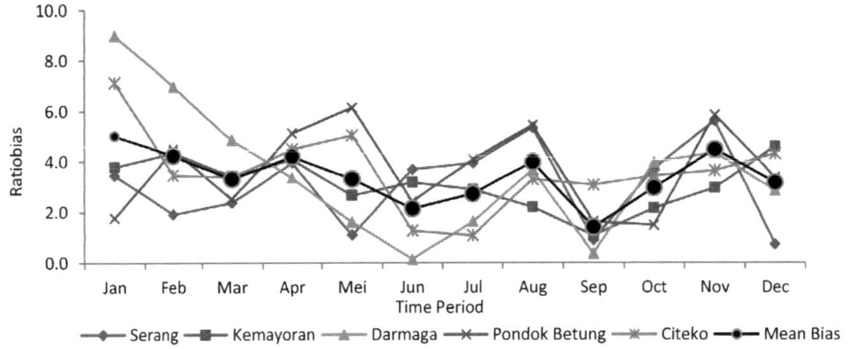

Figure 5. Monthly ratio bias for QPE radar data coincident with each rain-gauge locations over the period June 2013 to June 2016 . The ratio bias averaged over all rain-gauge is denoted by black solid dots.

Table 2. Rain-gauges used for the case study and ratio bias of radar data at rain-gauge location.

Location	Elevation (meter above sea level)	Time Period												Mean Bias
		Jan	Feb	Mar	Apr	Mei	Jun	Jul	Aug	Sep	Oct	Nov	Dec	
Serang	37	3.4	1.9	2.4	4.0	1.1	3.7	3.9	5.3	0.9	3.7	5.6	0.7	3.1
Kemayoran	5	3.8	4.4	3.4	4.1	2.7	3.2	2.9	2.2	1.1	2.2	3.0	4.6	3.1
Darmaga	29	9.0	7.0	4.9	3.4	1.6	0.1	1.6	3.7	0.4	4.0	4.3	2.8	3.6
Pondok Betung	169	1.8	4.5	2.5	5.1	6.1	2.4	4.1	5.4	1.6	1.5	5.8	3.4	3.7
Citeko	973	7.1	3.5	3.4	4.5	5.0	1.3	1.1	3.3	3.1	3.4	3.6	4.3	3.6
Mean Bias		5.0	4.2	3.3	4.2	3.3	2.1	2.7	4.0	1.4	3.0	4.5	3.2	3.4

The RMSE results for the static bias adjustment method show that there are high fluctuation for the five locations with the highest in Darmaga at August and the smallest is ini Kemayoran at September.

Table 3. Root mean square error (RMSE) with *Static gauge-adjustment* method for five locations.

Location	**Time Period**											
	Jan	Feb	Mar	Apr	May	Jun	Jul	Aug	Sep	Oct	Nov	Dec
Serang	5.9	3.8	15.9	12.7	1.8	5.8	9.3	15.6	4.5	8.0	17.1	0.4
Kemayoran	12.2	11.9	13.1	20.7	11.8	10.0	8.8	8.6	0.5	7.5	9.2	13.1
Darmaga	19.2	20.3	8.0	5.8	0.9	8.0	11.6	24.5	8.4	5.7	16.4	5.2
Pondok Betung	4.3	7.0	15.4	18.3	13.6	10.2	9.4	12.2	3.2	2.8	15.7	3.5
Citeko	10.6	4.6	5.1	9.2	8.1	1.7	1.4	7.5	2.7	6.9	9.3	10.6

5. Conclusion

Two methods combining rainfall estimation from a QPE radar product and an automatic rain gauge network have been implemented. A 3 (three) years QPE radar was carried out against 5 rain gauges location of hourly measurement. Several statistics have been computed to evaluate the performance of the radar-gauge merging methods. The result pointed out that mean field bias correction can significantly reduce the error of the radar estimation. Nevertheless, there is a clear benefit of using a spatial correction factor. Based upon this study the best method is the mean field bias correction which makes use of the radar as secondary information to improve the spatial interpolation of gauge values.

References

[1] Cole S J and Moore R J 2008 *Journal of Hydrology* **358** 159–181
[2] Zhang X and Srinivasan R 2010 *Journal of Environmental Modelling & Software* **25** 1781-88
[3] Bianchi B, Van L P J, Hogan R J and Berne A 2013 *Journal of Hydrometeorology* **14**(6) 1897-1909
[4] Wood S J, Jones D A and Moore R J 2000 *Hydrology and Earth System Sciences* **4**(4) 545–554
[5] Löwe R, Thorndahl S, Mikkelsen P S, Rasmussen M R and Madsen H 2014 *Journal of Hydrology* **512** 397–407
[6] Vanaja J 2013 *Rainfall-runoff modeling using Doppler weather radar data* (Chennai (IN). Anna University)
[7] Chumcheana S, Seedb A and Sharmac A 2005 *Journal of Hydrology* **317** 123–137

Data Cubes Integration in Spatial OLAP for Agricultural Commodities

A I Putri[1] and I S Sitanggang[1]

[1]Department of Computer Science, Bogor Agricultural University, Jalan Meranti Wing 20 level 5 Kampus IPB, Bogor, West Java, Indonesia

Email: ananda.putri152@gmail.com, imas.sitanggang@ipb.ac.id

Abstract. Ministry of Agriculture Indonesia collects data of agricultural commodities in Indonesia in the annual period. Agricultural commodities data include food crops, horticulture, plantations, and livestock. The data are available in the spreadsheet format. This study developed data cubes for food crops, plantations, and livestock using the galaxy schema of data warehouse and integrated the data cubes into the SOLAP Horticulture using SpagoBI. SOLAP is useful for data analysis and data visualization. The application displays agricultural commodities data in form of crosstab and chart. This study also developed the location intelligence module that visualizes agricultural commodities data on the map. The system was tested using the black box approach. The result showed that main functions including roll up, drill down, slice, dice, and pivot work properly. This application is expected to enable users to easily obtain data summaries of agricultural commodities.

1. Introduction

Agricultural commodities in Indonesia are very diverse which are divided into several sectors: food crops, horticulture, plantations, and livestock. Ministry of Agriculture Indonesia collects data of agricultural commodities in Indonesia in the annual period from the national to the district level. The data are available in the spreadsheet format so that it is not easy for users to get historical data summaries of agricultural commodities. Every year the data of agricultural commodities continue growing. The growth of data will generate datasets in very large quantities. Therefore, we need a tool that can automatically summarize information and transform data into a useful knowledge [1]. Data warehouse is a long-term storage space to store a collection of data from multiple sources in a scheme to help users in decision making [1]. A data warehouse can store data in a multidimensional space to make a data cube. A data warehouse uses the multidimensional data model and Online Analytical Processing (OLAP) to explore data in detail [2].

Spatial data are data that contain geographic information of a location on the earth's surface. A spatial data warehouse is a collection of spatial data which is subject oriented, integrated, time variant, and non-volatile to make decisions [1]. Development of a spatial data warehouse to integrate thematic and geographic data produce a spatial data cube for analyzing multidimensional spatial data [1]. Nowadays, there is a system which is called Spatial Online Analysis Processing (SOLAP) that can be used for data analysis and data visualization. A SOLAP is integrated with a spatial data warehouse to help users in obtaining summaries of historical data easily and quickly.

This study developed spatial data cubes for food crops, plantations, and livestock using the galaxy schema of data warehouse and integrated the data cubes into the SOLAP Horticulture using SpagoBI. SpagoBI was utilized to create data cubes. SpagoBI is a complete, open source, and flexible business intelligence tool. SpagoBI is complete because it covers all the needs of business intelligence with innovative solutions and provides a variety of analysis tools [3]. This study developed a SOLAP and a location intelligence module for agricultural commodities. The application provides data summaries represented in crosstab, chart or dynamic maps. We expect that the application can facilitate users to get data summaries of food crops, plantations or livestock appropriately.

2. Data and Method

2.1. Data

The data of agricultural commodities used in this study are food crops, plantations, and livestocks from 2000 to 2014 at the district level in Indonesia. The data were obtained from the website of Ministry of Agriculture Indonesia at http://aplikasi.pertanian.go.id/bdsp/newkom.asp in Microsoft Excel format (.xls) [4]. Attributes of the data include types of commodity, location, time, and status of the data. The measures in food crops data are harvested area (ha), production (ton), and productivity (kg/ha). The measures in plantations data are acreage (ha), production (ton), and productivity (kg/ha) and the measures in livestocks data are population (tail) and production (ton). Figure 1 shows the data of harvested area of food crop in Majalengka, West Java Province in the period of 2010-2019.

Hasil Pencarian Berdasarkan Komoditi

Sub Sektor	:	Tanaman Pangan
Indikator	:	Luas Panen
Level	:	Kabupaten/Kota
Provinsi	:	Jawa Barat
Kabupaten/Kota	:	Kab. Majalengka
Status Angka	:	Angka tetap
Tahun	:	2010-2019

Komoditi	Satuan	2010	2011	2012	2013	2014	2015	2016	2017	2018	2019
Jagung	Ha	18,577.00	16,062.00	19,738.00	17,149.00	15,943.00	0.00	0.00	0.00	0.00	0.00
Kacang Hijau	Ha	1,473.00	1,601.00	965.00	811.00	1,134.00	0.00	0.00	0.00	0.00	0.00
Kacang Tanah	Ha	1,112.00	962.00	696.00	679.00	553.00	0.00	0.00	0.00	0.00	0.00
Kedelai	Ha	2,347.00	1,514.00	1,428.00	507.00	1,344.00	0.00	0.00	0.00	0.00	0.00
Padi	Ha	103,392.00	98,289.00	101,279.00	108,008.00	105,558.00	0.00	0.00	0.00	0.00	0.00
Padi Ladang	Ha	2,284.00	1,534.00	1,818.00	1,887.00	2,758.00	0.00	0.00	0.00	0.00	0.00
Padi Sawah	Ha	101,108.00	96,755.00	99,461.00	106,121.00	102,800.00	0.00	0.00	0.00	0.00	0.00
Ubijalar	Ha	732.00	671.00	727.00	621.00	485.00	0.00	0.00	0.00	0.00	0.00
Ubikayu / Ketela Pohon	Ha	1,538.00	1,169.00	1,123.00	920.00	581.00	0.00	0.00	0.00	0.00	0.00

Sumber Data : Kementerian Pertanian

Figure 1. Food crop data.

This study was conducted in several steps including design of data warehouse, implementation of data warehouse, design of SOLAP, new data cubes integration, implementation of SOLAP for new data cubes, implementation of location intelligence for new data cubes, as well as SOLAP and location intelligence testing.

2.2. Design of data warehouse

A data warehouse is usually modeled by a multidimensional data structure called a data cube. The dimension is a collection of attributes in the scheme and the fact is value of several measurements [1]. The multidimensional data scheme used in this study is the galaxy scheme. In this scheme, there are some dimension tables and fact tables that can share dimension tables so the scheme contains some star schemas that are connected each other. In the galaxy scheme, one dimension table can be used simultaneously by multiple fact tables. The fact table contains some measures and some foreign keys that connect with the primary key in a dimension table [5]. A dimension table contains detailed data and descriptive text that describes a foreign key in the fact table.

2.3. Implementation of data warehouse
In implementation of the data warehouse, data are managed using the DBMS PostgreSQL. Multidimensional data consist of dimensions, measure, and facts. The data are stored in fact tables and dimension tables. Data cubes are implemented in SpagoBI Studio.

2.4. Design of SOLAP and new data cubes integration
SpagoBI Studio was used to design and to construct data cubes of food crops, plantations and live stocks. The scheme of data warehouse adopted is the galaxy scheme. The galaxy scheme can integrate new data cubes with dimension tables that have been built in the SOLAP for horticultural crops.

2.5. Implementation of SOLAP for new data cubes
SOLAP for agricultural commodities was built on the web application using SpagoBI as the OLAP server. The new data cubes stored in SpagoBI Studio are connected with SpagoBI Server to perform OLAP operations.

2.6. Implementation of location intelligence for new data cubes
Implementation of location intelligence was carried in SpagoBI Server. Maps for visualization of agriculture data are stored in GeoServer. SpagoBI has two types of dynamic map visualizations namely map zone and map point. A map zone is a polygon map. Each polygon has a different color representing different value of measures. The higher of measure value, the darker color of the polygon. A map point is a map showed by dots. Each point has a different size. The higher of measure value, the greater size of the dots on a map point.

2.7. SOLAP and location intelligence testing
Data verification was done before system testing to evaluate the data provided by the SOLAP. In addition, SOLAP testing was performed on OLAP operations including roll up, drill down, slice, dice, and pivot on each data cube. SOLAP testing was also conducted to evaluate data visualization in crosstab and chart for selected dimensions and measure. Some functions in the location intelligence were tested including displaying a map in a polygon or a point, displaying a map according to the selected indicators, and data filter on a map. System testing was done to find the functions that could not work properly. Moreover, testing results are important for further system development.

3. Result and Discussion

3.1. Preprocessing of data
Data of food crops, plantations, and livestock were obtained from the website of Ministry of Agriculture Indonesia. The data include agriculture commodities from 2000 to 2014 at the district level. Status number of the data used is fixed numbers. Measures of food crops, plantations and livestock have different units so data conversion on the measures is required. Number of records in food crops, plantations crops and livestock are 59,997, 80,140 and 199,610 respectively.

3.2. Design of data warehouse
The data warehouse scheme adopted in this study is the galaxy scheme. Dimension tables that are connected to the new fact tables are time, location, and status of data. This study created three new dimension tables and three fact tables. Three dimensions are food crop commodities, plantation commodities, and livestock commodities. Three fact tables are food crop fact, plantations fact, and livestock fact. Therefore the data warehouse of agriculture commodities has seven dimensions and four fact tables. The SOLAP for agricultural commodities has four data cubes including horticultural crops, food crops, plantations and livestock. Figure 2 shows the galaxy schema for data cubes of food crops, plantations and livestock.

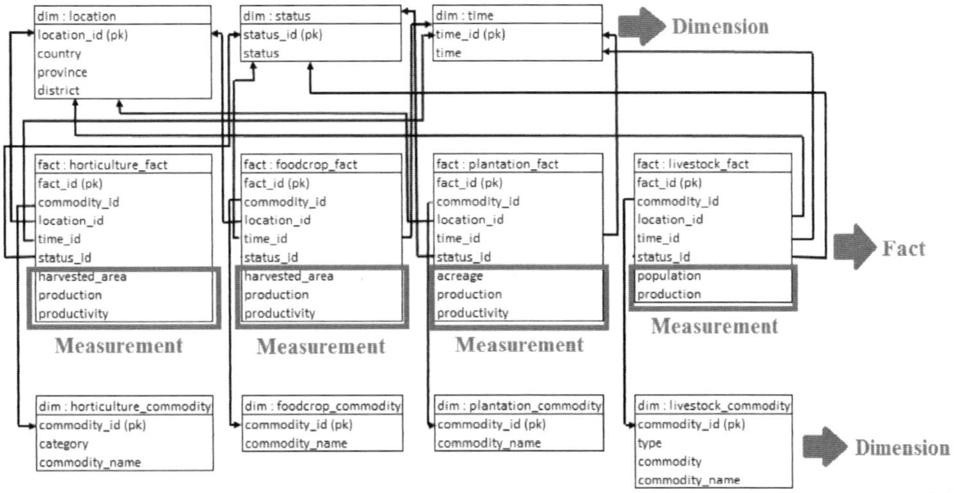

Figure 2. Galaxy scheme for data cubes of horticulture, food crop, plantation, and livestock.

3.3. Implementation of data warehouse

Fact tables and dimension tables in the data warehouse of agriculture commodities are stored on DBMS PostgreSQL. The data warehouse consists of seven dimension tables and four facts tables. Seven dimension tables are location, status, time, horticulture commodities (horticulture_commodity), food crop commodities (foodcrop_commodity), plantation commodities (plantation_commodity), and livestock commodities (livestock_commodity). Four fact tables are horticultures (horticulture_fact), food crops (foodcrop_fact), plantations (plantation_fact), and livestocks (livestock_fact). Dimension tables were implemented before the fact tables were created because each dimension table contains a primary key that is used as a foreign key in the fact table.

3.4. Design of SOLAP and new data cubes integration

This study designs SOLAP and data cubes using SpagoBI Studio. The dimension tables and fact tables stored in the DBMS PostgreSQL are connected to SpagoBI Studio using the PostgreSQL JDBC driver. Data cubes were created using SpagoBI Studio by assigning the type of dimension to dimension tables, fact tables to cubes, and measures in fact tables to measure.

3.5. Implementation of SOLAP for new data cubes

SOLAP was implemented by creating a server in SpagoBI Studio and then the server was connected to the data warehouse of agriculture commodities using SpagoBI Server. Furthermore, the OLAP template was made in SpagoBI Studio with the xml format for the multidimensional data model. The OLAP template is deployed into SpagoBI Server.

The result of SOLAP implementation is a spatial data warehouse of agriculture commodities which is available in the SpagoBI Server. OLAP operations such as roll up, drill down, slice, dice, and pivot are performed on the agriculture crops data and the results are displayed in crosstab and chart. The crosstab and chart can be downloaded into Microsoft Excel or PDF formats. Figure 3 shows the setting for OLAP operations.

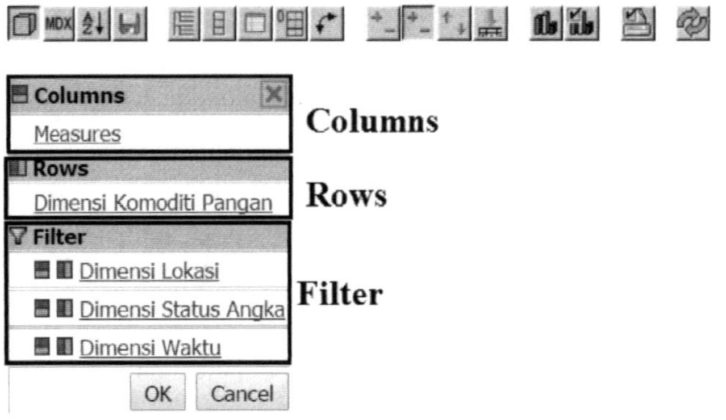

Figure 3. OLAP navigator.

Figure 3 shows the view of OLAP navigator. In the OLAP navigator, users can select measures and dimensions including the dimension of the location, commodity, status of data, and time. The dimensions and measure can be displayed as columns or rows in crosstab according to user's needs. The filter is used to select contents of the selected dimension. Figure 4 provides the result of SOLAP operation for food crops with two dimensions namely commodity and location from 2000 to 2014. Moreover, an example of SOLAP operation in harvested area of food crops in West Java from 2000 to 2014 is provided in figure 5.

		Measures		
Dimensi Komoditi Pangan	Dimensi Lokasi	Luas Panen (Ha)	Produktivitas (Ku/Ha)	Produksi (Ton)
−All Dimensi Komoditi Pangan	+All Dimensi Lokasi	452,369,204	3,250,682	2,142,393,875
[12] Jagung	+All Dimensi Lokasi	1,202,300	1,901	5,737,382
[12] Kacang Hijau	+All Dimensi Lokasi	90,016	421	96,465
[12] Kacang Tanah	+All Dimensi Lokasi	140,134	705	188,738
[12] Kedelai	+All Dimensi Lokasi	157,208	7,362	3,635,550
[12] Padi	+All Dimensi Lokasi	2,104,631	25,623	12,512,298
[12] Padi Ladang	+All Dimensi Lokasi	150,527	8,121	640,584
[12] Padi Sawah	+All Dimensi Lokasi	1,954,175	19,188	11,872,086
[12] Ubijalar	+All Dimensi Lokasi	13,823	7,337	315,477
[12] Ubikayu / Ketela Pohon	+All Dimensi Lokasi	159,117	9,491	3,677,400
Jagung	−All Dimensi Lokasi	47,413,416	210,446	173,757,133
	−jawa	24,353,778	96,863	98,326,616
	+banten	79,139	2,432	218,516
	+daerah istimewa yogyakarta	926,529	2,723	3,020,064
	+daerah khusus ibukota jakarta	1,634	361	529
	−jawa barat	1,672,351	54,583	7,013,854
	Kab. Bandung	132,940	702	491,688
	Kab. Bandung Barat	24,445	321	140,233
Hierarchy	Kab. Bekasi	510	321	737
	Kab. Bogor	2,847	212	11,426
	Kab. Ciamis	66,797	712	314,222
	Kab. Cianjur	88,098	569	286,874
	Kab. Cirebon	3,092	614	8,135
	Kab. Garut	705,315	785	3,327,015
	Kab. Indramayu	3,262	548	8,741
	Kab. Karawang	3,079	41,539	9,847
	Kab. Kuningan	60,278	601	194,605

Figure 4. Food crops data summary in crosstab.

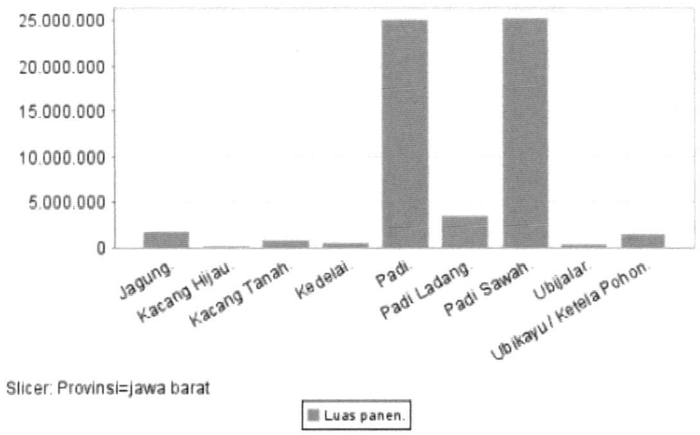

Figure 5. Food crops data summary in chart.

3.6. Implementation of location intelligence for new data cubes

The initial step in location intelligence implementation is preparing and storing the map of Indonesia in GeoServer in the shape file format (shp). The map of Indonesia was obtained from Geospatial Information Agency (BIG). Furthermore, the map is displayed in the SpagoBI server. The second step is creating datasets for each data cube in SpagoBI Server to be visualized on the map of Indonesia. Data retrieved from the data warehouse of agriculture commodities are commodity names, id locations, provinces, countries, times, harvested area, crop production and crop productivity. This study utilizes the dynamic map for data visualization. The dynamic map displays data in polygon maps (map zone) and point maps. The location intelligence requires a template to connect GeoServer to SpagoBI Server. The template is represented in the JSON format. In the template there are indicators to be presented on the map. The template is used to match the location id in GeoServer with the location id in the dataset. Figure 6 shows the result of the location intelligence module for harvested area of food crops that is represented in the map zone for the period of 2000-2014. The darker color of the polygon, the greater value of the harvested area in the region. In data visualization, there are some functions that can be used by users to explore the map. These functions include indicators, filters, and legends. The indicator has a function to show the measure. Filter is used to filter the data to be displayed on the map. Legend provides information related to different colors on the map.

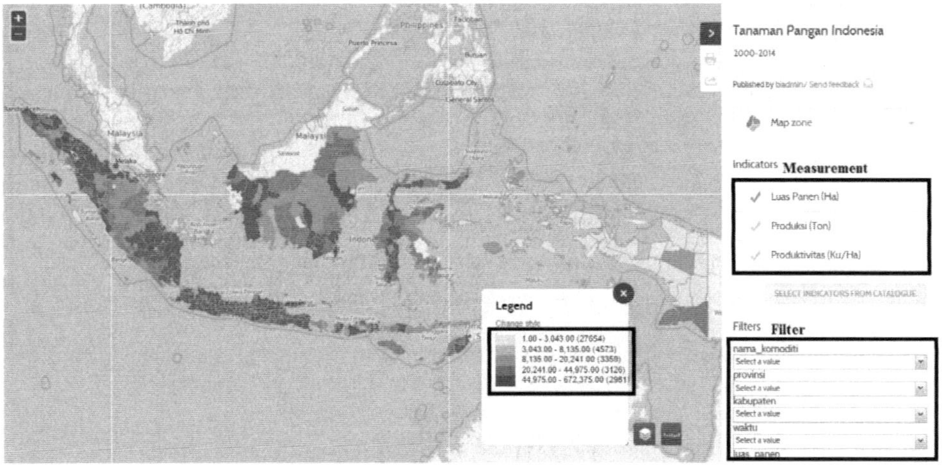

Figure 6. Representation of food crops data summary in a map zone.

Figure 7 shows the result of the location intelligence module for harvested area of food crops that is represented in the map point for the period of 2000-2014. In the figure 7, size of dots relates to the value of the harvested area in the region.

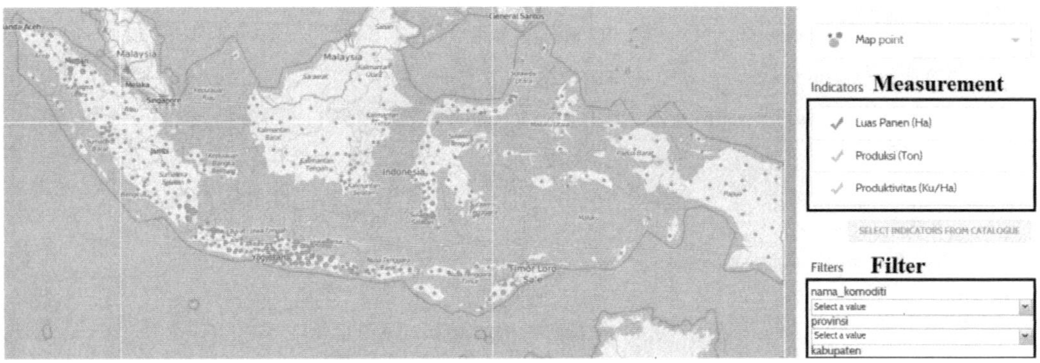

Figure 7. Food crops data in a map point.

3.7. SOLAP and location intelligence testing

Data verification was done before SOLAP and location intelligence testing. The result shows that data summaries provided by the SOLAP are valid and correspond to the original data that are available in website of Ministry of Agriculture. The testing of SOLAP for agricultural commodities and the location intelligence module was performed using the black box approach. Table 1 shows the test scenarios for SOLAP for agricultural commodities, whereas table 2 shows the test scenarios for the location intelligence module. The testing results on the SOLAP for agricultural commodities and the location intelligence module show that all functions on the system work properly.

Table 1. Testing results of the SOLAP for agricultural commodities.

No	Task	Scenario	Expected result	Result
1	Selecting data for certain dimension and measure	Examine each dimension of data cube	Data are displayed according to selected dimensions	The task works properly
2	Displaying data in crosstab	Selecting one or more dimension(s) in OLAP Navigator and click "ok" to display the crosstab	Data are represented in crosstab	The task works properly
3	Displaying data in chart	Selecting each dimension of data cube and choose "chart"	Data are represented in chart	The task works properly
4	Displaying data as the result of OLAP operations on each data cube	Selecting dimensions according to OLAP operations	Data are displayed according to selected OLAP operations	The task works properly

Table 2. Testing results of the location intelligence module.

No	Task	Scenario	Expected Result	Result
1	Displaying map zone for each data cube	Selecting map zone	Different color of polygons according to measure values for each district	The task works properly
2	Displaying map point for each data cube	Selecting map point	Different size of points according to measure values for each district	The task works properly
3	Filtering in map	Selecting one or more filter(s)	A map is displayed according to the filtering result	The task works properly
4	Displaying map according to indicator	Selecting one indicator	A map is displayed according to the indicator	The task works properly
5	Zoom in and zoom out on the map	Selecting zoom in or zoom out features on the map	A map shown to be larger or smaller than the previous map	The task works properly

4. Summary

Development and integration of new data cubes for horticulture crops, food crops, plantations and livestock in the SOLAP for agricultural commodities were successfully performed using SpagoBI. The SOLAP for agricultural commodities is integrated to the data warehouse developed based on the galaxy scheme. The SOLAP provides agriculture data summaries represented in crosstab and chart. The location intelligence module for each data cube was also successfully implemented to visualize agriculture data summaries in dynamic maps. The testing results of SOLAP and the location intelligence module for data cubes of horticultural crops, food crops, plantations and livestock indicate that all functions in the SOLAP and the location intelligence work properly.

References

[1] Han J, Kamber M and Pei J 2012 *Data Mining: Concept and Techniques* 3rd ed (San Francisco (US): Elsevier)

[2] Boulil K, Ber L F, Bimonte S, Grac C and Cernesson F 2014 Multidimensional modeling and analysis of large and complex watercourse data: an OLAP-based solution, *Ecological Informatics* (Elsevier) pp 30

[3] Cazzin G 2012 *Business Intelligence with SpagoBI* (Padua (IT): SpagoBI Competency Center)

[4] [Ministry of Agriculture] Ministry of Agriculture 2016 The database agricultural statistics retrieved from http://aplikasi.pertanian.go.id/bdsp/ newkom.asp.

[5] Kimball R and Ross M 2013 *The Data Warehouse Toolkit: The Definitive Guide to Dimensional Modeling* (Indianapolis (IN-US): Wiley)

North American Multi Model Ensemble (NMME) Performance of Monthly Precipitation Forecast over South Sulawesi, Indonesia

A M Setiawan [1, 2*], Y Koesmaryono[1], A Faqih[1] and D Gunawan[2]

[1] Department of Geophysics and Meteorology, Bogor Agricultural University (IPB), Bogor, Indonesia
[2] Centres for Climate Change Information, Indonesia Agency for Meteorology Climatology and Geophysics (BMKG), Jakarta, Indonesia

Email: amsari.setiawan@gmail.com

Abstract. The North American Multi-Model Ensemble (NMME) as one of the multi-model seasonal forecasting system, regularly generate monthly precipitation forecast for all globe with 0.5 – 11.5 months lead time. This useful information can be used as general input to regional and local precipitation forecast. This study quantifies monthly precipitation hindcast data performance in South Sulawesi provided by seven coupled models from the NMME during 29 years (1982 – 2010) period. Climate Hazards Group InfraRed Precipitation with Stations (CHIRPS) dataset and the Standardized Verification System (SVS) for long-range forecasts (LRF) analysis applied to asses monthly precipitation prediction. Almost all individual model shows relatively high skill when used to make June – November monthly precipitation prediction and low skill when used to forecast monthly precipitation in December – May periods. Multi-Model Ensemble (MME) performed using Simple Composite Method (SCM) increased performance of monthly precipitation prediction. Nevertheless, this improvement only applies at short lead time (< 3.3 months). This result also shows that NMME is more promising when it used to make precipitation forecast application during dry period (i.e. drought prediction) rather than wet period over South Sulawesi region.

Keywords: NMME performance; Monthly Forecast Verification; South Sulawesi; SCM-MME; SVS-LRF

1. Introduction

The North American Multi-Model Ensemble (NMME) is one of the multi-model seasonal forecasting system, consisting of coupled models from US modeling centers, including the NOAA National Centers for Environmental Prediction (NOAA/NCEP), Center for Ocean-Land-Atmosphere Studies (COLA), NOAA's Geophysical Fluid Dynamics Laboratory (NOAA/GFDL), National Aeronautics and Space Administration/ Global Modeling and Assimilation Office (NASA/GMAO) and Canadian modeling centers [1]. The routine generation of global monthly precipitation forecasts of each individual model are freely available from and easily downloaded from International Research Institute for Climate and Society/ Lamont-Doherty Earth Observatory (IRI/LDEO) website.

This dataset has been widely used in previous studies to verify monthly mean precipitation [2] and 2-meter temperature forecasts[3], monthly prediction of Sea Surface temperature as ENSO indices [4,5], seasonal precipitation [6,7], seasonal global drought onset [8] and interannual prediction [1]. Information about forecasting skill for each model needed before it can be used to improve operational routine monthly deterministic forecast provided by National Meteorological and Hydrological Services (NMHS).

Precipitation over South Sulawesi as one of the primary national rice production centers strongly affected by El Niño event [4,9,10], indicated by wider drought coverage area [11]. Precipitation information become very important to this region. The objectives of this study are to quantifies monthly precipitation hindcast data performance in South Sulawesi provided by seven individual coupled models from the North American Multi-Model Ensemble (NMME) and evaluate the forecasting skill of Simple Composite Method Multi Model Ensemble (SCM-MME) formed from NMME. This study investigates model performance for all time period (1982 – 2010) and stratified according to each calendar month and lead time.

2. Data and Methodology

2.1. NMME

The gridded monthly precipitation predictions of each individual model were downloaded from the International Research Institute for Climate and Society/ Lamont-Doherty Earth Observatory (IRI/LDEO) collection of climate data (http://iridl.ldeo.columbia.edu/SOURCES/.Models/.NMME/). All time coverage with all lead and ensemble mean [12] of each NMME member used and averaged over South Sulawesi. Table 1 provides a brief description of the models used to in this study.

Table 1. Description of monthly precipitation prediction model used in this study.

Model	Lead Time (Months)	Time Coverage	Ensemble Members	Modeling Centre	References
CFSv2	0.5 – 9.5	1982 – 2010	24	National Centers for Environmental Prediction (NOAA/NCEP)	[13]
CMC1-CanCM3	0.5 – 11.5	1981 – 2010	10	Environment Canada's Canadian Meteorological Centre (CMC)	[14]
CMC2-CanCM4	0.5 – 11.5	1981 – 2010	10	Environment Canada's Canadian Meteorological Centre (CMC)	[14]
CCSM3	0.5 – 11.5	1982 – Present	6	The Center for Ocean-Land-Atmosphere Studies (COLA)	[15]
CCSM4	0.5 – 11.5	1982 – Present	6	The Center for Ocean-Land-Atmosphere Studies (COLA)	[16]
GFDL-CM2p5-FLOR-B01	0.5 – 11.5	1982 – 2012	10	NOAA's Geophysical Fluid Dynamics Laboratory (GFDL)	[17]
GEOS5 GMAO-062012	0.5 – 8.5	1981 – Present	12	National Aeronautics and Space Administration/ The Global Modeling and Assimilation Office (NASA/ GMAO)	[18]

An MME of seven individual coupled models was formed using the simple composite method [19], with equal weights assigned to ensemble mean predictions of each model, assuming that each model is independent. The mean bias from each model was removed by forming anomalies with respect to each model's own seasonal climatology. Because the lead times and hindcast periods were different in each model, a common range of 0.5–8.5 months lead time and the 1982–2010 hindcast period were selected for the MME.

2.2. Observation

This study quantifies NMME monthly precipitation hindcast performance using 0.05° x 0.05° blended gauge-satellite of daily and monthly precipitation estimates of the Climate Hazards Group InfraRed Precipitation with Stations (CHIRPS) dataset [20]. This grided dataset were obtained from Climate Hazards Group/ The Department of Geography, University of California Santa Barbara (ftp://ftp.chg.ucsb.edu/pub/org/chg/products/CHIRPS-2.0/) for the 30-year period from 1981 – 2010. South Sulawesi CHIRPS dataset averaged over the study area (119.25E – 121.75E and 7S – 2S) (figure 1). General topographical characteristics of study area described with gridded elevation data were obtained from National Geophysical Data Center, NOAA, with 2 – minute Gridded Global Relief Data (ETOPO2)v2 [21] .

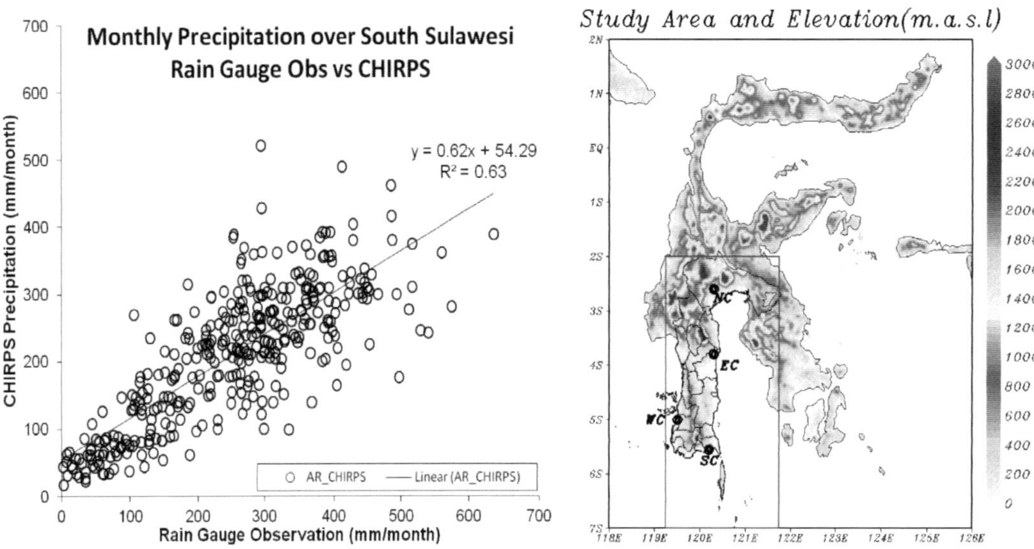

Figure 1. Scatter plot between all rain-gauged monthly precipitation observation and CHIRPS averaged over study area (left); and elevation characteristic (right) over study area (blue box) (meter above sea level/ m. a. s. l) with station location (black circle).

Observational-based stations are distribute over 4 region and district in South Sulawesi (Table 2) with various altitude between 14 – 81 meter above sea level (m. a. s. l.). Various altitude and station type such as rain gauge (Obs Gauge), Agricultural Meteorological station (AgriMet), and BMKG weather station (BMKG). This observational dataset provided by The Indonesia Agency for Meteorology Climatology and Geophysics (BMKG). Station form four different districts located at West Coast (WC), South Coast (SC), East Coast (EC) and North Coast (NC) South Sulawesi Province are used to determine their representation in CHIRPS dataset. Good correlation between CHIRPS values and recorded insitu precipitation over study area also shown (Figure 1).

Table 2. Observational station used for comparison with CHIRPS monthly precipitation dataset. Missing data information calculated from 30-years period from January 1981 to December 2010.

No	StationName	District	Latitude (dms)	Longitude (dms)	Altitude (m.a.s.l)	Station Type	Missing Data (%)
1	STAMET HASANUDDIN	MAROS (WC)	05 04 15.6 S	119 33 07.7 E	14	BMKG	3.61
2	BATUKAROPA	BULUKUMBA (SC)	05 28 09.1 S	120 12 29.2 E	81	AgriMet	0.28
3	SIWA	WAJO (EC)	03 44 59.1 S	120 21 50.2 E	25	Obs Gauge	8.61
4	STAMET MASAMBA	LUWU UTARA (NC)	02 33 16.0 S	120 19 27.0 E	50	BMKG	11.39

2.3. Verification

The Standardized Verification System (SVS) for long-range forecasts (LRF) analysis [22] was used on the hindcast data and monthly precipitation observation from the Climate Hazards Group InfraRed Precipitation with Stations (CHIRPS) dataset [20] to evaluate model skills for predicting South Sulawesi precipitation. The SVS consists of four parts including diagnostic measures, key parameters, verification data sets, and details of forecast systems [22]. The monthly precipitation anomaly verification was stratified according to each calendar month and lead time. The Mean square skill score (MSSS) is defined as one minus the ratio of the mean square error (MSE) of the forecasts to the MSE of climatology forecast. The values can be used to compare forecast performance to climatology forecast. MSSS decomposition provides information on phase errors through forecast/observation correlations, amplitude errors through the ratio of the forecast to observed variances, and overall bias error. Ratio of the square roots of the variance explains how well the variance of model can explain the variance of observation. The information can be used to adjust or weight forecasts when they are used as an input into regional and local forecasts. If model performance is good with a strong correlation and small amplitude error, the model can be considered suitable for use in forecasts. The MSSS will not exceed zero if forecasts have the same amplitude with the observations and no overall bias unless the correlation exceeds approximately 0.5 [22]. A good forecast is achieved when the ratio of the forecasted to observed variances is close to one, overall bias (the difference between the forecast and observation) is near zero, and the MSSS is close to one. Negative/zero MSSS values indicate that deterministic forecasts are worse than/the same as climatology forecasts.

3. Result and Discussion

3.1. Monthly Performance of Individual Model

The monthly climatological mean precipitation calculated for each station with 30-years period (1981-2010). CHIRPS monthly precipitation data averaged over study area and monthly climatological mean calculated with same period with station data. In general, lower precipitation amount in study area occurred in August – October. Higher precipitation with high variation value between each region (especially between west and east coast) observed during December – June period (figure2).

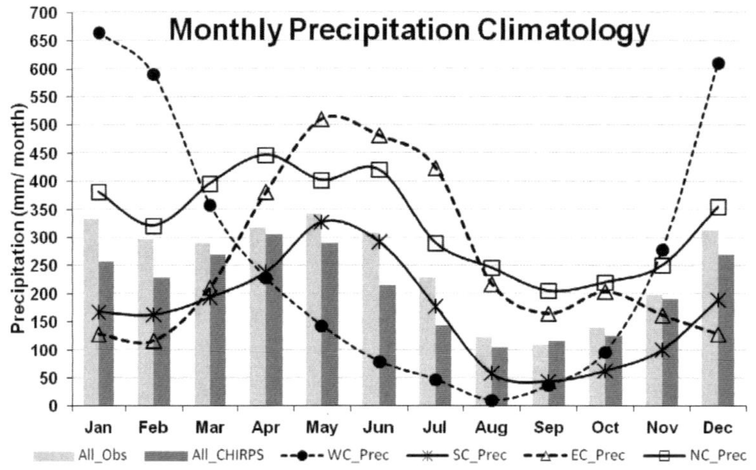

Figure 2. Monthly South Sulawesi climatological mean (1981-2010) of precipitation based on regional averaged CHIRPS data (blue bar) and all gauged-based data (green bar). Line chart represent precipitation in west coast (WC_prec/ circle dashed line), south coast (SC_prec/ cross line), east coast (EC_prec/ triangle dashed line) and north coast (NC_prec/ square line).

Monthly climatological mean pattern obtained from station are similar and slightly underestimates compared to gauged based observation. Furthermore, relatively highly significant relationship with monthly precipitation from observation ($R^2 = 0.63$, $\alpha = 0.05$) makes CHIRPS data is considered to represent the rainfall observation in the study area. This precipitation dataset has been used to evaluate NMME skill [23] and verified in various past studies with good correlation and small error between CHIRPS values and recorded insitu precipitation [24–26].

The monthly precipitation verification was stratified according to each calendar month and lead time. A lead time of 0.5 month implies a prediction made at the very beginning of the target month. Comparisons of model performance based on anomaly correlation values are presented in figure 3.

The correlation coefficient values are plotted for each month and for each individual model lead time. Noticeable skill differences based on anomaly correlations are founded in monthly precipitation forecasts for June - November at medium lead times (4.5 – 9.5). The NASA/GMAO has relatively high skill (anomaly correlations > 0.6) while other individual models show lower skill (0.3 – 0.8). These correlations are statistically significant at 95 % confidence level. Almost all individual model shows low skill (anomaly correlations < 0.4) when used to forecast monthly precipitation in January – May periods. MME performed using simple composite method increased performance of monthly precipitation prediction. Nevertheless, this improvement only applies at short lead time (< 3 months) and not seen in January – May prediction.

Model performance comparison based on overall bias values between each model output and observed precipitation are presented in figure 4. Similar with figure 3, overall bias values are plotted for each month and for each individual model lead time. Almost all model show overestimates prediction during dry period (June – August) and underestimates during wet period (January – March), except for CFSv2 which show overestimates prediction for all month. Significant MME improvement related to smaller overall bias value applies at three month forecast and shorter lead time.

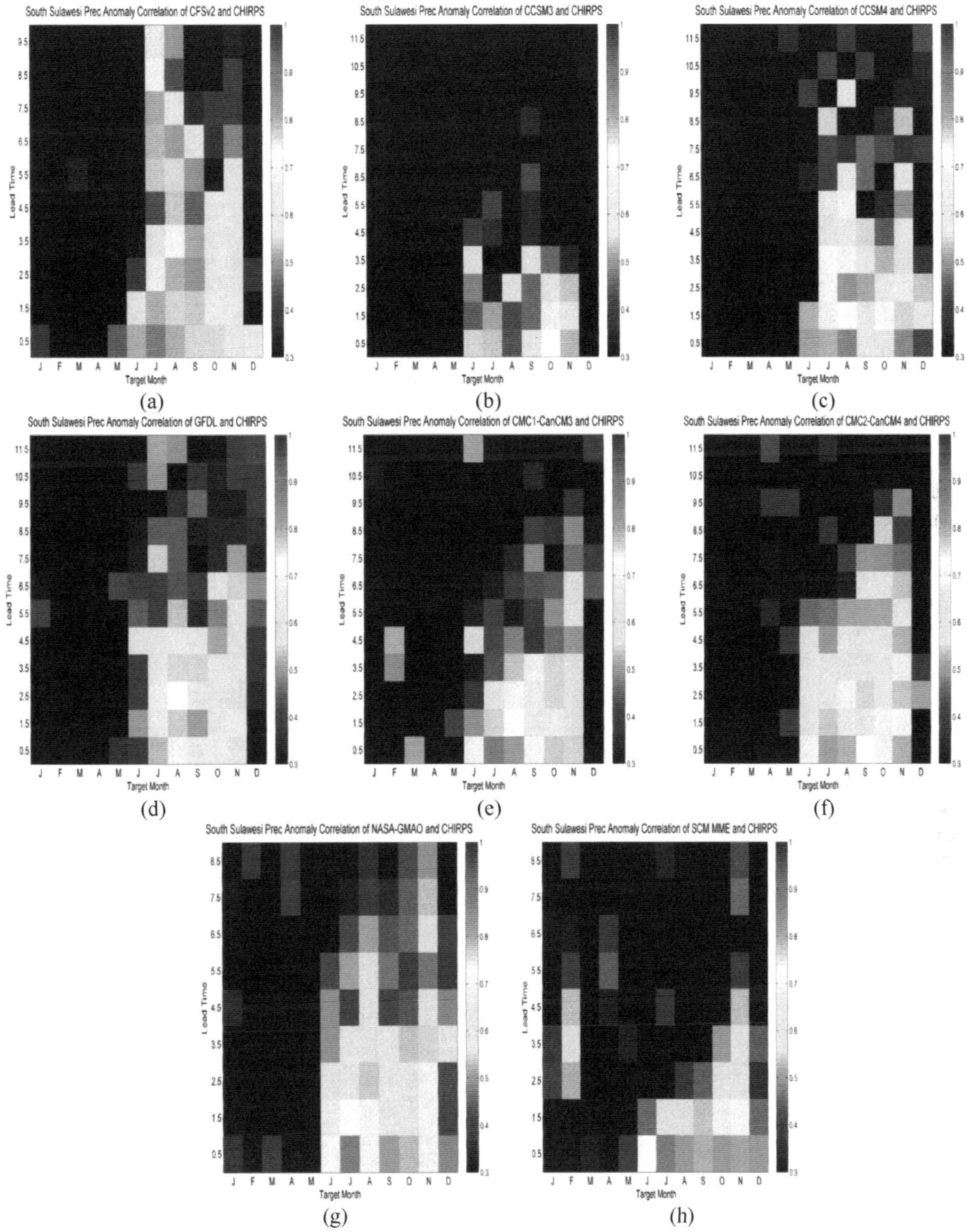

Figure 3. Performance of monthly precipitation individual model and Multi Model Ensembles (MME) forecast (1982 – 2010) based on anomaly correlation coefficient value of monthly precipitation over South Sulawesi with CFSv2 (a), COLA-CCSM3 (b), COLA-CCSM4 (c), GFDL-CM2p5-FLOR-B01 (d), CMC1-CanCM3 (e), CMC2-CanCM4 (f), NASA-GMAO-062012 (g) and Simple Composite Method (SCM)-MME (h). Target month indicated by horizontal axis and lead time on the vertical axis.

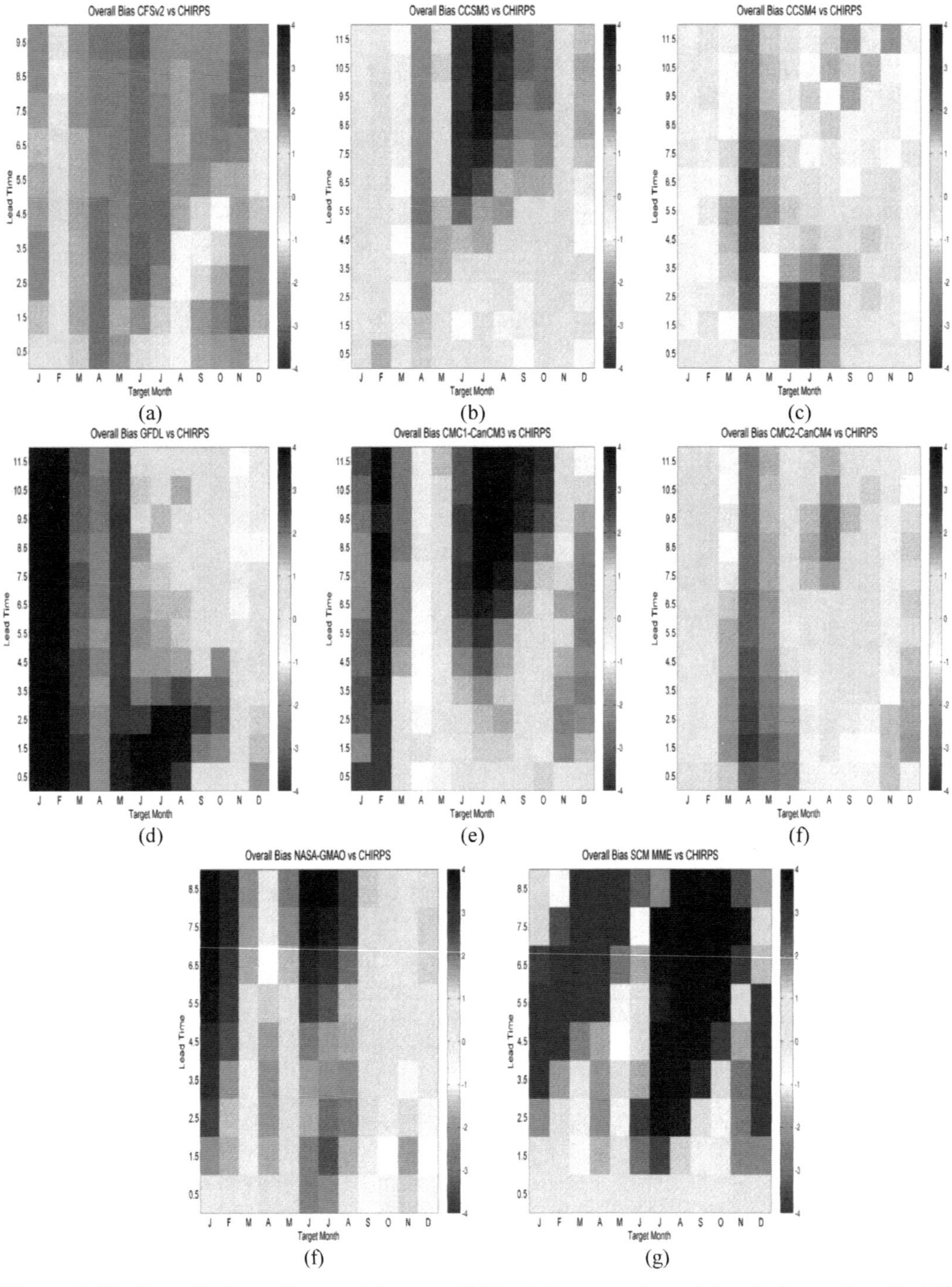

Figure 4. Similar with figure 2, except for overall bias between each model and observation. Positif (negatif) overall bias values indicated by color ramp, associated with overestimates (underestimates) of model prediction.

IOP Conf. Series: Earth and Environmental Science **58** (2017) 012035 doi:10.1088/1755-1315/58/1/012035

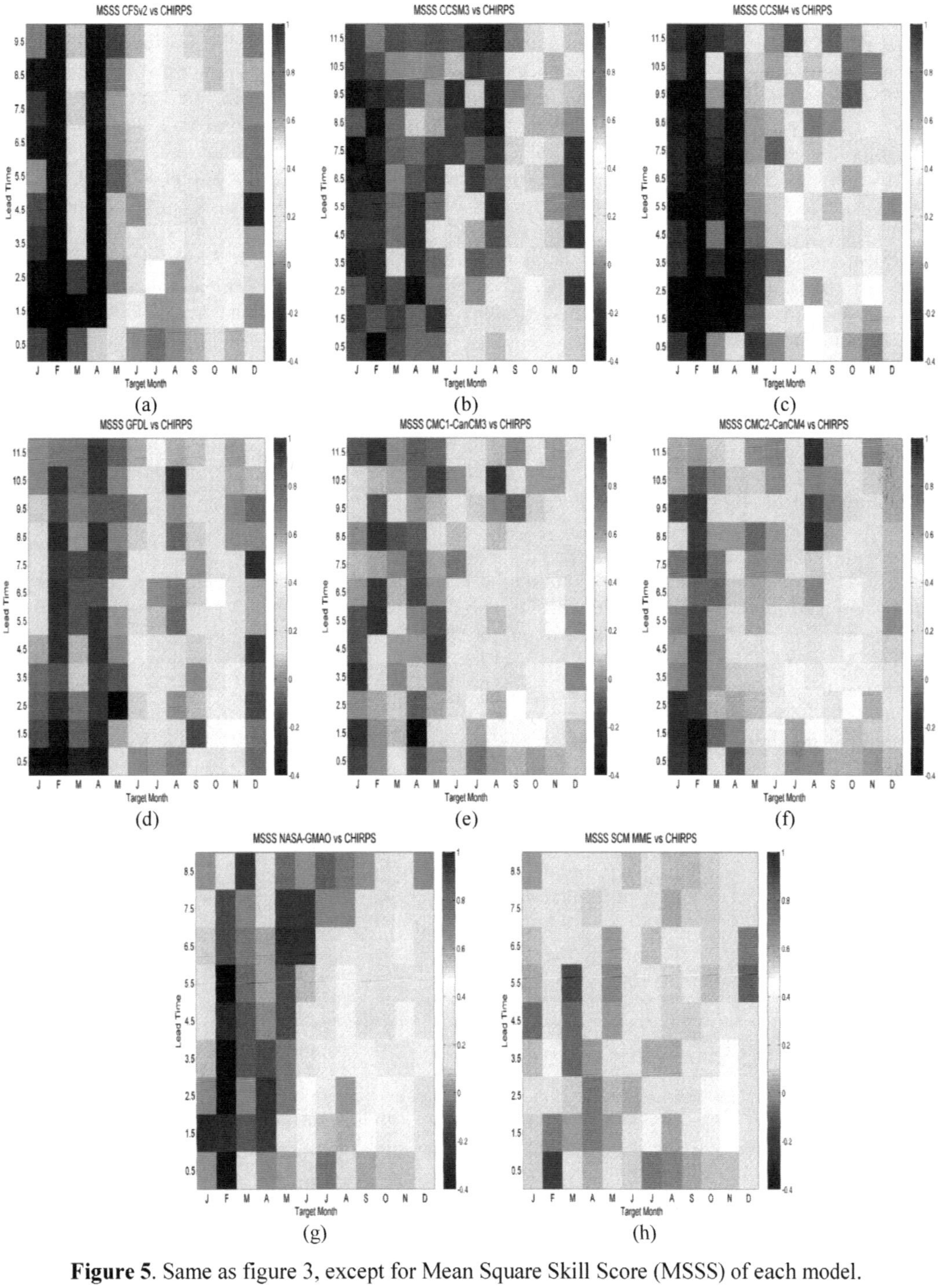

Figure 5. Same as figure 3, except for Mean Square Skill Score (MSSS) of each model.

Figure 5 describes the Mean Square Skill Score (MSSS) of each individual model as a function of target month and lead time. The most noticeable skill difference based on MSSS is found in forecasts for June - November at 2.5 to 5.5 month lead times. The NASA-GMAO has relatively high skill (MSSS of 0.4 or higher) while other individual models show lower skill. This result is slightly different from previous results that reviewed the model performance only based on overall bias. The MSSS not only is based on correlations, but also takes into account the phase errors, amplitude errors and overall bias error of the forecasts. Worse forecast skills than climatology were found in forecast for January to March (CFSv2 at all lead time), January to May (COLA-CCSM3 at all lead time), January to April (COLA-CCSM4 at 1.5 months lead time or longer), December to April (GFDL-CM2p5-FLOR-B01 at all lead time), January to May (CMC1-CanCM3 at 1.5 months lead time or longer) and January to February (CMC2-CanCM4 at all lead time). Simple Composite Method (SCM)-MME make significant performance improvement especially during June – November target month with less than 4.5 month lead time.

Climatological monthly precipitation variability between each region during December – May are suspected to be reason of lower skill prediction in study area. There are opposite monthly rainfall pattern in this period especially between east coast and west coast region (figure 2), although general rainfall condition well captured by NMME prediction during June – November. Therefore, rainfall prediction for December – May should not generalized and further downscaling procedure may apply to capture regional precipitation characteristic.

3.2. Multi Years Performance of NMME

Further analysis was conducted to determine the performance comparisons between each individual model and multi-models ensemble forecast for monthly precipitation forecast during multi years (1982 – 2010) period. This comparison presented in figure 6.

The MME, COLA-CCSM4 and NASA/GMAO have relatively high skill (correlations of 0.6 or higher) while GFDL-CM2p5-FLOR-B01 models show lowest skill (correlation < 0.2) if model performance only stratified by coefficient value. A CFSv2 and CMC2-CanCM4 show underestimates prediction at almost all lead time, while GFDL-CM2p5-FLOR-B01, CMC1-CanCM3 and NASA/GMAO show overestimates prediction. These overall bias issues for each individual model are well addressed in MME and improvements showed by near zero overall bias for all lead time.

COLA-CCSM4 shows good performance according to model sensitivity explained by near one ratio of the square roots of the variance. Based on the MSSS, the most noticeable skill difference was found in 4.5 to 8.5 month lead times forecast. The GFDL has relatively low skill (MSSS of 0.1 or lower) while other models show higher skill (0.2 – 0.5) due to overall bias and the ratio of square root of the variance value.

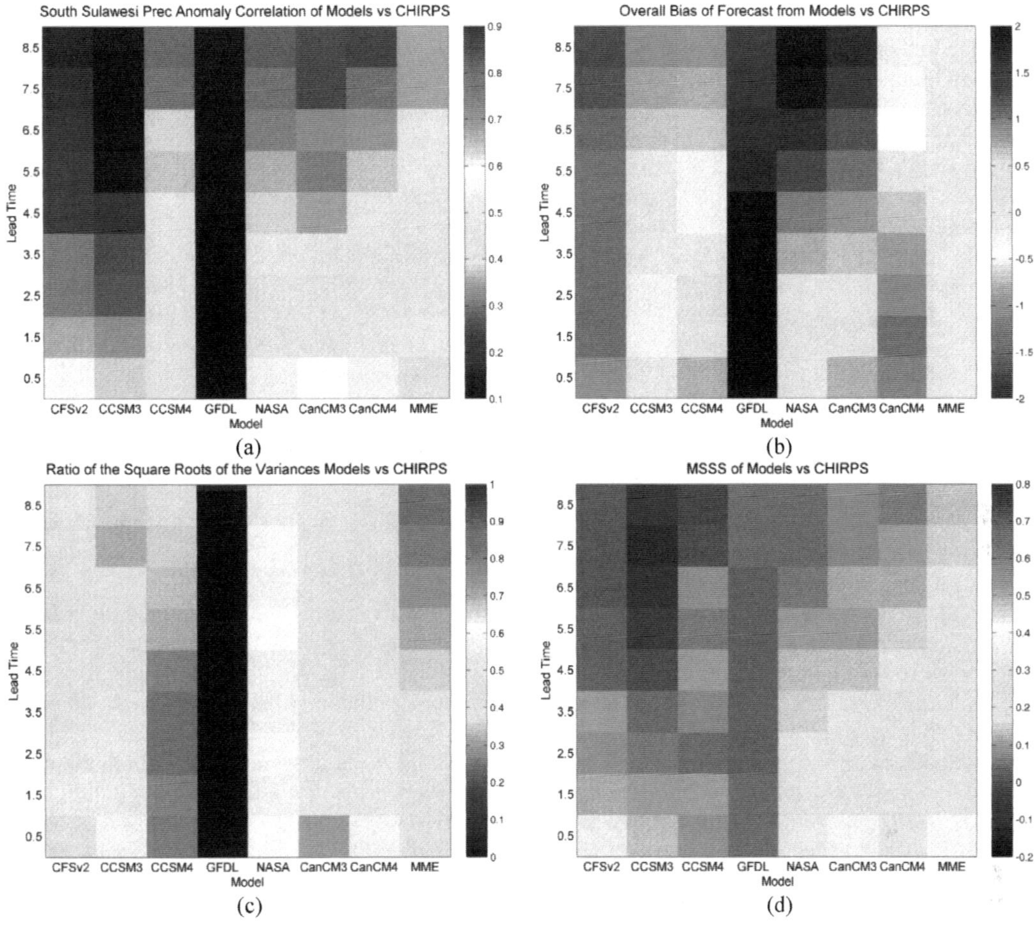

Figure 6. Performance comparison of multi years monthly precipitation from Multi Model Ensembles (MME) forecast (1982 – 2010) over South Sulawesi based on Standardized Verification System (SVS) for Long-Range Forecasts (LRF) analysis. Each figure indicates anomaly correlation value (a), overall bias (b) and ratio of square roots of the variances (c) as decomposition of Mean Squared Skill Score (MSSS) (d). The individual model name is indicated on the horizontal axis, and lead time prediction on the vertical axis.

4. Conclusion

Noticeable skill differences in monthly precipitation prediction were found between each individual model, especially at medium to long lead times. The performance of monthly precipitation MME forecasts for all time period (1982–2010) using the Simple Composite Method (SCM)-MME method show model improvement. Significant performance improvement by SCM-MME especially during June – November target month with less than 4.5 month lead time. Climatological monthly precipitation variability between each region during December – May are suspected to be reason of lower skill prediction in study area. Therefore, rainfall prediction for December – May should not generalized and further downscaling procedure may apply to capture regional precipitation characteristic. This result also shows that NMME is more promising when it used to make precipitation forecast during dry period rather than wet period over South Sulawesi region.

Acknowledgements
The first author thanks the Indonesia Agency for Meteorology Climatology and Geophysics (BMKG) for providing a scholarship for his PhD program. This research is funded by Department of Geophysics and Meteorology, Bogor Agricultural University (IPB) as part of BMKG – IPB scholarship program. The NMME project and data dissemination was supported by NOAA, NSF, NASA and DOE.

References
[1] Kirtman B P, Min D, Infanti J M, Kinter J L, Paolino D A, Zhang Q, van den Dool H, Saha S, Mendez M P, Becker E, Peng P, Tripp P, Huang J, DeWitt D G, Tippett M K, Barnston A G, Li S, Rosati A, Schubert S D, Rienecker M, Suarez M, Li Z E, Marshak J, Lim Y-K, Tribbia J, Pegion K, Merryfield W J, Denis B and Wood E F 2014 *Bull. Am. Meteorol. Soc.* **95** 585–601
[2] Wang H 2014 *Hydrol. Process.* **28** 4472–86
[3] Becker E, van den Dool H, Zhang Q and Peña M 2013 *Sci. Technol. Infus. Clim. Bull.*
[4] Setiawan A M 2014 *Drought Characteristics in Indonesia Related to Warm ENSO Episodes* (Busan)
[5] Barnston A G, Tippett M K, Van Den Dool H M and Unger D A 2015 *J. Appl. Meteorol. Climatol.* **54** 1579–95
[6] Infanti J M and Kirtman B P 2014 *J. Hydrometeorol.* **15** 529–50
[7] Ma F, Ye A, Deng X, Zhou Z, Liu X and Duan Q 2016 Evaluating the skill of NMME seasonal precipitation ensemble predictions for 17 hydroclimatic regions in continental China **144** 132–44
[8] Yuan X and Wood E F 2013 Multimodel seasonal forecasting of global drought onset **40** 4900–5
[9] Setiawan A M 2011 *Determination of Reference ENSO Index for Indonesian Region Based on Correlation Analysis of Spatial and Temporal Pattern with Standardized Precipitation Index (SPI)* (Bandung: Institut Teknologi Bandung (ITB))
[10] Hidayat R, Ando K, Masumoto Y and Luo J J 2016 *IOP Conf. Ser. Earth Environ. Sci.* **31** 12043
[11] Surmaini E, Hadi T W, Subagyono K and Puspito N T 2015 *Theor. Appl. Climatol.* **121** 669–84
[12] Ishizaki Y, Nakaegawa T and Takayabu I 2012 *Clim. Dyn.* **39** 185–206
[13] Saha S, Moorthi S, Wu X, Wang J, Nadiga S, Tripp P, Behringer D, Hou Y-T, Chuang H, Iredell M, Ek M, Meng J, Yang R, Mendez M P, van den Dool H, Zhang Q, Wang W, Chen M and Becker E 2014 *J. Clim.* **27** 2185–208
[14] Merryfield W J, Lee W-S, Boer G J, Kharin V V., Scinocca J F, Flato G M, Ajayamohan R S, Fyfe J C, Tang Y and Polavarapu S 2013 *Mon. Weather Rev.* **141** 2910–45
[15] Collins W D, Bitz C M, Blackmon M L, Bonan G B, Bretherton C S, Carton J A, Chang P, Doney S C, Hack, James J, Henderson T B, Kiehl J T, Large W G, McKenna, Daniel S, Santer B D and Smith R D 2006 *J. Clim.* **19** 2122–43
[16] Gent P R, Danabasoglu G, Donner L J, Holland M M, Hunke E C, Jayne S R, Lawrence D M, Neale R B, Rasch P J, Vertenstein M, Worley P H, Yang Z L and Zhang M 2011 *J. Clim.* **24** 4973–91
[17] Delworth T L, Broccoli A J, Rosati A, Stouffer R J, Balaji V, Beesley J A, Cooke W F, Dixon K W, Dunne J and Dunne K A 2006 *J. Clim.* **19**
[18] Vernieres G, Rienecker M M, Kovach R and Keppenne C L 2012 *The GEOS-iODAS: Description and Evaluation* vol 30(Greenbelt, Maryland: National Aeronautics and Space Administration)
[19] Jeong H-I, Lee D Y, Ashok K, Ahn J-B, Lee J-Y, Luo J-J, Schemm J-K E, Hendon H H, Braganza K and Ham Y-G 2012 *Clim. Dyn.* **39** 475–93
[20] Funk C C, Peterson P J, Landsfeld M F, Pedreros D H, Verdin J P, Rowland J D, Romero B E,

Husak G J, Michaelsen J C and Verdin a. P 2014 *U.S. Geol. Surv. Data Ser.* **832** 4
[21] National Geophysical Data Center 2006 ETOPO2v2 2-minute Global Relief Model
[22] World Meteorological Organization 2010 Standardized Verification System (SVS) for Long-Range Forecasts (LRF) *Manual on the Global Data Processing and Forecasting System* WMO-No. 485 vol I-Global (Geneva: World Meteorological Organization) p II.8.1-II.8.17
[23] Wanders N, Wood E F, Lott F C, Gordon M, Graham R J, Macleod D A, Jones A, Di F, Shukla S, Funk C and Hoell A 2014 *Environ. Res. Lett.* **9**
[24] Duan Z, Liu J, Tuo Y, Chiogna G and Disse M 2016 *Sci. Total Environ.*
[25] Katsanos D, Retalis A and Michaelides S 2015 *Atmos. Res.* **Volume 169** 459–464
[26] Hessels T M 2015 *Comparison and Validation of Several Open Access Remotely Sensed Rainfall Products for the Nile Basin* (Delft University of Technology)

Use of Drought Index and Crop Modelling for Drought Impacts Analysis on Maize (*Zea mays* L.) Yield Loss in Bandung District

E Kurniasih[1,2*], Impron[1] and Perdinan[1]

[1] Department of Geophysics and Meteorology, Faculty of Mathematics and Natural Sciences, Bogor Agricultural University (IPB), Jl. Meranti Kampus IPB Dramaga, Bogor, Indonesia 16680
[2] Stasiun Klimatologi Lasiana-Kupang, Indonesian Agency for Meteorology Climatology and Geophysics (BMKG), Jl. Timor Raya Km 10.7 Lasiana, Kupang, Nusa Tenggara Timur, Indonesia 85311

Email: endahkurniasih222@gmail.com

Abstract. Drought impacts on crop yield loss depend on drought magnitude and duration and on plant genotype at every plant growth stages when droughts occur. This research aims to assess the difference calculation results of 2 drought index methods and to study the maize yield loss variability impacted by drought magnitude and duration during maize growth stages in Bandung district, province of West Java, Indonesia. Droughts were quantified by the Standardized Precipitation Index (SPI) and the Standardized Precipitation Evapotranspiration Index (SPEI) at 1- to 3-month lags for the January1986-December 2015 period data. Maize yield responses to droughts were simulated by AquaCrop for the January 1986-May 2016 period of growing season. The analysis showed that the SPI and SPEI methods provided similar results in quantifying drought event. Droughts during maize reproductive stages caused the highest maize yield loss.

Keywords : Aquacrop, drought, maize, standardized precipitation index, standardized precipitation evapotranspiration index

1. Introduction

Drought is the consequence of a natural reduction in amount of precipitation received over an extended period of time. Other climatic factors such as high temperatures, high winds, and low relative humidity are often associated with drought and can significantly aggravate the severity of the event. This reduction of precipitation may lead to a condition where water supply is insufficient to meet water demands of human activities and the environment. This situation induce a series of cascading impacts in a wide range of economic sectors and the environment [1]. Base on degree of severity, time span of the event, spatial extent, total death toll, economic consequences, social disruption and long-term impact criteria, the most important global hazard is drought [2].

Agriculture is one of the most economic sectors which sensitive to drought because crop productivity is determined by access to water resources and soil moisture reserves [3]. In Indonesia, drought decrease planting area, decrease harvested area and increase crop failure in the food crop agricultural sector. The total area of food crops affected by drought based on the average 5-year data (2011-2015) recorded an area of 236.257 hectares of the total planting area of 13,293,109 hectares. West Java was the province with the highest drought affected area, followed by South Sulawesi, Central Java and East Java [4].

Effective drought monitoring which provide drought early warnings is a critical component of national drought strategies. Information of the onset, spacial extend, intensity and duration of drought contributed to better drought preparedness and management in agriculture sector to maintain national food security. Cop loss due to drougth can be influenced by drougth magnitude-duration and growing stages of the plants when drought occurs [5], understanding of this information might helps national drought watch and monitoring agencies to fit the right time to give drougth early warning information for goverment policy and farmers in order avoid or minimize the crop production loss.

The essential tools to monitor and assess drought events are drought index. Drought index are, typically a single number, proxies based on climatic information and it is assumed that they adequately quantify drought intensity [5]. Many drought index were developed and used by meteorologists and climatologists to quantify and monitor drought [6-11]. The standardized precipitation index (SPI) [6] is the most used index to quantify drought and/or wet events all over the world since it is recomended by WMO [12, 13]. Precipitation is the only input parameter of the index and it can be computed for different time scales. The SPI is based on the probability of precipitation for any time scale that transformed into an index [6, 14]. Another index recomended by WMO is The Standardized Precipitation-Evapotranspiration Index (SPEI) [8]. The original SPEI input parameters are precipitation and temperature data. Mathematically, it is similar to SPI, but it includes the effect of temperatur variability. Monthly-mean temperature data are used to calculate the simplest potential evapotranspiration, PET [15]. The difference between the precipitation and PET for the month provides a simple measure of the water surplus or deficit for the analyzed month named the climatic water balance. The index values are obtained by normalization of the climatic water balance into a log-logistic probability distribution to transform the original values to standardized units that are inter-comparable in space and time. The SPEI has significant potential as a world meteorological drought index that suits agricultural system needed because it uses a simple climatic water balance [16] and appropriate for drought analysis under climate change scenarios because it has been considering temperature parameter [8].

Drought and/or wet intensity defined by the SPI and the SPEI methods can be categorized into 7 groups. The detail information shown in table 1. A drought event begins when the SPI reach a value of -1.0 or less. The drought event ends when SPI or SPEI becomes positive. It has been determined that SPI or SPEI is in normal condition 65% of the time, in moderate drought 10% of the time, in severe drought 5% of the time, and in extreme drought 2% of the time. The 2 % of SPI or SPEI values within the "extreme drought" category is a percentage that is typically expected for an "extreme" event.

Table 1. SPI and SPEI categories of drought.

SPI/SPEI	Moisture category	Frequency (%)
≥2.0	Extreme wet	2
1.50 to 1.99	Severe wet	6
1.49 to 1.00	Moderate wet	10
0.99 to -0.99	Normal	65
-1.00 to -1.49	Moderate drought	10
-1.5 to - 1.99	Severe drought	5
≤-2.00	Extreme drought	2

Drought impacts on crops production can be estimated by crop models. Models are mathematical equations which describe the relationships between crop growth, development, yields, technology and climate. One of the main aims of constructing operational crop models is to estimate the harvestable (economic) yield. Crop simulation models are useful tool in separating yield gain into climate variability, new management practices and technologies factors [17]. By minimizing non-climatic factors, crop model users could isolate the climate variability effect on crop yield to assess drought impact on every crop growing stages. Crop simulation model had been well reported to determine the patterns of drought stress for short- and medium-duration upland rice and maize around flowering and early grain-filling on 12 locations and more than 30 environments (6 years x 5 to 6 planting dates) in Brazil [18].

It is impportant to assess drought event on a spesific crop in a spesific area to build an effective drought early warning system. Maize planting area in West Java province often affected by drought. Ministry of Agriculture report, it is observe specifically maize planting area in West Java province, showed that maize planting area in Bandung is the most frequently affected by drought [19].

The main goal of this paper is to assess the difference calculation results of SPI and SPEI and to study the maize yield loss variability impacted by drought intensity and duration during maize growth stages in Bandung district, province of West Java, Indonesia. Section 2 describes material and methods. Section 3 shows the results and discussion. Section 4 concludes the paper.

2. Material and Methods

2.1 Meteorological Data and Drought Identification

The monthly precipitation and average temperature (T_{ave}) data of Bandung area (06° 55' S, 107° 36' E and 791 above sea level) were collected from Dramaga Climatological Station for the 1986-2015 period. These data have been used to calculate drought index using the SPI [6] and the SPEI [8] methods at 1- to 3-month lag. The potential evapotranspiration (PET) needed by the original SPEI method were calculate using the PET Thornthwaite calculation method [15]. Both index were calculated in the SPEI package build under R version 3.2.5. The performance of the index were evaluated using two statistical indicators, Homogeneity test and Pearson correlation test.

2.2 Analysis of Maize Yield during Drought condition

Prediction of maize yield has been carried out by the crop model named AquaCrop [20-22] version 5 build by Food and Agriculture Organization (FAO). In order to evaluate the effect of drought on maize production, the simulation ran on two planting dates scenarios (1st and 15th day) for each month and repeated in each experimental year (30 years x 12 months x 2 planting dates). Each growing cycles ran with the assumption that other factors (soil and management) don't have negative effect on crop growth stages unless the climate condition. The daily temperatures (average temperature,T_{ave}, maximum temperature, T_{max}, and minimum tempereture, T_{min}) and precipitation data of Bandung for the January 1986-May 2016 period were used to identified the climate effect during the crop life cycle. Conservative (constant) parameters of maize, MaizeGDD.CRO file from AquaCrop data bank as described by *Hsiao et al.* [21], were choosen for this study. The detail information shown in table 3. The soil was set constantly as sandy loam (Andosol Soil) based on the typical soil of Bandung area.

3 Results and Discussion

3.1 Analysis of Drought Driving Factor and Drought Index

Driving factors for drought during maize growing cycles in this experiment are precipitation, temperatures (T_{max}, T_{ave} and T_{min}) and PET. The long-term climatic data analysis of normal precipitation, water balance and temperatures during period of 1986-2015 are presented on figure 1. Annual average precipitatian and PET of Bandung are about 2179 mm and 1148 mm. The highest temperature occur in September-October. It can be concluded that the driest months occur in June-October when the precipitation is below normal, the temperature is hight and the water are deficit.

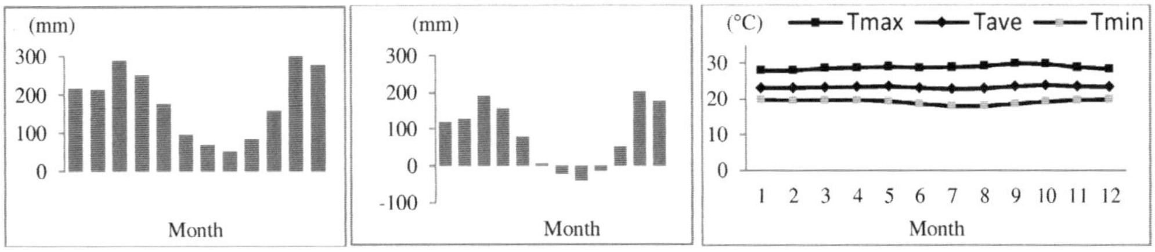

(a) (b) (c)

Figure 1. Normal precipitation (a), water balance (b) and temperatures (c) during period of 1986-2015.

The long-term fluctuation of SPI and SPEI values at different time scale lag in Bandung are presented in Figure 2. Homogeneity test of SPI and of SPEI values showed that there is no significant differences between the two index for every time scale lag. The correlation between the SPI and the SPEI was high for the different series, independently of the time scale analyzed (Pearson correlation R-values between 0.98 and 0.99). As mentioned in Vicente-Serrano *et al.* [9] for the calculation at the Helsinki observatory (1910-2007), these results are consistent with the hypothesis that the main variable for droughts is precipitation. This result shows that under climate conditions in which low interannual variability of temperature dominates, both drought index respond mainly to the variability of precipitation. Furthermore, they also said that drought could be determined only from precipitation data when temporal trends in temperature are not apparent. The inclusion of PET only affects the index when PET differs from average conditions under global warming scenarios.

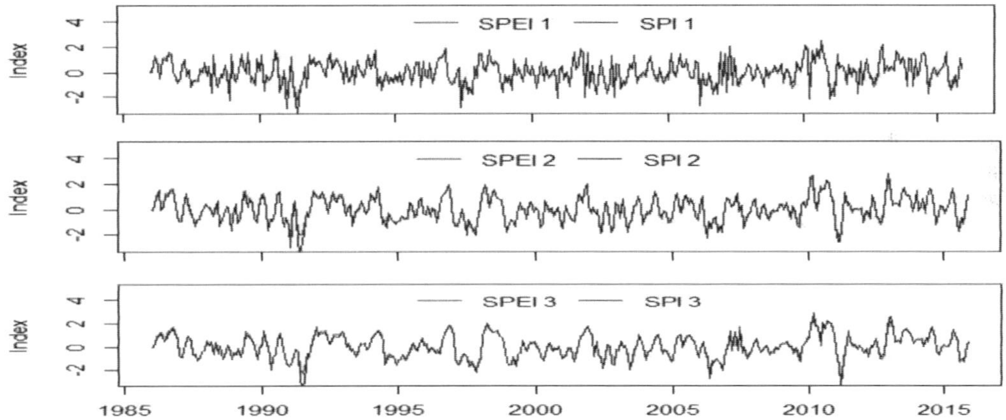

Figure 2. Drought index quantified by the SPEI and the SPI in different time scales; 1-month (above), 2-month (middle) and 3-month (bottom) refer to the 1986-2015 period.

Figure 3 presents the distribution of potential drought intensity in Bandung calculated by the SPI and SPEI refer to the 1986-2015 period. Based on index calculated result, droughts event occur almost every year, except in 1986 and 1992. The drought index analysis showed that the most persistent drought was in 2006 (8 months for SPI and SPE 3-months lag, 3 months for SPI and SPEI 2 month lag, and 4 months for SPEI 1-month lag), 1997 (5 months for SPI and SPE 3-months lag, 3 months for SPI and SPEI 2 month lag, and 3 months for SPI and SPEI 1-month lag) and 1991 (3 months for SPI and SPEI 3-months lag). The longest severe to extreme drought occurred in 1991 (2-3 months).

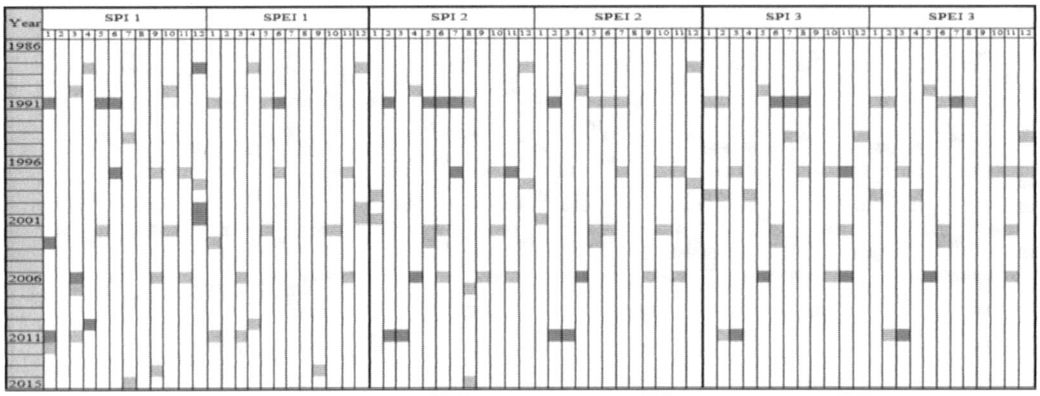

Drought categories: ☐ normal-wet condition; ☐ moderate drought; ☐ severe drought; ☐ extreme drought

Figure 3. Multiannual distribution of moderate, severe and extreme drought in Bandung as quantified by SPI and SPEI at 1- to 3-month lag refer to the 1986-2015 period.

3.2 Maize Yield Simulation

Environmental abiotic stresses such as water and temperature have major negative impacts on canopy development, biomass production and yield, depending on timing of occurrence, severity and duration. AquaCrop is designed to simulate crop responses to water as the main factor effected crop yield. This models also concern about temperature and carbon dioxide concentration that required for the simulation under climate change scenarios. Water stress effects on productivity and water use processes are simulated by impacting: (1) canopy growth; (2) stomata conductance; (3) canopy senescence; (4) root deepening, and (5) harvest index [20].

Maize simulation results by AquaCrop presented in Figure 4. The Best maize production occurred during 1986, followed by 1992 and 1989. While the worst maize production occurred during 1997, followed by 2006, 1991 and 2002. The best years of the maize production happened when the drought index doesn't indicate a drought event. Vice versa, the worst year of maize production occurs when the drought index indicates the incidence of droughts.

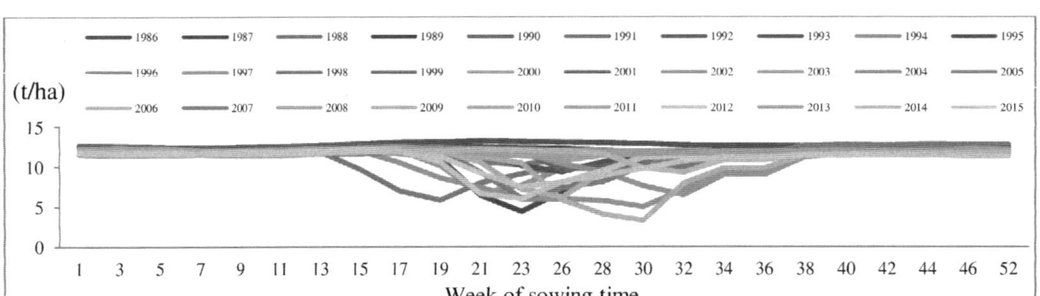

Figure 4. Yields of maize base on crop sowing time.

Maize sowing time during June to early July 1986 period gained the maximum yields (13.3 t/ha). The difference between the highest yield and anothers maize yields asumed based on the climatic conditions of Bandung. Thus, the entire crop yields below the maximum value is considered as the impacts of dryness/wetness conditions. The simulation shown that the lowest yield occurs when sowing in early July 2006(3.3 t/ ha). Hardest drought occurred in all phases of the plant lifecycle on 1 July to 15 November 2006. The crop model application noted that the harvest index (HI) of the crop decreased by up to 25% (maximum HI for maize is 48%) caused by canopy expansion decreased by 33% of the total maximum biomass.

The lower production is commonlly the result of sowing in the 11th to the 18th week (May-September). This is reasonable considering crop growth stages at these planting times pass the moments in which the water balance are deficit (see Figur 1(b)). The example of the life cycle of maize and the duration of each cycle, calculated by Aquacrop for 1700 growing degree days, can be seen in table 2.

Table 2. Maize growing stages.

Stage	Emergence	Vegetative	Flowering	Yield formation and ripening
Time	0-6st day	7th - 55th day	56th - 67th day	68th - 105th day

3.3 Drought Impact on Maize Yield Loss

The fluctuations in maize yields over time were calculated based on the agro-meteorological conditions (i.e dryness/wetness conditions) during the growing season from 1986-2016 period. In this study, we assumed that climatic influences (precipitation, PET and temperature) on maize yields generally occurred at a higher frequency than non-climatic influences (soil types, genetic and technological improvements, etc).

The correlation between drought magnitude and maize yield loss showed in Table 3. The drought intensity and frequency were calculate for SPEI 3 $\leq$ -1.0. The maize yield loss were calculated as the difference between the maximum yield value and each maize yields (t/ha). Detail information about maize showing time and maize production loss data shown in table 4. This information combined with crop growth stage in table 2 will help us to understand about the time when the growth phase of maize are affected by drought events. As an example can be seen during maize cultivation time in 1987. Maize yield loss still quite low when the crop being planted at the 1st to 19th week because the plants did not experience with drought stress. But when the showing time had been done on the 20th to 26th week, maize yield loss becomes very hight. This is due to drought affected crops in the generative phase. Maize showing time during 28th to 32th week experience drought in the vegetative phase, but the yield loss quite low.

It can be concluded from Table 4 that impacts of drought on maize yield loss depend on drought magnitude and duration and on plant development stages when droughts occur. The highest corn yield loss, 10 t/ha (75%), in Bandung happened during drought ongoing since the vegetative phase to harvest. Yield losses due to drought in the vegetative phase of lowering the yield to 40%, while when the generative phase (flowering to the formation of the kernel) lowering the yield to 56%. The percentage of yield losses was calculated by divided the difference between potential and actual yields (in Table 3) and yield potential and multiplied by 100.

4 Conclusion

This study represents the use of drought index to assess the drought magnitude and duration impacts on maize yield loss. Because of the lack of metadata of maize yields actual data, we used crop simulation to represent them. A detailed analysis of drought-yield relations under optimal agro-technological condition had been showed. Both SPI and SPEI have the sama result to described drought onset, duration and intensity in Bandung. Drought impacts on maize yield loss depend on drought magnitude and duration and on plant growth stages when droughts occur. Drought events during maize reproductive stages caused the highest maize yield loss, than drought during vegetative stages.

Some suggestions for the further research may arise from this study. Improved accuracy of the drought index can be done by extending the period of data used (over 30 years) in different climatic zone area. Each plant has a different tolerance to drought so it is needed to do research on various types of plants to determine the differences reduction of the crop yields.

Acknowledgements
We appreciate the insightful suggestions of reviewers. The study was supported by Indonesian Agency for Meteorology Climatology and Geophysics (BMKG).

References

[1] Wilhite DA 2011 Quantification of agricultural drought for effective drought mitigation and preparedness : key issues and challenges *Agricultural drought index : Proceeding of An Expert Meeting* WMO 13-21
[2] Bryant E 2005 *Natural hazard (2nd edition)* Cambridge University Press : Chapter Introduction 10-13
[3] Sivakumar MVK 2011 Agricultural drought-WMO perspectives *Agricultural drought index : Proceeding of An Expert Meeting* WMO 22-30
[4] *Nota dinas laporan perkembangan serangan OPT, banjir dan kekeringan mei 2016* Direktur Perlindungan Tanaman Pangan 16
[5] Potova V, Boroneant C, Boincean B and Soukup J 2015 *Int. J. of Climatol.* 10.1002/joc.4481
[6] McKee TB, Doesken NJ and Kleist J 1993 The relationship of drought frequency and duration to time scales proceedings of the 8th *Conference on Applied Climatology, Anaheim-California American Society Boston*
[7] Heim RR 2002 *Bull Amer Meteor Soc* **83** 1149–1165
[8] Vicente-Serrano SM, Beguería S and López-Moreno JI 2010 *Journal of Climate* **23** 1696-1718
[9] Cloppet E 2011 Agricultural drought index in france and europe : strengths, weakness and limitation *Agricultural drought index : Proceeding of An Expert Meeting* WMO 83-94
[10] Sivakumar MVK, Stone R, Sentelhas PC, Svoboda M, Omondi P, Sarkar J and Wardlow B 2011 Agricultural drought index : summary and recomendations *Agricultural drought index : proceeding of an expert meeting* WMO 172-197
[11] Stone RC 2011 Agricultural drought index in current use in australia : strengths, weakness and limitation *Agricultural drought index : Proceeding of an expert meeting* WMO 72-82
[12] WMO 2012 *Standardized precipitation index user guide* (Geneva, Switzerland)
[13] Hayes M, Svoboda M, Wall N and Widhalm M 2010 *Bulletin of the American Meteorological Society* **92** 485-488
[14] Bordi I, Frigio S, Parend P, Speranza A and Sutera A 2001 *Annali di Geofisica* **44(5/6)** 979-993
[15] Thornthwaite CW 1948 *Geographical Review* **38** 55–94
[16] Stagge JH, Tallaksen LM, Chong YX and Van Lanen HAJ 2014 Standardized precipitation-evapotranspiration index (SPEI): sensitivity to potential evapotranspiration model and parameters *Hydrology in a Changing World: Environmental and Human Dimensions Proceedings of FRIEND-Water,* Montpellier (France) 367-373
[17] Motha RP 2011 Use of crop models for drought analysis *Agricultural drought index : Proceeding of an expert meeting* WMO 138-148
[18] Heinemann AB, Dingkuhn M and Delphine 2008 *Euphytica* **162(3)** 395-410
[19] Indonesia Ministry of Agriculture ____ Monthly report of drought impact on maize planting area in West Java province during 2005-2015 period of observation [Unpublished]
[20] Steduto P, Hsiao TC, Fereres E and Raes D 2012 *Crop yield response to water* FAO Irrigation and Drainage Paper 66 (Rome)
[21] Hsiao TC, Heng L, Steduto P, Rojas-Lara B, Raes D and Fereres E *Agron. J.* **101** 448-459 doi: 10.2134/agronj2008.0218s
[22] Heng LK, Hsiao T, Evett S, Howell T and Steduto P 2009 *Agron. J* **101(3)** 488-498

IOP Conf. Series: Earth and Environmental Science **58** (2017) 012036 doi:10.1088/1755-1315/58/1/012036

Appendix

Table 3. Conservative (constant) parameter of Aquacrop taken from Hsiao *et al.* [21] and used for the simulation of maize productivity

Base temperature	: 8.0 °C
Cut-off temperature	: 30.0 °C
Canopy cover per seedling at 90% emergence (cc_o)	: 6.5 cm^2
Canopy growth coeffi cient (CGC)	:1.3%
Maximum canopy cover (CC_x)	: function of plant density
Crop coeffi cient for transpiration at CC = 100%	: 1.03
Decline in crop coef. after reaching CC_x	: 0.3%)
Canopy decline coefficient (CDC) at senescence	: 1.06%
Water productivity, normalized to year 2000 (WP*)	: 33.7 gm (biomass) m^{-2}
Leaf growth threshold (p_{upper})	: 0.14
Leaf growth threshold (p_{lower})	: 0.72
Leaf growth stress coeffi cient curve shape	: 2.9
Stomatal conductance threshold (pupper)	: 0.69
Stomata stress coeffi cient curve shape	: 6.0
Senescence stress coeffi cient (pupper)	: 0.69
Senescence stress coeffi cient curve shape	: 2.7
Reference harvest index (HI_o)	: 48%
GDD from 90% emergence to start of anthesis	: 800
Duration of anthesis, in GDD	: 190
Coefficient, inhibition of leaf growth on HI	: 7.0
Coefficient, inhibition of stomata on HI	: 3.0

Table 4. Impact of drought magnitude (quantified by SPEI 3-month lag) on maize yields loss (according to AquaCrop simulation based on difference sowing time). The numbers here show the value of the difference between the maximum yield and the yield at different sowing time (t/ha). The greater value, the greater the production loss. Maximum maize yield in this study was 13.3 t/ha.

Year	Week of sowing time																							
	1	3	5	7	9	11	13	15	17	19	21	23	26	28	30	32	34	36	38	40	42	44	46	52
1986	0.7	0.7	0.8	0.8	0.8	0.7	0.6	0.4	0.2	0.2	0	0.1	0.2	0.3	0.4	0.5	0.6	0.6	0.6	0.7	0.8	0.7	0.8	0.8
1987	0.9	0.9	0.9	1	0.9	0.8	0.8	0.6	0.7	2.2	6.8	8.9	6.4	3.6	2.2	1.4	1.3	1.3	1.2	1.2	1	1.2	1.2	1.1
1988	1	1.1	1.2	1.3	1.3	1.3	1.1	0.9	0.7	1.3	2.7	3	2.3	1.7	1.4	1.5	1.5	1.5	1.3	1.3	1.3	1.1	1.1	1.2
1989	1.1	1.3	1.3	1.4	1.2	1.3	1.2	1.1	1	0.7	0.8	0.9	1	1.2	2.2	1.7	1.6	1.6	1.3	1.1	1	1.3	1.4	1.4
1990	1.5	1.7	1.7	1.5	1.5	1.4	1.3	1.1	1.2	1.2	1.2	1.3	1.3	1.4	1.6	1.6	1.8	1.8	1.6	1.5	1.4	1.4	1.4	1.3
1991	1	1.1	1.2	1.2	1.2	1.2	1.5	3.6	6.2	7.4	5.3	4.2	3.3	3.6	3.2	3.3	1.6	1.6	0.6	0.5	0.5	0.4	0.5	0.5
1992	0.6	0.7	0.8	0.9	0.8	0.8	0.6	0.5	0.6	0.8	1	1.3	1.5	1.6	1.6	1.7	1.7	1.7	1.8	1.6	1.4	1.2	1	0.9
1993	1	1.3	1.3	1.4	1.6	1.5	1.4	1.4	1.4	1.2	0.9	1.2	1.2	1.2	1.4	1.5	1.6	1.6	1.7	1.6	1.5	1.4	1.3	1.2
1994	1.2	1.4	1.4	1.3	1.3	1.2	1	0.9	1.2	3.2	6.6	5.5	3.5	3.3	2.4	1.7	1.6	1.6	1.8	1.7	1.7	1.6	1.6	1.5
1995	1.7	1.7	1.6	1.6	1.7	1.7	1.4	1.4	1.1	1.1	1.5	3.1	3.9	2.7	1.9	1.7	1.6	1.6	1.5	1.4	1.4	1.5	1.4	1.5
1996	1.7	1.7	1.6	1.6	1.5	1.5	1.6	1.5	1.4	1.5	1.3	1.3	1.5	1.3	1.4	1.5	1.3	1.3	1.2	0.9	1.1	1	1.2	1.3
1997	1.4	1.4	1.5	1.5	1.3	1.2	1.2	1.2	2.7	4.7	5.9	7.3	7.3	7.5	8.3	6.1	3.6	3.6	1.8	1.9	1.7	1.7	1.7	1.7
1998	1.6	1.6	1.7	1.7	1.9	1.7	1.7	1.7	1.5	1.4	1.4	1.4	1.5	1.5	1.6	1.7	1.5	1.5	1.3	1.3	1.4	1.3	1.3	1.3
1999	1.2	1.3	1.4	1.3	1.2	1.1	1.1	0.9	0.9	0.8	0.9	1.3	3.6	3.3	2.6	2	1.5	1.5	1.2	1.3	1.1	1.1	1.1	1.1
2000	1.1	1.3	1.4	1.5	1.6	1.4	1.3	1.2	1	0.9	1.1	1.9	2.2	1.6	1.6	1.6	1.5	1.5	1.5	1.4	1.4	1.2	1.2	1.2
2001	1.2	1.4	1.6	1.6	1.4	1.5	1.3	1.2	1.1	1.1	1.1	1.4	1.3	1.4	1.5	1.5	1.4	1.4	1.6	1.6	1.6	1.5	1.5	1.4
2002	1.3	1.6	1.7	1.7	1.6	1.7	1.5	1.4	1.3	1.2	1.2	1.6	2.9	3.9	5.7	6.8	4.2	4.2	2	1.8	1.6	1.6	1.4	1.5
2003	1.6	1.5	1.6	1.5	1.5	1.5	1.2	1.1	1.7	2	1.9	1.2	1.2	1.3	1.3	1.3	1.4	1.4	1.4	1.3	1.3	1.3	1.5	1.4
2004	1.3	1.3	1.6	1.5	1.3	1.2	1.2	0.9	0.7	1.1	1.7	1.8	1.5	1.5	1.7	1.7	1.7	1.7	1.7	1.7	1.5	1.5	1.6	1.6
2005	1.6	1.7	1.6	1.6	1.6	1.6	1.5	1.4	1.3	1.2	1.2	1.4	1.5	1.6	1.6	1.5	1.4	1.4	1.4	1.4	1.3	1.4	1.4	1.6
2006	1.6	1.7	1.6	1.5	1.4	1.3	1.2	1	0.9	1.3	2	2.8	7.4	9.2	10	5.3	3.4	3.4	1.8	1.7	1.6	1.5	1.3	1.4
2007	1.3	1.3	1.4	1.5	1.5	1.3	1.2	1.2	1.1	1.1	3.9	7.2	5.9	5.2	3.2	1.8	1.6	1.6	1.4	1.3	1.3	1	1	1.1
2008	1	0.9	1	1.2	1.2	1.1	1.1	1	0.9	1.2	1.6	1.8	1.9	1.8	1.6	1.7	1.6	1.6	1.3	1.3	1.4	1.3	1.2	1.2
2009	1.2	1.4	1.4	1.3	1.4	1.5	1.6	1.3	1.2	1.2	2.7	6.1	7.1	3.9	2	1.7	1.5	1.5	1.5	1.3	1.4	1.5	1.3	1.3
2010	1.4	1.6	1.8	1.7	1.9	1.9	1.6	1.4	1.2	1.1	1.3	1.4	1.5	1.6	1.6	1.6	1.4	1.4	1.3	1.2	1.3	1.3	1.4	1.5
2011	1.6	1.7	1.6	1.5	1.5	1.4	1.3	1.2	1.2	1.2	1.1	1.6	2.5	1.9	1.7	1.7	1.6	1.6	1.7	1.7	1.9	1.7	1.6	1.5
2012	1.6	1.5	1.5	1.6	1.4	1.4	1.2	1.1	1	2	4.1	5.8	5.2	3.7	2.1	1.9	1.8	1.8	1.7	1.6	1.6	1.6	1.6	1.6
2013	1.7	1.8	1.8	1.7	1.7	1.6	1.5	1.3	1.2	1.1	1.1	1.1	1.4	1.5	1.7	1.6	1.6	1.6	1.5	1.4	1.2	1.3	1.1	1.1
2014	1.4	1.5	1.7	1.7	1.7	1.7	1.5	1.4	1.3	1.2	1.2	1.2	1.3	1.6	3.4	4	2.5	2.5	1.7	1.6	1.6	1.6	1.5	1.5
2015	1.5	1.7	1.6	1.6	1.6	1.5	1.4	1.3	1.3	2.6	6.6	7.3	5.1	4.4	3.5	2.6	2.1	2.1	1.8	1.8	1.8	1.8	1.9	1.9

Categories of drought : normal-wet condition; moderate drought; severe drought; xtreme drought

Preliminary Study of Nocturnal Rainfall Mechanism over Semi Closed Equatorial Oceans as observed by TRMM-PR

F Alfahmi[1*], R Hidayat[1], R Boer[1] and Perdinan[1]

[1] Faculty of Mathematics and Natural Sciences, Bogor Agricultural University, Bogor, Indonesia

Email: furqonalfahmi@gmail.com

Abstract. Relation of shape and size of coastal line to coastal rainfall mechanism over Indonesian Maritime Continent (IMC) is not clearly understood. Cendrawasih Bay is a unique coastal line because it is the biggest curvature type in IMC, near equatorial lines and semi-closed oceans. The mechanism of nocturnal rainfall over Cendrawasih Bay has been studied using data from Tropical Rainfall Measuring Mission Precipitation Radar (TRMM-PR) 3B42 product over during 1998–2015. Correlation coefficient of TRMM-PR and observational data from Biak Meteorological Station in 2014 is 0.6. Composite TRMM-PR data showed that nocturnal rainfall over the Cendrawasih Bay began to form at midnight and has peak intensity after midnight. Rain rate intensity over the Cendrawasih Bay is bigger than over inland and adjacent waters with speed of rainfall migration is approximately 5 m/s. Northern monsoon flows and land breeze along curvature coast induced low level convergence over Cendrawasih Bay during late evening until morning.

1. Introduction

Indonesian Maritime Continent (IMC) is the largest archipelago with the greatest convective activity and the longest coastal line in Asia. Convective activity in this area is very complex [1]. Mountainous islands, such as Sumatra, Java, Kalimantan (Borneo), Sulawesi, Papua and many other islands of various sizes are located in the region. The complicated topography induced complex convective activity can significantly affect the formation and distribution of rainfall [2].

Diurnal rainfall over IMC is one of the most significant variations of convection and has close relation to coastal area. Rainfall occurs mostly at boundary of sea land [3]. Many previous studies have described the characteristics and propagation of the diurnal convection over IMC [4]–[8]. Peak of rainfall usually occurs at land on daytime and over oceans during evening until early morning [9]. Peak of rainfall in both Sumatera and Borneo Island show a movement of rainfall from inland to oceans [5], [7].

IMC has a long coastal line with unique shape and size. Relationship of shape and size of coastal line to rainfall mechanism was never discussed on previous studies. This paper specially will discuss about nocturnal rainfall mechanism over the biggest curvature coastal line in IMC, Cendrawasih Bay. There is a very unique location because it is directly facing Pacific Oceans, the biggest curvature coastline and located near equatorial line.

2. Data

The data sets used in the present study are the Tropical Rainfall Measuring Mission Precipitation Radar (TRMM-PR) 3B42 product and National Centers for Environmental Prediction-Final Analysis (NCEP-FNL). Some previous study examined diurnal variation using TRMM-PR 3B42 product at IMC [9]–[13]. To examine diurnal rainfall, we made a composite 3-hourly dataset of rain rate based only on the PR sensor for 18 years (1998–2015) and also calculated total average rain rate for analysis over Cendrawasih bay. Migration of rainfall from inland to oceans was detected by time-spatial slice methods [5]. Correlation coefficient of TRMM-PR and observational data from Biak Meteorological Station in 2014 was 0.6.

Simulation of dynamical atmosphere over Cendrawasih bay was calculated by Weather Research and Forecasting – Advanced Research WRF (WRF-ARW). The initial and boundary conditions were from NCEP-FNL. In these simulations, 5-layer thermal diffusion and Yonsei University scheme boundary layer schemes were employed. The cloud microphysics and radiation schemes used were the WRF Grell-Devenyi (GD) ensemble scheme and RRTMG scheme, respectively. The WRF model was run at 39 vertical levels with the model top at 50 hPa. Parameters of WRF-ARW product used to analyze diurnal cycle were wind to calculate Low Level Convergence Filed, Convective Available Potential Energy (CAPE) and lifted condensation level (LCL).

3. Result and Discussion

3.1. Total Mean of Rainfall

Figure 1a below is mean of rain rate base on TRMM-PR 3B42 during 1998-2015. Fig. 1a shows that peak of rainfall over inland and adjacent water has integrated rainfall system. High rainfall intensity over inland will be followed with high rainfall intensity over adjacent oceans. Rainfall intensity over oceans near coastal line is lower than two locations before. Land-sea rainfall system at Sumatera had been examined by Mori [5]. Peak of rainfall seemed to continue from island to oceans. Migration of rainfall from land to oceans about 11.1 m/s. Base on peak of rainfall, land-sea rainfall system over Papua Island seemed to consist of three land-sea rainfall system; Sorong land-sea rainfall system, Nabire land-sea rainfall system and Asmat land-sea rainfall system.

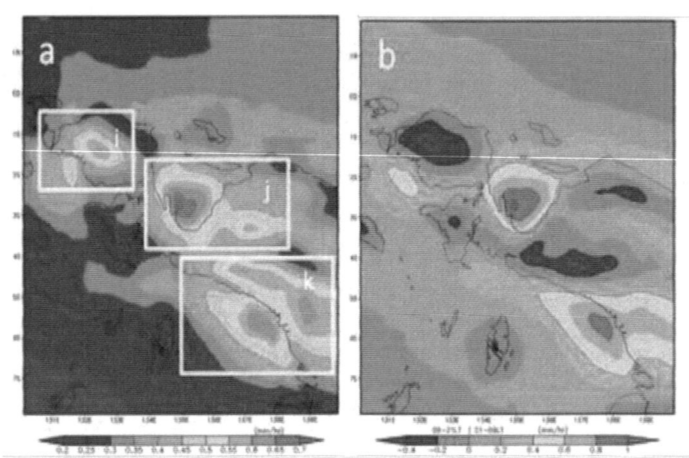

Figure 1(a) Total mean of rain rate during 1998 – 2015, index-*i* is Sorong system, *j* is Nabire system, *k* is Asmat system **(b)** Total mean of daytime and nocturnal rain rate during 1998 – 2015. Minus mean composite of daytime rainfall and plus mean composite of nocturnal rainfall.

Diurnal rainfall divided into two part, nocturnal rainfall is rainfall that occurred from 20.00-08.00LT and daytime rainfall is rainfall that occurred during 08.00-20.00LT [14]. Fig. 1b shows the location of nocturnal and daytime rainfall. Nocturnal rainfall mostly occurred at offshore area (Cendrawasih Bay, Sorong Waters and Asmat waters), and daytime rainfall occurred at inland area (Nabire, Sorong and Asmat). Neumann [15] suggested that land breeze front on curvature coastal line have a significant effect as a trigger midnight convection at oceans, while daytime rainfall convection over island was triggered by sea breeze front [16] and merger cumulus [2], [9].

Cendrawasih Bay has semi-annual variability of rainfall and have strong influence El Niño–Southern Oscillation (ENSO) during normally dry season [17]. Peak of rainfall over Cendrawasih Bay base on TRMM-PR during 1998- 2010 occurred on March, June, and September. Those conditions likely correlated with annual apparent motions of the sun. March 21st and September 23rd are the date on which the annual apparent motion of the sun is located at the equatorial line, while June 21st is the date on which the annual apparent motion of the sun is located at the vernal equinox. Impact of annual apparent motion of the sun is variability of monsoon wind. March and September are month of transition and June is month of peak of dry seasons.

3.2. Distribution and Diurnal of Rainfall over Semi Closed Equatorial Oceans

Diurnal distribution and variation of rainfall that had been detected by the Precipitation Radar (PR) onboard the TRMM satellite for 18-year period from 1998 to 2015 is displayed in Fig. 2. In the early afternoon (1030–1330 LT) most areas of Papua Island and Cendrawasih Bay are nearly rain-free. During the late afternoon and early evening hours (1330–1930 LT), rainfall occurs mostly over the island along the coast and close to the mountainous area of the island. Previous research described that rainfall occurred along coastal line may be triggered by sea breeze front [16], [18], [19] . During these times of the day, rainfall coastal zone continues to grow and spread even across the top of the mountain.

Rainfall on valley began to intense at evening time (1930 – 2230 LT). Local mountain breeze drove rainfall over valley on western of Sarmi. In another part, some of the peak rainfall moved to west and began to intense at oceans.

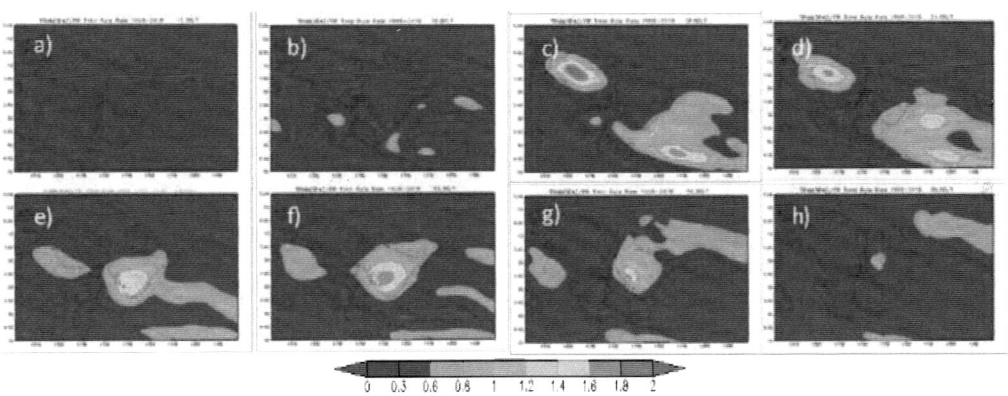

Figure 2 Diurnal rainfall variation composited 3-hourly TRMM-PR 3B42 product during 1998 – 2015 **(a)** 1030 – 1330 LT **(b)** 1330 – 1630 LT **(c)** 1630 – 1930 LT **(d)** 1930 – 2230 LT **(e)** 2230 – 0130 LT **(f)** 0130 – 0430 LT **(g)** 0430 – 0730 LT **(h)** 0730 – 1030 LT.

At nighttime until late morning (2230–1030 LT) rain occurred mainly over Cendrawasih Bay. This was main period of nocturnal rainfall. Peak of nocturnal rainfall occurred at after midnight until 0300 LT. Peak rainfall over western ocean of Sumatera [5], [20], western ocean of Borneo [7] and Cendrawasih Bay were almost same time.

Nocturnal rainfall intensity over Cendrawasih Bay was higher than adjacent sea. Nocturnal rainfall intensity at north of Papua was 0.6 mm/hr, meanwhile, rainfall intensity over Cendrawasih Bay was 1.8 mm/hr. Shape of curvature coastline type created effect of cumulus merger. Rainfall from east, south and west migrated together to middle of oceans. Predominant cloud type at north of Cendrawasih Bay (Biak Island) during nocturnal rainfall was stratiform. Nocturnal rainfall over Biak

Island was due to the northward propagation of a cloud system that appeared at midnight in the northern coastal sea region of Papua Island [6].

3.3. *Migration of Rainfall*

To examine migration of rainfall, we created meridional-time slice and zonal-time slice during diurnal rainfall. Migration of meridional and zonal rainfall to and from around Semi Closed Equatorial Oceans are shown in Fig. 3. Arrow α and β in Figure 3 indicates direction of movements from onshore to inland and land to oceans. From Fig. 3 we can show that there is rainfall migration from land to oceans for all direction, except from open oceans directions.

Analysis of zonal movement (BB') shows that rainfall began to form on coast on the east side of Cendrawasih Bay at afternoon until early evening. Rainfall moved form on coast on the east side to westward with speed 5.2 m/s (β_2). This movement had peak intensity reach 1 mm/hr. During shifting, the peak intensity of rainfall decreased at distance 10-30 km from coastline. Same as with coast on the east side, coast on the west side had movement rainfall to oceans at evening time (β_1). β_1 had the same speed with β_2 but had different direction. β_2 had easterly movement, but β_1 had westerly movement. During shifting, the peak intensity of westerly movement has more decreased than easterly movement. This condition occurred because westerly movement runs though small water between lands.

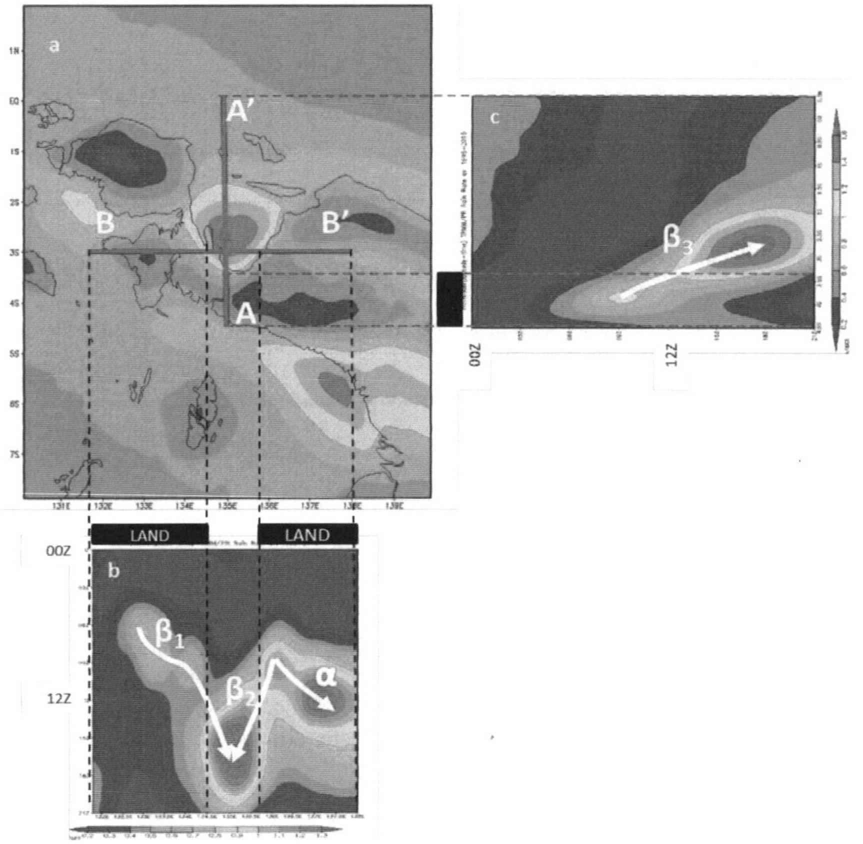

Figure 3 (a) Cross section positions for zonal-time slice and meridional-time slice **(b)** Zonal-time slice cross section. β_1 and β_2 is westerly and easterly movement rainfall from land to oceans area. α is westerly movement rainfall from onshore to inland area. **(c)** Meridional-time slice cross section. β3 is southerly movement rainfall from land to oceans.

Speed movement of α (7.9 m/s) was faster than β1 and β2. Rainfall of α was disappeared at late evening. Characteristic of α rainfall's at this region was the same as with α rainfall at Sumatera, predominant on the deep convective-type peak rain rate [5], [6].

Analysis of meridional movement (AA') showed that rainfall move from south to the middle of the oceans (β_3), but there was no migration rainfall from north. Speed of rainfall β_3 was 4.6 m/s. Rainfall at inland relatively low, but increased after reach the oceans. Average speed of rainfall β for all direction was 5 m/s.

Merger of rainfall system (β_1, β_2, β_3) looks occurred in the middle of the oceans. This merger condition was induced by local wind circulation. Different temperature between oceans and inland create wind circulation to convergence in the middle of the oceans at land breeze conditions.

3.4. Diurnal Variations of Low Level Convergence Field, CAPE and LCL

Diurnal variation of rainfall was likely affected by low level convergence (shown in Fig. 2). Low level convergence began to form at 0730 LT near mountainous area. It continued to develop until 1930 LT, but it had peak magnitude at 1200 LT. There was no rainfall over mountain at 0730. Rainfall began to form at late afternoon. Approximately, delayed three hours between peak of low level convergence and rainfall occurred at mountain.

Low level convergence was induced by local circulation. Low level convergence at mountainous area was formed by sea breeze convergence at middle of island. Sea breeze convergence zones related to the shape of coastline and orographic effects. This mechanism at Java island can create merger cumulus [21], but at Papua island cannot create cumulus merger at convergence zone. This condition may be because the size of island is too large [22] and mountainous area is too high [2].

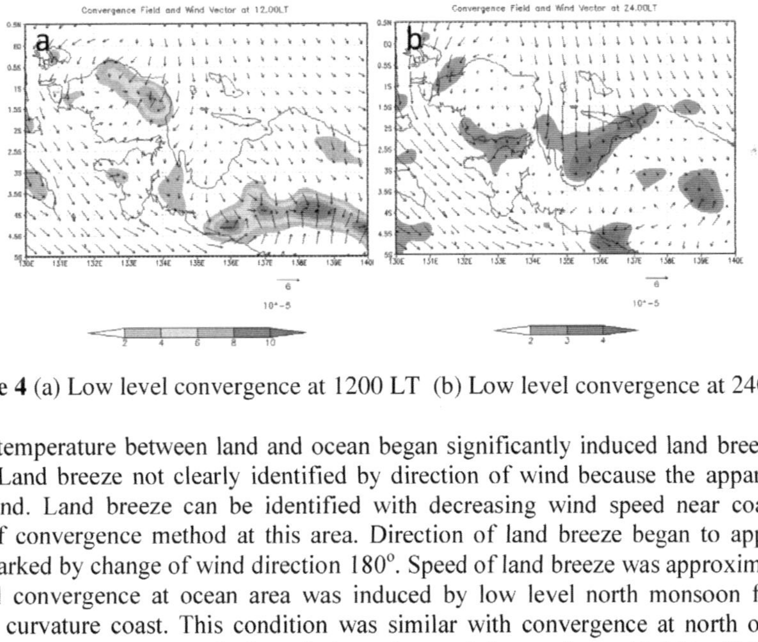

Figure 4 (a) Low level convergence at 1200 LT (b) Low level convergence at 2400 LT.

Different temperature between land and ocean began significantly induced land breeze circulation at 1930 LT. Land breeze not clearly identified by direction of wind because the apparent wind still move to inland. Land breeze can be identified with decreasing wind speed near coastal line and calculation of convergence method at this area. Direction of land breeze began to appear 2230 LT which was marked by change of wind direction 180°. Speed of land breeze was approximately 5 m/s.

Low level convergence at ocean area was induced by low level north monsoon flow and land breeze along curvature coast. This condition was similar with convergence at north of Borneo [7], [23]. Land breeze convergence had peak intensity at 3-hour after midnight. This gradient of temperature made strong land breeze caused increasing intensity of convergence.

Diurnal variations of Convective Available Potential Energy (CAPE) showed that dynamical atmosphere was very unstable at midnight. CAPE is effectively the positive buoyancy of an air parcel and is an indicator of atmospheric instability. The higher value of CAPE means unstable atmospheric condition. CAPE was highest magnitude at 0300 LT (1700 J/kg).

Lifting Condensation Level (LCL) is the height at which the relative humidity (RH) of an air parcel will reach 100% when it is cooled by dry adiabatic. The LCL is a good approximation of the height of the cloud base when air is lifted mechanically from the surface such as convergence area. Fig. 5 shows that base cloud at midnight over Cendrawasih Bay was lower than at daytime. Base cloud at midnight approximately 350 m, while at daytime 500 m. Cloud base was significantly associated with cloud types producing heavy rainfall. Cumulonimbus usually have base cloud approximately 200 – 400 m. Cumulonimbus clouds can produce high winds, heavy rainfall, lightning, gust fronts, waterspouts, funnel clouds, and tornadoes.

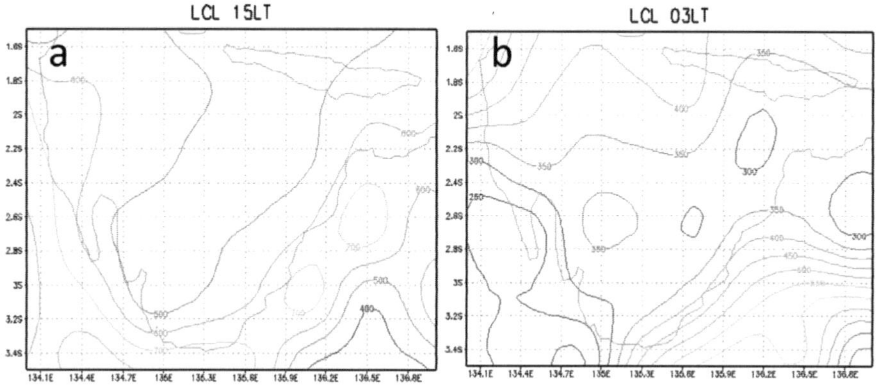

Figure 5 (a) LCL (m) at 1500LT **(b)** LCL (m) at 0300 LT.

4. Summary and Concluding Remarks

Diurnal rainfall over Cendrawasih Bay has different characteristics with adjacent water and another open sea. Curvature coastal line has significant effect for diurnal variations. Merger rainfall and convergence field over bay induced rainfall has high intensity at nocturnal rainfall. Nocturnal rainfall over the Cendrawasih Bay begins to form at midnight and has peak intensity after midnight (0300LT).

There is migration of nocturnal rainfall from all direction except open sea. Different temperature land and oceans induce land breeze that affects migration of rainfall. Speed of rainfall movement at nocturnal system is 5 m/s.

Northern monsoon flow and land breeze along curvature coastal line induce low level convergence at evening over oceans. Atmosphere is unstable which marked by high magnitude of CAPE. Cumulonimbus is very easily formed due to lower LCL.

This paper just focuses on TRMM data and simple parameter for analysis. More work is needed to examine more clearly the relationship of size of curvature of coastal line and mechanism of nocturnal rainfall. More intense observation is also needed because meteorological station observation over Nabire is just 12-hr operational.

References

[1] CS Ramage 1968 *Mon. Wea. Rev.* **96** 365–370
[2] T Qian, CC Epifanio and F Zhang 2011 *J. Atmos. Sci.* **69(1)** 130–149
[3] GY Yang and J Slingo 2001 *Mon. Wea. Rev.* **129**784–801
[4] H Ichikawa and T Yasunari 2006 *J. Clim.* **19(7)** 1238–1260
[5] S Mori *et al.* 2004 *Mon. Wea. Rev.* **132(8)** 2021–2039
[6] Y Tabata *et al.* 2011 *J. Atmos. Solar-Terrestrial Phys.* **73(9)** 1031–1042
[7] P Wu, DY Manabu and J Matsumoto 2008 *J. Meteorol. Soc. Japan* **86A** 187–203
[8] P Wu, M Hara, JI Hamada, MD Yamanaka and F Kimura 2009 *J. Appl. Meteorol. Climatol,* **48(7)** 1345–1361

[9] A Pribadi, P Wongwises, U Humphries, A Limsakul and A Wangwongchai 2012 *Journal of Sustainable Energy & Environment* **3** 81–86
[10] K Kikuchi and B Wang 2007 *Journal of Climate* **21** 2680–2696
[11] BC Bhatt, S Sobolowski and A Higuchi 2016 *J. Meteorol. Soc. Japan.* **94A** 89–103
[12] Trismidianto, TW Hadi, S Ishida, Q Moteki, A Manda and S Iizuka 2016 *Sola* **12** 6–11
[13] R Hidayat and S Kizu 2010 *Int. J. Climatol.* **30** 1816–1825
[14] S Yin, D Chen and Y Xie 2008 *Int. J. Climatol.* **29(8)** 1154–1170
[15] J Neumann 1951 *Journal of Meteorology* **8** 60–67
[16] C Azorin-Molina, BH Connell and R Baena-Calatrava 2009 *J. Appl. Meteorol. Climatol.* **48(10)** 2069–2085
[17] E Aldrian and RD Susanto 2003 *Int. J. Climatol.* **23(12)** 1435–1452
[18] DE Kingsmill 1995 *Mon. Wea. Rev.* **123(10)** 2913–2933
[19] RG Fovell 2005 *Mon. Wea.Rev.* **133(1)** 264–278
[20] N Sakurai *et al.* 2005 *J. Meteorol. Soc. Japan* **83(5)** 835–850
[21] JH Qian, AW Robertson and V Moron 2010 *J. Atmos. Sci.* **67(11)** 3509–3524
[22] Z Xian and RA Pielke 1991 *Journal of Applied Meteorology* **30(9)** 1280–1304
[23] RH Johnson and DC Kriete 1982 *Mon. Wea. Rev.* **110** 1898–1911

Trajectory and Concentration PM_{10} on Forest and Vegetation Peat-Fire HYSPLIT Model Outputs and Observations (Period: September – October 2015)

Khairullah[1,2], S Effendy[1]* and E E S Makmur[2]

[1]Departement of Geophysics and Meteorology, Faculty of Mathematics and Natural Sciences, Bogor Agricultural University (IPB), Indonesia
[2]Meteorology, Climatology and Geophysics Agency, Indonesia

Email: khairullah.bmkg@gmail.com

Abstract. Forest and vegetation peat-fire is one of the main sources of air pollution in Kalimantan, predominantly during the dry period. In 2015, forest and vegetation fire in Central Kalimantan and South Kalimantan emit large quantities of smoke leading to poor air quality. Haze is a phenomenon characterized by high concentration of particulate matter. This research objective is to analyze trajectory and dispersion of concentration particulate matter, PM_{10} in Banjarbaru and Palangka Raya. Dynamics of PM_{10} (Particulate matter less than or 10 μm in size) on vegetation peat-fire is done using GDAS (Global Data Assimilation System) output with a horizontal resolution 1° which corresponds to 100 km x 100 km for input model. Climate conditions in the period September to October 2015 at generally during dry season of El Nino year. The Hybrid-single Langrangian Integrated Trajectory (HYSPLIT) model was used to investigate concentration and long-range movement of this pollutant from the source to the receptor area. We used time-series data on PM_{10} readings obtained from two stations Banjarbaru (South Kalimantan) and Palangka Raya (Central Kalimantan) belonging to Meteorology Climatology and Geophysics Agency (BMKG). We also used weather parameter such as wind speed and direction. We investigated trajectory run from hotspots information MoF (Sipongi Output Programs) and HYSPLIT. We compared concentration obtained from PM_{10} observation and its concentrations trend. The dispersion pattern, as simulated by HYSPLIT showed that the distribution of PM_{10} was greatly influenced by the wind direction and topography. There is a large difference between the concentration of PM_{10} Palangka Raya and Banjarbaru.

1. Introduction

Forest and vegetation peat-fire (biomass burning) affects significant impacts on human health and climate. This issue was a source smoke haze. This problem has become an emerging disaster issue over last few years, predominantly during the dry period [1]. Smoke haze is a source of disasters such as poor visibility, degradation of local air quality and economic loss [2].

Biomass burning which consists of carbon emission is responsible for increasing particulate matter in El Nino-Southern Oscillation (ENSO) period [3]. During period of smoke haze,the existence of particles of PM_{10}, $PM_{2.5}$ and TSP in air could exceeds the health-safe threshold. Particulate matter emitted by the fires is the dominant pollutant causing exceedances of ambient air quality thresholds on a regional scale [4]. Particulate PM_{10} was a major driver of the pollutant in the atmosphere from biomass burning [2].

In the El Nino year 2015, Indonesia experienced abnormal drought conditions. The 2015 fire season by September, in much of Kalimantan were blanketed in thick smoke that lasted through October. The thick smoke that blanketed the equator in September-October 2015 was a worst since 1997 with the worst in Palangka Raya [5].

Banjarbaru and Palangka Raya is an important observation point PM_{10} BMKG in Kalimantan. Smoke haze with increasing particulate that affected Banjarbaru 2015 probably derived source of suburban South Kalimantan until the surround of Palangka Raya. In order to reduce these losses, we need a model that can describe the impact of the pollutant distribution area. Dispersion model is very useful to design strategies to reduce emissions and manage ambient air quality [6]. The concentration of pollutans varies by wind direction at a receptor [7].

HYSPLIT (Hybrid Single-Particle Langrangian Trajectory) is useful to predict the trajectory, dispersion and concentration of pollutants from a point source, line and area [8]. HYSPLIT uses meteorological data to track the movement and concentration either forward or backward in the period and at regular intervals [9].

This research objective is to analyze trajectory and dispersion of concentration particulate matter, PM_{10} in Banjarbaru and Palangka Raya.

2. Materials and Methods

This study uses weather observations data and concentration of particulate matter PM_{10} from meteorological station observations BMKG Palangka Raya and Banjarbaru the period from September to October 2015. We also use weather data model from GDAS (Global Data Assimilation System). Data GDAS1 is downloaded with a horizontal resolution of 1 ° corresponding to 100 km × 100 km.

The tool that used for running the simulation is HYSPLIT application (Hybrid Single-Particle Langrangian Trajectory) version 4.9 to track the trajectory and concentration. Rstudio was running by R version 3.2.3 and Openair to display weather data. GoogleEarth was used to display the location of stations.

The methodology used for this study is divided into these following steps:

2.1. Inventory data

Inventory observation weather data, model GDAS data, observation concentration PM_{10} data and hotspots data. Synoptic weather observation data on the surface of an hourly scale obtained from 2 locations Banjarbaru and Palangka Raya meteorological observation stations period September-October 2015 (Figure 1). GDAS model data was downloaded at the website https://ready.arl.noaa.gov/ready2-bin/extract/extract.pl/. Data GDAS1 has a resolution of 1 degree every data network measuring 111.1984 km for every latitude and longitude (Table 1). Data in the form of weekly data that is stored in a bin format consisting of meteorological parameters [10]. PM_{10} observation data was obtained from measurement the automatic instruments. Hotspots data was obtained from the hotspot MoF (Sipongi Output Programs) by the ministry of forestry and environment Indonesia. Hotspots data was given by the website http://sipongi.menlhk.go.id/hotspot/main [11]

Table 1. List of BMKG stations measuring air quality in South Kalimantan and Central Kalimantan Province

Province	Stasion	Longitude	Latitude	Altitude	Code Station WMO
South Kalimantan	Banjarbaru	114,841	-3,462	55 m	96687
Central Kalimantan	Palangka Raya	113,944	-2,226	27 m	96655

IOP Conf. Series: Earth and Environmental Science **58** (2017) 012038 doi:10.1088/1755-1315/58/1/012038

Figure 1. Location weather observation and air quality stations BMKG Palangka Raya and Banjarbaru

2.2. Plotting data

Plotting observation concentration PM_{10} data. Plotting the observation data for PM_{10} in Banjarbaru and Palangka Raya period September - October 2015. This is to determine the peak concentration of PM_{10} during the smoke haze because of forest and vegetation peat-fire. Determining the observation time by selecting multiple time when the peak concentration of PM_{10} were selected by determining the time range of the simulation.

2.3. Simulation of the trajectory and concentration with HYSPLIT

Simulation of the trajectory and concentration HYSPLIT time from September to October, 2015 using the model data GDAS air quality dispersion HYSPLIT performed to estimate the value of meteorological elements such as wind speed and direction to determine its trajectory. Simulation trajectory and concentration HYSPLIT done using the starting point of the source area (hotspots) with the forward trajectory and from the observation point PM_{10} as the area affected.

2.4. Analysis variance

Analysis variance was conducted by quantitative and descriptive methods. This analysis aims to determine patterns and distribution concentrations of pollutants. This analysis was done by looked at the figures of the simulation results (trajectories and concentrations). Analysis variance of concentration was done by quantitative method [7]. We also used trend linear regression for PM_{10} to see forecasting concentration and cumulative frequency distribution concentrations of PM_{10}.

3. Results and Discussion

3.1. Inventory Data

The results of measurement PM_{10} concentrations and meteorological elements such as wind direction and speed shown by both stations BMKG Banjarbaru (Figure 2) and Palangka raya (Figure 3). Figure 2 and Figure 3 shows the concentration of PM_{10} reach the peak. The average result observations for PM_{10} in Banjarbaru and Palangka Raya September-October 2015 shows a quiet different values (Table

2). Banjarbaru average value of PM_{10} concentrations significantly lower than in Palangka Raya. Banjarbaru value is yet below the threshold of 150 ug / m^3. In contrast, the concentration of that particulate in Palangka Raya was a lot above the threshold. It means that during September-October 2015 the average concentration of PM_{10} in Palangka Raya was quite extreme. In fact, during September - October 2015 it was very less rainy in southern part of Borneo because of a strong El Nino. The haze from forest and land fires in Central Kalimantan was the worst incident since 1997, released a lot of particulate matter from the peatlands into the atmosphere.

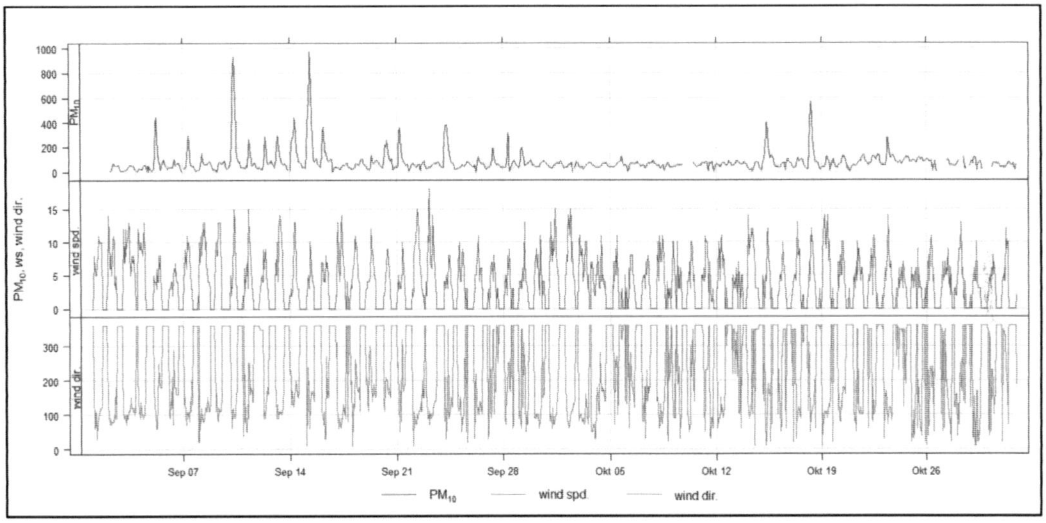

Figure 2. Time series data of concentration $PM_{10,}$ direction and wind speed in Banjarbaru

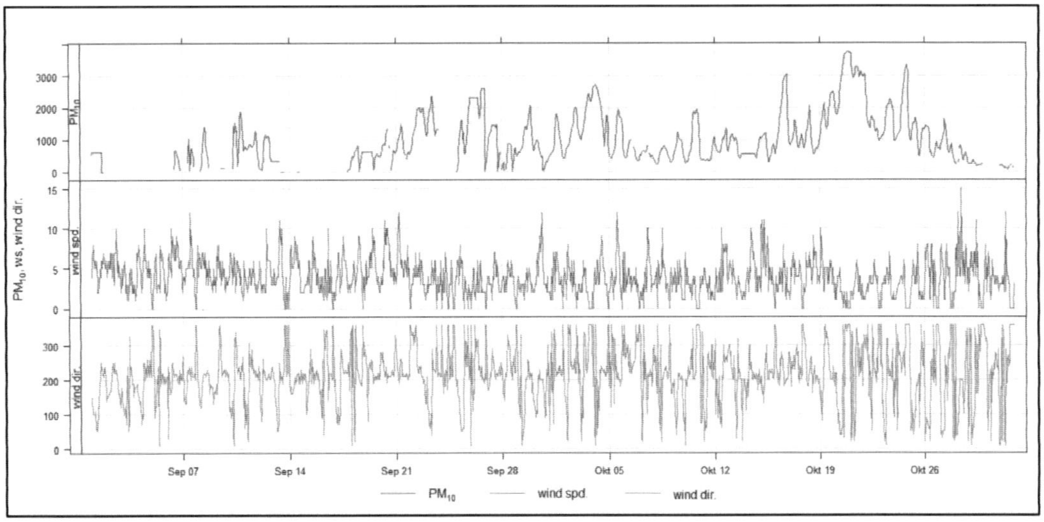

Figure 3. Time series data of concentration PM_{10}, direction and wind speed in Palangka Raya

Table 2.Average concentrations of PM_{10} observation Banjarbaru and Palangka Raya September-October 2015

Parameter	Stasion	Month/ Year ($\mu g/ m^3$)		Indonesian Standard
		09 2015	10 2015	
PM_{10}	Banjarbaru	76.82 ± 75.17	70.43 ± 54.5	150 (24 hours)
	Palangka Raya	868.80 ± 625.5	1167.9 ± 836.2	

Table 3. The area forest and vegetation fire 2015

Province	The area forest and vegetation fire (ha)
South Kalimantan	1714.89
Central Kalimantan	122882.9

Source : Sipongi - The ministry of forestry and environment Indonesia, 2016 [11]

South Kalimantan represented by Banjarbaru and Central Kalimantan represented by Palangka Raya. The area of forest and land burned in 2015 Central Kalimantan has larger the area forest and vegetation fire than South Kalimantan (Table 3), so the emissions are released larger in Central Kalimantan. This causes the pollutant concentrations of PM_{10} in Central Kalimantan larger than South Kalimantan.

3.2. Plotting Data

The result of the calculation in Banjarbaru boxplot (Figure 4) and boxplot in Palangka Raya (Figure 5), could observant will recognize some of the peak point values for PM_{10}. It is presumed at the time an incidence of the peak concentration PM_{10} in Palangka Raya. Palangka Raya was both of as source pollutants and the receptor.

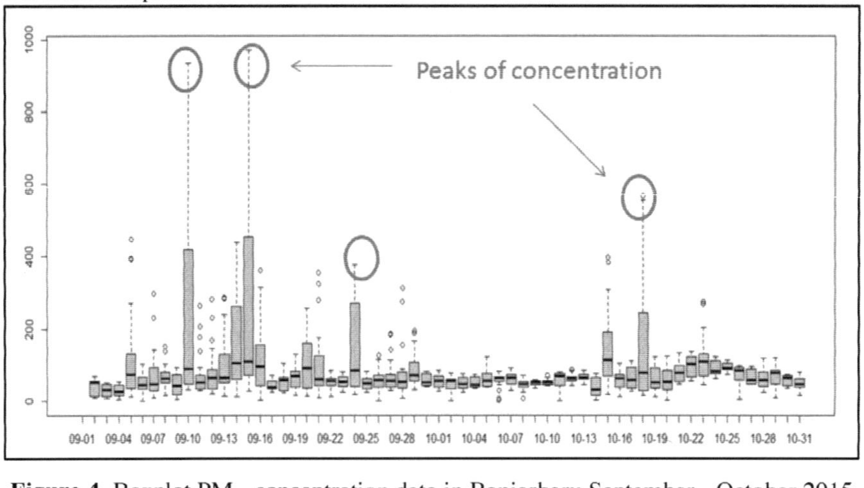

Figure 4. Boxplot PM_{10} concentration data in Banjarbaru September - October 2015

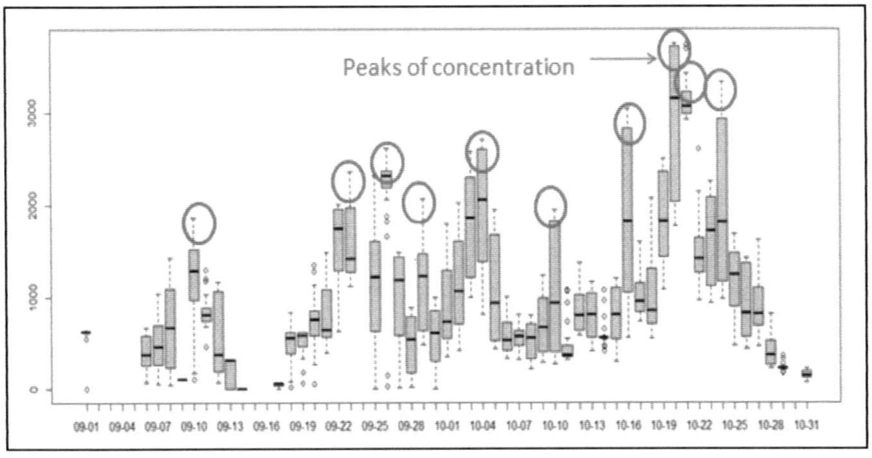

Figure 5 Boxplot PM_{10} concentration data in Palangka Raya September - October 2015

After we find peaks of the PM_{10} concentration, we select some of them to be simulated with HYSPLIT applications. HYSPLIT application could simulate trajectories and dispersion of pollutants PM_{10}. Here are some of peaks concentrations PM_{10} in Banjarbaru and Palangka Raya (Table 4.):

Table. 4 Peaks concentrations of PM_{10}

Location	Peaks concentrations of PM_{10}	
	Concentrations (μg/ m^3)	Time
Banjarbaru	934.9	09 September 2015 09:00:00
	970.9	15 September 2015 08:00:00
	375.6	24 September 2015 07:00:00
	567.8	18 October 2015 07:00:00
Palangka Raya	1839.5	10 September 2015 20:00:00
	2357.1	23 September 2015 07:00:00
	2613.5	26 September 2015 21:00:00
	2072.3	29 September 2015 20:00:00
	2708.1	04 October 2015 02:00:00
	1945.1	10 October 2015 20:00:00
	3049.3	16 October 2015 18:00:00
	3760.8	20 October 2015 20:00:00
	3743.1	21 October 2015 00:00:00
	3334.2	24 October 2015 19:00:00

3.3. Simulation trajectories PM_{10}

The simulation result with some sample data peak points concentration by HYSPLIT application has been carried out. Figures 6 and 7 show the simulation HYSPLIT Banjarbaru and Palangka Raya.

HYSPLIT trajectory simulations use a moving frame of reference for the advection and diffusion calculations as the air trajectories [8]. Figure 6 and 7 shows a comparison of the trajectory of pollutants PM_{10} descriptively. Those figure shows us HYSPLIT trajectory results in Banjarbaru on

September (Figure 6) and in Palangka Raya on October (Figure 7). HYSPLIT will track the movement of pollutants to see which direction it moves. Instead, HYSPLIT use GDAS1, weather model data with a period of one week. The Global Data Assimilation System (GDAS) is the system used by the National Center for Environmental Prediction (NCEP) Global Forecast System (GFS) model to place observations into a gridded model space for the purpose of starting, or initializing, weather forecasts with observed data [10].

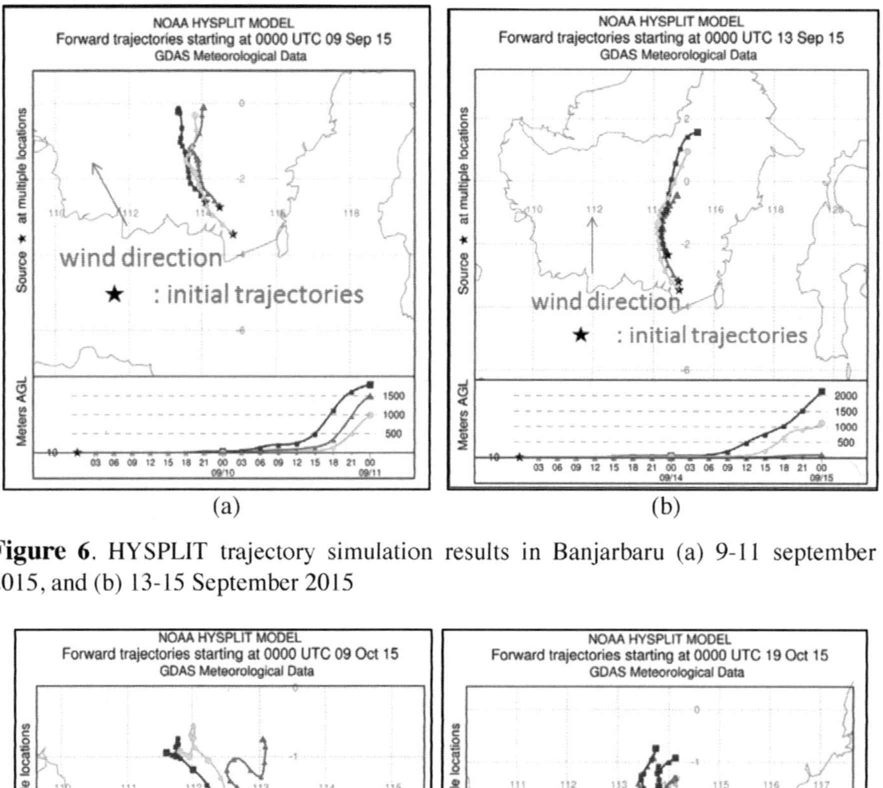

Figure 6. HYSPLIT trajectory simulation results in Banjarbaru (a) 9-11 september 2015, and (b) 13-15 September 2015

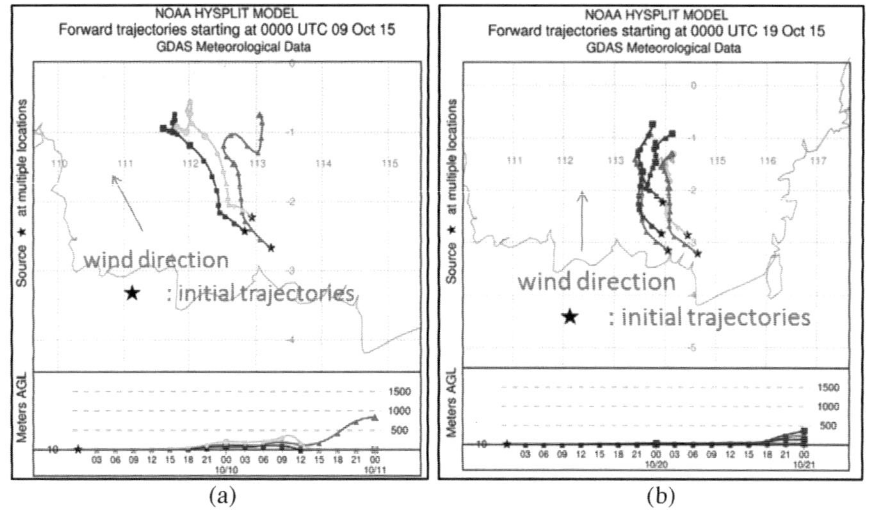

Figure 7. HYSPLIT trajectory simulation results in Palangka Raya (a) 9-11 October 2015, and (b) 19-21 October 2015

Figure 6 to 7 shows us the source of pollutants from some of hotspots bring pollutants PM_{10} from HYSPLIT model tends toward the north and northwest, or in other words, comes from the south and southeast. The stars show the initial trajectories, sources of pollutants. The direction of the wind is the same as the direction of movement of the synoptic wind through Borneo (Figure 8). Banjarbaru dan Palangka Raya has the same trajectory simulation results. Both of them the directions tend toward the north and northwest. In addition, HYSPLIT trajectory is greatly influenced by wind direction. HYSPLIT is useful to investigate where wind direction that bring pollutants PM_{10} move.

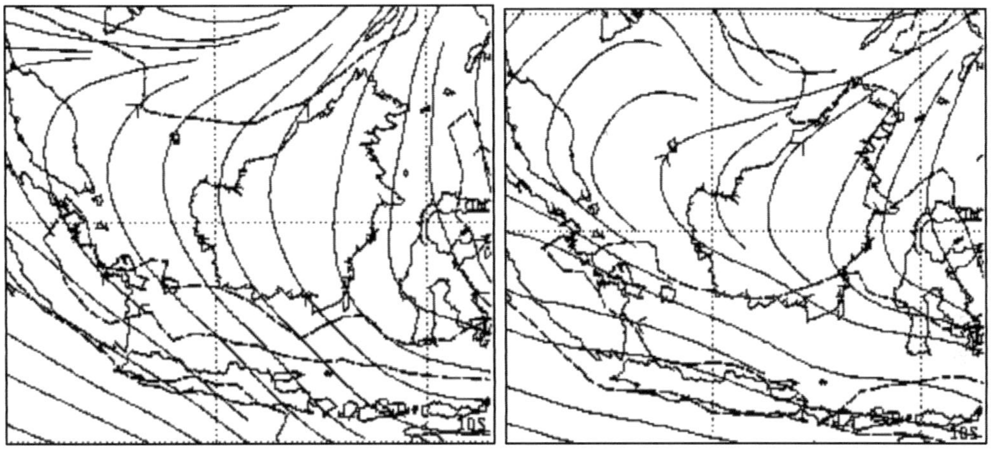

(a) (b)

Figure 8. Wind direction through Borneo on September 13, 2015 00 UTC (a) and October 19, 2015 00 UTC (b). The wind moved from the southeast and turn on the equator from the southwest. (Source : http://www.bom.gov.au/australia/charts/archive/) [12]

Another factor that must be considered in the model trajectory or direction of the wind is topography. Swachner Muller mountain in the north of Central Kalimantan is an important factor that can deflect the wind in the area(Figure 9). Topography may lead to the spread of pollutants in an area that is not flat (large topographic variations) making it difficult to disperse. So the three factors of weather, topography, and the emissions are important factors in air quality models [6]. HYSPLIT model simulation is also influenced by the local topography.

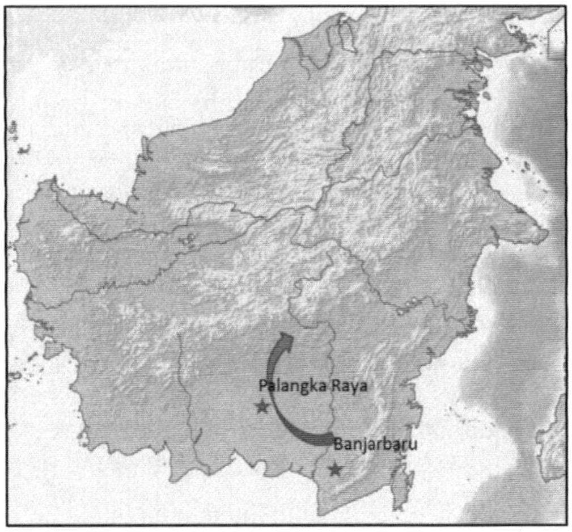

Figure 9. Topography of Central Kalimantan can make the wind turn wind

3.4. Simulation concentrations of PM_{10}

Simulation pollutants concentration uses a fixed three-dimensional grid as a frame of reference to compute pollutant air concentrations [8]. Simulation of PM_{10} concentration can indicate how much pollutants are dispersed. Through this simulation we can see where pollutants such as particulate matter will be retained as indicated by an area with a high concentration of pollutants. From some of multiple hotspots we can see the sources of the emissions that lead to the emergence of pollutants PM_{10} (Figure 10). We can trace concentrations of PM_{10} where dispersed from the source since the starting point until in the receptor areas where the pollution concentrations are retained.

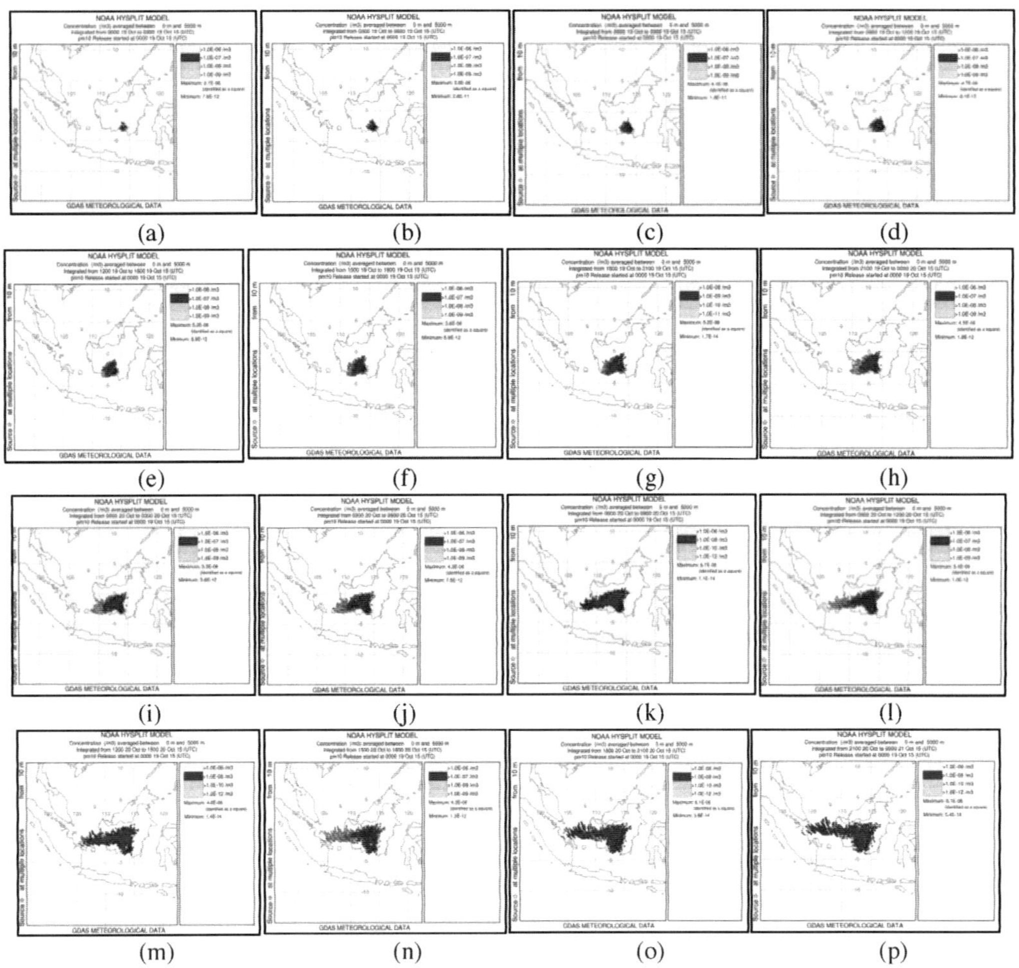

Figure 10. Simulation concentrations of PM_{10} Palangka Raya 19-21 October 2015, 19 October 2015 : (a) 00-03UTC, (b) 03-06UTC, (c) 06-09UTC, (d) 09-12UTC, (e) 12-15UTC, (f) 15-18UTC, (g) 18-21UTC, (h) 21-00UTC, 20 October 2015 : (i) 00-03UTC, (j) 03-06UTC, (k) 06-09UTC, (l) 09-12UTC, (m) 12-15UTC, (n) 18-21UTC, (o) 21-00UTC,21 October 2015: (p) 00-03UTC

Figure 10 shows the simulated concentration of PM_{10} in Palangka Raya. Simulation is used on 19-21 October 2015 with a three-hour period. Simulations begins on 9 October 2015 00 UTC until 11 October 2015 00 UTC. Simulation of PM_{10} concentration can indicate how much pollutants are dispersed. Through this simulation we can see where pollutants such as particulate matter will be dispersed to the eastern of the source. From some of multiple hotspots we can see the sources of the emissions. Concentration with HYSPLIT simulation will be useful with a quantitative analysis pollutants PM_{10} (Figure 10).

3.5. Analysis variance

Figure 11 and 12 show that there is a trend of increasing concentrations of pollutants both in Palangka Raya and Banjarbaru in the period September- October 2015 on the data hourly, three-hourly and monthly. In Palangka Raya has larger trend of increase compared with Banjarbaru. In fact, at concentrations Banjarbaru three hours in a slight decrease. Wind direction is also a factor that affects will bring many pollutants out of the affected area.

The wind carries pollutants PM_{10} move. Due to the large particle size of 10 microns, in the strong wind it will not carry over all. However, pollutants will also be a lot of stuck in the source area. Distance from Palangka Raya to Banjarbaru also quite far around 220 km, so the pollutants PM_{10} will be spend more time in Palangka Raya than in Banjarbaru. Thus, the concentration of pollutants in Palangka Raya higher than Banjarbaru.

Figure 13 shows the cumulative frequency distribution for PM_{10} for Banjarbaru and Palangka Raya during September – October 2015 (based on real-time measurements). Both of the periods Palangka Raya concentrations of PM_{10} (September and October 2015) was significantly higher than Banjarbaru. The cumulative frequency distribution of Palangka Raya higher than Banjarbaru.

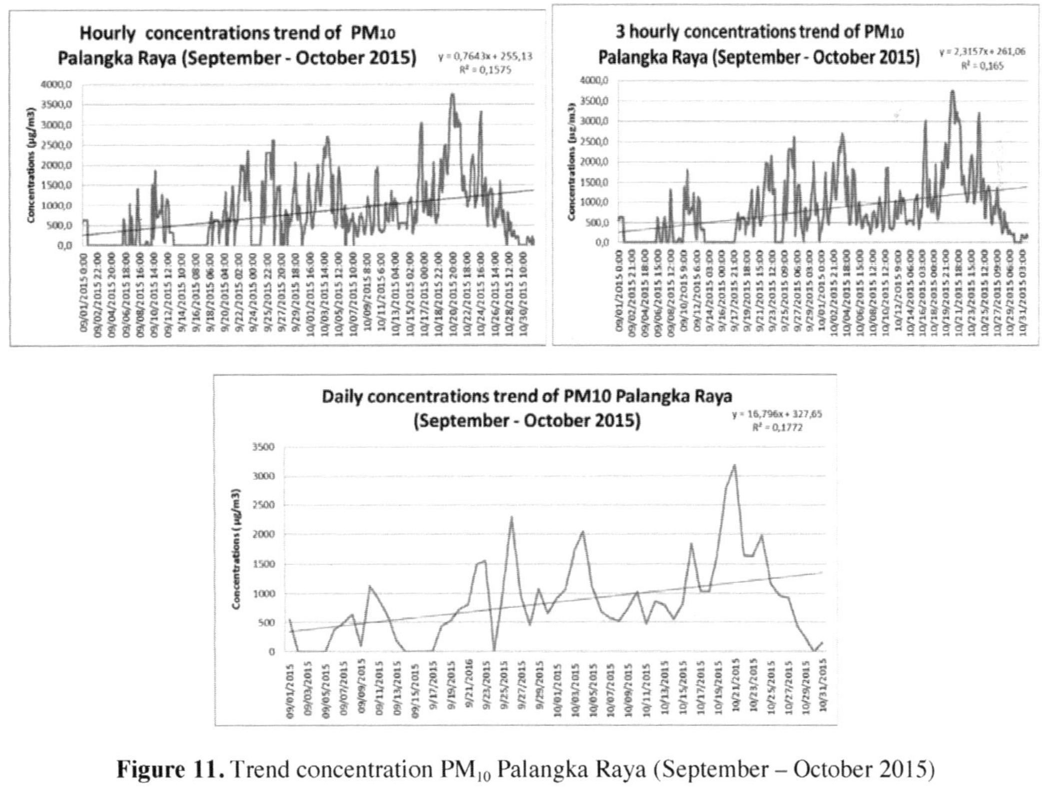

Figure 11. Trend concentration PM_{10} Palangka Raya (September – October 2015)

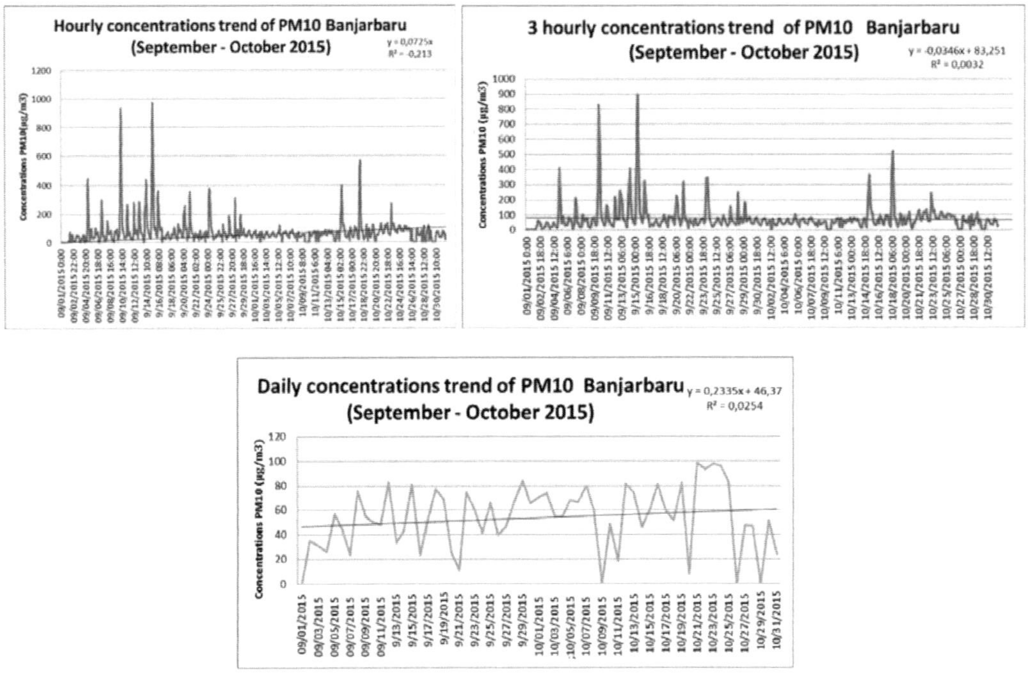

Figure 12. Trend concentration PM_{10} Banjarbaru (September – October 2015)

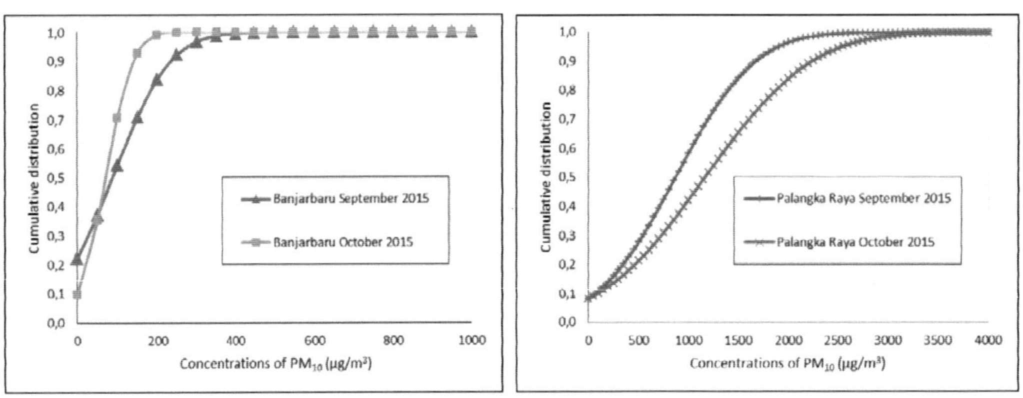

(a) (b)

Figure 13. Cumulative frequency distribution concentrations of PM_{10} (a) Banjarbaru (b) Palangka Raya

The assumptions are that the population mean for each period are normally distributed with common but unknown variances. Assuming a larger diameter PM_{10} and weak wind speeds of less than 3 m/s, it can not happen dispersion resulting in the accumulation of pollutants. A cumulative analysis aims to find a cumulation of pollutants due to the measurement of pollutants, with the number of measurements of pollutants measured not only on that day alone but also from pollutants in the days before. Cumulative frequency distribution concentrations of PM_{10} is useful to know the differences cumulative frequency distribution concentrations between Banjarbaru and Palangka Raya.

Hotspots could be useful to determine which became a source of pollutants and areas exposed to pollutants. There are two case studies on the current September-October 2015. In the first case, 13 September 2015 showed the number of hotspots more in Central Kalimantan (Figure 14). The number of hotspots in Central Kalimantan greater than South Kalimantan (Table 5). By contrast, in the second case October 19, 2015 there are more hotspots in South Kalimantan than Central Kalimantan (Figure 15). Hotspots were clustered and numerous is an indication of the affected region. However, from simulation (Figures 6 and 8) the wind was moving from South Kalimantan to Central Kalimantan. Thus, Central Kalimantan is both the source and exposed area, because the wind came from South Kalimantan. Then, the Central Kalimantan PM_{10} concentrations are higher as well get pollutants from South Kalimantan is carried by the wind direction.

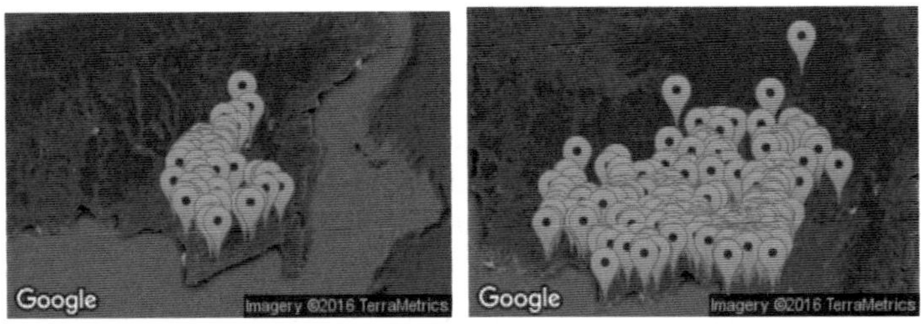

(a) (b)

Figure 14. Hotspots in South Kalimantan (a) and Central Kalimantan (b) 13 September 2015[11]

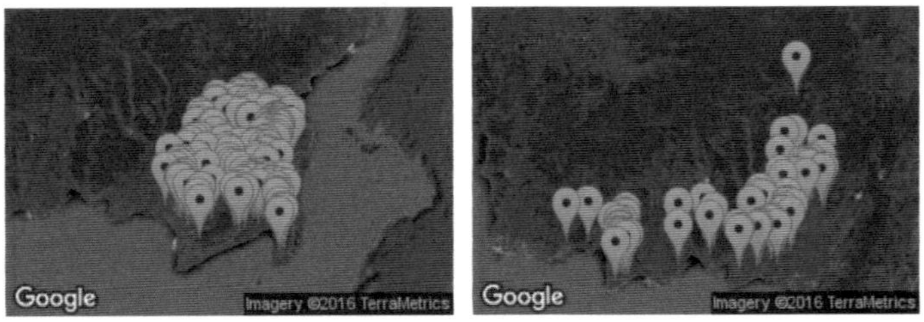

(a) (b)

Figure 15. Hotspots in South Kalimantan (a) and Central Kalimantan (b) 19 October 2015 [11]

Table 5. Summary hotspots data

Province	Date	Hotspots
South Kalimantan	13 September 2015	53
Central Kalimantan		240
South Kalimantan	19 October 2015	141
Central Kalimantan		63

4. Conclusions

PM_{10} concentration data from observations are very useful for tracking extreme events by looking at its peak. Simulations using HYSPLIT similar to those seen in the synoptic observations. The dispersion pattern, as simulated by HYSPLIT showed that the distribution of PM_{10} is greatly influenced by the wind direction and topography. Concentration with HYSPLIT simulation will be useful with a quantitative analysis and trend. There is a large difference between the concentration of

PM_{10} Banjarbaru and Palangka Raya. Area sources of PM_{10} and areas of exposure is an important area for the dispersion of pollutants and its concentration. Cumulative frequency distribution concentrations of PM_{10} is useful to know the differences cumulative frequency distribution concentrations between Banjarbaru and Palangka Raya.

References

[1] Sirimongkonlertkul N, Upayokhin, P, and Phonekeo, V 2013 *Kasetsart Journal* **47** 768-780
[2] Anwar A, Liew J N, Othman M and Latif M T 2010 *Sains Malaysiana* **39** (2) 169-174.
[3] Perez I A, Artuso F, Mastura M, Kulshrestha U, Sanchez M L and Garcia M A 2015 *Hindawi Publishing Corporation Advance in Meteorology* **284213** : 1-10.
[4] Heil A, Langmann B, Aldrian E 2005 *Journal Mitigation and Adaptation Strategies for Global Change* **12** 113-133
[5] Thinking ASEAN 2015 Mobilizing Mass Efforts to Fight the Haze in Southeast Asia. Issue 5 November 2015 4-6.
[6] Aggarwal, A, Haritash, A. K and Kansal, G 2014 *International Journal of Advanced Technology in Engineering and Science* **02** 06 355-365
[7] Carslaw, D 2015 *Journal Environmental Modelling and Software* **27-28** 52-61
[8] Stein A F, Draxler R R, Rolph G D, Stunder B J B, Cohen M D and Ngan F 2015 *American Meteorological Society* 2059-2074
[9] Draxler A A and Hess G D 1998 *Aust. Meteor. Mag.* **47** 295-308
[10] Su L, Yuan Z, Fung J C H and Lau A K H 2015 *Science of the Total Environment* **506-507** 527–537.
[11] Sipongi- The Ministry of Forestry and Environment Indonesia 2016 Sebaran Titik Panas [internet]. available at : http://sipongi.menlhk.go.id/hotspot/sebaran/ [Accessed on 2016 November 10].
[12] Bureau of Meteorology Australia [BoM] 2016 Analysis Chart Archive [internet]. available at : http://www.bom.gov.au/australia/charts/archive/ [Accessed on 2016 November 10].

Descriptive Statistics and Cluster Analysis for Extreme Rainfall in Java Island

K E Komalasari[1*], H Pawitan[2] and A Faqih[2]

[1]Postgraduate student at Bogor Agricultural University, Indonesia
[2] Department of Geophysics and Meteorology, Bogor Agricultural University, Indonesia

Email : kurnia.bmkg@gmail.com

Abstract. This study aims to describe regional pattern of extreme rainfall based on maximum daily rainfall for period 1983 to 2012 in Java Island. Descriptive statistics analysis was performed to obtain centralization, variation and distribution of maximum precipitation data. Mean and median are utilized to measure central tendency data while Inter Quartile Range (IQR) and standard deviation are utilized to measure variation of data. In addition, skewness and kurtosis used to obtain shape the distribution of rainfall data. Cluster analysis using squared euclidean distance and ward method is applied to perform regional grouping. Result of this study show that mean (average) of maximum daily rainfall in Java Region during period 1983-2012 is around 80-181mm with median between 75-160mm and standard deviation between 17 to 82. Cluster analysis produces four clusters and show that western area of Java tent to have a higher annual maxima of daily rainfall than northern area, and have more variety of annual maximum value.

Keywords: extreme rainfall, maximum daily rainfall, descriptive statistics, cluster analysis

1. Introduction

Rainfall is one element of climate has a major influence on human life, but in extreme conditions of rainfall often causes damages to human life and the natural environment. Extreme weather (including extreme rainfall) as a rare event based on the statistical distribution on a certain area, and will be specifics for different place [1].

There are two ways to identifying extreme rainfall from real data. The first approach considers the maximum or minimum of variable taking in successive periods. The second approach focuses in the appearance of value exceeding a given threshold [2]. In this study will using block maxima method for one day period to identify extreme rainfall and conducted 40 rain stations in Java Island corresponding daily rainfall observation data for period 1983-2012.

Descriptive statistics analysis was performed to analysis condition of centralization data, variation and distribution patterns of maximum daily rainfall data in Java Region. Regional clustering will generated to obtain geographical distribution of maximum of daily rainfall in Java Island.

2. Data and Methodology

Data in this study is daily rainfall data in Java Region ($5^{o}25'$-$8^{o}48'$S and $105^{o}1'$-$114^{o}4'$E) which consists of 40 rain stations that spread across six provinces in Java Region during the period 1983-2012. Distribution map of the Station a can be seen in Figure 1.

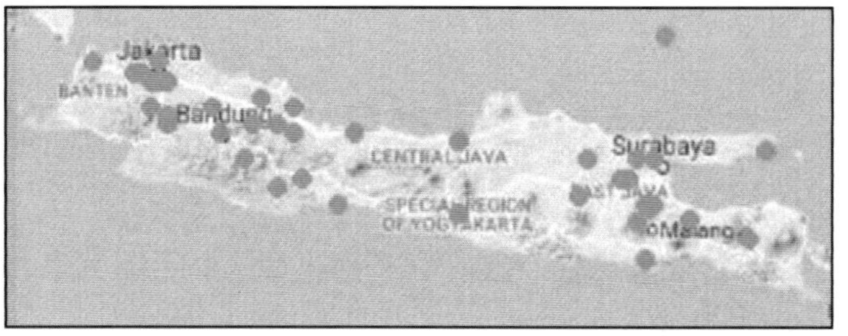

Figure 1. Distribution map of rain station for this study.

2.1. *Identification of Maximum Daily Rainfall*

WMO has recommended 11 indices to identified extreme rainfall [3], and one of them is maximum daily rainfall that used in this study. Analysis are based on maximum daily rainfall from 40 stations over Java regions that identified from maximum daily rainfall amount in a year during period 1983-2012.

2.2. *Descriptive Statistics Analysis*

Descriptive statistics analysis performed for maximum daily rainfall during period 1983-2012. Statistical analysis used to describe the condition of centralization data, variation and distribution patterns of extreme rainfall data in Java Region. Measures of central tendency data that will used is mean (average) and median (middle value). Variation of data will measured by Inter Quartile Range (IQR) and obtained from the difference between third quartile to first quartile, standard deviation which indicates the deviation of actual data against the average formulated with:

$$Sd = \sqrt{\frac{\sum_{i=1}^{n}(x_i - \bar{x})^2}{n-1}} \quad (1)$$

The pattern of distribution data will measured by skewness and kurtosis, skewness value of normal distribution is zero, when data have extends distribution to right with a long tail to the higher value that mean the data have positive value skewness, in another hand if the data have distribution with a long tail to the lower value that mean the data has a value skewness negative, skewness value calculated by:

$$Sk = \frac{n}{(n-1)(n-2)}\sum\frac{(x_j - \bar{x})^3}{s} \quad (2)$$

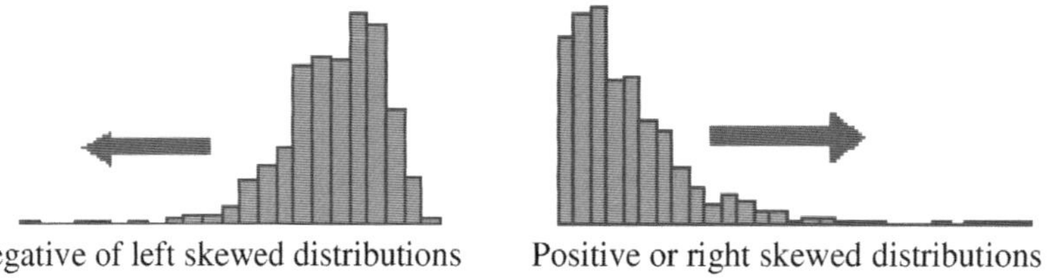

Negative of left skewed distributions Positive or right skewed distributions

Figure 2. Two type of distribution pattern with different skewness.

Kurtosis value indicates high or low data distribution curve with reference to the normal curve. Kurtosis value of normal distribution is 0, curve that higher than normal distribution has a positive kurtosis value, while the lower curve of the normal distribution have negative kurtosis value. Kurtosis value is calculated by the following formula.

$$Kr = \left\{ \frac{n(n+1)}{(n-1)(n-2)(n-3)} \sum \frac{(x_j - \bar{x})^4}{s} \right\} - \frac{3(n-1)^2}{(n-2)(n-3)} \tag{3}$$

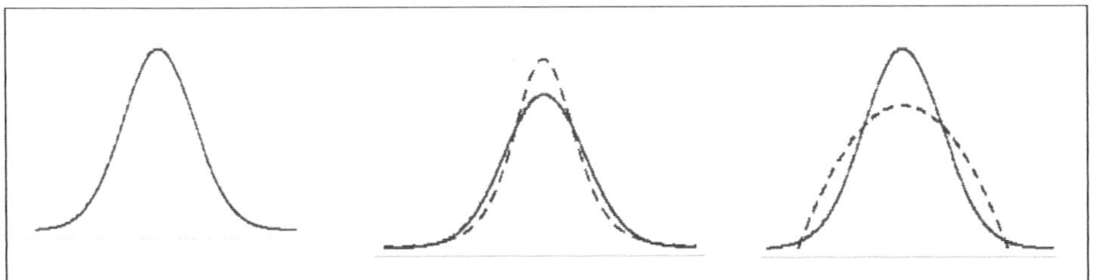

Figure 3. Three type of distribution pattern with different kurtosis

2.3. Principal Component Analysis

Principal component analysis aims to avoid any multicollinearity between variables before cluster analysis. The principal component is combination of these variable that orthogonal (perpendicular) and independent each other. The principal component is formed by a sequence variance from the biggest to the smallest. If principal component y is a combination from $x_1, x_2, x_3, \ldots, x_5$ so

$$y = w_1x_1 + w_2x_2 + \ldots + w_px_p \tag{4}$$

w_i is weights or coefficients for the variables i, xi is the variable to i and y are linear combinations of the variables x. Percentage of total diversity is considered sufficient if 75% or more is able to be explained by 4 or 5 first principal components [4].

2.4. Cluster Analysis

Cluster analysis is a technique for classifying objects into groups according to their particular characteristics. Each observation must have a high homogeneity in a group but high heterogeneity with other groups [5]. Cluster analysis carried out for seven first component from principal component analysis result. Squared euclidean distance is used to measure distance of each object in this study. The distance measure used in this study is squared euclidean distance is formulated as below[6]:

$$d_{ij} = \sqrt{(x_{i1} - x_{j1})^2 + (x_{i2} - x_{j2})^2 + \cdots + (x_{ip} - x_{jp})^2} \tag{5}$$

Cluster technique that used in this study is a ward method or also known as minimum variance method. Ward's linkage was the most robust hierarchical clustering method in handling the wrong grouping due to noise in data [7, 8].

3. Result

3.1. Descriptive Statistics

Result of statistics descriptive analysis for 40 stations show that average of maximum daily rainfall in Java Region during period 1983-2012 is around 79,7 -181,1mm with median between 74,5-159,8mm and standard deviation between 17,2 to 81,8. Most of rainfall stations have positive kurtosis value that means distribution of data have heavy tails, and outliers. All of data have positive skewness that indicate skewness distribution curve have tail at value above the median, and mean is on the right of the peak value. List for detail can be seen in Table 1, box plots for 40 rainfall stations in Java in Figure 4.

Table 1. Station name and statistics of maximum daily rainfall for all the stations in Java Island.

Name Station	Mean	Median	Kurtosis	Skewness	Standard Deviation	IQR
Stageof Bandung	79.7	76.0	1.9	1.2	17.6	19.8
Cibukamanah	115.6	112.5	1.3	0.8	31.5	37.5
Sumurwatu	86.7	75.0	4.7	1.9	31.2	26.0
Staklim Bogor	127.9	119.1	2.6	1.4	34.4	38.8
Cicurug	97.9	91.0	11.8	2.8	45.4	33.8
Karangkendal	85.3	82.5	0.5	0.8	24.4	38.3
Bantardewa	113.4	100.0	2.6	1.5	46.5	32.0
Pacet	90.2	92.2	0.3	0.5	17.6	23.5
Leles	93.6	79.0	20.8	4.3	66.6	34.0
Nariewattie	181.1	159.0	6.3	2.3	81.8	45.5
Jatiwangi	126.3	122.7	-0.5	0.3	26.0	38.9
Cimalaka	80.3	76.5	2.2	1.3	18.4	21.5
Bd.Bekasi	91.3	89.5	0.3	0.6	22.0	31.0
Linggarjati	112.3	74.5	8.4	2.8	77.0	44.5
Tnjg Priok	122.3	106.7	0.1	1.0	40.0	47.1
Halim	130.2	108.8	4.6	2.1	52.0	42.8
Stageof Tanggerang	112.5	97.9	5.8	2.2	46.8	40.3
Pndk Betung	128.3	109.1	6.3	2.5	63.7	33.2
Budiarto Curug	102.2	89.2	6.5	2.4	34.6	25.1
Serang	86.7	84.0	3.6	1.4	21.9	24.8
Semarang Klim	134.5	116.0	2.1	1.6	49.0	38.0
Tegal	112.8	99.4	1.5	1.4	38.8	34.8
Cilacap	173.3	159.8	0.7	1.0	64.4	59.4
Adisucipto	123.2	111.6	7.8	2.7	58.9	37.2
Bojonegoro	102.4	98.5	-0.2	0.4	26.3	41.0
Banyuwangi	98.3	93.0	-0.5	0.6	33.6	55.5
Sangkapura	138.5	121.7	3.0	1.7	53.4	54.8
Kalianget	99.9	86.8	14.9	3.6	59.4	26.8
Juanda	103.7	100.9	1.8	1.2	19.8	25.2
Stamet Perak I	94.5	92.0	3.4	1.5	30.1	30.2
Stamet Perak II	88.2	81.1	0.2	0.8	22.2	27.5
Stageof Tretes	136.3	128.5	0.5	0.6	26.3	29.5
Tlekung	87.9	84.5	1.2	1.3	23.4	17.3
Gandekan	103.1	92.0	9.5	2.6	57.3	50.8
Sukowono	98.7	96.0	-0.1	0.3	17.2	24.3
Tapen	98.2	92.0	0.6	1.0	21.4	20.0
Saradan	129.4	117.5	-1.1	0.1	44.3	76.5
bantur	120.8	97.0	2.4	1.8	59.1	54.0
Telebuk	102.4	93.5	-0.6	0.7	23.0	31.5
Botogardu	103.2	97.0	6.0	2.0	28.6	33.0

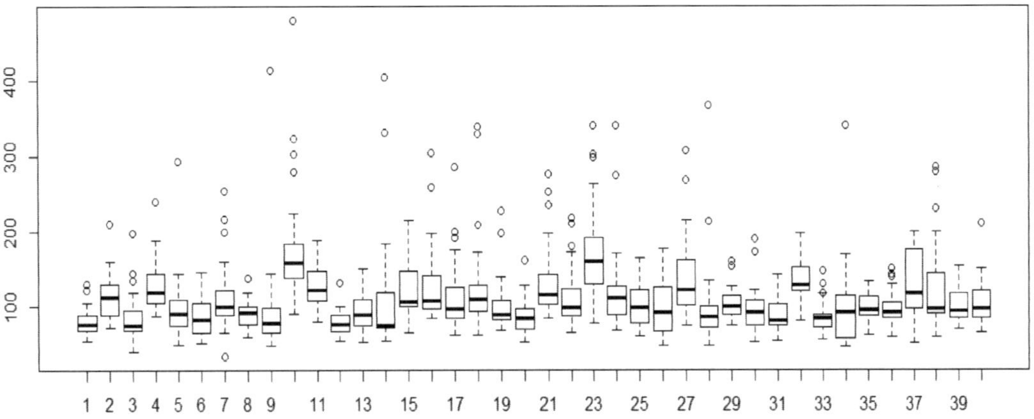

Figure 4. Box plot for maximum daily rainfall in 1983-2012.

3.2. Cluster Analysis for Maximum of Daily Rainfall

In this study we use 30 years maximum daily rainfall of 40 rainfall station period 1983-2012 to obtain clustering of rainfall station. Grouping of station was started with Principal Component Analysis (PCA). Principal component analysis aims to avoid multicollinearity between variables. Scree plot components based on variations can be seen in Figure 5 and a summary of each component are presented in Table 2.

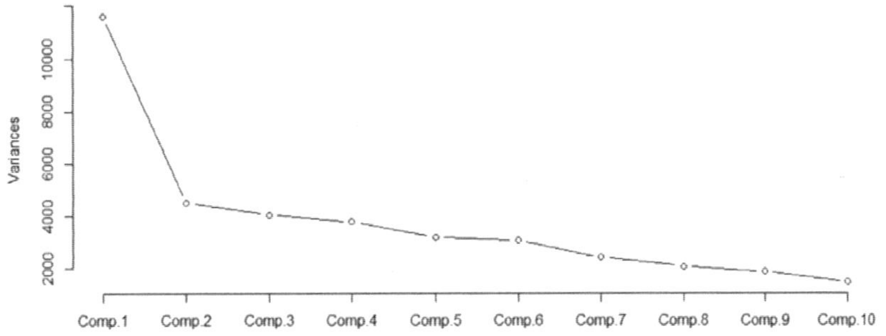

Figure 5. Scree plot of principal component based on variance.

Table 2. Summary for each principal component.

	PC.1	PC.2	PC.3	PC.4	PC.5	PC.6	PC.7	PC.8	PC.9	PC.10
Standard deviation	107.72	67.12	63.55	61.32	56.28	55.31	49.05	45.23	42.95	38.10
Proportion of Variance	0.27	0.10	0.09	0.09	0.07	0.07	0.06	0.05	0.04	0.03
Cumulative Proportion	0.27	0.37	0.47	0.56	0.63	0.70	0.76	0.80	0.85	0.88

According to the table 2, the first component can explain 27% variance of data, and then by adding a second component can explain until 37% variance of data. Following Morrison [4] that percentage of total diversity is considered sufficient if first 4 or 5 principal components account 75% or more of the total variance, thus in this study will using first 7 principal component that account 76% of total variance.

These seven component are formed from a combination of maximum daily rainfall for period 1983-2012. The equations for seven principal component is presented below:

PC1 = -0.13*t1984-0.13*t1985-0.32*t1986-0.19*t1987-0.13*t1988-0.15*t1989-0.18*t1990 -0.20*t1991-0.3*t1992-0.31*t1993-0.16*t1995-0.26*t1996-0.14*t1997-0.11*t1998-0.24*t1999 -0.15*t2000-0.24*t2001-0.15*t2003-0.18*t2005-0.33*t2007-0.18*t2008-0.17*t2009-0.18*t2011

PC2 = 0.13*t1983+0.14*t1985-0.42*t1986-0.29*t1988-0.23*t1989+0.14*t1990+0.13* t1991-0.37*t1993+ 0.16*t1997-0.50*t1999+0.14*t2000-0.34*t2007-0.16*t2008

PC3 = 0.13*t1983+0.27*t1985+0.33*t1986+0.4*t1989-0.15*t1991-0.31*t1992+0.29*t1993 -0.2*t1995-0.34*t1996+0.23*t1999-0.13*t2001-0.16*t2002-0.12*t2003- 0.18*t2005 +0.27*t2006-0.133*t2007+0.1*t2011+0.11*t2012

PC4 = 0.22*t1987+0.54*t1988-0.22*t1992-0.13*t1993+0.38*t1995-0.19*t1996+0.32 *t1999-0.24*t2000-0.25*t2002-0.22*t2004-0.12*t2007+0.14*t2008-0.12*t2011

PC5 = 0.12*t1984+0.2*t1986-0.33*t1988-0.38*t1990+0.46*t1995+0.21*t1997-0.12*t2000 +0.14*t2001 + 0.17*t2006-0.51*t2007+0.2*t2008

PC6 = -0.15*t1985-0.16*t1986-0.12*t1987-0.31*t1988+0.29*t1990+0.12*t1991-0.46* t1995-0.10*t1996-0.14*t1997+0.32*t1998+0.16*t1999-0.28*t2000+0.17*t2001-0.21*t2006-0.15*t2008-0.26*t2009

PC7 = -0.27*t1986+0.19*t1987-0.64*t1991-0.18*t1992-0.16*t1993-0.10*t2000+0.24* t2001+ 0.3*t2002 +0.27*t2003-0.34*t2004+0.12*t2006-0.14*t2008

Cluster analysis by ward method n squared euclidean distance against these 7 first principal components of these stations. Dendogram shows the process of grouping of each object separated into larger groups based on their distance. Dendogram of this cluster analysis can be seen in Figure 6, list of cluster member and location of cluster member can be seen in Figure 7 and Table 3.

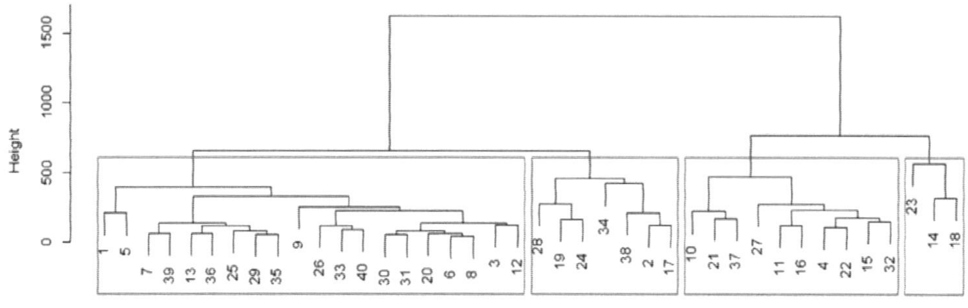

Figure 6. Cluster Dendogram with 4 groups.

IOP Conf. Series: Earth and Environmental Science **58** (2017) 012039 doi:10.1088/1755-1315/58/1/012039

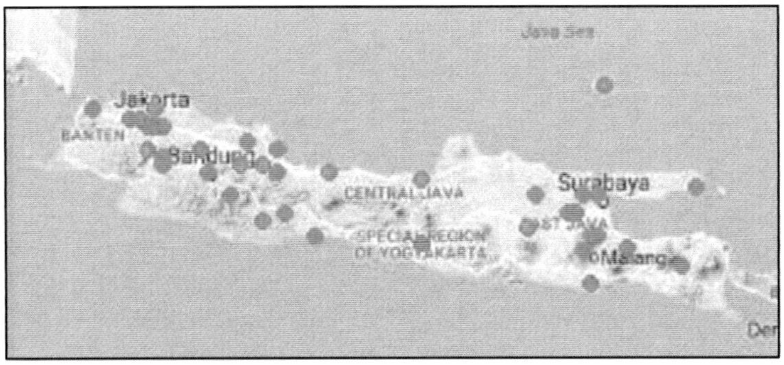

Figure 7. Location of the cluster members, cluster A (red), cluster B (green), cluster C (blue), cluster D (purple).

Table 3. Member of each group.

Group A		**Group B**	**Group C**	**Klaster D**
Stageof Bandung	Bojonegoro	Cibukamanah	Staklim Bogor	Linggarjati
Sumurwatu	Banyuwangi	Stageof Tanggerang	Nariewattie	Pndk Betung
Cicurug	Juanda	Budiarto Curug	Jatiwangi	Cilacap
Karangkendal	Stamet Perak I	Adisucipto	Tnjg Priok	
Bantardewa	Stamet Perak II	Kalianget	Halim	
Pacet	Tlekung	Gandekan	Semarang Klim	
Leles	Sukowono	bantur	Tegal	
Cimalaka	Tapen		Sangkapura	
Bd.Bekasi	Telebuk		Stageof Tretes	
Serang	Botogardu		Saradan	

From Figure 7 can be seen that cluster A, and B was spread all over Java Island except cluster C and D. Cluster C and D tent to spread in western area of Java Island. It's mean that in western area of Java Island have more complex type of annual maximum of daily rainfall group.

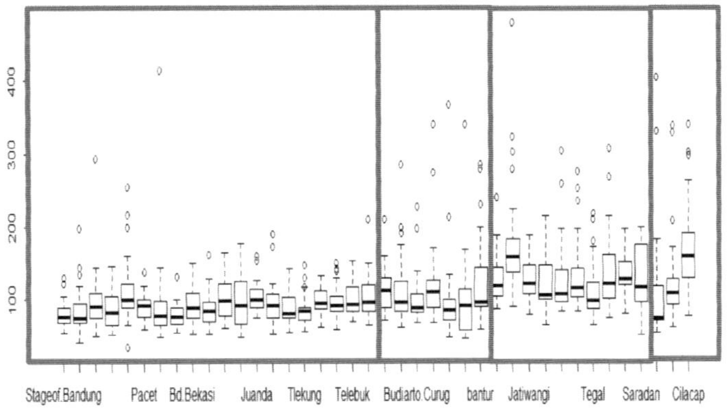

Figure 8. Box plot for annual maximum data of each station based on his group.

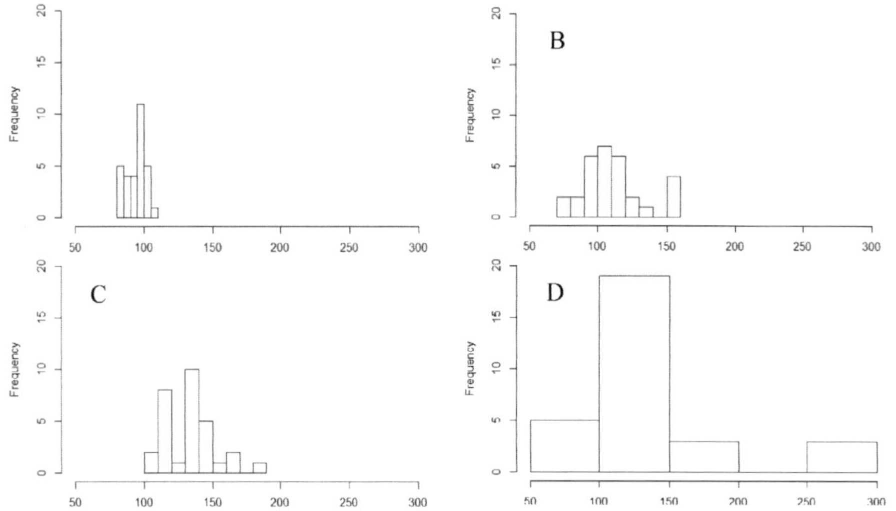

Figure 9. Histogram for each group

Picture 8 and 9 show that cluster A has a lowest annual maximum value than the other clusters with small variations over period 1983-2012. Cluster B and C tent to have a higher value than Group A and have a larger variations, cluster C have tent to have a higher value than Group B. Cluster D just spread in western area in Java Island that have a highest variation value over period 1983-2012.

4. Conclusion

This study investigated descriptive statistics and grouping of rainfall station in Java Island for period 1983-2012. Most of rainfall stations have positive kurtosis value means the distribution of data have heavy tails, and outliers. All of data have positive skewness that indicate skewness distribution curve have tail at value above of the median, and mean is on the right of the peak value. Statistics descriptive analysis for 40 stations show that average of daily maximum rainfall in Java Region during period 1983-2012 were around 80 -181mm with median between 75-160mm and standard deviation between 17 to 82.

Cluster analysis produce 4 cluster that cluster A has a lowest annual maximum value than the other clusters with small variations over period 1983-2012. Cluster B and C tent to have a higher value than Group A and have a larger variations, cluster C have tent to have a higher value than Group B. Cluster D just spread in western area in Java Island that have a highest variation value over period 1983-2012. It is mean that western area of Java tent to have a higher annual maxima of daily rainfall than northern area, and have more variety of annual maximum value.

References

[1] Intergovernmental Panel on Climate Change (IPCC) 4th Assessment Report 2007 Contribution of Working Group I to the Fourth Assessment Report of the Intergovernmental Panel on Climate Change http://www.ipcc.ch/pdf/assessment-report/ar4/wg1/ar4-wg1-annexes.pdf

[2] Gilli M and Këllezi E 2006 An application of extreme value theory for measuring financial risk *Comput. Econ.* **27** 1-23

[3] Klein Tank A M G, Zwiers F W and Zhang X 2009 Guidelines on analysis of extremes in a changing climate in support of informed decisions for adaptation *WCDMP No.72 WMO-TD No. 1500* http://eca.knmi.nl/documents/WCDMP_72_TD_1500_en_1.pdf

[4] Morisson D F 1990 *Multivariate Statistical Method* 3rd Edition (New York: McGrawHill Publishing Company)

[5] Sharma S 1996 *Applied Multivariate Techniques* (New York: John Wiley & Sons, Inc)

[6] Johnson R A dan Wichern D W 2007 *Applied Multivariate Statistical Analysis* (New Jersey: Pearson Prentice Hall, Inc)
[7] Singh W 2008 *Robustness of three hierarchical agglomerative clustering techniques for ecological data* (Iceland: University Of Iceland) http://www.unuftp.is/static/files/rannsoknarritegrdir/WarshaSingh_Masters Thesis.pdf
[8] Balcan M F, Liang Y and Gupta P 2014 Robust hierarchical clustering *J.Machine Learning Research* **15** 1401-4051

Exposure to Drought: Duration, Severity and Intensity (Java, Bali and Nusa Tenggara)

N L Adhyani[1], T June[2] and A Sopaheluwakan[1]

[1] Meteorological Climatological and Geophysical Agency, INDONESIA
[2] Department of Geophysics and Meteorology, Faculty of Mathematics and Natural Sciences, Bogor Agricultural University, INDONESIA

Email: nadhyani@gmail.com, tania_june@yahoo.com, ardhasena@gmail.com

Abstract. Occurrence of drought is a slow process lasted for a long time until the rainy season come. This natural disaster has broad and severe impact. This research was conducted to examine the level of severity and intensity of drought in Java, Bali and Nusa Tenggara using Standardized Precipitation Evapotranspiration Index (SPEI). SPEI is drought index used for quantifying drought occurrence, and can be used to analyze duration, severity and intensity. Drought is a climatological phenomenon which is difficult how to determine its onset, duration, magnitude, intensity, spatial extent, etc. Therefore, this information of exposure to drought could be describe the characteristics of drought events. To summarise the calculation, for 30 years (1985-2014) from 22 stations BMKG obtained the longest and strongest drought occurred at meteorology Serang – Banten, with long duration is 11 months and severity -16.816 at October 2002 until August 2003. In one period of drought event, not always the longest it's mean the strongest.

1. Introduction

Occurrence of drought is a slow process lasting for a period longer than a season. This natural hazard has broad and severe impact which often lead to disaster. Drought caused by anomaly weather conditions such as the decreasing of the intensity of rainfall as compared to normal conditions [1,2,3], and the rising of the temperature significantly affect the severity of the drought [4]. It is also affected by conditions of the low soil moisture and surface water availability, which is insufficient [5,6,7]. Drought can be categorized into three types of drought, namely meteorological drought, agricultural drought and hydrological drought [2]. This natural hazard has broad and severe impact which often lead to disaster. Understanding drought characteristic is, an essential element in well-prepared drought management plans.

In representing the dryness of a region, a lot of research using a drought index, which can provide a quantitative assessment of climatic conditions in terms of intensity, duration and severity of drought. There are two main objectives of the drought index, assessing the vulnerability of various systems to drought, monitoring and early warning. Various indices have been developed to detect and monitor droughts, and have been used to analyze meteorological drought. A drought index is a main variable in order to assess the effect of a drought and to determine various drought characteristics, such as duration, intensity and severity [8]. The most common used meteorological drought indicator is the standardized precipitation index (SPI) whichhas key advantage that it can be calculated with different timescales. McKee *et al.* [9] proposed the concept of standardized precipitation index (SPI) based on the long-term precipitation or stream flow record for a chosen period, and adaptable for the analysis of drought at variable time scales; it can be used for monitoring agricultural and hydrological purposes.

Vicente *et al.* [10] proposed a new climatic drought index: the standardized precipitation evapotranspiration index (SPEI). The SPEI is based on a climatic water balance (precipitation – potential evapotranspiration) anomalies. Under global warming conditions, SPEI can identified an increase in drought severity associated with higher water demand as a result of evapotranspiration. It can be calculated at different time scales, allowing exploring the vulnerability of various systems to drought, and based upon the original SPI calculation procedure. With SPEI method can provide information of drought exposure, which can describe the characteristics of drought events by characterizing its severity, duration, interval time, intensity, probability, and return period of drought.

The areas of Java, Bali and Nusa Tenggara is an agricultural center which will be affected by drought. This research is conducted to examine the level of severity, duration and intensity of drought to know the characteristics of drought in Java, Bali and Nusa Tenggara using Standardized Precipitation Evapotranspiration Index (SPEI) with time scale 1 and 3 months.

2. Study Area and Data Used

The location of research focused on the islands of Java, Bali and Nusa Tenggara which are located approximately between 5.7° – 11° S and 105° – 126° E. The stations are selected based on length recording data and completeness of data (empty data minimum). Because of SPEI calculation based on the calculation of the SPI. Calculating SPEI using precipitation and temperature data which data recorded during a period of 30 years.

Data from the Center for Database Meteorology Climatology and Geophysics (BMKG) represent of Java, Bali and Nusa Tenggara BMKG there are 22 stations with a data length of 30 years (1985-2014). There are 17 stations in Java, one station in Bali, and 4 stations in Nusa Tenggara. Data used is climate parameter data, there are monthly precipitation totals, and monthly mean temperature, with a unit of precipitation is millimeter, and temperature is degrees Celsius. To support data processing, required supporting data such as topography data and elevation each station the list of station shown by Table 3, while the position of the station is shown in Figure 1.

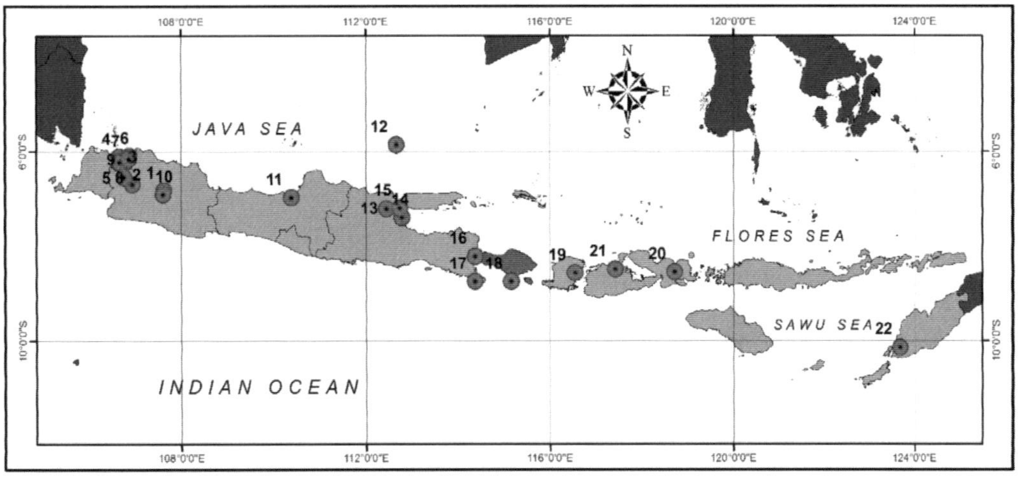

Figure 1. Location map of the 22 stations BMKG in Java, Bali and Nusa Tenggrara.

Table 1. List of 22 Stations BMKG in Java, Bali and Nusa Tenggara.

No	No. Sta	Name of Station	Longitude	Latitude	Elevation
1	88882	Stageof Lembang	107.6167	-6.8333	1200 M
2	96733	Sta. Klim. Pondok Betung - Tangerang	106.7500	-6.2558	26.2 M
3	96735	Sta. Geof. Tangerang	106.6667	-6.2667	14 M
4	97637	Sta. Met. Serang	106.6465	-6.1167	14 M
5	96739	Sta. Met. Budiarto - Curug	106.6500	-6.2333	46 M
6	96741	Sta. Met. Maritim Tanjung Priok	106.8667	-6.1000	2 M
7	96745	Sta. Met. Kemayoran - Jakarta	106.8533	-6.1600	4 M
8	96751	Sta. Met. Citeko - Bogor	106.9333	-6.7000	920 M
9	96753	Sta. Klim. Darmaga - Bogor	106.7498	-6.5536	190 M
10	96783	Sta. Geof. Bandung	107.6000	-6.9167	791 M
11	96835	Sta. Klim. Semarang	110.3833	-6.9833	4 M
12	96925	Sta. Met. Sangkapura - Bawean	112.6633	-5.8675	3 M
13	96933	Sta. Met. Perak I - Surabaya	112.4461	-7.2236	3 M
14	96935	Sta. Met. Juanda - Surabaya	112.7839	-7.4028	2.8 M
15	96937	Sta. Met. Maritim Perak II - Surabaya	112.7356	-7.2056	3 M
16	96945	Sta. Geof. Tretes	114.3833	-8.2167	50 M
17	96987	Sta. Met. Banyuwangi	114.3833	-8.7486	50 M
18	97230	Sta. Met. Ngurah Rai - Denpasar	115.16917	-8.74583	3 M
19	97240	Sta. Met. Selaparang - Mataram	116.5333	-8.56667	16 M
20	97270	Sta. Met. M. Salahuddin - Bima	118.69292	-8.54275	2 M
21	97260	Sta. Met. Sumbawa Besar	117.41367	-8.48833	3.8 M
22	97374	Sta. Klim. Lasiana - Kupang	123.66722	-10.13861	19 M

3. Methodology

The first step is the quality control data by checking the data used. Terms SPEI calculation is data to be filled in complete or without any missing data during the period of use. If there are empty data then do imputation to complete the data. After missing data problem solved, the next step calculated SPEI value on different time scale one-month (SPEI-1) and 3 months (SPEI-3) for each of the BMKG station.

3.1 Method of Completing the Missing Data

Ideally estimating missing data require comparison data from multiple stations are close and correlated with data from the test station [11]. This study used the method of average - arithmetic average [12], a simple method for filling missing rainfall data. Measurements were carried out at several stations at the same time summed and then divided by the number of stations, and if the magnitude of the difference between the average annual rainfall from each stations with the average annual rainfall that will estimated is less than 10%. The formula of missing data (monthly, seasonal or yearly) in station *(p)* is:

$$p = \frac{p_1 + p_2 + p_3 + \cdots + p_n}{n}$$

where p is missing rainfall data; p_1, p_2, p_3, ..., p_n are rainfall data at the station 1,2,3, ..., n; and n is number of stations.

Another method used is linear regression method which is based on a linear relationship between the data of the test station (station A) and data from station B, and C. The formula is:

$$pA = a + b * pB + c * pC$$

where a, b, and c are constants regression.

3.2 Calculating of SPEI Value

The SPEI calculation was done using simple multiscale drought index that combines precipitation and temperature data with the capacity to include the effects of temperature variability on drought assessment. The SPEI is based on the original SPI calculation procedure. The SPI is calculated using monthly or weekly rainfall as input data. While SPEI uses the monthly (or weekly) difference between precipitation and potential evapotranspiration (PET). The first step to estimate of potential evapotranspiration (PET), this represents a simple climatic water balance uses Thornthwaite [13] method, for estimate of PET is not focus only one method.

In this study, SPEI calculation refers to the calculations carried out by Vicente *et al.* [10], the steps calculation is as follows

3.2.1 Computation of the climatic balance. The simplest approach to calculate PET [13], which has the advantage of only requiring data on monthly-mean temperature and supporting data such as the number of days of the month, the maximum numbers of sun hours, the latitude from each stations. Following this method, the monthly PET (mm) is obtained by

$$PET = 16K\left(\frac{10T}{I}\right)^m$$

where T is the monthly-mean temperature (°C); I is a heat index, which is calculated as the sum of 12 monthly index values i, the latter being derived from mean monthly temperature using the formula

$$i = \left(\frac{T}{5}\right)^{1.514}$$

m is a coefficient depending on I: $m = 6.75 \times 10^{-7} I^3 - 7.71 \times 10^{-5} I^2 + 1.79 \times 10^{-2} I + 0.492$; and K is a correction coefficient computed as a function of the latitude and month,

$$K = \left(\frac{N}{12}\right)\left(\frac{NDM}{30}\right)$$

Here NDM is the number of days of the month and N is the maximum number of sun hours, which is calculated using

$$N = \left(\frac{24}{\pi}\right)\varpi_s$$

where ϖ_s is the hourly angle of sun rising, which is calculated using $\varpi_s = \arccos(-\tan\varphi \tan\delta)$, where φ is latitude in radians and δ is the solar declination in radians, calculated using

$$\delta = 0.4093\left(\frac{2\pi J}{365} - 1.405\right),$$

where J is the average Julian day of the month.

The next step with the values obtained ETP, then to get the difference between precipitation (P) and potential evapotranspiration (PET) month i is calculated using the equation climatic water balance,

$$D_i = P_i - PET_i,$$

it is a simple measure of the water surplus or deficit for one month. The ratio of P to PET as a suitable parameter for obtaining a drought index that accounts for global warming processes [14]. This approach has some disadvantages: the parameter is not defined when PET 5 0 (which is common in many regions of the world during winter), and the P/PET quotient reduces dramatically the range of variability and the role of temperature in droughts.

3.2.2 Creation of cumulative series at desired time scale, fitting the data to an adequate distribution function (LogLogistic) and transforming the data int0 (standardized) z-values. Calculated D_i at different time scales, following the same as with the SPI procedure. The difference $D_{i,j}^k$ in the specific month j and year i depends on the selected time scale k. For example, accumulated difference for one month in a given year i with 12-month time scale is calculated using

$$X_{i,j}^k = \sum_{l=13-k+j}^{12} D_{i-1,l} + \sum_{l=1}^{j} D_{ij}$$

if $j < k$ and

$$X_{i,j}^k = \sum_{l=j-k+1}^{j} D_{i,l}$$

if $j \geq k$ and

where $D_{i,l}$ is the *P - PET* difference in the first month of year *i*, in millimeters.

For calculation of the SPI on different time scales, the probability distribution of the gamma family is used (the two-parameter gamma or three-parameter Pearson III distributions), because the frequencies of precipitation accumulated at different time scales are well modeled using these statistical distributions. Although the SPI can be calculated using two-parameter distribution, such as the gamma distribution, the distribution of the three-parameters needed to calculate the SPEI. In the distribution of two-parameter, the variable x has a lower limit of zero ($0 > x < \infty$), while the distribution of the three-parameters, x can take values in the range ($\gamma > x < \infty$), where γ is the parameter origin of the distribution; consequently, x can have a negative value, which are common in *D* series.

Vicente *et al.* [10] tested the most suitable distribution to model the values of *Di* calculated at different time scales. For this purpose, L-moment ratio diagrams used to allow for comparison of the empirical frequency distribution of *D* calculated at different time scales with the number of theoretical distributions. The L moments are times analogous to a conventional central moment, able to characterize a wider range of distribution functions and strengthen in relation to the outliers in the data.

To create the L-moment ratio diagrams, L-moment ratios (L skewness τ_3 and L kurtosis τ_4) must be calculated. Here τ_3 and τ_4 are calculated as follows:

$$\tau_3 = \frac{\lambda_3}{\lambda_2} \quad \text{and} \quad \tau_4 = \frac{\lambda_4}{\lambda_2}$$

where λ_2, λ_3, and λ_4 are L-moments of the *D* series, obtained from probability-weighted moments (PW Ms) using the formulas

$$\lambda_1 = w_0,$$
$$\lambda_2 = w_0 - 2w_1,$$
$$\lambda_3 = w_0 - 12w_1 + 30w_2 - 20w_3,$$

The PWMs of order s are calculated as

$$w_s = \frac{1}{N}\sum_{i-1}^{N}(1-F_i)^s D_i$$

Where Fi is frequency estimator calculated following the approach of Hosking [15]:

$$F_i = \frac{i-0.35}{N}$$

where i is the range of observations arranged in increasing order and N is the number of data points. Vicente et al. [10], get the values τ_3 and τ_4 calculated from the D series of 11 observation points between 1910 and 2007 in different regions of the world, tropical (Tampa, Florida; Sao Paulo, Brazil), monsoon (Indore), Mediterranean (Valencia, Spain), semiarid (Albuquerque), continental (Wien, Austria), cold (Punta Arenas, Chile) and marine (Abashiri, Japan). The dataset was obtained from the GHCN-monthly database.

The probability density function of three-parameters log-logistic distributed variable is expressed as

$$f(x) = \frac{\beta}{\alpha}\left(\frac{x-y}{\alpha}\right)^{\beta-1}\left[1+\left(\frac{x-y}{\alpha}\right)^{\beta}\right]^{-2}$$

where α, β, *dan* γ are scale, shape, and origin parameters, respectively, for D values in the range ($\gamma > D < \infty$).

Parameters of the log-logistic distribution can be obtained following different procedures. Among them, the L-moment procedure is the most robust and easy approach [16]. When L-moment are calculated, the parameters of the Pearson III distribution can be obtained following Singh *et al.* [17]:

$$\beta = \frac{2w_1 - w_0}{6w_1 - w_0 - 6w_2}$$

$$\alpha = \frac{(w_0 - 2w_1)\beta}{\Gamma\left(1+\frac{1}{\beta}\right)\Gamma\left(1-\frac{1}{\beta}\right)}$$

$$\gamma = w_0 - \alpha\Gamma\left(\frac{1+1}{\beta}\right)\Gamma\left(\frac{1-1}{\beta}\right)$$

where $\Gamma(\beta)$ is the gamma function of β.

The log-logistic distribution adapted very well to the D series for all time scales. The probability distribution function of the D series, according to the log-logistic distribution, is given by

$$F(x) = \left[1+\left(\frac{\alpha}{x-y}\right)^{\beta}\right]^{-1}$$

The $F(x)$ values for the D series at different time scales adapt very well to the empirical $F(x)$ values at the different observations, independently of the climate characteristics and the time scale of the analysis. This demonstrates the suitability of the log-logistic distribution to model $F(x)$ values from the D series in each region of the world. With $F(x)$ the SPEI can easily be obtained as the standardized values of $F(x)$. For example, following the classical approximation of Abramowitz and Stegun [18],

$$SPEI = W - \frac{C_0 + C_1W + C_2W^2}{1 + d_1W + d_2W^2 + d_3W^3}$$

where

$$W = \sqrt{-2\ln(P)} \qquad for \qquad P \leq 0.5$$

and P is the probability of exceeding a determined D value, $P = 1 - F(x)$. If $P > 0.5$, then P is replaced by $1 - P$ and the sign of the resultant SPEI is reversed. The constants are $C_0 = 2.515517$, $C_1 = 0.802853$, $C_2 = 0.010328$, $d_1 = 1.432788$, $d_2 = 0.189269$, and $d_3 = 0.001308$.

The average value of SPEI is 0, and the standard deviation is 1. The SPEI is a standardized variable, and it can therefore be compared with other SPEI values over time and space. An SPEI of 0 indicates a value corresponding to 50% of the cumulative probability of D, according to a log-logistic distribution.

At the 3-month time scale of the output SPEI-3 value series, if the data were used 30 years (360 months) there will be a 358 of SPEI-3 values. Furthermore, based on the value of SPEI-1 and SPEI-3, drought analysis can be done in Java, Bali and Nusa Tenggara.

3.3 Duration, severity and intensity of drought

A drought index is main variable in order to assess the effect of a drought and to determine various drought characteristics, such as duration, intensity and severity [8]. In this research, used meteorological drought indicator is SPEI [10] and calculated for the different time scale, 1 month and 3 months.

Figure 2, is used to determine drought characteristics. The negative and positive values of SPI are considered as the drought and non-drought event. As drought is defined when the values of SPI fall below zero, a drought event is considered a period with negative SPEI values. In order to measure length of drought duration and magnitude of drought severity, a threshold value must be defined. The drought duration *(D)* is the period length in which the SPEI is continuous negative, started from the SPEI values is equal to -1 and ends when the SPEI values turn out to be positive. The drought severity *(S)* is the cumulated SPEI values within the drought duration, which is defined by

$$S = -\sum_{i=1}^{D} SPEI_i$$

and intensity of drought is the ratio of severity of drought to its duration.

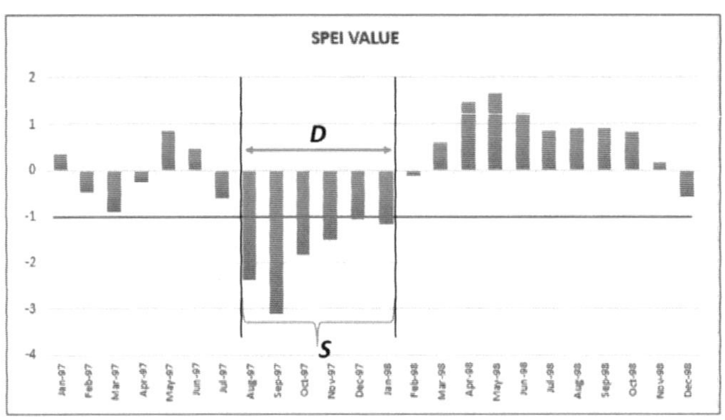

Figure 2. Definition of drought characteristics.

Many drought planners appreciate the flexibility SPI as part of the monitoring and early warning efforts, because the SPEI calculation based on the calculation of the SPI then the classification level of dryness follow SPI classification proposed by McKee *et al.* [9], as follows:

Table 2. Classification Value of SPI/SPEI.

SPI/SPEI Value	
2.0+	extremely wet
1.5 to 1.99	very wet
1.0 to 1.49	moderately wet
-0.99 to 0.99	near normal
-1.0 to -1.49	moderately dry
-1.5 to -1.99	severely dry
-2 and less	extremely dry

4. Results and Discussion

4.1 Climatology condition

The pattern of rainfall on Java, Bali and Nusa Tenggara is a monsoonal as described by Aldrian *et al.* [19]. High rainfall occurred at the beginning and end of the year, in the middle of the year tends to be low, see at figure 3. Based on Oldeman climate classification, which differentiates wet months, humid months and dry months. criteria used limits of monthly rainfall, which the wet months is the monthly rainfall > 200 mm, the humid months is about 100 - 200 mm, and dry months with monthly rainfall < 100 mm [12]. That periods of dry months is apparent in Java, Bali and Nusa Tenggara start from May until October. More details see Table 3, show recaps dry months in Java, Bali and Nusa Tenggara, represented by BMKG station.

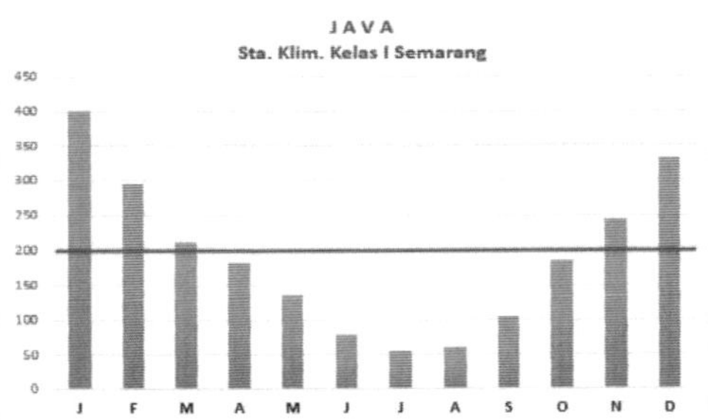

(a). Climatology Station Semarang

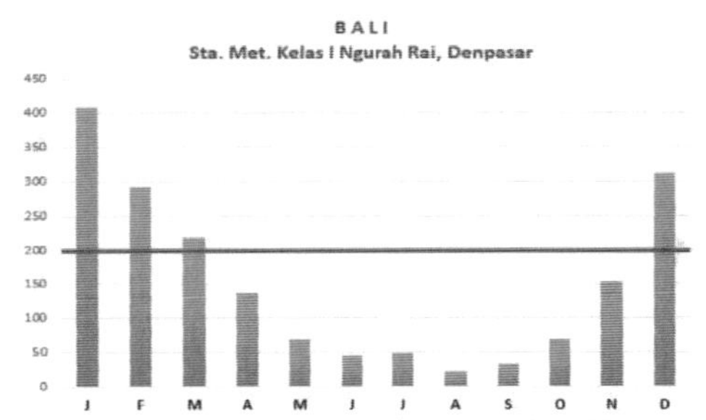

(b). Meteorology Station Ngurah Rai, Denpasar

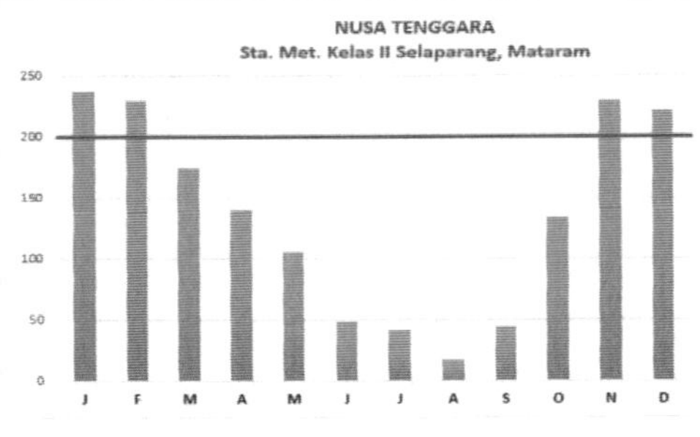

(c). Meteorology Station Selaparang, Mataram

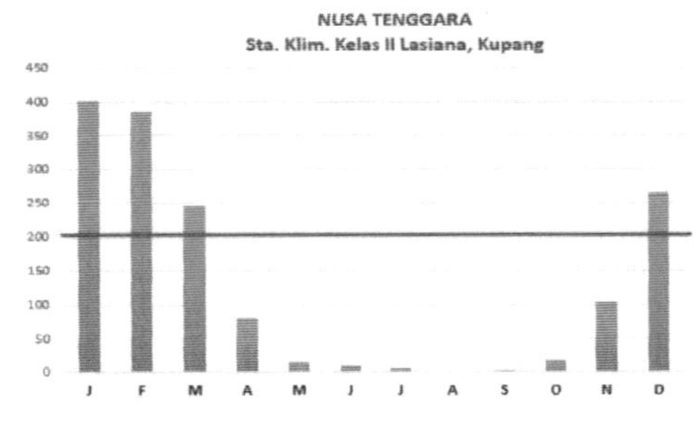

(d). Climatology Station Lasiana, Kupang

Figure 3. Average Monthly Rainfall of 30 years (1985 – 2014) in Java, Bali and Nusa Tenggara

4.2 Analysis of drought

During the period 1985 - 2014, the results of SPEI-1 at 22 stations BMKG there are 37-54 drought events. Most drought events recorded in Meteorology Station Budiarto, Curug is as much as 54 times of drought events. Which have a length of more than 2 months that occurred in 1991, 1994, 1997 dan 2006. The longest drought occurred about 4 months in 1997, while the lowest incidence of droughts recorded in Meteorology Station M. Salahudin – Bima as much as 37 times of droughts.

SPEI-3 values showed there are 22 to 36 of drought events, with the highest incidence records in Meteorology station Sumbawa Besar, and lowest in Meteorology station Kemayoran – Jakarta. In Sumbawa Besar, the drought events Which have a length of more than 2 months that occurred in 1987-1988, 1996, 2001, 2003, 2004, 2012 and 2014. The longest drought occurred about 6 months in 2003. Figure 4 shows the number of occurrences of droughts for each station BMKG for SPEI-1 and SPEI-3. The results showed that the longer time scale of SPEI calculation then the number of occurrences of drought will be decrease however length of drought will be increase.

4.3 Duration, severity and intensity of drought

Results of SPEI-1 calculation, obtained maximum duration of each station, the maximum value is shown in Figure 5.a ranges between 2 – 6 months or very short until short duration. The longest recorded at three stations, there are meteorology maritime Tanjung Priuk-Jakarta, meteorology station Citeko – Bogor, and meteorology station M. Salahudin – Bima, into the category short.

While the results of SPEI-3 shows the duration of the drought that occurred longer than SPEI-1. The maximum duration of SPEI-3 is shown in Figure 5.b ranges between 5-11 months in category short until medium duration. There are 2 stations with medium duration which 11 months duration, and the longest recorded at the geophysics station Tangerang and meteorology station Serang.

SPEI-1 calculation, obtained maximum severity of each station, the maximum value is shown in Figure 6.a ranges between very low and low category. The highest severity recorded at the meteorological station Citeko – Bogor occurred at 1985 which a peak of drought in February 1985 with index value -2.338 (extremely dry).

The maximum severity of SPEI-3 calculation is shown in Figure 6.a ranges between low and high category. The highest severity recorded at the meteorology station Serang occurred at 2002 – 2003 which a peak of drought in March 2003 with index value -1.802 (severely dry).

Table 3. Recapitulation of Dry Months in Java, Bali and Nusa Tenggara.

NO	NOSTA	STATION	DRY MONTHS	
			Total	Description
1	88882	Stageof Lembang	4	June - September
2	96733	Sta. Klim. Kelas II Pondok Betung - Tangerang	1	August
3	96735	Sta. Geof. Kelas I Tangerang	5	June - October
4	97637	Sta. Met. Kelas I Serang	5	June - October
5	96739	Sta. Met. Kelas III Budiarto - Curug	3	July - September
6	96741	Sta. Met. Kelas I Maritim Tanjung Priok - Jakarta	6	May - October
7	96745	Sta. Met. Kelas III Kemayoran - Jakarta	4	June - September
8	96751	Sta. Met. Kelas III Citeko - Bogor	2	July - August
9	96753	Sta. Klim. Kelas I Darmaga - Bogor	0	-
10	96783	Sta. Geof. Kelas I Bandung	4	June - September
11	96835	Sta. Klim. Kelas I Semarang	3	June - August
12	96925	Sta. Met. Kelas III Sangkapura - Bawean	4	July - October
13	96933	Sta. Met. Kelas III Perak I - Surabaya	6	May - October
14	96935	Sta. Met. Kelas I Juanda - Surabaya	4	July - October
15	96937	Sta. Met. Kelas II Maritim Perak II - Surabaya	6	May - October
16	96945	Sta. Geof. Kelas II Tretes	4	July - October
17	96987	Sta. Met. Kelas III Banyuwangi	6	May - October

18	97230	Sta. Met. Kelas I Ngurah Rai - Denpasar	6	May - October
19	97240	Sta. Met. Kelas II Selaparang - Mataram	4	June - October
20	97270	Sta. Met. Kelas III M. Salahuddin - Bima	7	April - October
21	97260	Sta. Met. Kelas III Sumbawa Besar	6	May - October
22	97374	Sta. Klim. Kelas II Lasiana - Kupang	7	April - October

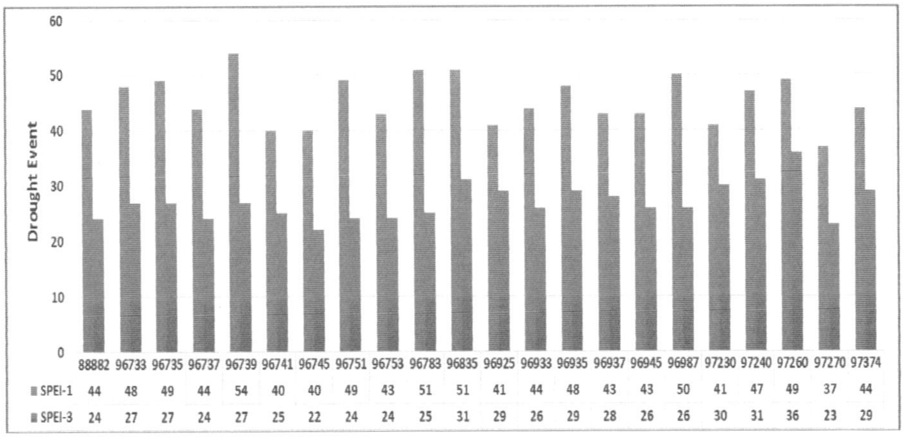

Figure 4. The numbers of drought events for 1985 – 2014 in Java, Bali and Nusa Tenggara.

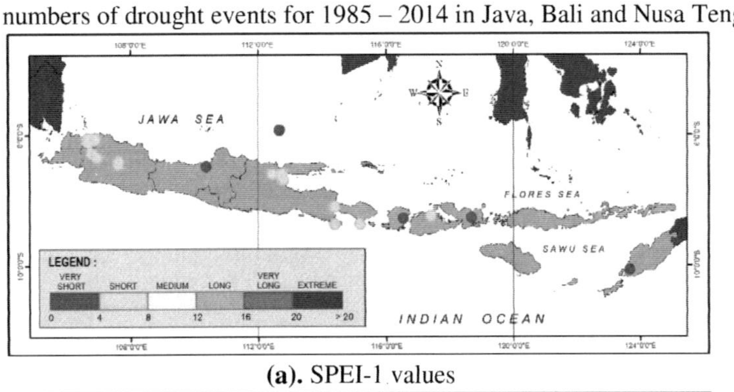

(a). SPEI-1 values

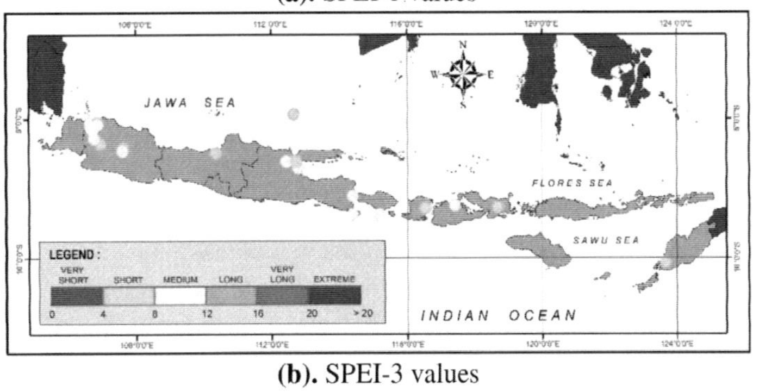

(b). SPEI-3 values

Figure 5. Maximum duration of drought in Java, Bali and Nusa Tenggara.

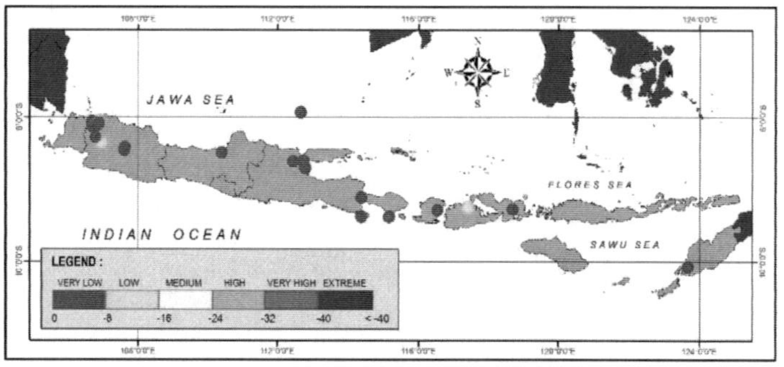

(a). SPEI-1.

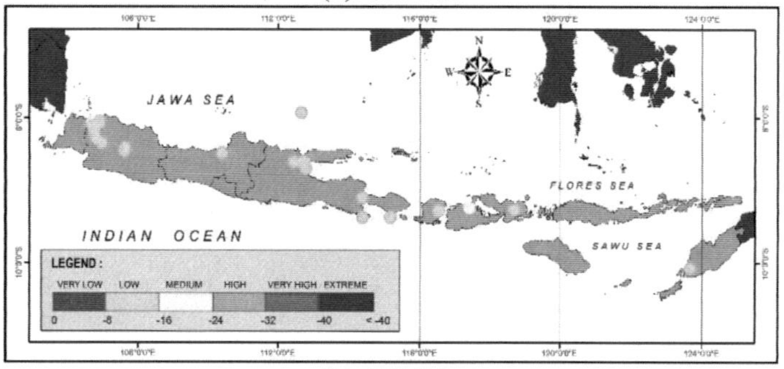

(b). SPEI-3.

Figure 6. Maximum severity of drought in Java, Bali and Nusa Tenggara.

The maximum intensity for SPEI with time scales 1 (SPEI-1) and 3 (SPEI-3) months shown in Figure 7. The magnitude of the intensity its mean in one period of drought has a high severity and short duration. For SPEI-1, the highest intensity that occurred in geophysics station Lembang – Bandung about -2.931 in October 2012 with a value of severity -2.931 with a month duration. And the highest intensity for SPEI-3 was recorded in meteorology station Selaparang – Mataram about -2.485 with severity -7.455 and duration of 3 months occurred in June – August 2009.

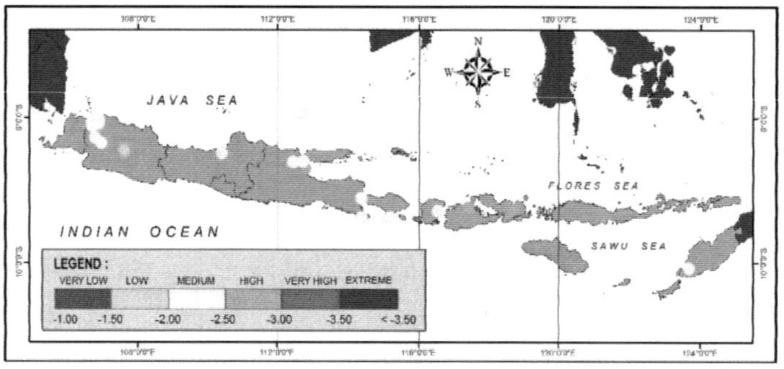

(a). SPEI-1

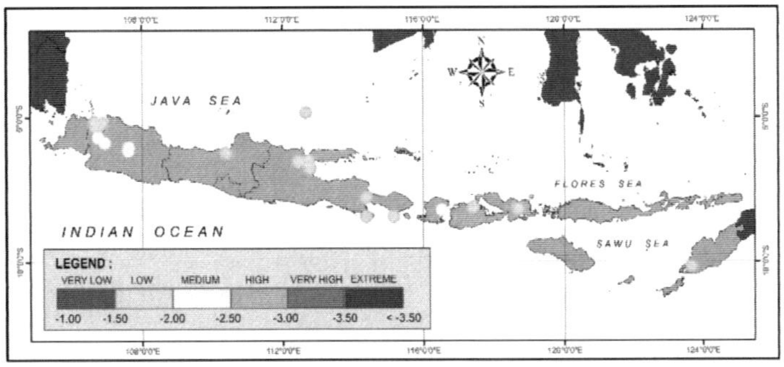

(b). SPEI-3.

Figure 7. Maximum intensity of drought in Java, Bali and Nusa Tenggara.

Based on the statement above, that the longer of the time scale of SPEI then the number occurrences of drought event will decrease but the duration of the drought will increase. From the calculations, obtained the longest and strongest drought occurred at meteorology station Serang – Banten, with long duration is 11 months and severity -16.816 at October 2002 until August 2003. In one period of drought event, the longest it doesn't mean the strongest, and vice versa.

Table 4 shows the summary results of duration, severity and intensity calculation for each station with the time scales 1 and 3 months. Recapitulation shown apparent difference in duration, severity and intensity of each region, it describe the characteristics of drought in one regions had a difference with other regions.

Table 4. Recapitulation of drought events the longest and strongest for SPEI-1 and SPEI-3.

NO	STATIONS	TIME SCALE	LONGEST		STRONGEST		HIGHEST	
			YEAR	D	YEAR	S	YEAR	I
1	88882	SPEI-1	1991	4	1991	-6.062	2012	-2.931
		SPEI-3	2012	4	2012	-9.575	2012	-2.394
2	96733	SPEI-1	2011	3	2011	-5.131	2007	-1.969
		SPEI-3	1994 - 1995	8	2011	-10.353	2003	-1.753
3	96735	SPEI-1	2012	3	2012	-4.374	2002	-1.852
		SPEI-3	1997 - 1998	11	1997 - 1998	-14.994	2011	-1.787
4	96737	SPEI-1	2009	5	2009	-6.520	2003	-2.170
		SPEI-3	2002 - 2003	11	2002 - 2003	-16.816	1985	-1.888
5	96739	SPEI-1	1991	4	1991	-5.303	1985	-2.156
		SPEI-3	1997 - 1998	8	1997 - 1998	-12.451	1999	-1.642
6	96741	SPEI-1	2003	6	2003	-8.843	2010	-2.608
		SPEI-3	1997 - 1998	7	1997 - 1998	-9.481	2003	-2.322
7	96745	SPEI-1	2002	5	2002	-6.891	2014	-2.494
		SPEI-3	1997 - 1998	10	1997 - 1998	-13.056	2003	-1.760
8	96751	SPEI-1	1985	6	1985	-11.121	2003	-2.035

		SPEI-3	2002 - 2003	6	1985	-11.168	1985	-2.234
9	96753	SPEI-1	2010 - 2011	4	2010 - 2011	-6.460	2010	-2.135
		SPEI-3	2006	9	2006	-13.297	2001	-2.128
10	96783	SPEI-1	1991	4	1991	-6.249	2010	-1.906
		SPEI-3	2006	8	2006	-11.645	1991	-2.026
11	96835	SPEI-1	2014	3	2014	-3.688	1998	-2.003
		SPEI-3	1997 - 1998	7	1997 - 1998	-9.034	2007	-1.822
12	96925	SPEI-1	2014	3	2014	-4.589	2004	-2.129
		SPEI-3	2005	7	2005	-10.808	1997 - 1998	-1.997
13	96933	SPEI-1	2009	4	2009	-7.134	2014	-2.289
		SPEI-3	2009	8	2009	-11.083	2004	-1.899
14	96935	SPEI-1	2009	4	2009	-6.532	2010	-2.628
		SPEI-3	2008	9	2008	-12.434	2009	-1.675
15	96937	SPEI-1	2009	4	2009	-7.413	1991	-2.118
		SPEI-3	2006 - 2007	6	2006 - 2007	-8.779	2009	-1.898
16	96945	SPEI-1	2008	4	2008	-5.124	2007	-2.001
		SPEI-3	2006 - 2007	8	2006 - 2007	-11.647	1997	-1.699
17	96987	SPEI-1	1992	5	1992	-6.739	2005	-2.067
		SPEI-3	1991 - 1992	9	1991 - 1992	-14.909	1998	-1.708
18	97230	SPEI-1	1997 - 1998	5	1997 - 1998	-7.164	2001	-2.334
		SPEI-3	1997 - 1998	9	1997 - 1998	-14.934	1995	-1.969
19	97240	SPEI-1	2010	3	2010	-3.934	2000	-2.177
		SPEI-3	1997 - 1998	6	1997 - 1998	-10.283	2009	-2.485
20	97260	SPEI-1	2014	2	2014	-3.927	2005	-2.771
		SPEI-3	2004	6	2004	-8.090	2009	-1.916
21	97270	SPEI-1	2009	6	2009	-8.606	2010	-2.639
		SPEI-3	2009 - 2010	9	2009 - 2010	-16.509	1997	-1.856
22	97374	SPEI-1	2013	3	2014	-4.015	2002	-2.456
		SPEI-3	2006 - 2007	5	2006 - 2007	-9.212	2012	-1.930

Information :
D : Length of drought (Duration) in the month; S : Severity; I : Intensity

5. Summary and Conclusion

Analysis of duration, severity, and intensity of drought in Java, Bali and Nusa Tenggara had been conducted using SPEI with time scales 1 and 3 months. Twenty two meteorology station were used for a period of 30 years (1985 – 2014). The results shows that duration, severity and intensity of drought increase from western to eastern part of Indonesia. The Longest of drought, the strongest severity of drought in each stations been identified. The longer of the time scale of SPEI then the number of occurrences of drought will decrease but the dryness will be increase. Duration, severity and intensity were calculated for Java, Bali and Nusa Tenggara with the following the longest and the strongest drought of SPEI-1 occurred in meteorology station Citeko which range duration 6 months at 1985. And SPEI-3 give the results that meteorology station had the longest and strongest drought event with 11 months duration from October 2002 until August 2003.

By calculating the duration, severity and intensity of each region can describe the characteristics of drought events, and the difference for one region to another. In the future needs to calculate the duration, severity and intensity with long time scales such as 6 and 12 months, and next analysis it helps the incidence of droughts relationship with the global phenomena such as El-nino.

References

[1] Mutjtahiddin M I 2014 *Jurnal Meteorologi dan Geofisika* **15** 2 99-107
[2] NOAA 2008 *Drought National Oceanic and Atmosphere Administration National Weather*

Service

[3] Pramudia A 2002 *Analisis Sensitivitas Tingkat Kerawanan Produksi Padi di Pantai Utara Jawa Barat Terhadap Kekeringan dan El-Nino* [Tesis] (Bogor (ID): Institut Pertanian Bogor)

[4] Vicente-Serrano S M 2006 *Water Resources Management* **20** 37-60.

[5] Suryanti I 2008 *Analisis Hubungan Antara Sebaran Kekeringan Menggunakan Indeks Palmer dengan Karakteristik Kekeringan (Studi Kasus : Provinsi Banten)* [Skripsi] (Bogor (ID): Institut Pertanian Bogor)

[6] Tosunoglu F and Can I 2016 *Nat Hazards* **82** 1457-1477.

[7] World Meteorological Organization. SPI User Guide WMO-No.10-90. Chair, Publications Board. WMO, 2012.

[8] Mishra A K and Singh V P 2010 *Journal of Hydrology* **391**(1-2) 202-216.

[9] McKee T B, Doesken N J and Kleist J 1993 *The Relationship of Drought Frequency and duration to Time Scales*, Proceding of the 8th Conference on Applied Climatology.

[10] Vicente-Serrano S M, Begueria S and Lopez-Moreno J I 2010 *J. Climate*. **23** (7) 1696–1718.

[11] Saputro D R dkk., 2011, *Pendugaan data tidak lengkap curah hujan di Kabupaten Indramayu (berdasarkan data tahun 1980-2000). Sains*. IPB Press. Bogor.

[12] Asdak C 2002 *Hidrologi dan Pengelolaan DAS* (Yogyakarta: UGM Press)

[13] Thornthwaite C W 1948 *The Geographical Review* **38**

[14] Tsakiris G, Pangalou D and Vangelis H 2007 *Water Resour. Manage*. **21** 821–833.

[15] Hosking J R M 1990 *J. Roy. Stat. Soc.* **52B** 105–124.

[16] Ahmad M I, Sinclair C D, and Werritty A 1988 *J. Hydrol.* **98** 205–224.

[17] Singh V P, Guo H, and Yu F X 1993 *Stochastic Hydrol. Hydraul.* **7** 163–177.

[18] Abramowitz M and Stegun I A 1965 *Handbook of Mathematical Functions, with Formulas, Graphs, and Mathematical Tables* (Washington: Dover Publications) pp 1046.

[19] Aldrian E, Gates L D and Widodo F H 2003 *Variability of Indonesian rainfall and the influence of ENSO and resolution in ECHAM4 simulations and the reanalyses*, Max-Planck-Institut fur Meteorologie. May 2003. MPI Report 346,ISSN 0937-1060

Mapping Extreme Rain Conditions in Sumatra by Influence Global Conditions

S Supriyadi[1,2*], R Hidayati[1], R Hidayat[1] and A Sopaheluwakan[2]

[1]Bogor Agricultural University, Faculty of Mathematics and Natural Sciences, Building Wing 19 Level 4, Bogor
[2]Indonesian Agency for Meteorology Climatology and Geophysics (BMKG)

Email: lamet.1980@gmail.com

Abstract. Determination of the extreme value category with a specific threshold value for a region with diverse climate characteritics may lead to incorrect extreme characterization. This study aims to determine thresholds of extreme rainfall in Sumatra, identify its spatial patterns and occurence in Sumatra and examine the link between extreme weather events to global climate phenomena which are the ENSO and the Dipole Mode. The data used in this study are daily rainfall data from 38 stations in Sumatra and indices of global climate phenomena, the dipole mode index (DMI) and the Southern Oscillation Index (SOI). The method used to determine the threshold for extreme value is with a probability of 95% based on their corresponding distribution, cluster analysis and correlation analysis. The expected result is an extreme value for all parts of Sumatra, the extreme character of all clusters by time and region, as well as its relationship with DMI and SOI. Preliminary results obtained that extreme daily rainfall values ranged from 42.4 to 114.7 mm per day. Extreme daily rainfall for western region of Sumatra has an average value higher and more varied than the eastern region. Extreme daily rainfall over northern region has an average value higher and more varied than the southern region. Clustering analysis of 38 observation stations with ward method gets 5 (five) clusters. At least 20 years long periods of ideal data are needed to find the threshold of extreme rainfall over Sumatra region. Threshold of extreme rainfall with long periods of different data showing different values for each station are sometimes larger or smaller that may be caused by different spread of the data for each station.

1. Introduction

Sumatra Island is geographically an island which is above equatorial region surrounded by ocean. Rainfall in Indonesia in general is influenced by the interaction of land, atmosphere and the oceans around it. The phenomenon of positive anomaly of sea surface temperatures in the Indian Ocean to the west and negative anomalies in sea surface temperatures occurred west of Sumatra, resulting in increased rainfall in Sumatra's west coast region known as the Indian Ocean dipole mode (IOD) [1]. There is a negative IOD in most areas of Sumatra causing increased rainfall. In Indonesia the events of floods and severe droughts are always associated with El Niño-Southern Oscillation (ENSO) [2]. The results of research related to extreme rainfall with cumulative distribution function method for Sumatra conducted by Marpaung et al [3] using rainfall data from TRMM and observations indicate that extreme rainfall events with the number of incidence days of 1 to 2 days per annum dominant place. In 1998 and 1999 it appears to increase in extreme rainfall events for a number of days per year incidence greater than 2 days, especially in the water of the West and East Sumatra. This increase is allegedly due to the influence of global phenomenon La Nina and the Dipole Mode. The threshold value of extreme rainfall for Sumatra is 60 to 130 mm / day. Several studies of extreme rainfall also done abroad including Zhang et al [4] that did a study in the valley of the Yangtze River, China. They analyzed daily rainfall data of 150 stations from the 1960-2002 period in the Yangtze River valley. They use extreme climate index of the highest daily rainfall amount of 5-day and 10-day (R5D and R10D) and the maximum daily rainfall at the 95th percentile (prec95p) and the 99th percentile (prec99p). Frequency of R5D and R10D show a downward trend. The phenomenon is more noticeable on the middle of the Yangtze River valley. Moberg [5] analyzed temperature and precipitation for stations in the Western Europe region. They calculated the daily extreme climate index of the data series in the period 1901-2000. They use criteria of extreme climates percentile of 90, 95 and 98 to calculate the index of extreme temperature and precipitation in Western Europe during the winter and summer. The study showed that in winter a trend is warmed (~ 1.0 °C / 100 years) than summer (~ 0.8 °C). In the winter, total rainfall average in 121 stations in Europe showed a significant trend of increased ~ 12% per 100 years. The trend of daily rainfall in the winter with the calculation of the 90th percentile, 95 and 98 also show the same thing. Other studies calculate extreme values using the median quantile regression that was first introduced by Roger Koenker and Gilbert Basset [6]. Some researchers determine the extreme value distributions that use the Generalized Extreme Value (GEV) developed by Jenkinson [7] to model extreme precipitation. This approach is referred to as block maxima using distribution modeling for maximum data at each time interval called the block. Other researchers use the approach of Peak Over Threshold (POT), which was originally introduced in the field of hydrology to define extreme values. POT method uses a threshold value for defining the precipitation with a value exceeding the threshold as extreme precipitation. POT overall uses observational data in modeling extreme precipitation, this is what distinguishes it from the GEV approach. Extreme rainfall events from time to time in Indonesia are more frequent that often leads to loss of property and loss of life. The extreme value issued by BMKG under regulation KBMKG No: Kep.009 Year 2010 is rainfall intensity of 50 mm/day or 20 mm/hour. The determination of threshold values with the extreme categories that limit values for a region as diverse character of the climate was not able to provide characterization of the extreme right. Therefore, this study aims to determine the threshold of extreme rainfall in Sumatra and group them by similarity.

2. Data and Methods

2.1. Research Locations

Location of the study is the island of Sumatra in the form of 18 points stations in Sumatra and surrounding areas.

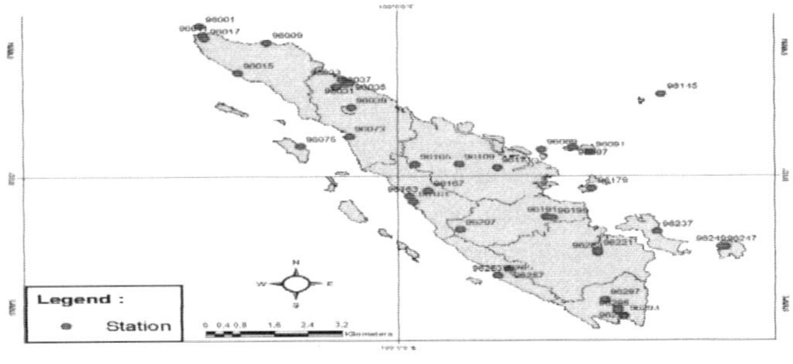

Figure 1. Map of research site.

2.2. Data

The data used in this study are the daily rainfall data of observation stations in the island of Sumatra in the period of 1983-2012.

Table 1. Inventory of daily rainfall stations data in Sumatra Island.

No.	Station Id	Name of station	Station Location	Latitude	Longitude	Mean Sea Level	Period
1	96291	Masgar	Pesawar, Lampung	-5.172	105.18	71	1996-2012
2	96295	Raden Inten II	South Lampung, Lampung	-5.27	105.18	81	1981-2012
3	96293	Maritim Panjang	Bandar Lampung,	-5.47	105.33	1	1998-2012
4	96297	Kota bumi	North Lampung, Lampung	-4.836	104.87	60	1994-2012
5	96253	Fatmawati Airport	Bengkulu City, Bengkulu	-3.88	102.33	16	1981-2012
6	96257	Kepahiyang	Kepahiyang, Bengkulu	-3.63	102.59	517	1985-2012
7	96255	Baai Island	Baai Island, Bengkulu	-3.87	102.32	10	1985-2012
8	96223	Kenten	Palembang, South Sumatra	-3.00	104.70	8	1981-2012
9	96221	Sutan Mahmud Badarudin II	Palembang, South Sumatra	-2.90	104.70	11	1981-2012
10	96237	Depati Amir Airport	Pangkalpinang, Bangka Belitung	-2.17	106.13	33	1981-2012
11	96249	H.A.S Hanandjoedin	Belitung, Bangka Belitung	-2.757	107.75	47	1981-2012
12	96247	Pilang	Belitung, Bangka Belitung	-2.757	107.65	22	1995-2012
13	96191	Sei Durian	Muaro Jambi, Jambi	-1.60	103.49	34	1998-2012

14	96207	Depati Parbo	Kerinci, Jambi	-2.083	101.45	450	1985-2012
15	96195	Sultan Thoha	Sultan Thoha, Jambi	-1.634	103.64	26	1983-2012
16	96171	Japura	Indra Giri Hulu, Riau	00.33	102.32	19	1982-2012
17	96109	Sultan Syarif Kasim	Pekanbaru, Riau	0,324306	101.43	27	1981-2012
18	96087	Hang Nadim	Batam, Riau Islands	1.117	104.12	28	1993-2012
19	96089	Raja Haji Abdullah Airport	Tanjung Balai Karimun, Riau Islands	01.03	103.38	1	1994-2012
20	96091	Kijang Airport	Tanjung Pinang, Riau Islands	0,6375	104.53	18	1983-2012
21	96145	Tarempa Airport	Kepulauan Anambas, Riau Islands	03.20	106.25	2	1991-2012
22	96179	Dabo Airport	Lingga, Riau Islands	-0.48	104.58	29	1991-2012
23	96167	Sicincin	Sicincin, West Sumatra	-0.575	100.71	137	1985-2012
24	96163	Minangkabau International Airport	Padang, West Sumatra	-0.793	100.29	6	1982-2012
25	96165	Silaing Bawah	Padang Panjang, West Sumatra	0,317	100.40	773	1996-2012
26	96161	Teluk Bayur	Kota Padang, West Sumatra	-0.996	100.37	2	1994-2012
27	96031	Sampali	Deli Serdang, West Sumatra	3.617	98.78	25	1980-2012
28	96033	Belawan	Medan, West Sumatra	3.788	98.71	3	1982-2012
29	96035	Kualanamu Airport	Deli Serdang, North Sumatra	3.642	98.89	23	1981-2012

30	96037	Tuntungan	Deli Serdang, North Sumatra	3.501	98.56	86	1991-2012
31	96039	Parapat	Simalungun, North Sumatra	2.694	98.92	1061	1997-2012
32	96073	F.L Tobing Sibolga	Tapanuli Tengah, North Sumatra	01.55	98.88	10	1981-2012
33	96075	Binaka	Gunung Sitoli - Nias, North Sumatra	1.165	97.71	175	1982-2012
34	96001	Cut Bau Maimun Saleh Airport	Sabang, Nanggroe Aceh Darussalam	0,269	95.35	126	1981-2012
35	96009	Malikussaleh Airport	Aceh Utara, Nanggroe Aceh Darussalam	5.228	96.93	28	1983-2012
36	96011	Sultan Iskandar Muda Airport	Aceh Besar, Nanggroe Aceh Darussalam	5.521	95.42	20	1982-2012
37	96015	Tjut Nyak Dien Meulaboh Airport	Nagan Raya, Nanggroe Aceh Darussalam	4.049	96.25	3	1982-2012
38	96017	Indrapuri	Aceh Besar, Nanggroe Aceh Darussalam	05.40	95.46	51	1995-2012

2.3. Method

2.3.1. Cluster analysis

Cluster analysis is part of a multivariate analysis with the purpose to categorize objects based on the characteristics of its observations, where each object in the same group have a level of closeness and similarity between one another [8]. The result of cluster analysis is the formation of groups that have a degree of similarity and heterogeneity of high inter-cluster. Nazaripour and Khosravi [9] classify the intensity of daily rainfall in Zahedan, Iran with Ward cluster analysis method. Techniques within the Ward's method calculate the sum of the squares between the two groups for all variables. Ward's equation is as follows:

$$d\,(r,s) = \frac{n_r n_s d_{rs}^2}{(n_r + n_s)} \tag{1}$$

d (r, s) : distance
drs : distance between *r* and *s*
nr : group-r
ns : group-s

2.3.2. Analysis of Opportunities

In determining the threshold of extreme rainfall that contains zero millimeters of rainfall, the chances are calculated by the method of distribution of mixture [10], namely:

$$H\,(x) = q + pP_x\,(X) \tag{2}$$

where :

$H\,(x)$: probability distribution for a mixture of rain recesses value that exceeds x for $x \geq 0$
q : zero chance of rain events
p : probability of rain events greater than zero, $p = 1 - q$
$Px\,(X)$: probability of rain events by gamma cumulative distribution opportunities ($x > 0$)

2.3.3. Data Length Determination

The steps in determining the length of the data that is used to find the value extreme are:

1. Find a station that has a data length more than 30 years
2. Divide data into several periods where each period is a multiple of five years
3. Determine the extreme value of each station
4. Create scatter plot graph
5. Determine the value that is closest to the ticket

3. Result and Discussion

3.1. Analysis Cluster

Results of grouping rainfall stations in Sumatra based extreme value with threshold of each month are five clusters (figure 2).

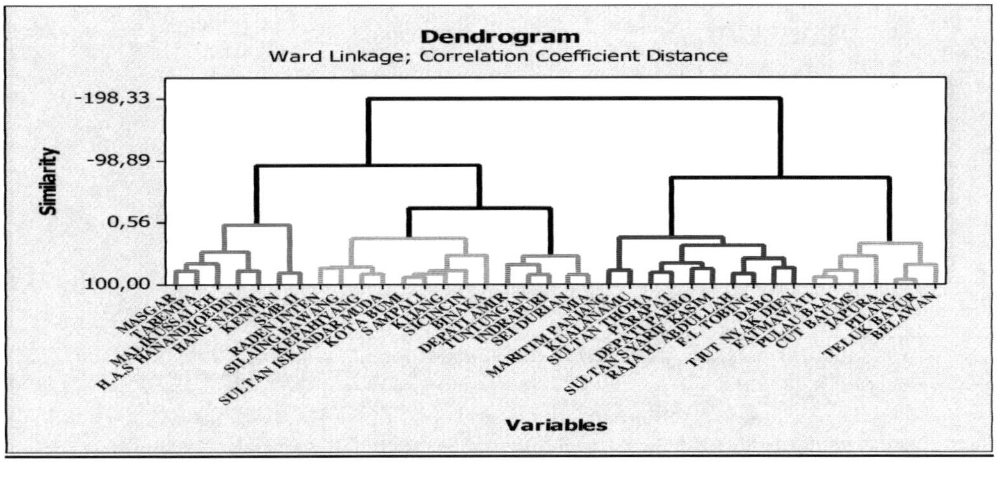

Figure 2. Dendogram of rainfall station groups in Sumatra Island.

Based on the analysis results it is obtained that in cluster 1 there are 7 stations, 9 stations in cluster 2, 10 stations in cluster 3, 7 stations in cluster 4 and 5 stations in cluster 5. Groups of rain stations are shown in table 2.

Table 2. Station groups based on the extreme threshold.

No.	Station Id	Name of station	Cluster	No.	Station Id	Name of station	Cluster
1	96291	Masgar	1	20	96091	Kijang Airport	2
2	96295	Raden Inten II	2	21	96145	Tarempa Airport	1
3	96293	Maritim Panjang	3	22	96179	Dabo Airport	3
4	96297	Kota bumi	2	23	96167	Sicincin	2
5	96253	Fatmawati Airport	4	24	96163	Minangkabau International Airport	5
6	96257	Kepahiyang	2	25	96165	Silaing Bawah	2
7	96255	Baai Island	4	26	96161	Bayur Bay	4
8	96223	Kenten	2	27	96031	Sampali	2
9	96221	Sutan Mahmud Badarudin II	1	28	96033	Belawan	4
10	96237	Depati Amir Airport	5	29	96035	Kualanamu Airport	3
11	96249	H.A.S Hanandjoedin	1	30	96037	Tuntungan	5
12	96247	Pilang	4	31	96039	Parapat	3
13	96191	Sei Durian	5	32	96073	F.L Tobing Sibolga	3
14	96207	Depati Parbo	3	33	96075	Binaka	2
15	96195	Sultan Thoha	3	34	96001	Cut Bau M. Saleh Airport	4
16	96171	Japura	4	35	96009	Malikussaleh Airport	1
17	96109	Sultan Syarif Kasim	3	36	96011	Sultan Iskandar Muda Airport	2
18	96087	Hang Nadim	1	37	96015	Tjut Nyak Dien Airport	3
19	96089	Raja Haji Abdullah Airport	3	38	96017	Indrapuri	5

The pattern of rainfall threshold in cluster 1 shows that the threshold value is the highest rainfall in January. The pattern of rainfall threshold in cluster 2 in figure 4 shows that the threshold value is the highest precipitation in February. The pattern of rainfall threshold in cluster 3 shows that the threshold values are the high rainfall in August in contrast to cluster 1 and cluster 2. The pattern of rainfall threshold in cluster 4 shows that the limit values are high in June and August and the lowest is in April. The pattern of rainfall threshold in cluster 5 shows that the extreme rainfall has two peaks, where the first peak occurs in February and the second peak occurs in August (figure 3).

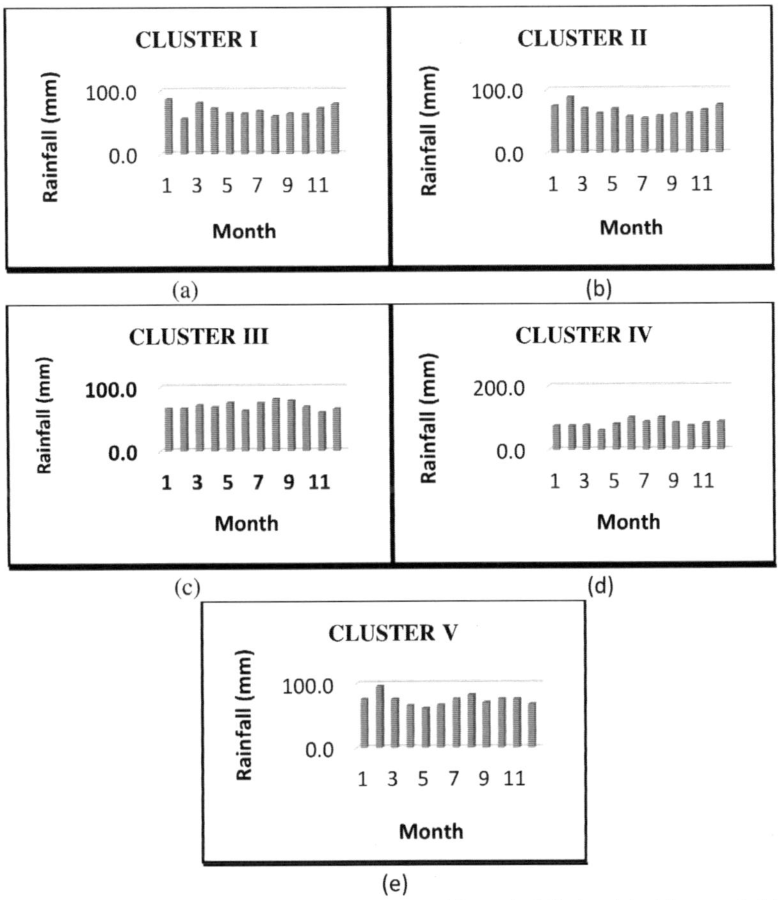

Figure 3. The pattern of extreme monthly rainfall in (a) Cluster I;(b) ClusterII;(c) Cluster III; (d) Cluster IV (e) Cluster V.

3.2. Analysis of Opportunities

Figure 4 shows that the extreme daily rainfall values range from 42.4 to 114.7 mm. Extreme value of daily rainfall in western region of Sumatra has an average value higher and more varied than the eastern region. While the northern region of Sumatra has an extreme value of daily rainfall is higher than in the south.

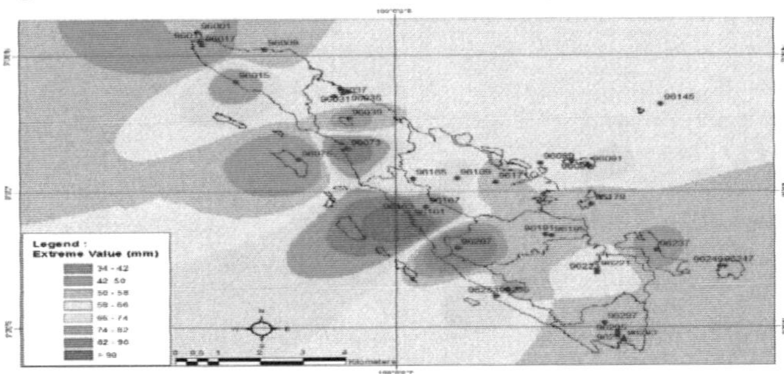

Figure 4. Threshold of extreme rainfall in Sumatra island (1998-2012).

3.3. Analysis of Data Period

Minimum length of data period is very important as a reference in determining the length of data that is ideal for calculating the thresholds of extreme rainfall. Determination of a minimum length of the data period was done by taking a sample of rainfall observation data over 30 years. After sorting the data from 38 stations in Sumatra, seven stations that have long data period for 35 years are Raden Inten II Lampung, Fatmawati Bengkulu, Kenten South Sumatra, Sutan Mahmud Badarudin II South Sumatra, Depati Amir Bangka Belitung, Sampali North Sumatra, Kualanamu North Sumatra.

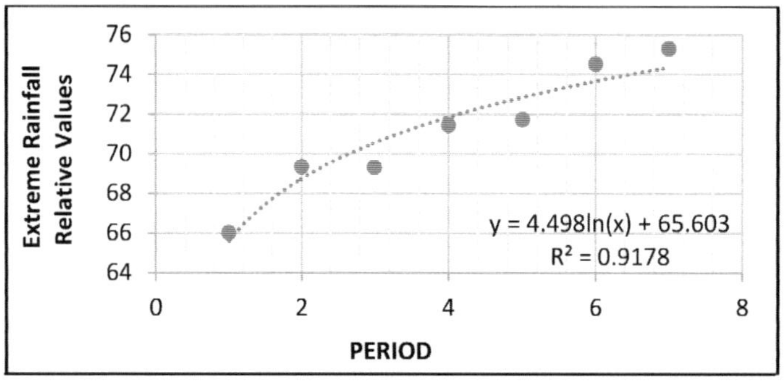

Figure 5. Graph of the average threshold of each period.

Figure 5 shows that the period approaching the intersection line is the 4th period (1978-1997) so that it can be concluded that long periods of minimum rainfall data taken in seeking the minimum threshold require extreme data over 20 years. After sorting, as many 18 stations have data more than 20 years including Raden Inten II, Fatmawati, Kenten, Sutan Mahmud Badarudin II, Depati Amir, HAS Hanandjoedin, Sultan Thoha, Japura, Minangkabau International Airport, Sampali, Belawan, Kualanamu F.L Tobing Sibolga, Binaka, Kijang, Cut Bau Maimun Saleh, Sultan Iskandar Muda and Tjut Nyak Dien Meulaboh.

As many 28 stations have data at least 30 years. These stations include Raden Inten II, Fatmawati, Kepahiyang, Baai Island, Kenten, Sutan Mahmud Badarudin II, Depati Amir, HAS Hanandjoedin, Depati Parbo, Sultan Thoha, Japura, Sultan Syarif Kasim, Hang Nadim, Kijang, Tarempa, Dabo, Sicincin, Minangkabau International Airport, Sampali, Belawan, Kualanamu, Tuntungan, F.L Tobing, Sibolga, Binaka, Cut Bau Maimun Saleh, Malikussaleh, Sultan Iskandar Muda and Tjut Nyak Dien Meulaboh.

Table 3. Calculation of minimum data length using the 7 observation stations.

PeriodTEN		Extreme Values							
		Raden inten	Fatmawati	Kenten	SMB II	Depati amir	Sampali	Kualanamu	Average
Period 1	1978-1982	64,57	77,00	71,52	58,04	59,88	64,37	66,63	66,00
Period 2	1978-1987	70,93	81,92	70,00	65,65	59,77	64,74	72,27	69,33
Period 3	1978-1992	73,28	83,77	71,16	62,21	59,35	66,81	68,52	69,30
Period 4	1978-1997	73,30	98,00	72,00	63,73	59,40	66,15	67,40	71,43
Period 5	1978-2002	73,30	84,00	74,00	75,16	59,11	67,28	69,17	71,72
Period 6	1978-2007	72,33	98,00	75,00	75,80	58,40	71,71	70,38	74,52
Period 7	1978-2012	72,05	96,81	77,00	75,60	58,30	77,00	69,21	75,28

4. Summary and Concluding Remarks

Clustering analysis of 38 observation stations with the ward method results 5 (five) clusters. Each cluster has a different pattern every month. Extreme daily rainfall in the region of Sumatra has a value ranging from 42.4 to 114.7 mm. Extreme daily rainfall over western region of Sumatra has an average value higher and more varied than the eastern region. Extreme daily rainfall over the northern region has an average value higher and more varied than the southern region. Threshold of extreme rainfall with long periods of different data showing different values for each station are sometimes larger or smaller that may be caused by different spread of the data for each station.

References

[1] Saji N, Goswami B, Vinayachandran P and Yamagata T 1999 *Nature* **401** 360–363
[2] Rosanne D'A, Rob A, Rob W, Jonathan P, John S, Jason E S, Satria B and La Ode Ngkoimani 2008 *Jurnal Climatolology*
[3] Marpaung, Satiadi and Harjana 2012 *Analisis Kejadian Curah Hujan Ekstrim di Pulau Sumatera Berbasis Data Satelit TRMM dan Observasi Permukaan* Jurnal Sains Dirgantara **9(2)**
[4] Q Zhang, X Chen and B Stefan 2011 *Atmospheric and Climate Sciences* **1** 1-8
[5] Moberg A, Jones, Lister, Walther, Brunet, Jacobeit, Alexander, Della Marta, Luterbacher, Yiou, Chen, Klein Tank, Saladie, Javier Sigro, Aguilar, Alexandersson, Almarza, Auer, Barriendos, Begert, Bergstro, Bo¨hm, C. J. Butler, Caesar, Drebs, Founda, Gerstengarbe, Micela, Maugeri, Hermann, Pandzic, Petrakis, Srnec, Tolasz, Tuomenvirta, Werner, Linderholm, Philipp, Wanner, and Elena Xoplaki 2006 *J. Geophys. Res* **111** D22106
[6] R. Koenker and G Basset 1978 *Econometrica* **46(1)** pp. 33-50
[7] Jenkinson 1955 *Q J Meteorol. Soc* **81**
[8] Everitte S B and Dunn G 1998 *Applied Multivariate Data Analysis* (New York USA: Halsted Print)
[9] H Nazaripour and M Khosravi 2011 *Geography and Environmental 30 Planning, 21th Year* **40(4)**
[10] I G Darmaputra, Y Koesmaryono, I Santosa 2006 *Jurnal Agromet Indonesia* **20(2)** 14 – 24

Influence of Climate Variability on Brown Planthopper Population Dynamics and Development Time

S Romadhon[1,2], Y Koesmaryono[1] and R Hidayati[1]

[1] Department of Geophysics and Meteorology, Bogor Agricultural University, Indonesia
[2] Meteorological, Climatological, and Geophysical Agency, Jakarta, Indonesia

Email: sy.romadhon@gmail.com

Abstract. Brown planthopper or *Nilaparvata lugens* (BPH) is one of the rice major pest in Indonesia. BPH can cause extensive damage and almost always appear in each planting season, frequent explosions attack (outbreaks) resulting in very high economic losses. Outbreaks of BPH were often occurred in paddy fields in Indramayu regency and several endemic regency in Java island, where rice is cultivated twice to three times a year both in the rainy and dry cropping seasons. The output of simulation shows the BPH population starts increasing from December to February (rainy season) and from June to August (dry season). The result relatively had same pattern with light trap observation data, but overestimate to predict BPH population. Therefore, the output of simulation had adequately close pattern if it is compares to BPH attacked area observation data. The development time taken by different stages of BPH varied at different temperatures. BPH development time at eggs and adults stage from the simulation output is suitable with BPH real lifestage, but at nymphs stage the result is different with the concept of development time.

Keywords: brown planthopper, simulation, population, development time

1. Introduction

BPH is still serious pest of rice across the world, mainly in tropical climates. Outbreak frequency of BPH has been increasing in Asian rice growing countries in recent years. BPH is well known for its status as a typical outbreak-type pest, which is characterized by great fluctuations in population size from year to year and from field to field [1]. The pest feeds directly on the growing plant, reducing its yield potential. If the pest density is high, the plant will die and results a condition called hopperburn. BPH also transmit the grassy stunt and ragged stunt virus, which can further reduce yield. The alternative host for BPH are *Eleusine coracana* (wild millet), *Leersia hexandra* (southern cutgrass), *Saccharum officinarum*, *Zea mays*, and *Zizania latifolia* (Manchurian wildrice) [2].

BPH population always fluctuates according to the dynamic conditions of its environment. Both biotic and abiotic factors are believed to be the factor responsible for the change in population. In agricultural ecosystems, climate variability leads to changes in the biodiversity of insect pests as a result of rising temperature, changes in precipitation pattern and an increase in the frequency and intensity of some extreme climate events that cause an increase in the frequency of occurrence of pests and diseases. Climate, mainly the temperature is known to exert a strong influence on the distribution and abundance of species, often through effects on mortality. Insects in general are poikilotherms, the temperature of their bodies is approximately the same as that of the environment. Therefore, temperature is probably the most important environmental factor influencing insect behavior, distribution, development, survival, and reproduction [3]. Higher temperatures will cause predictions of developmental period from the mean to be too short, and lower temperatures will make them too long. The developmental response of insects to temperature is important in understanding the ecology of insect life histories. Understanding the relationship between temperature and various biological characteristics is needed for the development of reliable pest population prediction systems and management strategies.

During the dry season, when evaporation always exceeds precipitation, so that rainfall becomes a key factor determining BPH populations. However, in the wet season extra rainfall often appears, heavy rain and floods will wash nymphs and adults off the host plant, and eggs laid in the rice sheath cannot hatch after submergence. Dry weather enhances emigration but rainfall inhibits the planthoppers from taking off. BPH prefer lowland to upland rice and since thick vegetation is a better habitat for them, since various species have distinct preferences for plants at different growth stages, they proliferate when rice plants of different ages are available. The shortage of host plants results in overcrowding, which adversely affects the population buildup. It reduces the rate of nymphal development, increases the percentage of macropterous adults, lengthens the preoviposition period, and decreases the number of eggs laid [4].

BPH adults have two forms of wing, macropterous (long winged) and brachypterous (short winged). Macropterous (long winged) can fly which allows them to escape adverse habitats and track changing resources. This is the migration type. Brachypterous (short winged) are flightless but usually possess higher fecundity. This type will dominant if the host plant is abundant. Under normal conditions, the field population decreases sharply about 70 days after the rice is transplanted, because the nutritional conditions of the rice sheath on which the BPH feeds tend to deteriorate [5]. As the rice plant ages, most of the BPH population becomes macropterous and emigrates. During the first 4-5 weeks after transplantation, macropterous adults that apparently had immigrated from the outside were dominant, though some brachypterous adults probably originating from the eggs laid in the nursery bed [6].

The objective of this research is to learn the BPH characteristics including its population dynamics and development time according to climate variability using simulation model as the anticipation and adaptation efforts aimed at developing agriculture that is adaptive to climate variability and pest attacks today and in the future.

2. Materials

Area of this research is Indramayu regency which is situated in coordinate 107"51'-108"36' east longitude and 6"15' - 6"40' south latitude. Materials presented are those data which were collected from various sources. The climatological data are acquired from BMKG climatology station Dramaga from year 2007 to 2015, and data of BPH that infected in paddy fields were provided by Indramayu installation pest laboratory as input for initial BPH population and validation. The main tools utilized are CSIRO DYMEX^{TM} Builder and Simulator version 3, and Microsoft Office 2010. The flowchart of this research is shown in figure 1.

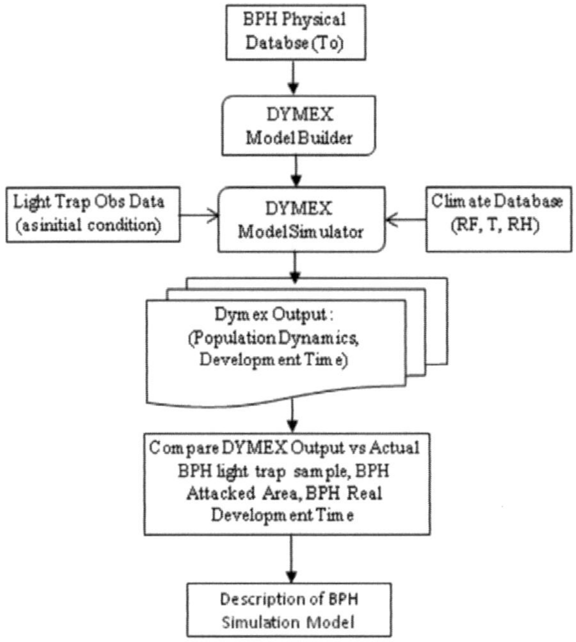

Figure 1. Research Flowchart.

3. Results And Discussion

During period of January 2007 to December 2015, the climate variables in Indramayu varied from years to years. Minimum temperature range is between 18°C and 27°C, maximum temperature range is between 25°C and 39°C, range of maximum RH is 55%-99%, minimum RH range from 17% to 97%, and the yearly rainfall is from 1955 mm to 3521 mm. Figure 2 shows climate parameters in Indramayu regency.

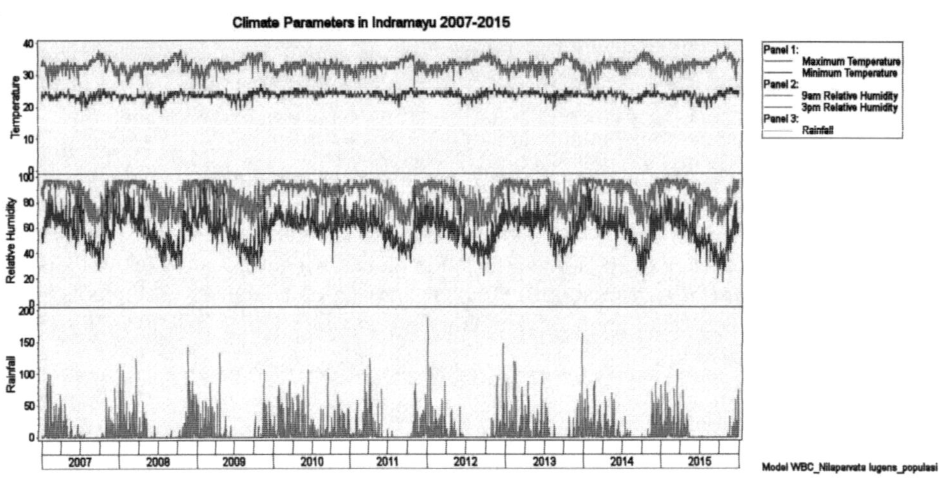

Figure 2. Climate parameters in Indramayu regency.

Those climate conditions are favorable for BPH to grow and multiply rapidly when the host plant is available. The climate phenomena of La Nina in late 2008 and early 2009 was followed by El Nino in late 2009 and continuous weak to moderate La Nina from April 2010 to 2012 were producing unusual climate conditions which were very conducive to the outbreaks of particular pests such as BPH [7]. Each BPH stage has their requirements to stay comfortably and active as shown on the table 1.

Table 1. Temperature and relative humidity of BPH requirements. (Source: [4])

Stage	Temperature range	Optimum temperature	RH range
Eggs	15°-32° C	25°-27° C	70-85%
Nymphs	15°-32° C	25°-30° C	70-85%
Adults	10°-32° C	28°-30° C	70-85%

According the range on table 1, sometimes (maximum) temperature in Indramayu was higher than the range, then BPH still alive but in an aestivation mode. When an insect is progressively warmed to higher temperature, a sequence of distinct observable or measureable events occurs. Firstly, the BPH moves in an increasingly uncoordinated way and becomes immobile called 'critical thermal temperature' (CTmax). As the temperature is further increased, all small-scale movement of appendages (legs, antennae) ceases as the organisms enter a state of 'heat coma' (HCT), after which, at a higher temperature, the insect dies at its upper lethal temperature (ULT). The differences in body size and volume also affect heat tolerance; thus the CTmax, heat coma temperature and ULT of nymphs were consistently and significantly lower than that of adults, and for one of these indices (heat coma), adult males were less heat tolerant than females [8].

From the Indramayu installation pest laboratory data, we obtained the information that BPH first appearance is on the early to mid of December while the wet cropping season is about to begin and the peak population occurred on early to mid of January. In dry cropping season, BPH will appear again on mid to late of June and become highest population on about mid to late of July.

The output of simulation model for BPH population dynamics had relatively near to the pattern of the light trap observation in Indramayu installation pest laboratory from 2007-2009, but the model's result is overestimate to predict the BPH population (figure 3). This case could explain because the BPH light trap observation data are only captured in one place and often happened the light trap tools were out of order so the monitoring of BPH population is not continuous. The model also has limitation that may hinder possible influential factors other than ones included. Meanwhile, if we compares the model's result with the attacked area data (assumption the population has linearity with attacked area data) from 2007-2014, they shared the adequately close pattern. So, those results are sufficient to decide that the simulation model can be used for anticipation of the upcoming outbreaks of BPH and other pests.

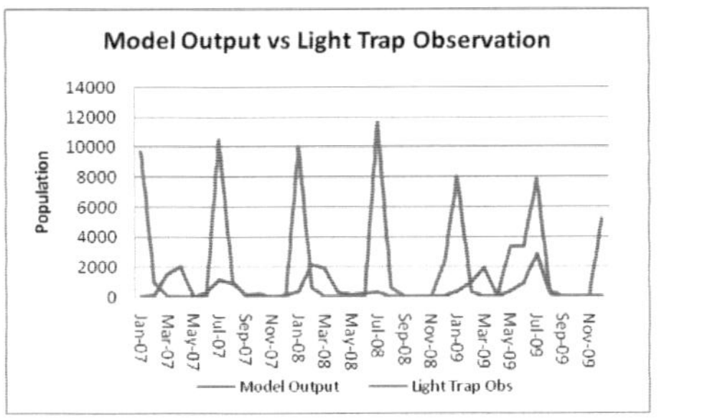

(a)

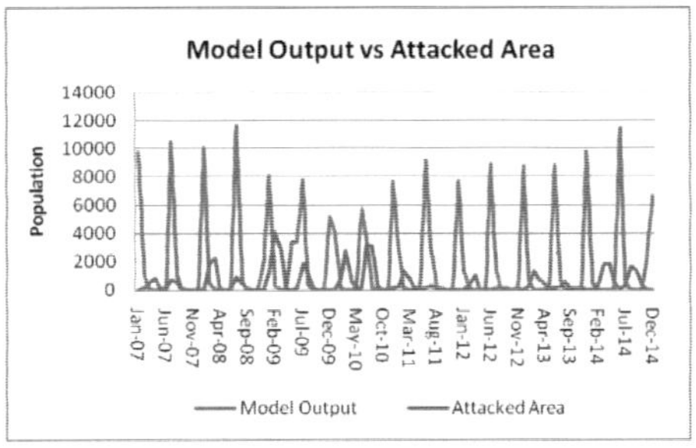

(b)

Figure 3. Model Output versus Observation; a) versus Light Trap; b) versus Attacked Area.

Since insects including BPH are cold-blooded animals (poikilotherms), their rate of growth and development is proportional to ambient temperature. At very low temperatures, there is no development at all. As temperature increases, there are some points (different for various species) at which development begin to occur. This low-temperature "gate" is called the Developmental Threshold. As temperatures increase above the developmental threshold, the rate of insect development gets faster. At high temperature, development rate levels off and then drops quickly near the upper limit of survival. Similar insects reared at different temperatures will require different amounts of time to complete development. Cooler temperatures retard development, warmer temperatures stimulate development. Since the relationship between temperature and development rate is linear through a wide range of temperature, it is possible to predict how long development will take at each different temperature. The concept of Day-Degrees is a convenient way to track development at different temperatures. One day-degree is the amount of development that occurs in one day (24 hours) when the temperature is one degree above the development threshold. Day degrees accumulate over time, each day the temperature is above the developmental threshold, more day-degrees are added to the accumulation. Each insect species requires a certain total number of day-degrees to complete its development. This is also known as the physiological time of development [9].

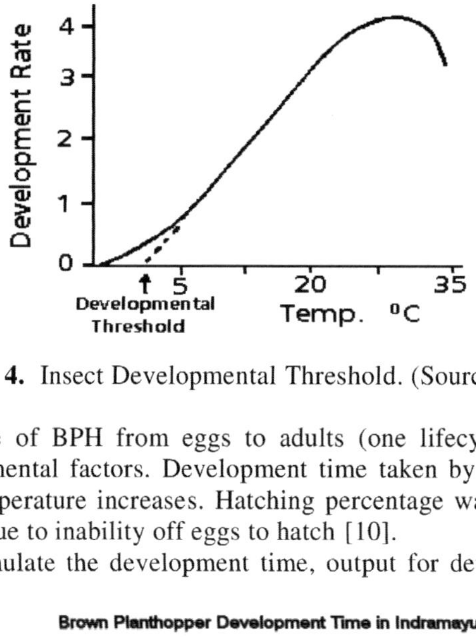

Figure 4. Insect Developmental Threshold. (Source: [9])

Total development time of BPH from eggs to adults (one lifecycle) is about 25-55 days, depending on the environmental factors. Development time taken by different stages decreases considerably when the temperature increases. Hatching percentage was drastically reduced with increasing in temperature due to inability off eggs to hatch [10].

The model also can simulate the development time, output for development time of BPH is given in figure 5.

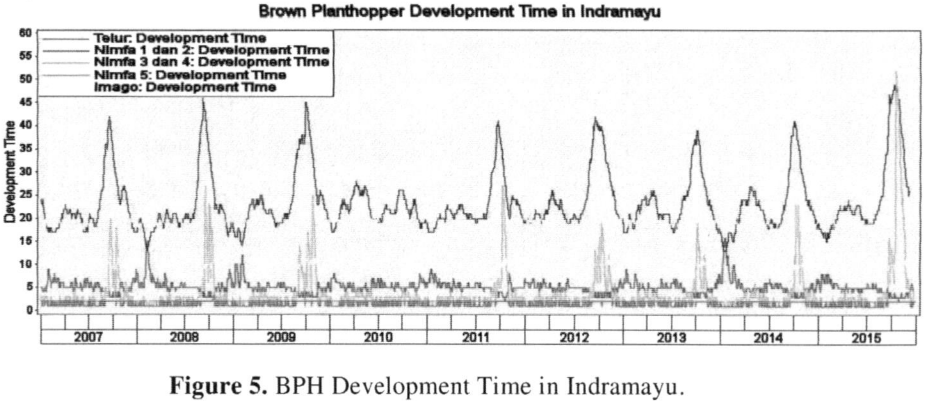

Figure 5. BPH Development Time in Indramayu.

From the figure 5, we can notice the development time at all stages in late 2009 until late 2011 is shorter than any period on the graph. We know in almost through 2010, the La Nina event occurred and made the excess precipitation in most of area in Indonesia which is favorable to BPH. Therefore, the shorter time development of BPH means the BPH grows and regenerates more quickly. Meanwhile, the occurrence of El Nino event on late 2014 was not significantly extending the BPH development time based on the graph.

The longest development time spent by first and second instars of BPH nymph stage (blue line) that took 23 days in average, while the shortest development time is fifth instar of BPH nymph stage (purple line) that only took 1-3 days because it will enter the imago (adults) stage soon.

a) b) c)

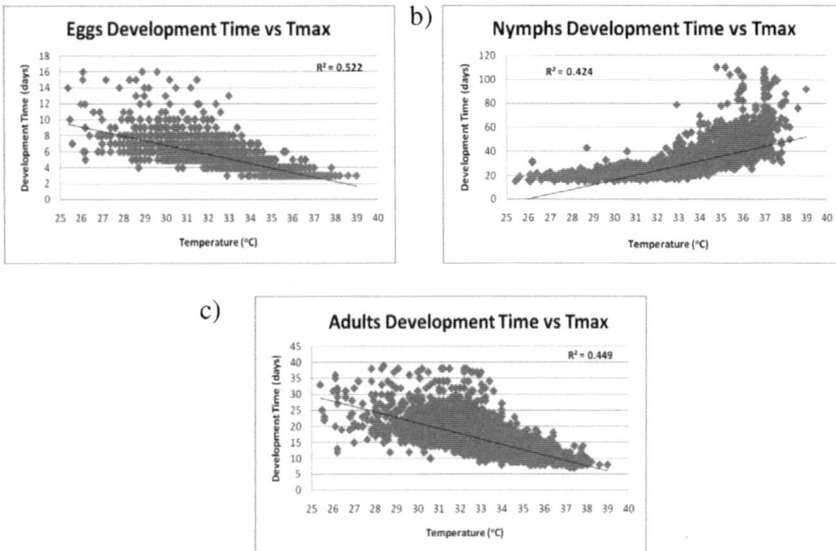

Figure 6. Plot of BPH Development Time versus maximum temperature; a) Eggs ; b) Nymphs; c) Adults.

Based on the graphs at figure 6, we obtain the relationships between development time and (maximum) temperature that are suitable to the development time concept for eggs and adults stage with high correlation. But at the nymphs stage, there are not suitable with the concept of development time. The relationships are reverse with the eggs and adults stage. This case happened because the BPH real lifestage of nymphs has five instars, but the simulation only count three instars and can occur miscounting of development time of nymphs stage. Another assumption that the nymphs stage is different with the others. BPH did the aestivation mode dominantly at this stage, so their growth rate becomes slower and automatically the development time becomes longer.

Lastly, besides using the simulation models to predict the outbreaks population of BPH in relation with climate variability as prevention to minimize risks of BPH attack, we need to do comprehensive and integrated anticipation on pest management such as using crop rotation, using pest resistant crop varieties, using the natural enemies, and using the pesticide only if necessary.

4. Conclusions

Climate variability, mainly temperatures factor plays an important role in the different biological activities on poikilothermic organisms including BPH such as population dynamics, developmental time and lifecycle. The output of simulation model for BPH population dynamics is relatively close to the pattern of both light trap observation data and the attacked area data in Indramayu installation pest laboratory, but overestimate to predict the population. Meanwhile, the output for eggs and adults development time are match with the concept which explain the higher temperature the shorther development time but at the nymphals stage is not suitable with such concept due to the limitation of the model.

Acknowledgement
Many thanks to the Meteorological, Climatological, and Geophysical Agency (BMKG) for the opportunity and provides the climate data.

References

[1] Kisimoto R 1981 *Rev. Plant Prolec. Res.* **14** 26-58.
[2] Dyck V A and Thomas B 1977 *International Rice Research Institute*
[3] Bale J S, G J Masters, I D Hodkinson, C Awmach, T M Bezemer, V K Brown, J Butterfield, A Buse, J C Coulson, J Farrar, et al 2002 *Global Change. Biol* **8** 1-16
[4] Pathak M D, Khan Z R 1994 *International Rice Research Institute* 20-23
[5] Wu G R, Yu X P, Tao L Y and Ren Z J 1994 *Wing dimorphism and migration in the Brown planthopper, Nilaparvata lugens Stal.* H.V Danks (ed). Insect Life Cycle Polymorphism, p.263-275
[6] Sawada H, Kusmayadi A, S W Gaib Subroto, Sumardiwijaya E and Mustaghfirin 1993 *Population Ecology* **35** 113-137
[7] Stigter, C (Kees) J 2012 *Unusual Climate Conditions of 2010/2011 and Pest & Diseases Outbreaks* Work In Progress Seminar University of Indonesia
[8] Phiyopungkul J, Jeremy P and Jeffrey B 2012 *PlosOne* **7(1)**
[9] Meyer J 2003 *Tutorial page of General Entomology* (North Carolina State University)
[10] Manikandan N and J S Kennedy 2013 *International Journal of plant Protection* **6(2)** 376-378

Dry spell frequency in West Java, Indonesia

V I Purnaningtyas[1,2*], R Boer[1] and A Faqih[1]

[1]Department of Geophysics and Meteorology, Faculty of Mathematics and Natural Sciences, Bogor Agricultural University (IPB), Indonesia
[2]Meteorological Climatological and Geophysical Agency (BMKG), Indonesia

Email: vivi_indhira@yahoo.com

Abstract. The climatology and variability of dry spells are valuable information for scientists, engineers, planners, and managers working in water-related sectors such as agriculture, ecology, hydrology, and water resources. The dry spell concerns consecutive dry days which are the largest number of consecutive days with less than 1 mm of daily precipitation within a year. The objective of this study is to analyse the spatial and temporal characteristics of dry spells and also specifically investigate the frequency patterns of the dry spell distribution based on historical observed daily precipitation from 1981 to 2010. The longest dry spell occured at Pasirukem with the length of 252 days. The northern coast of West Java have higher probabilities of dry spells more than 5, 10, 15, and 20 days.

1. Introduction

Changes in climate indicators such as surface temperature, sea surface temperature, sea levels, rainfall, extreme weather and climate events, resulted in broad changes of the global climate which impacts various aspects of life [1]. According to Surmaini et al. [2], agriculture is one of the areas most vulnerable to climate change, particularly food crops that are mainly annual crops and relatively sensitive to stress associated with the excess and shortage of water. Three major factors associated with global climate change which also impacts the agricultural sector is the change of rainfall patterns, increased extreme climate events (floods and droughts), as well as increased air temperature and sea levels [3]. Changes in rainfall patterns and the rise of temperatures caused significant decline in agricultural production, furthermore, extreme climate events such as floods and droughts lead to increased number of parched crops [2].

Zhang et al. [4] mentioned some properties of extreme climates which have great impact on agriculture of which is dry spells or series of dry days. Dikshit et al. [5] conducted research on twenty varieties of rice and found that with the dry spell occuring for sixteen days, caused delays in the harvest time about two to twenty-seven days and a decline in yields about 10-91%. Furthermore, Niewolt [6] stated that in the tropics, the occurrence of dry spell for seven days or more have serious impacts on crops.

The objective of this study is to analyse the spatial and temporal characteristics of dry spells, by investigating the frequency patterns of its distribution and the probability for some given length of dry spells. The results of this study are expected to provide improved climate information which can be used to reduce the vulnerability of the climate sensitive agricultural activities towards extreme weather/climate events in West Java.

2. Data and Methods

The data used in this study consisted of daily rainfall records obtained from Indonesian Agency for Meteorology Climatology and Geophysics (BMKG) for the period 1981 to 2010. There are 54 rain gauge stations used and the distribution is shown in Figure 1.

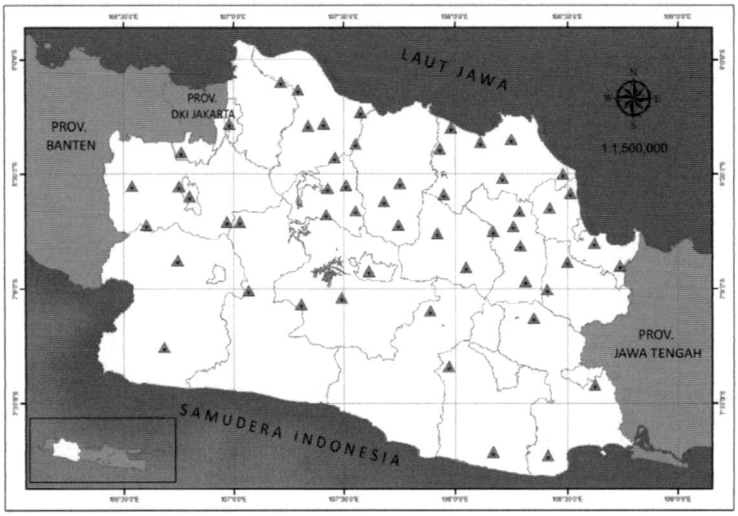

Figure 1. The area and the distribution of rain gauges of this study.

The analysis of dry spell uses the definition given by the *Expert Team on Climate Change Detection and Indices* (ETCCDI) which concerns consecutive dry days in which are the largest number of consecutive days with less than 1 mm of daily precipitation within a year [7]. Frequencies of dry spells were derived from daily data obtained from certain locations during the period of study using the following equation:

$$F_{ijk} = \sum_{i=1}^{m} F_{ijk}$$

where i is dry day occurences in a month (j) and year (k), with (m) the total number of years [8]. The probability density function (PDF) is used to calculate the probability of dry spells which has values more than 5, 10, 15, and 20 days at each stations using the following equation:

$$P(X > x) = 1 - P(X < x) = 1 - F(x)$$

3. Results and Discussions

The observed frequency of dry spells during the period of study showed that a one-day dry spell has the highest frequency of occurence at every location, see the example in Figure 2. The picture shows dry spell distribution of two different stations. It can be seen that the incidence of dry spell in Darmaga is no more than 30 days, while in Pasirukem a dry spell of more than 200 days have occurred. This shows that there are sizeable differences in the incidence of dry spell between stations due to differences in rainfall patterns.

The Meteorology, Climatology and Geophysics Agency divides the West Java region into 36 rainfall patterns that consists of 34 Seasonal Zones (Zona Musim/ZOM), and two Non Seasonal Zones (Non Zona Musim/Non ZOM). ZOM areas have a clear distinction between the dry season and the rainy period, while Non ZOM areas generally do not have a clear distinction between the dry season and the rainy season, with rainfall mainly all high or all low throughout the year. Non ZOM areas in

West Java are Bogor, North Sukabumi and North Cianjur where maximum incidences of dry spell are less than 50 days.

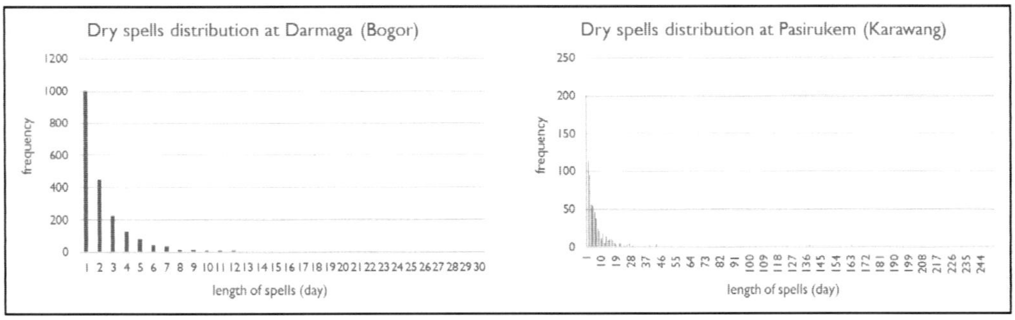

Figure 2. The observed frequency of dry spells at Darmaga (Bogor) and Pasirukem (Karawang).

The spatial observed distribution of maximum runs of dry spells are presented in Figure 3. The longest run of dry spell of 252-days was observed during mid-April to mid-December 1982 at Pasirukem. It can be seen that most stations in West Java have experienced dry spells of more than 150 days. Dry spells with durations of less than 100-days were dominant over high lands and close to large water bodies.

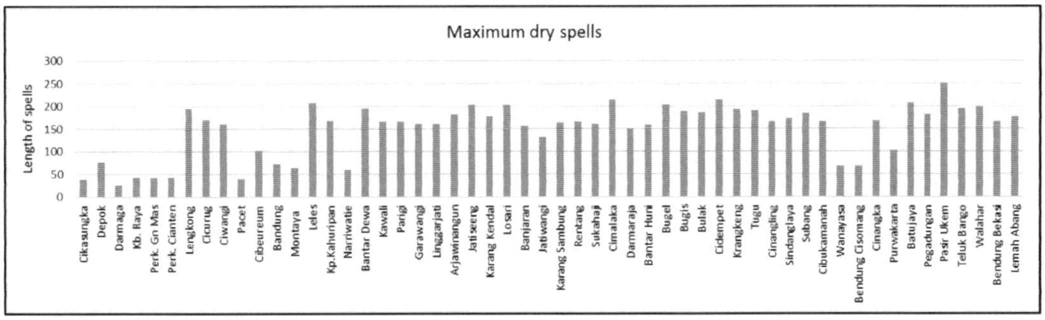

Figure 3. The spasial distribution of maximum dry spells.

Most of the maximum incidence of dry spells at each station occurs during the El Nino phenomenon. The maximum incidence of dry spell mostly occured in 1982, 1991 and 1997 in which the El Nino intensities are moderate and strong.

Figure 4 shows the example of PDF graphs at Darmaga and Pasirukem. From these graphs we can see the probable period occurence of dry spells. The probability of dry spells between 1-3 days at Darmaga is 82% and above. While the probability of dry spells between 1-25 days in Pasirukem is 94% and above. This shows that in Darmaga, dry spells only lasts for one to three days, while in Pasirukem dry spells lasting for one to twenty-five days occur frequently. It can be said that the Pasirukem region is drier than the Darmaga region. For other stations, dry spells occurences vary between 1 to 20 days.

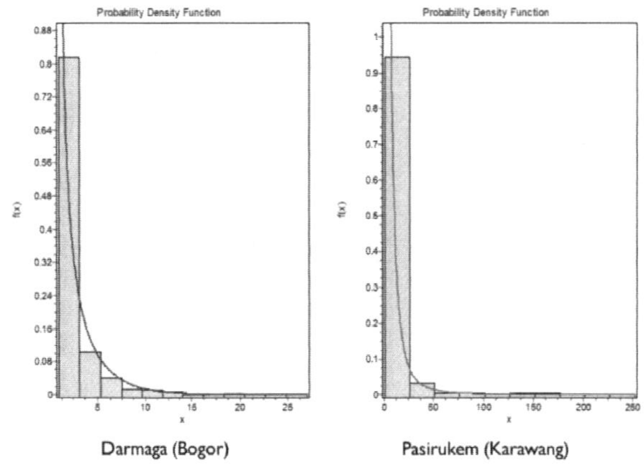

Figure 4. The PDF graphs of dry spell at Darmaga and Pasirukem.

Figure 5 shows the probability of dry spell occurance which takes on a value of more than 5, 10, 15, and 20 days over West Java. The probability of dry spell of more than 5 days ranged between 10% - 50% with the highest probability of occurrence is in the northern coast of West Java and the lowest odds is in the western regions of mid West Java. The probability of dry spell events over 10 days ranged from 0-40% and the probability of dry spell events over 15 days and 20 days ranged from 0-20%.

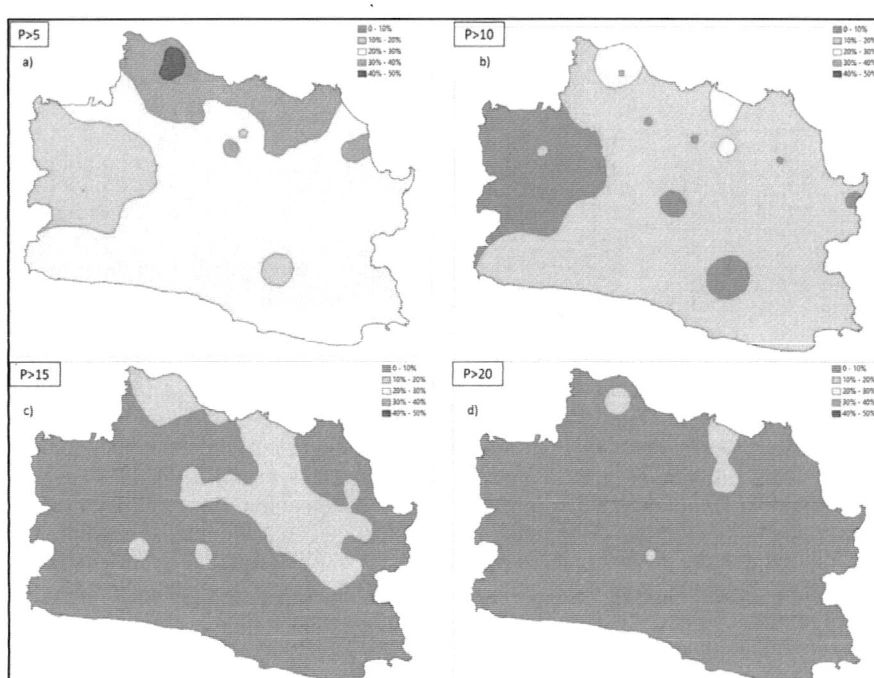

Figure 5. Probability of dry spells more than 5 days (a), more than 10 days (b), more than 15 days (a), and more than 20 days (d).

Based on the analysis results, it can be said that the area around the northern coast of West Java tend to be more prone to dryness which could disrupt agricultural activities in the region. Therefore, efforts should be made to reduce or avoid the impact of the drought events by arranging the planting calendar in which time and cropping patterns are adjusted.

4. Conclusions

Based on the frequency distribution of dry spell events, almost the entire region of West Java have experienced dry spells of more than 150 days. The regions which experienced dry spells event of less than 100 days were in mountainous areas or near reservoirs. Based on the analysis of probability occurrences, the north coast of West Java is more prone to the incidence of dry spells, so it is necessary to regulate irrigation and cropping pattern settings so that agricultural activities can be ran well.

Acknowledgments

This research was supported by Bogor Agricultural University and Indonesian Agency for Meteorology Climatology and Geophysics.

References

[1] Kementrian PPN/Bappenas 2014 Rencana Aksi Nasional Adaptasi Perubahan Iklim.
[2] Surmaini E, Runtunuwu E, Las I 2011 *J. Litbang Pertanian* **30(1)** 1-7
[3] Salinger MJ 2005 *J. Climate Change* **70** 9-29
[4] Zhang X and Yang F 2004 RClimdex(1.0) User manual. Ontario (CA): Climate research Branch Environment Canada
[5] Dikshit UN, Parida D, Satpathy D 1987 *IRRN* **12** 6-7
[6] Niewolt S 1989 *J Agriculture and Forest Meteorology* **45** 251-263
[7] Nastos PT and Zerefos CS 2009 *J. Atm Research* **94** 616-628
[8] Tilya FF and Mhita MS 2007 Frequency of Wet and Dry Spells in Tanzania *Climate and land degradation* (Berlin: Springer Berlin Heidelberg) 197-204

Developing Vulnerability Analysis Method for Climate Change Adaptation on Agropolitan Region in Malang District

Y Sugiarto[1,2,3*], Perdinan[1,2,3], T Atmaja[2,3] and A Wibowo[4]

[1]Department of Geophysics and Meteorology, Bogor Agricultural University
[2]Prima Intelektual Area (PI AREA)
[3]Generasi Hijau Indonesia
[4]Ministry of Environment and Forestry

Email: yon.ipb@gmail.com

Abstract. Agriculture plays a strategic role in strengthening sustainable development. Based on agropolitan concept, the village becomes the center of economic activities by combining agriculture, agro-industry, agribusiness and tourism that able to create high value-added economy. The impact of climate change on agriculture and water resources may increase the pressure on agropolitan development. The assessment method is required to measure the vulnerability of area-based communities in the agropolitan to climate change impact. An analysis of agropolitan vulnerability was conducted in Malang district based on four aspects and considering the availability and distribution of water as the problem. The indicators used to measure was vulnerability component which consisted of sensitivity and adaptive capacity and exposure component. The studies earned 21 indicators derived from the 115 village-based data. The results of vulnerability assessments showed that most of the villages were categorised at a moderate level. Around 20% of 388 villages were categorized at high to very high level of vulnerability due to low level of agricultural economic. In agropolitan region within the sub-district of Poncokusumo, the vulnerability of the villages varies between very low to very high. The most villages were vulnerable due to lower adaptive capacity, eventhough the level of sensitivity and exposure of all villages were relatively similar. The existence of water resources was the biggest contributor to the high exposure of the villages in Malang district, while the reception of credit facilities and source of family income were among the indicators that lead to high sensitivity component.

Keyword: agriculture, economy, industry, Podes, Poncokusumo, tourism, village, water

1. Introduction

Agriculture is an important sector ensuring the sustainability of human life on earth through its function in the provision of food. Agriculture becomes one of the important pillars of the village to improve livelihood and welfare of the community. By strengthening the agricultural sector, there are opportunities to expand employment, improve social welfare and increase the sustainability to reduce dependency on imported goods [1]. Because the farm is identical to the village, the sustainable agriculture means to support the existence of the village as a base area of agricultural land, labor availability in enough quantities, and the presence of environmental resources, including water sources as one of the main prerequisites agricultural activities.

Total food requirement of agricultural products continuously increases along with the increase of world population which reached more than 7 billion. Many countries realize how agriculture plays a strategic role in sustainable development, but the data showed a lowering trend in the agricultural sector's contribution to Gross Domestic Product (GDP) globally. There was a significant reduction of agricultural sector's contribution to world GDP from 8.1 percent in 1995 to 3.9 in 2014. As the country with the greatest economic value in Southeast Asia, Indonesia is an agricultural based country which also experienced a lowering trend in GDP contribution coming from the agricultural sector. Although the contribution of the agricultural sector in Indonesia was quite high (around 17.1% of GDP) in 1995, but there trend continuously decreased to became 14% in 2007 and 13.3% in 2014. This trend was in line with the magnitude of agricultural land conversion and decrease the number of workers and family farmers who are the backbone of the agricultural sector.

Friedmann and Douglass [2] has introduced a new concept of Agropolitan as one of the regional development planning concept. It was quite popular in many decades in developing countries such as Indonesia. Agropolitan has been defined as a centers of economic activity based on agriculture, where the agricultural sector will be able to grow and thrive because it is able to serve, encourage, and transporting activities (agribusiness) in center of farming villages and surrounding villages. According to Rustiadi and Hadi [3], agropolitan is a concept of regional development that arises because of the problems of development gaps between city as a center of economic activity and the growth of rural areas as a center of agricultural activities. Agropolitan development is a solution to reduce the flow of urbanization as well as to empower the rural economy. Through the concept of Agropolitan, agriculture can be developed as part of regional planning and development that can make the village as the center of economic activities based on agriculture and agro-industries that have high added values to encourage people in the villages to live in their respective areas. Agropolitan can be developed in rural areas with the population between 50,000 and 150,000 people [2].

Mercado [4] mentions five characteristics of agropolitan area, which includes a) a relatively small geographical scale; b) a high level of self-sufficiency and independence in decision making and planning, based on popular participation and cooperative action at the local level; c) the diversification of employment in rural areas to include agricultural and non-agricultural activities, emphasizing the growth of small-scale rural industrialization; d) the function of urban-rural industries and their relationship to local resources and economic structures; and e) the use and evaluation of local resources and technology. The idea of agropolitan introduced by Friedmann and Douglass [2] will basically empower the village for not only maximizing production and local trade, but also can increase social knowledge to the wider community, thereby increasing the capacity of communities to address difficult and complex problems.

The goals of agropolitan development can be achieved if some factors such as social, economic, spatial planning and the environmental are completed. One component of the environment is the management of sustainable water resources. Water is one of the resources that most vulnerable to climate change. Changes in the climatic parameters such as precipitation and air temperature and other parameters directly or indirectly affect the condition of water resources in the region. In the development of agricultural areas, water availability for agricultural land is very important. Water availability can be met either through rainfall or irrigation, but water provision and arrangements are important to ensure the availability of water in agricultural activities.

As a whole model of development, agropolitan composed some key aspects in achieving its goals, i.e., agriculture itself, economic aspects, aspects of the industry (agro-industry) and aspects of tourism (agro-tourism). However, behind the excellent concept of agropolitan model, there are also many challenges.

Water resources became one of the main problems is quite important to note. Because the source of the water is very vulnerable to the impacts of climate change. Based on the results of the survey in 2012, Malang regency is one area that has a high degree of vulnerability to climate change, especially in the sector of water resources and agriculture. However, so far there was no models that has been developed especially in the focus of agropolitan vulnerability in Malang. Therefore, it is very important to develop assessment methods of vulnerability with the focus on Agropolitan region to be used as part of climate change adaptation analysis in Malang[5].

2. Related works

Model of agriculture vulnerability now has evolved. But the development model of vulnerability to climate change with a focus on the agropolitan and some considered aspects of water resources has not been developed. Developing indicators of vulnerability has been proposed by Perdinan *et al.* [6] based on the aspect of supply chain i.e. production, post-harvest and storage, distribution and consumption. Meanwhile, Agustian [7] analyzed the factors that affected the production of maize cultivation in Garut. Agustian [7] concluded that the scale (land area) and seed investment significantly and positively related to corn production. In addition, Boer e*t al* [8] has divided indicators of vulnerability with a focus on aspects such as ecosystem development, poverty, education, health, infrastructure, livelihoods and governance. Meanwhile, the Indonesian government through the Ministry of Agriculture especially Balitklimat has developed indicators of food security which is in accordance with the definition of food security of the FAO [9] and Law No. 7 1996. According Estiningtyas *et al.* [10], there are four components in this focus: 1) the availability of food; 2) the stability of food supply; 3) accessibility / affordability of food, and 4) the quality/food safety.

3. Methodology

The differences in meaning between the components, indicators, sub indicators and data/variables is important to note before developing vulnerability assessment method of agropolitan region. The components are among the largest part of the division. While the indicators are part of the components and the data/variables are a constituent part of the indicators. The development of vulnerability assessment in this research is based on IPCC concept (AR-5) where the climate-related disaster risk (R) is a function of the hazards (H), vulnerability (V) and exposure (E) [11].

The vulnerability and exposure were assessed using data and socio-economic information. Therefore, data selection for each indicator was determined based on the availability of data to the village level. BPS data was a major source of data usage in this study. Among them are BPS Kecamatan Dalam Angka 2015 from each subdistrict in Malang District, Population Census of 2010, Potensi Desa (Podes), Data of 2014 and Agricultural Census 2013, and the baseline set was in 2014.

Understanding variation or range of values for socio-economic data in aimportant, an therefore it is necessary to normalize socio-economic data. Normalization data was done so that any data that was used has a range of zero to one (0-1). Normalization step aimed to equalize the value of the variable to be calculated by other variables that have been normalized as well. In this guide, normalization techniques were caculated based on Mechanical Normalization Median proposed by Perdinan *et al.* [12]. Then the next step was to calculate a combined score between variables normalized into sub-indicators in each component. Combination of the indicators into Components was using the following equation:

$$\text{Component } A\ (KK)_i = w * (\text{K1+K2+K3+K4}) \tag{1}$$

where: *i* represent the village-i and w is weights for each indicator. Selection of the appropriate amount of weight carried the relation between the indicator and sensitivity to agriculture and climate change. The maximum value of the weight is 1.

The use of the same value range for measuring the level of vulnerability on each component is required as a standard value. Each component has a value range by 0-1. Furthermore, the each component was grouped by the value/index. Classification was done by dividing the proportional value of the degree of vulnerability (0-1) into five classes (table 1). Component-level vulnerabilities categorized as "Very High" if the index above 0.8 to 1, and categorized as "Very Low" when vulnerability index below 0.2 to 0. There were five groups, namely VL (Very Low), L (Low), M (moderate), H (High), and VH (Very High). Staining of each category of components was made for mapping purposes vulnerability index agropolitan area in Malang.

Table 1. Criteria of index.

Components Value	Staining Exposures; Sensitivity and Vulnerability	Staining Adaptive Capacity
0.0 – 0.2	Very Low (VL)	Very Low (VL)
0.2 – 0.4	Low (L)	Low (L)
0.4 – 0.6	Mediun (M)	Mediun (M)
0.6 – 0.8	High (H)	High (H)
0.8 – 1.0	Very High (VH)	Very High (VH)

4. Result and Discussion

4.1 Determination of Component Agropolitan

Regional vulnerability assessment method of Agropolitan was prepared with basic consideration that agropolitan as a single model of development that consisted of several main aspects in achieving its goal of agricultural aspects, economic aspects, aspects of the industry (agro-industry) and aspects of tourism (agro-tourism). Each aspects was composed of supporters of each component. Agropolitan is a model of development that relies on centralized infrastructure development similar cities in rural areas which aims to boost agricultural production, support the growth of agro-processing small and medium scale, and to encourage the diversity of economic activities in rural areas [13]. Referring to the notion of agropolitan and water resource issues that occur in Malang agropolitan, it is important to consider in the development of the vulnerability assessment of agropolitan region should consider the availability and distribution of water resources as one of the main problems of climate change impacts.

4.1.1. Aspects of Agriculture. Agriculture becomes the main focus in the development of the concept Master Plan Agropolitan, so that all constituent components of Agropolitan consider this aspect. In the preparation of the agricultural component, the various indicators that support agricultural water resources as affected by climate change impacts should also be considered. Various indicators constituent aspects of agriculture including the density of agricultural workers, the existence of the affected buildings, agricultural pollution, the existence of the affected farm, and decreased quality of land (the exposure component) also need to be included. Non-productive age population, agricultural workers, labor agriculture industry and sources of income (the sensitivity component) and access to the agricultural also important component for adaptation capacity.

4.1.2. Aspects of Economy. Economic aspects and agriculture in Malang was very closely related. The district's main economy was highly dependent on agriculture. The agricultural was a primary sector in the economy of Malang. Therefore, most of the economic indicators was very closely linked to the concept of agropolitan. This indicator was developed by the density of the component exposure of agricultural workers, the productive age population, agricultural workers, level of poverty, and income sources and public acceptance of credit facilities for component sensitivity. In the adaptive capacity component, constituent indicators consisted of economic levels were compiled based on the availability of venture capital agriculture, agricultural markets, supporting agricultural enterprises, cooperatives and local revenue (Pendapatan Asli Daerah or PAD). The adaptive capacity was also composed by indicator of regional tourism.

4.1.3. Aspects of Industry. Malang had prospective industries, but they were not able to develop progressively. This can be seen from the difference between production and their potential. Not all areas suitable and conducive to be used as an industrial area. Many factors still needed to develop the industrial sector, such as: the availability of qualified human resources, natural resources, and considerable investment. However, referring to the availability of data, the drafters of this indicator consisted of industrial pollution, agricultural and industrial workforce proportion of micro and small industries.

4.1.4. Aspects of Tourism. Tourism in Malang Regency is determined by a variety of indicators that strongly influence including tourist location, traders inside and outside of the tourist attractions, facilities of tourist attractions, cleanliness sights, human resources (managers, traders, visitors), the availability of water resources, and ticket prices to enter tourist sites. However, referring to the availability of data, the indicator is organized by the indicators of the presence of tourist attractions and the ability of excursions.

4.2 Selected Indicators for Vulnerability Assessment Method of Agropolitan

Considering four aspects which were mentioned above and water resources as the main issue then drafted vulnerability concepts matrix constituent agropolitan was established with reference to the third component, namely the IPCC vulnerability exposure, susceptibility consisting of sensitivity and adaptive capacity. Table arrangement with a rationale for the data on each indicator is shown in Appendix 1 and Appendix 2. The selection of these indicators also referred to the availability of data at the village level throughout Malang. After selection of indicators, selection of data at the village level was carried out as a material forming indicator. This selection puts 115 village-level data and information to be used to compile the 51 variables derived. From the 51 variables, these derivatives were combined to form 36 sub-indicators used to compile the 21 indicators in all components of the degree of vulnerability (table 2).

Table 2. Total use of data and information constituent indicators.

Komponen	Indicator	Sub Indicator	Variabel	Data
Exposure	8	12	15	115 Data and information
Sensitivity	7	10	13	
Adaptive capacity	6	14	23	
Total	21	36	51	

4.3 Analysis of Vulnerability of Malang District and Sub-District Poncokusumo

The map of exposure, sensitivity, adaptive capacity and vulnerability of the entire village in Malang based on Podes Data 2014 is presented in figure 1. From the total number of 388 villages in the district of Malang, for exposure map, there were 341 villages or 88% in the category of "medium" while 41 villages or 11% were in category of "low". Specifically for agropolitan region in District Poncokusumo, village Ngadas was the only village from 17 villages with "low" level of exposure whereas the remaining of 16 villages were in the category of "medium".

Most villages including Pujon considered to have high exposure caused by several factors. The first was the poor condition of water sources. This indicator was determined by the existence of water source for bathing/washing and drinking/cooking. Some villages were only using one source of water for various purposes (rivers, irrigation canals, lake/reservoirs/dam). The existence of natural resources in a region will be exposed to the increasing threat of climate change. If a community activity depends only on one/less water source then the area will be categorized as highly exposed. Meanwhile, the presence of natural attractions also greatly contributed to determine the area dealing with high exposure. Nature tourism was heavily affected by the impact of climate change, therefore, the more places that have natural attractions, the more potentially to be exposed. Ngijo village, Kalisongo, Bendosari and Pandesari have potentially affected by nature of climate change.

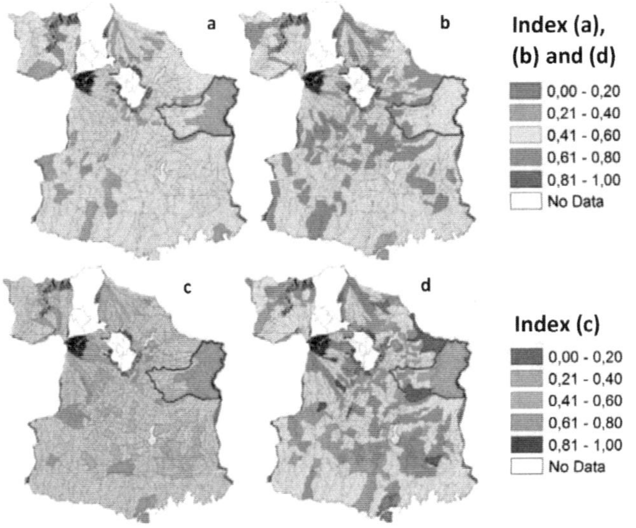

Figure 1. Map of Exposure (a), Sensitivity (b), Adaptive Capacity (c) and Vulnerability (d) of Malang District to climate change.

For sensitivity component, the distribution was more varied throughout the villages (figure 1). However, the majority of the villages are still located on the criterion of "medium" which comprised of 269 villages, or 69% of the villages. In contrast to the exposure, sensitivity component tended to have more vulnerable, because there were 84 villages (22%) categorized as "high" in this criteria. In Sub Poncokusumo, from 17 villages there were three villages that were included in the category of "high" while the rest were classified as "medium". The villages with high category were spread more in the central and southwestern districts and only a small portion was in the north. The villages with the category of "high" had an index interval of 0.60 - 0.80. The name of villages with the value of index that more than 0.65 were Tulungrejo, Wonorejo, Corner, Karangrejo, Maguan, Plandi, Plaosan, Jeru, Sukolilo and Pait. While in Sub-District Poncokusumo there were three villages in the category "High" including Dawuhan, Pajaran and Ngebruk. The factors contributing to the high sensitivity were the main indicators of public acceptance to credit facility and the family income. While the other factors were the dependence of the water source followed by family agricultural workers.

The majority of villagers were still rely their livelihood on staple crop agriculture such as rice. For example, in the Sub-District of Pajaran, The Pajaran village had 6429 inhabitants, and approximately 4.8% (310) of the people worked in agriculture (crop cultivation, fishery, and livestock), while 13.6% (875) people worked as farm labors. More and more resident's livelihood was depend on agriculture, the region will be more sensitive, especially if their agricultural products was in the form of staple crops (rice). This because agriculture is one of the most vulnerable sectors affected by climate change mainly related to rainfall conditions which have high implication to the presence of water for agriculture.

Analysis of the adaptive capacity of Malang shows that the capacity of most of the villages in the district classified as "medium" (figure 1). This data suggested that the adaptability of the villages to the climate change in the space of Agropolitan was good enough. Based on the recapitulation, the majority of the village which consisted of 291 villages (75%) belong to the criterion of "medium" and 77 villages (20%) included in the criteria of "low". Only 20 villages (5%), which had capacity of adaptation of "high". While in District Poncokusumo, the majority of villages varied from the level of adaptation of "low" to "high", and the majority tended to have the level of "medium".

Villages with low capacity needs to get more attention. Some of these areas were Ringinsari, Sumbersuko, Kepatihan, Sukorejo, Petungsewu, Ngadirejo, Taji, Argosari and Ampeldento. While the village with low capacity in District Poncokusumo were Dawuhan, Sumberejo and Ngadireso. The majority of the villages graded as low capacity mainly caused by some factors including lack of access to agricultural (0175), the economic level (0175) and the industrial level (0.175), followed by lack of disaster preparedness (0100). The values showed the weighting applied when calculations. The analysis also showed that the respective Index dominant indicator worth very low (<0.333). This indicated the ability of the village to adapt, such as inadequate infrastructure which was very low.

Based on calculations, the majority of villages were included in the category of "medium", but its spread fairly distributed throughout the area categorized as "low" and "high". From 388 villages surveyed, around 222 villages (57%) were included in the criteria of "medium" while 81 villages (21%) were counted as "low" and 68 villages (18%) were categorized as "high". Similarly, the Sub-District Poncokusumo, the majority of villages were likely categorized at the level of vulnerability of "low" and "medium". Only two villages were drop into category of "high" and "very high" (figure 1). Although only a view villages that fell into this category but these villages need serious attention. Villages that enter into the category of "very high" were the village Kepatihan, Sukorejo, Dawuhan, Sumberejo, Kranggan, Petungsewu, Ngadirejo, Taji and Argosari. As mentioned before the high vulnerability was caused by low adaptive capacity and/or high sensitivity. For example, with the interval of 0-1 for each component of vulnerability, Dawuhan village in Sub-District Poncokusumo had a sensitivity of 0.60 while the index of adaptive capacity was only 0.30.

In summary, the distribution criteria for each component shows that more than 50% of the villages entered in the category of "medium" based on the calculation of each component. Only a few villages were included in the category of "low" or "high". That is, for the majority of the exposure and sensitivity village is quite vulnerable to the effects of climate change, even though 80% of villagers have enough adaptability above (figure 2).

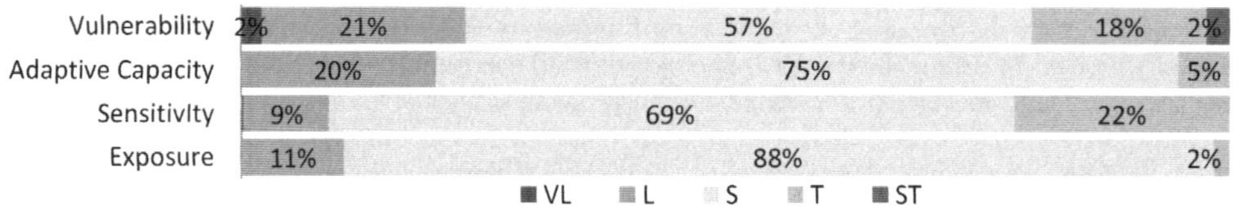

Figure 2. Percentage of villages based on criterya. Note: VL, very low; L. low; S. medium, T, high; ST, very high.

5. Conclusions

Vulnerability assessment method of the impact of climate change on the agropolitan region can reflect the level of vulnerability and exposure as components to a risk assessment. The results of vulnerability assessments to all villages in the district showed that most of the villages were categorized at the moderate level, but there were 20% of 388 villages that has high to very high vulnerability. One of the causes was the average of economic level for villages in Malang was still low. In the case of agropolitan region in the sub-district of Poncokusumo, the vulnerability of the village varied between very low to very high. Assessment of the factors that contribute to high vulnerability showed that most villages were vulnerable due to the low adaptive capacity, even though the level of sensitivity was not much different and the level of exposure of all villages were relatively similar. Based on the average of the villages in the district of Malang, access to basic facilities contributed the most excellent indicators of adaptive capacity, while the low level of economic contributed to low adaptive capacity. The existence of water resources was the biggest contributor to the high exposure of the villages in the district of Malang. The acceptance of credit facilities and sources of family income indicators led to high sensitivity component at the district level. In the area that was developed specifically as agropolitan region which is in sub-district Poncokusumo, the vulnerability was relatively higher than the average village in the district. One of the villages, which is Sumberejo have very high levels of vulnerability caused by low levels of adaptive capacity. Compared with other villages in the same district, the sensitivity and exposure levels were not too different and have a similar pattern for all indicators.

References

[1] Rosdiana H, Inayati and Murwendah 2014 *Procedia Environmental Sciences* **20** 563 – 572

[2] Friedmann and J M Douglass 1975 *"Agropolitan Development: Towards a New Strategy for Regional Planning in Asia." In: United Nations Centre for Regional Development, Growth Pole Strategy and Regional Development Planning in Asia* (Nagoya) pp 333-387

[3] Rustiadi E, dan Hadi S 2006 *Kawasan Agropolitan Konsep Pembangunan Desa-Kota Berimbang* (Bogor: Crestpent Press P4W-LPPM IPB) pp 1-31

[4] Ruben G and Mercado *2002 Regional development in the Philippines: a review of experience, state of the art and agenda for research and action. Philippine Institute for Development Studies Suriansamga Pag-aaral Pangkaunlaranng Pilipinas, February* 2002 http://unpan1.un.org/intradoc/groups/public/documents/apcity/unpan005113.pdf

[5] [Bappeda] 2015 *Pedoman Umum Pengembangan Kawasan Agropolitan dan Minapolitan Provinsi Jawa Timur 2015* Bappeda Provinsi Jawa Timur. Surabaya.

[6] Perdinan, Tri A, Ujang S, Yon S, Lina F and Ryco F A 2016 Deriving Vulnerability Indicators for Crop Production Regions in Indonesia *3rd International Symposium on LAPAN-IPB Satellite FSEM*

[7] Agustian and Adang 2015 J Dinamika Produksi dan Penerapan Teknologi Pertanian - Jawa Barat

[8] Boer, Rizaldi, Adi R, Akhmad F, Perdinan and Anter P S 2015 SPARC - Indeks Kerentanan Dan Risiko Iklim Desa Di Propinsi Nusa Tenggara Timur *KLHK - CCROM SEAP*

[9] [FAO] (Food and Agriculture Organization of the United Nations) 2005 *1983 World Food security: a reappraisal of the concepts and approaches*. Director General's Report. Rome.Estiningtyas

[10] Estiningtyas, Woro, Aris P and Yayan A 2014 *Buletin Hasil Peneitian Aroklimat dan Hidrologi* **11** Balai Penelitian Agroklimat dan Hidrologi.

[11] [IPCC] 2014 *Climate Change 2014: Synthesis Report. Contribution of Working Groups I, II and III to the Fifth Assessment Report of the Intergovernmental Panel on Climate Change* [Core Writing Team, R.K. Pachauri and L.A. Meyer (eds.)]. IPCC, Geneva, Switzerland, 151 p

[12] Perdinan, Yon S, Tri A, Ujang S, Ryco F A and Enggar Y A 2015 *Climate Change Clid Centered Risk Assessment in Surabaya* (FMIPA IPB – UNICEF)

[13] Rustiadi et al. 2011 *Perencanaan dan Pengembangan Wilayah* (Jakarta: Crestpent Press dan Yayasan Pustaka Obor Indonesia)

Study of the effectiveness of several tree canopy types on roadside green belt in influencing the distribution of NO_2 gas emitted from transportation

R D Desyana[1], B Sulistyantara[1], N Nasrullah[1] and I S Fatimah[1]

[1] Department of Landscape Architecture, Bogor Agricultural University, Kampus IPB Darmaga Jl. Meranti, Wing 9/Level 4, Bogor 16680 Indonesia

E-mail: dwicadesyana@gmail.com, bbsulistyantara@yahoo.co.id, nizarnasrullah@yahoo.com, indungsittifatimah@gmail.com

Abstract. Transportation is one significant factor which contributes to urban air pollution. One of the pollutants emitted from transportation which affect human's health is NO_2. Plants, especially trees, have high potential in reducing air pollutants from transportation through diffusion, absorbtion, adsorption and deposition. Purpose of this study was to analyze the effectiveness of several tree canopy types on roadside green belt in influencing distribution of NO_2 gas emitted from transportation. The study conducted in three plots of tree canopy in Jagorawi Highway: Bungur (*Lagerstroemia speciosa*), Gmelina (*Gmelina arborea*) and Tanjung (*Mimusops elengi*). The tree canopy ability in absorbing pollutant is derived by comparing air quality on vegetated area with ambience air quality at control area (open field). Air sampling was conducted to measure NO_2 concentration at elevation 1.5m, 5m and 10m at distance 0m, 10m and 30m, using Air Sampler Impinger. Concentration of NO_2 was analyzed with Griess-Saltzman method. From this research, the result of ANOVA showed that tree plot (vegetated area) affected significantly to NO_2 concentration. However the effect of distance from road and elevation was not significant. Among the plots, the highest NO_2 concentration was found on Control plot (area without tree canopy), while the lowest NO_2 concentration was found in Tanjung plot. Tanjung plot with round shape and high density canopy performed better in reducing NO_2 than Bungur plot with round shape and medium density canopy, regardless the sampling elevation and distance. Gmelina plot performed the best in reducing horizontal distribution of NO_2 concentration at elevation 1.5 and 5m, but the result at elevation 10m was not significant.

1. Introduction

Human activity is one of the largest contributors of urban pollution. Since the second half of the 20th century, measuring pollution levels widely practiced in developing countries as a result of increasing urban pollution [1]. One of the causes of urban pollution is the increasing number of people which directly proportional to the increase in the number of motor vehicles as a means of transportation. Based on data from the Central Bureau of Statistics [2], the population of Bogor City in 2012 increased by 3.87% from the previous year. While the number of public transport in Bogor City increased by 3.61% from the previous year. The volume of the higher motor vehicle is directly proportional to the high level of pollutants released to the air pollution in big cities in Indonesia, including Bogor.

Pollutants can be defined as chemicals that are added or inserted into the atmosphere through human activity causing increased concentrations of these substances [3]. There is a lot of air pollutants in the atmosphere, but there are nine types of air pollutants that are considered important, namely carbon oxides, sulfur oxides, nitrogen oxides, volatile organic components, the suspension of particles, photochemical oxides, radioactive substances, heat and sound [4].

Pollutants generated in the combustion of fuel is gas emissions of motor vehicles in the form of CO, hydrocarbons (HC), SOx, NOx and particles. Nitrogen dioxide (NO_2) along with nitrous oxide is a gas group as the most common air pollutants compared with other forms of nitrogen oxides in the atmosphere. The nature of the toxicity of NO_2 gas four times higher than the NO gas. Most sensitive organs to NO_2 gas pollution is the lung. Lungs contaminated with NO_2 gas will swell, making it difficult to breathe, and even cause death [3].

To reduce the level of pollutants in the air, there are several ways can be done, either reduce pollutants at source (effluent) or reduce pollutants in ambient air. To reduce the level of pollutants in the air source, the most widely used is the biological mechanism, such as the use of plants and vegetation. Plants as the landscape elements have the potential and important role in influencing the distribution of pollutant including NO_2 gas through some processes. The process of gas absorption by plants occurs mainly in the leaves. Most of the gas exchange occurs in the leaf through the stomata [5]. The ability of plants to gas absorption can be seen from the air mainly from plant metabolic processes such as photosynthesis, respiration, and transpiration [6]. Tree with higher leaf surface area ratio than projected broad canopy (leaf area index) can act as a biological filter to reduce pollutants and improve air quality [7]. According to Grey and Deneke [8], stomatal density species with moderate to high are potential in reducing gas pollutants.

Pollutants emitted from transportation that spreads to the roadside can be absorbed by trees and bushes planted in the green belt. Hermawan [9] has examined the ability of the highway green belt in reducing pollutant particles horizontally shaped, while Sulistijorini [10] has examined the effectiveness and tolerance of plants in absorbing NO_2. Pollutants can be vertically spread by diffusion and join other types of pollutants in ambient air. The condition of the tree canopy above the road as the field of the roof can absorb gaseous pollutants, so it is necessary to investigate the absorption of NO_2 gas by various types of tree canopy.

2. Literature Review

Pollutants can be defined as chemicals that are added or inserted into the atmosphere through human activity causing increased concentrations of these substances [3]. Pollutants can be classified into primary and secondary pollutants. Primary pollutants are pollutants that are directly released into ambient air by the natural processes and human-related activities. Primary pollutants can be deposited onto the surface of the ground and the wind and fused with other primary pollutants to form secondary pollutants. Pollutants can be in gaseous or solid particles.

Krupa [3] also suggests that pollutants can be produced from a static source (stationary) or dynamic sources (moving). Static source is a source of pollutants that have a stationary position or stationary, such as a volcano or a factory chimney. Dynamic source is a source of pollutants that can switch positions, eg motor vehicles. Darmono [4] mentioned that there is a lot of air pollutants in the atmosphere, but there are nine types of air pollutants that are considered important, namely carbon oxides, sulfur oxides, nitrogen oxides, volatile organic components, the suspension of particles, photochemical oxides, radioactive substances, heat and sound.

Pollutants generated in the combustion of fuel is gas emissions of motor vehicles in the form of CO, hydrocarbons (HC), SOx, NOx and particles. Nitrogen dioxide (NO_2) along with nitrous oxide is a gas group as the most common air pollutants compared with other forms of nitrogen oxides in the atmosphere. The nature of the toxicity of NO_2 gas four times higher than the NO gas. Most sensitive organs to NO_2 gas pollution is the lungs. Lungs contaminated with NO_2 gas will swell, making it difficult to breathe, and even cause death [3].

Nasrullah [11] stated that the amount of pollutant that has been released in the environment can be reduced by the presence of vegetation. The mechanism of plants in reducing pollutants consists of:

1. Diffusion, the dispersion of pollutants into the atmosphere wider. Tall tree canopies can deflect wind gusts into the atmosphere.
2. Absorption, absorption of gaseous pollutants through stomata and into the leaf tissue.
3. Adsorption, adsorption of pollutant particles from the surface of the leaves, stems, twigs. Plant parts adsorb dust and metal particles contained in the air.
4. Deposition of large particles by leaves and other plant parts.

3. Objectives of Research

This research has a common goal to study differences in the ability of different tree canopy to absorb NO_2 gas pollutant, while the specific objectives are to determine which tree canopy that effectively reduce NO_2 gas and to determine the relation between tree canopy types and pollutant distribution vertically and horizontally

3.1. Expected Output

The results of this research are expected to give an idea of the importance of plant selection with tree canopy which is effective in reducing gaseous pollutants on roadside green belt, and it can be useful for guidelines in the selection of plants, especially tree, in the preparation of planting plan on roadside green belt.

3.2. Activity Overview

Similar studies have been conducted in the research related to the function of vegetation in reducing the levels of pollutant in ambient air. Aspects that have been studied include species with high capacity to adsorb pollutant particles of lead (Pb), nitrogen oxides (NOx), carbon dioxide (CO) as well as the width of the green belt relationships which are effective in reducing pollutants. The new aspect had been examined is the relationship between the type of tree canopy and its effectivity in reducing NO_2 gas. The result of this research activity is expected to complete the preparation of guidelines for plant selection, especially trees that effectively reduce NO_2 gas for roadside green belt.

4. Research Methods

4.1. Location and Time of Research

To conduct research regarding the relation between canopy types and reduction of pollutant concentration, field study had been conducted in the end of June to early July 2016 on roadside green belt of Jagorawi Highway, Cibubur segment from km 7-13 on Jakarta-Bogor track. Air sampling was conducted for measurement of NO2 gas on June 30, 2016. It was assumed that more vehicles crossing highway, so the gaseous pollutant level is higher than other roads in Bogor City.

Preparation of tools and materials for NO2 measurement of air quality parameters as well as the analysis was conducted in the laboratory of Environmental Research Center (PPLH) LPPM IPB. There were 3 plots of trees as follow: Bungur (*Lagerstroemia speciosa*) with round canopy, Gmelina (*Gmelina arborea*) with columnar canopy, and Tanjung (*Mimusops elengi*) with round canopy. Selected plots consisted of at least two rows of trees of same species, with average thickness 10m. Sampling also conducted at control plot (open field area without tree canopy). Table 1 shows the canopy shape, location and sampling time for each plot.

Table 1. Canopy shape of green belt, location at Jagorawi Highway and sampling time.

Canopy Shape	Tree Species	Position (km)	Sampling Time
Round	Bungur (*Lagerstroemia speciosa*)	10.400	11-12am, 02-03pm, 05-06pm
Columnar	Gmelina (*Gmelina arborea*)	11.400	11-12am, 02-03pm, 05-06pm
Round	Tanjung (*Mimusops elengi*)	12.200	11-12am, 02-03pm, 05-06pm
Control	Open area without tree canopy	12.800	11-12am, 02-03pm, 05-06pm

4.2. Implementation of Research

4.2.1. Measurement of NO_2 Gas Concentration
This study conducted by measuring NO_2 in ambient air using Air Sample Impinger. Measurement was done at 3 elevations above road surface (1.5m, 5m, 10m) and 3 distances (0m, 10m, 30m) from road body, as stated in Figure 1. Distance 0 m is the edge of roadside green belt that close to road curb or at point 2 m from the edge of the road (asphalt surface).

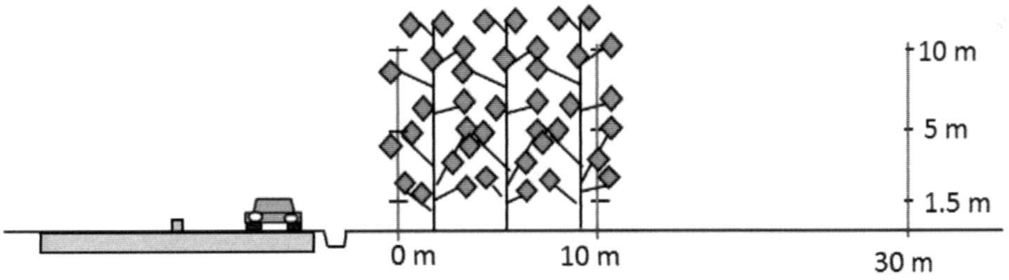

Figure 1. Sampling elevation of air quality measurement

4.2.2. Measurement of NO_2 Level in The Laboratory
Data of NO_2 analyzed using spectrophotometer following Griess-Saltzman method in PPLH IPB Laboratory.

4.2.3. Data Analysis
The effectiveness of the reduction of NO_2 gas concentration by each type of canopy is found by comparing air quality from the results of measurement on green belt to the air quality on control plot area. The effect of roadside green belt to concentration reduction of pollutant would be formulated through experimental design involving three factors: type of green belt canopy, the distance of the area from the road, and the elevation above surface. To determine the influence of experimental factors, the data of NO_2 gas concentration was analyzed using ANOVA. If the experiment factor was significant, test of Duncan Multiple Range at the level of $\alpha \leq 5\%$ would be applied. To assess which canopy type is better in reducing concentration, the value of concentration reduction rates were compared between plots.

4.2.4. Formulation of Recommendation
Recommendations for the selection of plant type which can be proposed in planting plan for roadside green belt based on canopy types of plant which has the highest effectivity of pollutant reduction. Other factor to consider is the relation between the shape of the canopy and the reduction of pollutant concentration.

5. Results and Discussion

5.1. Source of Pollutant

Vehicles passing on Jagorawi Highways are categorized as dynamic source of pollutant (moving source). At the time of NO2 gas sampling, volume of vehicles ranged 9,264-9,891 units. Table 2 shows sampling time, total vehicles, estimation of NO_2 emission and average wind velocity at each plot. Highest emission was reached at 05-06 pm (rush hour).

Table 2. Sampling time, total vehicles, estimation of NO_2 emission and average wind velocity.

Sampling Time	Total Vehicles	Estimation of NO2 Emission/km (g)	Wind Velocity (m/s)			
			Control	Bungur	Gmelina	Tanjung
11 – 12 am	9,264	19,195.20	2.07	1.50	1.82	0.18
02 – 03 pm	9,404	19,861.92	2.30	1.03	1.30	0.15
05 – 06 pm	9,891	21,156.76	2.27	1.30	1.63	0.22

Description : Emission of gasoline vehicle 2.2 g NO_2/km, emission of diesel vehicle 0.68 g NO_2/km based on Strauss and Mainwaring in Sulistijorini [10]. Data of total vehicles was obtained from Jasa Marga.

5.2. Concentration of NO_2 in Each Plot

Results of analysis of variance (ANOVA) showed that tree plots were significantly differ among plots. However the effect of distance from the road and elevation of sampling were not significant to NO_2 concentration. The highest NO_2 concentration was found on Control plot (area without tree canopy). This result indicates that tree canopy has significant effect in influencing concentration of NO_2. Average value in Control plot was not significantly different with Gmelina plot, while value in Bungur plot was not significantly different with Tanjung plot. Thus, average value in Bungur plot and Tanjung plot were significantly different with Control plot. The lowest concentration of NO_2 was found in Tanjung plot. Tanjung trees have round canopy with high density which can effectively block the wind movement and absorb more gaseous pollutant. This theory is suitable with the wind velocity data, where the wind velocity in Tanjung plot was lowest among the four plots (0.15-0.22 m/s). Average of NO_2 concentration in each plot presented in Table 3.

Table 3. Average of NO_2 concentration on the sampling plot in several distance from the road.

Distance (m)	Plot				Average
	Control	Bungur	Gmelina	Tanjung	
0	64.39	43.87	65.36	35.06	52.17
10	58.88	41.71	45.83	36.63	45.76
30	78.96	46.84	45.43	42.86	53.52
Average	67.41^a	44.14^b	52.21^a	38.18^b	

Description : Value which is followed by different alphabet character is significantly different by Duncan Multiple Range Test at the level of $\alpha \leq 5\%$.

5.3. The Distribution of NO_2 Gas Concentration

Concentration of NO_2 in three plots with tree canopy were lower compared to area without tree canopy (Control plot). Highest average concentration of NO_2 (99.8 g/km) was found at Gmelina plot with distance 0m and elevation 5m. The concentration of NO_2 emission in the sample plots presented in Figure 2.

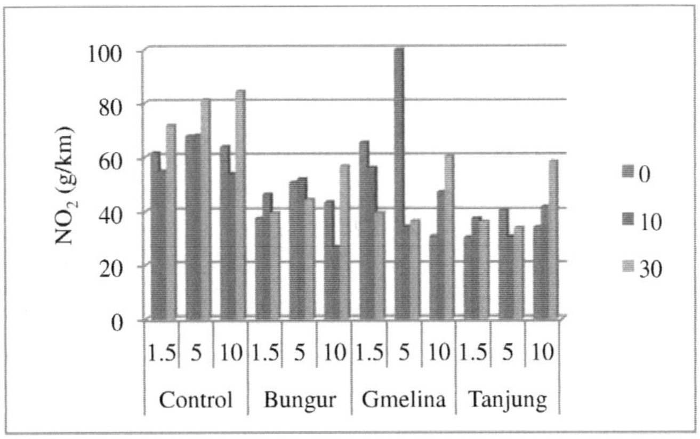

Figure 2. Distribution of NO_2 concentration at each plot

In Control plot, the concentration of NO_2 was slightly decreasing at distance 10m and increasing at distance 30m. The wind velocity in Control plot was highest among other plots (2.07-2.30 m/s), thus enhance the distribution of NO_2 horizontally. Vertical distribution of NO_2 was also not significant due to the absence of tree canopy. The concentration of NO_2 in the Control plot presented in Figure 3.

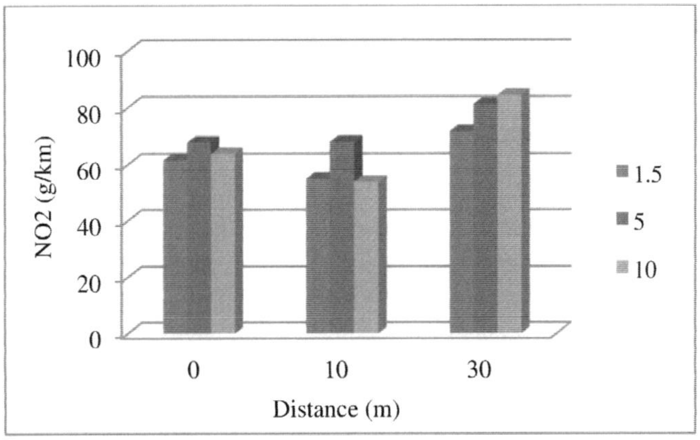

Figure 3. NO_2 concentration in Control plot

The concentration of NO_2 in Bungur plot is shown in Figure 4. As stated earlier, analysis of variance performed effect of sampling distance and elevation and their interaction were not significant to NO_2 concentration. Vertical distribution of NO_2 shows higher value at elevation 5m, at distance 0m and 10m. The lowest value found at elevation 10m and distance 10m. In this plot, Bungur trees only existed until distance 10m. The highest canopy of Bungur trees were lower than 10m. Area between 10m and 30m was considered open area with only small plants existed. Compared to Control plot, Bungur plot was considered better in reducing the concentration of NO_2.

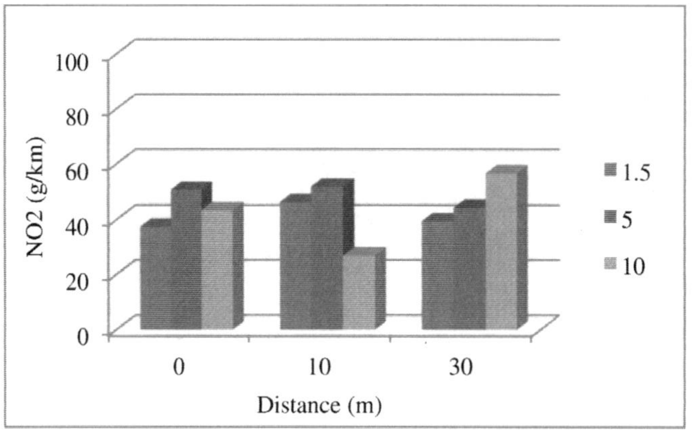

Figure 4. NO_2 concentration in Bungur plot

Highest NO_2 concentration was found in Gmelina plot, at distance 0m and elevation 5m. It is assumed that NO_2 gas was trapped right below the Gmelina tree canopy before absorbed by the tree. Concentration of NO_2 at elevation 10m at the same distance was lower than any sampling points in this plot. The horizontal distribution of NO_2 also decreasing following the distance at elevation 1.5m and 5m. This indicated that Gmelina tree canopy is effective in reducing NO_2 concentration horizontally. In contrary, NO_2 concentration at elevation 10m was increasing following the distance. The columnar-shaped Gmelina canopy can reach higher than 10m height, but the higher part of the canopy has less density than the lower part of the canopy. This indicated that Gmelina tree canopy at elevation 10m was not dense enough to reduce the NO_2 effectively. It can be assummed that beside the tree canopy shape, the density of leaves also has significant effect in influencing NO_2 concentration. The NO_2 concentration in Gmelina plot was shown in Figure 5.

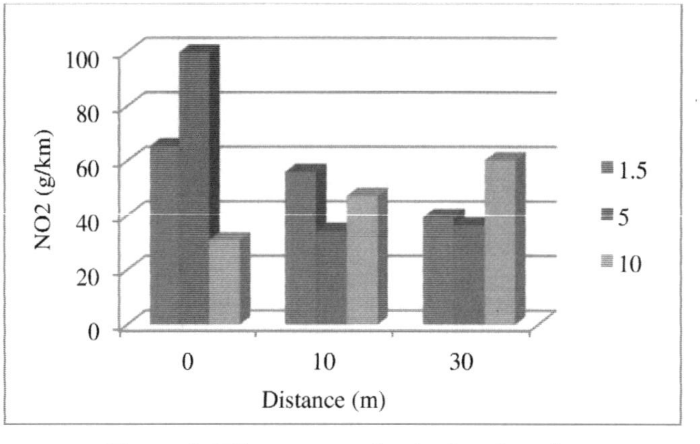

Figure 5. NO_2 concentration in Gmelina plot

The NO_2 concentration in Tanjung plot was shown in Figure 6. Average concentration of NO_2 in Tanjung plot was lowest among all plots. The vertical and horizontal distribution of NO_2 showed no significance between sampling elevation and distance except for one sampling point at elevation 10m and distance 30m. It can be assumed that Tanjung plot performed the best in reducing vertical and horizontal distribution of NO_2 based on the canopy type of the trees. The round canopy with high density of leaves can block the wind movement effectively, thus decreasing the distribution of NO_2

gas in the plot. As found in Gmelina plot, it can be assummed that beside the tree canopy shape, the density of leaves also has significant effect in influencing NO_2 concentration.

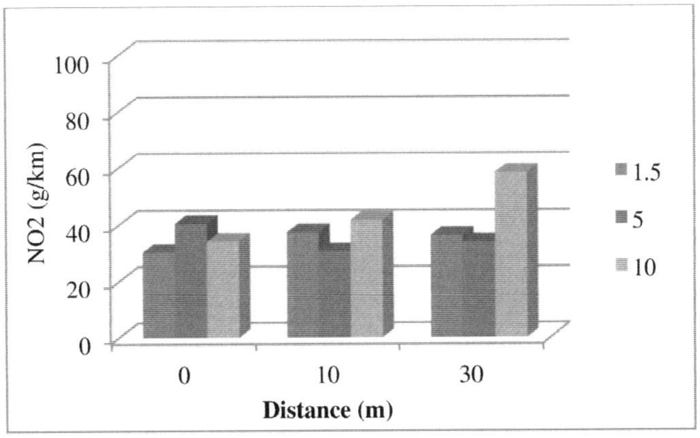

Figure 6. NO_2 concentration in Tanjung plot

Table 4 shows the NO_2 concentration in distance 10m and 30m if concentration at the distance 0m used as reference in each plot of sampling. At elevation 1.5m in Bungur and Tanjung plot, the NO_2 concentration increased at distance 10m and decreased at distance 30m (123.75 % to 105.50 % and 123.26 % to 119.41 % respectively). Most significant decreasing of NO_2 concentration occured at Gmelina plot, as follow 85.62 % (distance 10m) to 60.38 % (distance 30m) at elevation 1.5m and 34.51 % (distance 10m) to 36.60 % (distance 30m) at elevation 5m. This result indicated that Gmelina plot was the most effective in reducing NO_2 concentration horizontally, within the canopy area (below 10m elevation).

Table 4. NO_2 concentration in distance 10 m and 30 m if concentration at the distance 0m used as reference in each plot of sampling.

Elevation (m)	Distance								
	0	10				30			
		Bungur	Gmelina	Tanjung	Control	Bungur	Gmelina	Tanjung	Control
	%								
1.5	100	123.75	85.62	123.26	89.11	105.50	60.38	119.41	116.64
5	100	102.37	34.51	75.67	100.29	87.70	36.60	83.81	119.76
10	100	61.91	152.75	122.03	84.28	130.11	195.15	170.32	132.40

6. Conclusions and Recommendations

From this research, the result of the analysis of variance showed that tree plot has significant effect to NO_2 concentration. However the effect of distance from the road and elevation of sampling were not significant. The highest NO_2 concentration was found on control plot (area without tree canopy).

Roadside green belt with the lowest average of NO_2 concentration was found in Tanjung plot. Tanjung trees and Bungur trees have the same round shape of canopy with difference in leaves density. Tanjung tree with round shape and high density canopy performed better in reducing NO_2 than Bungur tree with round shape and medium density canopy, regardless the sampling elevation and distance.

Gmelina plot performed best in reducing horizontal distribution of NO_2 at elevation 1.5 and 5m, but the result at elevation 10m was not significant. Results indicated that beside of tree canopy shape, leaves density also takes part in reducing NO_2 concentration. Therefore, research regarding factor of leaves density of the canopy is suggested.

Acknowledgements
We would like to thank to DIKTI (Directorate General of Higher Education) who support this research through BOPTN (State Higher Education Operational Aid). We also would like to thank PT JASA MARGA who has permitted us to conduct this research.

References

[1] Brunekreef B 2008 *Procedia Social and Behavioral Sciences*, Beijing For. (China) p 6661-69
[2] Badan Pusat Statistik Kota Bogor 2013 *Bogor dalam Angka* (Bogor: BPS)
[3] Krupa SV 1997 *Air Pollution, People, and Plants* (Minnesota: The American Phytopathological Society
[4] Darmono 2001 *Lingkungan Hidup dan Pencemaran: Hubungannya dengan Toksikologi Senyawa Logam* (Jakarta: UI Press)
[5] Gardner F P, Pearce R B, Mitchell R L, 1991 *Fisiologi Tanaman Budidaya*, translator Susilo H translation from: Crop Physiology (Jakarta: UI Press)
[6] Fujii S, Cha H, Kagi N, Miyamura H, Kim YS 2005 *Building and Environment* **40** 105-112
[7] Hirabayashi S, Kroll CN, Nowak DJ 2012 *Env. Poll.* **171** 9-17
[8] Grey G W and Deneke F J 1978 *Urban Forestry* (New York: John Wiley and Sons Inc.)
[9] Hermawan R 2012 *Efektivitas struktur jalur hijau jalan dalam mereduksi partikel timbal dari emisi kendaraan bermotor* [dissertation] (Bogor : Institut Pertanian Bogor)
[10] Sulistijorini 2009 *Keefektifan dan toleransi jenis tanaman jalur hijau jalan dalam mereduksi pencemaran akibat aktivitas transportasi* [dissertation] (Bogor : Institut Pertanian Bogor)
[11] Nasrullah N, Gandanegara S, Suharsono H 2001. *Buletin Taman dan Lanskap Indonesia* **4** 1-5

Enzymatic activity of Glucose Oxidase from *Aspergillus niger* IPBCC.08.610 On Modified Carbon Paste Electrode as Glucose Biosensor

T Rohmayanti[1], L Ambarsari[1*] and A Maddu[2]

[1]Department of Biochemistry, Faculty of Science, Bogor Agricultural University, Indonesia
[2]Department of Physics, Faculty of Science, Bogor Agricultural University, Indonesia

Email: ami_icha@yahoo.com

Abstract. Glucose oxidase (GOx) has been developed as glucose sensor for measuring blood glucose level because of its specificity to glucose oxidation. This research aimed to determine kinetic parameters of GOx activity voltametrically and further test its potential as a glucose biosensor. GOx, in this research, was produced by local fungi *Aspergillus niger* IPBCC.08.610 which was isolated from local vine in Tarakan, East Borneo, Indonesia. GOx was immobilized with glutaraldehyde, which cross-linked onto modified carbon paste electrode (MCPE) nanofiber polyaniline. Intracellular GOx activity was higher than extracellular ones. Immobilized GOx used glutaraldehyde 2.5% and dripped on the surface of MCPE nanofiber polyaniline. MCPE have a high conductance in copper with the diameter of 3 mm. The concentration of glucose in the lowest concentration of 0.2 mM generated a current value of 0.413 mA while 2 mM of glucose induced a current of 3,869 mA value. Km and Imax of GOx in MCPE activities polyaniline nanofiber were 2.88 mM and 3.869 mA,respectively, with turnover (Kcat) of 13 s^{-1}. Sensitivity was 1.09 mA/mM and response time to produce a maximum peak current was 25 seconds. Km value was then converted into units of mg/dL and obtained 56.4 mg/dL. $GOx_{immo\text{-}IPB}$|MCPE electrode is potential to be able to detect blood glucose level in a normal condition and hypoglycemia conditions

Keywords: *Glucose oxidase, Glucose Biosensor, Electrode Paste Carbon*

1. Introduction

Indonesia is a tropical country, which consists of diverse microorganism. Microorganisms may be valuable and potential sources for various purposes, which can be revealed through comprehensive research. Production of various enzymes using microorganisms is one thing that can be developed. Ironically, to date Indonesia has not been able to produce enzymes industrially yet still fully dependent on imported products [1]. One of widely used enzymes is glucose oxidase (GOx). In this research, GOx was produced directly from local *Aspergillus niger* which was isolated from local vine *Dryobalanops* in Tarakan, East Borneo, Indonesia. Previous study showed that this enzyme has activity and yield of 27.77 U/mg and 72.86%, respectively.

GOx is an enzyme that catalyzes the oxidation of β-D-glucose into D-gluconolactone and hydrogen peroxide using molecular oxygen as an electron acceptor [2]. Nowadays, GOx is generally applied to biosensor and other commercial applications. In addition, the enzyme is also used in biomedicine for determining blood glucose levels. Determination of blood glucose levels using GOx through either colorimetric or amperometric assay is more effective than that non-enzymatic methods because GOx has a high substrate specificity towards β-D-glucose [3]. Thus, biosensor based on GOx could be used in analyzing glucose content of a blood. GOx can oxidize glucose in blood to gluconic acid and hydrogen peroxide by using oxygen molecule as the electron acceptor which will be taken as the signal, and it is then decoded in computer software [4].

Application of GOx onto electrode in biosensor system needs immobilization process [5]. GOx immobilization into polyaniline-nanofiber at carbon pasta electrode surface is presumably to rise electrode conductance in biosensor. Nanometer-sized polyaniline widens electrode surface so that its application may be more economically efficient. GOx immobilized into carbon graphite can be repeatedly used so that it is cheaper and more efficient.

This research has a novelty as a biosensor because of GOx produced from local fungi *A. niger* IPBCC.08.610. This research aimed to determine kinetic parameters of GOx activity voltametrically, to further test its potential as glucose biosensor in blood. It can be developed potential diversity of microorganisms to make high value product.

2. Materials and Methods

2.1. *Materials*

Local fungi *A. niger* IPBCC.08.610 purchased from the fungal collection of Biology Department of Bogor Agricultural University (BAU), was isolated from Tarakan East Borneo, polyaniline-nanofiber, glutaraldehyde, carbon graphite, ortho-dianisidine and hydrogen peroxida (H_2O_2). The other materials used such as Potato Dextrose Agar media, sucrose, peptone, $(NH_4)_2HPO_4$, $MgSO_4.7H_2O$, KH_2PO_4, $CaCO_3$, ammonium sulfate, glucose, Bradford reagent, Bovine Serum Albumin (BSA) standard, paraffin wax, acetic buffer 0.1 M, KCl, ($K_3Fe(CN)_6$).

2.1.1. Glucose oxidase production, isolation, and purification. This assay was done based on Rohmayanti 2016 with modifications. Production media of A. niger contained 0.4 g/L $(NH_4)_2HPO_4$; 0.2 g/L KH_2PO_4; 0.2 g/L $MgSO_4.7H_2O$; 40 g/L $CaCO_3$; 3.3 % sucrose; and 0.35 % glucose. Culture was then incubated at 37^0C on 120 rpm for 40 hours. After being incubated, mycelium was filterred by using gauze. The supernatant contained the crude extract of extracelullar glucose oxidase. Filtered mycelium was homogenized using quartz sand and sodium phosphate buffer 0.1 M pH 6.0, then centrifuged at 12,000 x g temperature of 4^0C for 20 minutes. The supernatant obtained was crude extract of intracellular. The purification of both extracellular glucose oxidase and intracellular crude extract was done through precipitation with saturated 80% $(NH_4)_2SO_4$. Protein content was measured according to Bradford method (1976) and its activity based on Bergmeyer method (1988).

2.1.2. Measurement of Carbon Paste Electrode (CPE) and Modified Carbon Pasta Electrode (MCPE) Cyclic Voltammogram. This experiment was conducted based on Colak *et al.* [6] with modifications. CPE was prepared by mixing 100 μL paraffin and 0.15 gram graphite powder into mortar until pasta shaped. Furthermore, the carbon paste electrodes was inserted into the prepared tube of glass capillary size variation copper diameter of 1 mm and 3 mm. The best diameter copper size of EPK was then modified by the addition of polyaniline nanofiber by 2 mg polyaniline. Whereas, MCPE was made of 2 mg polyaniline which mixed into mortar with 100 μL paraffin and 0.15 g graphite powder. Furthermore, each graphite pasta was added into electrode tube which prepared from capillary glass with its diameter and height were respectively 3 mm and 10 mm. CPE and MCPE (as working electrode alternately), Ag|AgCl electrode (reference electrode) and platina electrode (counter

IOP Conf. Series: Earth and Environmental Science **58** (2017) 012046 doi:10.1088/1755-1315/58/1/012046

electrode) were placed into solution test contained 3 mL potassium chlorid (KCl) solution. These electrodes then linked to three distinct wires that had been connected to potentiostat.

2.1.3. Immobilization of Glucose Oxidase into MCPE. This assay was done based on Ambarsari 2016 with modifications. About 50 μL from each GOX extract was mixed with 1 mg BSA, 50 μL of 0.1 M acetic buffer (pH 4.5), and 30 μL of 2.5% glutaraldehyde. The mixture was placed into micro tubes and then shaked slowly. It was then dropped down (4 μL) to the upper surface of MCPE. . Electrode utilized immobilized-GOx from local isolate was called $GOx_{immo\text{-}BAU}$|MCPE. Furthermore, the $GOx_{immo\text{-}BAU}$|MCPE was dried at 4°C. Then, it was washed several times with acetic buffer pH 4.5 to get rid of unimmobilized enzyme. Electrode was then kept in refrigerator at 4°C in 0.1 M acetic buffer (pH 4.5). The prototype electrode is showed at figure 1.

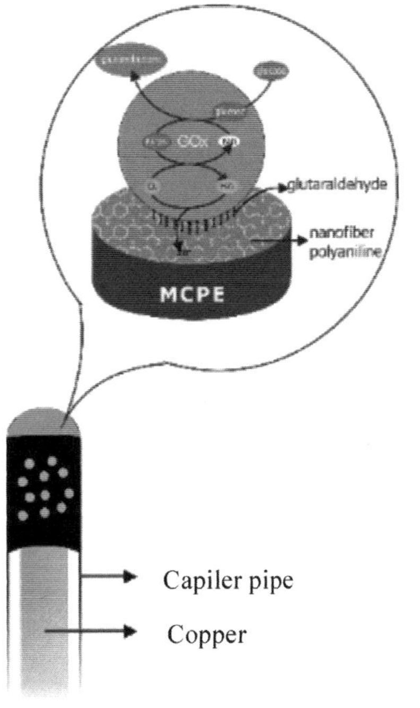

Figure 1. MCPE Performance and GOD reaction

2.1.4. Immobilized Glucosa Oxidase Kinetics in MCPE. Cyclic Voltammogram of $GOx_{immo\text{-}BAU}$|MCPE was measured in 1 mL 0.1 M acetic buffer (pH 4.5) supplemented with 1 mL 0.1 M potassium ferisianida and 180 μL glucose standard solution with concentration of 0.2-8 mM. Cyclic Voltammogram analyses was then subjected for designing the Michaelis-Menten curve. Electrode kinetics parameter determination of $GOx_{immo\text{-}BAU}$|MCPE required linier line of Michaelis-Menten chart, then plotted into Lineweaver-Burk equation to obtain K_m and I_{maks} value.

3. Result and Discussion

3.1. GOx Fraction of Intracellular and Extracellular Crude Extract and Saturated $(NH_4)_2SO_4$ 80% Precipitate

GOx was produced by *A. niger* IPBCC.08.610 in intracellular and extracellular manner. Intracellular GOx activity was higher than extracellular, about 1.3084 U/mL. Intracellular and extracelullar GOD activity of crude extract was 1.31 U / mL and 0.17 U/ mL, respectively. Biomass produced by A.niger be large enough round cells and white. Then GOx intracellular crude extract was purified by saturated $(NH_4)_2SO_4$ 80%.

Precipitation of intracelullar GOx crude extract was the initial step to purify enzyme that significantly increased protein concentration, reduced the volume, and separate the protein target from partial undesired contaminant. Indeed, the purity level of precipitated GOx was 20 times higher than GOx crude extract, with specific activity and yield of 27.77 U/mg and 72.86%, respectively.

Based on the chart Lineweaver-Burk, the activity of the GOX is glucose concentration dependent. Enzymatic activity at low glucose concentration showed linear pattern in corresponds to the increased glucose concentrations. However, enzymatic activity at more than 0.06 M glucose was reduced V_{maks} and Km value of precipitated GOx was 20.747 U/mg and 56 mM, respectively. Moreover, the GOx k_{cat} value of ammonium sulfate 80% fraction was predicted about 53 s^{-1}.

3.2. Cyclic Voltammogram of CPE and MCPE

Cyclic Voltammogram may visualize the produced current by electrochemical cell as current value. Conductance change of this electrode was measured by using cyclic voltammetry method because it could evaluate electrochemical reaction on the surface of electrolite solution and electrode [7]. Figure 2 shows that cyclic voltammogram of CPE and MCPE has higher current evolving than cyclic voltammogram of MCPE. The increasing current is conceived by the change of the width area of cyclic voltammogram. This is due to an improvement of polyaniline into nanofiber that had role to raise its conduction capacity. Moreover, the variation of the size would affect the mass and thickness of carbon graphite which added to electrode capillary, so that it causes distinct electron transfer that could be analyzed from its current and potential.

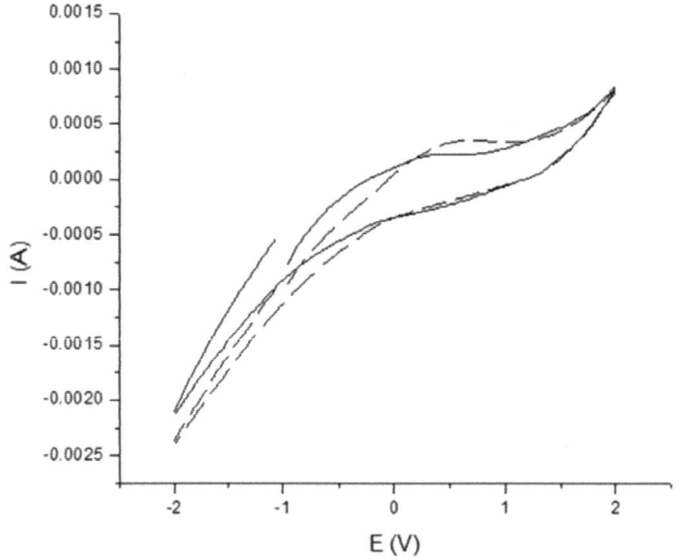

Figure 2. Cyclic voltamogram diameter of CPE ——— 1mm, ---------- 3mm.

3.3. Immobilized Enzyme Kinetics Parameter in MCPE

Immobilized GOx was dripped onto the surface of modified carbon paste electrodes (MCPE) nanofiber polyaniline. MCPE have a high conductance in copper with the diameter of 3 mm. GOx which was immobilized on MCPE was called as $GOx_{immo-IPB}$|MCPE. GOx kinetic parameters (Km and Imax value) was determined using cyclic voltammetry. within range of glucose concentration of 0.2 - 8.0 mM

According to Michaelis-Menten curve, below 2 mM of glucose concentration the $GOx_{immo-BAU}$|MCPE electrode performance to its subtrate would evolve by the increasing concentration of glucose, on the other hand, enzymatic activity increased following elevated concentration of glucose.However, when glucose concentration was about 2 mM, GOx activity has reached its maximum activity, thus additional glucose concentration had no effects to the oxidation current value of $GOx_{immo-BAU}$|MCPE electrode (figure 3).

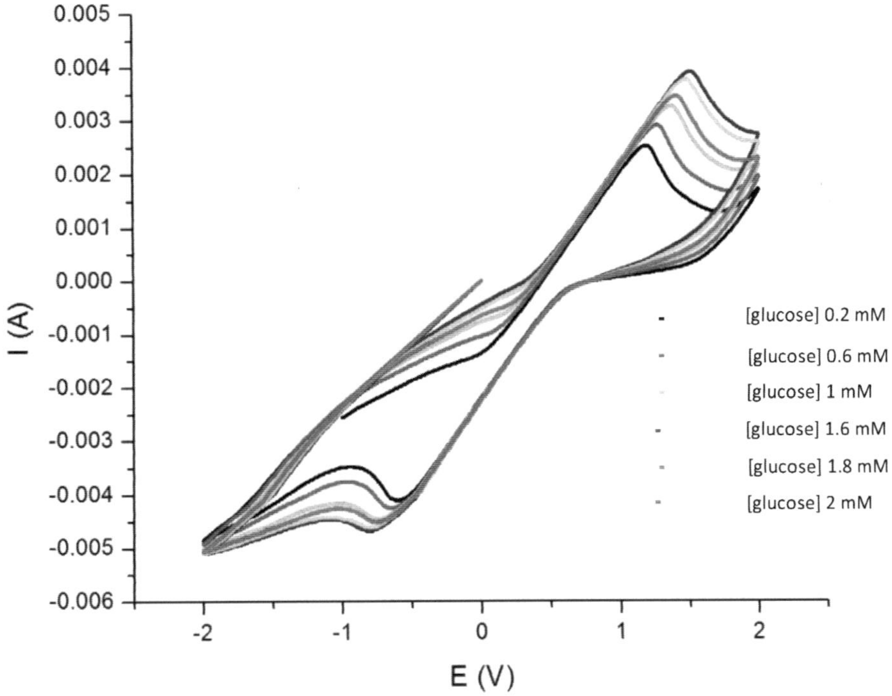

Figure 3. Cyclic Voltamogram performance of $GOx_{immo-BAU}$|MCPE.

Linier chart shows glucose substrate concentration to electrode current value of $GOx_{immo-BAU}$|MCPE determined with Michaelis-Menten equation (figure 4). The range of electrode performace was in the concentration of 0.2 – 2 mM, with regression value (R^2) of 0.986. Km and I_{maks} value of GOD electrode in this research is better than previous study by Colak et al., 2012, which was 0.61 mM and 1.22 μA, respectively. On the other hand, Ambarsari (2016) result indicated lower Km value of 0.70 mM and higher I_{maks} value of 4.24 mA which was probably due to the utilization of GOx commercial with high thickness and high activitity (5204.3 U/mL). In addition, $GOD_{immo-BAU}$|MCPE sensitivity was 1.099 $mAmM^{-1}$.

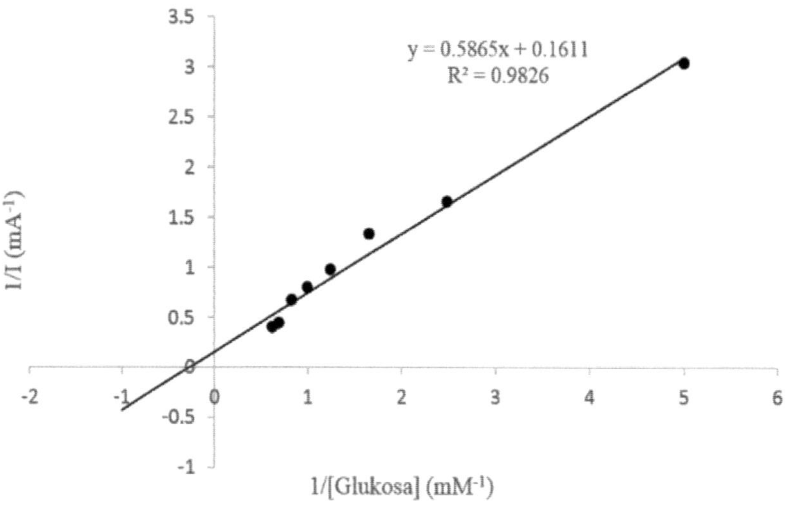

Figure 4. Electrode kinetics parameter of $GOx_{immo-BAU}$|MCPE.

Km value of $GOx_{immo-IPB}$|MCPE was 56.4 mg/dL, thus it is potential for this electrode to be used as glucose biosensor. . According to the Km values, $GOx_{immo-IPB}$|MCPE electrode is able to detect blood glucose level in a normal condition, hyperglycemia, and hypoglycemia conditions. Human normal blood glucose level is about 70-110 mg/dL, and hypoglycemia blood glucose condition to be in a state if glucose level under 60 mg/dL and in a state of hyperglycemia when glucose level above 270 mg/dL.

Table 1. Kinetics parameter of $GOx_{immo-IPB}$|MCPE electrode.

Electrode	Kcat (s^{-1})	Range [glukosa] (mM)	Sensitivity (mA mM^{-1})	Km (mM)	Imax (mA^{-1})
$GOx_{immo-IPB}$|MCPE	13	0.2-1.6	1.09	2.88	3.869
GOD $(NH_4)_2SO_4$	53	2-100	-	30	-

4. Conclusion

Biochemical characteristics of GOx of *A. niger* (IPBCC.08.610) consists of enzymatic activity, kinetics parameter and its immobilization in carbon pasta electrode. The enzymatic activity of intracellular GOx was higher than extracellular one. MCPE have a high conductance value in copper with the diameter of 3 mm. Linear region of the electrode performance $GOx_{immo-IPB}$|MCPE, which is half of the maximum speed, is at a concentration of 1.6 mM or 56.4 mg/dL. $GOx_{immo-IPB}$|MCPE electrode is hihgly potential to be used as glucose biosensor to to detect blood glucose level in a normal condition and hyperglycemia.

References

[1] Saryono 2008 *J Natur Indonesia* **11** 19-23
[2] Sabir S, Bhatti H N, Zia M A and Sheikh MA 2007 **45(4)** 443-446
[3] Ahmad A, Syaiful A, Firman A, and Patong AR 2007 *Indo J Chem* **7** 97-104
[4] Simpson C, Jordaan J, Gardiner N S and Whiteley C 2007 *Protein Express Purif* **51** 260-266
[5] Ambarsari L, Setyawati I, Kurniasih R, Kurniatin P A and Maddu A 2016 *Indo J Chem* **1** 1-16
[6] Colak O, Arslan H, Zengin H and Zengin G 2012 *Int J Electrochem Sci* **7** 6988-6997
[7] Zhu J, Chen M, Qu H, Zhang X, Wei H, Luo Z, Colorado HA, Wei S and Guo Z 2012 *Polymer* **53** 5953-5964

Effect of Manganese Addition on the Structure, Magnetic Properties and Microwave Absorption of $La_{0.8}Ba_{0.2}Mn_xFe_{½(1-x)}Ti_{½(1-x)}O_3$

W A Adi[1], M N Indro[2] and A A Kusumastuti[2]

[1]Center for Science and Technology of Advanced Materials, National Nuclear Energy Agency, Indonesia.
[2]Departement of Physics, Bogor Agricultural University, Indonesia.

Email : dwisnuaa@batan.go.id, indromn@gmail.com, aanggun@gmail.com

Abstract. We have carried out modification of $La_{0.8}Ba_{0.2}Mn_xFe_{½(1-x)}Ti_{½(1-x)}O_3$ (x = 0.1 – 0.8) magnetic materials by wet milling method. Raw materials of La_2O_3, $BaCO_3$, Fe_2O_3, TiO_2 and $MnCO_3$ were mixed according to stoichiometry calculation for each composition. The mixture was milled for 5 hours and then sintered at 1000 °C for 5 hours. The refinement results by X-ray diffraction pattern shows that the increasing Mn composition enhances the mass fraction of $La_{0.8}Ba_{0.2}Mn_xFe_{½(1-x)}Ti_{½(1-x)}O_3$ phase which has the same structure as $LaMnO_3$. For x = 0.8 a single phase of $LaMnO_3$ was formed. The single phase has a crystal monoclinic crystal structure with space group of I 1 2 / a 1, with lattice parameters given by a = 5.519(5) Å, b = 5.5537(5) Å and c = 7.8176(9) Å, $\alpha = \gamma = 90^o$ and $\beta = 90.345(6)^o$, V = 239.64(3) $Å^3$, ρ = 6.463 $gr.cm^{-3}$, wRp = 5.96, and χ^2 (chi-squared) = 1.17. The hysteresis curve shows that the sample with composition x = 0.8 produces ferromagnetic behaviour at room temperature. The ferromagnetic properties arise due to the mixed valence of Mn^{3+} and Mn^{4+} ions through a double exchange mechanism. The results of the microwave absorption indicated that there was a broadening of absorption peak frequency at 9.9 GHz. The reflection loss (RL) increases with the increasing of $LaMnO_3$ phase. For x = 0.8 we have the best of RL where the microwave absorption was calculated reaching 95% at the highest peak frequency with a thickness of 1.5 mm. Thus we have been successful in creating a single phase of $La_{0.8}Ba_{0.2}Mn_xFe_{½(1-x)}Ti_{½(1-x)}O_3$ with application as a microwave absorber.

Keywords: Perovskite, $La_{0.8}Ba_{0.2}Mn_xFe_{½(1-x)}Ti_{½(1-x)}O_3$, composition, phase, ferromagnetic, microwave absorber

1. Introduction

One research topic which attracts great interest nowadays is a material that can absorb microwaves with applications in the field of electronics, telecommunications and defense. There are two important factors that must be met so that the material can absorb microwaves, namely that the material should have magnetic loss (component permeability) and dielectric loss (permittivity component) is high [1].

Recently, some of the magnetic material successfully demonstrated absorption in the microwave region. The magnetic material is hexa-ferrite system [2-5] which has proven capable of absorbing microwave because they have high permeability values. Meanwhile other interesting magnetic material called perovskite ABO_3 are manganite-based [6-9] because they have a high permittivity value has also attracted lots of interest.

The magnetic structure of $LaMnO_3$ is anti-ferromagnetic. Magnetic interactions in this system belong to the indirect exchange interaction, called superexchange mechanism. Superexchange mechanism is the magnetic interaction between the adjacent Mn^{3+} ions mediated by nonmagnetic ions O^{2-} with electron spin pairing. The hysteresis curve of $LaMnO_3$ measured at room temperature has a linear magnetic pattern as a function of the applied magnetic field. Meanwhile the presence of Ba substitution into La on this system causes a magnetic phase transformation. In contrary, the magnetic structure of $La_{0.8}Ba_{0.2}MnO_3$ is ferromagnetic. This magnetic phase transformation occurs due to the presence in the mixed valence Mn through double-exchange mechanism. The mechanism of double exchange is the magnetic interaction in which the displacement of the electron spin is parallel to the nearest neighbor by doing twice hopping simultaneously from Mn^{3+} ions to Mn^{4+} ions through O^{2-} ions. The hysteresis curve of $La_{0.8}Ba_{0.2}MnO_3$ measured at room temperature has a nonlinear magnetic pattern as a function the applied magnetic field [10].

In this study, we will present structural engineering materials called perovskite $La_{0.8}Ba_{0.2}MnO_3$, where the element manganese is substituted with iron (Fe) to increase its permeability and titanium (Ti) to increase the permittivity in the material. The focus of discussion is to determine the material structure, magnetic properties and electromagnetic wave absorption capability due to addition of manganese and titanium.

2. Material and Method

The raw materials used were lanthanum oxide (La_2O_3), barium carbonate ($BaCO_3$), manganese carbonate ($MnCO_3$), iron oxide (Fe_2O_3), and titanium dioxide (TiO_2) from Merck product with high purity. Each of raw materials was weighed suitable with the compositions stoichiometry as x function (x = 0 – 0.8) according to reaction as follows:

$$0.8La_2O_3 + 0.4BaCO_3 + 2xMnCO_3 + ½(1-x)Fe_2O_3 + (1-x)TiO_2 \rightarrow 2La_{0.8}Ba_{0.2}Mn_xFe_{½(1-x)}Ti_{½(1-x)}O_3 + zCO_2$$

The samples consist of eight compositions, namely, $La_{0.8}Ba_{0.2}Mn_{0.1}Fe_{0.45}Ti_{0.45}O_3$, $La_{0.8}Ba_{0.2}Mn_{0.2}Fe_{0.4}Ti_{0.4}O_3$, $La_{0.8}Ba_{0.2}Mn_{0.3}Fe_{0.35}Ti_{0.35}O_3$, $La_{0.8}Ba_{0.2}Mn_{0.4}Fe_{0.3}Ti_{0.3}O_3$, $La_{0.8}Ba_{0.2}Mn_{0.5}Fe_{0.25}Ti_{0.25}O_3$, $La_{0.8}Ba_{0.2}Mn_{0.6}Fe_{0.2}Ti_{0.2}O_3$, $La_{0.8}Ba_{0.2}Mn_{0.7}Fe_{0.15}Ti_{0.15}O_3$, and $La_{0.8}Ba_{0.2}Mn_{0.8}Fe_{0.1}Ti_{0.1}O_3$ furthermore are respective called by x = 0.1, x = 0.2, x = 0.3, x = 0.4, x = 0.5, x = 0.6, x = 0.7, and x = 0.8. Methods used are a solid state reaction use mechanical milling technique of wet. This method is simple and inexpensive. The each of mixtures were milled by HEM (high energy milling SPEX8000 mixer/mil) for 5 hours and dried in oven at 100 °C for 5 hours. The samples made were sintered in Furnace at 1000 oC for 5 hours, and then the each of samples were characterized by using X-ray diffraction (XRD) to know phases formed.

The samples with compositions selected were characterized furthermore by using scanning electron microscope (SEM), energy dispersive spectroscopy (EDS), particle size analyzer (PSA), vibrating simple magnetometer (VSM), and vector network analyzer (VNA) respective to know morphology of particle, elemental composition analysis, distribution of particle size, magnetic properties, and performance of microwave absorption. Especially for the phase qualitative and quantitative of analysis were carried out using Rietveld method with applying GSAS program [11].

3. Results and Discussions

Figure 1 shows X-ray diffraction profiles of synthesized material which respective profiles of system $La_{0.8}Ba_{0.2}Mn_xFe_{½(1-x)}Ti_{½(1-x)}O_3$ (x = 0.1 – 0.8) are compared.

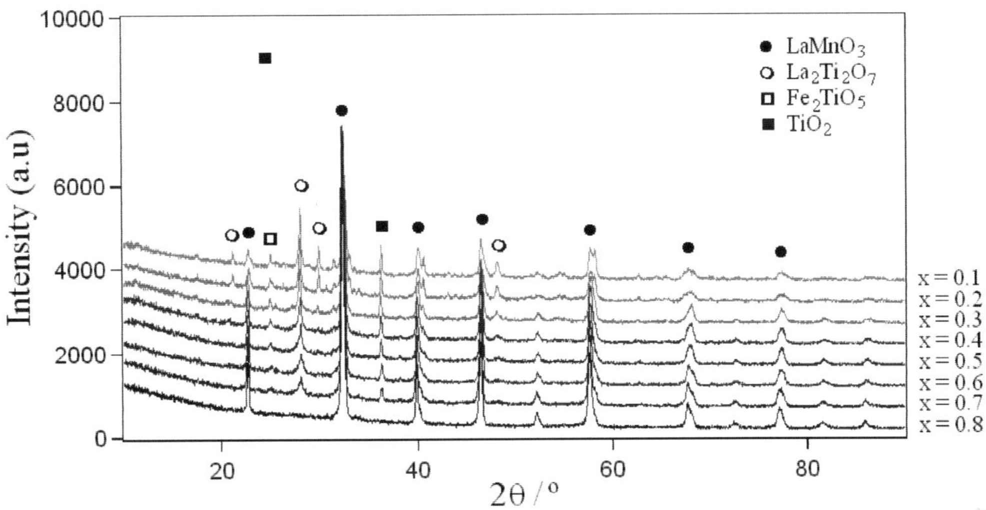

Figure 1. X-ray difraction profile of magnetic system $La_{0.8}Ba_{0.2}Mn_xFe_{½(1-x)}Ti_{½(1-x)}O_3$.

The results of phase identification shows that sample with composition x = 0.1 consists of four phases, namely phase of $LaMnO_3$ (COD : 1001820), $La_2Ti_2O_7$ (COD: 2002196), Fe_2TiO_5 (COD: 2002318), and TiO_2 (COD: 9015662). In this composition appears that the substitution of Fe and Ti excess caused a reaction becomes imperfect so that Fe and Ti atoms form another phase. Samples with composition x = 0.2 - 0.4 appears that phase of $LaMnO_3$ increased while others decreased. Fe_2TiO_5 phase to be lost on the composition x = 0.5, so that the sample only consists of three phases. Phase of $LaMnO_3$ growing increases with increasing manganese in the sample composition and phase $La_2Ti_2O_7$ and TiO_2 are getting down (composition x = 0.6 - 0.7) and disappeared on the composition x = 0.8. Single phase of $LaMnO_3$ obtained when the composition x = 0.8. Thus the change in the number of phases which are formed on the composition x = 0.1, x = 0.5 and x = 0.8.

Results refinement X-ray diffraction pattern and morphology of particle for a third of the composition is shown in figure 2. While the crystal structure parameters and mass fraction of phase formed is shown in table 1.

Table 1. The value of structure parameters, criteria of fit (R_{wp}), goodness of fit ($\chi 2$) and the mass fraction of phase formed with variations in composition.

Sample (x)	Phase	Lattice parameter (Å) a	b	c	V (Å³)	ρ (g/cm³)	Fraction wt%	R_{wp} (%)	χ^2
0.1	$LaMnO_3$	5.525(1)	5.536(1)	7.769(1)	237.6(1)	6.518	42.66	9.20	1.41
	$La_2Ti_2O_7$	25.571(7)	7.812(3)	5.584(3)	1115.6(8)	5.782	45.21		
	Fe_2TiO_5	3.780(5)	11.432(1)	9.505(5)	410.7(3)	3.745	2.92		
	TiO_2	4.508(1)	4.508(1)	2.964(1)	60.24(6)	4.403	9.21		
0.5	$LaMnO_3$	5.5038(7)	5.5427(7)	7.8074(9)	238.17(4)	6.503	69.45	7.30	1.27
	$La_2Ti_2O_7$	25.605(1)	5.031(1)	13.695(7)	1134.0(2)	5.694	24.08		
	TiO_2	4.496(2)	4.496(2)	2.965(1)	59.97(7)	4.424	6.46		
0.8	$LaMnO_3$	5.519(5)	5.5537(5)	7.8176(9)	239.64(3)	6.463	100.00	5.96	1.17

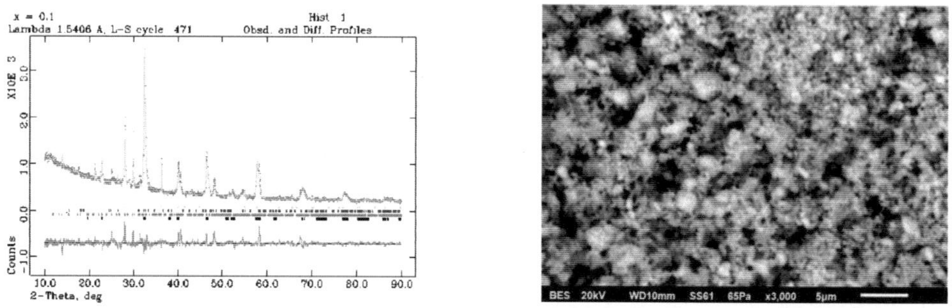

(a) Composition x = 0.1

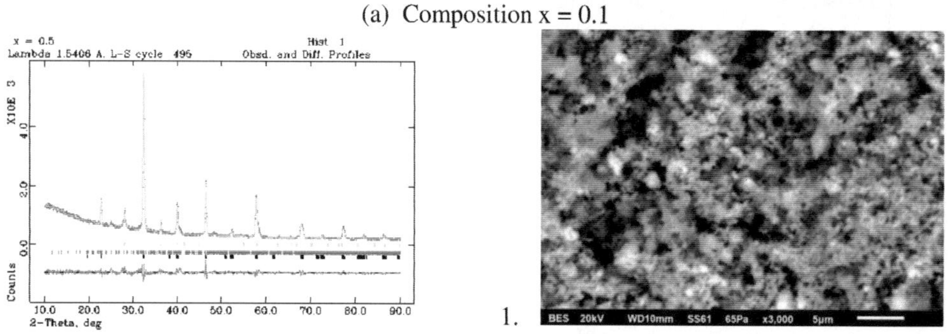

1.

(b) Composition x = 0.5

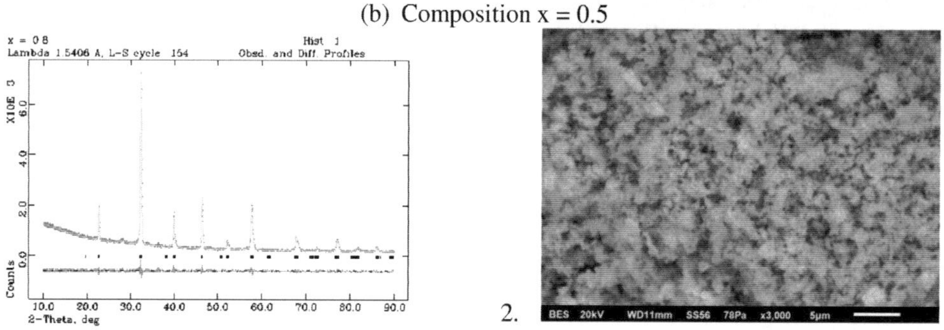

2.

(c) Composition x = 0.8

Figure 2. Refinement result of XRD and SEM image on the $La_{0.8}Ba_{0.2}Mn_xFe_{½(1-x)}Ti_{½(1-x)}O_3$.

Figure 2 and table 1 shows a correlation between the phases formed and the morphology of the particles of each composition. In composition x = 1 and x = 0.5 has been shown that the samples consist of multi-phase and is supported by the results of particle morphology is very diverse. While the composition x = 0.8, sample is single-phase so that it seemed at the SEM image shown morphology uniform and homogeneous across the surface. Single phase of system $La_{0.8}Ba_{0.2}Mn_xFe_{½(1-x)}Ti_{½(1-x)}O_3$ particles were successfully synthesized by solid state method appears at the manganese composition of x = 0.8.

In addition, the hysteresis curve shows that the sample is ferromagnetic behavior as shown figure 2. The ferromagnetic behavior for the samples because the presence of mix valence from manganese ions between Mn^{3+} and Mn^{4+} ions, giving rise magnetic interactions. The magnetic interactions are due double exchange and superexchange which occur respective between Mn^{3+}/Mn^{3+} and Mn^{3+}/Mn^{4+} ions [10].

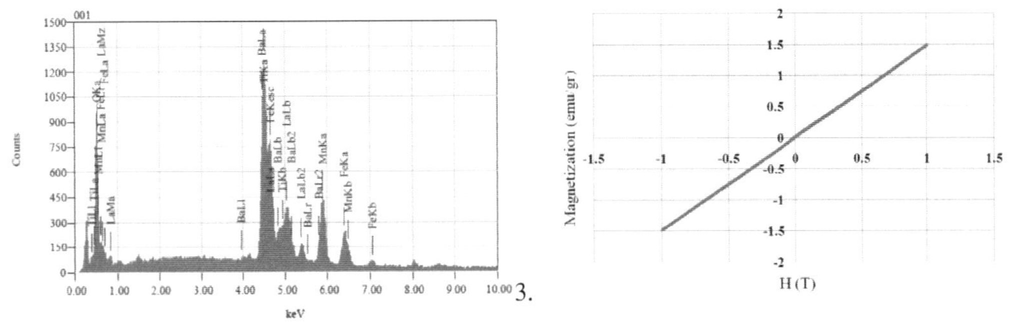

3.

(a) Composition x = 0.1

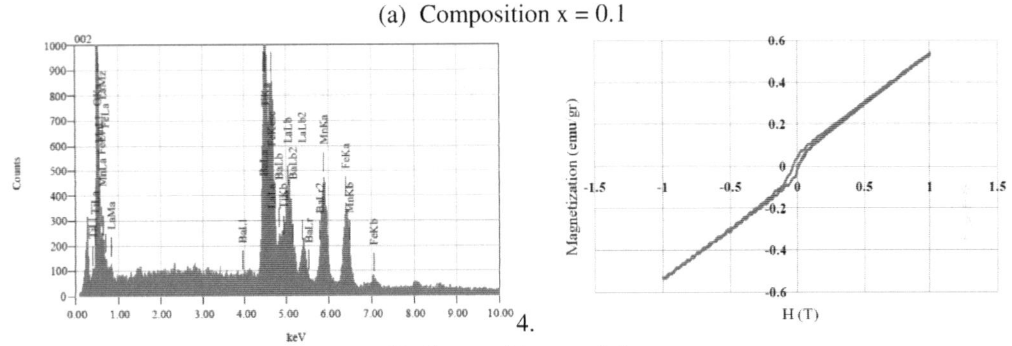

4.

(b) Composition x = 0.5

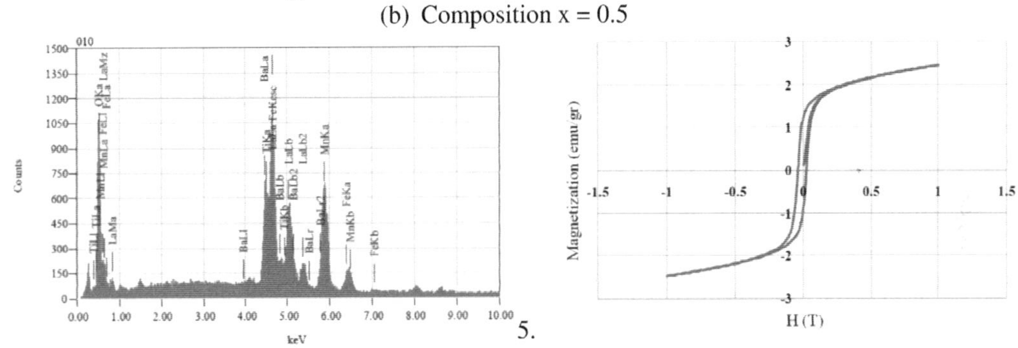

5.

(c) Composition x = 0.8

Figure 3. The element analysis and magnetic properties on the $La_{0.8}Ba_{0.2}Mn_xFe_{½(1-x)}Ti_{½(1-x)}O_3$.

In figure 3 is shown that the magnetic properties of the composition x = 0.1 is paramagnetic because the content of Mn in the composition is so low so that the $LaMnO_3$ phase is relatively small as shown in figure 4. Then in the composition x = 0.5, Mn content greater so present ferromagnetic properties of materials this is because the mass fraction of $LaMnO_3$ phase greater. While the composition x = 0.8 magnetic properties change into ferromagnetic because the composition is already forming a single phase of $LaMnO_3$.

The ferromagnetic properties of system $La_{0.8}Ba_{0.2}Mn_xFe_{½(1-x)}Ti_{½(1-x)}O_3$ depends on the presence of Mn ions with different valence (mixed-valence). The presence of Mn^{3+} and Mn^{4+} is closely linked with ferromagnetic behavior to this system where this phenomenon can be explained by the double-exchange theory of Zener [12]. Mn^{3+} ion has the electron configuration $3d^4$, so that the orbital t_{2g} fully charged while there is a single electron spin in orbital e_g. However Mn^{4+} ion has the electron configuration $3d^3$ in which the orbital t_{2g} also fully charged while the e_g orbital empty. Therefore, the electron spin configuration Mn^{3+} is able to hop to fill the void in the configuration Mn^{4+} through the intermediation anion O^{2-} without changing the direction of their spin.

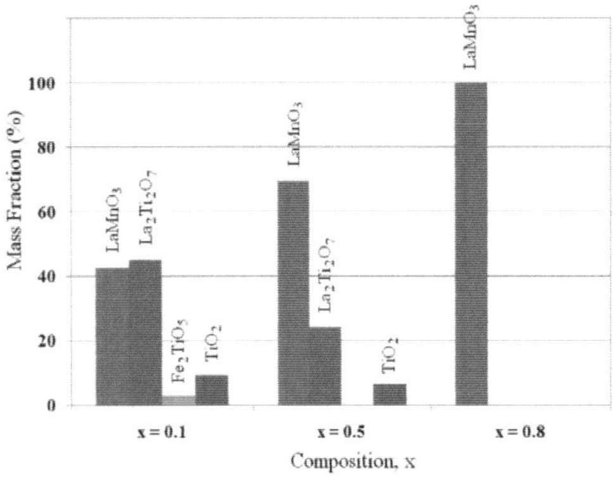

Figure 4. Mass fraction of phase formed.

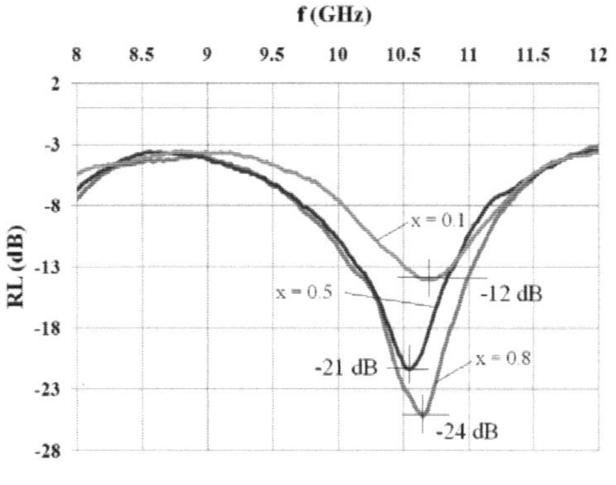

Figure 5. Reflection loss of the samples.

Measurement results of VNA shows that there are one of absorption peaks with reflection loss values of higher at range frequency of 8 – 12 GHz as shown figure 5. It appears that the greater the mass fraction of $LaMnO_3$ phase formed the larger RL possessed by the material. This indicates that the manganese composition of x = 0.8 has the best of microwave absorbing properties in the frequency range of 8- 12 GHz, whose value of the absorbing peaks are -24 dB at 10.5 GHz. The study concluded that a single phase $La_{0.8}Ba_{0.2}Mn_{0.8}Fe_{0.1}Ti_{0.1}O_3$ material has been successfully synthesized showing a good candidate as microwave absorbing materials.

4. Summary

Single phase of system $La_{0.8}Ba_{0.2}Mn_xFe_{½(1-x)}Ti_{½(1-x)}O_3$ particles were successfully synthesized by solid state method. The refinement of x-ray diffraction for the manganese composition of x = 0.8 has been confirmed a single phase. The hysteresis curve shows that the sample with the composition of x = 0.8 is ferromagnetic behavior at room temperature. The ferromagnetic properties arise because of the mixed valence of Mn^{3+} and Mn^{4+} ions through a double exchange of mechanism. The manganese composition of x = 0.8 has the best of microwave absorbing properties in the frequency range of 8- 12 GHz, whose value of the absorbing peaks are -24 dB at 10.5 GHz.

Acknowledgement

The authors gratefully acknowledge the support of the Center for Science and Technology of Advanced Materials, National Nuclear Energy Agency for the research facilities

References

[1]. S. Duggal, G D Aul 2014 *International Journal of Engineering and Advanced Technology* **3(5)** 12-19
[2]. Wisnu A A and Azwar M 2012 *J. Basic. Appl. Sci. Res* **2(8)** 7826-7834
[3]. Z Zhang, X Liu, X Wang, Y Wu and R Li 2012 *J. Alloys Compd* **525** 114-119
[4]. C A Herme, P G. Bercoff and S E Jacobo 2012 *Mat. Res. Bull* **47** 3881-3887
[5]. E Kiani, A S H Rozatian, M H Yousefi 2014 *J. Magn. Magn. Mater* **361** 25-29
[6]. S Zhang and Q Cao 2012 *Materials Science and Engineering B* **177** 678– 684

[7]. K S Zhou, H Xia, K. L Huang, L W Deng, D Wang, Y P Zhou, S H Gao 2009 *Physica B* **404** 175–179
[8]. A. A. Wisnu, A. Manaf 2013 *Advanced Materials Research* **789** 97-100
[9]. Zhang S and Cao Q 2012 *Materials Science and Engineering B* **177** 678– 684
[10]. P Sardjono and A A 2014 Wisnu *Journal of Advanced Materials Research* **896** 381-384
[11]. B H Toby 2000 *Journal of Applied Crystallography*
[12]. C Zener 1951 *Phys. Review* **82(3)** 403-405

Secretory Structure, Histochemistry and Phytochemistry Analyses of Stimulant Plant

C Umah[1], Dorly[1*] and Y C Sulistyaningsih[1]

[1] Department of Biology, Bogor Agricultural University, Indonesia.

Email: dorly_ipb@yahoo.com

Abstract. Plants that are used as stimulant supposed to contains various metabolit compounds that are produced or secreted by secretory structures. This study aimed to identify the secretory structure of plant used as stimulant and chemical compounds accumulated in it. The secretory structure and its histochemistry were observed on plant material that are used as herbal ingredient. Phytochemical content was analyzed by using a qualitative test. The result showed that the idioblast cells and secretory cavities were found in the leaves of *Decaspermum fruticosum,* and *Polyalthia rumphii.* Most idioblast cells contained lipophilic substances and terpenoids or alkaloids, while secretory cavity contained alkaloid. Phytochemical analysis for *D. fruticosum,* and *P. rumphii* contain terpenoids, phenols, steroids, and flavonoids

Keywords: phytochemistry, histochemistry, secretory structure, stimulant.

1. Introduction

Tonic is a medicine that can strengthen or give additional energy to the body. It can play an important role tightening or strengthening body physiological system. Tonic plants are kinds of plants that can raise and keeping body stamina againts disease attack and protect body vitality to keep in perfect healt [1]. Plants that are potential as tonic are often used to keep health by traditional society, for example Anak Dalam Tribe who live in in the area of Bukit Duabelas National Park.

Bukit National Park is one of tropical rain forest Jambi in lowland in that area were found 110 kind of plants used as medicinal. From 110 plants spend as medicine, there are 22 spesies have been examined their chemical contents including their availability of alkaloid, saponin, flavonoid, tannin and polyphenol [2-3].

Anak Dalam Tribe utilize a traditional medicines to maintain health based on knowledge to maintain health based on knowledge, belief experience and direct practices which have been conducted from generation to generation from their ancestor. Knowlegde about traditional medicines was the foundation of new medicine invention on the modern theraphy system [4]. Plants that are used as medicinal material contain metabolite compounds which are potential as active ingredient among others are alkaloid, terpenoid, polyphenol, saponin, flavonoid, tannin, triterpen, and steroid. Compounds contents found on the plants utilized as tonic are among others alkaloid compounds terpenoid, quinon, phenol, polysaccharide, and glycoprotein [5].

Scientific study on secretory structure producing metabolite compounds and histochemical test on tonic plants used by Anak Dalam Tribe have never been carried out. The scientific study enabled to develop information about metabolite compounds content on plants used as tonic material as well as developing through tissue culture or cell culture.

The objective of this research was to indentify plant secretory structure used as tonic and accumulated compounds in the structure.

2. Materials and Methods

2.1. Plant Collection and Identification

Plant collection was conducted in Bukit Duabelas National Park area, Jambi. Plant material used were leaf of saledemo (*Decaspermum fruticosum* and ganja sayur (*Polyalthia rumpii*) (Table 1). For light microscope observation, leaves were fixed using 70% ethanol. Fresh leaves were collected for the histochemistry test, while for qualitative phytochemical content analysis the samples were dried under the sun light for two days, and then dried using oven at 50°C for five days. Identification of plant local name originated from Bukit Duabelas National Park was helped by the head of Anak Dalam Tribe (Tumenggung Tarip). Further plant identification was carried in Herbarium Bogoriense, LIPI, Bogor.

Table 1. Kinds of tonic plant, usage and used methods.

Local name	Scientific name (Family)	Part of plant used	Usage*	Usage methods*
Saledemo	D. fruticosum	Leaf	cold medicinal	Boiled leaves, drunk and used for taking a bath
Ganja sayur	P. rumphii (Annonaceae)	Leaf	Appetizer	Raw eaten

* Tumenggung Tarip, Anak Dalam Tribe pers.comm, 2014

2.2. Histological Analysis

Leaves paradermal section were prepared as a semi permanent preparation by using wholemount method [6]. Fixative leaves in ethanol 70% were washed with distilled water and soaked in 50% HNO_3 solution, rinsed in the water and then the adaxial and abaxial of the leaves were peeled by using razor blade. The section result was soaked in the clorox solution for 3-5 minutes, then was coloured by safranin 1%. The samples were observed by using a light microscope (Olympus CX21).

2.3. Histochemistry Test

Leaf samples of *D fruticosum* and *P.rumphii* were sliced transversally as thick as 15-20 µm using freezing microtome (Yamato RV-240). The sliced result were tested by using several reagent. Alkaloid compounds test on cell or tissue. Lamina leaf section was soaked in Wagner reagent for 24 hours. A positive result was indicated by presence of reddis-brown deposit. As negative control alkaloid, the leaf section pre-treated with 5% tartaric acid for 48 hours to dissolve alkaloid [7]. Lipophilic compounds content test by using sliced samples washed in 70% ethanol for one minute, then coloured with 0.03% sudan IV and heated in a water bath at 40 ^{0}C for 30 minute. The sliced sample were then washed with ethanol 70%. The presence of lipophilic compounds was indicated by colour orange [8]. Terpenoid compounds test in the leaf tissue was identified by soaking samples slices in 5% cupric acetate reagent for 24 hours. Terpenoid compound content will be shown if there is yellow or brownish-yellow color [9].

2.4. *Qualitative Phytochemical Analyses*

Leaf samples that had been air dried will be made into powder by using blender. The powder is qualitatively tested to detect the existence of terpenoid, steroid, flavonoid, alkaloid, and phenol compounds according to the methods of Harborne [9].

2.4.1. *Triterpen and Steroid Compounds Test.* Approximately 3 gram of powder sample will be added 5 ml concentrated ethanol while being heated then it will be filtered. The obtained filtrate will be dropped on the drop plate and will be heated until dry, then it will be added with 1 ml diethyl ether, and Lieberman-Burchard reagent (3 drops of unhydrous acetic acid + 1 drop of sulphuric acid). If the colour of red or purple is obtained it shows triterpen compounds is positive, but if the colour of green or blue appears it shows steroid compounds is positive.

2.4.2. *Alkaloid Compounds Test.* Approximately 3 gram of sample powder will be extracted by a little chloroform , then it was added with 3-5 drops of ammonia and was filtered. Filtrate was added with 1-2 ml H_2SO_4 2M, then was shaken until two layers will be formed. Acid layer (no colour) will be dropped on three test plates, the first plate will be dropped with Dragendorf reagent, the second plate will be dropped with Mayer reagent and the third plate will be dropped with Wagner reagent. If there is or orange, white and brown precipitation consecuently against the three reagents mentiones above it means the result is positive against alkaloid compounds.

2.4.3. *Flavonoid Compounds Test.* Approximately 3 gram of sample powder will be given a little distilled water and was heated for 5 minutes, then it will be given filtered. The filtrate was added a little Mg powder, several drops of concentrated chloric acid and 2 ml amyl ethanol, if the colour of orange appears it means flavonoid compounds is positive.

2.4.4. *Phenol Compounds Test.* Approximately 3 gram of sample powder was added with hot 70% methanol , then it will be filtered. The filtrate was dropped with 10% NaOH. If there is yellow until red colour it means phenol compounds is positive.

3. Results and Discussion

3.1. *Secretory Structure*

Secretory structure was plant cell or tissue plant function as secretory place of metabolite compounds such as essential oils, gum, resin, latex, alkaloid, glycoside, and mineral salts. Secretory structure were differentiated in to two based on location, namely external and internal secretory structure. External secretory structure comprises trichome, nectary or honey gland, hydathode and stigma, while internal secretory structure can consist of idioblast, secretory cavity, secretory duct and laticifer [10].

Secretory structure found in the two studied tonic plants, consist of idioblast cell and secretory cavity (Figure 1). *D. fruticosum* plant as included in the Myrtaceae family [11]. In *D. fruticosum* leaves were found secretory structure as idioblast cells and secretory cavities which are spread in palisade and sponge tisues (Figure 1).

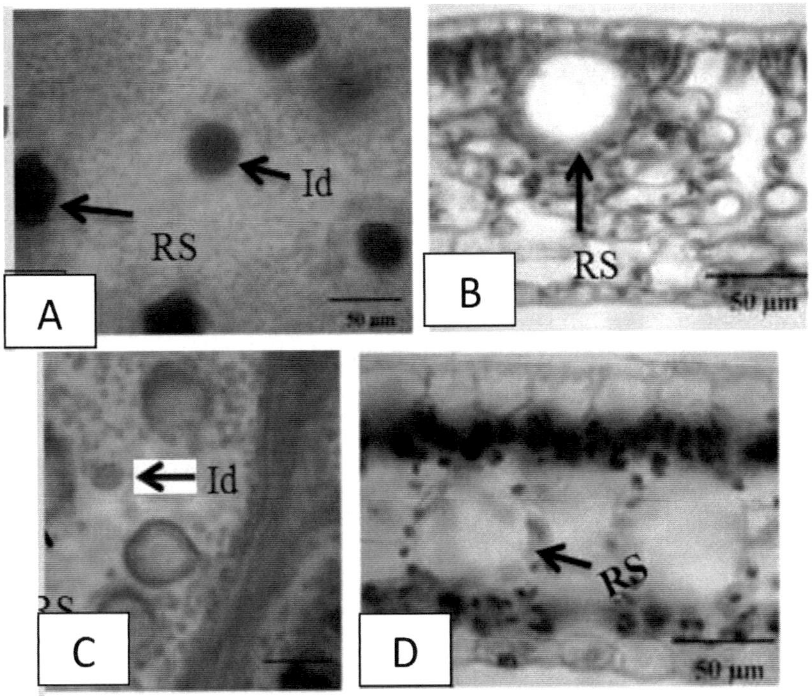

Figure 1. Secretory structure on tonic plant *D fruticosum* leaf (A, B) *P. rumphii* leaf (C, D), idioblast cell (Id), RS) secretory cavity.

D. fruticosum leaf idioblast cell was round with various measurements. The measurements of idioblast cell in palisade and sponge tissues are not different but the higher density was in sponge tissues (Table 2). Secretory cavity on this plant are round with epithel cell surround it. Secretory cavity measurement was bigger than idioblast cell do not the same as in the palisade and sponge tissue, while density was higher in the palisade tissue (Table 2). On the leaves of *Ugni molinae* Turcz as included in the Myrtaceae family there is also a secretory cavities. Metcalfe and Chalk [12] reported the availability of secretory cavity is a general characteristic of Myrtaceae family, even though the structure is also found in other families.

Table 2. Measurement and density of various tonic plant structure secretory.

Plant name	Organ	Secretory structure	Diameter measurement (µm)		Density (mm-2)	
			Palisade	Sponge	Palisade	Sponge
D. Fruticosum (Myrtaceae)	Leaf	Idioblas cell	26.3-36.1	31.6-39.6	16.8-18.6	20.4-33.4
		Secretory cavity	46.1-58.3	52.7-56.1	26.3-35.5	16.6-20.2
P. Rumphii (Annonaceae)	Leaf	Idioblas cell	21.4-25.4	17.7-26.1	20.4-33.4	25.5-48.0
		Secretory cavity	30.4-63.0	43.4-69.6	41.8-43.6	30.2-48.8

P. Rumphii plant which is the member of the Annonaceae family [13] has the secretory structure as idioblast cells spread in the whole mesophilic tissues beside there are also secretory cavity spread among palisade and sponge tissue (Figure 1). Secretory cavity size is bigger than idioblast cell. Idioblast cells round with the size and density that are not different between palisade and sponge tissue. Secretory cavity is round with the size and density which are really varies (Table 2).

3.2. Histochemistry Analysis

Idioblast cell *in D. fruticosum* contain lipophilic compounds and terpenoids, whereas the lipophilic compounds containing in the *P. rumphii.* Secretory cavity in *D. fruticosum* and *P. rumphii* leaves contain alkaloid compounds. On *Peganum harmala* leaves idioblast cell is specific cell which accumulate alkaloid compounds [14]. On *Mandevilla guanabarica* leaf (Apocynaceae) idioblast cell produces lipophilic compounds [15].

The result of *D. fruticosum* leaf histochemistry test showed the availability of alkaloids, terpenoids and lipophilic compounds (Table 3). Alkaloid compounds was detected in the secretory cavity, lipophilic and terpenoid compounds were found in the different idioblast cells (Figure 2). From several study results, besides producing alkaloid compounds, secretory cavity also produce other secretes, such as on *Ugni molinae* Turcz (Myrtaceae) produces lipophilic compounds [16], while in leaf of Myrrhinium atropurpureum Schott var. atropurpureum (Myrtaceae) produces lipophilic and terpenoid compounds [17]. *P. rumphii* leaf contains alkaloid compounds which was detected on the secretory cavity and lipophilic compounds was detected in the idioblast cell (Figure. 3)

Table 3. The results of histochemistry tests of tonic plants.

Secretory structure	Plant name	Test result		
		Alkaloids	Terpenoids	Lipophilic compounds
Idioblast cell	*D. fruticosum*	-	+	+
	P. rumphii	-	-	+
Secretory cavity	*D. fruticosum*	+	-	-
	P. rumphii	+	-	-

\+ = detected secondary metabolite compounds
\- = undetected secondary metabolite compounds

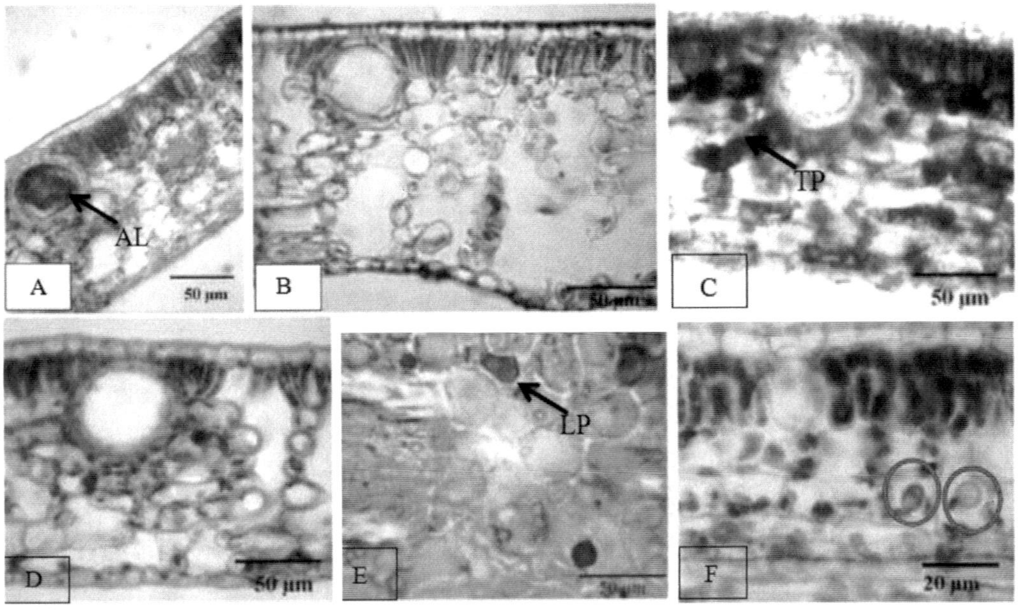

Figure 2. Histochemistry test result on *D fruticosum* leaf: compound (A) alkaloids control (B) terpenoids (C), terpenoids control (D), lipophilic compounds (E), lipophilic compounds control (F): Al) alkaloid compounds, TP) terpenoid compounds, LP) lipophilic compounds.

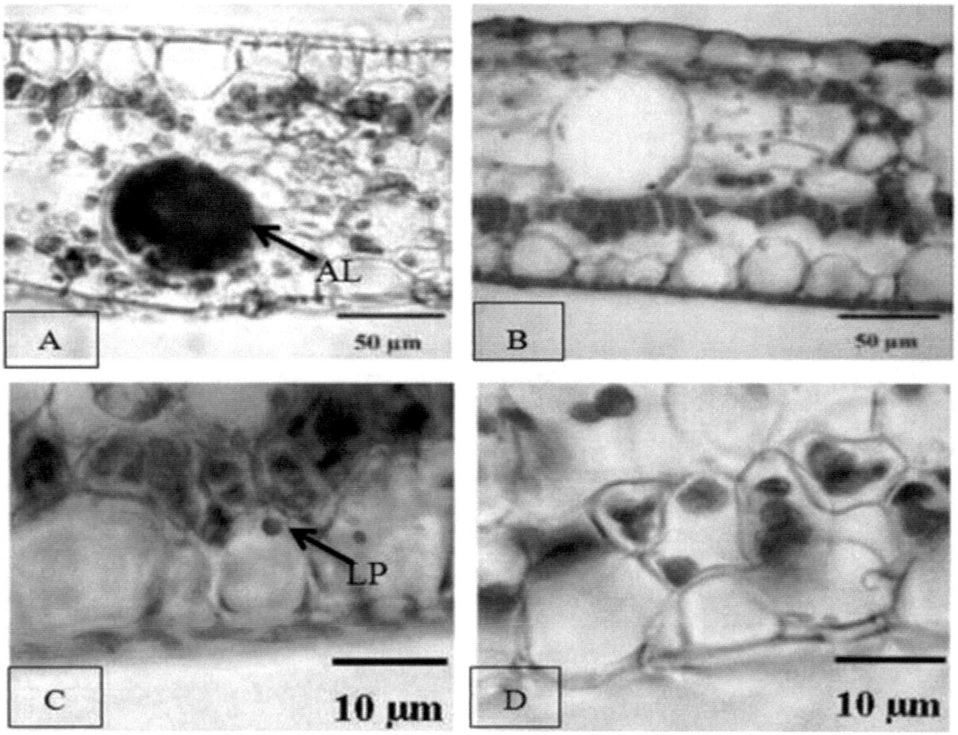

Figure 3. Histochemistry test result *P. rumphii* leaf: Alkaloid compounds (A), alkaloids control (B), lipophilic compounds (C), lipophilic compounds control (D):Al) alkaloid compounds, TP) terpenoid compounds, LP) lipophilic compounds.

3.3. Phytocemical Analysis

Qualitative phytochemical analysis result of *D.fruticosum* and *P. rumphii* leaf showed there phenolic, flavonoid, steroid and triterpene compounds, but alkaloid compounds was undetected (Table 4). Alkaloid compounds was undetected on phytocemistry test was shown by no precipitation occur when it was given Mayer reagent, Wagner reagent and Dragendrof reagent. Different from the phytochemical result, histochemistry result on *D.fruticosum* and *P. rumphii* showed there was alkaloid compounds. Alkaloid compounds was undetected on phytochemical test because the test was carried out against powder originated from the whole plant organ tissue, so that metabolic compounds concentration was low. Histochemistry test showed positive result because it was conducted against tissues obtained from organ slices so that the test was directly observed on the secretory structure.

The highest triterpenes content was found on *D. fruticosum* leaf. The highest steroids content was on the *P. rumphii* leaf. On the *D. fruticosum* leaf, triterpenes was the main component on metabolite compounds, while other compounds were available in low number (Table 4). According to Wagner [5] alkaloid, terpenoid, quinonic, phenol, polysaccharide, glycoprotein compounds are potential as tonic, as well as flavonoids group such as flavones and flavonols [18]. Phytochemical test result of *P. rumphii* showed the existence of phenol, flavonoid, and steroid compounds (Table 4). Paarakh and Khosa [19] reported that several species from genus *Polyalthia* also contain alkaloid compounds.

Table 4. Qualitative analysis result of phytochemistry compounds on tonic plants.

Plant	Test result				
	Alkaloids	Phenols	Flavonoids	Steroids	Triterpenes
D. fruticosum	-	+	+	++	++++
P. rumphii	-	++	++	+++	+++

Explanation:

- : no compounds
+ : a little bit
++ : moderate
+++ : a lot
++++ : very much

4. Conclusion

D. fruticosum and *P. rumphii* used has tonic have secretory structure in the form idioblast cell and secretory cavity. Idioblast cell on *D. fruticosum* leaves contain lipophilic and terpenoid compounds, while on *P. rumphii* only contain lipophilic compounds. In the secretorial cavity structure on *D. fruticosum and P. rumphii* contain alkaloid compounds. Phytochemical compounds content produced by *D. fruticosum* and *P. rumphii* are phenol, flavonoid, steroid and triterpene compounds.

References

[1] Suriawiria 2000 *Obat Mujarab dari Pekarangan Rumah*. (Jakarta (ID): Papas Sinar Sinarti)
[2] [Deptan] Departemen Kehutanan 2010 *Informasi Taman Nasional Bukit Duabelas*. (Jambi (ID): Balai Taman Nasional Bukit Duabelas Jambi)
[3] Sasmita K, Mulyani W, Priyantoro B, David, Marpaung JP and Algopeng Z 2011 *Pengenalan Tumbuhan Obat Taman Nasional Bukit Duabelas Jambi, Indonesia*. (Jambi (ID): Balai Taman Nasional Bukit Duabelas)
[4] Ramawat KG. 2009 *Herbal Drugs: Ethnomedicine to Modern Medicine* (Heidelberg (DE): Springer Verlag Press)
[5] Wagner H 1985 *Immunostimulants From Medicinal Plants* In: Chang HM, Yeung HW, Tso WW, and Koo A *Chinese Medicinal Materials Research* Singapore (SG):World Scientific Publisher
[6] Sass JE 1951 *Botanical Microtehnique* Iowa (US): Iowa State Coll Press
[7] Furr Y and Mahlberg PG 1981 *J. Nat. Prod.* **44(2)**153-159
[8] Boix YF, Victorio CP, Defaveri ACA, Arruda R do carmo de oliveirea, Sato A and Lage CLS 2011 *Plant Biosystems* **145(4)** 848–856
[9] Harborne JB 1987 *Metode Fitokimia Penuntun Cara Modern Menganalisis Tumbuhan*. Pasmawinata K and Soediro I, Penerjemah; Niksolihin S, editor (Bandung (ID): Penerbit ITB Press) Terjemahan dari: *Phytochemical Methods*. Ed ke 2.
[10] Dickison WC 2000 *Integrative Plant Anatomy*. 1st ed. (New York: Academic Press)
[11] Backer CA and Van Den Brink RCB 1963 *Flora of Java*. Volume ke-1 (Groningen: NVP Noordhoff)
[12] Metcalfe C and Chalk L 1979 *Anatomy of The Dicotyledons*. (Oxford (GB): Clarendon Press)
[13] Backer CA and Van Den Brink RCB 1968 *Flora of Java*. Volume ke-3 (Groningen: NVP Noordhoff)
[14] Khafagi IK. 2007 *Amer. J.Plant Physiol.* **2(1)** 17-26
[15] Cordeiro SZ, Sato A, do Carmo de Oliveira Arruda R and Simas NK 2012 *Biochem System and Ecol.* **45** 102– 107
[16] Retamales HA, Scherson R and Scharaschkin T 2014 *Rev. Chil. de Historia Nat.* **87(1)** 27

[17] Victorio CP, Moreira CP, Souzac M da Costa, Sato A and Arruda R do carmo de oliveira 2011 *NPC* **6(7)** 1045–1050
[18] Hollman, PCH, Hertog MGL and Katan MB 1996 *Food Chem.* **57(1)** 43-46
[19] Paarakh PM and Khosa RL 2009 *JP* **2(4)** 594-605

Study of Structural and Electrical Conductivity of Sugarcane Bagasse-Carbon with Hydrothermal Carbonization

M Kurniati[1], D Nurhayati[1] and A Maddu[1]

[1] Biophysics Division, Department of Physics, Faculty of Mathematics and Natural Sciences, Bogor Agricultural University, Bogor, Indonesia.

Email : mkurniati@apps.ipb.ac.id, dedehn49@gmail.com, maddu3@gmail.com

Abstract. The important part of fuel cell is the gas diffusion layer who made from carbon based material porous and conductive. The main goal of this research is to obtain carbon material from sugarcane bagasse with hydrothermal carbonization and chemical-physics activation. There were two step methods in this research. The first step was sample preparation which consisted of prepare the materials, hydrothermal carbonization and chemical-physics activation. The second one was analyze character of carbon using EDS, SEM, XRD, and LCR meter. The amount of carbon in sugarcane bagasse-carbon was about 85%-91.47% with pore morphology that already form. The degree of crystallinity of sugarcane bagasse carbon was about 13.06%-20.89%, leaving the remain as the amorphous phase. Electrical conductivity was about 5.36 x 10^{-2} Sm^{-1} - 1.11 Sm^{-1}. Sugarcane bagasse-carbon has porous characteristic with electrical conductivity property as semiconductor. Sugarcane bagasse-carbon with hydrothermal carbonization potentially can be used as based material for fuel cell if only time of hydrothermal carbonization hold is increased.

1. Introduction

Fuel cell is a system to converse chemical energy from hydrogen and oxygen directly into electrical energy. It has some advantages compared with another conventional electrical generator, such as high efficiency, does not make air pollution, and having flexible construction that can harness fuel which has been available. One of fuel cell which has been developed is Polymer Electrolyte Membrane Fuel Cell (PEMFC). It has some main components which have important function, one of them is gas diffusion layer. Gas diffusion layer has function to diffuse Hydrogen (anode) and Oxygen (cathode), as catalytic support and as a conductor of electron movement. Gas diffusion layer is commonly made from a porous and conductive carbon based material. This type of carbon is actually similar to that of carbon that commonly used as adsorbent[1].

One of agricultural waste of biomass which nowadays developed for carbon based material is sugarcane bagasse. Sugarcane bagasse is a waste material which has fibrous from rod of cane and been extracted from its roomie. Sugarcane bagasse consist of cellulose 37.65 % with carbon can be up to 24.7%. This high number put forward sugarcane bagasse as potential waste for carbon resource through carbonization.

Carbonization process is a conversion of organics compound into charcoal with no oxygen hence complex compounds inside the organic material separated into charcoal with high content of carbon [2]. One of carbonization method that has been concerned recently is hydrothermal carbonization with hydro charcoal as result. Hydrothermal carbonization is biomass conversion through thermos chemical process and use water on low temperature. Hydrothermal carbonization has advantage compared with pyrolysis carbonization, such as it is use low temperature hence the energy is low waste and air pollution can be reduced [3]. The purpose of this research is to know the structure of pore and conductivity material carbon from sugarcane bagasse with hydrothermal carbonization process use various temperature over 200 °C, 250 °C and 300 °C.

2. Method

Material used in this research was sugarcane bagasse with no roomie and very dry. Sugarcane bagasse was mashed up until became sugarcane bagasse powder with very soft condition to be moved into a digester tube for hydrothermal carbonization. Sugarcane bagasse and water was placed in tube with 15 % from volume of water. Water volume was 1/3 from digester tube. Digester tube is over 20 cm in length and 6.35 cm in diameter completed with electrical heater.

Sugarcane baggase with good shape (powder) moved into digester tube and mixed until it smooth. The material and aquades was put into digester at room temperature then closed tightly while heater turned on. The temperatures were set at 200 °C, 250 °C, and 300 °C with three times of repetition for getting an appropriate sample weight for being activated and tested in the next step. When the temperature is reached, heating was hold over six hours, afterwards the heating was ended. Charcoal from this hydrothermal carbonization (hydro-charcoal) was cleaned on paper filter to be separated between filtrate and residue. After hydrothermal carbonization, sugarcane bagasse carbon was activated with chemical and physical activation. Chemical activation was conducted using potassium hydroxide (KOH) with ratio of KOH: charcoal was set 3:1. For this, 2 grams of charcoal marinated in 26 ml aquades was reconstituted with 6 gram of KOH in 24 hours of marinated. After chemical activation, the product of charcoal been filtered for filtrate and residue separation. The residue was then heated at 60 °C for 24 hours. The charcoal product of chemical activation was then moved into atmospheric fixed bed reactor and been activated with temperature up to 700 °C. When it was reached, temperature was hold for 30 minutes.

The result of activation was then marinated with HCL 10 % for 1 hour and then cleaned up with hot water to remove impurities and neutralizing pH. Next, it was oven dried at 105 °C until reaching constant weight. The dried carbon material was analysed for the content of carbon, structure of atom, and conductivity use SEM-EDS, XRD, and LCR-meter.

3. Result and Discussion

Table 1 indicates the weight of the dregs of mashed sugarcane bagasse before hydrothermal carbonization and the formed charcoal following hydrothermal carbonization. The mass charcoal decreases with increase in temperature. This is because the volatile compounds contained in sugarcane bagasse evaporate. Increasing temperatures cause the more evaporated compounds so that the remaining charcoal mass decreases.

Table 1. Comparison of mass sugar cane husks before and after the hydrothermal carbonization.

Hydrothermal Carbonization Temperature (°C)	Sugarcane Bagasse Mass (grams)	Charcoal Mass (grams)	Missing Mass (grams)	Missing Mass Percentage (%)	Charcoal Mass Percentage (%)
200		4.02	7.38	64.74	35.26
250	11.4	3.53	7.87	69.04	30.96
300		2.56	8.84	77.54	22.46

After hydrothermal carbonization, the charcoal is activated in either physics or chemical activation. The data of comparison of charcoal mass before and after activation can be seen on Table 2. The average of charcoal mass is approximately 21.67 % from prior to the activation process. Based on these data, the mass loss indicates that carbon materials and volatile materials as well as ash materials degrade during the process.

Table 2. Comparison of charcoal mass before and after activation.

Hydrothermal Carbonization Temperature (°C)	Charcoal Mass (grams)		Missing Mass (gram)	Missing Mass Percentage (%)	Charcoal Mass Percentage (%)
	Before Activation	After Activation			
200	2	0.25	1.75	87.5	12.5
250	2	0.33	1.67	83.5	16.5
300	2	0.72	1.28	64	36

Based on the table, the smallest charcoal mass is 0.25 grams which was obtained from the charcoal of hydrothermal carbonization process at 200 °C (charcoal HC 200 °C). The next smallest one is charcoal HC 250 °C with 0.33 grams and the largest one charcoal HC 300 °C with 0.72 grams. The differences of obtained results show that there are still a volatile and ash material specifically inside charcoal HC 200 °C so that the loss mass is approximately 1.75 grams. It also works for charcoal HC 250 °C and HC 300 °C. After activation, along with the increase in temperature of carbonization thermal, the loss mass decreases. It is due to the slight remaining of volatile and ash material on charcoal HC 300 °C before activation.

3.1. Analysis of Carbon Content

The active carbon is a porous material with activated carbon content of 87 % - 97 %. Table 3 showed that the carbon content in sugarcane bagasse charcoal which is produced by hydrothermal carbonization has reached the value of 85.02 % - 91.47 %. The highest result is 91.47 % which is obtained from carbon charcoal hydrothermal carbonization results with temperature 200 °C (charcoal HC 200 °C). The next one are charcoal HC 250 °C with 88.39 % carbon content and charcoal HC 300 °C with 85.02% carbon content.

Table 3. The content of carbon in charcoal.

Temperature charcoal HC (°C)	Carbon content (%)
200	91.47 ± 9.74
250	88.39 ± 9.32
300	85.02 ± 8.99

Carbon content inside sugarcane bagasse charcoal seems decreasing along with the increase in hydrothermal carbonization temperature. As for the carbon content of charcoal HC 250 °C and 300 °C HC, the measured content carbon is only for the part that got shot X ray. Each part of charcoal has the different carbon content so we can assume that it could be the area with the less optimum content of carbons. This is also supported by data electrical conductivity which gets bigger as the increasing of hydrothermal carbonization temperature. One of the factors that affects the increased electrical conductivity is the carbon content, as the more carbon content will increase the electrical conductivity of the charcoal itself [4]. Therefor it cannot be concluded that the decreasing carbon content is along with the increasing temperature.

3.2. Analysis of the Structure of Charcoal

In this research, *Scanning Electron Microscope* (SEM) is used to observe the structure of the charcoal characters. It was used to know the morphology appearance of the sugar cane husks charcoal as the result of hydrothermal carbonization process. The character structure of the desired sugarcane bagasse charcoal is a porous structure with a diameter of 1-30 μm, because one of the carbon properties which is used in *gas diffusion layers* is porous carbon with a diameter of 1-30 μm.

Figure 1a, 1b, and 1c showed the results of morphology observations of the sugar cane husks charcoal with 1000 times magnification. Based on the results, the pore morphology on sugar cane husks charcoal has been formed with a diameter range 4-11 μm. Morphology of porous charcoal looks broken and irregular. Porous structure of coals HC 200 °C (Figure 1a) is more evident than charcoal HC 250 °C (Figure 1b) and charcoal HC 300 °C (Figure 1c).

(a) (b) (c)

Figure 1. Observations of the morphology, SEM magnification 1000 x (a) HC 200 °C (b) HC 250 °C and (c) HC 300 °C.

The porous of charcoal HC 250 °C and charcoal HC 300 °C appears more broken and irregular which is caused by the rising temperature of hydrothermal carbonization. It is also caused by stirring charcoal when leached with aquades to neutralize the pH. Charcoal HC 200 °C is the first charcoal leached by using aquades. At first, the process stirring is only done in the late minute when approaching a neutral pH. But for the charcoal HC 250 °C and 300 °C, stirring starts from beginning to end of the leaching. Charcoal HC 300 °C is the last leached charcoal by using aquades after previously soaked with HCl. Solution. Because of that matter, neutralizing pH charcoal takes a longer time. The neutralization time charcoal affects stirring. The more frequent stirring will damage porous structure of coals.

The next structure observation is done by using X-ray Diffraction. This observation is aimed to inform the change of the charcoal structure after passing stages of hydrothermal carbonization and activation as well as knowing the degree of crystallinity of coals. After the match with no. 22 Graphite JCPDS- 1069, X-ray diffraction pattern of sugar cane husks charcoal approached the X-ray diffraction patterns of graphite (Figure 2).

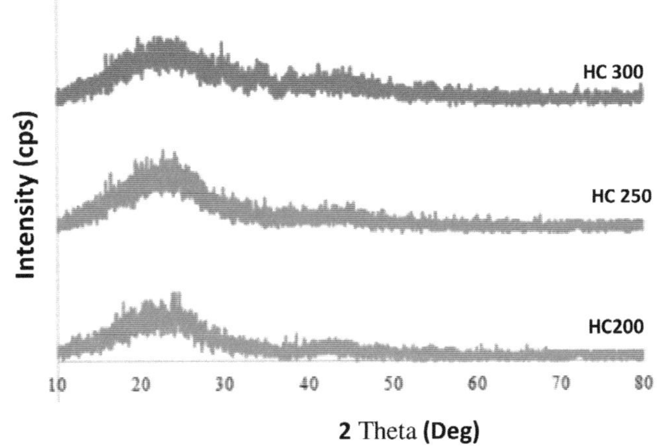

Figure 2. X-ray Diffraction pattern Charcoal HC 200 °C, 250 °C and 300 °C.

Figure 2 shows the X-ray diffraction patterns of charcoal cane dregs for the entire hydrothermal carbonization temperature variations. X-ray diffraction pattern of charcoal HC 200 °C, 250 °C and 300 °C indicates that carbon peak specific respectively appears at an angle 2Θ; 23.61°,23,61 ° and 23.93 °. The overall x-ray diffraction pattern shows the hill with a large area and the blunt end. It indicates that the crystal structure of charcoal tends to be amorphous.

The present of amorphous phase causes the low degree of crystallinity. The degree of crystallinity is ratio comparison of crystal phase against amorphous phase and phase crystal. It represents the regularity of crystal structures on a material. The degree of crystallinity of commercial carbon which can be used for the base material of gas diffusion layer components is 27.79 %. Table 4 shows that the highest degree of crystallinity is obtained from charcoal HC 200 °C with 20.89 % and the lowest one is on charcoal HC 300 °C with 13 %. The degree of crystallinity successively decreases along with the rising of hydrothermal carbonization temperature. It is influenced by stirring factor when leaching charcoal after activation. Leaching charcoal aims to rid the pollutant which sticks to the charcoal after soaked in acid neutralizer and activator. It is performed by using charcoal until it reaches a pH close to neutral pH. The needed time is various so because of that, it can affect speed of neutralization. The more frequent stirring will damage porous structure of coals so that the crystal structure of charcoal becomes more amorphous.

Table 4. Crystallinity index of charcoal.

The temperature of the charcoal HC (°C)	Crystallinity index (%)
200	20.89
250	18.31
300	13.06

3.3. Analysis of Electrical Conductivity

Electrical conductivity is a measure of a material ability in the conduct an electric current. Beside porous carbon structure condition, electrical conductivity of charcoal is also one of the important terms in the use of carbon for gas diffusion layer [5]. Electrical conductivity of charcoal is measured using LCRmeter. Electrical conductivity data can be seen in Table 3. The data is measured at 136 Hz frequency because with this frequency the pattern of electrical conductivity of charcoal is clearly visible. The electrical conductivity of sugar cane husks charcoal increases along with the rising of hydrothermal carbonization temperature. The electrical conductivity of the activated charcoal was from the lowest to highest HC 200 °C, HC 250 °C and 300 °C with values as shown in Table 5.

Table 5. Electrical conductivity of charcoal.

Temperature of the charcoal HC (°C)	Electrical conductivity (Sm^{-1})
200	5.36×10^{-2}
250	1.18×10^{-1}
300	1.11

From Table 5, we can conclude that electrical conductivity of charcoal cane dregs is categorized into semiconductors. It is because the range of electrical conductivity of semiconductor material is 10^{-8} Sm^{-1} - 10^{3} Sm^{-1}, while the electrical conductivity of a conductor material is more than 10^{3} Sm^{-1}. This value is very low in comparison with the graphite which has 3.4×10^{3} Sm^{-1} in value of the electrical conductivity.

References

[1] Han M, Xu JH, Chan SH, Jiang SP 2008 *Electrochimica Acta* **53** 5361-5367
[2] Kalderis D, Bethanis S, Paraskeva P, Diamadopoulos 2008 *Bioresc technol* **99** 6809-6816
[3] Kannan AM, Kanagala P, Veedu V 2009 *Journal of Power Sources* **192** 297-303
[4] Nishimiya K, Hata Ishihara S 1995 *Wood Research* **82** 34-36
[5] Mochidzuki K, Soutric F, Tadokoro, Antal MJ 2003 *Industrial & Engineering Chemistry Research* **42(21)** 5140-5151

Detection of vitamin b$_1$ (thiamine) using modified carbon paste electrodes with polypyrrole

N M Mupparíqoh[1], W T Wahyuni[1,2] and B R Putra[1]

[1] Chemistry Department, Bogor Agricultural University, Bogor, Indonesia
[2] Pusat Studi Biofarmaka Tropika, Lembaga Penelitian dan Pengabdian kepada Masyarakat, Bogor, Indonesia

Abstract. Vitamin B_1 (thiamine) is oxidized in alkaline medium and can be detected by cyclic voltammetry technique using carbon paste electrode (CPE) as a working electrode. polypyrrole-modified CPE were used in this study to increase sensitivity and selectivity measurement of thiamine. Molecularly imprinted polymers (MIP) of the modified CPE was prepared through electrodeposition of pyrrole. Measurement of thiamine performed in KCl 0.05 M (pH 10, tris buffer) using CPE and the modified CPE gave an optimum condition anodic current of thiamine at 0.3 V, potential range (-1.6-1 V), and scan rate of 100 mV/s. Measurement of thiamine using polypyrrole modified CPE (CPE-MIPpy) showed better result than CPE itself with detection limit of 6.9×10^{-5} M and quantitation limit 2.1×10^{-4} M. CPE-MIPpy is selective to vita min B_1. In conclusion, CPE-MIPpy as a working electrode showed better performance of thiamine measurement than that of CPE.

1. Introduction

Vitamin B_1 or thiamine (3-[(4-amino-2-methylpyrimidin-5-yl)methyl]-5-(2hydroxyethyl)-4_methylthiazol-3-ium chloride) (figure 1) have a role in the development of brain and neurons [1]. There are at least 30 g Stock of vitamin B_1 in human body [2]. Lack of vit B1 could cause beriberi deseas, which could deffect the neuron system. It could be prevented by consuming food containing vit B_1 from any resource or suplement.

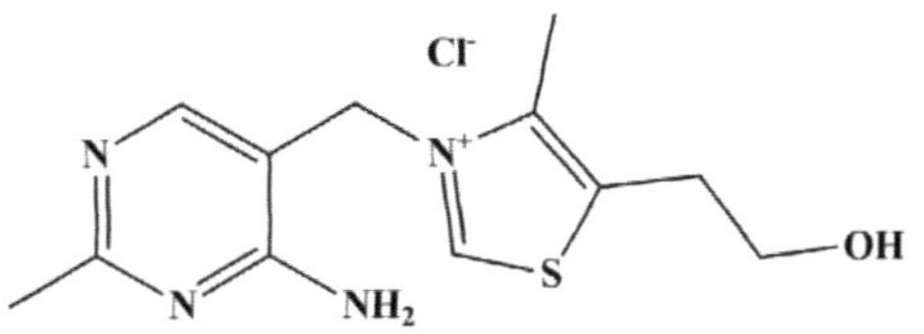

Figure 1. Structure of Vitamin B_1.

The importance of vitamin B_1 roles in human body, lead to development of methods to detect vit B_1 in foods or certain suplements. Some of the developed methods including high performance liquid chromatography [1], chemiluminesence [3], dan fluoresence [4]. The eminence of those methods are high accuracy and coefficient of determination, also low detection limit. However, those methods also have some flaws including complicated sample preparation, requiring high tech instrument, and costly. Those problems could be avoided with electrochemical method, particulary voltammetry. The advantages of these technique including sensitivity, and the ability to produce interpretable data even in a low concentration level, fast, easy, and low cost anlysis [5,6]. Vitamin B_1 could be detected electrochemically due to oxidizable character into thiochrome in basic [6].

Carbon paste electrode (CPE) producing low current and selectivity, thus need to be modified to upgrade the performance. CPE Modifier used in this study is moleculary imprinted polymers (MIP) polypyrola. The previlege of using MIP polypyrola has been proved by Koirala et al. [7]. MIP polypyrola could be electropolymerized in various material with electro conductivity and give selectivity measurement. Molecules print formed as the result of interaction between polymers and molecule filters that makes MIP plypyrola selective. Pyrrole coating on CPE used electropolymerization method with eminence including, ability to coat narrow and uneven surface of CPE and thickness of polypyrrole coat could be adjusted with variation of time in polymerization [8]. This study intended to create CPE mofied by MIP polypyrrole (CPE-MIPpy) that could detect vitamin B_1 using voltammetry method. Modification expected to increase sensitivity and selectivity of CPE compared to the unmodified CPE.

2. Experimental

2.1 Materials

The materials needed are graphite, paraffin, standard vitamin B_1-HCl (Himedia), KCl, $K_3Fe(CN)_6$, $CoCl_2.6H_2O$, pyrrole (Sigma Aldrich), NaOH, H_2SO_4, HCl, KOH, H_3PO_4, H_3BO_3, CH_3COOH, CH_3OH, N_2, tris buffer, copper wire, waxed paper, Pt electrode, an electrode Ag / AgCl, deionized water, filter paper, and distilled water.

2.2 Apparatus

The tools used are glass tools common in laboratories, potentiostat EDAQ, Sonicator 42 Hz (As One), analytical balance (Sartorius), pH meter (Hanna Instruments), scanning electron microscope JEOL JSM-6360LA, glass tube diameter 2.5 mm, mortar and pestle, and software E-CHEM, Origin Pro v 2.1.0 and 7.0.

2.3 Preparation of carbon paste electrode

Graphite and paraffin mixed with a ratio of 7: 3 (w/w). The mixture was homogenized using a sonicator for 15 minutes, then crushed and compacted using a mortar and pestle for 30 minutes to form a paste. Then the mixture was put in a 2.5 mm diameter glass tube that has been inserted copper until the remaining vacant space of about 3 mm from the tip of the tube and the tube base glued. Once solid, the electrode surface is flattened by means of unidirectional rubbed oil paper. Rubbing suspended when the black color of the carbon is no longer attached to the waxed paper. Carbon paste electrodes can be measured after being stored ± 2 days at room temperature.

2.4 Determination of the measurement conditions

The determination of the optimum range potential for measuring vitamin B_1 uses a variation of the range potential -1.6-1 V each 0.2 V. While the hose scan rate varied from 50 to 250 mV/s each multiple of 50 mV / s. For the determination of pH, vitamin B_1 in KCl solution mixed with a solution of NaOH 0.1 M Tris buffer pH 10 and pH 10 separately. Each mixture scaning on the range potential and optimum scan rate.

2.5 Preparation of polypyrrole modified CPE (CPE-MIPpy)

MIP polipyrrole made through electropolymerization using the potentiostat. The working electrode that has been made is connected with a potentiostat along with Ag/AgCl as a reference electrode and Pt as counter electrode. The third electrode is dipped into a solution of vitamin B_1 0.01 M pyrrole monomer and 0.1 M in Britton-Robinson buffer of pH 3, and then do electrodeposition at a potential of 0.9 V for 180 seconds. After electropolymerization completed, a layer that is formed is washed and soaked with deionized water for 24 hours. Deionized water used to extract vitamin B_1 that is embedded in the film. The electrode is hereinafter referred CPE-MIPpy.

2.6 Analytical procedure
Techniques used for the voltammetric determination of vitamin B_1 are cyclic voltammetry (CV). The measurements were performed in electrolyte (KCl) with tris buffer (pH 10) at laboratory temperature. The 0.001M stock solution was prepared by diluting vitamin B1 in deionized water. The calibration curves were measured in triplicate and their statistical parameters (e.g., slope, intercept, correlation coefficient, and limit of detection) were calculated. The detection limits were calculated as the concentration of an analyte using 3s/m where s is the standard deviation of intercept and m is slope.

3. Result and Discussion

3.1 Optimization conditions for the determination of vitamin B_1

3.1.1 *Electrolyte*. One of the factors that need to be considered when measurements voltammetry is an electrolyte. Electrolyte serves as an electron transfer medium so that electrons move to the electrode surface and unreadable as current. In this study KCl selected as an electrolyte because it does not provide background currents that influence the reaction of vitamin B_1. KCl has a wide potential range on the CPE [9], the redox reaction between K^+ and Cl occur in a very positive potential, that is 2.93 V and 1.36 V [10].

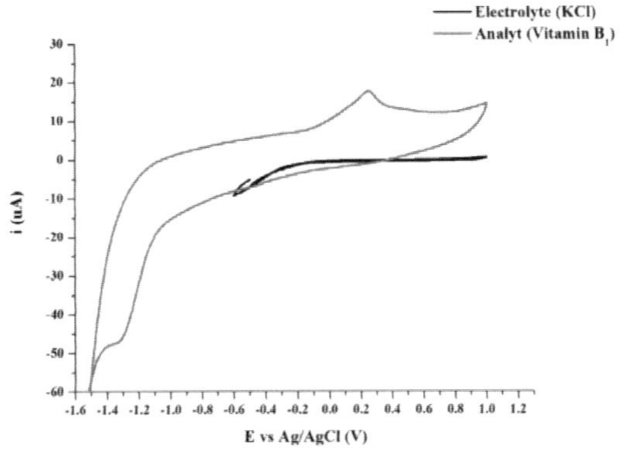

Figure 2. Cyclic voltammogram of vitamin B1 5 mM in KCl 50 mM and tris buffer pH 10. Scan rate 100 mV/s.

KCl as electrolyte concentration varied at 3, which is 50, 100, and 500 mM. The results showed that the concentration electrolyte of 50 mM KCl most stable because background current low value. At a concentration of 100 mM KCl, and 500 mM background currents higher, probably derived from the non-Faraday current form of charging current. Figure 2 shows the cyclic voltammogram (CV) of vitamin B_1 5 mM in 50 mM KCl. It can be observed that the peak of vitamin B_1 (0.3 V) is not disturbed by the presence of background current of 50 mM KCl electrolyte.

3.1.2 Optimun pH. Oxidation of vitamin B_1 to tiokrom (Figure 3) is influenced by the pH of the electrolyte solution. According to Oni et al. [6]. Vitamin B1 can be oxidized to tiokrom in alkaline medium (pH 8-10). In this study measured vitamin B_1 measured in KCl electrolyte and electrolyte KCl pH conditioned using NaOH and Tris. The results indicate that there is a difference of current and peak potential of oxidation of vitamin B_1. Response vitamin B_1 in the KCl and KCl-tris buffer pH 10 have the same value of oxidation potential (0.25 V), whereas the oxidation potential of vitamin B_1 in 0.1 M KCl-NaOH pH 10 was observed at 0.72 V (Figure 4). Oxidation potential shifts toward more positive potential indicates the oxidation reaction occurs more difficult.

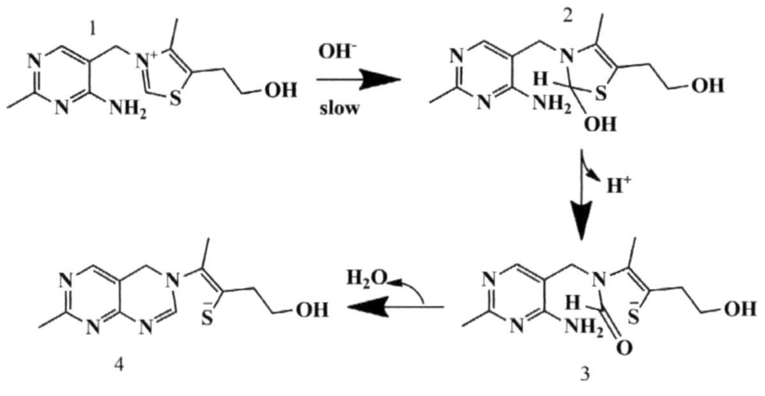

Figure 3. The mechanism of the oxidation reaction of vitamin B_1 to tiokrom.

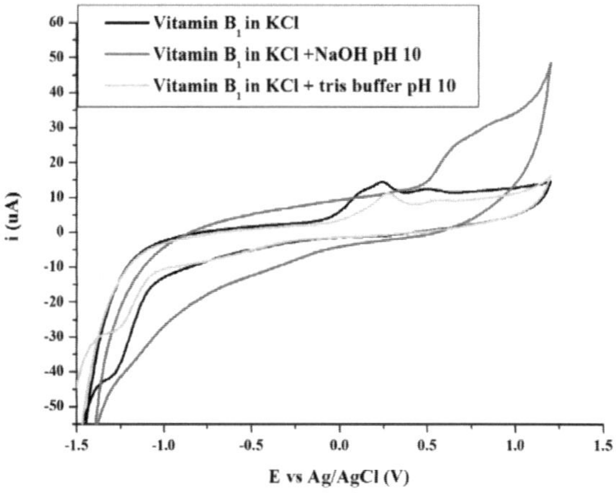

Figure 4. Cyclic voltammogram of vitamin B1 5 mM in KCl 50 mM, adding NaOH pH 10 dan tris buffer pH 10. scan rate 100 mV/s.

Vitamin B_1 in KCl give the current 1.6 times greater compared with KCl-tris buffer pH 10. However, in repeat measurements made, the intensity of the oxidation peak current inconsistent (data not shown). Based on these results, KCl-tris buffer pH 10 is used as the electrolyte in the next measurement.

3.1.3 *Potential Range*. Potential window illustrates the potential range of the redox reaction of the analyte. Good potential range can describe analyte peak clearly without interference peak of the redox electrolyte. Figure 5 shows the cyclic voltammogram vitamin B_1 measured at various potential window. In -1.2-1 potential -1-1 V and V have not seen the peak oxidation of vitamin B_1. Oxidation of vitamin B_1 peak of the new look as a potential window widened from oxidation peak -1.4-1 V. Vitamin B_1 is also observed during use a potential window -1.6-1 V. Range potential selected for subsequent measurement is -1.6-1 V, because it generates a current the highest (17.9 uA at a potential of 0.3 V).

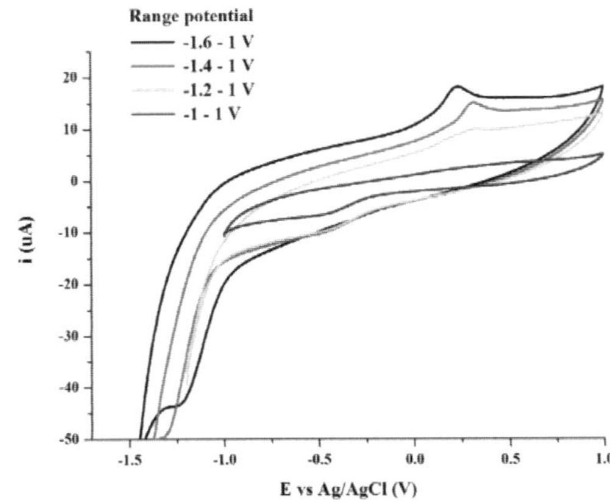

Figure 5. Cyclic voltammogram of vitamin B_1 5 mM in KCl 50 mM with tris buffer pH 10. Scan rate 100 mV/s on a variety of potential range.

3.1.4 *Scan Rate*. Scan rate effect on the peak intensity of the oxidation of vitamin B_1. Selection of the optimum speed Payar need to notice the easy nature or absence of vitamin B_1 to undergo oxidation reactions and oxidation of high peak currents generated. Increased scan rate current proportional to the intensity of oxidation of vitamin B_1 (Figure 6). scan selected speed is 100 mV / s. It is considering the speed of the oxidation reaction of vitamin B_1 to tiokrom which is slow. At higher scan rate, feared oxidation of vitamin B_1 to tiokrom can not happen perfectly.

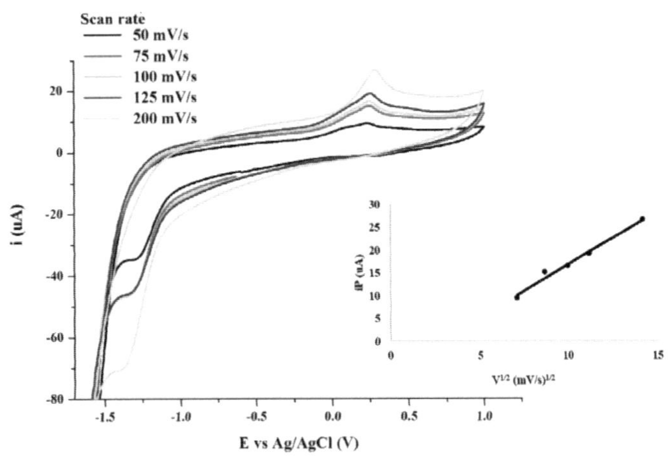

Figure 6. Cyclic voltammogram of vitamin B1 5 mM in KCl 50 mM with tris buffer pH 10 potential range –1.6–1 V on a variety of scan rate. Inset: scan rate root relationship between oxidation current peak curve

Based on the Randles-Sevcik equation, if the peak current (Ip) is proportional to the square root scan rate $V^{1/2}$ is proportional, then the charge transfer that occurs from the solution to the surface of the electrode under the influence of diffusion. Meanwhile, if the relationship between the peak current (Ip) and scan velocity V is proportional to the charge transfer that occurs from the solution to the surface of the electrode under the influence of adsorption [11]. The peak current generated by vitamin B_1 is directly proportional to the square root of the speed scan given to the value of $R^2 = 0.9832$ (inset in Figure 6). This indicates that the charge transfer is under the influence of diffusion. Diffusion occurs from the bulk solution to the electrode surface due to the concentration gradient and flux increases on the electrode with increasing scan rate [12].

3.2 CPE modified polymers molecularly imprinted polypyrrole (CPE-MIPPy)

Polypyrrole selected as an CPE modifier because it has good electrical conductivity and can be synthesized electrochemically on the surface of CPE through electropolymerysation [7]. Vitamin B_1 is mixed with pyrrole monomer before electropolymerysation process that vitamin B_1 is at polipirola matrix. Trapped vitamin B_1 in polipirola occur through hydrogen interaction between the N-H group at the pyrrole with O-H and N-H group in vitamin B_1. Following that, vitamin B_1 is extracted with a porogenic solvent in order to obtain moleculary imprinted polymer (MIP) on the surface of CPE. MIPpy existence on the CPE's surface is expected to improve the selectivity of CPE on the measurement of vitamin B_1. MIPpy role as mold that can only be passed by a molecule of vitamin B_1. So only the vitamin B_1 can reach the electrode surface and produce a response in the form of oxidation current.

3.2.1 Oxidation of vitamin B_1 at CPE-MIPPy. Cyclic voltammograms of vitamin B_1 which measured by CPE and CPE-MIPpy. show the increase of peak oxidation intensity of vitamin B_1 of 1.5 when measured with CPE-MIPpy (Figure 7). The process of electron transfer on the surface of CPE-MIPpy more easily occur because the electrical conductivity character owned by polypyrrole.

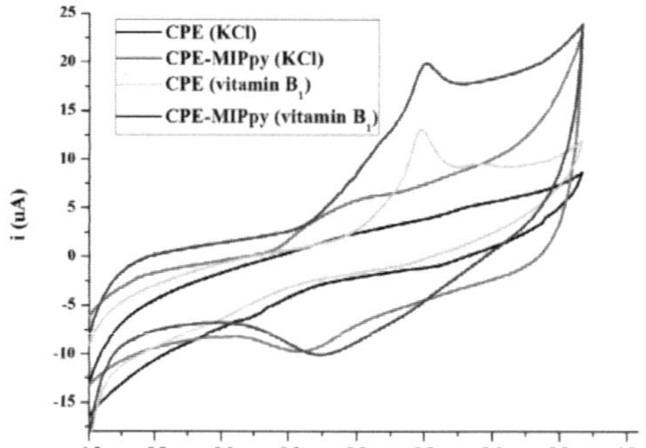

Figure 7. Cyclic voltammogram of vitamin B_1 5 mM in KCl 50 mM with tris buffer pH 10 at CPE-MIPpy. Scan rate 100 mV/s with bubbling with N2(g).

3.2.2 *Selectivity CPE-MIPpy on measurement vitamin B_1.* The selectivity indicates the ability of a method to distinguish the analyte from the matrix or other nuisance. Selectivity CPE-MIPpy on the measurement of vitamin B_1 was evaluated using vitamin B_6 (pyridoxine) as a bully. CPE-MIPpy can detect vitamin B_1 and vitamin B_6 simultaneously (Figure 8). Peak oxidation of vitamin B_6 can be detected because vitamin B_6 molecules can pass through the mold vitamin B_1 and reaches the surface of the CPE. Escape of vitamin B_6 possible because the molecule size smaller than vitamin B_1. Even

 IOP Publishing
doi:10.1088/1755-1315/58/1/012050

CPE-MIPpy can be bypassed by vitamin B_6, vitamin B_1 measurement is not disturbed due to oxidation peak vitamin B_1 and vitamin B_6 are at a different potential so that its peak can be distinguished. This opens the opportunity to do the simultaneous measurement of vitamin B_1 and vitamin B_6 use CPE-MIPpy.

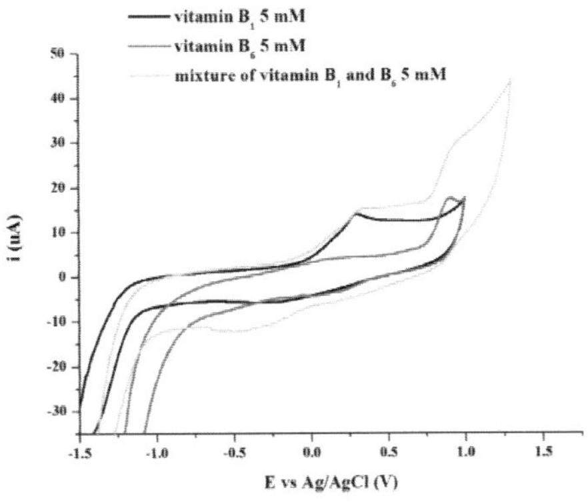

Figure 8. Cyclic voltammogram of vitamin B1 5 mM, Vitamin B6 5 Mm, and mixture at CPE-MIPpy. Scan rate 100 mV/s and bubbling with $N_{2\,(g)}$.

3.3 Surface morphology of CPE and CPE modified polipyrrole

CPE surface morphology, CPE-MIPpy, and CPE-py observed using SEM. CPE has the smoothest surface structure while the CPE-MIPpy among the most rugged and irregular. Irregular structure of the CPE-MIPpy formed from the interaction of vitamin B1 with polipyrole when elektropolimerisasi then vitamin B1 is extracted, leaving a mold. CPE-py more irregular than CPE-MIPpy as contained in the surface just polipyrrole.

Figure 9. The surface morphology CPE (above), CPE-MIPpy (bottom left) and CPE-py (bottom right) were characterized using SEM with a magnification of 5.000 times.

3.4 Performance evaluation of CPE and CPE-MIPpy on measurement vitamin B_1

3.4.1 *Linearity*. Linearity showed the relation between current response to variation of concentration at optimum measurement conditions. Linearity CPE and CPE-MIPpy measured by variations concentration of vitamin B_1 0.256-10 mM with 6 times repetition. The results showed that higher concentration of vitamin B_1 oxidation, increasing current intensity. This indicates that more molecules of vitamin B_1 are oxidized (Figure 10). Linearity is evaluated based on the value of the coefficient of determination (R^2) obtained from concentration curve with peak current of oxidation. The linear regression equation was obtained for measurement of vitamin B_1 with CPE and CPE-MIPpy has ipa (uA) = 1336.9 x + 1.66 (R^2 = 0.9866) and ipa (uA) = 2805.8 x + 3:40 (R^2 = 0.9900). Relationship between concentration and oxidation current intensity of vitamin B_1 in measurements with CPE-MIPpy more linear than with CPE. Besides the slope value measurement of vitamin B_1 with CPE-MIPpy showed higher sensitivity CPE-MIPpy on the measurement of vitamin B_1 is better than CPE.

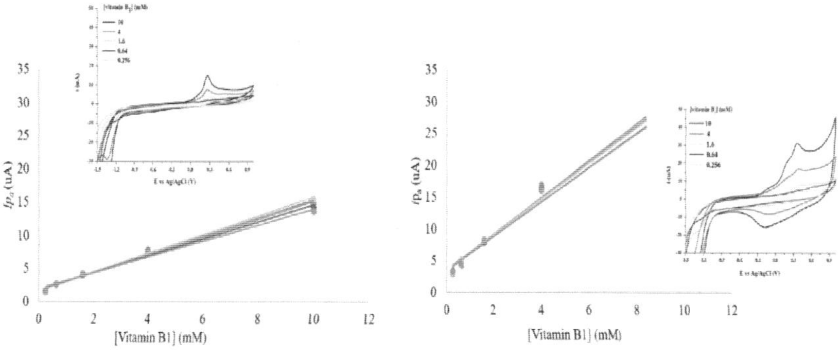

Figure 10. Curve relationship between the concentration of vitamin B_1 and vitamin B1 oxidation peak currents. Inset: cyclic voltammograms of various concentrations of vitamin B_1.

3.4.2 *Limit of detection and limit of quantitation.* Limit of detection (LOD) indicates the lowest concentration of an analyte that can be detected by the instruments or methode [14], in this case carbon paste electrodes, but not for quantitation. LOD for CPE-MIPpy of 6.9 x 10^{-5} M. This value is 9.7 more lower than the CPE, meaning the sensitivity of CPE-MIPpy in the measurement of vitamin B_1 is better than the CPE. Limit of quantitation (LOQ) indicates the lowest concentration of an analyte that can be determined by a method on the level of good precision and good accuracy. LOQ CPE-MIPpy on the measurement of vitamin B_1 11.42 more lower than CPE. These results indicate that the CPE-MIPpy can be used well for measurement of vitamin B_1 at concentrations low enough.

3.4.3 *Precision.* Precision indicates the value of measurement accuracy based on the percent relative standard deviation (%RSD). The smaller %RSD increasingly rigorous methods / techniques used. Precision measured at concentrations of 5 series and each performed 6 repetitions. %SBR to CPE and CPE-MIPpy row by 3.68% and 3.90%, which means cyclic voltammetry technique has good accuracy for CPE and CPE-MIPpy.

3.4.4 *Stability and Reproducibility.* The stability test is required to determine the consistency of the response from the working electrode. Stability and reproducibility were evaluated based on %RSD from current and potential of oxidation peak vitamin B1. The stability oxidation peak of CPE better than CPE-MIPpy. CPE has stable current response to the measurement until day 7, while the current response of CPE-MIPpy decreased by 32% after using for 2 days (Figure 11). Stability oxidation potential of CPE-MIPpy lower than CPE, but still within the limits of tolerance with the value of %RSD is less than 5% [13].

Reproducibility is required to evaluate the uniformity of CPE and CPE-MIPpy. Reproducibility was evaluated based on %RSD current and oxidation peak potential of vitamin B1. Reproducibility of oxidation peak current of CPE-MIPpy better than the CPE because the current of three electrodes is uniform by 21 ± 0.7 uA, whereas CPE has fluctuate currents by 20 ± 6.3 uA (Figure 12).

Repeatability potential oxidation peak of CPE is bad with the value of %RSD is 5.40%. Less uniformity between the CPE to the others CPE for the manufacture are still conventional. Reproducibility potential oxidaton of CPE-MIPpy is very good because the polymerization of pyrrole on the surface of CPE done using the potentiostat. Overall, CPE-MIPpy has the sensitivity and reproducibility is more higher but still less stable than the CPE.

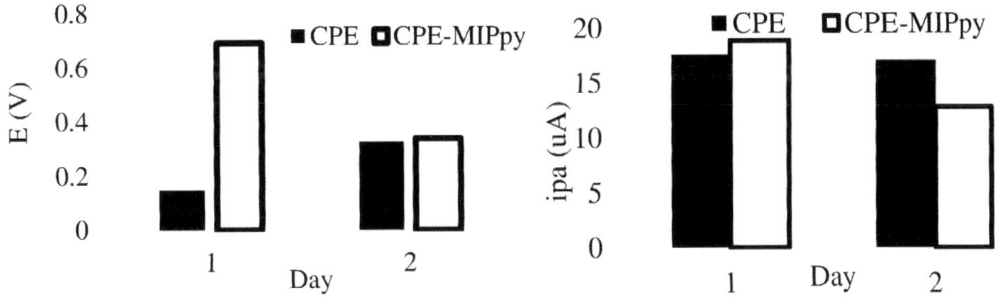

Figure 11. Stability current measurement (right) and potential (left) uses vitamin B_1 and CPE CPE MIPpy.

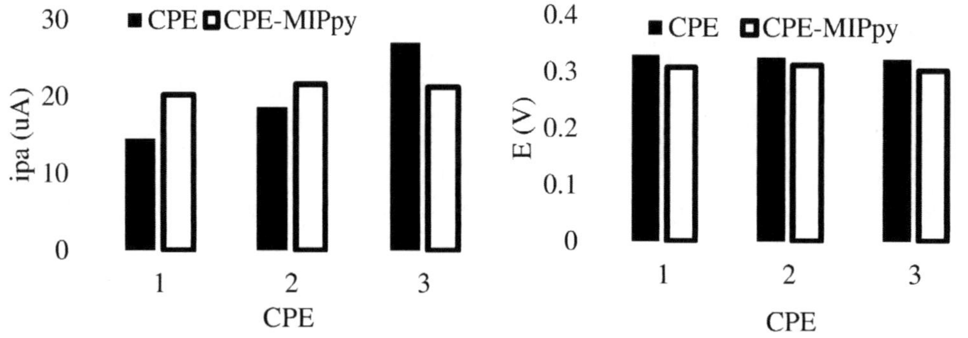

Figure 12. reproducibility of the current measurement (left) and potential (right) uses vitamin B1 CPE and CPE-MIPpy.

Table 1. Regression data of the calibration line for quantitative determination of vitamin B1 using Cyclic voltammetry.

Parameters	**CPE**	**CPE-MIPpy**
Measured peak potential (V)	0.32	0.32
Linearity range (M)	2.56 × 10-4 - 1 × 10-2	
Slope (uA/M)	1 336.9	2 805.8
Intercept (uA)	1.66	3.4
Determination coefficient (R^2)	0.9866	0.99
Limit of detection (M)	6.7×10^{-4}	6.9×10^{-5}
Limit of quantification (M)	2.4×10^{-3}	2.1×10^{-4}
Stability of peak current (%RSD)	1.64	8.31
Stability of peak potential (%RSD)	1.41	2.10
Reproducibility of peak current (%RSD)	1.89	1.56
Reproducibility of peak potential (%RSD)	5.41	0.26

4. Conclusion

Vitamin B_1 can be detected using methods voltammetry with carbon paste electrode as the working electrode scanning at of -1.6-1 V, scan rate at 100 mV/s and the addition of tris buffer pH 10 signal is generated in the form of oxidation peak currents that are potentially ± 0.32 V. measurement of vitamin B_1 with the modified polipirola CPE (CPE-MIPpy) showed better results than the CPE. CPE-MIPpy selective measurement of vitamin B_1. CPE-MIPpy as the working electrode shows better performance for the measurement of vitamin B_1 from the CPE.

References

[1] Suh JH, Junghyun K, Juhee J, Kyunghyun K, Suel GL, Hyun-Deok C, Yura J, Sang Beom H 2013 *Bull Korean Chem Sos.* **34** 1745-1750
[2] Jain A, Mehta R, Al-Ani M, Hill JA, Winchester DE 2015 *J. Cardfail.* **21** 1000-1007
[3] Ruiz TP, Lozano CM, Martinez MDG 2009 *JPBA* **50** 315-319
[4] Purbia R dan Paria S 2016 *Biosensor and Bioelctronic* **79** 467-475
[5] Brahman PK, Dar RH, Pitre KS 2012 *AJC* **016** 1-8
[6] Oni J, Philippe W, Tebello N 2002 *J. Electroanal.* **14** 1165-1168
[7] Koirala K, Sevilla FB, Santos JH 2015 *J.SNB*

[8] Panasyuk T, Dall'Orto VC, Marrazza G, El'skaya A, Piletsky S, Rezzano I, Mascini M 1998 *J. Analitycal L.* **31** 1809-1824
[9] Skoog DA, Holler FJ, Crouch SR. 2007. *Principles of Instrumental Analysis sixth edition.* Canada (CD): Thomson books/cole
[10] Chang R. 2003. *General Chemistry: The Essential Concepts Third Edition.* New York (US): The McGraw-Hill Companies
[11] Scholz. 2010. *Electroanalytical Methods: Guide to Experiments and Applications.* Ed ke-2. Berlin (DE): Springer
[13] Wang J. 2000. *Analytical Electrochemistry second edition.* New York (US): Wiley-VCH.
[14] [ICH] International Conference on Harmonisation. 2005. *Validation of Analytical Procedures*

IOP Conf. Series: Earth and Environmental Science **58** (2017) 012051 doi:10.1088/1755-1315/58/1/012051

A Statistical Bias Correction Tool for Generating Climate Change Scenarios in Indonesia based on CMIP5 Datasets

A Faqih

Department of Geophysics and Meteorology, Faculty of Mathematics and Natural Sciences, Bogor Agricultural University (IPB). IPB Dramaga Campus, FMIPA Building, Wing 19, Level 4, Jl. Raya Dramaga, Jawa Barat, Indonesia 16680. Phone/Fax: +62-251-8623850.

Email: akhmadfa@apps.ipb.ac.id

Abstract. Providing information regarding future climate scenarios is very important in climate change study. The climate scenario can be used as basic information to support adaptation and mitigation studies. In order to deliver future climate scenarios over specific region, baseline and projection data from the outputs of global climate models (GCM) is needed. However, due to its coarse resolution, the data have to be downscaled and bias corrected in order to get scenario data with better spatial resolution that match the characteristics of the observed data. Generating this downscaled data is mostly difficult for scientist who do not have specific background, experience and skill in dealing with the complex data from the GCM outputs. In this regards, it is necessary to develop a tool that can be used to simplify the downscaling processes in order to help scientist, especially in Indonesia, for generating future climate scenario data that can be used for their climate change-related studies. In this paper, we introduce a tool called as "Statistical Bias Correction for Climate Scenarios (SiBiaS)". The tool is specially designed to facilitate the use of CMIP5 GCM data outputs and process their statistical bias corrections relative to the reference data from observations. It is prepared for supporting capacity building in climate modeling in Indonesia as part of the Indonesia 3rd National Communication (TNC) project activities.

1. Introduction

Developing climate change scenarios require baseline and projection data from the outputs of global climate models (GCM). The data are gridded and usually stored in specific file, such as netCDF or Grib, where not many people familiar with the format. This can be a considerable constrain for the researcher who would like to use the data for their climate change related studies but do not have enough experience or skill in dealing with the data. The outputs of GCMs from the Climate Model Inter-comparison Project Phase 5 (CMIP5) are the main data prepared for climate change studies, particularly for the assessment of future climate scenarios. The Intergovernmental Panel on Climate Change (IPCC) used these datasets in the Fifth Assessment Report (AR5) published in 2013 [1]. The data are not only complex in term of their huge file size and uncommon file format, but they also commonly contain biases. These biases need to be statistically corrected before we can use them for further climate-change related studies. Statistical bias correction is one of many methods that can be selected to perform statistical downscaling of GCM outputs.

Processing climate data in order to generate climate scenarios for a selected studied region are complicated and require special skills. Therefore, it is necessary to develop a tool that can be used to simplify the downscaling processes, starting from reading the complex data, doing interpolation, performing bias correction and preparing downscaling results that can be easily used for further analysis. In this study, we develop a tool or software that can be used for the preparation of climate change scenarios in Indonesia based on the outputs of CMIP5 GCMs. The tool called as "Statistical Bias Correction for Climate Scenarios (SiBiaS)", and now has been developed to version 1.1. The tool is prepared for supporting capacity building in climate modeling in Indonesia as part of the Indonesia 3rd National Communication (TNC) project activities.

This paper aims to provide general description of SiBiaS as the tool for generating downscaled future climate scenarios in Indonesia. It also describes potential uses of the tool in supporting climate change vulnerability and impact studies over different domain location in Indonesia.

2. Methodology

2.1. Data

SiBiaS is specially designed to facilitate the use of CMIP5 GCM data outputs and process their statistical bias corrections relative to the reference data from observations. There are a number of GCMs where their data outputs can be taken from the Climate Models Inter-comparison Project version 5 (CMIP5) database. The data can be obtained directly from the PCMDI/ESGF data portal (http://www-pcmdi.llnl.gov or http://pcmdi9.llnl.gov/esgf-web-fe/), or from other sources. Especially for SiBiaS v1.1, we downloaded the CMIP5 data from a more user-friendly website called as the KNMI climate explorer (http://climexp.knmi.nl). There are 24 models selected and stored for the SiBiaS database. The list of the models where their output used in the software is shown in Table 1.

Table 1. Name of GCMs and their references used in SiBiaS v1.1.

No	GCM	Reference
1	BCC-CSM1-1	[2-4]
2	CCSM4	[5]
3	CESM1-CAM5	[6]
4	CSIRO-Mk3-6-0	[7]
5	FIO-ESM	[8]
6	GFDL-CM3	[9, 10]
7	GFDL-ESM2G	[11, 12]
8	GFDL-ESM2M	[11, 12]
9-11	GISS-E2-H*	[13]
12-14	GISS-E2-R*	[13]
15	HadGEM2-AO	[14, 15]
16	HadGEM2-ES	[14, 15]
17	IPSL-CM5A-LR	[16]
18	IPSL-CM5A-MR	[16]
19	MIROC5	[17]
20	MIROC-ESM	[17]
21	MIROC-ESM-CHEM	[17]
22	MRI-CGCM3	[18, 19]
23	NorESM1-M	[20]
24	NorESM1-ME	[21]

* *Three outputs are used for each of these two models (GISS-E2-H and GISS-E2-R): r1i1p1, r1i1p2, and r1i1p3. Therefore the total is 24 GCM outputs for each scenario used in SiBiaS v1.1*

The program consists of rainfall and surface temperature data outputs from 24 CMIP5 GCMs under 4 RCP scenarios. RCP is the most recent scenario used by the IPCC for their new climate projection studies for the Fifth Assessment Report (AR5), which is published at the end of 2013 [1]. The RCP scenarios consists of four climate change scenarios defined by radiative forcing pathways until 2100, i.e. RCP2.6, RCP4.5, RCP6.0 and RCP8.5 [22, 23]. Those four scenarios represent different pathways of radiative forcing in the future. RCP2.6 is the lowest scenario with radiative forcing peak at ~3 Wm-2 before 2100 and then declines ("peak and decline" pathway). Meanwhile, the highest scenario is RCP8.5. It has "rising" pathway with radiative forcing is expected to be greater than 8.5 Wm-2 in 2100.

The main data used as reference are taken from the gridded observation of rainfall and surface temperatures data, which is already cropped only for Indonesia domain. The Climate Hazards Group InfraRed Precipitation with Station Data (CHIRPS) version 2.0 [24, 25] is used as the observed reference file of rainfall data in the system. The data has two selections of spatial grid resolutions, i.e. 0.25° (low) and 0.05° (high) grid resolutions. For surface temperature data, the Climate Research Unit TS3.22 [26] data is used for reference file. This data has a relatively coarse 0.5° grid resolution. In addition to this data that already automatically available in the program, users can also add reference data from observed stations by entering the data files into the browse file menu. To be able for the program to read the user-defined data, the data has to be prepared by following the file format as shown in the user manual [27].

2.2. Method

There are two choices of statistical bias correction techniques implemented in this program, i.e. i) Delta method, and ii) Statistical intensity distribution method. A previous study provides more details information regarding the Statistical intensity distribution method [28]. While, information about the comparison between the strengths and weaknesses of various bias correction or statistical downscaling method is summarized in another study [29].

Delta method [e.g. 30] is a simple downscaling method widely used for preparing climate change scenarios at local level. It uses either through addition or multiplication of delta ($\Delta\mu$) and the data in the baseline periods. In general, SiBias follow the equations below:

$$x_{cor,i} = x_{mp,i} + \mu_{ob} - \mu_{mb} \tag{1}$$

$$x_{cor,i} = x_{mp,i} \text{ x } \frac{\mu_{ob}}{\mu_{mb}} \tag{2}$$

Addition method of the delta is mostly used for temperature (Equation 1), while multiplication is used for rainfall data (Equation 2). In order to maintain the variability of the model when calculating the time series of the future rainfall or temperature, $\Delta\mu$ is calculated from the differences or from the ratio between the mean value of the observed (μ_{ob}) and the modelled baseline (μ_{mb}).

Meanwhile, the Statistical intensity distribution method is based on the correction of distribution of climate data from the model, which is transformed in order to match with the distribution of observed data [28]. In the process of performing bias correction methods, the observation data and simulation assumed to have a specific distribution pattern, where for rainfall data commonly follows a Gamma distribution pattern [28]. Gamma-based PDF follows Equation 3 below:

$$pdf(x) = \frac{e^{(-\frac{x}{\theta})} x^{(k-1)}}{\Gamma(k)\theta^k} \tag{3}$$

Where x is the rainfall data and k is the shape parameter and θ is the scale parameter. In general, the correction process is conducted by calculating the value of the inverse cumulative distribution function (CDF) curve for observed (y) and modeled (x) data. Data from the inverse CDF is then used to develop a regression equation where y = f (x). The equations obtained from the regression is considered as correction factor that can be used to correct current baseline and future projection data from the models. Figure 1 provides an example of steps in performing bias correction based on the statistical intensity distribution method. We have tried several regression approaches in order to find the best fit of the regression model. It is found that the third order polynomial regression that is forced the fitted line to go through the origin point (0,0), has better result in providing correction factors for the rainfall data. Dissimilar to the distribution function of rainfall data that follows Gamma distribution, we modify the fitting of distribution for temperature data. The pattern of temperature distribution generally differs from rainfall data due to its Gaussian-shaped that tend to follow normal distribution patterns.

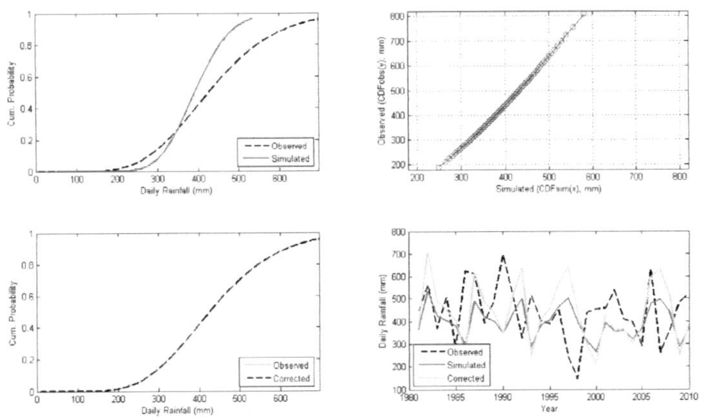

Figure 1. An example describing steps of bias correction method based on Statistical intensity distribution method.

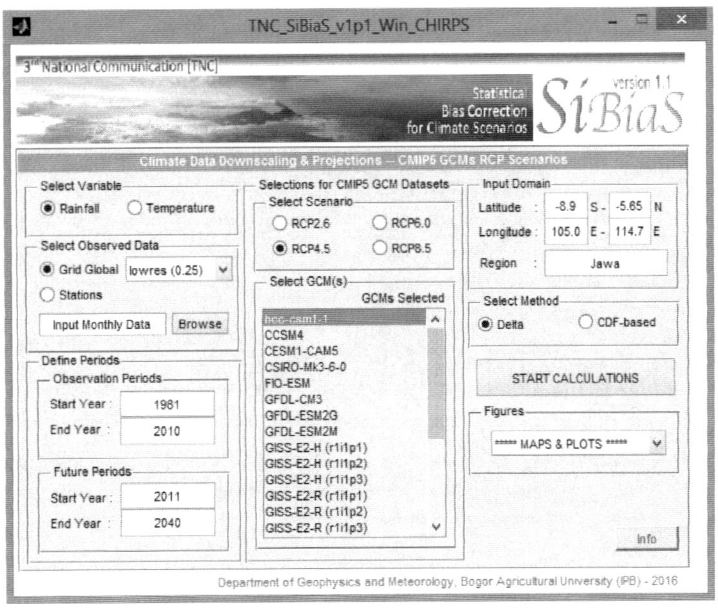

Figure 2. SiBiaS v1.1 interface.

3. Results and Discussion

SiBiaS is built to process baseline and projection data from the CMIP5 GCMs outputs into downscaled climate change scenarios for various locations in Indonesia. The program consists of rainfall and surface temperature data outputs from 24 CMIP5 GCMs under 4 RCP scenarios. This data can be selected and processed for the calculation of downscaling within the selected region in Indonesia. The tool performs correction of rainfall or temperature data for every single grid location within single month to consider seasonality in climate data, meaning if the data has 10 grid location then it will have 10 grid x 12 month (=120) correction factors. The result of the calculation in the program will deliver two types of outputs, i.e. i) graphic plots from the results of the area-averaged monthly climatology, monthly time series, anomalies of monthly time series and probability density function (PDF), and ii) downscaled data in simple ASCII format. SiBiaS v1.1 interface is shown in figure 2.

3.1. Graphic plots resulted by SiBiaS

The tool offers simple graphical results for different purposes in delivering downscaled future climate scenarios. It also can be used for assessing agreement between model in simulating rainfall or temperature in the selected area of study. Figure 3a shows an example of the uncorrected temperature climatology from the model baseline compared to the observed data during the same baseline period in 1981-2005. It shows the model performance in reproducing surface temperature climatology in Indonesia. This kind of figures can also be used to identify the best model that matches the climatology of the observed data.

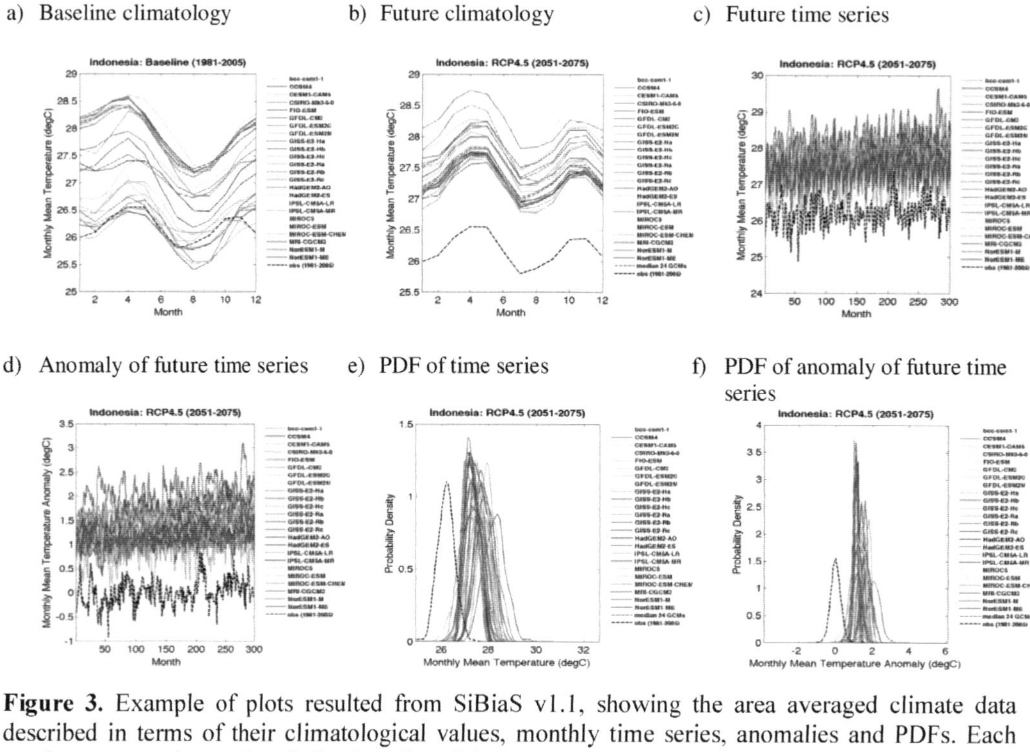

Figure 3. Example of plots resulted from SiBiaS v1.1, showing the area averaged climate data described in terms of their climatological values, monthly time series, anomalies and PDFs. Each graph compares the results of all selected models.

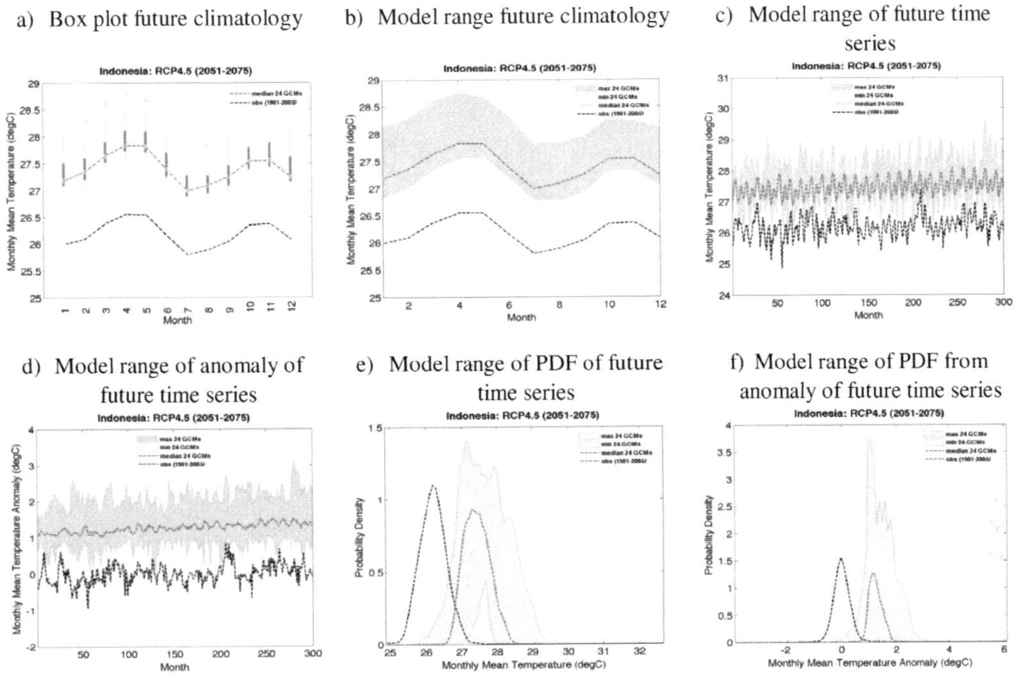

Figure 4. Example of plots resulted from SiBiaS v1.1, showing the area averaged climate data described in terms of their climatological values, monthly time series, anomalies and PDFs. Each graph compares the results of all selected models.

The important aspect in performing the bias correction or downscaling of climate data from GCM outputs is to obtain the corrected data for future climate scenarios. Figure 3b-3d provides the examples of the area-averaged of the bias-corrected data by using delta approach. The figures show the scenarios of future rainfall climatology and monthly time series and their anomalies resulted from each individual models. Those monthly time series and their anomalies are also used to display and compare the PDFs resulted from every models as shown in figure 3e and 3f. The individual model results combined into this specific single figure, delivering the median value and the range between minimum and maximum of all the selected models performed in figure 3. The example of the simplified figure showing the range across different models is shown in figure 4. The figure is also can be used to get information about the range of uncertainty in the future climate scenarios resulted from many models.

3.2. Comparing downscaled outputs for different regions in Indonesia

SiBiaS is designed for producing bias-corrected data for generating future climate scenarios within user-defined domain in Indonesia. Therefore, the users can select various area of interest within the country to deliver future climate scenarios. Figure 5 demonstrates an example of further uses of the SiBiaS outputs for comparing future scenarios of rainfall climatology over different areas in Indonesia. It shows the box plots resulted from the outputs of 24 GCMs, depicting uncertainty of rainfall climatology in the future. The median value provide information regarding the tendency of the rainfall climatology might unfold in the future over different selected regions in Indonesia, i.e. Sumatra, Jawa, Kalimantan, Sulawesi, Bali and Nusa Tenggara, Maluku, and Papua. It is shown that some areas in Indonesia are expected to experience less rainfall in the dry season and more rainfall in the rainy season in the future, as shown by their scenario of future rainfall climatology.

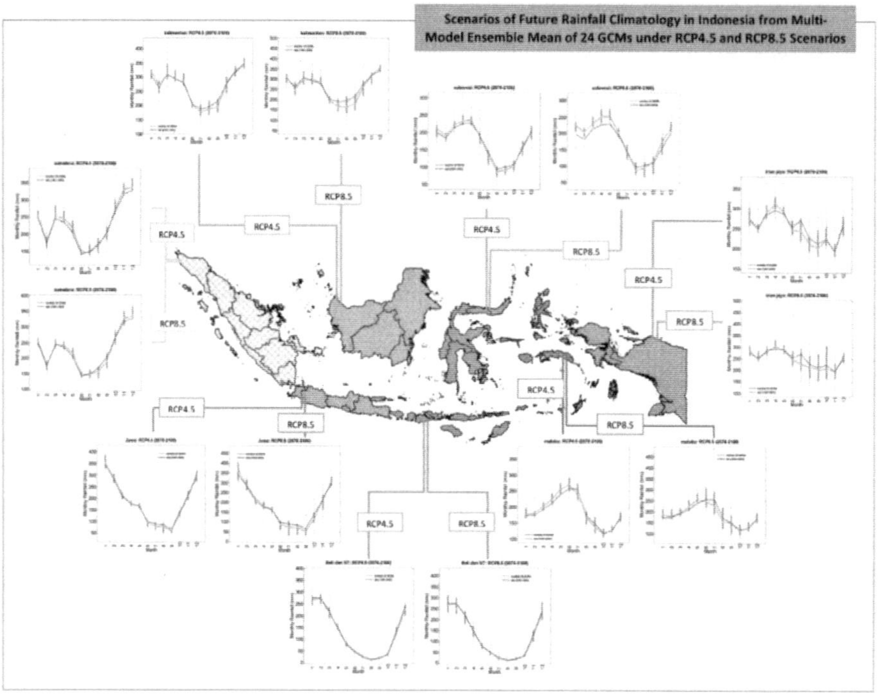

Figure 5. Example of plots resulted from SiBiaS v1.1 for comparing the result of future scenarios of rainfall climatology from different regions in Indonesia.

3.3. Delivering more analysis from the downscaled data outputs

More analysis can be delivered by using the ASCII data outputs of SiBiaS. The data is automatically stored in the output folder of the program. It consists not only the data of area-averaged values used for creating default graphical plots, but also spatial data from every grid locations within the selected region. In order to perform further result from this data outputs, third party software should be used for creating maps and graphical plots. The data outputs resulted by SiBiaS consist information such as, monthly climatology, monthly time series, anomalies of monthly time series, and PDF values from raw and anomaly data. The data is available for every grid locations and for the area-averaged value of the selected domain. In addition, the data output also provides information regarding latitude and longitude locations of every grid points within the selected domain. This information is very important for creating spatial mapping from the data outputs.

Figure 6 provides an example of the map created from the output data resulted by SiBiaS. The maps compare the future scenarios of seasonal rainfall in September-October-November (SON) period over Sumatera during three different periods under 4 RCP scenarios. This type of maps is very useful for comparing the result of climate scenarios in different periods in the future as resulted by using different climate change scenarios. Figure 6 is depicted from the median value of the selected 24 GCMs. Many other different maps and graphical plots can be explored and created from the outputs of SiBiaS. Further analysis related to future climate scenarios and impact analysis can also be done by using the data outputs.

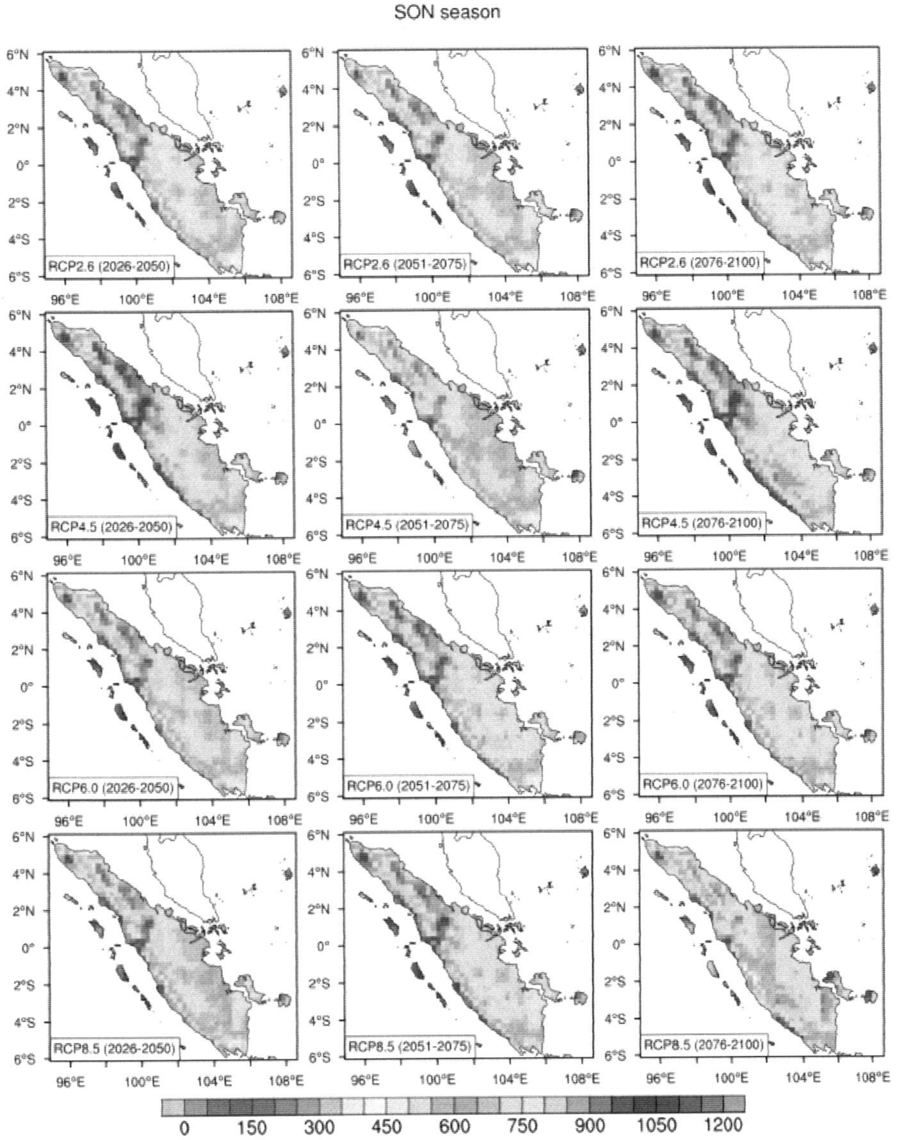

Figure 6. Spatial maps representing the future scenarios of seasonal rainfall in September-October-November (SON) period over Sumatera during three different periods (2026-2050, 2051-2075, and 2076-2100) under 4 RCP scenarios (RCP2.6, RCP4.5, RCP6.0 and RCP8.5).

4. Conclusion

SiBias is built to perform user defined climate change scenarios for various locations in Indonesia. It processes baseline and projection data from the CMIP5 GCMs outputs under different Representative Concentration Pathways (RCP) scenarios. The analysis result from SiBiaS can be used as important information for providing climate change scenarios to support climate change impacts, adaptation and mitigation studies in Indonesia.

Acknowledgments
We acknowledge the United Nation Development Programme (UNDP) and the Ministry of Environment and Forestry of Indonesia for supporting this climate modeling work under the Indonesia Third National Communication. We acknowledge the World Climate Research Programme's Working Group on Coupled Modelling, which is responsible for CMIP, and we thank the climate modeling groups for producing and making available their model output.

References

[1] IPCC 2013 Climate Change 2013: The Physical Science Basis. Contribution of Working Group I to the Fifth Assessment Report of the Intergovernmental Panel on Climate Change ed T F [Stocker, D. Qin, G.-K. Plattner, M. Tignor, S. K. Allen, J. Boschung, A. Nauels, Y. Xia, V. Bex and P.M. Midgley (eds.)] (Cambridge, United Kingdom and New York, NY, USA: Cambridge University Press)
[2] Xin X, Zhang L, Zhang J, Wu T and Fang Y 2013 *J. Meteorol. Soc. Jpn.* **91** 413–29
[3] Wu T 2012 *Clim. Dyn.* **38** 725–44
[4] Xin X, Wu T, Li J, Wang Z, Li W and Wu F 2012 *Atmos. Ocean. Sci. Lett.* **6** 21–6
[5] Gent P R and Danabasoglu G 2011 *J. Clim.* **24** 4992–8
[6] Hurrell J, et al. 2013 *Bull. Am. Meteorol. Soc.*
[7] Rotstayn L D, Jeffrey S J, Collier M A, Dravitzki S M, Hirst A C, Syktus J I and Wong K K 2012 *Atmos. Chem. Phys.* **12** 6377–404
[8] Collins W D, Rasch P J, Boville B A, Hack J J, McCaa J R, Williamson D L, Briegleb B P, Bitz C M, Lin S-J and Zhang M 2006 *Journal of Climate* **19** 2144-61
[9] Delworth T L, Broccoli A J, Rosati A, Stouffer R J, Balaji V, Beesley J A, Cooke W F, Dixon K W, Dunne J, Dunne K A, Durachta J W, Findell K L, Ginoux P, Gnanadesikan A, Gordon C T, Griffies S M, Rich Gudgel, Harrison M J, Held I M, Hemler R S, Horowitz L W, Klein S A, Knutson T R, Kushner P J, Langenhorst A R, Lee H-C, Lin S-J, Lu J, Malyshev S L, Milly P C D, Ramaswamy V, Russell J, M. Daniel Schwarzkopf, Shevliakova E, Sirutis J J, Spelman M J, Stern W F, Winton M, Wittenberg A T, Wyman B, Zeng F and Zhang R 2006 *Journal of Climate* **19** 643–74
[10] Donner L J, Wyman B L, Hemler R S, Horowitz L W, Ming Y, Zhao M, Golaz J-C, Ginoux P, Lin S-J, Schwarzkopf M D, Austin J, Alaka G, Cooke W F, Delworth T L, Freidenreich S M, Gordon C T, Griffies S M, Held I M, Hurlin W J, Klein S A, Knutson T R, Langenhorst A R, Lee H-C, Lin Y, Magi B I, Malyshev S L, Milly P C, Naik V, Nath M J, Pincus R, Ploshay J J, Ramaswamy V, Seman C J, Shevliakova E, Sirutis J J, Stern W F, Stouffer R J, Wilson R J, Winton M, Wittenberg A T and Zeng F 2011 *Journal of Climate* **24** 3484-519
[11] Dunne J P, John J G, Shevliakova E, Stouffer R J, Krasting J P, Malyshev S L, Milly P C D, Sentman L T, Adcroft A J, Cooke W, Dunne K A, Griffies S M, Hallberg R W, Harrison M J, Levy H, Wittenberg A T, Phillips P J and Zadeh N 2013 *J. Climate* **26** 2247–67
[12] Dunne J P, John J G, Adcroft A J, Griffies S M, Hallberg R W, Shevliakova E N, Stouffer R J, Cooke W, Dunne K A, Harrison M J, Krasting J P, Malyshev S L, Milly P C D P, P. J. , Sentman L T, Samuels B L, Spelman M, Winton M, Wittenberg A T and Zadeh N 2012 *Journal of Climate* **25**
[13] Schmidt G A, Ruedy R, Hansen J E, Aleinov I, Bell N, Bauer M, Bauer S, Cairns B, Canuto V, Cheng Y, Del Genio A, Faluvegi G, Friend A D, Hall T M, Hu Y, Kelley M, Kiang N Y, Koch D, Lacis A A, Lerner J, Lo K K, Miller R L, Nazarenko L, Oinas V, Perlwitz J, Perlwitz J, Rind D, Romanou A, Russell G L, Sato M, Shindell D T, Stone P H, Sun S, Tausnev N, Thresher D and Yao M-S 2006 *Journal of Climate* **19** 153-92
[14] Collins W J, Bellouin N, Doutriaux-Boucher M, Gedney N, Halloran P, Hinton T, Hughes J, Jones C D, Joshi M, Liddicoat S, Martin G, O'Connor F, Rae J, Senior C, Sitch S, Totterdell I, Wiltshire A and Woodward S 2011 *Geosci. Model Dev.* **4** 1051-75

[15] Martin G M, Bellouin N, Collins W J, Culverwell I D, Halloran P R, Hardiman S C, Hinton T J, Jones C D, McDonald R E, McLaren A J, O'Connor F M, Roberts M J, Rodriguez J M, Woodward S, Best M J, Brooks M E, Brown A R, Butchart N, Dearden C, Derbyshire S H, Dharssi I, Doutriaux-Boucher M, Edwards J M, Falloon P D, Gedney N, Gray L J, Hewitt H T, Hobson M, Huddleston M R, Hughes J, Ineson S, Ingram W J, James P M, Johns T C, Johnson C E, Jones A, Jones C P, Joshi M M, Keen A B, Liddicoat S, Lock A P, Maidens A V, Manners J C, Milton S F, Rae J G L, Ridley J K, Sellar A, Senior C A, Totterdell I J, Verhoef A, Vidale P L and Wiltshire A 2011 *Geosci. Model Dev.* **4** 723-57

[16] Dufresne J L, Foujols M A, Denvil S, Caubel A, Marti O, Aumont O, Balkanski Y, Bekki S, Bellenger H, Benshila R, Bony S, Bopp L, Braconnot P, Brockmann P, Cadule P, Cheruy F, Codron F, Cozic A, Cugnet D, Noblet N, Duvel J P, Ethé C, Fairhead L, Fichefet T, Flavoni S, Friedlingstein P, Grandpeix J Y, Guez L, Guilyardi E, Hauglustaine D, Hourdin F, Idelkadi A, Ghattas J, Joussaume S, Kageyama M, Krinner G, Labetoulle S, Lahellec A, Lefebvre M P, Lefevre F, Levy C, Li Z X, Lloyd J, Lott F, Madec G, Mancip M, Marchand M, Masson S, Meurdesoif Y, Mignot J, Musat I, Parouty S, Polcher J, Rio C, Schulz M, Swingedouw D, Szopa S, Talandier C, Terray P, Viovy N and Vuichard N 2013 *Climate Dynamics* **40** 2123-65

[17] Watanabe M, Suzuki T, O'ishi R, Komuro Y, Watanabe S, Emori S, Takemura T, Chikira M, Ogura T, Sekiguchi M, Takata K, Yamazaki D, Yokohata T, Nozawa T, Hasumi H, Tatebe H and Kimoto M 2010 *Journal of Climate* **23** 6312-35

[18] Yukimoto S, Adachi Y, Hosaka M, Sakami T, Yoshimura H, Hirabara M, Tanaka T Y, Shindo E, Tsujino H, Deushi M, Mizuta R, Yabu S, Obata A, Nakano H, Koshiro T, Ose T and Kitoh A 2012 *Journal of the Meteorological Society of Japan. Ser. II* **90A** 23-64

[19] Yukimoto S, Yoshimura H, Hosaka M, Sakami T, Tsujino H, Hirabara M, Tanaka T Y, Deushi M, Obata A, Nakano H, Adachi Y, Shindo E, Yabu S, Ose T and Kitoh A 2011 Meteorological Research Institute - Earth System Model v1 (MRI-ESM1) - Model Description. ed T R o MRI: Ibaraki, Japan) p 83 pp

[20] Iversen T, Bentsen M, Bethke I, Debernard J B, Kirkevåg A, Seland Ø, Drange H, Kristjansson J E, Medhaug I, Sand M and Seierstad I A 2013 *Geosci. Model Dev.* **6** 389-415

[21] Tjiputra J F, Roelandt C, Bentsen M, Lawrence D M, Lorentzen T, Schwinger J, Seland Ø and Heinze C 2013 *Geosci. Model Dev.* **6** 301-25

[22] Moss R H, Edmonds J A, Hibbard K A, Manning M R, Rose S K, Vuuren D P v, Carter T R, Emori S, Kainuma M, Kram T, Meehl G A, Mitchell J F B, Nakicenovic N, Riahi K, Smith S J, Stouffer R J, Thomson A M, Weyant J P and Wilbanks T J 2010 *Nature* **463**

[23] Moss R, Babiker M, Brinkman S, Calvo E, Carter T, Edmonds J, Elgizouli I, Emori S, Erda L, Hibbard K, Jones R, Kainuma M, Kelleher J, Lamarque J F, Manning M, Matthews B, Meehl J, Meyer L, Mitchell J, Nakicenovic N, O'Neill B, Pichs R, Riahi K, Rose S, Runci P, Stouffer R, Vuuren D v, Weyant J, Wilbanks T, Ypersele J P v and Zurek a M 2008 Towards New Scenarios for Analysis of Emissions, Climate Change, Impacts, and Response Strategies. (Geneva: Intergovernmental Panel on Climate Change) p 132 pp

[24] Funk C, Peterson P, Landsfeld M, Pedreros D, Verdin J, Shukla S, Husak G, Rowland J, Harrison L, Hoell A and Michaelsen J 2015 The climate hazards infrared precipitation with stations-a new environmental record for monitoring extremes *Scientific Data* **2**

[25] Funk C C, Peterson P J, Landsfeld M F, Pedreros D H, Verdin J P, Rowland J D, Romero B E, Husak G J, Michaelsen J C and Verdin A P 2014 A quasi-global precipitation time series for drought monitoring *U.S. Geological Survey Data Series* **832, 4 p**

[26] Harris I, Jones P D, Osborn T J and Lister D H 2014 *Int. J. Climatol.* **34** 623–42

[27] Faqih A 2016 Proyeksi Iklim menggunakan Luaran GCM CMIP5: Statistical Bias Correction for Climate Scenarios (SiBiaS) versi 1.1 [Panduan Pengguna]. Departemen Geofisika dan Meteorologi, Fakultas Matematika dan Ilmu Pengetahuan Alam, Institut Pertanian Bogor)

[28] Piani C, Haerter J O and Coppola E 2010 *Theor. Appl. Climatol.* **99** 187–92

[29] Watanabe S, Kanae S, Seto S, Yeh P J-F, Hirabayashi Y and Oki T 2012 *Journal of geophysicsal research* **117**

[30] Graham L P, Andréasson J and Carlsson B 2007 *Clim. Change* **81(S1)** 293–307

IOP Conf. Series: Earth and Environmental Science **58** (2017) 012052 doi:10.1088/1755-1315/58/1/012052

Synthesis of a Carbon-activated Microfiber from Spider Webs Silk

E Taer[1], W S Mustika[1] and R Taslim[2]

[1] Authors' Department of Physics, Faculty of Mathematics and Natural Sciences, University of Riau, INDONESIA.
[2]Authors' Department of Industrial Engineering, Faculty of Science and Technology, State Islamic University of Sultan Syarif Kasim, INDONESIA.

Email: erman_taer@yahoo.co.id , widyasintamustika@ymail.com

Abstract. Carbon fiber of spider web silk has been produced through the simple carbonization process. Cobwebs are a source of strong natural fiber, flexible and micrometer in size. Preparation of micro carbon fiber from spider webs that consist of carbonization and activation processes. Carbonization was performed in N_2 gas environment by multi step heating profile up to temperature of 400 ^{o}C, while the activation process was done by using chemical activation with KOH activating agent assistance. Measurement of physical properties was conducted on the surface morphology, element content and the degree of crystallinity. The measurement results found that micro carbon fiber from spider webs has a diameter in the range of 0.5 -25 micrometers. It is found that the carbon-activated microfiber takes the amorphous form with the carbon content of 84 % .

1. Introduction

Carbon material is one of the functional materials that are intensive studied in the recent decades [1]. Various aspects of life can be attributed to the use of carbon such as, health, beauty, water purification and gas, as well as the supply of energy storage device [2,3]. The wide range of carbon application because its excellent properties such as high surface area and good electrical conductivity. The availability of carbon in various forms such as: particle, fiber and granules as well as various sizes in the nanometer, micrometer and millimeter into its own advantages in the preparation for the use of a unique carbon. Carbon fiber itself can still be categorized further into carbon fiber hollow and solid [4]. Besides, the source of carbon material origin becomes a major factor in the production of carbon. One step to produce carbon fiber is to perform the carbonization process to transform it into carbon fiber material. Preparation the fiber material is a research process that requires a lot of development.

The supply process of original carbon fiber material is relatively expensive that can become a problem in the development of carbon fiber on a large scale [5]. The availability of fiber in nature may be one of the choices in the development of carbon fiber. Some natural fiber material which had been made into a carbon fiber including cottons and cocoons [6,7,8]. Spider web silk is one of the fiber materials that are naturally available in a microsize. The spider web properties were strong and relatively flexible that was certainly a special consideration in the selection of this material as a carbon source of natural fiber. In this study will be focused on the production process of carbon fiber from spider web silk that be emphasized at adjustment of pyrolisis condition. This study was concentrate on observation of the morphological structure to see the size of the carbon fibers, the supporting element content and the degree of crystallinity of the carbon fiber of cobwebs.

2. Experiment

A Common houses spider web (CHSW) (Parasteatoda tepidariorum) was selected as a raw materials for production of carbon activated microfiber. The preparation of the carbon activated microfiber materials consist of some stages i.e collecting, washing, drying, molding pellets, carbonization and activation as shown in Figure 2. Sample CHSW weighing ± 0.13 g was immersed in distilled water for 48 hours and continued with the process washing at 80 ° C with rotation speed of 300 rpm for 15 minutes using a hot plate stirrer and then dried for 36 hours using an oven. The CHSW sample is formed into the pellets using a Hydraulic Press at a pressure of 6 tons. Pellet CHSW carbonized up to a temperature of 400 ° C in an environment of nitrogen gas (N2). The process was performed by multistep heating profile with N2 gas flow rate of 1.5 L / min that had been reported previously [9,10]. CHSW carbonization profile shown in Figure 1. The heating process begun at room temperature 30 °C up to soaking temperatures of 280 ° C for 4 hours 17 minutes with a temperature increase rate of 1 ° C / min. Next, the sample was heated at a temperature of 280 ° C for 1 hour, then heating is continued up to a temperature of 400 ° C at a rate of temperature rise of 3 ° C / min for 40 minutes. This process had been successfully transformed into activated carbon CHSW. After the carbonization process, carbon CHSW was activated by using 0.7 M KOH.

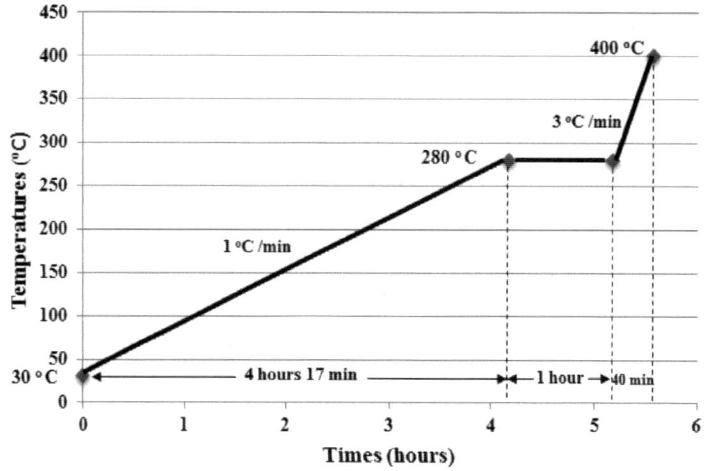

Figure 1. Carbonization process of CHSW.

Structure CHSW samples were analyzed using X-Ray Diffraction instrument. This analysis used the instrument X 'Pert Pro in diffractometer cubic system, using a light source and wavelength anode Cu Kα was to 15.406 nm in the range of diffraction angle (2θ) is 10-100°. The surface morphology and chemical composition of the samples was measured using a Hitachi S-3400N. SEM measurements was performed at a potential of 10,000 V used a detector of back scatter electron (BSE).

3. Result and Discussion

CHSW main stages to transform into carbon microfiber (CMF) were the washing process, carbonization and activation. Overall treatment have a purpose to remove other materials than carbon, so these processes lead to loss of mass in the sample. The sample mass shrinkage data are presented following in Table 1.

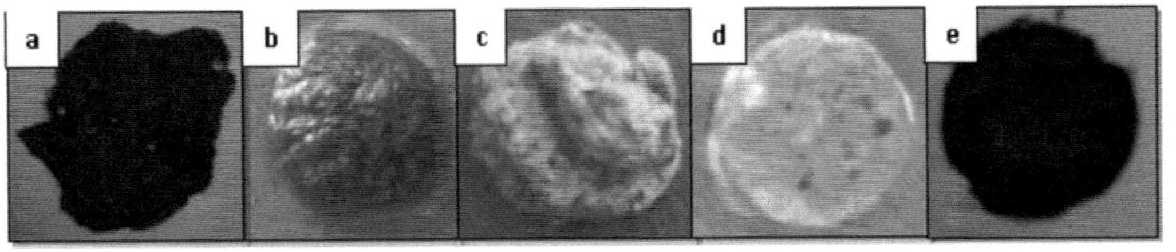

Figure 2. The sample for each step in preparation process: a) a spider web once collected, (b) after washing, (c) after drying, (d) after compression, (e) after carbonization.

Table1. Mass shrinkage of the sample.

Process	Initial mass (g)	Final mass (g)	Mass lost (%)
Washing	1.298	0.356	72.57
Carbonization	0.334	0.157	52.99
Activation	0.157	0.154	1.91

The mass loss occured when washing the sample were equal to 72.5 7%, this process removed dust and dirt on CHSW. Carbonization process caused mass loss of 52.99%. This carbonization process caused evaporation elements other than carbon, which can evaporate at a temperature ≤ 400 °C. The activation process caused a small mass loss of 1.91%. This mass loss was caused by breaking the carbon chains and carbon material that binded to the elements or compounds in addition to carbon.

CHSW sample XRD pattern shown in Figure 3. The peak of carbon for dhkl field (002) and (100) appeared at a diffraction angle 2θ in the range of 15 to 35°. Sharp peaks that appearing are impurities other than carbon, namely SiO_2, the same result had also been reported in a previous article [11]. Characterization of samples CHSW before carbonization indicated that CHSW was carbon in the form of compounds. CHSW sample peak was detected at diffraction angles 2θ of 22.728° and 42.799°. Carbonization process caused the peak shift at diffraction angles 2θ of 26.399° and 43.637°, while the activation process brought out carbon peaks at diffraction angles 2θ of 26.065° and 43.976°.

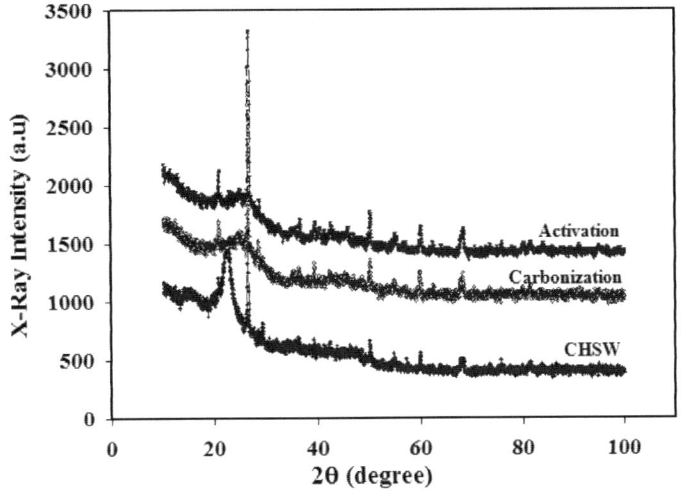

Figure 3. XRD pattern of CHSW sample.

SEM micrographs after washing, carbonization and activation are shown in Figure 4. This figure showed that the morphology CHSW was a form of fiber. After carbonization the carbon morphology CHSW like fine long fibers. Diameter of carbon fiber were in the range 0.5 to 25 μm.. The activation process showed a clearer shape fiber with an average size of 8.72 μm.

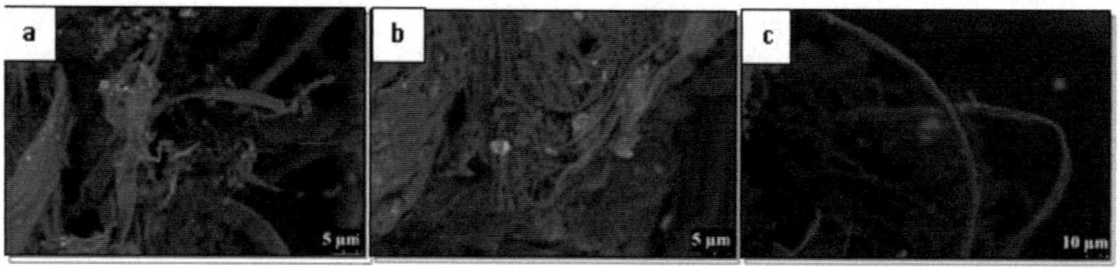

Figure 4. Figure of sipeder web fiber: (a) raw materials as prepared, b) after carbonization, c) after activation process.

Figure 5 and Table 2 respectively presents a test EDX spectrum and composition of constituent elements of CHSW and activation CHSW. Raw material showed carbon element content of 50.68%. Carbonization process that had been carried out successfully increases the carbon content in CHSW which amounted to 77.34%. In addition to the element carbon in the samples indicated the presence of other elements such as Al, Si and Ca. The EDX test results correlated with the XRD data. XRD data also indicate the presence of materials Si, Al and Ca are characterized by the presence of a sharp peak at two-theta angle 28.78°, 21.065° and 50.302 °, 60.152°, respectively [12-13]. Chemical activation process managed to reduce impurities in this reality was supported by data that the percentage of materials non-carbon apparent decrease as the data shown in Table 4. On the other hand, KOH activators increase the oxygen content (O) on carbon CHSW so that the composition of C was reduced quantitatively. The activation process caused KOH bound with sample group of oxygen. Washing step can eliminate K bond, but cannot sever the bond of O.

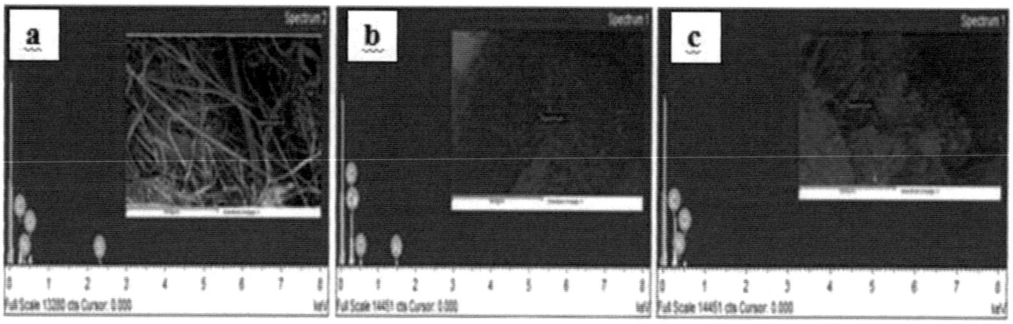

Figure 5. The EDX spectrum of CHSW sample, a) after washing process, b) after carbonization process and c) after activation process.

Table 2. EDX data of CHSW.

Unsure	Percentage of mass (%)			Percentage of atomic weight (%)		
	CHSW	Carbonization	activation	CHSW	carbonization	activation
C	50.68	77.34	72.49	57.95	83.47	77.51
N	11.70	17.63	4.12	11.48	14.28	3.77
O	33.60	-	23.25	28.85	-	18.66
S	4.02	-	0.15	1.72	-	0.06
Al	-	3.43	-	-	1.65	-
Si	-	0.62	-	-	0.28	-
Ca	-	0.99	-	-	0.32	-
Total	100.00	100.00	100.00	100.00	100.00	100.00

Some literature had been describing the micro carbon fiber synthesis of various types of raw materials. The following comparison of micro carbon fiber and research conducted are summarized in Table 3. Based on the data shown in Table 3 is clearly indicated that CHSW is an acceptable range for a micro carbon fiber. The other side of excellence CHSW is available in lengths of more than the other micro carbon. This fact is supported by the data that the natural fiber silk spider webs have a regular shape in the form of a network arrangement. Shape structures are at the network will also produce carbon micro-fiber with the same structure. Another advantage that can be written in this paper is certainly the cost of providing relatively low, so the micro carbon fiber of spider silk network's potential to be developed in a number of applications in the future [1].

Table 3. Micro carbon fibers from several precursors.

Precursor	Fiber size	Reference
Phenol-formaldehyde and tetraethyl orthosilicate	1 μm	[14]
Polyacrylonitrile (China Fabric)	12.7 μm	[15]
spider silk scaffold and CNT	1 μm	[16]
Carbon Microfiber Fabric ACC-5092-20	10 μm	[17]
Silk Cocoon	10-40 nm	[6]
Banana Peel	26 nm	[18]
CHSW	0.5-25 μm	Recent study

4. Conclusions
Based on all the data and analysis that has been done can be concluded that the carbon activated microfiber made from the spider web shows potential to be developed as a functional carbon material. The size of the diameter and length were very supportive and a high purity level cause the carbon fiber of spider webs easy to be activated, and was developed as a raw material for a variety of applications.

Acknowledgements
We say to thank to kemenristek dikti for funding payment that had been provided through research project on HIBAH KOMPETENSI to first researcher Dr, Erman Taer, M.Si with the title of the project is Penyediaan nano karbon sebagai inti elektroda komposit untuk aplikasi pada superkapasitor.

References
[1] Kötz R And Carlen M 2000 *Electrochimica Acta* **45** 2483-2498
[2] Taer E, Yusriwandi, Taslim R, Syam IDM, Deraman M 2015 *KnF Engineering 2016* 1-6
[3] Sun G, Wang X, Chen P 2015 *Materials Today* **18** 215–226
[4] D'Almeida ALFS, Barreto DW, Calado V, D'Almeida JRM 2006 *Polym Polym Compos* **14(1)** 73–80
[5] Monteiro S-N, Calado V, Margem F-M, Rodriguez R-JS 2012 *Journal of Materials Research an Technology* **1** 189-199
[6] Zheng J, Zhao Q, Ye Z 2014 *Applied Surface Science* **295** 86-91

[7] Ekrami E, Dadashian F, Soleimani M 2014 *Fiber and Polymers* **15** 1855-1864
[8] Liang Y, Wu D, Fu R 2013 *Sientific Reports* **3** 1119
[9] Taer E, Deraman M, Taslim R, Iwantono 2013 *AIP Conf. Proc* **1712** 33-37
[10] Taer E, Taslim R, Deraman M 2016 *AIP Conf. Proc* **1712** 050011
[11] Taer E, Deraman M, Talib IA, Awitdrus A, Hashmi SA, Umar AA 2011 *Int.J Electrochemical.* Sci **6** 3301-3315
[12] Rebhi A, Makhlouf T, Njah N 2009 *Physics Procedia* **2** 1263-1270
[13] Li X, Xu X, Zhou Q, Qi T, Liu G, Peng Z, Cui Y, Li J 2016 *Int. Journal of Refractory Metals and Hard Materials* **60** 82-91
[14] Zhou D, Dong Y, Cui L, Lin H, Qu F 2014 *J.Nanopart Res* **16** 2732
[15] Shi F, Lu C, Liang M 2010 *JMEPEG* **19** 643–649
[16] Steven E, Saleh WR, Lebedev V, Acquah SFA, Laukhin V, Alamo RG, Brooks JS 2013 *Nature Communication* **4** 2435
[17] Etacheri V, Sharon D, Garsuch A, Afri M, Frimer AA, Aurbach D 2013 *Journal of Materials Chemistry* A **1** 5021
[18] Phiriyawirut M and Maniaw P 2012 *Scientific Report* **2** 56-62

IOP Publishing
doi:10.1088/1755-1315/58/1/012053

Lead accumulation by jabon seedling (*Anthocephalus cadamba*) on tailing media with application of compost and arbuscular mycorrhizal fungi

L Setyaningsih[1*], Y Setiadi[2], S W Budi[2], Hamim[3] and D Sopandie[4]

[1]Department of Forestry, Nusa Bangsa University, Jl. Kh. Sholeh Iskandar Km 4, Bogor, Indonesia. 16166
[2]Department of Silviculture, Faculty of Forestry, Bogor Agricultural University (IPB), Jl. Bundaran Akademik. Campus IPB Darmaga, Bogor, Indonesia. 16680
[3]Department of Biology, Faculty of Mathematics and Natural Sciences, Bogor Agricultural University (IPB), Jl. Meranti, Campus IPB Darmaga, Bogor, Indonesia. 16680
[4]Department of Agronomy, Faculty of Agriculture, Bogor Agricultural University (IPB), Jl. Meranti. Campus IPB Darmaga, Bogor, Indonesia. 16680

Email: luluk.setya@gmail.com

Abstract. Lead (Pb) is one of the dangerous heavy metal contained in tailing that needs remediation activity. This study aimed to investigate the potency of jabon to take up and accumulate lead in its tissue by the application of compost and arbuscular mycorrhiza fungus (AMF) on pot observation. In Pb-containing tailing media, the average levels of Pb in roots seedling was 50% greater as compared to the levels of Pb in the stem and leaves of seedlings. Application of compost in tailings media significantly increased ($p \leq 0.5$) the average levels of Pb in the roots and stems, but decreased Pb levels in leaves. Applications AMF significantly decreased ($p \leq 0.5$) the average levels of Pb in the roots, stem and leaves of seedlings by approximately 18-33%. The combination applications of compost and AMF significantly ($p \leq 0.5$) increased the level of Pb in the roots, stems and leaves of seedlings at 6, 16 and 27 fold respectively than that in control plant (without compost and AMF). After 12 weeks exposure, lead bioconcentration factor varied from 0.1-1.6 in seedling tissue with transport factor varied from 0.1-1.0. The application of active compost and AMF increased 1-15 fold lead accumulation from control, and the biggest accumulation was 452.9 $x10^{-2}$ mg/plant with Pb concentration of 1.5 mM. Active compost and AMF application supported jabon seedling to act as lead phytostabilizer and to remove lead from the tailing to the above part of the plant.

1. Introduction

Tailing produced from mining activities has extreme physical and chemical properties, which in addition to lower carbon and macro nutrients, it also contains heavy metals, such as lead (Pb) that can be produced naturally or triggered by mining activity [1,2]. Some reports observed that lead content in the tailings was approximately 4.80 ppm with total dissolved Pb was 172 ppm [2], while from another observation [3] it was 110 mg Pb/kg, and it was already at the critical threshold according to Alloway (year of publication) [4]. Lead has been categorized as a second dangerous element after arsenic [5] and at a certain concentration can be very harmful to living things, if deposited in the food chain, so that the effort to remove this particular element from the environment is a main priority to do.

As other heavy metals, lead is not easy to degrade so that the cleaning efforts (remediation) can be done by stabilizing the soil that reduce the mobility and bioavaibility of contaminants from leaching by wind or water erosion, and or transfer from extracted soil. This activity can be carried out by employing a certain plants by a process known as phytoremediation [6,7]. Phytoremediation is based on the fact that the plant life activities as a pumping using solar energy to extract and concentrate certain heavy metals from the environment [8]. Phyto-stabilization and phyto-extraction are two among several mechanisms of phyto-remediation. Phyto-stabilization uses plants to reduce the availability (bioavailability) of contaminants in the environment [9]. It is mainly due to strong roots arrangement that resists the leaching and erosion of contaminated soil by root absorption of pollutants or precipitate them to the rhizosphere, so that it will be stable [10]. On the other hand, phyto-extraction uses plants that have the ability to accumulate pollutants and further remove them from soil while concentrate them in the plant tissue, for later harvested of its biomass [9]. Dry biomass of plant shoots is expected containing contaminants in higher concentrations than those found in the soil. Thus, the potential remediation of a plant can be determined by measuring the bioconcentration factor (BCF), the value of the transport factor (TF) and the value bioaccumulation [7,10,11,12].

Phytoremediation of lead, however, is often inhibited by the lower solubility of Pb in the soil, or lower Pb translocation to the plant that can be harvested, or Pb toxicity for the plants [7,13]. The selection of resistant plants to Pb and improvement of physical and chemical properties of the tailings using compost and mycorrhizal applications are expected to be suitable to overcome these obstacles. Compost application is important and strongly recommended in agriculture practices [14], and it becomes an important component in tailing reclamation to support phytostabilization [15,16]. In addition to compost, arbuscular mycorrhizal fungi (AMF) has also been recognized to be able to filter heavy metals such as aluminum, arsenic, boron, cadmium, copper, iron, lead, nickel, selenium and zinc to the tolerant concentration for plant growth, so it is less toxic to the plants [17]. On the other hand, AMF is also able to increase accumulation of heavy metals such as Cu, Ni, Pb, Zn, Fe and Co on root and shoot tissues of grass *Ehranta calycina* Sm [18,19,20], as well as accumulation arsenic on Brachiaria Grass *(Brachiaria decumbens)* [21]. An attractive system to advance plant based environmental clean-up is provided by AMF. Its associations are integral functioning parts of plant roots and are widely recognized to enhance plant growth on severely disturbed sites, including those contaminated with heavy metals [22].

In addition to its utilization as alternative industrial-forest tree, Jabon, one of the fast-growing timber species, [23], has been chosen as afforestation and reforestation tree in the post mining area. However, detailed information regarding Jabon seedling ability to perform remediation activities on lead-contaminated tailings has not been revealed. Likewise, activated compost utilization and AMF to support the ability of forest trees seedlings in phytoremediation of tailing media is still very limited.

This study aimed to determine the effect of compost and AMF to support Jabon seedlings ability to accumulate lead in the tailings contained lead.

2. Methods

Research was carried out at Greenhouse Laboratory of Silviculture, Faculty of Forestry and Research Center of Biological Resources and Biotechnology, Bogor Agricultural University (IPB), Bogor, Indonesia.

2.1. Research design

This study was carried out using completely randomized design (CRD) with three factors. The first factor was addition of Pb (NO3)2, consisted of 4 levels: 0, 0.5, and 1.5 mM of Pb (equivalent to 0, 150, 450 mg kg^{-1} of tailing); The second factor was compost, consisted of 2 levels (without and with the addition of compost (1: 20 w / w)); and the third factor was AMF inoculants consisted of 2 levels (with and without AMF Mycofer Inoculant). Mycofer Inoculant consisted mix of 4 species of AMF, i.e: *Glomus manihotis* INDO-1, *Glomus etunicatum* NPI 126, *Gigaspora margarita*, *Acaulospora tuberculata* INDO-2.

2.2. Preparation of Media

Tailing was collected from the tailing dam in the area of national gold mining company, owned by PT Aneka Tambang Tbk. UBPE Pongkor, Bogor, Indonesia. Dried tailing was characterized by the following properties: tend to be alkaline (pH 8) with very low carbon (0.24%) and nitrogen (of 0.02%) content, while the availability of other macro nutrients was (P = 6 ppm and Mg = 0.73 cmol/kg). Total Pb content was 114 ppm and available Pb was 17.3 ppm. The activated compost was made from cow manure that was combined with rock phosphate and rice husk with the ratio of 500:20:50 (weight). One liter of bio-activator solution that had been diluted with 100 liters of water was added and the mixture was incubated for a week. Chemical properties of compost was: having pH slightly acid (6), the carbon content was very high (17.78%), macro nutrients content (N, P, K, Mg, Ca) was very high (1:57%, 2765 ppm, 1768 ppm, 21:01 cmol / kg, 40.61 cmol / kg respectively). Pb content was 22 ppm and only 0.2 ppm was in the form of available. Tailings and compost were mixed with the ratio of 20: 1 (w / w).

2.3. Preparation of seedlings and AMF

Jabon seeds were obtained from the Tissue Culture Laboratory, SEAMEO BIOTROP, Bogor, Indonesia, which was in the form of plantlets that had been acclimatized. Jabon seedling were prepared into small pots for 3 weeks before transferring to the tailings for the experimental treatment. AMF inoculation was carried out while the seedling were prepared and after being transferred to tailing media. A total of 10 grams of Mycofer inoculant containing spores approximately 25-50 AMF was applied in the middle of the hole of seedlings during planting.

2.4. Pb Treatment

The solution of Pb $(NO_3)_2$ was prepared to provide Pb treatment with 3 levels (0, 0.5, and 1.5 mM of Pb). The 50 ml amount of Pb $(NO_3)_2$ was applied to the seedling by pouring the solution to the area around the roots of 1 week seedlings.

2.5. Biomass measurements

Jabon seedlings biomass was measuring the dry weight of shoots and roots of seedlings. Seedlings were harvested and dried in an oven at a temperature of 70°C to obtain a constant dry weight (estimated at between 20-28hours).

2.6. Lead content measurements

Pb content was analyzed using the American Public Health Association method [24] by Atomic Absorption Spectroscopy (AAS) which previously had been calibrated with a standard solution of Pb. Pb accumulation in seedling tissue was calculated by measuring the levels of Pb in the tissue. After the seedlings were harvested and separated into 3 part (roots, stems and leaves), then they were dried in the oven with a temperature of 45 ± 2°C until constant weight. Special for roots seedling, before being dried they were soaked first for 30 minutes in 1.0 mM H-EDTA and rinsed three times in H_2O in order to eliminate attachment Pb on the root surface [25]. The value of Pb then was used as the basis for calculating value of Factor Bioconcentration/BF (Bioconcentration Factor/BCF), Factor Transport/FT (Translocation Factor/TF) and bioaccumulation /B (bioaccumulation/B) [7], by the equation as follows:

$FB = C_{root} / C_{soil}$ ... (C_{root} is the concentration of Pb in roots and C_{soil} is Pb concentration in soil / media)

$FT = C_{aerial} / C_{root}$ .... (C_{aerial} is the concentration of Pb in the shoots)

$B = C_{aerial} \times DW$ ..... (DW is tissue dry weight)

The data of Pb, FB, FT and bioaccumulation of Pb in the plant tissue were analyzed using SPSS data processing version 15 by comparing mean and analysis of variance.

3. Results

3.1. Jabon Seedling biomass

After 12 weeks of growth, biomass of Jabon seedling was measured. Single factor compost (C) influenced significantly ($p \leq 0.05$) to the seedling biomass except the roots biomass. Compost application increased the average of stem, leaf and total biomass of Jabon seedling, respectively 50%, 200% and 110%. Single factor mycorrhiza also significantly influenced ($p \leq 0.05$) the average biomass of Jabon seedlings, except biomass of stem. Applications of AMF increased the biomass of roots, leaves and total biomass for 48%, 20% and 14%, respectively. While the single factor Pb gave significant effect to the leave's biomass, only. But in general, Pb exposure has reduced seedling biomass, as total plant biomass reached 49% when the seedlings were exposed to 0.5 mM Pb, and decreased 14% when exposed to 1.5 mM Pb. Double and third interaction factor, compost-mycorrhiza-Pb, did not significantly affect the Jabon biomass. However, application of compost and mycorrhizal increased the biomass of Jabon leaf by more than 200%, and 140% . The greatest total biomass of seedling was found in the treatment of compost-AMF-Pb 0, amounting to 228.9 g (table 1).

3.2. Pb levels in the tissue of Jabon seedlings

The measurements of lead content in Jabon tissue seedlings was carried out 12 weeks after planting. The result showed that lead was found in all parts of the seedling tissues, either in the roots, stems and leaves with the value varied between 0.2 - 86 ppm. The average levels of lead in the roots was larger up to 50% (30.2 ppm) than in the stem (16.2 ppm) and leaves (15.4 ppm) of the seedlings. The application of compost on the media tailing significantly ($p \leq 0.05$) increased average level of Pb in the roots and stems, but it was decrease in leaves.

Applications of AMF significantly lowered ($p \leq 0.05$) average levels of Pb in roots, stems and leaves of seedlings, with the decrease of 18-33%. Exposure to Pb $(NO_3)_2$ by 0.5 mM (150 mg kg^{-1} tailings) resulted in an increase of lead content on the whole seedling tissues, but it decreased significantly when the Pb treatment in the tailings reached 1.5 mM (Table 2). The combination of compost together with AMF significantly ($p \leq 0.05$) increased the level of Pb in the roots, stems and leaves of control seedling (without Pb treatment) up to 6, 16 and 27 fold respectively. However, on the seedling exposed to lead tailings by 0.5 mM, the application caused the level of Pb in the seedling was lower (46.8 ppm in the roots, 30.2 ppm in the stem and 16.5 ppm in the leaves. When seedlings were exposed to 1.5 mM of Pb, the effect of compost and AMF application was inconsistent, since Pb content was decreased in the roots and leaves, while it was increased in the Jabon stem (table 2).

3.3. Bioconcentration factor of Pb (BF)

Bioconcentration factor (BF) shows the comparison of lead level which is transported into the seedling tissues to that present in the tailing media. After 12 weeks of lead exposure in the tailing media, lead bioconcentration factors varied widely, from 0.1-1.6. The highest BF presented by roots tissue followed by the stems, and then the leaves. The higher value of FB (exceeded 1) was shown only by the seedlings applied with compost without any exposure to lead (Pb Compost-0). The effect of AMF to BF was inconsistent among all applications so that there was no particular pattern on the value of BF of lead (figure 1).

3.4. Transport Factors of Pb (TF)

Transport Factors (TF) illustrates the comparison of Pb in the shoot (stems and leaves) to that in the roots of seedlings. The TF value was divers in Jabon seedling after 12 week application in media tailing exposed by different lead concentration and compost application in combination with mycorrhizal fungi (Figure 2). TF value of seedling Jabon was ranged between 0.1 - 1.1 for stem and

0.1 - 1.0 for leaves. The application of compost enhanced TF of lead to an average of 0.7 on the stem and 0.4 on the leaves. Applications of AMF did not change the average of TF (0.4 - 0.5). Increased lead exposure lowered TF on the stem to an average of 0.3, but increased TF of leaves up to 0.6. The TF value ≥ 1 was found only in seedlings without compost, without AMF and at the Pb level of 0.5 mMol, with compost, without AMF and Pb 0, and with compost and AMF and Pb 1.5 (Figure 2).

Table 1. Dry biomass of Jabon seedlings grown on tailing media exposed to different Pb concentrations and different applications of compost and AMF at 12 weeks after planting.

Treatment			Root		Stem		Leaves		Total	
			================== gram ==================							
Without Compost	-AMF	Pb 0	24.0	ab	70.0	bc	82.9	bc	176.8	bcd
		Pb 0.5	12.0	ab	2.0	a	15.5	ab	29.2	ab
		Pb 1.5	5.8	a	6.3	a	14.3	ab	26.2	ab
	+AMF	Pb 0	25.0	ab	24.6	ab	54.6	bc	104.0	bcd
		Pb 0.5	18.2	ab	16.3	ab	40.2	bc	74.5	bc
		Pb 1.5	25.3	ab	12.8	ab	50.3	bc	88.2	cd
Compost	+AMF	Pb 0	8.3	a	24.5	ab	111.8	bcd	144.4	bcd
		Pb 0.5	15.4	c	39.2	bc	68.6	bc	122.9	bcd
		Pb 1.5	14.6	bc	40.0	ef	168.9	bcd	223.3	de
	+AMF	Pb 0	20.7	ab	33.5	bc	174.8	bcd	228.9	de
		Pb 0.5	10.8	a	24.2	ab	72.1	bc	106.8	bcd
		Pb 1.5	18.8	ef	42.2	bc	162.5	bcd	223.3	de
F-Level		C	0.09	ns	36.68	s	42.04	s	32.93	s
F- Level		M	22.54	s	2.04	ns	3.09	s	4.95	s
F- Level		P	0.37	ns	0.51	ns	2.19	s	1.44	ns
F- Level		C*M	9.1	s	1.03	ns	0.01	ns	0.41	ns
F- Level		C*P	0.69	ns	0.75	ns	2.21	s	1.67	ns
F- Level		M*P	2.87	s	1.41	ns	1.05	ns	1.2	ns
F- Level		C*M*P	0.29	ns	0.26	ns	0.98	ns	0.48	ns

Note: Average data followed by similar letter in the same column means not significantly different based on 5% of DMRT test.
ns : p > 0.05, not significantly different based on 95 % confidence interval of DMRT test
s : p ≤ 0.05, significantly different based on 95% confidence interval of DMRT test
C : Compost, M : Micorrhiza, P : Pb, AMF : Arbuscular Mycorrhizal Fungi

3.5. Bioaccumulation of lead in tissue seedlings

Bioaccumulation of lead describes the amount of lead absorbed and accumulated in the seedling tissues. The result showed the diversity of bioaccumulation values in the seedling tissue which was influenced by Pb treatment and application of compost in combination with AMF as shown in Table 3. The entirely treatments either single, double and triple interaction of the factors including Pb treatments, compost and AMF significantly influenced bioaccumulation of Pb ($p \leq 0.05$) in the Jabon seedlings.

Part of Jabon shoots (stems and leaves) seem to accumulate more lead than the roots, reaching an average of 137.6 x 10^{-2} mg/plant at the tip. While in the roots it was only 43.3 x 10-2 mg/plant. The application of compost provided a significant increase ($p \leq 0.05$) on Pb accumulation, and in the shoots it was more than three-fold, but the increase was not significant in the root seedlings. The AMF application also increased Pb accumulation up to 50% in the shoots and 22% in the roots. Pb treatment also increased significantly Pb accumulation in the tissue seedlings, the largest accumulation when exposed to 0.5 mM Pb was in the root (56.4 x 10-2 mg / plants) and stems (45.4 x 10^{-2} mg / plant), and when it was exposed to 1.5 mM of Pb the largest accumulation was in the leaves (152.1 x 10^{-2} mg / plants) and shoots (190.3x10^{-2} mg / plant) (Table 3).

Table 2. The Pb content in the tissue of Jabon seedlings grown on tailing media exposed to different Pb concentrations and different applications of compost and AMF 12 weeks after planting.

Treament			Root		Stem		Leave	
			----------------------- ppm ----------------					
Without Compost	-AMF	Pb 0	3.5	a	1.1	a	0.3	a
		Pb 0.5	60.8	f	60.2	b	60.5	d
		Pb 1.5	43	e	1.9	a	17.7	bc
	+AMF	Pb 0	2.5	a	1.3	a	0.2	a
		Pb 0.5	26.4	c	10.5	ab	12.6	abc
		Pb 1.5	33.3	cd	3.9	ab	12.6	abc
Compost	-AMF	Pb 0	21.4	b	15.7	ab	6.8	ab
		Pb 0.5	42.2	de	20.3	ab	12.7	abc
		Pb 1.5	39.2	de	7.7	ab	10.7	abc
	+AMF	Pb 0	26.4	c	17.1	ab	8.4	ab
		Pb 0.5	46.8	e	30.2	ab	16.5	ab
		Pb 1.5	27.6	c	27.4	ab	20.7	c
F-level		C	6.66	s	1.75	ns	3.14	s
F-level		M	22.57	s	0.58	ns	7.32	s
F-level		P	163.30	s	10.62	s	25.08	s
F-level		C*M	27.23	s	7.68	s	18.05	s
F-level		C*P	20.38	s	3.47	s	15.66	s
F-level		M*P	13.57	s	4.04	s	11.49	s
F-level		C*M*P	16.15	s	3.95	s	8.18	s

Note: Average data followed by similar letter in the same column means not significantly different based on 5% of DMRT test.
ns : $p > 0.05$, not significantly different based on 95 % confidence interval of DMRT test
s : $p \leq 0.05$, significantly different based on 95% confidence interval of DMRT test
C : Compost, M : Micorrhiza, P : Pb, AMF : Arbuscular Mycorrhizal Fungi

4. Discussion

Jabon seedlings (12 weeks after planting) were still capable to grow following exposure to Pb for up to 1.5 mM However, plant biomass was declined for nearly 50%. It indicates that Pb can affect the metabolism of Jabon seedling or even Pb can be toxic at certain concentrations [8, 26]. It was reported that up to 1500 ppm, Pb exposure has caused death of *Paraserianthes falcataria* as indicated by clorosis and necrosis [27]. The existence of Pb ion can take over Mg ions in chlorophyll that may inhibit enzymatic activity of those enzymes responsible in the synthesis of chlorophyll [28,29], thus resulting metabolic constraints that reduces plant growth.

The application of compost and AMF, both in single and double treatments, has increased the total biomass of Jabon seedlings for more than 140% as well as improving seedlings vegetative performance when exposed to lead. The role of compost in improving seedling tolerance to lead exposure presumably due to improvements in the chemical, physical and biological tailing media that support metabolic activities of the plants Production of enough carbohydrates via photosynthesis will provide energy to improve plant growth through the activity of cell division, as well as to produce compounds (i.e organic acids) to chelate and efflux lead toward cells that do not perform active metabolism [30,31,32,33]. This compartment may reduce the toxic effects of lead to plant cells although level of Pb in tissue is increased, as observed in our study (table 2).

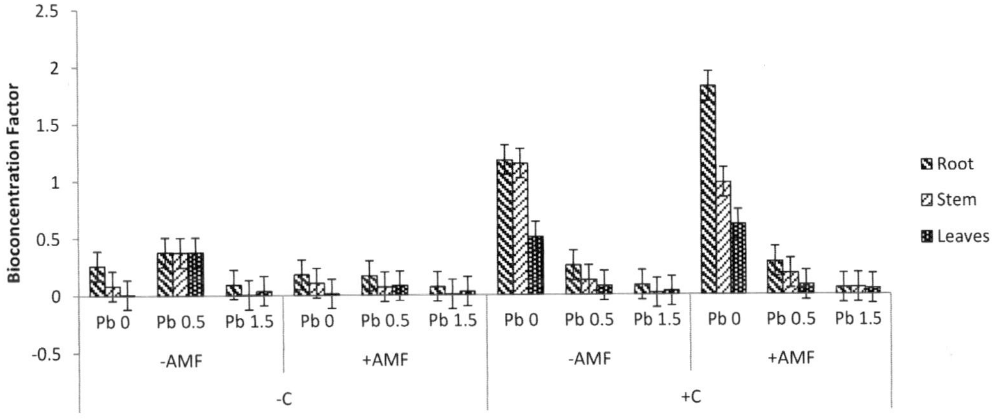

Figure 1. Bioconcentration factor (BF) of Pb in the tissue of Jabon seedlings grown at tailing media exposed to different Pb concentration and different application of compost (C) and AMF, 12 weeks after planting.

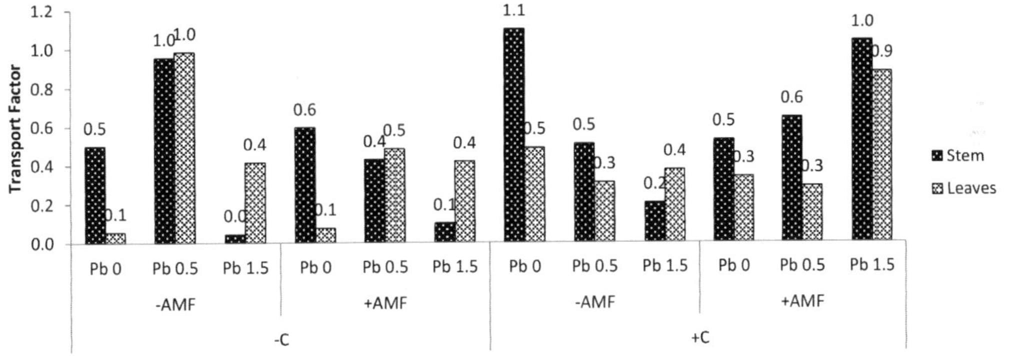

Figure 2. Transport Factor (TF) of Pb in the tissue of Jabon seedlings grown at tailing media exposed to different Pb concentration and different application of compost (C) and AMF, 12 weeks after planting.

AMF may contribute in improving seedling biotmass through direct and indirect ways. AMF may reduce the negative effects of Pb or increase seedlings tolerance by reducing level of Pb in tissue seedlings (table 2). Mycorrhizae may collect the heavy metals in the extrametrical mycelium, thus decreases heavy metals concentration in soil-root area as well as roots and shoots in the network [34, 35]. Mycorrhizae may help plants to escape from the toxic level of specific pollutants through the production of detoxification compounds, i.e such as organic acids or by binding pollutants in the tissues of fungi in the roots and form a physical barrier thaht inhibit the translocation of pollutants into the stem of the host plant [35]. Moreover, mycorrhizae can indirectly improve the ability of plants to survive on contaminated soil by improving plant nutrition status especially in soil with low phosphorus bioavailability, protecting plants from root pathogens, drought stress, improving soil aggregation and increasing the retention against xenobiotics [35].

Table 3. Bioaccumulation of Pb in the tissue of Jabon seedlings grown at tailing media exposed to different Pb concentration and different application of compost and AMF 12 weeks after planting.

Treatment			Root	Stem	Leaves	Shoot
			------------- x 10^{-2} mg/plant			
Without Compost	-AMF	Pb 0	8.2 a	7.7 ab	2.5 ab	10.2 ab
		Pb 0.5	69.4 bc	11.8 ab	96.0 cde	107.8 bcd
		Pb 1.5	22.3 a	1.2 a	26.1 ab	27.3 ab
	+AMF	Pb 0	6.1 a	3.2 a	1.1 a	4.3 a
		Pb 0.5	46.5 b	17.1 abc	51.1 abc	68.2 abc
		Pb 1.5	82.3 c	5.0 a	63.9 bcd	68.9 abc
Compost	-AMF	Pb 0	16.5 a	38.5 cd	76.3 bcd	114.7 bcd
		Pb 0.5	62.3 bc	79.5 e	87.6 cd	167.1 cd
		Pb 1.5	54.8 b	30.8 bc	181.2 f	212.1 de
	+AMF	Pb 0	53.17 b	57.3 de	147.2 ef	204.5 d
		Pb 0.5	47.53 b	73.0 e	119.6 de	192.6 cd
		Pb 1.5	50.33 b	115.6 g	337.2 g	452.9 e
F-level	C		622.5 tn	30409 n	118427 n	
F-level	M		682.1 tn	2585.7 n	13271 n	
F-level	P		9035.7 n	2137.7 n	51679 n	
F-level	C*M		75.9 tn	2132.9 n	20659 n	
F-level	C*P		853.7 n	605.2 n	29531 n	
F-level	M*P		1791.1 n	1728.1 n	8431.7 n	
F-level	C*M*P		21110 n	1650.9 n	326.1 n	

Note: Average data followed by similar letter in the same column means not significantly different based on 5% of DMRT test.
ns : $p > 0.05$, not significantly different based on 95 % confidence interval of DMRT test
s : $p \leq 0.05$, significantly different based on 95% confidence interval of DMRT test
C : Compost, M : Micorrhiza, P : Pb, AMF : Arbuscular Mycorrhizal Fungi

Jabon seedlings grown on the tailing media for 12 weeks which were exposed with different concentrations of lead were able to absorb and distribute Pb in all parts of their tissues, where the lead was mostly accumulated in root and less accumulated in stem and leaves. Bioconcentration factor values ranged from 0.1 to 1.6, demonstrated that the level of lead in the roots was widely varied, starting from the lowest for only one-tenth to the level of even exceeded from the concentration in the tailings media. The average levels of lead in roots was still 50% greater than in the stems and leaves, suggesting that Pb accumulation pattern was root> stem> leaf. This pattern is similar with previous study in other plants, including *Jatropha* [36] and *Chlorophytum comosum* [28]. The Application of Pb 0.5 mM significantly increased Pb level in the tissues, however when 1.5 mM of Pb was applied, the Pb level in the tissues was decreased (table 2) showing that application of high concentration of may inhibit Pb accumulation in the seedling tissues.

The application of compost markedly increased the levels of Pb in root and stem for 14% and 46%, repspectively, but to decreased by 30% in the leaves. The increased levels of Pb on the tissues in response to compost application was likely triggered by humic acid which is is able to bind metal as well as the normal pH value of compost (pH 6-7) which increased metal absorption by plant roots

[28,37]. It is reported that complex compound of humic acids consisted of carboxyl group and phenolic is able to form a complex binding with metal ions. Indeed, chelating formation is an important aspect of humic acid's role in regulating the availability (bioavaibility) of ion metals [38, 39]. Utilization of humic acid to increase heavy metal availability in the soil such as cadmium and lead has been demonstrated in wheat (*Triticum aestivum*). It is reported that higher concentration of humic acid (0, 140, 280 and 560 mg / kg) in the pot, leads to the increased concentration of heavy metals in plant biomass, while the plant biomass decreased with the increasing concentrations of heavy metals in wheat [40].

The AMF application, on the other hand, generally decreased levels of Pb in the tissues by 15% in the roots, 17% in the stem, and 33% in the leaves. Similar results were exhibited in previous study as lower levels of metals were detected in the tissues of infected Mindi (*Melia azedarach* Linn) seedlings with AMF in the tailing media applied with compost [2] and in maize infected with *Glomus intraradices* [41]. Meanwhile, opposite result was found in *G. intaradices*-inoculated tobacco plant [42] as well as in Brachiaria Grass [16]. It seems likely that AMF exhibits diverse effects on the levels of metal absorption by plants which is highly associated with one of the mechanisms of mycorrhizal fungi to protect the plant from heavy metal toxicity [22,43,44].

Based on the value of transport factor (TF) Pb in stem and leaves which were 0.5 and 0.4, respectively, it is likely that Pb was not transported from the roots to the shoots for more than 50%. The application of compost and FMA, although it increased the value of TF (the range of 0.4-0.7), however, the contribution was not significant because the range was still below one. This condition indicated that Jabon seedling grown in the tailing media applied with compost and AMF still categorized as metal excluder group or the plant prefer to save metal in the root [11]. The types of plants that are tolerant to heavy metals with high value of BF and lower TF has been suggested to be used as phytostabilizer in the contaminated area together with cover crop vegetations [45]. Jabon may have such properties and therefore has potential function as phytostabilizer for tailing area in order to minimize the occurrence of Pb migration into underground water and to reduce potential risk of entering Pb to the food chain.

Bioaccumulation becomes an important indicator to estimate potential crop in remediation activity. The application of compost and AMF provide a significant increase to the accumulation of lead in all parts of the seedling tissues, with the increase of 16-fold by compost, and 4-fold by AMF application. Consequently, a significant increase in biomass by compost and AMF application (Table 1) may cause even higher Pb accumulation. Such result has also been reported in the kenaf seedlings which treated with chicken manure application [7]. This fact indicates that despite the fact that Jabon seedling is categorized as metal excluder (TF<1), however, its capability to accumulate high amount of Pb in the shoots support the application of this plant as phytoextractor [42]. In addition, application of AMF and compost markedly promotes the Jabon seedling to conduct remediation activity.

5. Conclusion

After 12 weeks of lead exposure in the tailing media, the average levels of Pb in the roots of Jabon seedlings was up to 50% higher than that in the stem leaves. Compost increased the level of Pb in the roots and stems. Application of AMF decreased Pb levels for about 18-33%. Application of compost and AMF increased the accumulation of lead with the highest accumulation of lead was detected at 452.9 $\times 10^{-2}$ mg/plant in the shoot of the seedlings. Compost and AMF play an important role in phytoremediation activity mediated by jabo*n* seedling.

Acknowledgements

Acknowledgements to the Governments of Indonesia C.q. Directorate General of Higher Education for the research fund. Thanks are also conveyed to the Forest Biotechnology Laboratory, Research Center for Biological Resources and Biotechnology, Bogor Agricultural University, and Silviculture Laboratory, Nusa Bangsa University, Bogor, Indonesia.

References

[1] Ang L H and Ang T B 2000 *Proceedings of the Fourth Conference on Forestry and Forest Products Research* (Kuala Lumpur, Malaysian: Forest Research Institute Malaysia) pp 195–205
[2] Setyaningsih L 2007 *Pemanfaatan cendawan mikoriza arbuskula dan kompos untuk meningkatkan pertumbuhan semai Mindi (Melia azedarach Linn) pada media tailing tambang emas Pongkor* (Bogor: Sekolah Pascasarjana, Institut Pertanian Bogor)
[3] Pierzynski G M and Schwab A P 1993 *Journal Environ Qual* **22** 247-254
[4] Alloway B J 1995 *Heavy Metals in Soil. 2nd Edition* (London: Blackie Academic and Profesional)
[5] ATSDR Information Center 2015 *The ATSDR 2015 Substance Priority List*, https://www.atsdr.cdc.gov/spl/, Friday, 21 Oct. 2016, 5.55 am
[6] Clistenes W A and Baoshan X 2006 *Scientia Agricola* **63(3)**
[7] Mun H W, Ang L H and Lee D K 2008 *Journal of Environmental Science* **20** 1341-1347
[8] Kopittke P M, Asher C J, Kopittke R A and Menzies N 2007 *Environmental Pollution* **150** 280-287
[9] Salt D E, Smith R D and Raskin I 1998 *Annu. Rev. Plant Physiol Plant Mol Biol* **49** 643–68
[10] Kramer U 2005 *Biotechnology* **16** 133-141
[11] Baker A J M 1981 *Journal Plant Nutr* **3** 643-654
[12] Kim I S, Kang K H, Jhonson-Green P and Lee E J 2003 *Environmental Pollut* **126** 235-243
[13] Cunningham S D and Berti W R 2000 *Phytoextraction and phytostabilization: Technical, economic and regulatory considerations of the soil lead issue* In phytoremediation of contaminated soil and water. (Terry N, Banuelos G. eds) (Florida Lewis Publication) 359-376
[14] Altuhaish A, Hamim and Tjahjoleksono A 2014 *Emir. J. Food Agric* **26**(8) 716-722
[15] Ernst WHO 2005 *Chemie der Erde*, **65**(S1) 29-42
[16] Adriano D C, Wenzel W W, Vangronsveld J and Bolan N S 2004 *Geoderma* **122** 121-142
[17] Norland M 1993 *Soil Factors Affecting Mycorrhizal Use in Surface Mine Reclamation* (Bureu of Mines Information Circular: United States Departement of Interior)
[18] Killham K and Firestone M K 1983 *Plant Soil* **72** 39-48
[19] Smith S E, Read D J 1997 *Mycorrhizal Symbiosis* (London: Academic Press Limited)
[20] Chen B D, Zhu Y G and Smith F A 2006 *Chemosphere* **62** 1464-1473
[21] Gomes M P, P Silva M and C Nascentes 2015 *Bioremediation Journal* **19(2)** 151-159
[22] Shaker-Koohi and Sajjad 2014 *Int J Adv Biol Biom Res* **2(5)** 1854-1864
[23] Krisnawati H, Maarit K and Markku K 2011 *Anthocephalus cadamba Miq.Ecology, silviculture and productivity* (Bogor: Center for International Forestry Research)
[24] American Public Health Association 1998 *Standar methods for the examination of water and waterwaster* (USA: APHA, AWWA, WEF, Maryland, USA 20 the edition) p 3.56 & 4.178
[25] Jarvis M D and Leung D W M 2002 *Environmental and Experimental Botany* **48** 21-32
[26] Kabir M, Zafar I M and Shafiq M 2009 *Advances in Environmental Biology* **3(2)** 184-190
[27] Setyaningsih L, Y Setiadi, D Sopandie and B S Wilarso 2012 *JMHT* **XVIII (3)** 177-183
[28] Wang Y, Tao J and Dai J 2011 *African Journal of Biotechnology* **10** 14516-14521
[29] Xiao W, Hao H, Liu X Q, Liang C, Chao L, Su M Y and Hong F H 2008 *Biology of Trace Elements Research* **126** 257-68
[30] Taylor G J 1991 *Plant Biochemistry and Physiology* **10** 57-93
[31] Tong Y P, Kneer R and Zhu Y G 2004 *Trends Plants Sci* **9(1)** 7-9
[32] Reichman S M 2002 *The Responses of Plants to Metal Toxicity: A review focusing on Copper, Manganese and Zinc* (Melbourne: Australian Minerals & Energy Environment Foundation)
[33] Yang X, Ying F, Zhenli H and Peter J S 2005 *Journal of Trace Elements in Medicine and Biology* **18** 339–353
[34] Marschner H 1995 *Mineral Nutrition of Higher Plant* (London: Academic Press 2nd Ed)

[35] Vosatka M, Jana R, Sudova R and Martin V 2006 *Soils Phytoremediation Rhizoremediation* **9A** 237-257
[36] Shu, Xiao, Yin L Y, Zhang Q F and Wang W B 2011 *Environ Sci Pollut Res* **(9)** 1-10
[37] Lee S Z, Chang L, Yang H H, Chen C M and Liu M C 1998 *J Hazand Mater* **63** 37-49
[38] Stevenson F J 1994 *Humus Chemistry, Genesis, Composition, Reactions* (New York: Wiley) p 496
[39] Ghabbour E A and Davies, G 2001 *Humic Substances: Structures, Models and Functions* (Cambridge, U.K.: RSC publishing) ISBN 978-0-85404-811-3
[40] Baodong C and Yong-Guan Z 2006 *Journal of Soils and Sediments* **6(4)** 236-242
[41] Malcova R, Vostka M and Gryndler M 2003 *Applied Soil Ecology* **23(1)** 55-67
[42] Sudova R, Daniela P, Tomas M and Miroslav V 2007 *Applied Soil Ecology* **35** 163–173
[43] Leyval C, Turnau K and Haselwander K 1997 *Mycorrhiza* **7** 139-153
[44] Karimi A, Habib K, Mozhgan S and Mirhassan R S 2011 *African Journal of Microbiology research* **5(13)** 1571-1576
[45] Yoon J, Cao X D, Zhou X Q and Ma L Q 2006 *Science Total Environment* **368** 456-464

Predictive Modeling of Rice Yellow Stem Borer Population Dynamics under Climate Change Scenarios in Indramayu

E Nurhayati [1,2*], Y Koesmaryono[1] and Impron[1]

[1]Department of Geophysics and Meteorology, Faculty of Mathematics and Natural Sciences, Bogor Agricultural University, Jl. Meranti Kampus IPB Dramaga, Bogor – Indonesia 16680
[2]Stamet Radin Inten II Bandar Lampung, Meteorology Climatology and Geophysics Agency Jl. Alamsyah Ratu Prawira Negara KM. 28 Branti, Lampung Selatan – Indonesia 35364

Email : evanurhayati1012@gmail.com

Abstract. Rice Yellow Stem Borer (YSB) is one of the major insect pests in rice plants that has high attack intensity in rice production center areas, especially in West Java. This pest is consider as holometabola insects that causes rice damage in the vegetative phase (deadheart) as well as generative phase (whitehead). Climatic factor is one of the environmental factors influence the pattern of dynamics population. The purpose of this study was to develop a predictive modeling of YSB pest dynamics population under climate change scenarios (2016-2035 period) using Dymex Model in Indramayu area, West Java. YSB modeling required two main components, namely climate parameters and YSB development lower threshold of temperature (To) to describe YSB life cycle in every phase. Calibration and validation test of models showed the coefficient of determination (R^2) between the predicted results and observations of the study area were 0.74 and 0.88 respectively, which was able to illustrate the development, mortality, transfer of individuals from one stage to the next life also fecundity and YSB reproduction. On baseline climate condition, there was a tendency of population abundance peak (outbreak) occured when a change of rainfall intensity in the rainy season transition to dry season or the opposite conditions was happen. In both of application of climate change scenarios, the model outputs were generated well and able to predict the pattern of YSB population dynamics with a the increasing trend of specific population numbers, generation numbers per season and also shifting pattern of populations abundance peak in the future climatic conditions. These results can be adopted as a tool to predict outbreak and to give early warning to control YSB pest more effectively.

Keywords: DYMEX, rice yellow stem borer, climate change scenario.

1. Introduction

Agriculture is one of sector which is very vulnerable to climate variability or climate change that adversely affects plant productivity either directly or indirectly among others through pest organism's attacks. The presence of pest certainly becomes a limiting factor of crop production. One of the major pests of rice plants that has high level intensity of attacks and wider spread of the population in some Asian countries and Indonesia in particular, is *Scirpophaga incertulas* (Walker) (Lepidoptera:Pyralidae) which is famously called *Rice Yellow Stem Borer* (YSB). The development of YSB pest is very dominantly influenced by climatic factors that is very significant affected the pattern of population such as physiology, abundance, phenology and geographical distribution [1]. Furthermore, the cause of YSB pest spread was suggested as a result of climate change [2,3]. Therefore, it was predicted that the climate change would also affect the pest dimensional changes [4].

In Indonesia, YSB is a major insect pest in rice plants, especially in the northern part of West Java Province. Indramayu District is one of endemic regions of YSB pests which is located and is renowned as an area of rice production centers that have a large enough contribution in national rice production. In general, this area is tropical regions that have a clearly dry season and the rainy season (*monsoonal*). The area also has lowland tophografy that is influenced by the sea breeze from the North Coast (Pantura), so it has a climate characteristics that tends to dry. Even though with that climatic conditions, YSB population was noted to be able to survive and thrive (adaptative) in the region, where the conditions may suitability for the pest with the minimum threshold temperature range of 10-15°C and maximum of 35-40°C [1].

The general assessment of climate change projections have been done to the long-term climate change scenarios based on emission scenarios, while the near-term climate projections was based on the results of climate prediction on decadal or interdecadal time-scale [5]. The study of near-term climate change projections at present time is very important, especially in relation to the efforts and government consideration for public policy in national development such as the development of agricultural sector, infrastructure and others. With regarding this issue, the information of climate change is very important, for example, in defining and preparing management planning and control strategies to anticipate pest attacks in the future (±10-20 years). The *5th Assessment Report* of the *Intergovernmental Panel on Climate Change* (IPCC AR5) has prepared the latest climate change projection data based on a scenario of near-term projections from 2016-2035 period [6]. We need new adapted-method of implementation of early warning systems of pest attacks in the future, particularly to support the implementation of YSB pest control measures to be more effective. Therefore, DYMEX Model may be able to help prediction of population dynamics which include the development and mortality process, abundance and YSB population peak time in climatic conditions which tend to fluctuate [7]. This research aimed to develop a predictive model of YSB pest population dynamics under climate change scenarios (2016-2035 period) using a DYMEX Model in Indramayu area, West Java.

2. Materials and Method

2.1. Data Analysis and Tools

Daily climate data of Jatiwangi region (6°45'S - 108°06'E, 50 meters above sea level) was obtained from Climatological Station Dramaga-Bogor (January 1st 1996 - December 31st 2015) contained minimum and maximum temperature, minimum and maximum humidity and rainfall data. For daily imago light trap populations data was obtained from the Installation of The Agency of Plant-Pests Organisms Observation and Forecasting (*Pengamat dan Peramalan Organisme Pengganggu Tanaman* (PPOPT) in Indramayu District (January 1st 2003 - December 31st 2009). YSB imago populations data were used to do calibration and validation test of imago population model outcomes. This study was performed using DYMEX 3.0 software to analyze YSB population dynamics and Microsoft Excel 2010 for data processing and statistical analysis.

2.2. YSB Population Dynamics Modeling

DYMEX model was used to analyze the relationship between climatic parameters and organism population changes in a certain region [8]. DYMEX consists of two parts, namely Builder (consists of various modules that are integrated for building models) and Simulator (to run the model that has been built in the builder). YSB modeling inputs in Builder was constructed from two main components, namely climate parameters and YSB development lower threshold of temperature (*To*). Climate parameters function was used to estimate soil moisture value that would be a main input models through the lifecycle process (*lifecycle module*). The lifecycle process in DYMEX described the process of development and mortality, the transfer of individuals to the next stage and the fecundity and reproductive [9]. In the lifecycle process, *timer* and *latitude module* are needed to determine *daylenght* value and then with the addition input variables of temperature and humidity the *evaporation* value can be estimated. Furthermore, the evaporation together with input variable of precipitation would estimate the *soil moisture* value. Because YSB is kind of *holometabola insect*, so it could be modeled from the eggs, larvae, pupae to adult (imago) phase (figure 1).

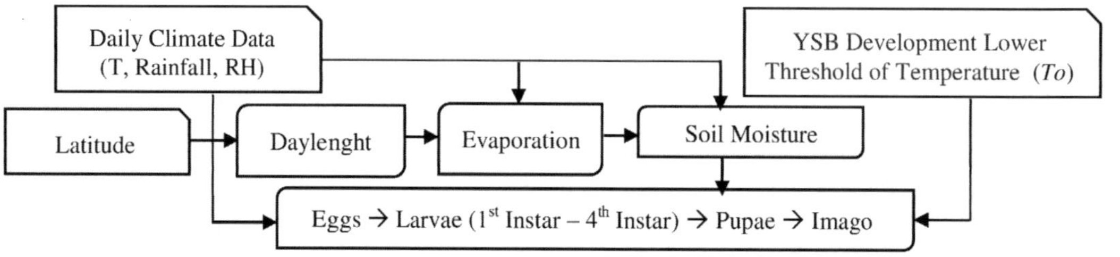

Figure 1. Diagram of YSB population dynamics in DYMEX Model.

The YSB development lower threshold of temperature value (*To*) was obtained from the value of the Development Rate (*DR*). *To* is a lower threshold temperature that could be tolerated by each phase of YSB development and has a relationship linear functions between temperature and *DR* value [10,9]. The *DR* value in each phase was determined using the equation $DR = 1/d$ [10a], where *d* was development period value (days) in egss, larvae and pupae phase at a constant temperature of 15-35°C which could be obtained from the table 1 [1a]. Linear regression model between the *DR* and *To* value could be obtained by the equation $y = a + bx$, with $y = DR$; $x = To$; *a* and $b = constants$ [2,9]. Determination of *To* phase imago was obtained from the equation of $DD = d\ (T\text{-}To)$, where *DD* is *Degree Days* and *T* is maximum temperature which is needed for the imago development at 35°C [9,13].

Table 1. The development period in YSB phase of eggs, larvae and pupae at a constant temperature.

Constant Temperature (°C)	Development Period (days)		
	Eggs	Larvae	Pupae
15	24.77	75.56	15.83
20	10.50	55.98	12.66
25	9.60	22.50	9.51
30	5.56	15.94	5.73
35	2.73	15.41	4.03

Source: Rahman and Khalequzzaman (2004)

The next process was setting mortality parameters in the model that was affected by the minimum temperature and soil moisture [8]. The transfer stages where the entire transfer phase after achieving full development, which reached physiological age is 1. If the physiological age threshold is set to a value of 1, the physiological age at any time would be the same as the accumulated degree days above threshold temperature of development [11,12,13]. The last stage was to determine the fecundity parameters and YSB reproduction. Fecundity is the ability of insects to reproduce that is modeled in the form of a YSB eggs potential parameter [14, 15]. Furthermore, in YSB reproduction process is assumed and controlled by the daily temperature cycles and precipitation [11,15].

DYMEX Simulator would be run after the Builder was formed to initialize the parameter inputs. Initialization models included setting values that is needed to run the model [8] and standard initialization implemented to ensure impersonation for simulation results [11]. The calibration and validation process of models could be achieved when value of accuracy level was over than 37%, and the model could be considered as a good result [11]. The evaluation of fitted models carried out based on the index of statistical coefficient of determination (R^2) between 0 – 1. The simulation model was applied using daily climate data with the baseline period of 20 years from January 1st 1996 - December 31st 2015 to analyze dynamic pattern of YSB population in Indramayu area at every stage of its development phase.

2.3. Climate Change Scenarios

Climate change scenarios (changes in air temperature and rainfall) based on a report AR5 [6] described the near-term climate change projections along 20 years period to the future (2016-2035). It referenced the baseline period (1996-2015) in sub region of Southeast Asia (10S-12N; 100E-150E) where Indramayu area is located. It was assumed that each value of climate change occurred in Southeast Asia region would represent the future climate change condition in Indramayu. It could be a new values to run the climate change simulation in DYMEX Model (table 2).

In this research the emissions increasing scenarios of *Representative Concentration Pathway* (RCP) was used. It consisted of RCP 2.6 model (*aggressive mitigation strategies*) and RCP 8.5 (*business as usual*). The different responses in dynamic population of YSB in each climate change scenarios on baseline climate conditions were analyzed using *t-test for two paired samples* (*paired two sample for means*) between the number of YSB populations before and after analysis of climate change scenarios.

Table 2. Temperature and rainfall changes under climate change scenarios (2016-2035 period).

Sub Region	Season	Climate Predictions in 2016-2035 Period					
		Tmin (°C)		Tmax (°C)		Rainfall (%)	
		RCP 2.6	RCP 8.5	RCP 2.6	RCP 8.5	RCP 2.6	RCP 8.5
Southeast Asia	Rainy (Winter)	0.3	0.3	1.2	1.2	0	2
	Dry (Summer)	0.3	0.4	1.2	1.3	1	0

Source: IPCC (2013); Rainy Season: October-March; Dry season: April-September; RCP 2.6: lower emissions; RCP 8.5: high emissions; Tmin: minimum temperature; Tmax: maximum temperature

3. Results and Discussion

3.1. Value Determination of YSB Development Lower Threshold of Temperature (To)

By using a linear regression model [10], *To* value of YSB could be obtained after calculating the value of *DR* in each phase of YSB development of amounted to 0.040, 0.095, 0.104, 0.180, 0.366 (for eggs); 0.013, 0.018, 0.044, 0.063, 0.065 (for larvae) and 0.063, 0.079, 0.105, 0.175, 0.248 (for pupae) at a constant temperature (15-35°C). *DR* value was also obtained, which was used to determine *To* value where no developmet in every phase of YSB (*DR = 0*). Thus *To* value was obtained of amounted to 14.39°C, 11.3°C and 10.65°C respectively for eggs, larvae and pupae phase. *To* value of imago phase was determained using the *Thermal Constant* or *Degree Days* (*DD*) and the average age (*days*) [1a], amounting to 75.16°days and 4.76 days. The value was 19.22°C for imago phase (table 3). These data then were used as a reference in preparing input data in Builder, in particular for using the YSB *lifecycle module*.

Table 3. Result of *To* value for the development egg, larva, pupa and imago of YSB phase.

Development Phase	Regression Equations [a]	R^2	*To* (°C)	*DD* (°day)	*d* (day)	*DR* (day^{-1})
Eggs	y = 0.0147x – 0.2115	0.84	14.39	56.27	2.73	0.367
Larvae	y = 0.003x – 0.0339	0.93	11.30	365.22	15.41	0.065
Pupae	y = 0.0093x – 0.099	0.92	10.65	98.13	4.03	0.248
Imago	-	-	19.22	75.16	4.76	0.224

[a]Temperature range for determination of *To* value in 15-35°C; *DR*: Development Rate; linear regression model $y = a + bx$, ($y = DR = 0$ if no development; $x = To$, a and b are constants); R^2: determination coefficient; *DD*: Degree Days; *d*: development period (Tmax=35°C)

Initialization parameter in Simulator was used to determine initial population numbers in each development phase. It assumed that the simulation model would be executed to satisfy the conditions in the field population numbers significantly, although the other factors outside climate was ignored. The time taken for each YSB development phase at a constant temperature of 28-36°C could influence the development period that was shorter. But when the temperature above 36°C was given, it would potentially reduce the populations of egg, larva, pupa and imago up to 50% (rate mortality of 0.5) [11]. It showed that the majority of YSB populations under higher temperature would have inactivity condition (*aestivation*), and their mortality even faster when the temperature was increased continuously up to >40°C [1]. This was occured because they were no longer able to tolerate these conditions. Therefore, based on the light trap data in Indramayu area gained an average of ±10 moths of YSB adult females per night. With the assumption that the potential eggs (fecundity) was 400 eggs per moth [15,14], so there would be found 4000 eggs, then a successful hatch into the first instar larvae were 2000 larvae, and the larvae became pupae were 1000 and finally a pupa became imago were 500 moths.

3.2. Calibration, Validation and Simulation Model

The calibration process was carried out based on the model of study area using daily YSB imago population observations data (light trap) that were available (January 1st 2003 – December 31st 2007) by testing the mortality parameters in the development phase of YSB imago. The model was built by ignoring factors of natural enemies (*biological agents*) in egg parasitism, application of pesticides and fertilizer, irrigation technique, that were used to improved varieties of rice and the influence of host plant to the YSB population that can cause mortality rates. Mortality parameters setting was done at each phase of development, particularly the imago. The parameters value was assumed to include the effects of which can lead to high or low mortality in every YSB phase outside climatic factors. Mortality assumption constant in eggs, larvae and pupae phase was 0.01 per day, whereas the mortality imago modeled using the rate of mortality of 0.1 per day. It reduced the population by 10% [8]. The best value of mortality was approaching the fluctuation pattern of observation data [7]. The variation values in imago mortality was tested from 0.1 to 0.25 per day. When the higher imago mortality parameters were applied, the model did not produce good results. Probably the model outputs of imago populations was smaller.

The best value on the calibration process was obtained at a mortality rate of 0.1 per day. It has been shown with R^2 value of 0.74. This condition could describe the prediction of peak abundance of YSB imago populations specifically population number per day. Although had a difference result, where the total predicted populations was higher than the observations number, however between them tended to have similar fluctuation pattern (figure 2a and 2b). In this regard, the observation data of light trap represented the presence or arrival indicators of YSB imago populations in a particular area. However, the data did not able to explain YSB imago populations number in the field related to the limitations of the scope area within the model. It was also associated to the YSB imago behavior (nocturnal insect pest) in response to the light where not all them attracted to the light traps depends on the power of the light or any changes occurred around the agroecosystem.

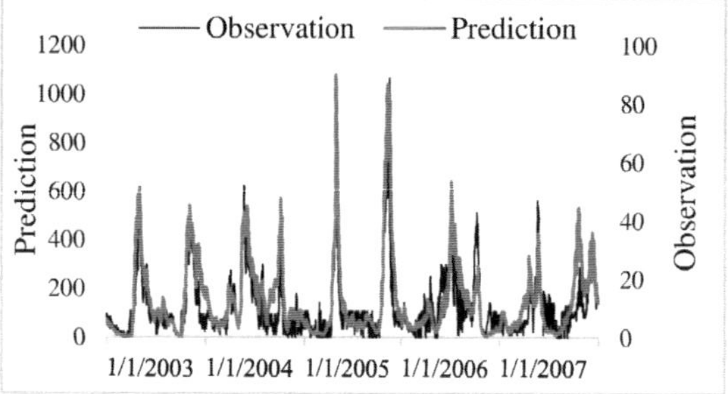

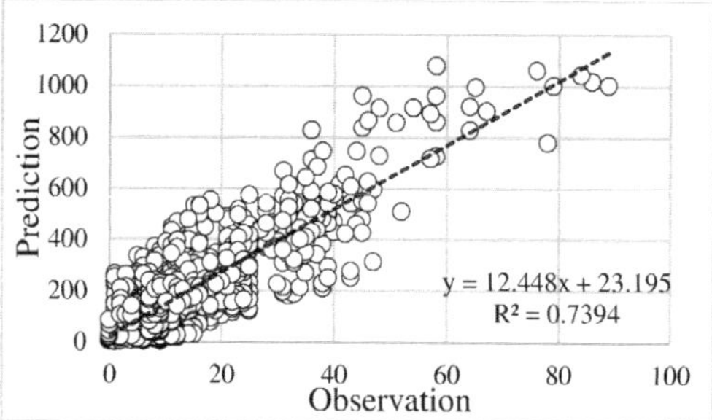

Figure 2. (a) The YSB imago population dynamics pattern of calibration results with use a mortality of 0.1 per day (01/01/2003 to 31/12/2007) (b) The coefficient of determination (R^2) between the populations of YSB imago prediction and observation of the calibration results.

Validation of the model was also carried out starting from January 1st 2008 - December 31st 2009. The results of validation model process was run very well so that it was able to predict the peak abundance of population during the simulation period with R^2 value of 0.88 (figure 3a and 3b). The accuracy level between calibration and validation results of model showed a high enough value that was above 37% [11]. This indicated that the model was considered to have good results, which was able to give an ilustration that YSB population dynamics in Indramayu area was strongly affected by climatic factors. Furthermore, the model was also able to simulate the response of YSB on baseline climate conditions in January 1st 1996 - December 31st 2015 (± 20 years).

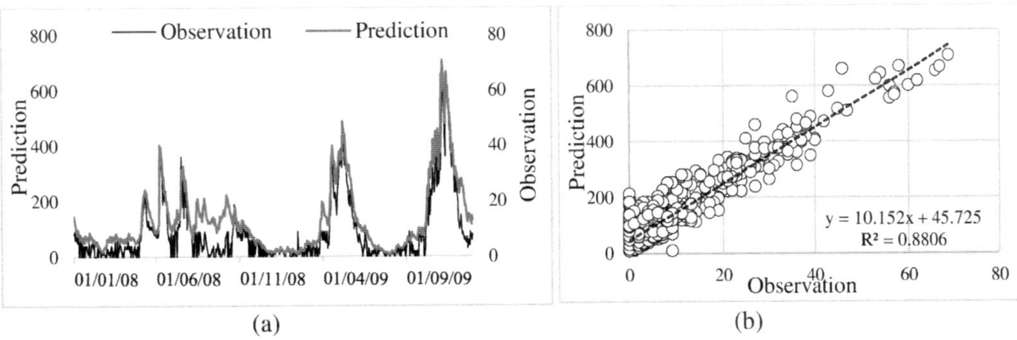

Figure 3. (a) The YSB imago population dynamics pattern of validation results (01/01/2008 to 31/12/2009) (b) The coefficient of determination (R^2) between the populations of YSB imago prediction and observation of validation results.

Simulation of the Model calculated monthly in the baseline of climate conditions in Indramayu area indicated that there was a tendency of abundance or imago population peak was occured in October when rainfall began to increase (>100 mm/month) and a decrease in rainfall intensity (<300 mm/month) that was in May. The YSB imago populations may reach >200 moths/month (figure 4a). Rainfall intensity between 0-30 mm/week that was ideal for YSB imago development, so they tended to be high and growing faster in these conditions [16]. Additionally, climate variables such as rainfall intensity between 5-40 mm/week or more was used as a reference of peak value of emergence YSB females imago to reproduce or the most ideal conditions for YSB imago to lay their eggs [11,15]. Optimum climatic conditions at the transitional season was the ideal conditions for YSB imago developmentwith the minimum and maximum temperature was about 23.3-24.2°C and 32.9-35.0°C and the minimum and maximum humidity of 44.5-64.9 % and 78.1-93.7%. On the contrary, YSB imago populations were lowest during January - March which was <60 moths/month. It was occur when an increase in rainfall intensity was ±300-400 mm/month (figure 4). High rainfall resulted in high mortality of YSB imago development. Furthermore, if the rainfall intensity >100 mm/week, YSB imago populations tended to decline dramatically [16].

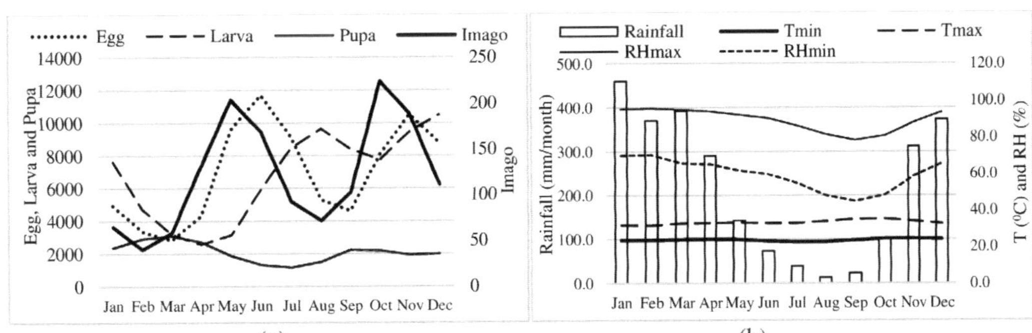

Figure 4. (a) The YSB population dynamics pattern of monthly simulation results average on baseline climate conditions (b) Climatic characteristics of Indramayu region on baseline climate conditions.

For Annual perspective (monthly simulation), the prediction of the peak abundance and populations number of YSB in Indramayu occurred on the beginning of October or at the First

Planting Season (PS. I) and on May or Second Planting Season (PS. II). During this season YSB reached more than 200 moths, then followed by emerging the eggs which reached the peak in June (PS. I) and November (PS. II) up to 12000 eggs. Subsequently, the larvae populations reached the peak (±10000 larvae) on August (PS. II) and December (PS. I). As a result, the abundance of pupae rose gradually from January to March (PS. I) and from September to November (transitional season) which achieved approximately 2000-3000 pupae. This proved that the climate cobditions greatly influenced the development and pattern of YSB population dynamics in Indramayu area.

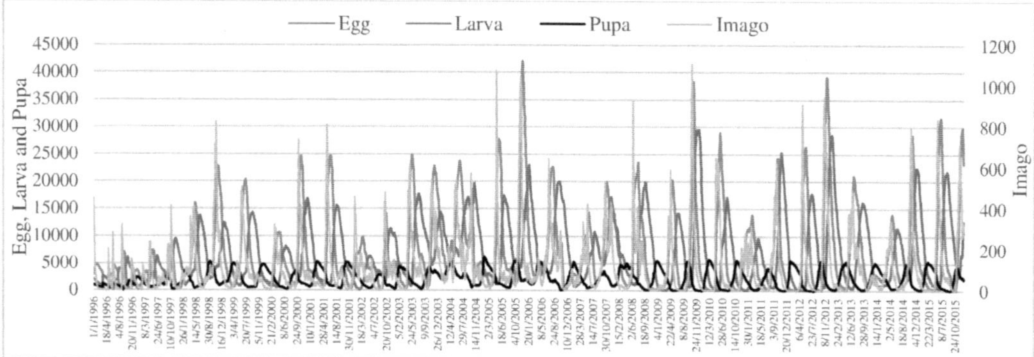

Figure 5. The YSB population dynamics pattern of daily simulation results on baseline climate conditions (period in 1996-2015).

For daily simulation, the model showed that YSB was continuously present, but the number was still relatively under control (below the threshold). In baseline climate conditions, the abundance of YSB imago populations tended to occur before rainy seson ended, and after dry season or during the early part of rainy season in every year. However, the potential explosion (outbreak) of YSB pests by climatic factors tended to occur when the phenomenon of extreme climate change was happened, such as after strong *El Nino* in 1997 followed by moderate *La Nina* which showed the explosion of YSB during the rainy season in 1998 (PS. 1998/1999). The other explosions occured when *Normal* or *La Nina* conditions and followed by weak to moderate *El Nino* were happened on PS. 2000/2001, PS. 2005 PS. 2005/2006, PS. 2008, PS. 2009/2010, PS. 2011/2012, PS. 2012, PS. 2013/2014, PS. 2014/2015, PS. 2015 and PS. 2015/2016 (figure 5). This data showed that the extreme environmental conditions such as *El Nino* and *La Nina*, which changes the distribution of rainfall could bring considerable influence to the surrounding organisms, including pests that tended to change drastically [17].

The high population of YSB has strong correlation with high or extreme temperature conditions, suggesting that high temperature triggers the development rate of YSB. Lower mortality rate also directly influenced the YSB developmental age which tended to make the cycle relatively shorter. In addition, the imago phase also reproduced faster under the high temperatures and low rainfall intensity conditions, which support the deployment of imago to be active until the laying of eggs. In normal conditions, each YSB female is able to produce about 100-600 eggs [14] and the age average of male and female are 2.5 and 5.09 day respectively [15]. It was assumed that the sex ratio between male and female was 1:2, and therefore one moth may produce approximately 400 eggs. It would also trigger the YSB development rate at the larva stages that have a lower mortality value. During the rainy season, larvae metamorphoses into pupae to become imago and imago reproduces quickly. If the temperature higher, the larvae development rate to the next stage would be more rapid. Therefore by the presence of host plants and water ability present every season or due to cropping pattern throughout the year (paddy-paddy-paddy) outbreak of YSB pest would be happened, where in a year might occur 2-3 generations of YSB populations.

3.3. Prediction of YSB Populations Under Climate Change Scenarios

The simulation models based on two climate change scenarios (RCP 2.6 and 8.5) in Indramayu area showed a significant difference in imago populations. The trend of populations for climate projection (2016-2035) for egg, larva and imago phase tended to increase, but pupa phase was decline from baseline conditions (table 4). The reason behind this was because pupa populations may be not able to tolerate high temperature which trigger their mortality. This may hamper the sustainability of pupae populations under those climate change scenarios. On the contrary, for the imago populations, it could support to its development. This probably due to YSB imago reached a comfort level at the optimum temperature of 33-35°C. The simulation model of two types of climate change scenarios used in Indramayu region showed a very high model sensitivity to climate change impacts to YSB population dynamics pattern in the future (figure 6).

Table 4. YSB population dynamics under climate change scenarios in 2016-2035 period.

Scenarios	Development Phase	Egg	Larva	Pupa	Imago
Baseline	Average Monthly Populations	9.637	6.796	2.087	122
RCP 2.6	t_{value}	-7.06	-35.82	4.41	-14.20
	P_{value} [a]	0.000	0.000	0.000	0.000
	S/TS	S	S	S	S
	Average Monthly Populations	10.438	7.806	1.980	263
RCP 8.5	t_{value}	-9.92	-41.77	5.44	-16.65
	P_{value} [a]	0.001	0.000	0.000	0.000
	S/TS	S	S	S	S
	Average Monthly Populations	10.813	8.092	1.963	268

[a] Confedence level by 5%; S: Significant; NS: Not Significant

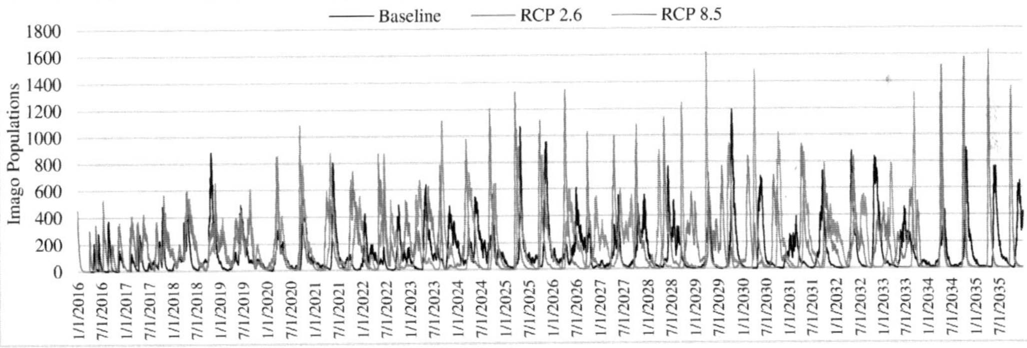

Figure 6. Prediction of YSB imago population dynamics in daily simulation (01/01/2016 to 31/12/2035) under climate change scenarios (RCP 2.6 and RCP 8.5).

The overall projection on application of two climate change scenarios described the rate of monthly YSB populations that was followed by shifting pattern of peak abundance in every phase. It occurred during 1-2 months ahead earlier than the baseline climate conditions. The increase of YSB populations in Indramayu area was due to highly comfort condition for YSB development or due to ideal growth conditions for each phase. The highest populations peak occurred during the changes of rainfall intensity, which from high to low or low to high (after rainy to dry season or after dry to rainy season). The monthly rainfall intensity that was changed from the baseline condition was 1.7 mm in dry season (RCP 2.6) and 3.3 mm in rainy season (RCP 8.5). In addition, increasing rainfall intensity, also followed by an increase in monthly minimum temperature which tended to be higher than the baseline climate condition which were 0.3°C (RCP 2.6) and 0.4°C (RCP 8.5) in dry season, while monthly maximum temperature raised up to 1.2°C (RCP 2.6) and 1.3°C (RCP 8.5). If during dry season it was supported by ideal humidity, so it was sufficient to provide optimum soil moisture levels, particularly in triggering the development of larvae and pupa phase. Both of the increase in temperature and precipitation significantly establish major influence to YSB imago development rate. From this, it can be calculated that the increasing abundance of specific YSB imago populations was more than 300 moths/month and generations number per year was reach more than 4 generations.

The prediction of monthly or seasonal YSB imago populations under the RCP 2.6 application scenario tended to be slightly higher than the RCP 8.5 before dry season (April-June) and ahead of rainy season (October-December). However, other conditions showed that the decreased back was happen ahead of rainy season peak (January-March) and in dry season peak (June-August) (figure 7). This might be due to YSB imago development rate in RCP 8.5 scenario that was highly influenced by the sensitivity of climate variables, particularly temperature and precipitation in every season differences. The mortality rate was 0.1 per day on RCP 8.5 application which tended to affect change in YSB imago populations. It was probably due to an increase in temperature and the rainfall intensity is too high. This factor could become a limiting factor of abundance which was slightly lowering the peak populations of YSB at the rainy or dry season.

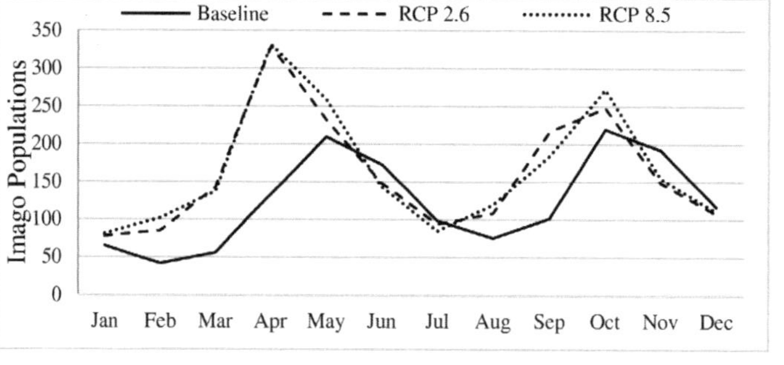

Figure 7. Prediction of monthly YSB imago population numbers under climate change scenarios.

In order to reduce the negative impact of the insecticides applied to control YSB pests, it is necessary to determine the economic threshold of insecticides application, so that the usage of insecticides is not excessive. The latest technology of YSB pest control was carried out immediately 4 days after the flight of moths (imago) both of the vegetative or generative phase, and it could be seen from the light trap catch [18]. The principle of this technology is not based on controlling intensity of larvae attack, but based on the effectiveness of YSB imago catches. The results of the study in West Java and Central Java showed that when the pest control was carried out after the attack, the rice yields went down. Therefore, the application of predictive YSB imago population dynamics under climate change scenarios as the output results from DYMEX Model become an alternative solution to determine the peak time of imago abundance.

4. Conclusions and Recommendations

The model DYMEX was able to predict the influence of climate on the dynamics of YSB population and produce a monthly or seasonal trend patterns, with the value of R^2 of 0.74 (calibration) and 0.88 (validation). The model was also able to predict the specific populations, generation number per season, and peak of populations in each season, and shifting pattern of peak abundance within climate change scenarios. The application of both climate change scenarios (2016-2035) of RCP 2.6 and RCP 8.5 was able to provide information required to assess the risk climate change impacts on YSB population densities for Indramayu farmers. Hence, the DYMEX Model could be used to formulate the management of YSB populations and control the pest in the future, especially on near-term projections. We recommend that for further assessments, the damage level analysis is required to estimate the potential yield losses due to the attack of YSB larva populations in the future, and to reduce the impact and improve control strategies more effectively.

Acknowledgment
We thank to the Pest Laboratorium PPOPT Installation in Indramayu District for daily YSB imago light trap data, and the Dramaga-Bogor Climatological Station for the daily climate data collection.

References
[1] Rahman M T and Khalequzzaman 2004 *Entomologia Sinica* **11** 47–60
[2] Litsinger J A, Alviola A L, Cruz C G D, Canapi B L, Batay-An E H and Barrion A T 2006 *International Journal of Pest Management* **52** 11–21
[3] Estay S A, Lima M and Labra F A 2009 *Journal of Applied* Entomology **133** 491–499
[4] Thomson L J, Macfadyen S and Hoffmann A A 2010 *Biological Control* **52** 296–306
[5] Candradijaya A, Kusmana C, Syaukat Y, Syaufina L and Faqih A 2014 *Informatika Pertanian* **23(2)** 159-168
[6] [IPCC] Intergovernmental Panel on Climate Change 2013 *Near-term climate change: The phisical science basis. Contribution of working group 1 to the fifth assessment report of the Intergovernmental Panel on Climate Change* (USA: Cambridge University Press)
[7] Koem S, Koesmaryono Y and Impron 2014 *J Entomol Indones* **11(1)** 1–10
[8] Maywald G F, Bottomley W and Sutherst R W 2007b *DYMEX model simulator version 3.* (Melbourne: CSIRO Publising)
[9] Nahrung H F, Schutze M K, Clarke A R, Duffy M P, Dunlop E A and Lawson S A 2008 *Forest Ecol Manag* **255** 3515–3523
[10] Lauziere I, Setamou M, Legaspi J and Jones W 2002 *J. Environmental Entomology* **31** 432–437
[11] Yonow T, Zalucki M P, Sutherst R W, Dominiak B C, Maywald G F, Maelzer D A and Kriticos D J 2004 *J. Ecological Modelling* **173** 9-30
[12] Manikandan N, Kennedy J S and Geethalakshmi V 2013 *Effect of elevated temperature on life-history parameters of rice yellow stem borer (Scirpophaga incertulas Walker)* Article *in* Current Science March 2016
[13] Shi P, Zhong L, Sandhu H S, Ge F, Xu X and Chen W 2011 *Ecologi and Evolution* **2** 58–64
[14] Suharto H and Usyati N 2008 *Pengendalian hama penggerek batang padi. In*: Padi inovasi teknologi produksi. BBPTP (Jakarta. LIPI Press) p 327-349
[15] Krishnaiah N V, Prasad A S R, Reddy C S, Pasalu I C, Mishra B, Krishna Y S R, Prasad Y G and Prabhakar M 2004 *Technical Bulletin* **7** 1-39
[16] Banjarnahor A R K 2013 *Analisis iklim dan dinamika populasi hama penggerek batang padi kuning di Indramayu dengan menggunakan model simulasi DYMEX 3.0* (Bogor (ID): Institut Pertanian Bogor)
[17] Koesmaryono Y, Hana F T and Yusmin 2005 *J Agromet* **19(2)** 13-23
[18] Baehaki S E 2013 *IPTEK Tanaman Pangan: Teknologi pengendalian hama penggerek batang padi* BBPTP **8(1)** 1-14

Risk Modelling of Agricultural Products

E H Nugrahani

Department of Mathematics, Bogor Agricultural University, Bogor, Indonesia

Email : e_nugrahani@ipb.ac.id

Abstract. In the real world market, agricultural commodity are imposed with fluctuating prices. This means that the price of agricultural products are relatively volatile, which means that agricultural business is a quite risky business for farmers. This paper presents some mathematical models to model such risks in the form of its volatility, based on certain assumptions. The proposed models are time varying volatility model, as well as time varying volatility with mean reversion and with seasonal mean equation models. Implementation on empirical data show that agricultural products are indeed risky.

1. Introduction

Agriculture is a source of employment and income, contributes to a country's food security, generates export revenues, sustains rural development, it provides nutrition and livelihoods for poor farmers, it is an important part of poverty eradication initiatives, and it supports human development of the poor [3]. Farming is typically a risky business, because facing a risk implies the possibility of losing property or income. Farm risks can be of financial and business nature. Financial risk refers to the method of financing. Business risk of a farmer is related to production, personal, price and institutional risk [1].

Agricultural products are different from other products manufactured in factories. Despite of the technologies advancements, the production depends heavily on the uncertainty of nature and climatic conditions. Being essential items for human consumption, demand of agricultural products remains constant, but supplies fluctuates widely, leading to price fluctuations, and occasional spikes. Methods of storage exist, but are added to the cost because agricultural products are perishable commodities. Such factors make agricultural products low margin, high volume, and fast moving items for traders.

The traditional pricing model for agricultural products is usually a cost based approach, where the selling price is a mark-up above the product cost, with the extent of mark-up depend on market conditions, such as demand and supply. Correct pricing is an important issue, because the yield of agricultural business that the farmers can enjoy depends largely on it. However, the actual condition is often not so promising. Instead, the farmers have to deal with fluctuating agricultural commodity prices in the world market. This leads to the fact that agricultural products' price has relatively high volatility parameters, which means that agricultural business is a quite risky business. This means, volatility process is an important element in pricing management [2].

This paper is intended to propose some risk models for some main agricultural products in Indonesia, in the form of some volatility models. Moreover, to gain some insights on its implementation, some empirical data are collected and their volatility are analyzed.

2. The Proposed Risk Models

First of all, it should be noted that in the market we can only observe the price of certain product. Therefore, let's start the model formulation by defining some mathematical process based on this price. Let P_t be the price of the agricultural commodity at time *t*. The return process is defined as follows.

$$y_t = \ln\left(\frac{P_t}{P_{t-1}}\right). \tag{1}$$

The variation of this return process will define its volatility.

The proposed models on risk of agricultural products include the model of time varying volatility, with its extension to the mean reversion model, and the seasonal model. The formulation of the proposed models is based on [2] and modified as follows.

2.1 Time varying volatility model

Under the assumption that the return follows the discrete time process

$$y_t = \mu + \varepsilon_t, \tag{2}$$

then the volatility satisfies the following model

$$\sigma_t^2 = \omega + \alpha(\varepsilon_{t-1})^2\,. \tag{3}$$

This model gives the measure of risk as the parameter σ_ε, which is the standard deviation of the error term ε_t in (2).

2.2 Time varying volatility model with mean reversion

Under the assumption that the return satisfies the following mean reversion process

$$y_t = \alpha(\mu - \ln P_{t-1}) + \varepsilon_t, \tag{4}$$

then the volatility model has the following form

$$\sigma_t^2 = \omega + \alpha(\varepsilon_{t-1})^2 + \beta(\sigma_{t-1}^2). \tag{5}$$

This model gives the measure of risk as the parameter σ_ε, which is the standard deviation of the error term ε_t in (4).

2.3 Time varying volatility model with seasonal mean equation

Assume that there exists weekly seasonal mean equation on the return process

$$y_t = \mu + \sum_{k=1}^{N}\left[\gamma_k \cos\left(\frac{2\pi kt}{5}\right) + \varphi_k \sin\left(\frac{2\pi kt}{5}\right)\right] + \varepsilon_t, \tag{6}$$

then the volatility model has the following form

$$\sigma_t^2 = \omega + \alpha\left(y_{t-1} - \left(\mu + \sum_{k=1}^{2}\left[\gamma_k \cos\left(\frac{2\pi kt}{5}\right) + \varphi_k \sin\left(\frac{2\pi kt}{5}\right)\right]\right)\right)^2 + \beta(\sigma_{t-1}^2)\,. \tag{7}$$

This model gives the measure of risk as the parameter σ_ε, which is the standard deviation of the error term ε_t in (6).

3. The Empirical Data of Indonesian Agricultural Products

The collected data are daily prices of some agricultural products in the time frame of June – October 2016, obtained from the Indonesian Ministry of Trade. The agricultural products under consideration are rice, sugar cane, beef and chicken meat. The graphical visualization of the data is given in the following figure 1.

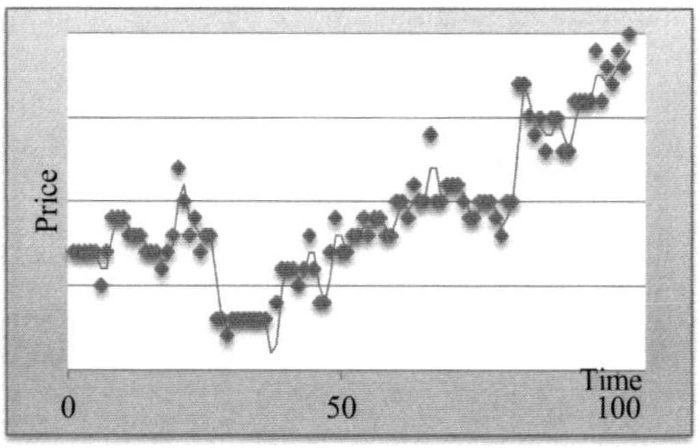

(a) Rice

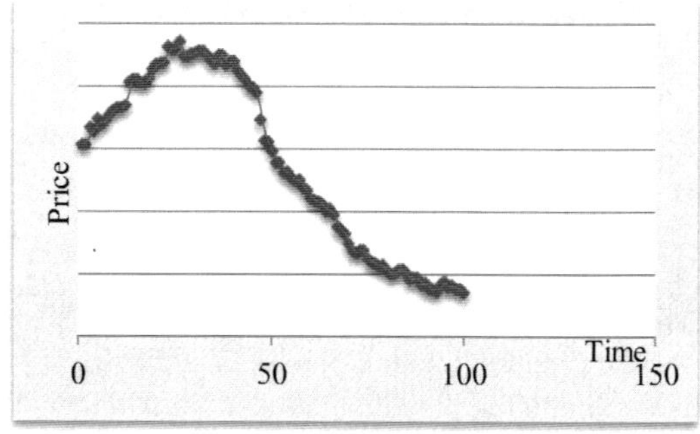

(b) Cane Sugar

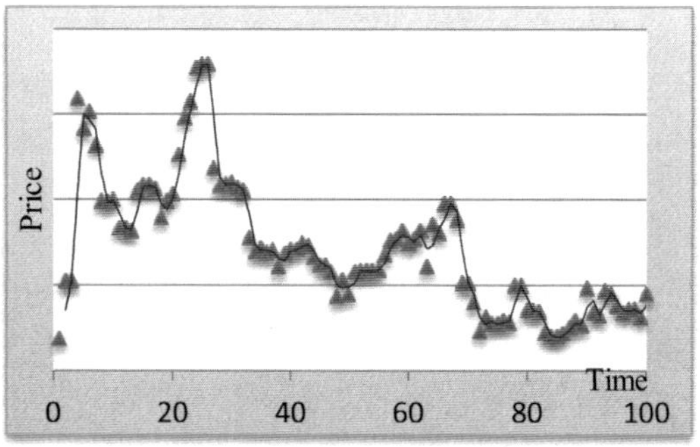

(c) Beef

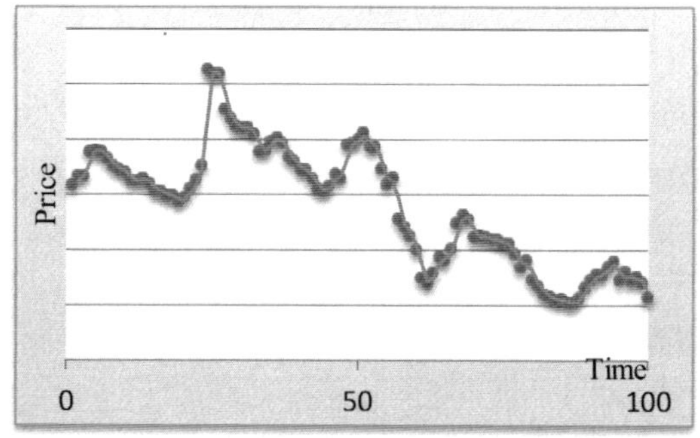

(d) Chicken

Figure 1. Daily prices of some agricultural products in Indonesia.

The price of all agricultural products under consideration in figure 1 show some significant fluctuations. Rice is the main source of food for most Indonesian, so the price of this strategic agricultural product is expected not to fluctuate widely. But it seems that this is not the case, the price of rice show quite significant fluctuation, similar to other agricultural products under consideration. For the case of chicken, it can be observed that there was significant price increase around beginning of the month July, which was the time of Ramadhan and Aidil Fitri celebration. It is widely known that, in this time of celebration, prices of almost all products in the market tend to increase significantly. On the other hand, the existing data show that rice and beef have the opposite trend, their price tend to decrease around this time. Nevertheless, both cases have similar impact, i.e. high fluctuation on the price of those agricultural products.

4. The Estimated Volatility

To obtain the idea of volatility, the following figure 2 gives graphical representation of the return according to (1) on the price of agricultural products under consideration. This graphical results show that there is significant fluctuation on the return process, which confirm that there are also fluctuations on the price of agricultural products. To be more specific, the fluctuation on the return of chicken price has the highest magnitude, so it can be estimated that this agricultural product has relatively high volatility. This will be discussed more closely in the next result of estimated volatility given in table 1.

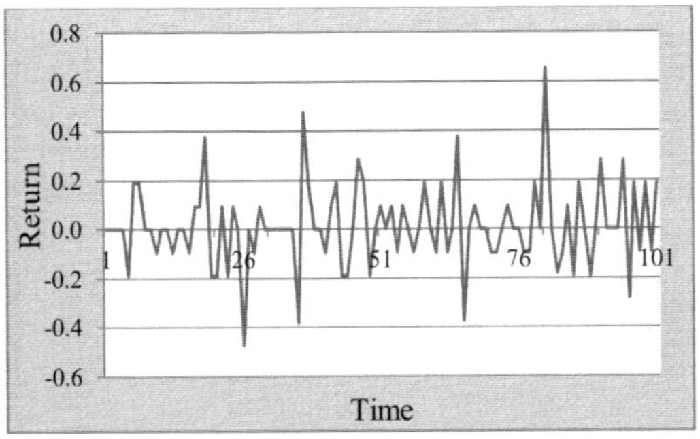

(a) Rice

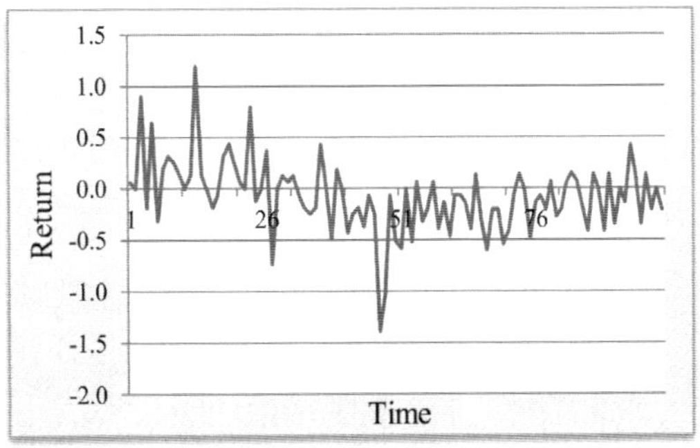

(b) Cane Sugar

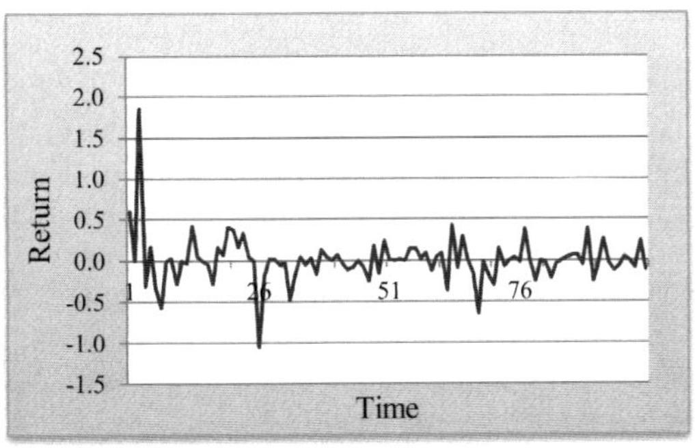

(c) Beef

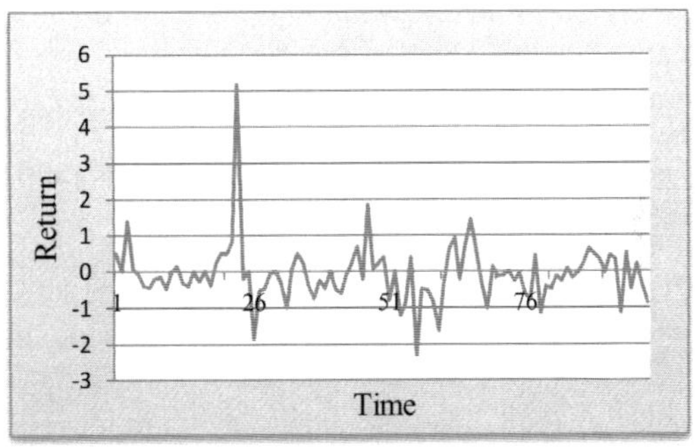

(d) Chicken

Figure 2. Return on price of agricultural products.

As a measure of risk, the volatility of each agricultural products was then computed according to each abovementioned model, i.e. the time varying, time varying with mean reversion and time varying with seasonal mean equation models. Model-1 is the model of time varying volatility as given in (3), while model-2 is the model of time varying volatility with mean reversion as given in (5), and model-3 is the model of time varying with seasonal mean equation volatility as given in (7). The values of estimated volatility are given in the following table1.

Table 1. Estimated volatility of the agricultural product's daily price.

Agricultural Product	Estimated Volatility		
	Model-1	Model-2	Model-3
Rice	16.70%	16.65%	74.85%
Cane Sugar	35.28%	36.05%	81.73%
Beef	29.48%	29.33%	75.73%
Chicken	82.54%	82.36%	105.65%

Table 1 gives the estimated volatility for all agricultural products, which show similar results for the first two models, i.e. the model of time varying volatility and the model of time varying volatility with mean reversion. On the other hand, the model of time varying volatility with seasonal mean equation show way too high volatility estimates, which is unreasonable. Therefore, it can be concluded that the model of weekly seasonal time varying volatility is not suitable for this case.

Furthermore, evaluating more closely the volatility among the agricultural products presented in table 1, chicken show the highest volatility, which confirms the fact that the fluctuation of its return has the highest magnitude as noted earlier. The next high volatile product is the cane sugar at the second place. Beef has relatively moderate volatility, despite of its high price in the market. Finally, the rice has relatively lowest volatility as expected. But nevertheless, the estimated volatility of almost 20% for such an important food product is rather alarming. This all at once concludes, that agricultural products contain a great deal of risk in its market price. Therefore, it also confirms that agricultural business is indeed a risky business.

5. Conclusions

Price of agricultural products show some fluctuations in the real world market, which implies that agricultural products can be considered as risky products. Some risk models in the form of volatility models are proposed. The high estimated volatilities on the prices confirm that agricultural products are risky.

References

[1] Ogurtsov VA, Van Asseldonk MPAM and Huirne RBM 2008 *NJAS* **56(1/2)** 39–58
[2] Ovararin K and Meade N 2011 Risk Modelling for Agriculture Commodity Prices The 31st Annual International Symposium on Forecasting (London)
[3] Stojanovski P and Muir-Wood R 2015 Comprehensive Disaster Risk Modeling for Agriculture *Planet @ Risk* Vol 3(1) pp 158–167, Davos: Global Risk Forum

Preliminary study : optimization of pH and salinity for biosurfactant production from *Pseudomonas aeruginosa* in diesel fuel and crude oil medium

A Z N Ikhwani, H S Nurlaila, F D K Ferdinand, R Fachria, A E Z Hasan[1*], M Yani[2], I Setyawati[1] and Suryani[1]

[1]Biochemistry Department, Mathematics and Natural Sciences Faculty, Bogor Agricultural University, Dramaga-Bogor, Indonesia
[2]Agroindustrial Technology Department, Agricultural Technology Faculty, Bogor Agricultural University, Dramaga-Bogor, Indonesia

Email : pakzainalhasan@gmail.com

Abstract. Biosurfactant is secondary metabolite surface active compound produced by microorganisms which is nontoxic and eco-friendly. Microorganism producing biosurfactant that is quite potential to use in many applications is from *Pseudomonas aeruginosa* strains. Good quality of biosurfactant production from microbes is supported by the suitable nutrients and environmental factors. The aim of this research was to obtain preliminary o data upon the optimum pH and salinity for the production of biosurfactant from *Pseudomonas aeruginosa* ATCC 15442 in diesel fuel and crude oil medium. *P. aeruginosa* ATCC 15442 cultured in diesel fuel and crude oil as carbon source showed biosurfactant activity. *P.aeruginosa*-derived biosurfactant was capable to form stable emulsion for 24 hours (EI24) in hydrocarbons n-hexane solutions. The particular biosurfactant showed EI_{24} highest value at pH 7 (31.02%) and 1% NaCl (24.00%) when *P. aeruginosa* was grown in 10% diesel fuel medium in mineral salt solution. As for the media crude oil, the highest EI_{24} value was at pH 6 (52.16%) and 1% NaCl (33.30%).

1. Introduction

Surfactants are compounds that can reduce the surface tension of a liquid, interfacial tension between two liquids or between a liquid and a solid. Surfactant is characterized as an organic compound consisting of a hydrophobic group (tail) and hydrophilic (head). This indicates that the surfactant molecules are not water soluble (oil-soluble components) as well as water-soluble components [1]. Meanwhile, biosurfactants is a surface-active compounds produced by microorganisms. Metabolites of these microorganisms are nontoxic and environmentally friendly. Characteristic of microbial surfactants is associated with surface activity, tolerance to pH, temperature and ionic strength, biodegradability, low toxicity, emulsion and de-emulsion ability, as well as antimicrobial activity. Those advantages make it a potential object of research for multiple applications such as in industrial cosmetics, pharmaceutical, food preservatives, detergents, petroleum, mining, metallurgy and many others [2].

Biosurfactants produced by various microorganisms, mainly bacteria, fungi, and yeast are vary in terms of their chemical compositions, as it depends on its natural environment, the type of microorganisms and nutrients availability. Environmental and nutrient conditions that may limit biosurfactant production including pH, temperature, agitation, aeration, dilution rate, metal ion concentration, as well as carbon and nitrogen sources [1]. Environmental factors such as pH is very important in the acquisition and characterization of biosurfactants. Quality of biosurfactant is determined by its emulsification activity. For example, a high emulsification activity of biosurfactant was produced by *P. aeruginosa* PBSC1 at pH 7 [3]. Other environmental factors that is quite important for some applications is salinity. Strain *Virgibacillus salarius* KSA-T was known to produce high concentration of biosurfactant at salinity of 4% (w/v) [4]. Meanwhile, production of surfactant by *P. aeruginosa* RS29 was decreased after the addition of NaCl> 0.8% (w/v), indicating that the particular strain in sensitive toward high salinity [5].

In this study, biosurfactant production by *P. aeruginosa* ATCC 15442 was optimized based on pH and salinity conditions. It was reported that *P. aeruginosa*-derived biosurfactant production is applicable to many purposes, including for the microbe-enhanced oil recovery (MEOR) and bioremediation. Biosurfactants produced from these bacteria is a type of glycolipid, particularly rhamnolipid, a secondary metabolite that is produced at the beginning of the stationary phase of the microbial growth [6].

This research aimed to optimize biosurfactant production from *P. aeruginosa* ATCC 15442 in laboratory scale. *P. aeruginosa* ATCC 15442 was grown in two types of fermentation medium, namely diesel oil and crude oil-enriched saline mineral (mineral salt solution (MSS)). The parameters used in this optimization were pH and salinity of the media. Biosurfactant obtained from each treatment were tested for emulsification index. The study was conducted in May 2016 to July 2016 at the Laboratory of Biochemistry, Department of Biochemistry, Faculty of Mathematics and Natural Sciences IPB.

1.1. Materials

The bacteria that used in this study was *P. aeruginosa* ATCC 15442 from Bogor Agricultural Institute Culture Collection (IPBCC) which routinely grown in nutrient broth (NB) or triptic soy agar (TSA) for starter culture. Media used in the growth of the bacteria producing biosurfactant was diesel oil and crude oil enriched with mineral salt solution (MSS). Mineral salts solution was composed by MSS composition used was 2.0 g/L KH_2PO_4, 5.0 g/L K_2HPO_4, 3.0 g/L $(NH_4)_2SO_4$, 0.1 g/L NaCl, 0,01 g/L $FeSO_4.7H_2O$, 0.2 g/L $MgSO_4.7H_2O$, 0.01 g/L $CaCl_2.2H_2O$, and 0.002 g/L $MnSO_4.H_2O$, with the addition glucose (0.03%) and yeast extract (0.03%). pH of the medium was adjusted by 4 N HCl and 3 M NaOH. The material used for biosurfactant activity response to test emulsification index (EI) is n-hexane. Other materials used are for refreshment bacterial liquid medium as well as starter medium and for a refresher on solid media.

1.2. Tools

Equipment used in for routine handling of bacteria including Petri dish, ose, Bunsen, Erlenmeyer flasks, beakers, laminar flow cabinets, autoclave, pH meter, incubator temperature, rotary shaker, and water bath shaker. Equipment used for biosurfactant analysisng via EI test including refrigerated centrifuge, test tubes, vortex mixer and a ruler. Other equipment used in this study was micropipette and 0.2 μm millipore filter, and other chemical glassware.

1.3. Procedures data analysis

1.3.1. Cultivation biosurfactant producing bacteria (modification Francy et al. 1991 [7]). P. aeruginosa ATCC 15442 was grown in 4% TSA (w/v) and then incubated at 37 °C for 18 hours. The single colonies of *P. aeruginosa* culture was further taken and grown in 5% NB (w/v) and incubated for 14 hours at room temperature using a rotary shaker (110 rpm). That broth culture was grown in 50 mL of a mixture diesel fuel or crude oil with MSS in 1% (v/v). Diesel fuel and cured oil were added to the media for for 10% and 5%, respectively. Culture was then incubated at 30 °C.

1.3.2. Effect of pH in biosurfactant production (modification Yakimov et al. 1995 [8]*).* Medium for the growth of *P. aeruginosa* was adjusted to pH value of 4, 5, 6, 7, 8, and 9. The medium was then incubated in water bath shaker at 110 rpm, 30 °C for an *optimum* time of biosurfactant production based on EI_{24} value. *P. aeruginosa* was incubated for 4 and 7 days in diesel oil and crude oil medium, respectively, which further were used for emulsification index analysis.

1.3.3. Effect of salinity on biosurfactant production (modification Prommachan et al. 2001 [9]*).* Effect of salinity on biosurfactant production was done by adding NaCl at different concentrations to the growth medium. NaCl concentration tested was 1%, 5%, 9%, 13%, and 17% (w/v) while pH was set up to 7. The culture was then incubated on a water bath shaker at 110 rpm, 30 °C for an optimum time of biosurfactant production based on EI_{24} in each medium which then was tested for emulsification index analysis.

1.3.4. Emulsification index analysis test (Pereira et al. 2013 modification [10]*).* A total of 2 mL of n-hexane was added to the 2 mL of the supernatant free of cell in a test tube. Supernatant was then centrifuged at 10000 g, 4 °C for 20 minutes which was followed with vortexing at high-speed for 2 minutes. Solution was then incubated for 24 hours at room temperature. Emulsification index was measured by the percentage of the emulsified layer height (cm) divided by the total height of the liquid in the tube (cm).

2. Result and discussion

P. aeruginosa biosurfactant-producing bacteria can be obtained from various sources, either directly from nature or culture collections that have been registered. *P. aeruginosa* used in this study was the collection of American Type Culture Collection (ATCC) strain number 15442 which is the re-culture from Institut Pertanian Bogor Culture Collection (IPBCC). This bacterial strain was isolated from water bottle animal with optimum growth temperature of 37 °C in TSA (tryptic soy agar) or TSB (tryptic soy broth) media under aerobic condition. Since the purpose of this study was to obtain good quality of biosurfactant, thus the *P. aeruginosa* was cultured in the presence of degradable hydrocarbon resources [12] and minimal nitrogen sources [13]. The carbon source in the media was either diesel oil or crude oil, while the nitrogen source was obtained from some mineral salts in the MSS and yeast extract.

Microbial fermentation process needs to be optimized to obtain a good quality of biosurfactant. In this study, optimization of pH and salinity of each media were conducted with further observation toward the emulsification index for 24 hours (EI_{24}).

2.1. Effect of pH in growth medium to emulsification index

Supernatant from the culture of *P. aeruginosa* used in this study showed biosurfactant activity on diesel fuel and crude oil medium at a slightly different pH conditions. Stable emulsion for 24 hours was seen at both neutral pH (7) and basic pH (8-9) on the diesel fuel medium, while in crude oil medium, it was found stable at acidic pH (5-6). On the other hand, at alkaline pH, the EI_{24} value was fairly low for less than 15%. The highest EI_{24} value in diesel medium and crude oil medium was found at pH 7 (27.35%) and pH 6 (52.16%), repectively (figure 1 and 2).

EI_{24} value optimum at pH 7 of diesel fuel medium in this study was also obtained by Saikia *et al.* (2012) with the same incubation time (4 days), but different carbon source, temperature and agitation speed. Saikia *et al.* (2012) get the value of EI_{24} about 70% at pH 7 and 8 with the incubation temperature was 35 °C and 110 rpm agitation speed—meanwhile in this study use 30 °C and agitation speed was 150 rpm—using glycerol as carbon source. Glycerol is used by Saikia *et al.* (2012) after previously tried with some other carbon sources. Their research showed that the strain of bacteria they used, namely *P. aeruginosa* RS29 produce maximum biosurfactant on the water-soluble substrate.
The highest EI_{24} using 10% diesel fuel medium in this research only get 27.35% at pH 7. A lot of data already reported mention that EI_{24} value > 50% is potential for biosurfactant production by

microorganisms (Rodríguez-Rodríguez *et al.*, 2012). Research of Prabakaran and Sumathi get 76% EI_{24} value on the carbon source media of 1% diesel fuel with pH 7, in the enrichment Bushnell Haas media, at 37 °C [14]. Furthermore, the enrichment media used in this study is MSS that being modified by Francy *et al.* (1991) which refered from Knetting and Zajic [15]. The difference EI_{24} result is fairly high value that can be caused by several factors, such as the bacterial strain used and the conditions of growth medium.

The enrichment compositions medium in this study and Prabakaran and Sumathi (2014) is almost the same. Both media are commonly used to study the utilization of hydrocarbons by microorganisms. The content of the medium is all important nutrients for bacterial growth except hydrocarbons as an energy source.

Magnesium sulphate ($MgSO_4$), calcium chloride ($CaCl_2$), and ferric chloride ($FeCl_3$) in Bushnell Haas media or iron phosphate ($FeSO_4$) on Knetting-Zajic media provide essential elements for the growth of the organism. Potassium phosphate is a buffer agent and ammonium nitrate (NH_4NO_3) on Bushnell Haas media or ammonium sulphate ($(NH_4)_2SO_4$) on Knetting-Zajic media is a source of nitrogen. Other elements present in the media Knetting-Zajic but none on Bushnell Haas medium are Mn, Na and Cl. In addition, the MSS which refers Francy *et al.* (1991) added 0.03% glucose as the carbon source and 0.03% yeast extract as nitrogen source.

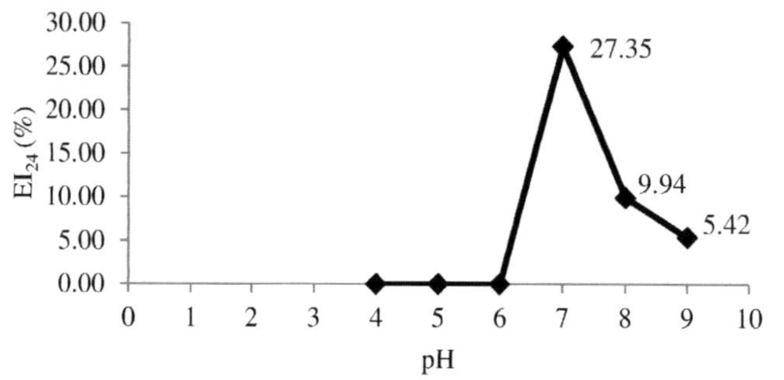

Figure 1. Effect of pH on the emulsification index (EI_{24}) of biosurfactant produced by *Pseudomonas aeruginosa* in diesel fuel medium.

Eraqi *et al.* research used 1% glycerol as a carbon source and 2% $NaNO_3$ as nitrogen source showed that rhamnolipid was produced at its maximum level at pH 6 [16]. Moreover *P. aeruginosa* OCD1 in showed the highest EI_{24} value at pH 6 with 2% n-octadecane as carbon source [17] which is similar to our result. Interestingly, it was reported that high EI_{24} (75.12%) value could be shown by *P. aeruginosa* PBSC1 at pH 7 in 1% crude oil as carbon source. Meanwhile, lower EI_{24} was exhibited by this particular bacterial strain when it was cultivated in pH 6. The bacteria used in their study was *P. aeruginosa* PBSC1 isolated from mangrove ecosystem.

Rhamnolipid which is the main type of bacterial biosurfactant strains derived from *P. aeruginosa* is known to reversibly change their morphology due to pH alterations. Their molecular aggregates changing from vesicles to lamella under acidic conditions and from lipid particles to micelle at slightly acidic conditions, in narrow pH range about 5-7. Rhamnolipid therefore is anionic surfactant at pH 6.8 and almost entirely protonated at pH 5, so that they are nonionic [18]. Rhamnosyl group structure is not affected in this narrow pH range, since their dissociation occurs at higher pH values is more than 11 [19].

In this study, the emulsion of biosurfactant in two different biosurfactant producing media at pH 4 was not detected. Similar results was shown by previous study with surfactant rhamnolipid [16]. This can be explained by rhamnolipid under acidic conditions which is strong enough to be in the protonated form, so that the water solubility decreases [20] [21].

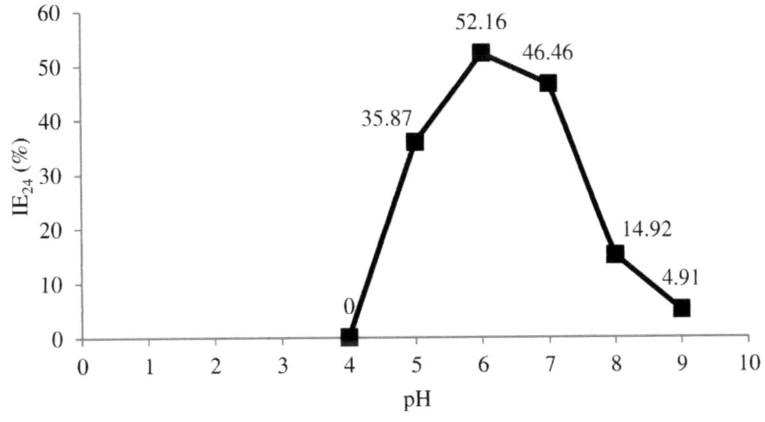

Figure 2. Effect of pH on the emulsification index (EI_{24}) of biosurfactant produced by *Pseudomonas aeruginosa* in crude oil medium.

2.2. Effect of salinity in growth medium on emulsification index

The concentration of salt in certain media can influence the production of biosurfactant from microorganisms. Indeed, in our study, the supplementation of salt for more than 1% did not induce any emulsion, both in diesel fuel and crude oil medium. The emulsion was formed only in 1% NaCl addition both in diesel fuel and crude oil medium for about 24.00% and 26.39%, repectively (figure 3 and 4). Previous study showed that 1.5% NaCl resulted an optimum production of biosurfactant [22]. *P. aeruginosa* used in this study was isolated from fishing harbor area. Other studies reported that *P. aeruginosa* DHT2 isolated from oil contaminated soil was tolerance to salinity for up to 10% due to biosurfactant production. Intrestingly, this *P. aeruginosa* DHT2 strain which were grown in diesel fuel medium as carbon source could yield 60% EI_{24} at 30 °C [23]. NaCl is known to activate the biosurfactant performance of many strains isolated from sea water or oil reservoir [8].

Previous study showed that EI_{24} value of biosurfactant produced by P. aeruginosa RS29 was decreased in NaCl supplementation for more than 0.8%, however, treatment of 1% NaCl could still maintain its EI_{24} for around 70%, significant reduction for about 40% was detected in 5% NaCl treatment. This indicated that *P. aeruginosa* strain RS29 was not quite tolerant toward salinity.In our study, activity of biosurfactant was not affected (EI_{24} value $< 50\%$) by various concentrations of NaCl tested.

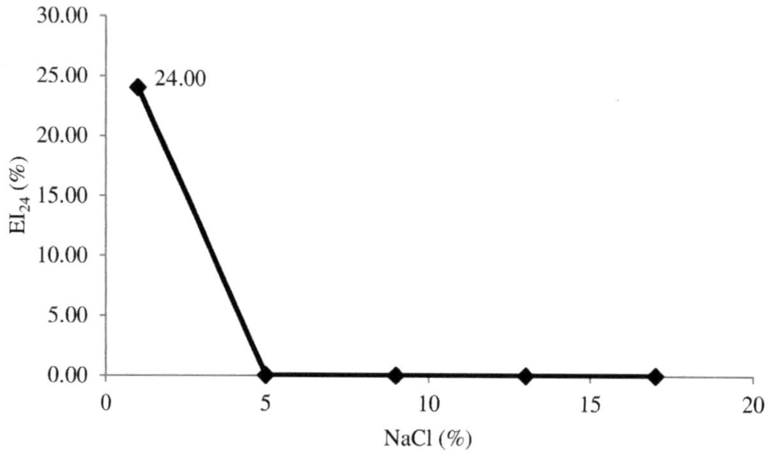

Figure 3. Effect of salt (NaCl) on the emulsification index (EI_{24}) of biosurfactant produced by *Pseudomonas aeruginosa* in diesel fuel medium.

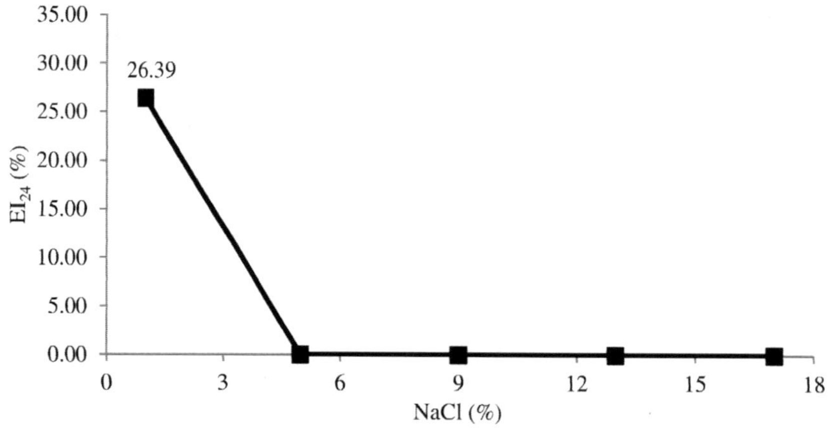

Figure 4. Effect of salt (NaCl) on the emulsification index (EI_{24}) of biosurfactant produced by *Pseudomonas aeruginosa* in crude oil medium.

3. Conclusion

Pseudomonas aeruginosa ATCC 15442, cultured in diesel fuel and crude oil as carbon source, showed biosurfactant activity. Biosurfactant activity is clear from stable emulsions for 24 hours (EI_{24}) in the emulsification test performed. Preliminary optimization showed EI_{24} highest value at pH 7 (27.35%) and 1% NaCl (24.00%) on 10% diesel fuel in mineral salt solution (MSS) medium. As for the crude oil media, the highest EI_{24} value was resulted at pH 6 (52.16%) and 1% NaCl (26.39%).

References

[1] Saharan B S, Sahu R K and Sharma D 2011 *Genet Engin Biotechnol J* **29**
[2] Rahman P K S M and Gakpe E 2008 *Biotechnol* **7(2)** 360-370
[3] Joice P A and Parthasarathi R 2014 *Int.J.Curr.Microbiol.App.Sci.* **3(9)** 140-151
[4] Elazzazy A M, Abdelmoneim T S and Almaghrabi O A 2015 *Saudi J Biologic Scien.* **22** 466–475
[5] Saikia R R, Deka S, Deka M and Banat I M 2012 *Ann Microbiol.* **62** 753-763

[6] Singh A, Hamme J D V and Ward O P 2007 *Biotechnol Adv* **25** 99–121
[7] Francy D S, Thomas J M, Raymond R L and Ward C H 1991 *J Ind Microbiol* **8(4)** 237-246
[8] Yakimov M M, Timmis K N, Wray V and Fredrickson H L 1995 *Applied Environm Microb.* **61(5)** 1706-1713
[9] Prommachan, H-Kittikun and Kawai F 2001 *The Annual Meeting of the Thai Society for Biotechnology*, Bio Thailand: From Research to Market; Bangkok, Thailand (TH).
[10] Pereira J F B, Gudiña E J, Costa R, Vitorino R, Teixeira J A, Coutinho J A P and Rodrigues L R 2013 *Fuel* **111** 259-268
[11] Müller M M, Kügler J H, Henkel M, Gerlitzki M, Hörmann B, Pöhnlein M, Syldatk C and Hausmann R 2012 *J Biotechnol* **162** 366-380
[12] Perfumo A, Smyth T J P, Marchant R, Banat I M 2010 *Production and roles of biosurfactants and bioemulsifiers in accessing hydrophobic substrates* (Berlin Heidelberg (DE): Springer) p 1501-1512
[13] Reis R S, Pereira A G, Neves B C, Freire D M G 2011 *Biores Tech* **102** 6377-6384
[14] Prabakaran M and Sumathi R 2014 *Int J Curr Res Chem Pharma Sci.* **1(1)** 69-78
[15] Knetting E and Zajic J E 1972 *Biotech Bioeng* **14** 179-390
[16] Eraqi W A, Yassin A S, Ali A E and Amin M A 2016 *Biotechnol Res Internat* **2016** 3464509
[17] Sahoo S, Datta S and Biswas D 2011 *J Adv Sci Res* **2(3)** 32-36
[18] Özdemir G, Peker S and Helvaci S S 2004 *Colloids Surfaces A: Physicochem Eng Aspects* **234** 135–143
[19] Ishigami Y, Gama Y, Fumiyoshi I and Choi H K 1993 *Langmuir* **9** 1634-1636
[20] Ishigami Y, Gama Y, Nagahora H, Yamaguchi M, Nakahara H, Kamata T 1987 *Chem Lett* **16** 763–766
[21] Schenk T, Schuphan I and Schmidt B 1995 *J Chromat A* **693** 7-13
[22] Gomathy C, Senthilkumar R 2013 *Internat J Res Environm Sci Technol* **3(3)** 86-91
[23] Kumar M, León V, Materano A D S, Ilzins O A and Luis L 2008 *World J Microbiol Biotechnol.* **24** 1047–1057

Atomic Orientation Detection using Nonlinear Optics and its Application as an Environmental Sensor

M R Tawakal[1], H Alatas[1] and H Hardhienata[1*]

[1]Theoretical Physics Division, Department of Physics, Bogor Agricultural University, Jl. Meranti, Gedung Wing S, Kampus IPB Darmaga, Bogor 16680, Jawa Barat, INDONESIA

Email: hendradi@ipb.ac.id

Abstract. Recent advances in theoretical nonlinear optics have enhanced our understanding regarding the origin of the higher harmonic generation in particular interpreting the higher harmonic intensity data from nonlinear spectroscopy. In this work we show that the atomic bond orientation at the surface as well as the bulk of an inversion symmetric material e.g. Silicon (111) facet can be determined from nonlinear spectroscopy alone without the need to apply the conventional X-Ray diffraction method. Our simulation shows that nonlinear spectroscopy data interpreted using bond models can reliably determine atomic orientation almost in a direct way because the model is based on a very simple assumption that the radiations are created from charges along the atomic bonds. In addition we present a brief description of possible application to use nonlinear spectroscopy as an environmental sensor.

Keywords: bond model, second harmonic generation, third harmonic generation, atomic orientation

1. Introduction

Material characterization using nonlinear spectroscopy method posses several advantages compared to its linear counterpart. The nonlinear technique applies femto or pico laser pulses as its incoming source on the sampled material which is nondestructive and does not require vacuum conditions as in linear optics. It has the advantage over the former in that more information can be obtained regarding the material symmetry due to the involvement of higher order (nonlinear) susceptibilities. It has been shown convincingly that linear spectroscopy will not reveal any changes in the output intensity for a Si(111) surface under azimuth rotation because it is linearly isotropic [1]. When using nonlinear spectroscopy however the second harmonic generation (SHG) intensity varies with respect to the rotating azimuth angle because unlike the linear part some elements in the tensor are anisotropic with regard to the azimuthal rotation. This feature provides valuable information regarding the atomic bond orientation at the nanoscale.

The possibility to determine atomic bond orientation using nonlinearity has been made available through the work of Powell *et. al.* [2] who introduced the bond hyperpolarizability model called simplified bond hyperpolarizability model (SBHM). Their key idea is that the driving nonlinear polarization inside the material creates anharmonic motion along the atomic bonds which produces dipole radiation that can be detected in the far field. This assumption is valid for linear optics and SHG and to a good approximation also to THG although the latter may (or may not) require further assumption of transverse bond oscillation in addition to the charge oscillation along its atomic bonds. Although the bond model validity has been questioned especially when dealing with vicinal semiconductor surfaces [3] it was shown by Alejo-Molina *et. al.* [4] that this model is agreeing very well with crystal symmetry principle e.g. Group Theory and withstanding a crucial test when applied to zincblende systems as shown by Hardhienata *et. al.* [5]. These findings have brought the confidence of SBHM validity into a higher level.

2. Theory of Bond Model

In this work we show that the molecular bond direction of a diamond crystal with (111) orientation can be predetermined both at the interface and bulk using nonlinear spectroscopy simulation taking the assumption of the bond model. When only a single input frequency is applied the general response of the material can be stated macroscopically as:

$$\vec{P}(\omega) = \chi^{(1)}(\omega)\vec{E}(\omega) + \chi^{(2)}(\omega)\vec{E}^2(\omega) + \chi^{(3)}(\omega)\vec{E}^3(\omega) + \ldots \qquad (1)$$

where $\vec{P}(\omega)$ is the total polarization produced by the linear (first term in eq. 1) and nonlinear polarizations (second, third, ect. terms) inside the material. The linear and nonlinear susceptibilities are respectfully given by $\chi^{(1)}$ and $\chi^{(2)}, \chi^{(3)}$ and so forth. Here $\chi^{(n)}$ is called the $n-$ th order nonlinier susceptibility and $E(\omega)$ is the incoming driving field. Because diamond crystal is centrosymmetric or non-inversion symmetric bulk dipolar contribution at the bulk is disallowed for even order higher harmonics due to parity symmetry when spatial dispersion is not considered. Hence the even order nonlinear susceptibility in eq. (1) is zero. In other words SHG can be used to determine the atomic bond orientation at the surface/ interface because the contribution from dipolar inside the bulk is zero although quadrupolar contribution may have to be taken into account [6-7]. Meanwhile, third harmonic generation (THG) is allowed at the bulk thus later we show how we can use THG to find out the molecular orientation inside the bulk. The advantage of using SHG and THG SBHM model simultaneously is that we can single out and identify different orientation at the interface and bulk.

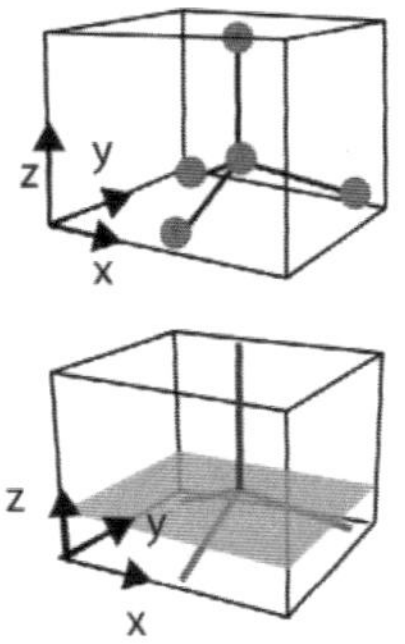

Figure 1. Atomic bond Orientation of a Si(111) facet (above, the upward bond is modelled by a different hyperpolarizability than the three down bonds (below).

The theory of the simplified bond hyperpolarizability model (SBHM) has been firmly established and explained lengthily in several research papers, particularly in Refs. [] and the curious reader is advised to read these works. In SBHM, particularly when doing second harmonic generation one first takes into consideration that the radiation comes from anharmonic oscillation of charges along the atomic bonds. According to the bond model the nonlinear far field for SHG and THG is calculated by:

$$\vec{E}_{g2\omega}(r) = \frac{k^2 \exp(ikr)}{4\pi\varepsilon_o rV}\left[(\bar{I} - \hat{k}\hat{k}) \cdot \sum_{j=1}^{4} \alpha_{2j}\hat{b}_j(\phi)(\hat{b}_j(\phi)\cdot\vec{E}_{loc})^2\right] \quad (2)$$

$$\vec{E}_{g3\omega}(r) = \frac{k^2 \exp(ikr)}{4\pi\varepsilon_o rV}\left[(\bar{I} - \hat{k}\hat{k}) \cdot \sum_{j=1}^{4} \alpha_{3j}\hat{b}_j(\phi)(\hat{b}_j(\phi)\cdot\vec{E}_{loc})^3\right] \quad (3)$$

where $\vec{E}_{loc}$ is the local field and $\hat{b}_j(\phi)$ is the onit vector of the $jth -$ bond. In principle for the Si case one can assume that the radiation is a sum of four bonds forming a tetrahedral structure ($sp - 3$ hybrid orbital) . The intensity of the nonlinear radiation can be obtained by multiplying the field in eq. (2-3) with its complex conjugate partner. In addition for the THG case in Si, the correct intensity must also consider Fresnel coefficients because THG dipolar nonlinear radiation comes mainly from within the bulk (it is not symmetry forbidden for odd order nonlinear susceptibilities).

Consider now the case of a diamond Silicon wafer for example, whose orientation is unknown, e.g. we want to determine without other spectroscopic methods methods whether the orientation is in the 111 or 100 direction. What is needed is to rotate the sample 360° about the vertical axis say the$-z$ axis and look at their intensity profile. The alerted reader may question the necessity to use nonlinear optics rather than just use linear rotational spectroscopy which is easier and cheaper to set up experimentally. The answer is that the linear radiation profile is isotropic with regard to rotation along the vertical axis for diamond and cubic lattices, the former belongs to Silicon which forms the backbone of modern day microelectronics. This can be traced back by looking at the linear susceptibility which is independent of the azimuthal rotation angle ϕ for an incoming p-polarized and outgoing p-polarized light as depicted in figure 1 taken from Ref. [1].

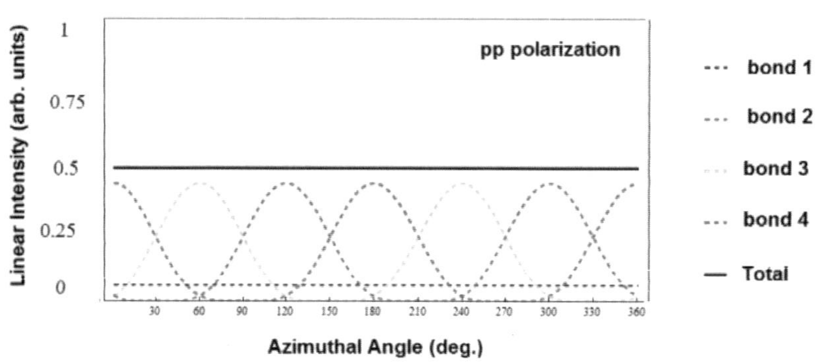

Fig 2. Linear intensity plot of a Si(111) orientation, although each bond produces peaks the total intensity is flat and reveal no dependence on the azhimuthal angle. [1]

In figure 2, the dashed lines depicts the linear radiation contribution from each of the Si(1111) bonds at the surface/interface. Interestingly although the linear susceptibility is anisotropic the field contribution from each bond is not flat but varies harmonically. However, the total sum of the field when squared (intensity) produces a flat intensity profile. Because the experimenter only sees the total intensity the information regarding the atomic bond orientation at the surface is lost.

In contrary the nonlinear radiation – even the first order nonlinear susceptibility already contains some elements in the tensor that is a function of the rotational azhimuth angle ϕ. One can straightforwardly see this by computing SHG susceptibility via:

$$\chi^{(2)}_{jkl} = \frac{N}{V}\sum_{j=1}^{4}\alpha_j b_j(\phi)\otimes b_j(\phi)\otimes b_j(\phi) \tag{4}$$

where N and V are respectively the amount of nonlinear dipoles per volume that radiates the nonlinear field and is just a scaling factor. If we have a Si111 facet the bond orientations are:

$$\hat{b}_1 = \begin{pmatrix} 0 \\ 0 \\ 1 \end{pmatrix} \quad \hat{b}_2 = \begin{pmatrix} \sin\beta \\ 0 \\ \cos\beta \end{pmatrix} \quad \hat{b}_3 = \begin{pmatrix} -\frac{\sin\beta}{2} \\ \frac{1}{2}\sqrt{3}\sin\beta \\ \cos\beta \end{pmatrix} \quad \hat{b}_4 = \begin{pmatrix} -\frac{\sin\beta}{2} \\ -\frac{1}{2}\sqrt{3}\sin\beta \\ \cos\beta \end{pmatrix} \tag{5}$$

Thus nonlinear optics reveals more information about the symmetry of the material than linear optics. In the following section we will disscuss the relevant results to determine the atomic orientation of a Si facet using SHG and THG.

3. Results and Discussion

The intensity for SHG is plotted for a Si(111) orientation can be obtained by squaring the far field in eqs. (2) and is presented in figure 3.

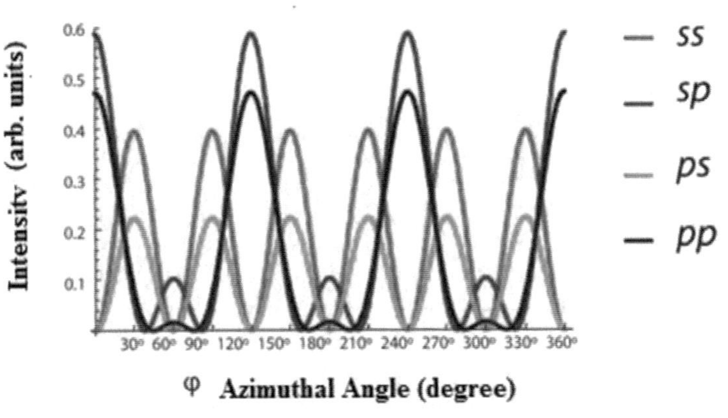

Figure 3. SHG intensity plot of a Si(111) surface orientation for the 4 polarizations.

The SHG intensity peaks can be explained as follows. When the Si(111) facet is shined by a *s* (fundamental) *–incoming* and *s* (SHG) *–outgoing* polarized light 6 intensity peaks are obtained. This can be explained from the basic assumption that each *down bonds* contribute twice to the peak, namely when the *s* incoming field is paralel to the bond direction which happens twice. Because there are 3 down bonds there are 6 peaks with equal height where each peak differ by a phase of $\pi/3$. Now assume that we do not know exactly how the atoms are oriented at the surface. By knowing experimentally that there are 6 equal peaks for the ss polarization we can guess using SBHM that the surface has C_{3v} symmetry so the orientation of the Si should be (111). If experimentally we have for the p –incoming and p-outgoing polarization 4 peaks instead it can analogously be shown using

SBHM that the orientation is Si(001) because the two bonds pointing upwards contribute 2 peaks whereas the two down bonds also contribute to 2 peaks this contrary to the case of a Si(111) *pp* SHG intensity because we expect only 3.

It is also notable that using SBHM we can determine the azhimuthal orientation of the atoms inside a material by looking at which azhimuthal angle the peaks occur. For the sake of clarity it is instructive to look at figure 4 which depicts a top view of the Si(111) surface. The blue line indicates the direction of the s-field. By careful analysis of the ss peaks it can be inferred that the first peak occurs after a rotation of $\frac{\pi}{6}$ so the down bond orientation at the surface before the rotation has to be as depicted. It must be stated that one cannot distinguish between any of the three down bonds so in this case the b_3 down bond can also be arbitrarily labelled as b_2 or b_4. Thus nonlinear optic RASHG can also serve as a direct method to access the molecular orientation especially for low symmetry structures such as Si.

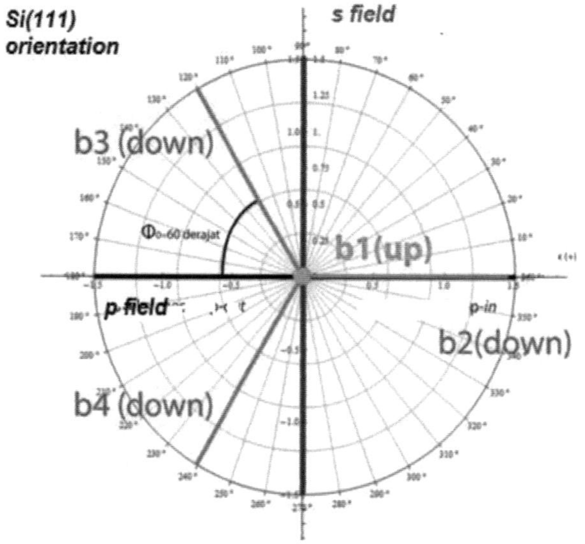

Figure 4. Top view orientation of the atomic bonds at the interface and bulk (in this case it is the same) for Si(111) (left).

The before mentioned discussion applies SHG to investigate surface atomic bond orientation which is valid because dipolar contribution are forbidden from within the bulk. Although this statement is not strictly correct because quadrupolar contribution from within the bulk also fairly contributes to the total SHG intensity, for certain cases such as Si(111) it does not produce significant change as to shift the phase of the peaks but only interferes with the dipolar contribution to increase or decrease the intensity peaks [6-7] as long as the atomic orientation at the surface and bulk are the same. Otherwise, one can still determine the atomic orientation but require a more complex model by assuming a different bond orientation for the quadrupolar SBHM model and add the fields with the dipolar surface radiation which is beyond the scope of our discussion.

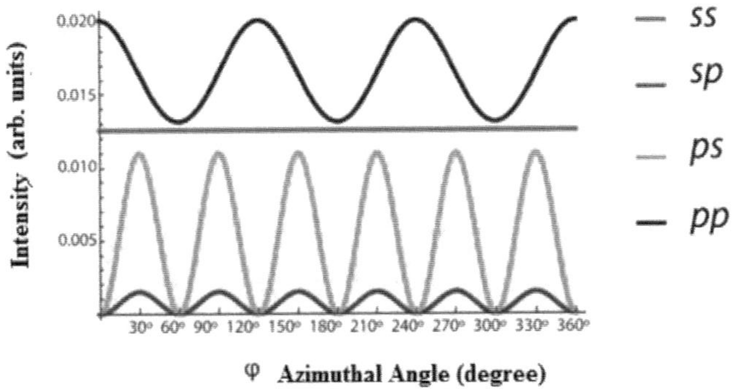

Figure 5. THG intensity peaks obtained using SBHM for a Si(111) bulk.

Another nice feature of nonlinear optics is the ability to use other harmonics in this case third harmonic generation (THG) to investigate the atomic orientation inside the bulk. It is well known that THG and other odd order nonlinear radiations are not dipole forbidden in centrosymmetric materials such as Si(111) so THG can be applied in analogy with SHG but in this case for guessing bulk atomic bond orientation. The THG intensity obtained by squaring eq. (3) is presented in figure 5. The simulation result that were obtained are in excellent agreement with the RASHG experimental data for a Si(111) orientation [8]. In a similar way, one can construct the SBHM model for THG and compare it with the experimental data to determine the atomic orientation as well as its position azimuthally before the rotation.

As a side note, we would also describe briefly the future potential of RASHG experiment backed by SBHM model to detect molecular deposition at the surface using nonlinear optics. Nonlinear optics offers many advantages in this regard because one can monitor atomic deposition such as oxygen gas in real time, without the need for ultra high vacuum, and in a non-destructive way. It has been relatively well understood that SHG can be applied to describe the surface kinetics of simple gas such as the dissociation of molecular oxygen at the Si/SiO_2 surface [9]. SBHM simulation can very well model SHG radiation from vicinal surfaces where it has also been shown by one of us recently using SBHM [10] that such configuration can sensitively detect atomic deposition and the vicinal angle. It is therefore suggested to modify this model accordingly e.g. include dimmer bonds for surface atoms opening the possibility to construct an environmental sensor especially using thin film semiconductors as substrate.

References

[1] H Hardhienata, A Prylepa, D Stifter and K Hingerl 2013 *J. Phys.* **423** 012046
[2] G D Powell, J F Wang and D E Aspnes 2002 *Phys. Rev. B* **65** 205320
[3] J. F. McGilp 2007 *J. Phys* **19** 016006
[4] Alejo-Molina A, Hardhienata H and Hingerl K 2014 *J. Opt. Soc. Am. B.* **31** 3
[5] H Hardhienata, A Alejo-Molina, C Reitböck, A Prylepa, D Stifter and K Hingerl 2016 *J. Opt. Soc. Am. B.* **33(2)** 195 - 201
[6] Cornelia R, David S, Adalberto A M, Kurt H and Hendradi H 2016 *Journal of optics* **18(3)**
[7] Cornelia R, David S, Adalberto A M, Hendradi H and Kurt H 2016 *Applied Surface Science*
[8] Tsang TYF 1995 *Physical Review A* **52(5)** 4116-4125
[9] T F Heinz, M M T Loy, WA Thompson and J Vac 1985 *Sci. Technol. B* **3** 1467
[10] Hendradi H 2016 *IOP Conference Series Earth and Environmental Science* **31(1)**

Simple conversion of fatty nitriles to primary fatty amines through iron(II)-tmphen catalyzed hydrosilylation

A Subekti[1], T T Irawadi[1*] and N Darmawan[1]

[1]Department of Chemistry, Bogor Agricultural University, Bogor, 16680, Indonesia

Email: tun_tedja@yahoo.com

Abstract. Hydrosilylation, a mild and save method, for reduction of fatty nitriles to primary fatty amines has been described. Using catalytic amount of iron(II)-tmphen complex, different long chain of fatty nitriles are reduced in good conversion number. The effect of the reaction times and the silane number toward the conversion was also studied in this work. The analysis by using gas chromatography showed that the the highest fatty nitriles conversion to its primary amines was achieved with nitrile:silane ratio 1:7 at 100°C for 24h under inert gas atmosphere. This commercially available and inexpensive catalyst and silane utilization offer a new method which is accessible for both laboratory and large scale primary fatty amines production.

1. Introduction

Fatty amine is a fatty acid derivative product which its demand is recently increasing. Amines is widely used as intermediate in many applications, such as surfactant, biocide, flotation agent, bitumen emulsifier, corrosion inhibitor, and also as solvent and polymer additive [1]. However, the increasing of fatty amine demand is not followed by its production. No contribution provided by the world major oil producer, Indonesia and Malaysia, is a factor which cause no significant increasing achieved in fatty amine number for 20 years [2]. In addition, a little development studied to find the most efficient method for production could be another factor.

There are some routes which can be applied for amine compound synthesis by using various methods, such as via amide [3-5], alcohol [6-8], nitro [9-11], and nitril [12-23]. The synthesis of amine via nitrile is widely applied because of its simplicity and less energy needed. A common method which has been applied for nitrile reduction is the use of metal hydride as the hydrogen source [17-20]. However, this method has some drawbacks, such as difficult handling and complicated post-reaction separation. Another method which could be applied is hydrogenation. However, this method commonly needs the presence of expensive metal catalyst, such as Pd [21], Ru [22], Ni, Co, and Pt [23], to increase the selectivity. In addition, the high temperature and pressure are often applied and this could increase the production cost.

In several years, hydrosilylation has attracted many reseachers because of its ability to reduce unsaturated compound in good chemo- and regioselectivity under milder conditions compared to the conventionally used reducing agents [24]. However, little is known about the hydrosilylation of C–N triple bonds, being the cyano group considered inert under the usual reaction conditions. The nitrile hydrosilylation had been investigated by using expensive cobalt complex [25], rhodium [26], ruthenium [27]. The study about nitrile hydrosilylation more focusing on aromatic nitriles conversion and there is no one concern with fatty nitrile. In addition, recent study about fatty nitrile reduction more often applied hydrogenation method [28-29].

Recently, Das et al. [30] reported that hydrosilylation catalyzed by inexpensive transition metal compounds, like iron, could convert nitrile to primary amine in good yield. In their trial experiment which tried to reduce benzamide by inexpensive dietoxymethylsilane in the presence of ammonium hydridotriironundecarbonylate benzonitrile was the major product instead of the desired benzyl amine. On the other side, the trial which used benzonitrile as substrate and different iron complex as catalyst gave benzyl amine as mayor product up to 85%. Although Das et al. [30] more focusing on amide hydrosilylation, but those trial give important information for fatty nitrile reduction with saver, milder, and more inexpensive method than the other previous common method.

2. Experimental

2.1. Materials

Lauronitrile (dodecanenitrile), Pamitonitrile (hexadecanenitrile), and stearonitrile (octadecanenitrile) as substrates and 3,4,7,8-tetramethyl-1,10-phenanthroline (tmphen) as ligand was obtained from Tokyo Chemicals Industries. Iron(II) acetate ($Fe(OAc)_2$), diethoxymethylsilane, toluene, triethylamine, potassium hydroxide and ethyl acetate was purchased from Sigma Aldrich.

2.2. Instrumentation

Reduction yield in this work was analyzed using gas chromatography/mass spectrophotometer (GC/MS) and uncoupled gas chromatography. Fatty amine was isolated by preparative thin layer chromatography (TLC-P) and the isolate was analyzed by using GC/MS and nuclear magnetic resonance (NMR) spectroscopy.

2.3. Procedure

2.3.1 Hydrosilylation of fatty nitriles. The hydrosilylation procedure was adapted from Das et al. [30] procedure. Iron(II) acetate (10%mol), tmphen ligand (20%mol), 3mmol diethoxymethylsilane ($(EtO)_2MeSiH$) and 2mL dry toluene was added to a 25mL round-bottom flask. After that, the mixture was stirred for 15minutes at ambient temperature. Furthermore, 1mmol nitrile substrate (lauronitrile) was added to the flask, the the mixture was stirred again. After 10minutes, the mixture was started to heat up at 100°C. The reaction was stirred at constant temperature (100°C) for 24h under nitrogen atmosphere. After that, the mixture was cooled and 5mL potassium hydroxide solution (25% in MeOH) was added into the flask. Then, the mixture was stirred again for 3h. Note that every step was carried out under nitrogen atmosphere. Furthermore, the mixture was completely dried and the yield was extracted with ether. The extract solution was then filtered off using celite pad and washed with ether again. After then, the filtrate was dried using rotary evaporator for analysis. The yield was then analyzed using GC/MS Shimadzu-Qp-505 equipped with column HP-5MS and MS as detector.

2.3.2 Investigation of time reaction and silane number effect. Reaction time was varied for 12h, 18h, and 24h, and another silane number (7mmol) was applied to investigate their effect toward the primary fatty amine yield. Every yield was then analyzed using gas chromatography equipped with column RTX-5 (Restec). Injection and column temperature was set 250°C and 100°C respectively and the mixture of nitrogen and air as a carrier. Column flow rate and linear velocity was set 1mL/minute and 27.7cm/s respectively. The best condition was then applied for other different long chain fatty nitriles.

2.3.3 Isolation of primary fatty amines. Fatty amine was isolated trough preparative thin layer chromatography (TLC) with the mixture of ethyl acetate-triethylamine as eluent. The isolation gave three spots on the TLC plate under ultraviolet 254nm. The isolate from each spots was gathered and redissolved into ether. After then, the isolates were analyzed using GC/MS.

3. Result and Discussion

3.1. First trial of fatty nitriles hydrosilylation

Hydrosilylation of fatty nitriles in this study was carried out in nitrogen atmosphere to prevent oxidation which was not expected. Iron metal ions Fe(II) in complex is easy to be oxidized to Fe(III). In addition, atmospheric nitrogen was also required to prevent moisture that can cause damage to the silane. GC/MS analysis on the first trial of hydrosilylation result that we carried out for 24h with nitrile:silane ratio 1:3 toward lauronitrile substrate gave three main peaks in the chromatogram (Figure 1). Based on the MS database, each peak with a retention time of about 13.0, 13.4, and 14.5 minutes respectively show represent the primary amine product (dodecylamine), lauronitrile (substrate), and the secondary amine product (didodecylamine).

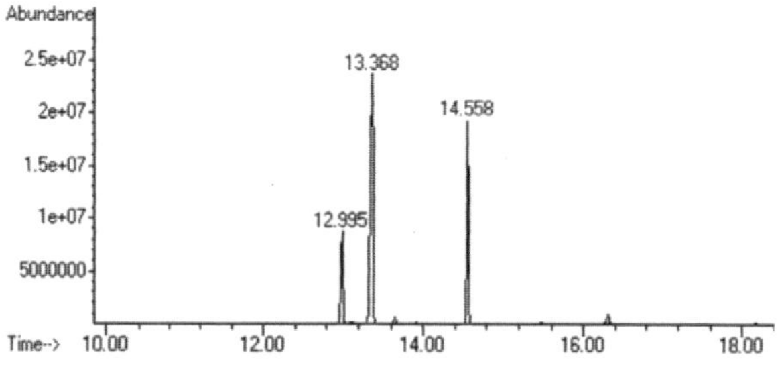

Figure 1. The chromatogram of the reaction mixture after hydrosilylation for 24h with nitrile:silane ratio 1:3.

The presence of dodecylamine compound was confirmed by fragmentation which let rearrangement resulting $CH_2{=}CHCH_2^+$ (m/z 41), $CH_2{=}CHCH_2CH_2^+$ (m/z 55) $(CH_2)_5NH_2^+$ (m/z 86), $(CH_2)_5CHNH_3^+$ (m/z 100) ion. Rearrangement has also occurred in lauronitrile fragmentation resulting $CH_2{=}C{=}NH^+$, $(CH_2)_3C{=}NH^+$, $(CH_2)_5C{=}NH^+$, and $CH_2{=}CH{-}(CH_2)_5C{\equiv}N^+$ ion with m/z 41, 69, 97, dan 124 respectively. In addition, the didodecylamine was confirmed by the fragment ion of $(CH_2)_4NH^+$ (m/z 71) and $(CH_3(CH_2)_{10}CH{=})(CH{=}CH_2)NH^+$ (m/z 210).

3.2. Investigation of time reaction and silane number effect

Further analysis on the result of reaction time and nitrile:silan ratio effect investigation by only using gas chromatography was showed in Figure 2. Although we used different coloumn from previous analysis but the chromatogram showed the same pattern. The peak with a retention time about 11.8, 12.5, and 14.5 minutes represent dodecylamine, lauronitrile, a didodecylamine respectively. Figure 2 showed that the reaction time gave a significant impact on the outcome of reduction through hydrosilylation. The longer reaction time, the higher primary amine yielded (Figure 2a-c). It could be explained as incomplete reaction occurred within 12h and 18h. This result was supported by the increasing of lauronitrile conversion number which is determined by comparing the area with the area of the standard (Table 1).

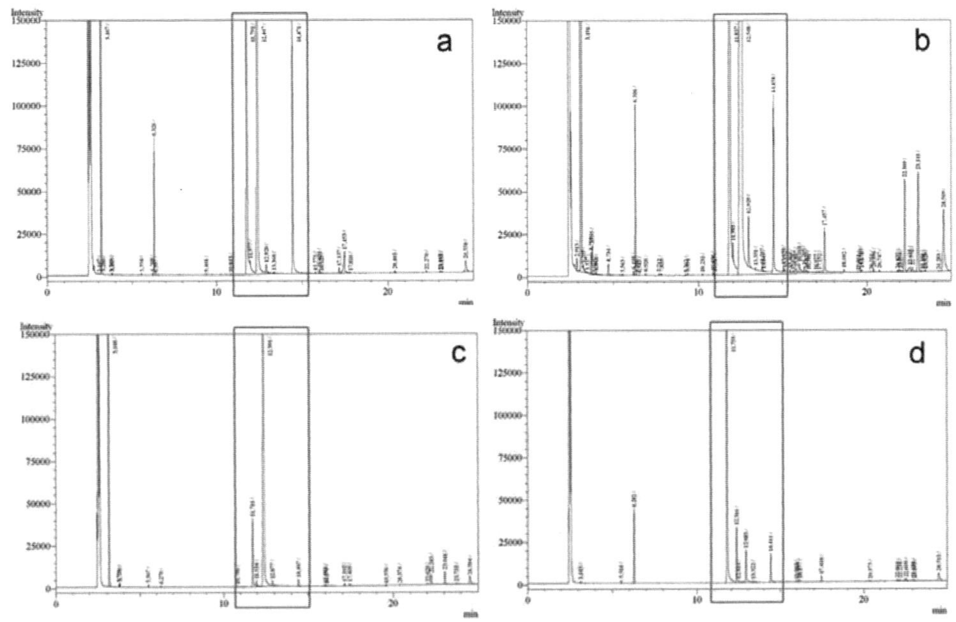

Figure 2. The chromatogram of lauronitrile hydrosilylation with nitrile:silane ratio 1:3 for 24h (a), 18h (b), 12h (c), and with nitrile:silane ratio 1:7 (d) for 24h.

Table 1. Fatty nitriles conversion data from GC analysis.

	Treatment		% Conversion
Substrate	Nitrile:silane ratio	Time (h)	
Lauronitrile (C12)	1:3	12	59.0
	1:3	18	60.0
	1:3	24	80.0
	1:7	24	99.4
Palmitonitrile (C16)	1:7	24	98.3
Stearonitrile (C18)	1:7	24	96.6

The conversion number in Table 1 showed no significant change for 12–18h but the chromatogram showed the increasing of didodecylamine peak (Figure 2b–c). It means that the formation of the secondary amine occurred over a period of 12–18h. Secondary amine is not an expected product in this study, but in some application it has been utilized since long ago. The sufficient availability of silane along the reaction was important for keeping lauronitrile hydrosilylation providing its primary amine. It was proven with the yield resulted from the reaction with nitrile:silane ratio 1:7. The hydrosilylation with that ratio gave more than 99% nitrile conversion and high primary amine selectivity.

The hydrosilylation of the longer chain fatty nitriles (C16 and C18) showed a slightly decreasing on the conversion number but it was rated highly (> 97%) (Table 1). We supposed that this decreasing occurred because the steric effect began to emerge. However, the conversion of other longer chain fatty niriles (> C18) is rarely carried out due to their little content number in palm oil. The conversion of the C16 and C18 gave very high primary hexadecylamine and octadecylamine with little secondary form.

The iron complex catalyst application in nitriles hydrosilylation showed the conversion number as good as the other previous catalyst [[24]]-[[26]]. However, by comparing the cost of the catalysts, this iron complex is still better than the other expensive catalysts. Comparing the methods for the same fatty nitrile substrates, the method in this work showed better conversion than the hydrogenation method

[[27]]-[[28]]. Besides, the addition of a large number of base, the high pressure and other special condition to achieve the best selectivity could raise the production cost which make these method disfavored. Thus, this research gave good information about an easier, saver, and milder method for the development of fatty amine production.

3.3. Isolation of primary fatty amine

The primary amines product in this work was isolated by using TLC preparative with ethyl acetate and triethylamine mixture (1:7) as a mobile phase. This separation of components in the post-hydrosilylation mixture, in this case lauronitrile as substrate, showed three spots under UV light with wavelength 254nm (Figure 3). Qualitative analysis was applied by comparing it with the spots of lauronitrile and the ligand. Thus, we recognized that the spot of isolate 1 and 3 represent the ligand and lauronitrile respectively. Although we could not ascertain what isolate 2 represent before but we supposed that it represented fatty amines.

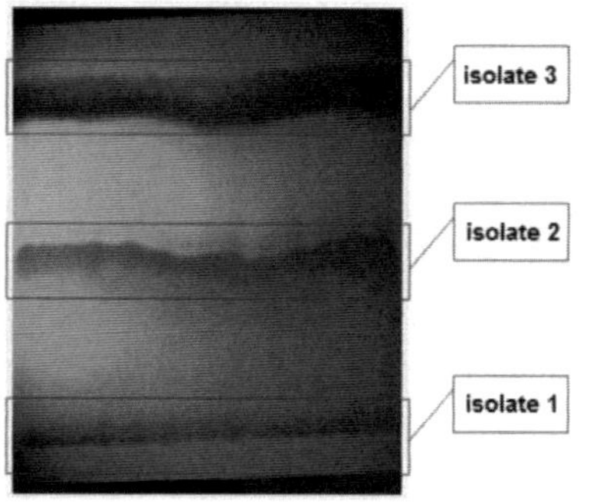

Figure 3. Preparative TLC plate of the hydrosilylation product after isolation under UV light 254nm.

Further analysis by using GC/MS on the isolate 2 and 3 gave certainty that those spots represented dodecylamine (primary lauroamine) and lauronitrile respectively. The didodecylamine (secondary lauroamine) was not further analyzed in this work because it was a byproduct. However, we supposed that didodecylamine was separated in the area between the isolate 2 and isolate 3. In addition, the ^1H NMR spectrum of the isolate 2 showed 5 signals at $\delta = 0.84$ (3H, $-CH_3$), 1.22 (18H, $-CH_2$), 1.40 (2H, $-CH_2$), 2.63 (2H, $-CH_2$), and 4.92 (2H, $-NH_2$).

4. Conclusion

Hydrosilylation catalyzed by iron(II)-tmphen complex could be applied as a saver, milder, and more inexpensive method in fatty nitriles reduction providing its analogue primary amines. Hydrosilylation with nitrile:silane ratio 1:7 for 24h at 100°C under nitrogen atmosphere gave high nitrile conversion and selectivity in producing primary fatty amines. The pure product could be obtained by chromatography separation using ethyl acetate:triethylamine mixture (1:7) as eluent.

References

[1] K Visek in: Watcher (Ed.), *Kirk Othmer Encyclopedia of Chemical Technology*, J Wiley. New York 2004 vol.2 217–225.
[2] W Rippilus 2011 *J Oil Palm Res.* **23** 1222–1226
[3] JT Reeves, Z Tan, MA Marsini, ZS Han, Y Xu, DC Reeves, H Lee, BZ Lu, CH Senanayake 2013 *Adv. Synth. Catal.* **355** 47–52
[4] M Stein, B. Breit 2013 *Ange. Chem.* **8** 2287–2290
[5] B Li, JB Sortais, C Darcel 2013 *Chem. Commun.* **49** 3691–3693

[6] C Gunanathan, D Milstein 2010 *Angew. Chem.* **45** 8661–8664
[7] S Imm, S Bahn, L Neubert, H Neumann, M Beller 2010 *Angew. Chem.* **44** 8126–8129
[8] Patil *J. Pharm. Res. Opinion.*
[9] L Pehlivan, E Metay, S Laval, W Dayoub, P Demonchaux, G Mignani, M Lemarie 2010 *Tetrahedron.* **10** 1971–1976
[10] L Pehlivan, E Metay, S Laval, W Dayoub, P Demonchaux, G Mignani, M Lemarie 2011 *Tetrahedron Lett.* **15** 1939–1941
[11] PL Gzikis, M Stratakis, IN Lyakakis 2013 *Catal. Commun.* **36** 48–51
[12] M Ochiai, H Hashimoto, H Tobita 2007 *Angew. Chem.* **119** 8340–8342
[13] M Chatterjee, H Kawanami, M Sato, T Ishizaka, T Yokoyama, T Suzuki 2010 *Green Chem.* **12** 87–93
[14] YM López-De Jesús, CE Johnson, JR Monnier, CT Williams 2010 *Top Catal.* **53** 1132–1137
[15] X Cui, Y Zhang, F Shi, Y Deng 2011 *Chem. Eur J.* **17** 2587–2591
[16] S Werkmeister, C Bornschein, K Junge, M Beller 2013 *Eur. J. Org. Chem.* Short Communication
[17] JM Khurana, G Kukreja 2002 *Synth. Commun.* **32** 1265–1269
[18] D Crain, S Armstrong, J Brunton, T Robben, SE Schmidt 2013 *Transactions of the Kansas Academic Science* **115** 139–144
[19] S Liu, Y Yang, X Zhen, J Li, H He, J Feng, A Whiting 2012 *Org. Biomol. Chem.* **10** 663–670
[20] J Way J, F Wuest 2013 *Nuclear Med. Biol.* **40** 430–436
[21] XF Guo, YS Kim, GJ Kim 2010 *Catal Today.* **150** 22–27
[22] X Miao, J Bidange, PH Dixneuf, C Fischmeister, C Bruneau, JL Dubois, JL Couturier 2012 *Chem. Cat. Chem.* **4** 1911–1916
[23] DJ Segobia, AF Trasarti, CR Apesteguia 2012 *Appl. Catal A: General.* **445–446** 69–75
[24] C Bornschein, S Werkmeister, C Junge, M Beller 2013 *New. J. Chem.* **37** 2061-2065
[25] T Murai, T Sakane, S Kato 1990 *J. Org. Chem.* **55** 449–453
[26] AM Caporusso, N Panziera, P Pertici, E Pitzalis, P Salvadori, G Vitulli, G Martra 1999 *J. Mol. Catal. A: Chemical.* **150** 275–285
[27] DV Gutsulyak, GI Nikonov 2010 *Angew. Chem. Int. Ed.* **49** 7553–7556
[28] DJ Ostgard, R Olindo, M Berweiler, S Roder, T Tacke 2007 *Catal. Today.* **121** 106–114
[29] H Chen, M Xue, J Shen 2010 *Catal. Lett.* **135** 246–255
[30] S Das, B Wendt, K Möller, K Junge, M Beller 2012 *Angew. Chem.* **124** 1694–1698

Clustering Haze Trajectory of Peatland Fires in Riau Province using K-Means Algorithm

H Khairat[1], I Sukaesih Sitanggang[1]* and D E Nuryanto[2]

[1]Department of Computer Science, Bogor Agriculture University, Indonesia.
[2]Meteorology Climatology and Geophysics Board, Indonesia.

E-mail: hamidatul.khairat@gmail.com, imas.sitanggang@ipb.ac.id, danang.eko@bmkg.go.id

Abstract. Haze is one of the impacts caused by forest and land fires. In order to determine how far the movement of haze from forest and land fires including in peatland, this research aims to determine the haze trajectory patterns from peatland fires in Riau Province in 2015 using HYSPLIT model. Furthermore, the haze trajectory patterns were grouped using the K-Means algorithm. The results show that the trajectory consists of 4887 positions of haze that moves to the northeast and northwest. The clustering results using K-Means show that the largest cluster has 3702 positions of haze and these haze movements are located at the average of height of 23.554 m AGL. Those haze positions were located in Riau, North Sumatra, Malaysia, and Malacca Strait.

1. Introduction

Peatland area has significant roles for biodiversity conservation and climate regulation, and provides important support for human welfare. The Province of Riau (Indonesia) has high deforestation rate in recent years and the majority of the deforestation has occurred on the peat soil. Deforestation and fire on the peatland area resulted in the increasing of CO and CO_2 emissions. The existence of haze is one of the impacts caused by forest and land fires, especially when it happens in peatland area. Haze caused by peatland fires will be blown away by the wind and can lead to a wide variety of health problems for the surroundinghuman. Therefore, peatland fire prevention is an important activity to reduce the damage and losses. In order to minimize the negative impacts of peatland fires, trajectory or movement pattern of haze as the results of peatland fires need to be identified and analyzed.

This work aims to analyze the haze trajectory of the peatland fires in the Province of Riau by implementing Hybrid Single-Particle Lagrangian Integrated Trajectory (HYSPLIT) model and K-Means algorithm. HYSPLIT model is designed to support a variety of simulations related to atmospheric transport and dispersion of pollutants [1]. K-Means clustering is one of methods that splits the data into several clusters [2]. In this work, HYSPLIT is used to model the meteorological data from peatland fires in Riau Province which aims to get the haze trajectory pattern. Furthermore, the haze trajectory patterns are grouped based on the movement direction of haze using K-Means algorithm. Based on the results, it is expected to be known how far the movement of haze from peatland fires.

2. Data and Method

2.1. Data

The data used in this research are the meteorological data in 2015. Data were obtained from the National Oceanic and Atmospheric Administration (NOAA), which can be accessed at this URL http://www.ready.noaa.gov/ready2-bin /extract/extracta.pl/. We focus the study to the area of Sumatra Island and its surrounding. The meteorological data are weekly data consisting of complex meteorological parameters which are stored in a bin-format file. The dataset includes the variables such as temperature (°C), relative humidity (%), wind direction (degree) and wind speed (m/s).

In addition to meteorological data, this work also utilized the sequence data of hotspots in Riau Province during 2015 of which occurrence's duration of 2 days. These sequences are obtained from the previous research [3], and consists of 86 hotspots. Attributes that are available in the hotspot sequences are the coordinates of the hotspots which are the longitude and the latitude.

Meteorological data in Sumatra were selected from global meteorological data. The hotspots in sequences are used as the initial location of haze trajectory. Meteorological data are selected based on locations of hotspots that occurred in the sequence. The selection was done by matching the date of hotspot occurrences with weekly meteorological data.

2.2. Determination of haze trajectory using HYSPLIT

Trajectory patterns contain information about spatial and temporal movements of objects. The spatial information indicates locations of points passed-through during the movement, while the temporal information indicates the time required for the movement from one point to another. The trajectory shows displacement or transition from one point to another in each transition time [4].

Hybrid Single-Particle Lagrangian Integrated Trajectory (HYSPLIT) is a model used to calculate the simple trajectory of air parcels to complex transport, dispersion, chemical transformation and deposition simulation. The initial development of HYSPLIT is a joint effort between NOAA and the Australian Bureau of Meteorology. HYSPLIT model is designed to support a variety of simulations related to atmospheric transport and dispersion of pollutants. Calculation method applies Lagrangian and Euler approaches. Lagrangian approach uses a moving object as a reference for calculating of air diffusion moved from their initial locations. Euler approach uses a 3-dimensional grid as a frame of reference for calculating the concentration of air pollutants [1].

2.3. Clustering haze trajectory using K-Means algorithm

Clustering is a process of grouping a set of objects into several clusters or groups so that objects in the same cluster have a high degree of similarity, however the objects are very different from objects in other clusters. Dissimilarities and similarities of the objects are calculated based on the attribute values that describe the objects. There are various methods that can be used to measure similarity value between object, one of them is the Euclidean distance [2].

K-Means clustering is a method that partitions data into one or more cluster(s). This method is efficient in large data processing with fast computation time [2]. The steps in K-Means algorithm are as follows:

1. Determine the value of k as the number of clusters to be formed along with the centre point of the cluster (centroid).
2. Calculate the distance of each object to each centroid using Euclidean distance.
3. Assign each object by the shortest distance between objects and the centroid.
4. Determine the new centroid by calculating the average value of the objects that exist in the same cluster.
5. Perform steps 2 through 4 until the centroid does not change.

In this work, clustering of haze trajectory patterns was done using K-Means algorithm. Attributes used in clustering are longitude and latitude coordinates of haze trajectory, as well as the height of haze positions. K-Means algorithm is available on 'fpc' package in R. Clustering results are analysed to determine how far the movement of haze in each cluster and regions affected by haze.

3. Results and Discussions

3.1. Data Selection

The hotspot sequence data consist of 6 sequence periods from July to October 2015. There are 86 hotspots in the sequences. Table 1 shows the selected hotspot coordinates of each sequence period of hotspots that are used as initial points of trajectory. Meteorological data were selected based on hotspot sequence data. Table 2 shows periods of meteorological data that are used in this study.

Table 1. Selected hotspot coordinates in sequences in 2015

Sequence period of hotspots	Longitude	Latitude	Duration (day)	District
July 9th – July 11th	100.445	2.080	3	Rokan Hilir
July 22nd – July 23rd	100.550	1.845	2	Rokan Hilir
July 26th – July 29th	102.662	-0.583	4	Indragiri Hulu
August 30th – September 1st	102.103	0.660	3	Siak
September 1st – September 2nd	101.281	0.432	2	Kampar
October 21st – October 22nd	102.493	-0.279	2	Indragiri Hulu

Table 2. Period of meteorological data based on the hotspot sequences

Sequence period of hotspots		Meteorological data
First day	Last day	
July 9th 2015	July 11th 2015	gdas1.jul15.w2
July 22nd 2015	July 23rd 2015	gdas1.jul15.w4
July 26th 2015	July 29th 2015	gdas1.jul15.w4 and gdas1.jul15.w5
August 30th 2015	September 1st 2015	gdas1.aug15.w5 and gdas1.sep15.w1
September 1st 2015	September 2nd 2015	gdas1.sep15.w1
October 21st 2015	October 22nd 2015	gdas1.oct15.w3 and gdas1.oct15.w4

3.2. Determination of Haze Trajectory using HYSPLIT

Haze trajectory patterns were generated using HYSPLIT with the input including meteorological data and hotspot sequences. Attributes of the hotspot sequence are the coordinates of hotspots that are used as starting points of haze trajectory. In this work, the hotspot sequence consists of 6 day periods of hotspot occurrences. Table 3 shows values of parameters for HYSPLIT in determining haze trajectory.

Table 3. Parameters for HYSPLIT to determine haze trajectory

Parameter	Value
Starting time	Initial date the sequence period of hotspots
Starting locations	Longitude and latitude coordinates of hotspots in the sequence
Height	10 m AGL
Total run time	Duration of hotspot occurrences (hour)
Direction	Forward trajectory
Top of model	10 km AGL
GIS-output	GIS-point
Vertical coordinate	m AGL (Above Ground Level)

The first step to get the haze trajectory in HYSPLIT is setting trajectory parameters. Trajectory parameters include time and location of the initial movement of haze, duration of simulation, and height of the haze source. Time, duration, and location of the initial movement of haze are customized based on the sequence data of hotspots. Height of the haze source depends on the vegetation in the location. This work assumed a tree with a height of 10 meters as height of the haze source. In addition, it is assumed that maximum height of haze (top of model) is the troposphere with the height of 10 km.

The results of haze trajectory in the period from July 9th until July 11th 2015 are shown in Figure 1. The haze trajectory contains time information with haze trajectory duration in hour, locations of haze trajectory, and height of haze in unit of m AGL (Above Ground Level). Figure 1 shows that in the period from July 9th until July 11th 2015 starting at 00:00 UTC or 07:00 am there were 18 trajectories or haze movements. The results show that the haze trajectory moves to the northeast toward Malaysia. The height of haze is between 0 m AGL and 1500 m AGL.

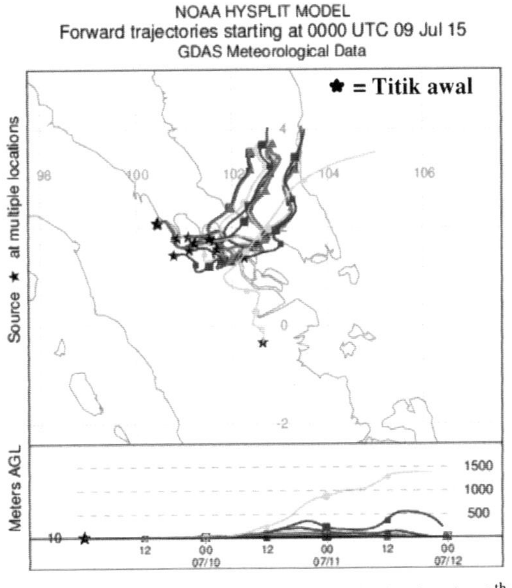

Figure 1. Haze trajectory in the period of July 9th to July 11th 2015

The haze trajectory in the period from July 9th until July 11th 2015 has 1241 positions of haze movements in which a maximum height of haze is 1395 m AGL. The entire period of the hotspot sequence in 2015 consists of 12 to 18 hotspots as initial points of haze trajectory. In this period, as many 4887 positions were obtained with the movement directions of haze to the northeast and northwest. Table 4 shows a summary of haze trajectory in 2015.

Table 4. Summary of haze trajectory in 2015

Date of haze trajectory	Movement direction of haze trajectory	Height level of haze (m AGL)	Height of haze in average (m AGL)
July 9th – July 11th	Johor and Pahang (Malaysia)	1395	54.817
July 22nd – July 23rd	Riau and North Sumatra	615	73.125
July 26th – July 29th	Nanggroe Aceh Darussalam	1300	387.189
August 30th – September 1st	North Sumatra and Malacca Strait	1070	103.221
September 1st – September 2nd	Riau and North Sumatra	665	57.286
October 21st – October 22nd	Riau	282	34.912

3.3. Clustering Haze Trajectory using K-Means Algorithm

Clustering haze trajectory was performed on the datasets consisting of latitude and longitude of haze trajectory and its height. Clustering was done using K-Means function that is available on 'fpc' package in R. In this research, number of clusters (*k*) was selected from 2 to 10 and the clustering results were evaluated by calculating the Sum Square Error (SSE). Figure 2 shows the SSE value and accuracy for each cluster.

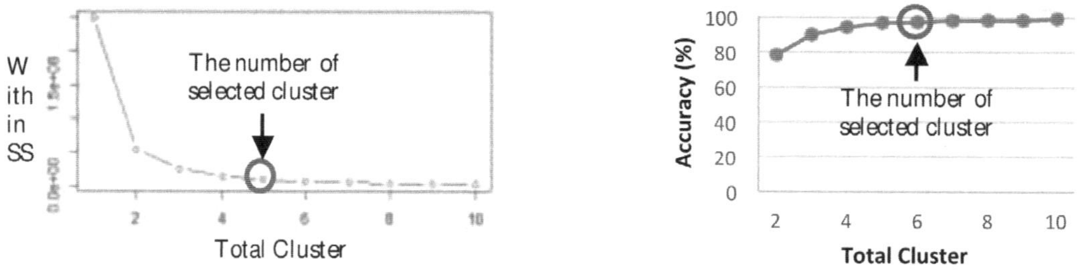

Figure 2. SSE value and clustering accuracy

The value of number of clusters (k) was selected such that percentage of clustering quality is high enough to represent the clustering results with the smallest value of SSE. Based on the experimental results, k of 5 was selected to cluster haze trajectory. The accuracy of clustering reaches 96.4%.

Table 5 shows the clustering results of haze trajectory at k of 5. Cluster 1 has the largest number of members which are positions of haze movement. As many 3702 positions (75.75% of overall positions) of haze trajectory are members of cluster 1. Cluster 2 has the smallest number of members which contains 89 positions of haze trajectory or about 1.82% of all data.

Table 5. Results of haze trajectory clustering with k of 5

Cluster	Number of members	Average (Center)			Percentage (%)
		Longitude	Latitude	Height (m AGL)	
1	3702	101.699	1.317	23.554	75.75
2	89	100.444	2.707	1138.831	1.82
3	169	99.210	2.821	797.639	3.46
4	287	99.886	2.342	467.498	5.87
5	640	100.744	1.598	180.673	13.10

Figure 3 shows a plot of five haze trajectory clusters. Locations of haze spread in Sumatra and a point of haze is located outside the territory of Sumatra that is around the Strait of Malacca and Malaysia.

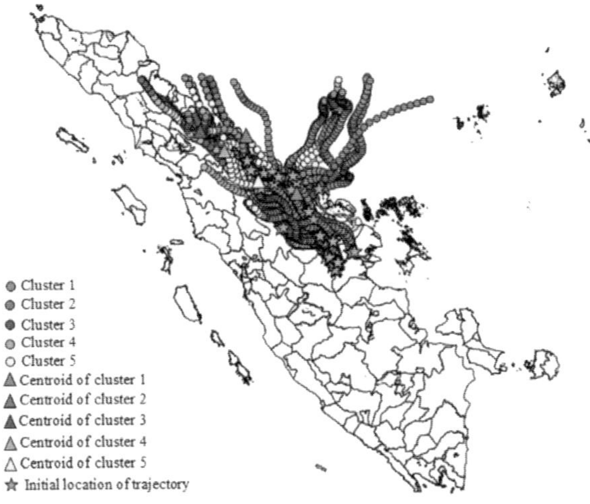

Figure 3. Haze trajectory clusters in Riau Province in 2015

Haze trajectory clusters were analysed based on height of haze in average on each cluster. According to Table 5, cluster 2 has the smallest number of haze trajectory positions with the average of haze height is 1138.831 m AGL. Haze positions move from the source of fire at height of 10 m AGL to 1138.831 m AGL. Cluster 1 has 3702 positions of haze trajectory in which the lowest height value in average is 23.554 m AGL. In this situation, the haze is dangerous for human health. The haze positions are located in 10 districts and cities in Riau province including Bengkalis, Indragiri Hilir, Indragiri Hulu, Kampar, Pelalawan, Rokan Hilir, Rokan Hulu, Siak, Pekan Baru City, and Dumai City, 4 districts in North Sumatra namely Asahan, Labuhan Batu, Serdang Bedagai, and Simalungun, 3 states in Malaysia namely Johor, Negeri Sembilan and Pahang. In addition, haze spread around the Strait of Malacca.

4. Summary

This study generated haze trajectory from peatland fires. The trajectory pattern mining results as many 4887 movement positions of haze that move towards the northeast and northwest of Riau Province. The results of haze trajectory clustering show that the largest cluster contains as many 3702 locations of haze movements in hotspot sequence period in which the lowest height of haze in average is 23.554 m AGL. The haze movements are located in 10 districts and cities in Riau Province, 4 districts in North Sumatra, 3 states in Malaysia, as well as around the Strait of Malacca.

References

[1] Stein A F, Draxler R R, Rolph G D, Stunder B J B, Cohen M D and Ngan F 2015 *Bull Amer Meteor Soc.* **96** 2059-2077

[2] Han J, Kamber M and Pei J 2011 *Data Mining: Concepts and Techniques* Ed 3 (US: Morgan Kaufmann)

[3] Abriantini G, Sitanggang I S, and Trisminingsih R 2016 Hotspot Sequential Pattern Visualization in Peatland of Sumatra and Kalimantan using Shiny Framework. *The 3rd International Symposium on LAPAN – IPB SATELLITE* (LISAT), October 25-26, 2016, Bogor, Indonesia.

[4] Giannotti F, Nanni M, Pedreschi D and Pinelli F 2007 *Trajectory Pattern Mining* (Italy: University of Pisa)

The Best Extraction Technique for Kaempferol and Quercetin Isolation from Guava Leaves (*Psidium guajava*)

I Batubara[1*], I H Suparto[1,2] and N S Wulandari[1]

[1,] Department of Chemistry, Bogor Agriculture University, INDONESIA.
[2] Tropical Biopharmaca Research Center, Bogor Agricultural University, INDONESIA

Email : ime@apps.ipb.ac.id

Abstract. Guava leaves contain various compounds that have biological activity such as kaempferol and quercetin as anticancer. Twelve extraction techniques were performed to obtain the best extraction technique to isolate kaempferol and quercetin from the guava leaves. Toxicity of extracts was tested against *Artemia salina* larvae. All extracts were toxic (LC_{50} value less than 1000 ppm) except extract of direct soxhletation on guava leaves, and extract of sonication and soxhletation using *n*-hexane. The extract with high content of total phenols and total flavonoids, low content of tannins, intense color of spot on thin layer chromatogram was selected for high performance liquid chromatography analysis. Direct sonication of guava leaves was chosen as the best extraction technique with kampferol and quercetin content of 0.02% and 2.15%, respectively. In addition to high content of kaempferol and quercetin, direct sonication was chosen due to the shortest extraction time, lesser impurities and high toxicity.

Keywords: guava leaves, high performance liquid chromatography, kaempferol, quercetin

1. Introduction

Guava (Psidium guajava) is a tropical fruit and guava leaves are often used as a traditional medicine to treat wounds, diarrhea, cough, mouth sores, and dengue fever. The guava leaf extract is reported to have antioxidant [1], anti-reaction nonenzimatik reducing sugar in patients with hyperglycemia [2], anti-inflammatory [3], and anticoagulants [4] activities. Some studies show that guava leaves extract contains phenolic compounds such as Ferulic acid, Gallic acid and flavonoids such as catechins, quercetin, and kaempferol [2].

Research on kaempferol and quercetin in guava has not been done, whereas kaempferol known to have many pharmacological activities such as antioxidant, antimicrobial [5], antidiabetic [6], anti-inflammatory and analgesic [7], whereas quercetin is known to have antioxidant activity[8], antibacterial[9], and antiviral [10]. Activities kaempferol and quercetin can be tested by Brine Shrimp Lethality Test (BSLT). This test is used as a preliminary test to determine the potential pharmacological activity of natural ingredients. It has also been perceived as an inexpensive alternative method for cytotoxicity assay. The combination of kaempferol and quercetin content of extracts reported a synergistic effect of antiproliferative in culture Human cancer cell lines [11].

A wide variety of extraction techniques to isolate kaempferol and quercetin has been developed. According to Tang et al.[12], kaempferol and quercetin have been isolated from the leaves of Ginkgo biloba with a yield respectively 0.0204 mg/g sample and 0.0371 mg/g sample, respectively [12].

Extraction technique used is macerated with the aid of sonication using methanol-water (85:15). Extraction of kaempferol (0.0007 mg/g sample) of *Sideroxylon foetidissimum* leaves according Erosa-Rejon et al. (2010) was performed by extracting the leaves with ethanol with maceration method for 1 week [13]. Extraction of kaempferol (613 mg/g sample) of soybean leaves by Zang et al. (2011) conducted by refluxing methanol leaves with 70% for 3 hours [6]. Loizzo et al. (2007) has succeeded in extracting kaempferol (12:12 mg / g sample) and quercetin (0.085 mg / g sample) from the leaves of *Ailanthus excelsa* with Soxhletation using 70% methanol [14]. The variety of extraction techniques result in yield kaempferol and quercetin produced also varies. Therefore, it is necessary to do research using a variety of extraction techniques to determine the best extraction technique of the most simple, cheap, and fast and produce maximum yield. This study aims to determine the best extraction techniques for the isolation of kaempferol and quercetin in guava leaf and determine its activity with the method BSLT.

2. Materials and Methods

The research was started by collecting the guava leaves from Conservation and Cultivation Station Unit of Tropical Biopharmaca Research Center at Darmaga Campus, Bogor. The leaves were powdered prior to extraction. The toxicity by BSLT toxicity test method, the total tannin content, total phenolic content, the total flavonoid content, and thin layer chromatography profile of extracts were determined. The kaempferol and quercetin content of the selected extract, then determined by high performance liquid chromatography to select the best extraction method.

2.1. Extractions

Guava leaves are divided into two parts; the first part, directly extracted with 4 kinds of techniques, namely maceration (M1), maceration by sonication (SN1), reflux (R1), and Soxhlet (S1). The second part was extracted with n-hexane (Soxhlet method), and the resulting residue is further divided into two parts. The first part, directly extracted with 4 kinds of techniques, namely maceration (M2), maceration by sonication (Sn2), reflux (R2), and Soxhlet (S2). The second part was extracted further with ethyl acetate (Soxhlet method); the residue was then extracted with 4 kinds of techniques, namely maceration (M3), maceration by sonication (SN3), reflux (R3), and Soxhlet (S3). The whole extract is obtained (M1, M2, M3, SN1, Sn2, SN3, R1, R2, R3, S1, S2, and S3) after concentrated by rotary evaporator

2.1.1. Maceration. The extraction process is based on Erosa-Rejon et al [13]. Leaf samples were extracted with ethanol at room temperature for 1 week. Extraction is done 3 times.

2.1.2. Maceration by sonication. The extraction process is based on Tang et al [12]. Leaf samples were extracted with methanol-water (85:15) with the aid of sonication for 3 hours. The extraction was repeated 3 times.

2.1.3. Reflux. The extraction process is based on Zang et al [6]. Leaf samples were added with 70% ethanol and then refluxing at a temperature of 60-70°C for 3 hours.

2.1.4. Soxhlet. The extraction process is based on Loizzo et al [14]. Leaf samples were added with methanol70% and the Soxhlet process was performed.

2.2. Total Phenolic Content Determination

Each about 25 mg diluted with methanol: water (1: 1) into a 25 ml flask. About 0.9 mL of standard solution (gallic acid) and extract was added by 4.5 mL reagent Folin-Ciocalteau, and shaken with a vortex. After 3 minutes, each solution was added to 3.6 mL of Na_2CO_3 7.5%, shaken and incubated for 1 hour. The absorbance of the standard solution and the sample was measured by UV-Vis spectrophotometer at a wavelength of 765 nm. The total phenol content of samples is determined using a regression equation of standard gallic acid.

2.3. *Tannin Content Determination*
Extract of 0.5 g was dissolved in DMSO, added to 10 ml of distilled water and heated at a temperature of 40-60 ° C for 30 minutes. The solution was filtered and diluted with distilled water until 50 mL and 5 mL of the solution was then added by 5 mL indigo carmine. The solution was subsequently titrated with 0.1 N $KMnO_4$.

2.4. *Total Flavonoid Content Determination*
The extracts about 200 mg was added with 1 mL hexamethylenetetramine (HMT) 0.5%, 20 mL acetone, and 2 ml of HCl. The mixture was hydrolyzed by refluxed for 30 minutes. The result was filtered and added with acetone to 100 ml. About 20 mL of solution was added to 20 ml of water and 15 ml ethyl acetate. Ethyl acetate fraction was collected in a 50 mL volumetric flask. Extraction was repeated by adding 10 mL ethyl acetate. The total ethyl acetate fraction was added with ethyl acetate until 50 ml. Furthermore, 10 ml of the mixture was added to 1 mL of 2% $AlCl_3$ and 5% glacial acetic acid in methanol. The mixture was homogenized and allowed to stand 15-30 minutes. The absorbance value at a wavelength of 425 nm was measured with a UV-VIS spectrophotometer

2.5. *Brine Shrimp Lethality Test*
Ten larvae of *Artemia salina* were added to 1 mL extract with varying concentrations of 200-14000 μg/ml. Control without the addition of extract was used as negative control. After 1 day (24 hours) the number of dead shrimp larvae was determined and the LC_{50} value was determined by probit analysis.

2.6. *Thin Layer Chromatography Profile*
Each extract was analyzed by Thin Layer Chromatography on silica gel $G_{60}F_{254}$ and *n*-hexane:ethyl acetate (2:98) as eluent. The detection was observed in the UV lamp 254 and 366 nm. An extract containing a fluorescent and Rf value the same with kaempferol and quercetin standards will be analyzed further.

2.7. *Kaempferol and Quercetin Content Determination*
About 30 mg of each extract was added to 2 ml of HCl 4 M, shaken and heated for 30 minutes and extracted by 2 mL ethyl acetate. Ethyl acetate fractions were separated and rinsed with 1 mL ethyl acetate 2 times. The collected ethyl acetate fraction evaporated and diluted with methanol till 10 ml. Standard solutions and the sample were then filtered with 0:45 micro μm membrane and 20 μl was injected in high performance liquid chromatography (HPLC) for analysis.

HPLC Conditions used is C18 column, UV detector at λ 370 nm, oven temperature 30 ° C, mobile phase acetonitrile 30% in KH_2PO_4 0.025 M buffer pH 2.5 with isocratic elution, and flow rate of 1 ml / min[14]. The kaempferol and quercetin content were determined by comparing the peak area with the peak area of standards.

2.8. *Statistical analysis*
All data obtained were analyzed using the analysis of variance at 95% confidence level ($p<0.05$) using the Duncan's test with SPSS 16.

3. Results and Discussion
The difference preparation and extraction technique used in this study is to find the highest content of kaempferol and quercetin. The separation process was Soxhlet with *n*-hexane and ethyl acetate and comparison with the leaves without separation step first. The aim of Soxhlet with *n*-hexane is to eliminate the fat attached to the leaf tissue so expect it easier to extract the polar compounds such as kaempferol and quercetin. The continue separation by Soxhlet with ethyl acetate aimed to eliminate

components that are semi-polar. Table 1 shows that the longer the separation step, the smaller % yield got. The highest yield is found in the extraction process of M2.

The highest yield of the extract is not guaranty the highest kaempferol and quercetin content. The screen which extract will be used for the next step, total phenolic, tannin and flavonoid content were determined to each extract. The result is shown in table 1. The highest total phenolic compound is found in extract of S1 while M1 has the highest tannin content. The high phenolic compounds in the extracts S1 indicates that phenolic compounds can be extracted optimally with direct extraction without separation step. According to Guimarães-Beelen et al. (2006) tannins can be optimally extracted by maceration using ethyl acetate and water [16]. M1 extracts were extracted with methanol: water so that the levels of tannins produced the highest. Other extracts have low levels of tannin extraction step because it has experienced a long and residue derived from ethyl acetate. Flavonoids are phenolic compounds that can be extracted using polar organic compounds such as methanol and ethanol [14]. The highest content of the total flavonoid extract is found in M1.

Table 1. Phenolic, Tannin, Flavonoid Contents and LC_{50} value against *A salina* of all extracts. Difference letter means significantly different at $p>0.05$

Extraction technique	Extract name	Yield (%)	Phenolic content (mg/g)	Tannin Content (%)	Flavonoid content (mg/g)	LC_{50} (ppm)
	M1	22.82^{d}	831.13^{e}	15.81^{a}	21.43^{a}	526.24^{f}
Maceration	M2	43.38^{a}	758.70^{g}	8.56^{c}	20.54^{a}	281.52^{e}
	M3	13.95^{e}	843.35^{d}	6.21^{i}	11.49^{c}	100.03^{d}
	Sn1	20.05^{d}	753.34^{g}	7.31^{f}	16.22^{b}	41.21^{b}
Sonication	Sn2	10.78^{e}	557.95^{i}	6.98^{g}	9.75^{d}	3862.62^{j}
	Sn3	6.60^{f}	868.19^{c}	6.47^{h}	5.09^{e}	745.54^{g}
	R1	14.22^{e}	837.68^{de}	8.91^{b}	11.24^{c}	31.22^{a}
Reflux	R2	13.49^{e}	877.82^{c}	7.63^{e}	4.44f	36.34^{a}
	R3	14.18^{e}	910.58^{b}	8.20^{d}	3.19^{i}	70.76^{c}
	S1	35.44^{b}	957.77^{a}	6.02^{j}	4.67^{f}	1340.81^{i}
Soxhlet	S2	24.57^{c}	807.49^{f}	7.78^{e}	3.85^{g}	1095.52^{h}
	S3	14.36^{e}	729.16^{h}	5.75^{k}	3.55^{h}	132.13^{d}

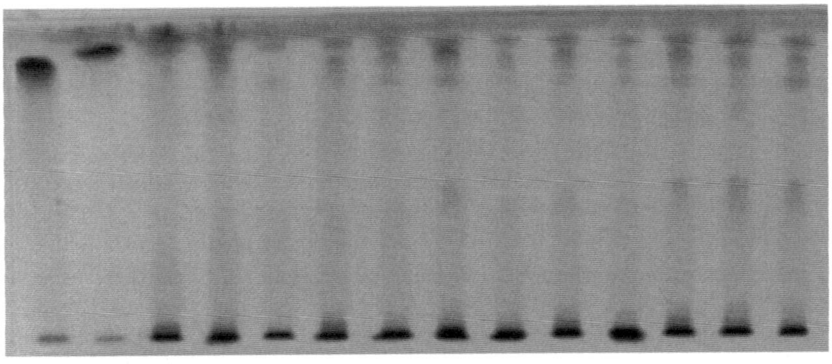

Figure 1. Chromatogram of extract and standard on UV 254 mm; from 1-14: quercetin standard, kaempferol standard, M1, M2, M3, Sn1, Sn2, Sn3, R1, R2, R3, S1, S2, and S3.

The combination of flavonols such as quercetin and kaempferol will give high potency as anticancer especially as antiproliperative and cytotoxic against cultured human cancer cell lines[11]. One of easy methods to determine the cytotoxic properties of the active compound is based on the number of deaths *Artemia salina* larvae at certain concentrations. The data are usually expressed in LC_{50} (Lethal Concentration 50%), which is a value that indicates the concentration of toxic substances that can cause 50% of the death test animals [17]. Based on the LC_{50} values shown in table 1, extract of Sn2, S1, and S2 which has a LC_{50} value is higher than 1000 ppm is less toxic, while other extracts is potent as a candidate for an anticancer agent.

In addition to the data of phenolic, tannin, and flavonoid content, TLC chromatogram is used to determine the extraction process to isolate kaempferol and quercetin. Figure 2 shows that all extracts generate almost the same bands. Kaempferol standard was found in Rf value of 0.87 whereas quercetin standard of 0.83. The structure of kaempferol and quercetin is shown in figure 2. Detection by UV 366 nm in figure 3 shows that both the standard kaempferol and quercetin had a green fluorescent. Different with the standard, the same Rf band in the extracts did not give green fluorescent but blue. The absence of green fluorescent at the same Rf can be assumed that kaempferol compounds in the extract were blocked by a blue fluorescent compound that has similar polarity with kaempferol.

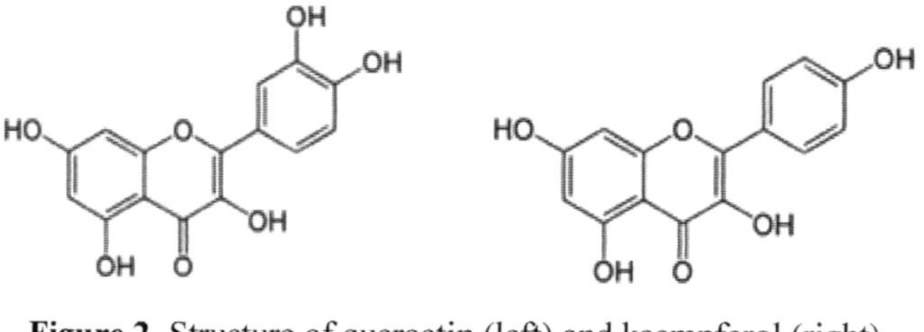

Figure 2. Structure of quercetin (left) and kaempferol (right).

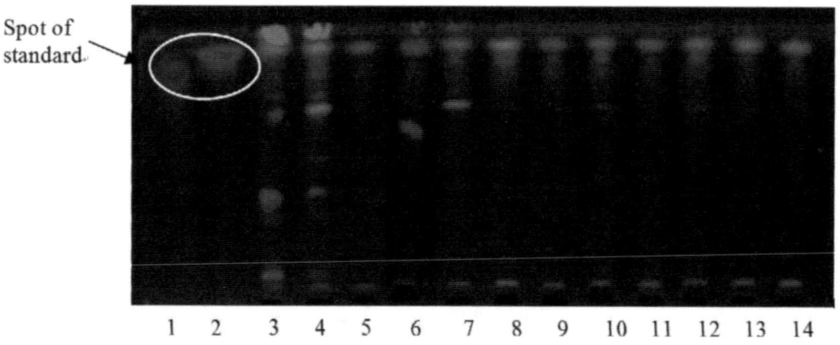

Figure 3. Chromatogram of extract and standard on UV 366 nm; from 1-14: quercetin standard, kaempferol standard, M1, M2, M3, Sn1, Sn2, Sn3, R1, R2, R3, S1, S2, and S3.

In fgure 2 showed that the extract of M1, M2, Sn1, and Sn2 has many bands with high intensity. The reflux and Soxhlet from original leaf and residues of *n*-hexane (R1, R2, S1, S2) showed a smaller number of bands and lower intensity of bands, while all extracts derived from residues of ethyl acetate (M3, SN3, R3, S3) show the lowest number and intensity of bands. It can be concluded that more separation step before extraction may be causing the loss of other compounds.

Table 2. Kaempferol and quercetin content on selected extract

Extract/standard name	Concentration (%)	
	Kaempferol	Quercetin
Sn1	0.02[b]	2.15[c]
Sn2	0.02[b]	1.97[b]
S1	0.03[b]	1.98[b]
S3	0.01[a]	1.14[a]

Difference letter means significantly different at $p>0.05$

Based on the fluorescent intensity on TLC, total phenolic content, tannin content, and flavonoid content, extract of Sn1, Sn2, S1 and S3 were selected to determine the kaempferol and quercetin content using HPLC. Selected extracts have high fluorescent intensity, low tannin and phenols content and high flavonoids content. Table 2 shows that the highest kaempferol content was found in extract of S1, while the highest quercetin content was found in Sn1 extract. Kaempferol content of the extract Sn1 is only small different with the S1, but has much higher LC_{50} value. Extract Sn1 had LC_{50} values of 41.21 ppm, whereas S1 extracts 1340.81 ppm. So Sn1 extract is considered better because it has a lower LC_{50} value. The results showed that the guava leaf sonication using methanol-water (85:15) directly from the sample was the best extraction techniques for the isolation of kaempferol and quercetin in guava leaves. The kaempferol and quercetin content on guava leaf extract with the Tang et al (2001) methods on this study is higher (0.04 mg/g sample and 4.30 mg/g sample respectively) compared with *Ginko biloba* leaf extracts reported by Tang et al [12].

4. Conclusion

Best extraction techniques to isolate kaempferol and quercetin from guava leaves are sonication using methanol-water (85:15) with a content of 0.02% by kaempferol and quercetin at 2:15%. This technique had yield about 20.05% and BSLT LD50 of 41.21ppm.

References

[1] Chen H Y and Yen G C 2007 *Food Chem.* **101** 689-694
[2] Wu J W, Hsieh C L, Wang H Y and Chen H Y 2009 *Food Chem.* **113** 78-84
[3] Jang M, Jeong S W, Cho S K, Ahn K S, Kim B K and Kim J C 2013 *Food Sci Biotechnol* **22**(5) 213-220
[4] Hsieh C L, Lin Y C, Yen G C and Chen H Y 2007 *Food Chem.* **103** 528-535
[5] Teffo L S, Aderogba M A and Eloff J N 2010 *S Afr J Bot* **76** 25-29
[6] Zang Y, Sato H and Igarashi K 2011 *Biosci Biotechnol Biochem.* **75**(9) 1677-1684
[7] Parveen Z, Deng Y, Saeed M K, Dai R, Ahamad W and Yu Y H 2007 *Yakugaku Zasshi* **127**(8) 1275-1279
[8] Duenas M, Manzano S O, Paramas A G and Buelga S C 2009 *J Pharm Biomed Anal.* **51** 443-449
[9] Arima H and Danno G 2002 *Biosci Biotechnol Biochem* **66**(8) 1727-1730
[10] Agustinus 2009 *Studi hematologis potensi metabolik jambu biji merah (Psidium guajava L.) pada penderita demam berdarah dengue* (Bogor(ID): Institut Pertanian Bogor)
[11] Ackland M L, Waarsenburg S V D and Jones R 2005 *In vivo* **19** 69-76
[12] Tang Y, Lou F, Wang J, Li Y and Zhuang S 2001 *Phytochemistry* **58** 1251-1256
[13] Erosa-Rejon G, Pena-Rodriguez L M, Sterner O 2010 *Rev Latinoamer Quim* **38**(1) 8-11
[14] Loizzo M R, Said A, Tundis R, Rashed K, Statti G A, Hufner A and Menichini F 2007 *Phytother Res* **21** 32-36

[15] Wang H and Helliwel K 2001 *Food Res Int* **34** 223-227
[16] Guimarães-Beelen P M, Berchielli T T, Beelen R, Filho J A and de Oliveira S G 2006 *Sci Agric* **6**(36) 522-528
[17] Meyer B N, Ferrigni N R, Putnam J E, Jacobson L B, Nichols D E and McLaughlin J L 1982 *Planta Med* **45** 31-34

IOP Conf. Series: Earth and Environmental Science **58** (2017) 012061 doi:10.1088/1755-1315/58/1/012061

Quality Evalution of Potato Clones as Processed Material Cultivated in Lembang

S T Rahayu[1*], T Handayani[1] and P S Levianny[1]

[1]Indonesian Vegetable Research Institute, indonesia, Jl.Tangkuban Perahu No.517 Lembang, West Java

Email: swarnit@yahoo.com

Abstract. Potatoes are widely grown in the temperate as well as tropical zones and are the fourth largest staple crop in the world after maize, wheat and rice. The study aimed to evaluate the quality of several potato clones as raw material on potato based products (chips and boiled). The study was conducted at Indonesian Vegetable Research Institute, Lembang about 1200 m asl height, in 2016. The design used was a randomized complete block design with three replications. The samples tested were 5 clones selection (clones number 1,2,3,4,10). In this study, variety Granola (Clone number 6) and Atlantic (Clone number 7) were used as a susceptible control, meanwhile the Katahdin (Clone number 8) and SP 951 (Clone number 9) were used as the resistant control. Chemical properties tested were starch, reduction sugar, water content, specific gravity, and Total Soluble Solute (TSS). The organoleptic assessment method used was hedonic test with scale of 1-5 (very like until very dislike) which had been done by 15 untrained panelists. Data was statisticaly analized by Duncan's test (5%). Clone 1 and 2 were preferred by panelist as raw material for potato chips, which got score of 'very like' until 'like' for color, size, taste, and texture parameters. Although there was no significant difference on color and size parameters for all samples of that boiled potato there, however, clone no 8 can be considered as the most favourite based on taste and texture parameters.

1. Introduction

Potatoes are the third largest food crop worldwide following rice and wheat, which are valued globally for their ease of cultivation, preparation and as a readily assimilated source of carbohydrate (CHO) energy [1]. Nutritionally, the role of carbohydrate metabolism is critical to the regulation of our energy intake and maintenance of body weight. Crop genetic resources are very important for crop-based industry. Heritage potatoes have a great diversity of skin and flesh color combined with unusual tuber shapes, textures and flavours. Starch granule sizes have been known to influence the digestibility and glycemic index of potato starch [2,3]. Bruise-free, high specific-gravity tubers are desirable, and the processing industry often pays incentives for each of those quality goals.

The physical maturity of tubers refers to skin set and the development of a mature periderm. Tubers that are physically immature have poor skin set, and a skin that is permeable to water and susceptible to skinning. Physiological maturity refers to the dry matter content and is achieved when tuber dry matter reach a maximum. Chemical maturity is critical for the processing industry. Maturity can be manipulated by various growth factors, such as seed treatment, planting date, nitrogen fertilizer management and harvest strategy [4].

Textural characteristics have been found to affect flavor perception in a wide variety of foods and beverages. The influence of textural characteristics on flavor perception has been shown to be dependent on the oral processing exhibited when consuming the food [5]. Maillard reaction between sugars and amino acids is considered to be the main mechanism for acrylamide formation, with asparagine and reducing sugars as the major reactants present in foods [6]. Since potatoes are high in both asparagine and reducing sugars such as glucose and fructose, frying potato chips generates optimum conditions for acrylamide formation [7]. Acrylamide is a chemical compound naturally formed during processes such as frying, roasting and baking in a wide variety of foods including French fries, potato chips, cereal, bread and coffee [8]. It has been reported to be a human neurotoxin, a rodent carcinogen and robable carcinogen to humans [9].

2. Materials and Methods

The experimental design used in this study was a randomized complete block design with three replications. The materials tested were 5 clones selection (clone number 1,2,3,4, and 10).Varieties for comparison are Granola, Atlantic, Katahdin, and SP 951 as control. Plant sources of inoculum were planted at least three rows around the field trial. Planting system was arranged in single row, with a spacing of 30 cm x 70 cm. Horse manure in dosage 30 ton/ha was applied a week before planting. Synthetic fertilizer (1000 kg/ha NPK with ratio 16:16:16, 100 kg urea/Ha, and 500 Kg Za /Ha) was applied twice on planting and 30 day after planting (DAP). Plant maintenance such as weeding was done for 2-3 times on 14, 45, and 75 DAP. Other maintenance, including insecticide and fungicide applications were applied at 14, 21, 28, and 35 DAP, with contact and systemic fungicides alternately. Chemical characteristic testing are starch, reduction sugar, water content, specific gravity, and Total Soluble Solute (TSS). Organoleptic tests were conducted by 15 untrained panelists who gave score of 1-5 ('strongly like' until 'strongly dislike'). Data was statisticaly analized followed by Duncan's test (5%).

3. Results and Discussion

Chemical characteristics of potatoes which were analyzed inthis study including starch content, reduction sugar, water content, specific gravity and total soluble solid (TSS, is shown in table 1. Since starch comprises the largest part of dry matter, it has direct influence on technological quality, especially with regard to the texture of the processed products [10]. Starch content showed a positive correlation with specific gravity, texture, pulp pH, and soluble solid, and was negatively correlated with reducing sugars[11]. High sugar accumulation and low starch content are unwanted since it will affect chips appearance and will not be preferred by consumers. Previous studies [12,13] verified a variation from 0.5 to 2.0% in reducing sugars contents in potato tubers. Tubers with reducinmg sugar contents higher than 2% are unacceptable for frying[14]. Based on our analysis, the starch content of selected clones values were ranging from 5.86%-10.61%. The highest starch content found in clone no.7 (control) and no 8. Meanwhile, the reducing sugar content of all clones were not significantly different and showed value below 2%. Among all selected clones, the control showed the lowest value, and clone no 1 showed the highest value.

Table 1. Chemical Characteristics of Potato Clones.

Clone	Starch (%)	Reduction sugar (%)	Water content(%)	Specific gravity (gcm^{-3})	TSS (° Brix)
1	6.88^{cd}	0.04	85.27^{b}	1.04	3.33^{ab}
2	7.25^{cd}	0.03	86.06^{b}	1.05	3.67^{a}
3	7.32^{cd}	0.03	89.02^{a}	1.03	2.67^{b}
4	7.19^{cd}	0.03	86.49^{ab}	1.05	3.40^{ab}
5	7.08^{cd}	0.03	86.82^{ab}	1.04	3.27^{ab}
6	7.20^{cd}	0.03	87.22^{ab}	1.05	3.47^{ab}
7	10.61^{a}	0.02	81.25^{c}	1.06	4.00^{a}
8	9.45^{ab}	0.03	84.93^{b}	1.05	3.60^{ab}
9	7.98^{bc}	0.03	85.55^{b}	1.03	3.47^{ab}
10	5.86^{d}	0.03	86.54^{ab}	1.02	3.13^{ab}

* The number followed by the same character on the same column is not significantly different, based on Duncan Test (5%).

Water content of selected clones varied from 81.25 to 89.02%. Clone no 7 (control) showed the lowest water content (81.25%), meanwhile the highest were found in clone no 3 (89.03%). Early maturing potatoes, which can be used such as for boiling, baking, frying, roasting or mashing usually have lower dry matter, higher water content and different starch structure [15].The specific gravity is important quality attributes because it can indirectly provide a dry matter content estimation, and is also related to industrial yield, oil absorption during frying, and final product quality. Studies showed that minimal value of specific gravity requirement for potato chips in national industry is 1.067[16]. Based on this research analysis, the specific gravity of the clones were not significantly different and ranging between 1.02 to 1.06. Clone no 7 (control) showed the highest value, which was 1.06, and followed by clone no 2, 4, and 8 which were 1.05. In this research, lowest TSS were found in clone 3 which was 2,67° Brix while the highest was found in clone no 7, 4,00 ° Brix. Clones no 7 used in this study is Atlantic cultivar which had been used as a control for processed potatoes, especially potato chips.

Figure 1. Potato Chips.

The results of organoleptic test on potato chips (table 2) showed that panelists gave score 'very like' until 'like' on color parameter of clone no 1, 2,4, 8, and 9. Meanwhile, on size parameter, panelist gave score 'like' for all clones except clone no 10. Clone no 1 had taste and texture that preferred by panelist who gave score 'very like' and showed to be significantly different with clone no 10.

Table 2. Organoleptic Test on Potato Chips

Clones	Color	Size	Taste	Texture
1	1.47^{d}	1.93^{b}	1.87^{c}	1.67^{d}
2	2.13^{cd}	2.20^{b}	2.53^{abc}	1.93^{cd}
3	3.67^{a}	2.80^{ab}	3.07^{ab}	2.47^{cd}
4	2.40^{bcd}	2.27^{ab}	2.27^{bc}	2.27^{cd}
5	2.67^{abc}	2.13^{b}	2.47^{abc}	2.93^{bc}
6	3.33^{ab}	2.87^{ab}	2.67^{abc}	3.07^{bc}
7	2.87^{abc}	2.40^{ab}	3.07^{ab}	3.73^{ac}
8	2.53^{bcd}	2.20^{b}	2.40^{abc}	3.00^{bc}
9	2.00^{cd}	2.13^{b}	2.13^{bc}	2.13^{cb}
10	3.33^{ab}	3.27^{a}	3.53^{a}	4.47^{a}

* 1= strongly like; 5= strongly dislike
** The number followed by the same character on the same column is not significantly different

Textural characteristics can influence flavor perception in liquid and soft foods to crisp/brittle foods. In addition, the effect of crispness level on flavor perception varies by flavor type, age group, and mastication pattern [17]. The colorless potato is more preferable as it also corresponds to the acrylamide contecnt. Acrylamide is correlated negatively with the Agtron chip color score. Agtron chip color represents a a score system in which 0 is equal to black and 90 is equal to white. Agton Score of 50 or greater is considered as color that is commercially acceptable. Potatoes that rich in reducing sugars will turn dark rapidly and produce high levels of acrylamide. In this study, the reducing sugar of two clones selected were lower (0,029 and 0,023) than that the reducing sugar content of the other potato clones which had dark color [18].

The dry matter from untransformed potato which contains high amylose during the various storage periods whereas significantly. Pyruvate decarboxylase in transgenic Snowden had increasing resistant starch and higher dry matter contents of phosphorus, and had decreasing acrylamide levels in potato chips made from these potatoes [19]. Recently, there has been increasing concern regarding the production of acrylamide during the frying process due to its potential carcinogenic properties. Acrylamide has been shown to be mainly formed in food by Maillard browning reactions of reducing sugars with asparagine at temperature above 12°C.

Table 3. Organoleptic Test on Boiled Potato

Clones	Color	Size	Taste	Texture
1	2.00 a	2.00 a	2.20 ab	2.27 b
2	2.60 a	2.60 a	2.33 ab	2.13 b
3	2.40 a	2.40 a	2.40 ab	3.40 ab
4	2.93 a	2.93 a	2.67 ab	2.87 ab
5	2.67 a	2.67 a	2.60 ab	2.80 ab
6	2.00 a	2.00 a	2.27 ab	2.87 ab
7	3.07 a	3.07 a	2.53 ab	3.80 a
8	2.87 a	2.87 a	2.07 b	2.27 b
9	2.07 a	2.07 a	2.20 ab	2.40 b
10	2.93 a	2.93 a	3.07 a	3.00 ab

* 1= strongly like; 5= strongly dislike
** The number followed by the same character on the same column is not significantly different

Figure 2. Boiled potato.

Organoleptic test on boiled potato clones had also been conducted to measure the clones as processed material (table 3). In general, clone no 10 were less accepted by panelists. Although, there was no significant difference between all clones for color and size parameters (table 3), however, a significant difference for taste and texture parameter was observed.

Textural characteristics have been found to affect flavor perception in a wide variety of foods and beverages. The influence of textural characteristics on flavor perception has been shown to be dependent on the oral processing exhibited when consuming the food. Changes in textural characteristics lead to variations in mastication that may change the physical properties of the food bolus altering how aromatic compounds are released from the food matrix [20]. The majority of work on texture–flavor interactions has been performed with a focus on rheological properties such as flow and deformation in liquid and soft food, however, crisp or brittle foods have received little attention. In the tasting parameter, panelists gave the highest score for clone no 8 which was significantly different than that clone no 10. Meanwhile, on texture parameter, panellists were preferred to clones no 1, 2, 8, and 9 which were significantly different to other clones.

4. Conclusion

Clone 1 and clone 2 were preferred by panelists as raw material for potato chips, which got score of 'very like' to 'like' for color, size, taste, and texture parameter. Although there was no significant difference on color and size parameters for all samples of that boiled potato there, however, clone no 8 can be considered as the most favourite based on taste and texture parameters.

Acknowledgment

We would like to gratefully thank Mrs. Enung Murtiningsih, Mr. Udin Samsudin and Mr. Mamat Rachmat from postharvest laboratory for helping our team during this study.

References

[1] Camire M E, Kubow S and Donnelly D J 2009 *Critical Reviews in Food Science and Nutrition* **49** 823-40.
[2] Noda T, Takigawa S, Matsuura-Endo C, Suzuki T, Hashimoto N, Kottearachchi NS, Yamauchi H and Zaidul ISM 2008 *Food Chem.* **110** 465-70
[3] Parada J, Aguilera J M 2009 *J. Food Sci.* **74** 34-38
[4] Sabba R P, Bussan J, Michaelis B A, Hughes R, Drilias M J and Glynn M T 2007 *Am. J. Potato Res* **84** 15
[5] Repoux H, Labouré H, Courcoux P, Andriot, I É. Sémon, C Yven and E. Guichard 2012 *Flavour and Fragrance Journal* **27** 414-23
[6] Friedman M 2003 *Journal of Agricultural and Food Chemistry* **51** 4504-16
[7] Zhang Y and Ren Y 2009 *Chemical Reviews* **109** 4375-85
[8] Ahn J S, Castle L, Clarke D B, Lloyd A S, Philo M R and Speck D R 2002 *Food Additives and Contaminants* **19** 1116-28
[9] Becalski A, Lau B P Y, Lewis D, Seaman S W 2003 *Journal of Agricultural and Food Chemistry* **51** 802-10
[10] Kadam S S, Wankier B N and Adsule N R 1991a *Potato production, processing and products* (Boca Raton: CRC Press) cap.5 p 9-35
[11] Feltran J, Carlos, Lemos, Borges L, Vieies and Lopes R 2004 *Scientia Agricola* **61(6)** 593-97
[12] Kusumiyati, Suradinata Y R, Sutari W and Sukarminah E 2015 *Laporan Akhir Penelitian Unggulan Perguruan Tinggi*
[13] Kadam S S, Wankier B N and Adsule N R 1991b *Potato production, processing and products* (Boca Raton: CRC Press) cap 5 p 111-54
[14] Pereira A S 1987 *Chemical composition, nutritional value and industrialization. In: Reifschneider, FJB coord* (Brasilia: Graphic line and Publisher) p 12-28
[15] Nayak B, Berrios J J and Tang J 2014 *Food Research International* **56** 35-46
[16] Budiman A 1999 *Seminar Kebutuhan dan Peluang untuk pengembangan PHT kentang*
[17] Luckett C R, Meullenet J F and Seo H S 2016 Food Quality and Preference **51** 8-19
[18] Asgar A, Rahayu S T, Kusmana M and Sofiari E 2011 *J.Hort.* **21(1)** 51-59
[19] Pinhero R G, Copp L J, Amaya C, Marangoni A G, Yada R Y 2007 *Physiologia Plantarum* **130(2)** 230-52
[20] Salles C, Chagnon MC, Feron G, Guichard E, Laboure H, Morzel M and YvenIn-mouth C 2011 *Critical Reviews Food Science and Nutrition* **51** 67-90

Determinant factors for Brebes shallot farmers in selecting shallot varieties (Case study in Brebes, Central Java Indonesia)

A Sembiring

Indonesian Vegetables Research Institute (IVegRI), IAARD, Ministry of Agriculture of Indonesia.

Email : rangkayoamah@gmail.com

Abstract. The using of a high quality seed is very important for a successfully cultivation and an increasing of vegetable production. The vegetable productions, including for shallot are much more affected by seed varieties. Previous studies showed that production rate, shape, colour, resistant to pest and disease, number of bulbs and demanding by market were factors that influence farmers to choose shallot variety. The aim of the study is to observe determinant factors that influence shallot farmers to change their shallot varieties and persist to use certain varieties. The study was conducted in June to December 2014 in Wanasari, Tanjung, Kemukten and Limbangan villages, Brebes District. Brebes was selected as it is the biggest shallot production in Indonesia. Data were collected from 35 shallot farmers through survey by using the structural questionnaire. The data then analysed with the descriptive statistic.

1. Introduction

The using of a high quality seed is very important for a successfully cultivation and an increasing of vegetable production including for shallot. Vegetables 'production are much more influence by seed. In selecting seed varieties, farmers consider various aspects such as seed produces high yield, resistant to pest and disease, and growth well. Moreover, it is fit with its environment, the price is affordable and it is favoured by market [1]-[4].

In the regard with the shallot, previous studies showed that farmers decided certain shallot varieties to be planted was affected by several factors such as production, time planting, a resistance of shallot to pest and disease, shape, colour, flavour (aroma), number of stems and market demand toward shallot varieties [4]-[7].

Indonesian government through Indonesian Vegetables Research Institute (IVegRI) have conducted some efforts to produce new and high qualities shallot varieties so that farmers could adopted the varieties. For shallot, until currently IVegRI has releases 11 shallot varieties, they are Bima Brebes, Maja Cipanas, Kramat 1, Kramat 2, Kuning, Sembrani, Katumi, Mentes, Pancasona, Pikatan dan Trisula.

Meanwhile, in shallot farmer itself develops various local shallot varieties for instance Lembah Palu from Central Sulawesi [8], Medan from Samosir-North Sumatera, Maja Cipanas from Cipanas-West Java, Sumenep from Madura, Kuning and Gombong from Brebes- Central Java [9]. Nevertheless until currently, majority of shallot farmers in Indonesia use Bima Brebes and Bima Curut varieties. Almost 95% of the farmers use the varieties [10].

The aim of the study is to observe determinant factors that influence shallot farmers to change their shallot varieties and persist to use certain varieties until currently. The study is expected could provide a contribution to shallot breeders, especially when they create new shallot varieties. These determinant factors should be considered by the breeders, hence, the farmers would like to adopt the varieties.

2. Materials and Methods

The study was conducted in June to December 2014 in Wanasari village, Tanjung, Kemukten and Limbangan Brebes district. Brebes was chosen because the area is the biggest central shallot production in Indonesia. Respondents were consisted of 35 shallot farmers. Data were collected through individual survey by using structural questionnaire and were analysed with the descriptive statistic.

3. Results and Discussion

3.1 The using of varieties at currently

The respondents are used Bima Brebes (12 respondents) and Bima Curut varieties (23 respondents) at currently. Before adopting Bima Brebes dan Bima Curut varieties, Brebes shallot farmers used various shallots varieties. For instance Kuning, Ampenan, Bangkok Warso, Mentes, Markonah and Philip varieties.

The first respondent planted Bima Brebes in 1995. Meanwhile, Bima Curut have been planted since 1987. As the general, the use of the varieties since the first time are described in table 1. The largest Bima Brebes' shallot planting was before 2000, they were 5 respondents (41.7%) and between 2006 to 2010 as much as 33.3%. Meanwhile, for Bima Curut, most of farmers had planted during 2006-2010 (43.5%) and after 2010 as much as 6 respondents (26.1%).

Table 1. The first time of respondents use Bima Brebes and Bima Curut varieties.

	Bima Brebes		Bima Curut	
Year	Number of respondents n=12	Percentage (%)	Number of respondents n=23	Percentage (%)
< 2000	5	41.7	2	8.7
2000-2005	2	16.7	5	21.7
2006-2010	4	33.3	10	43.5
> 2010	1	8.3	6	26.1
Total	12	100	23	100.0

3.2 Comparison of previous shallot varieties and current shallot varieties

Various shallot varieties are emerged in Indonesia and farmers have their own choice to select which varieties that they want to be planted. Indonesia government through IVegRI has released numerous shallot varieties with varied high qualities in order to be adopted by the farmers and provide benefit for them. The benefits include increasing the production and a resistance of shallot to pest and disease and decreasing cost of production. However, still, the farmers decide which variety to select for planting. Below are the respondents' description about the difference between their previous shallot varieties and the currently used varieties (Bima Brebes dan Bima Curut).

3.2.1 *Day of harvest*. In term of day of harvest, table 2 shows 5 respondents (41, 7%) said that their currently used variety (Bima Brebes) has shorter day of harvest compare to the previous one. As much as 41.7% respondents said that day of harvest of Bima Brebes was longer than the previous one. The rest, 16.7% respondents said that Bima Brebes has shorter than the previous varieties. Meanwhile, for

Bima Curut's respondents, 10 respondents (45.5%) said that day of harvest Bima Curut was shorter than the previous ones and 36.4% said that the currently used wathe same as the previous ones.

Table 2. Day of harvest of current shallot varieties an the previous varieties that are used by Respondents.

	Bima Brebes		Bima Curut	
Day of harvest	Number of respondents n=12	Percentage (%)	Number of respondents n=23	Percentage (%)
Shorter	2	16.7	10	45.5
Same	5	41.7	8	36.4
Longer	5	41.7	4	18.2
Total	12	100.0	22	100.0

3.2.2. *Yield.* In term of yield, both of the varieties' user said that their current varieties produced more yield than the previous ones (41.7% respondents of Bima Brebes and 42.9% respondents of Bima Curut). This is shown in table 3.

Table 3. Shallot's yield of current varieties compare to the previous ones.

	Bima Brebes		Bima Curut	
Yield	Number of respondents n=12	Percentage (%)	Number of respondents n=23	Percentage (%)
More	5	41.7	9	42.9
Same	4	33.3	5	23.8
Less	3	25.0	7	33.3
Total	12	100.0	23	100.0

3.2.3. *Time of planting*. About time of planting, Table 4 shows that all respondents Bima Brebes and Bima Curut users (100%) said that the current varieties could be planted both in dry and wet seasons. Meanwhile, the previous varieties could be planted only in dry season.

Table 4. Time of planting the currently used varieties compare to the previous ones.

	Bima Brebes		Bima Curut	
Time of planting	Number of respondents n=12	Percentage (%)	Number of respondents n=23	Percentage (%)
Dry and wet seasons	11	100.0	23	100.0
Dry season	0	0	0	0
Total	11	100.0	23	100.0

3.2.4. *Number of stems*. Table 5 shows majority of respondents (Bima Brebes 91.7% and Bima Curut as much as 73.9%) said that number of stems for currently used seed varieties more than the previous ones.

Table 5. Number of stems the current varieties compare to the previous ones.

Number of stems	Bima Brebes		Bima Curut	
	Number of respondents n=12	Percentage (%)	Number of respondents n=23	Percentage (%)
More	11	91.7	17	73.9
Same			3	13.0
Less	1	8.3	3	13.0
Total	12	100.0	23	100.0

3.2.5. *Bulb's size*. Meanwhile, most of the farmers said that bulb's size of the current varieties are bigger than the previous ones. This is answered by 66.7 % Bima Brebes's users and 82.6% Bima Curut's users as shown in table 6.

Table 6. Shallot bulb's size of current varieties compare to the previous ones.

Bulb's size	Bima Brebes		Bima Curut	
	Number of respondents n=12	Percentage (%)	Number of respondents n=23	Percentage (%)
Bigger	8	66.7	19	82.6
Same			3	13.0
Smaller	4	33.3	1	4.3
Total	12	100.0	23	100.0

3.2.6. *Bulb's shape*. From bulb's shape, majority of the farmers said that a bulb's shape of their currently used varieties were round shape. This was answered by 58.2% respondents of Bima Brebes and 60.9% respondents of Bima Curut as shown in Table 7.

Table 7. Bulb's shape of currently used varieties compare to the previous ones.

Bulb's shape	Bima Brebes		Bima Curut	
	Number of respondents n=12	Percentage (%)	Number of respondents n=23	Percentage (%)
Round	5	41.7	9	39.1
Oval	7	58.3	14	60.9
Total	12	100.0	23	100.0

3.2.7. *Bulb's colour*. Table 8 talking about bulb's colour. In term of colour, majority of respondents said that the bulb's colour of currently used varieties are dark red. This is answered by 100 % respondents of Bima Brebes and 82.6% respondents of Bima Curut.

Table 8. Bulb's colour.

Bulb's colour	Bima Brebes		Bima Curut	
	Number of respondents n=12	Percentage (%)	Number of respondents n=23	Percentage (%)
Light red			4	17.4
Dark red	12	100.0	19	82.6
Total	12	100.0	23	100.0

3.2.8. *Price selling*. About price selling, more than a half respondents said that both of the currently used varieties were more expensive than the previous ones as shown in Table 9.

Table 9. Price selling.

Price selling	Bima Brebes		Bima Curut	
	Number of respondents n=12	Percentage (%)	Number of respondents n=23	Percentage (%)
Cheaper			4	18.2
Same	4	33.3	5	22.7
More expensive	8	66.7	13	59.1
Total	12	100.0	22	100.0

3.2.9. *Resistance to pests and diseases*. In term of resistance of pests and diseases, as much as 8 respondents (66.7%) of Bima Brebes's users said that the variety was the same as the previous varieties. On the contrast, majority of Bima Curut's users (65.2%) said that their variety was more resistant than the previous ones as shown in table 10.

Table 10. Resistance to pest and disease of currently used varieties compare to the previous ones.

Resistance to pests and diseases	Bima Brebes		Bima Curut	
	Number of respondents n=12	Percentage (%)	Number of respondents n=23	Percentage (%)
More resistant	4	33.3	15	65.2
Same	8	66.7	6	26.1
Less resistant			2	8.7
Total	12	100.0	23	100.0

3.2.10. *Easy to seed*. Table 11 showed that almost 50% of the respondents agreed that both of the currently used varieties were easy to seed compare to the previous varieties. It was answered by 58.3% respondents of Bima Brebes and 60.9% respondents of Bima Curut.

Table 11. Easy to seed.

Ease to seed	Bima Brebes		Bima Curut	
	Number of respondents n=12	Percentage (%)	Number of respondents n=23	Percentage (%)
Easier	7	58.3	14	60.9
Same	3	25.0	7	30.4
More difficult	2	16.7	2	8.7
Total	12	100.0	23	100.0

3.2.11. *Age stored*. In term of age store the varieties, table 12 shows most of both respondents agreed that there was no difference of age store between the currently used varieties and the previous varieties. This was answered by 50% respondents of Bima Brebes and 43.5%) respondents of Bima Curut. However, there is 1/3 of the respondents (Bima Brebes 33.3% and Bima Curut 34.8%) who said that the age store of their currently used varieties was longer than the previous ones.

Table 12. Age stored of currently used varieties compare than the previous ones.

Age stored	Bima Brebes		Bima Curut	
	Number of respondents n=12	Percentage (%)	Number of respondents n=23	Percentage (%)
Shorter	2	16.7	5	21.7
Same	6	50.0	10	43.5
Longer	4	33.3	8	34.8
Total	12	100.0	23	100.0

3.2.12. *Production.* In term of production, majority of respondents' yield was on 7.5 ton to 10 ton/ha. This was answered by 41.7% respondents of Bima Brebes and 39.1% respondents of Bima Curut (Table 13)

Table 13. Production of currently used varieties.

Production	Bima Brebes		Bima Curut	
	Number of respondents n=12	Percentage (%)	Number of respondents n=23	Percentage (%)
5000-7500 kg	2	16.7	4	17.4
7501-10000 kg	5	41.7	9	39.1
10001-12500 kg	3	25.0	3	13.0
12501-15000 kg	1	8.3	3	13.0
> 15000 kg	1	8.3	4	17.4
Total	12	100.0	23	100.0

As the general, the production of Bima Brebes and Bima Curut as shown as the following; the highest production was 15.62 ton/ha and 18.75 ton/ha, the lowest production was 5.62 ton/ha and 5.4 ton/ha and the average yield of both the varieties was 10.4 ton/ha and 11.03 ton/ha. These information are shown in Table 14.

Table 14. Production rate of currently used varieties.

Production rate	Bima Brebes (ton/ha)	Bima Curut (ton/ha)
Higher	15.62	18.75
lowest	5.62	5.4
Average	10.4	11.03

Based on the numerous aspects above, farmers changed their shallot varieties because the new (currently used) varieties is better than the previous varieties. The excellency includes aspects : 1) the currently used varieties (Bima Brebes dan Bima Curut) have day of harvest shorter or at least same with the previous varieties 2) the currently used varieties produce more yield or at least same with the previous ones 3) currently used varieties could be planted both in dry and wet seasons, meanwhile the previous ones only could be planted in one season which is on dry season 4) currently used varieties have more number of stems than the previous ones. Those 4 factors above in line with the result of previous studies about farmers' preference on shallot varieties [1][2][3][4] 5) the shape of currently used varieties' bulb is oval and this shape is favoured by consumer 6) the bulb's colour of currently used varieties is dark red and this is also favoured by consumers [11]. These results is in line with previous studies about consumer's preference on shallot's shape and colour [4][5][6][7] 7) currently used varieties is more resistant to pest and desease compare to the previous ones 8) currently used varieties are easier to be seed 9) age store of currently used varieties' bulb is longer than the previous one.

All of these excellences currently used varieties (Bima Brebes and Bima Curut) make farmers are persist to use both of the varieties until currently, although from the production cost the current ones higher than the previous ones. Nevertheless, it still provides beneficial for the farmers compare to the shortages.

4. Conclusion

There are various excellency aspects of shallot variety such as high production, able to be planted during the year, ease to be seed, number of stems is higher, resistant to pest and disease and longer age store that cause farmers change their shallot variety to new ones and persist until today. For farmes in Wanasari, Tanjung, Kemukten and Limbangan Brebes, they choose Bima Brebes and Bima Curut varieties.

Based on the study, for the breeders who want to release new shallot varieties, it is needed to concern with those aspects. The breeders also are necessity to looking for other excellences of new shallot varieties. Hence, when they are introduced to farmers, the farmers would like to use the new shallot varieties.

Acknowledgments

Thank you very much for the Indonesian Centre for Horticulture Research and Development that had funded for the study and to Rofik Sinung Basuki, the senior researcher of IVegRI for the data.

References

[1] Adiyoga W, Suwandi W, Kartasih A 2014. *J. Hort.* **24(1)** 76-84.
[2] Joshi RK and Witcombe JR 2014 *Field Crops Research* **158** 15–23.
[3] Baswarsiati, Harnowo D and Pikukuh B 2014 *PERHORTI 2014* 548-554.
[4] Basuki RS 2010. *J. Hort.* **20(2)** 186-195.
[5] Sulastrini I, Hidayat IM dan Putrasamedja S 2014 *PERHORTI 2014* 341-346.
[6] Basuki RS 2009. *J. Hort.* **19(4)** 475-483.
[7] Basuki RS 2009 *J. Hort.* **19(3)** 344-355.
[8] Bahrudin and Ansar M 2015 *Proceeding of Seminar Nasional Perhimpunan Hortikultura Indonesia* 113-121.
[9] Putrasamedja S and Suwandi *1996 Monograf* 5, Balai Penelitian Tanaman Sayuran Lembang, Jawa Barat Indonesia
[10] Pusat Data dan Sistem Informasi Pertanian (Pusdatin) 2013, Jakarta, Kementan, 70.
[11] Adiyota W 2011 *J. Hort.* **21(3)** 280-294.

Molecular Dynamics Approach in Designing Thermostable Aspergillus niger Xylanase

N D Malau[1*] **and M Sianturi**[1]

[1]Department of Physics Education, Indonesia Christian University, Indonesia

Email : malaunyadaniaty@gmail.com

Abstract. Molecular dynamics methods we have applied as a tool in designing thermostable Aspergillus niger Xylanase, by examining Root Mean Square Deviation (RMSD) and The Stability of the Secondary Structure of enzymes structure at its optimum temperature and compare with its high temperature behavior. As RMSD represents structural fluctuation at a particular temperature, a better understanding of this factor will suggest approaches to bioengineer these enzymes to enhance their thermostability. In this work molecular dynamic simulations of Aspergillus niger xylanase (ANX) have been carried at 400K (optimum catalytic temperature) for 2.5 ns and 500K (ANX reported inactive temperature) for 2.5 ns. Analysis have shown that the Root Mean Square Deviation (RMSD) significant increase at higher temperatures compared at optimum temperature and some of the secondary structures of ANX that have been damaged at high temperature. Structural analysis revealed that the fluctuations of the α-helix and β-sheet regions are larger at higher temperatures compared to the fluctuations at optimum temperature.

1. Introduction

Enzyme applications in industrial processes have become very important as it is safe and environmental friendly, besides having cost advantages compared to the chemical reagents [1]. Enzymes are catalysts that are expected to reduce the impact of pollution and waste of energy because the reaction does not require high energy, be specific, and not toxic [2]. One types of enzymes that important role in the industry is the xylanase enzyme [3].

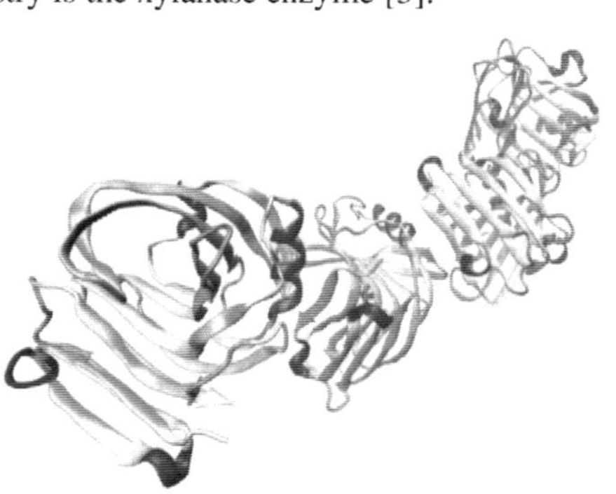

Figure 1. Representation of the secondary structure of *Aspergillus niger* Xilanase.

Xylanase (1,4-β-D-xylan xilanohidrolase, E.C 3.2.1.8) is a group of enzymes that have the ability to hydrolyze xylan. This enzyme is used in pre-bleaching process chemical pulp[4]. The use of xylanase can reduce the consumption of chlorine compounds up to 20-40% and can improve the quality of the paper that produced [5-6]. In addition, the use of enzymes in pulp bleaching process is lower cost than other process. But xylanases have to be functional at 60-70° C, which is the temperature of the incoming pulp for the bleaching operation.

Aspergillus niger xylanase has all the advantages in term of size, but the optimum temperature of this enzyme is 45°C which hampers it potential use in pulp bleaching industry. Thus there is a great need to design a new xylanase enzyme from Aspergillus niger that has the optimum temperature of 60 - 70° C, to take advantage of its ideal size.

Much work has been done to engineer enzymes to achieve the desirable state through mutations [7-8]. During past years many research groups have identified specific mutations that improve enzyme properties such as selectivity, activity, alternate catalytic route and thermal stability [9]. However, these random mutation approaches are too costly and time consuming. Computer aided design approaches offer an inexpensive means to narrow the set of possible solutions quickly, and then follow them up with experimental testing.

In this research we have carried out molecular dynamics [10] simulations and identified the region that has potential to be mutated by analyzing the root mean square deviation (RMSD) of backbone and secondary structure of Aspergillus niger xilanase at its optimum temperature (400 K) in comparison with its inactive temperature/ unfolding process (500 K).

2. Method

The crystal structure of Aspergillus niger Xylanase (pdb code : 1UKR) [11] used in the simulation was taken from the Protein Data Bank (PDB) [12]. This original file still contains of water molecules and 5.840 atoms. Protein must be prepared in a simulation box filled with the solvent atoms, TIP3P water box with the size of 92 x 92 x 92 $Å^3$ was used in the simulation (Figure 2). PSFgen program was used to apply a magnetic field in the system molecules. Force field that used to define the potential energy of the crystal structure is CHARMM22. To neutralize total charge system, four ions Cl^{-1} added to water box. All those preparations were done by using Virtual Molecular Dynamics program (VMD).

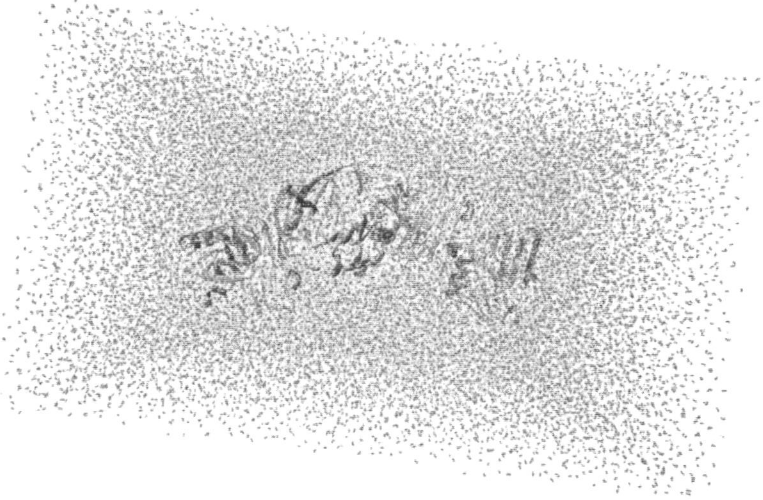

Figure 2. Protein in the water box.

Molecules that have solvated then minimized. This stage is the initial stage of MD simulation. The purpose of minimization is to avoid the Van Der Waals contacts that do not match (bad contact) and to minimize steric effects that high energy. Each simulations get started with minimization for 100 ps to make the proteins had the lowest energy (steady state). The second stage is the stage of heating and equilibration. Protein was heated from the temperature of 0 K up to 300 K, 400 K, 450 K and 500 K to the unfolding process for 40 ps with an increase every 25 K. Equilibration was done by the langevin protocol for 10 ps. Equilibration was meant to hold protein to stay stable in simulation system. Production run is the final stage of MD simulations where the constraints which applied to the equilibration process, then removed so that protein free to move. Production run performed for 1 ns with temperature at 300 K, and 2.5 ns with temperature at 400 K, 450 K and 500 K for the unfolding. The MD simulations were performed using NAMD v.2.9 simulator [9].

Overall simulations was using parameter integration time (time step) every 2 femto seconds (fs). The simulation was done by using periodic boundary condition (PBC) method to eliminate the effect of surface tension and to achieve density and pressure conditions were more uniform. Electrostatic energy system thoroughly calculated using the particle mesh ewald (PME) method, while the van der waals interaction was calculated using the Lennard – Jones potential with each cutoff was 12 Å. All simulations was performed in single CPU powered by 3.6 GHz Intel® core i7 processor with 16 GB of RAM and using Ubuntu 12.04 Linux platform.

To analyze the pathway of molecules unfolding and refolding which have been simulated, was done analyze using VMD to produce the output such as secondary structure and Root Mean Square Deviation (RMSD) using VMD.

3. Result and Discussion

3.1. Root Mean Square Deviation (RMSD)

Root Mean Square Deviation analysis is used to describe a change in the secondary structure of the protein during the simulation [13]. RMSD values that fluctuate indicate conformational change process from the beginning to the end of the simulation.

Figure 3.a and 3.b shows the fluctuations β-sheet and α-helix RMSD values at temperatures 400 K, 450 K and 500 K for 2.5 ns. From 400 K to 500 K simulation temperature, there was no significant change in the values of RMSD. Until 2 ns the values of RMSD was constant in 2 Angstrom. An increase temperature in the system is not affecting the protein stability. There is no significant change in the β-sheet and α-helix fluctuations of the whole enzyme.

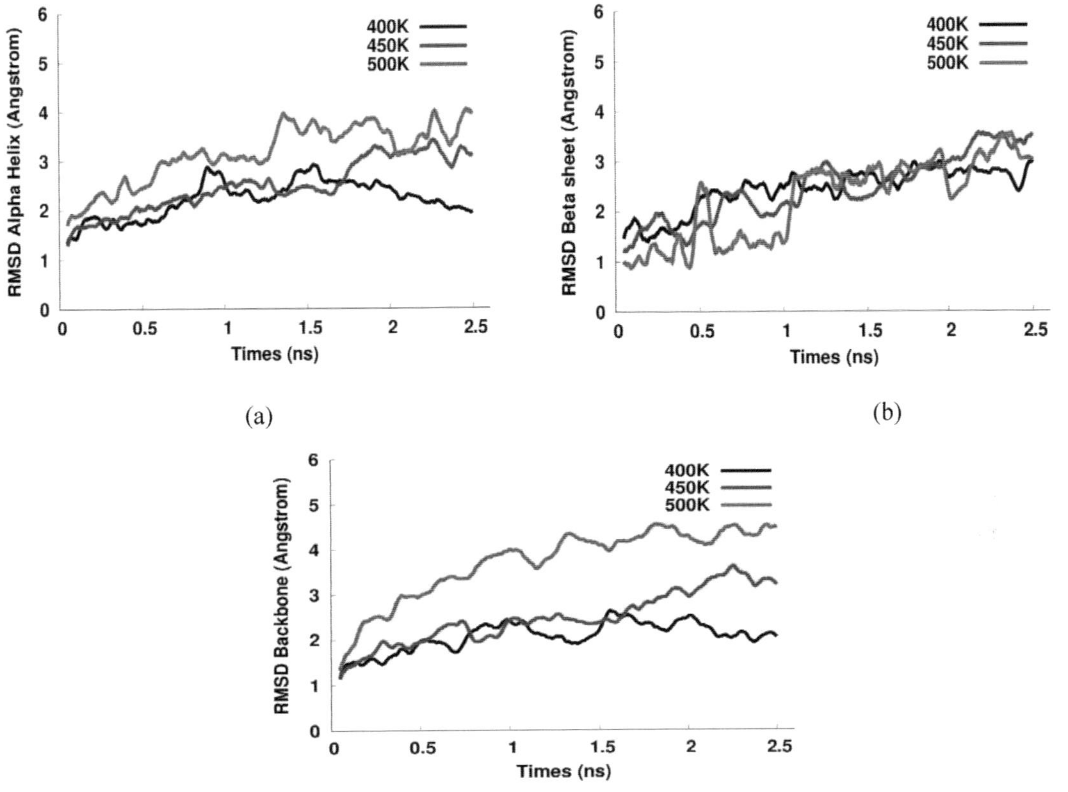

Figure 3. Root Mean Square Deviation (RMSD) as a function of simulation time for temperature 400 K, 450 K and 500 K (a) β-sheet structure (b) α-helix structure (c) backbone structure.

Figure 3.c shows the fluctuations Backbone RMSD values at temperatures 400 K, 450 K and 500 K for 2.5 ns. From 400 K to 450 K simulation temperature, there was no significant change in the values of RMSD. Until 2 ns the values of RMSD was constant in 2 Angstrom. An increase temperature in the system is not affecting the protein stability. Whereas the temperature 500 K, at this temperature RMSD enough constant in 2 Angstrom until 0.2 ns. The values of RMSD jumps drastically at 0.25 ns until 2.5 ns that is until 4.8 Angstrom. While still there is a need for detailed investigation of the reasons for this difference in the behavior of the fluctuations in the backbone region compared to the other region of the enzyme, we suggest that any attempt to design thermostable ANX should include mutation that could retain the backbone fluctuation of ANX at the similar level as that of fluctuations at the optimum temperature. We are currently undertaking in silico mutational studies to identify the mutations that will bring about this effect.

3.2. The Stability of the Secondary Structure

Analysis of changes in the secondary structure is an analysis that can be used as a marker has been the unfolding of proteins, especially the loss/ damage of a helix and sheet structure. Secondary structural analysis may also complete understanding of structural changes during the simulation [13].

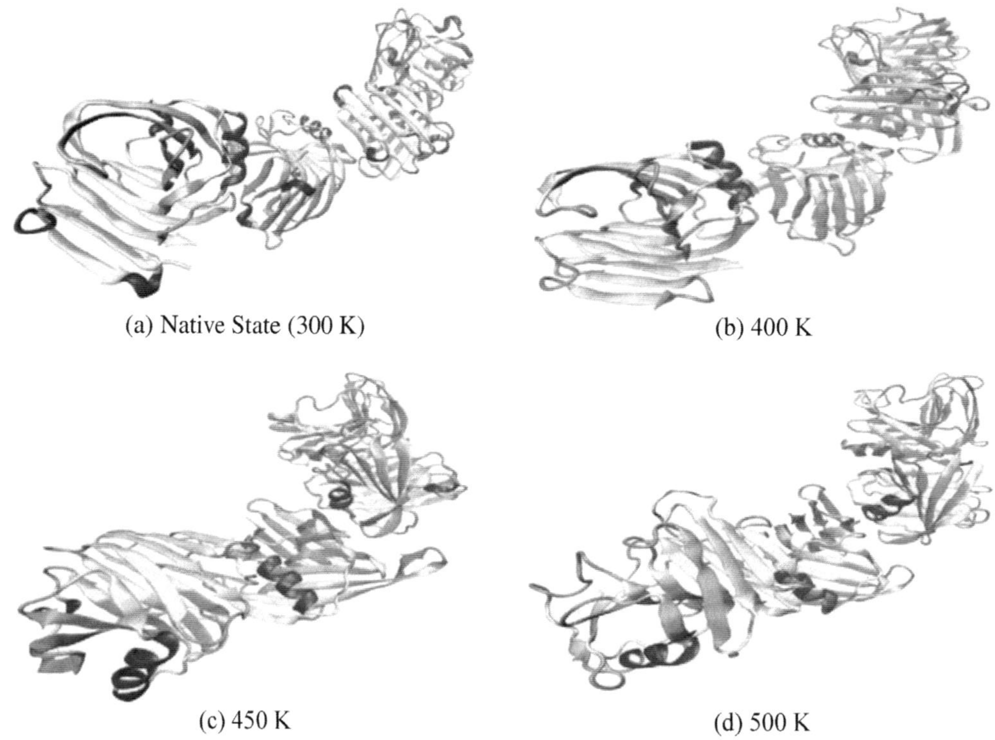

Figure 4. Protein secondary structure as a function of simulation time (a) for temperature 300 K, (b) for temperature 400 K (c) for temperature 450 K and (d) for temperature 500 K.

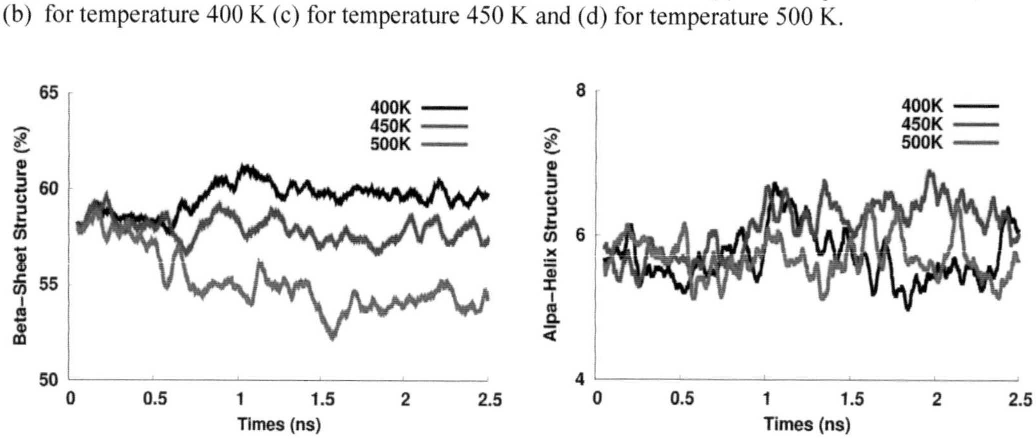

(a) (b)

Figure 5. The percentage change in the composition of secondary structure during the simulation at 400 K, 450 K and 500 K for 2.5 ns for (a) β-sheet structure. (b) α-helix structure.

The composition of secondary structures, especially α-helix or β-sheet at temperature 400 K and 450 K for 2.5 ns did not significantly changes (Figure 4.b and Figure 4.c). Meanwhile, at temperature 500 K composition of α-helix only slightly damaged while β-sheet had disappeared turned into coil and turn. This is indicates that proteins has been unfolding process. So, we suggest that any attempt to design thermostable ANX should include mutation that could retain the β-sheet had disappeared turned into coil and turn. We are currently undertaking in silico mutational studies to identify the mutations that will bring about this effect.

4. Conclusion

The analysis have shown that RMSD of backbone showed significant difference at higher temperatures compared other regions of the enzyme, we conclude that backbone region contribute significantly to the thermostability of ANX. Mutagenesis should be targeted to amino acid residues within backbone in ANX structure with the objective of maintaining the backbone regional fluctuation to the similar level as that of fluctuations at 400 K. If that can be achieved, we propose that such mutant ANX will be the thermostable and suitable for use in the pulp industry. Further work in this direction is in progress. Analysis of changes in the secondary structure is an analysis that can be used as a marker has been the unfolding of proteins, especially the loss/ damage of a helix and sheet structure. Secondary structural analysis may also complete understanding of structural changes during the simulation. So, we suggest that any attempt to design thermostable ANX should include mutation that could retain the β-sheet had disappeared turned into coil and turn. We are currently undertaking in silico mutational studies to identify the mutations that will bring about this effect.

Acknowledgments

The authors thank Research Computational Unit (RCU) Department of Physics Education FKIP UKI, for providing high performance computing to run all dynamics simulations. This research was supported by a research grant from Indonesia Christian University.

References

[1] Falch EA 1991 *Biotech Adv.* **9** 643-58
[2] Aunstrup KO, Andressen, Falch EA and Nielsen 1979 *Microbial Technology* **1** USA
[3] Marisa A, Lima 2013 *The Journal of Biological Chemistry* **288** 32991-3005
[4] Kirk TK and Jeffries TW 1996 *Enzymes for Pulp and Paper Processing* Washington USA
[5] Dhillon A, Gupta JK, Jauhari BM and Khanna S 2000 *Bioresource Technology* **73** 273 - 77
[6] Beg QK, Kapoor M, Mahajan L and Hoondal GS 2001 *Applied Microbiology and Biotechnology* **56** 326-38
[7] Otten LG, Sio CF, Van Der Sloot, Almer M, Cool RH and Quax WJ 2004 *Chem. Biochem.* **5** 820–25
[8] Schmidt DMZ, Mundorff EC, Dojka M, Bermudez E, Ness JE, Govindarajan S, Babbitt PC, Minshull J and Gerlt JA 2003 *Biochemistry* **42** 8387–393
[9] Morley KL and Kazlauskas RJ 2005 *Trends in Biotechnology* **23** 5
[10] Noorbatcha IA, Khan AM and Salleh HM 2010 *American Journal of Applied Sciences* **7** 823-28
[11] Krengel U and Dijkstra BW 1996 *J.Mol.Biol.* **263** 70-8
[12] Phillips JC, Braun S, Wang W, Gumbart J, Tajkhorshid E, Villa E, Chipot C, Skeel RD, Kale L and Schulten KJ 2005 *Comput. Chem.* **26** 1781
[13] Day R, Bennion BJ, Ham S and Dagget VJ 2002 *Mol. Biol.* **322** 189-203

Synthesis and Characterization of Composite Hydroxyapatite-Silver Nanoparticles

Charlena[1*], N A Nuzulia[2*] and Handika[1]
[1] Department of Chemistry, Bogor Agricultural University, INDONESIA.
[2] Department of Physics, Bogor Agricultural University, INDONESIA.

Email: charlena.ipb@gmail.com , aisyahnuzulia@apps.ipb.ac.id

Abstract. Hydroxyapatite (HAp) is commonly used as bone implant coating recently; however, the material has disadvantage such as lack of antibacterial properties, that can cause an bacterial infection. Addition of silver nanoparticles is expected to be able to provide antibacterial properties. Silver nanoparticles was obtained by reduction of $AgNO_3$ using glucose monohydrate with microwave heating at 100p for 4 minutes. The composite of hydroxyapatite-silver nanoparticles was synthesized using chemical methods by coprecipitation suspension of $Ca(OH)_2$ with $(NH_4)HPO_4$, followed by adding silver nanoparticles solution. The size of the synthesized silver nanoparticles was 30-50 nm and exhibited good antibacterial activity. Nevertheless, when it was composited with HAp to form HAp-AgNPs, there was no antibacterial activity due to very low concentration of silver nanoparticles. This was indicated by the absence of silver nanoparticles diffraction patterns. Infrared spectra indicated the presence of chemical shift and the results of scanning electron microscope showed size of the HAp-AgNPs composite was smaller than that of the HAp. This showed the interaction between HAp and the silver nanoparticles.

Keywords: antibacterial activity, composite, coprecipitation, nanoparticles

1. Introduction

Bone damage such as broken bones or fractures can be treated with the implants. Planting the implant will be able to replace the damaged bone or assist in the formation of new bone tissue. The implants can be made of metal, ceramic, polymer, or composites [1]. The field of medicine has been used as an implant metal as the metal has good mechanical properties, strong and malleable. Metals commonly used as an implant material of which is stainless steel, titanium alloys, and cobalt. But the form of metal implants when used in the long term would be harmful to the body. The release of ions from the implant-based metals such as iron ions, chromium, and nickel in the body led to allergy and demonstrated carcinogenic properties. Additionally, in the form of metal implants can cause swelling and pain around the bone. It is caused by a reaction of metal corrosion in body fluids and the low level of biocompatibility of these metals [2].

Prevention of corrosion and metal-based implants to increase of biocompatibility is necessary that successful to increases implantation of bone. This can be done by coating the implant with a metal-based material that is non-corrosive and have a good biocompatibility such as Hydroxyapatite [3]. biocompatibility of the hydroxyapatite due to the composition of hydroxyapatite has similarities with the composition of biological hard tissue such as bone and teeth [4]. Hydroxyapatite not only has good biocompatibility, in addition to hydroxyapatite is non-toxic, osteoconductive and bioactive. Therefore, the formation chemical bond between the implant and the bone tissue will be increases and accelerates the healing process of bone [5].

Hydroxyapatite is one type of bioceramics are widely applied in the fields of medical because of the similarity of its properties with chemical structure of the bone [6]. However hydroxyapatite has some disadvantages, among which it is fragile and antibacterial. This can reduce of the success of implantation and increased risk of death due to the infection by bacteria. Prevention of infection after bone implantation can be done by creating a composite of hydroxyapatite and material that is antibacterial. Nanoparticles of metals such as Zn, Cu, and Ag showed good antibacterial properties [7] One of the most studied nanoparticles are silver nanoparticles. Silver nanoparticles have stable properties and potential applications in various fields such as catalysts, optical sensors detectors and antibacterial agents [8].

Hydroxyapatite can be synthesized using waste blood shells, animal bones, stone gypsum, egg shells or *Bellamya javanica* shell [9]. *Bellamya javanica* shell of an abundant waste but has not been used commercially significant. This waste contains a variety of minerals one of which is calcium that can be used in the synthesis of hydroxyapatite [10]. In this study, hydroxyapatite synthesized from the utilization of waste *Bellamya javanica* shells through the wet method (precipitation). after obtained hydroxyapatite, hydroxyapatite-silver nanoparticle composite (HAp-AgNPs) synthesized by precipitation method using silver nitrate solution as the source of silver nanoparticles formed after reduced use glucose monohydrate. The research aims to synthesize silver nanoparticles as active substances that have antibacterial activity and synthesizing composite HAp-AgNPs potential as implant materials with antibacterial properties.

2. Mehods

2.1. Synthesis of Silver Nanoparticles

$AgNO_3$ solution 10^{-2} M is made by powder weighed as much as 0.1700 g $AgNO_3$, dissolved in 100 mL flask with distilled water and homogenized. A total of 16 mL of the solution was put in a beaker, then added 14 mL solution of glucose monohydrate 0.5 M. The solution was prepared by weighing 9.9000 g glucose monohydrate, dissolved in 100 mL flask with distilled water and then homogenized . After $AgNO_3$ 10^{-2} M and 0.5 M glucose monohydrate were mixed, added 10 mL of distilled water, and then homogenized. The mixture was heated in the microwave heating for 2 minutes then stirred using a magnetic stirrer for 20 minutes. Reheat for 1 minute and stir again for 20 minutes, repeat 2 times [11].

2.2 Synthesis of Hydroxyapatite

The powdered of calsium hydroxide from calcination *Bellamya javanica* shell used in the synthesis of HA as a source of calcium which is reacted with a solution of $(NH_4)_2HPO_4$ through the following reaction:

$$10Ca(OH)_2 + 6(NH_4)_2HPO_4 \rightarrow Ca_{10}(PO_4)_6(OH)_2 + 6H_2O + 12NH_4OH$$

The solution of $(NH_4)_2HPO_4$ 0.3 M dropped on the suspension of $Ca(OH)_2$ 0.5M using buret at 40±2 °C and stirred using a magnetic stirrer. In this synthesis, the pH is monitored and corrected to obtain a pH of 10. The solution was then sonicated for 3 hours, then decantation for 24 hours at room temperature. Furthermore, the solution was decanted and centrifuged at 4500 rpm for 15 minutes and then the precipitate is separated and rinsed with distilled water. Then, the precipitate is dried at 105 °C for 3 hours. After drying, the precipitated finely pulverized using a mortar and then sintered with a temperature of 500 °C for 3 hours. HA powder then cooled in the desiccator [12].

2.3 Synthesis of Composite HA-AgNPs
Powder $Ca(OH)_2$ from calcination *Bellamya javanica* shell used in the synthesis of composite HA-AgNPs as a source of calcium. Composite HA-AgNPs synthesized through the reaction between the suspension of $Ca(OH)_2$ solution $(NH_4)_2HPO_4$ and AgNPs solution. A total of 100 mL $(NH_4)_2HPO_4$ 0.3 M dropped to 100 mL suspension of $Ca(OH)_2$ 0.5 M using a buret at a temperature of 40±2 °C and stirred using a magnetic stirrer. Furthermore, as much as 50 mL AgNPs dripped on HA suspension using a buret. The solution was then sonicated for 3 hours and decantation for 24 hours at room temperature. The solution was decanted and centrifuged at 4500 rpm for 15 minutes and then the precipitate was separated and rinsed with deionized water. Furthermore, the precipitate is dried at 105 °C for 3 hours. After drying, the precipitated finely pulverized using a mortar and then sintered with a temperature of 500 °C for 3 hours. Powder HA-AgNPs then cooled in the desiccator [13].

2.2. Characterization
Silver nanoparticles has been synthesized and absorbance was measured with UV-Vis spectrophotometer at wavelength of 300-500 nm to estimate the size of AgNPs before being used in the synthesis of composite HA-AgNPs. Then HA and composite HA-AgNPs that has been synthesized are characterized using FTIR, XRD and SEM. FTIR characterization was performed to identify the functional groups on HA and the effect of adding AgNPs with the functional groups in the composite HA-AgNPs. XRD characterization was conducted to determine the phase contained in the composite HA and HA-AgNPs that has been synthesized. The results were compared with Joint Committee on Powder Diffraction Standards (JCPDS). Characterization using SEM was conducted to determine the morphology of HA and composite HA-AgNPs.

2.3. Antibacterial Activity Test
Antibacterial activity test performed on the active substance AgNPs, HA, and composite HA-AgNPs, the bacteria used were *Staphylococcus aureus*. Petri dishes are prepared and sterilized beforehand. Antibacterial activity test on AgNPs do with the disc method is to add 20 mL media Nutrient Agar (NA) into a sterile petri dish. Then, the bacteria is prepared at a concentration of 10-6 grown in media that have been made. Discs (diameter 6 mm) was added to the media. Furthermore, as many as 1 mL colloid AgNPs dropped into discs using a pipette. Then, incubated at 37 °C for 24 hours. Antibacterial activity test on HA and composite HA-AgNPs conducted using wells that the NA added 20 mL media into a sterile petri dish. Then, the bacteria is prepared at a concentration of 10-6 grown in media that has been made. A total of 0.2 grams of sample was mixed with 1 mL of distilled water, then put in the media. Furthermore, incubated at 37 °C for 24 hours. The antibacterial activity was calculated by the inhibitory zone diameter in mm. The test was done in duplicate. Value index of inhibition of bacteria can be calculated in the following equation:

$$\text{Index Inhibitory Control} = \frac{\text{diameter of inhibition zone - diameter discs}}{\text{diameter discs}}$$

3. Results and Discussion

3.1. Silver Nanoparticles Size Analysis
Silver nanoparticles sythesis in this research was done using chemical reaction method. $AgNO_3$ was reduced by monohydrate glucose. Synthesis of silver nanoparticle was done by adding monohydrate glucose into $AgNO_3$ solution, which is then heated gradually using microwave 100 W and after each heating phase finished, solution was stirred using magnetic stirrer. Gradual heating was done to prevent AgNPs agglomeration. Based on research by Sulistiawaty *et al.* (2015)[11], monohydrate glucose concentration was 0.5M and $AgNO_3$ concentration was 1×10^{-2} M. Formation of AgNPs was indicated by color changing from colorless to pale yellow. Pale yellow color showed that AgNPs started to form and when the heating phase ended, the solution will have yellow color (Figure1).

AgNPs

Figure 1. Synthesized silver nanoparticles.

Yellow color was caused by plasmon surface vibration excitation on AgNPs surface [14]. According to Solomon *et al.* (2007)[15], formation of nanoparticles was indicated by absorption peak at 400-500 nm, that peak involved electron cloud resonance on nanoparticles surface. Absorption appeared as result of inter bands transition in the system (4d-5sp) at higher energy level after exposed to light at UV-Vis wavelength [16]. Absorption peak at 420 nm indicated the formation of AgNPs. That absorption peak formed after $AgNO_3$ salt reacted with monohydrate glucose. Ag^+ ion from $AgNO_3$ solution could be reduced to Ag with aldehyde group in monohydrate glucose. During reduction, addition of electron transformed Ag^+ ion to become uncharged (Ag^0) [17]. According to Dankovich (2014)[18], reaction of AgNPs formation with monohydrate glucose reductor could be concluded as:

$$Ag^+_{(aq)} + C_5H_{11}O_5\text{-}CHO_{(aq)} + 2OH^- \longrightarrow Ag\text{-}NP + C_5H_{11}O_5\text{-}COOH_{(aq)} + H_2O_{(l)}$$

According to Solomon *et al.* (2007) [15], AgNPs size could be estimated based on wavelength of the absorption peak (Tabel 1). Synthesized silver nanoparticle from $AgNO_3$ precursor with monohydrate glucose reductor had maximum wavelength of 420 nm (Figure 2). Based on this, synthesized silver nanoparticles size was estimated ranging from 35-50 nm. Synthesized AgNPs using ascorbic acid as the reductor for $AgNO_3$. Characterization result using UV-Vis spectrophotometer and TEM showed that synthesized AgNPs had absorption peak at 414.5 nm and average size of 28 nm.

Table1. Silver nanoparticles size range based on spectrum pattern [15].

Particle size (nm)	Maximum wavelength (nm)	Spectrum band width (nm)
10-14	395-405	50-70
35-50	420	100-110
60-80	438	140-150

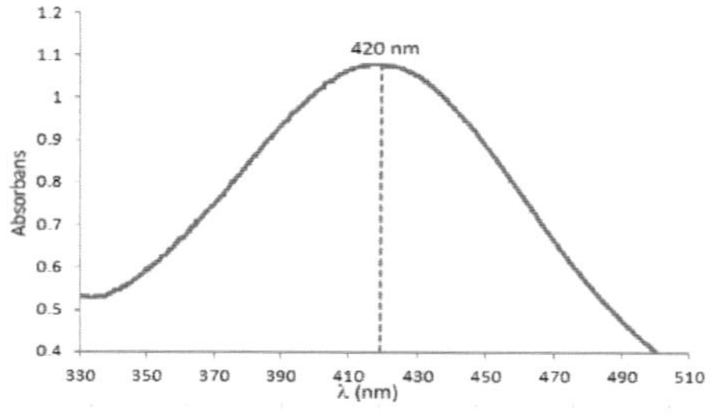

Figure 2. Silver nanoparticles UV-Vis Spectrum.

3.2. Hydroxyapatite-Silver nanoparticles Composite Synthesis

Hydroxyapatite (HA) is a type of bioceramic that have wide application in medical field, one of the is as metal based implant coating material because it have similar properties with bone chemical structure [6]. This bioceramic has good biocompatibility, non-toxic, osteoconductive, and also boactive. But HA doesn`t have antibacterial properties, which can lead to infection by bacteria during implantation process. Infection can cause implantation failure or even death. Addition of antibacterial material can provide ideal condition for bone regeneration process because it can prevent infection and increase patient tolerance [7]. One of the material with antibacterial properties is silver nanoparticles [8].

HA-AgNPs composite can be syntesized using presipitation method. This method is easy and simple, has high rendement, and relatively low cost, which is by reacting calcium and phosphate source with concentration ratio of 1.67. Then, AgNPs solution is added. Composite formation mechanism is Ag ion combines with HA then AgNPs particle growth occurs inside and on the surface of HA (Figure 3). AgNPs particle growth on the surface of HA happen when AgNPs interacts with OH^- ion of HA [13].

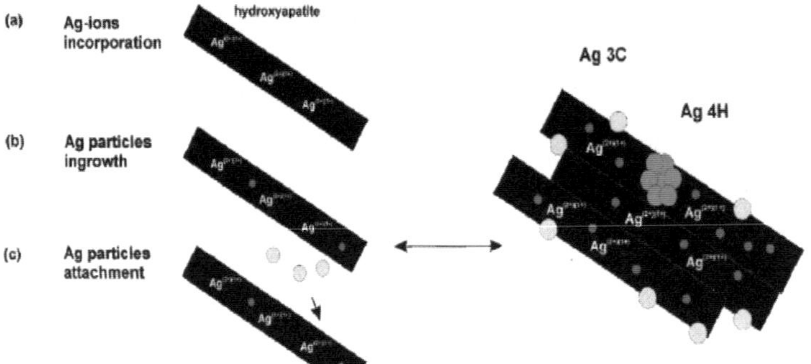

Figure 3. HA-AgNPs composite formation mechanism.

$(NH_4)_2HPO_4$ solution (Phosphate source) was added slowly into $Ca(OH)_2$ (calcium source) suspension under presipitation temperature of 40±2 °C and stirring for homogen mixing, then pH of the solution was measured until the pH was 10. Then AgNPs solution that has been synthesized before was added slowly into the suspension while stirred so that the mixture was homogen. That suspension was then sonicated. Sonication process was done to produce HA-AgNPs composite with high homogenity. Precipitate was seperated by sentrifugation. Precipitate was then filtered and washed with aquadest to get rid of unwanted solvent. Drying and sintering process could get rid of the water and other organic compound. Synthesized HA-AgNPs was in form of powder with yellow color. It had a different color from HA that was synthesized as control using presipitation method (Figure 4). HA and HA-AgNPs composite was characterized using FTIR, XRD, dan SEM.

Figure 4. Synthesized HA (a) and HA-AgNPs composite (b).

3.3. Analysis of Hydroxyapatite and Hydroxyapatite-Silver Nanoparticle Composite

Characterization using XRD, FTIR, and SEM was done to HA and HAp-AgNPs composite. Characterization using XRD supported in the determination of HAp-AgNPs composite structure, especially in determining crystalinity of hydroxyapatite and silver nanoparticles. XRD characterization on HAp and HAp-AgNPs composite showed specific diffraction pattern of hydroxyapatite (Figure 5). That was because the concentration of silver nanoparticles that was added, which was 0.4% b/b, was too low. Yan *et al.*(2015) [7] reported that Hydroxyapatite composite, chitosan, and silver nanoparticle that coated TiO_2 with addition of 0.3% b/b silver nanoparticle did not show specific diffraction pattern of silver nanoparticles. But the antibacterial activity increased by 20% and the test on cell showed that cell durability increased by 10%.

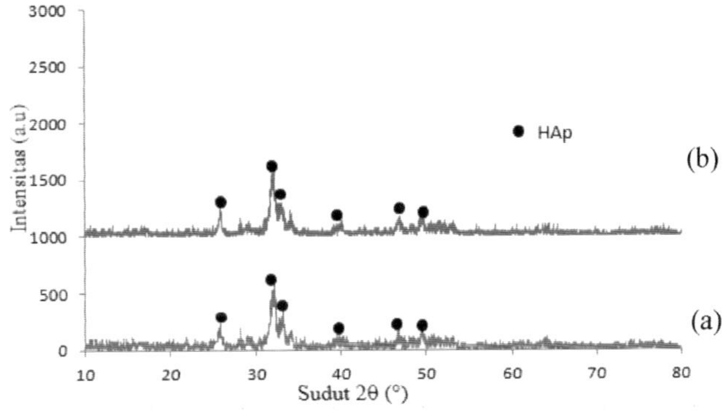

Figure 5. X-Ray diffraction pattern of HA (a) and HA-AgNPs composite (b).

Characterization using FTIR was done to determine functional group and interaction in HA-AgNPs composite. The result showed that the absorption bands of HA and HA-AgNPs composite was almost similar, that absorption band was dominated by apatite compound specific absorption but there was shift in wavenumber or the decrease of %transmitant of HA-AgNPs composite. Besides, there was absorption that showed the existence of aldehyde in HA-AgNPs composite (Figure 6). The shifting in wavenumber or decreasing of % transmitant could indicated the interaction between HA and AgNPs in the composite [7]. According to Vukamonovic *et al.* (2011)[13], shifting of OH^- ion strain that was in HA proved that there was interaction between HA and AgNPs. Besides, OH^- ion had an important role to control the size and stability of AgNPs in the composite. Decreasing of vibration from PO_4^{3-} indicated the decreasing of crystalinity that was caused by the addition of AgNPs.

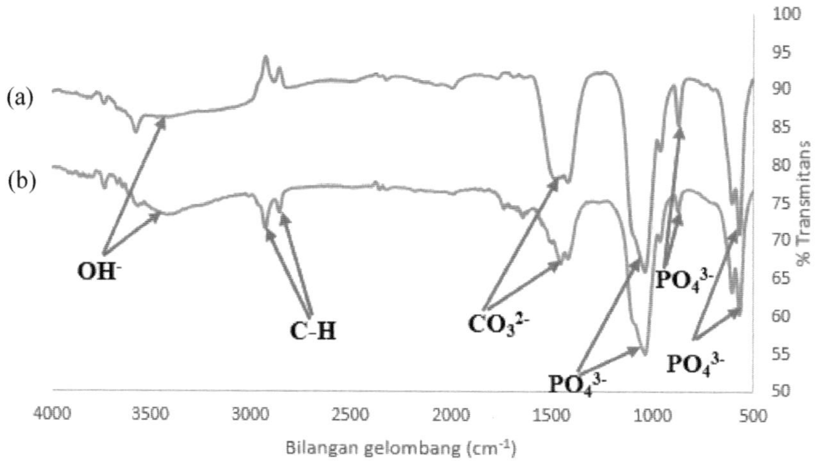

Figure 6. FTIR spectrum of HA (a) and HA-AgNPs Composite (b).

According to Meejo *at al.* (2006) [19], C-O stretching vibration for carbonate group usually appeared at wavenumber 1400-1600 cm^{-1}. Characterization using FTIR on HA and HA-AgNPs composite showed the strain at 1423 cm^{-1} which was caused by CO_3^{2-} ion. CO_3^{2-} ion could cause the formation of carbonate apatite type A if CO_3^{2-} replaced OH^- position or carbonate apatite type B if CO_3^{2-} replaced PO_4^{3-} position in hydroxyapatite. Besides, HA-AgNPs composite showed C-H strain at 2850 and 2924 cm^{-1} . That strain resulted from monohydrate glucose that was used as $AgNO_3$ reductor. This caused HA-AgNPs composite had yellow color as the result of caramelization reaction of monohydrate glucose.

According to Santos *et al.* (2004) [12], sythesis of hydroxyapatite using copresipitation method produced hydroxyapatite in the form of granule which size was not uniform. Characterization with SEM on HA and HA-AgNPs composite showed granules which was hydroxyapatite. But its size was still not homogen, granule size of HA was about 8 μm, meanwhile HA-AgNPs composite size was 6 μm. Surface morphology of HA-AgNPs composite was different from HA which was expected because the existence of silver nanoparticle on the surface of HA-AgNPs composite (Figure 7). Research by Vukamonovic *et al.* (2015) [20] showed that silver nanoparticle which was composited with hydroxyapatite caused hydroxyapatite size to become smaller and the silver nanoparticle was attached on the surface of hydroxyapatite.

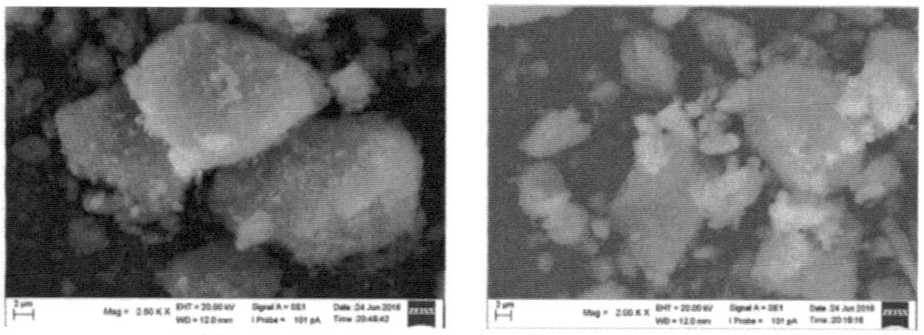

(a) (b)

Figure 7. Surface morphology of HA (a) and HA-AgNPs composite (b).

3.4. Result of Antibacterial Activity Test.
Antibacterial activity test on silver nanoparticle coloid was done using disk method, meanwhile test on HA and HA-AgNPs composite was done using well method. The difference on the method was because silver nanoperticle coloid had liquid form meanwhile HA-AgNPs composite had powder form. But those two method had same principle which was antibacterial activity was showed by transparent zone diameter around the disk or well that was made. Transparent zone would appeared after incubation for 24 hours at 37°C. Bacteria that was used in this test was *Staphylococcus aureus.* Measurement was done twice with ruler.

Staphylococcus aureus was chosen because this phatogenic bacteria was the cause of most infection on bone implant. *Staphylococcus aureus* was positive gram bacteria and coccus- shaped. Important substances in bacteria cell wall structure was polysaccharide and protein. That substances, with peptidoglycan and teichoic acid formed a rigid cell wall. The main chain of teichoic acid had a lot of phosphate group, so it had negative charges and could interact electrostaticly with cationic antibacterial compound. Silver nanoparticle as antibacterial agent had been researched for quite a long time but the mechanism of AgNPs was only partly known. Some estimated mechanism of AgNPs was that AgNPs could prevent ATP formation and DNA replication in bacteria, and it could pass through the cell wall, make hole ,and enter the cytoplasm. Then, AgNPs accumulation occured on cell membrane and increased permeability which could cause the death of bacteria [21]. Other mechanism was that Ag ion penetrated cell membrane and reacted with -SH group of cell protein. This would cause uncoupling of ATP synthesis during cell respiration, proton loss, and phosphate efflux system disruption [15].

Antibacterial test result observation showed that there was transparent zone when AgNPs was tested, but test on HA and HA-AgNPs composite there was no transparent zone (Figure 8). Calculation of bacterial inhibition index on AgNPs was 0.33, meanwhile HA did not show bacterial inhibition index. When HA was composited withAgNPs, there was potential of forming antibacterial implant coating material. But the result showed that HA-AgNPs composite showed that there was no bacterial inhibition index. This was because the concentration of AgNPs was too low. Low AgNPs concentration caused released Ag ion could not damage bacteria cell wall so the growth of bacteria was not inhibited. According to Vukamonović *et al.* (2015) [20], composite of hydroxyapatite and silver nanoparticles with 10% b/b addition of silver nanoparticle had antibacterial properties because silver nanoparticle was released and could damaged bacteria cell wall. This was proved by SEM characterization result on bacteria that showed the damaged cell wall.

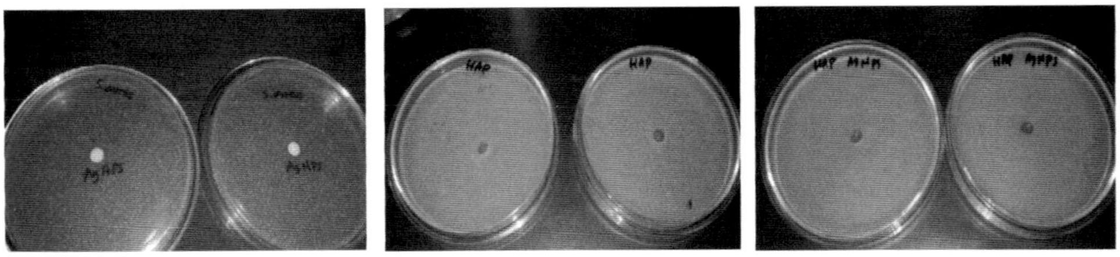

Figure 8. Antibacterial test result on AgNPs, HA, and HA-AgNPs.

4. Conclusion
Silver nanoparticles was successfully synthesized using reduction method with monohydrate glucose as reductor for $AgNO_3$. Synthesized AgNPs size was ranging from 35-50 nm and had antibacterial activity which was stated in inhibition index of 0.33. But when composited to form HA-AgNPs, antibacterial activity was not detected. This was caused by low concentration of AgNPs and proved by XRD analysis result that did not show the diffraction pattern of AgNPs. FTIR analysis showed the interaction between HA and AgNPs. Characterization using SEM showed that HA-AgNPs nanoparticles was granules and had smaller size compared to HA, with size range between 1-6 μm.

References

[1] Iwasaki T, Nakatsuka R, Murase K, Takata H, Nakamura H, Watano S 2013 *Journal Molecular Sciences* **14(5)** 9365-9378
[2] Gopi D, Sathishkumar S, Karthika A, Kavitha L 2014 *Industrial & Engineering Chemistry Research* **53(52)** 20145–20153
[3] Monmaturapoj N 2008 *Journal of Metals, Material and Minerals* **18(1)** 15-20
[4] Ramli RA, Adnan R, Bakar MA, Masudi SM 2011 *Journal of Physical Science* **22(1)** 25-37
[5] Sobczak A, Kowalski Z, Wzorek Z 2009 *Acta of Bioengineering and Biomechanics* **11(4)** 23-28
[6] Agrawal K, Singh G, Puri D, Prakash S 2011 *Journal of Minerals & Materials Characterization & Engineering* **10(8)** 727-734
[7] Yan Y, Zhang X, Li C, Huang Y, Ding Q, Pang X 2015 *Apllied Surface Science* **332** 62-69
[8] Haryono A, Dewi S, Harmami SB, Randy M 2008 *Jurnal Riset Industri* **2(3)** 156-163
[9] Salomoni R, Leo P, Ro drigues MFA 2015 *The Battle Against Microbial Pathogens: Basic Sciences, Technological Advances, and Educational Programs* 851-857
[10] Baby RL, Hasan I, Kabir KA, Naser MN 2010 *Journal of Scientific Research* **2(2)** 390–396
[11] Sulistiawaty L, Sugiarti S, Darmawan N 2015 *Indonesian Journal of Chemistry* **15(1)** 1-8
[12] Santos MH, de Oliveira M, SouzaL PF,Mansur HS, Vasconcelos WL 2004 *Materials Research.* **7(4)** 625-630
[13] Vukomanović M, Braćko I, Poljanšek I, Uskoković D, Škapin SD, Suvorov D 2011 *Crystal Growth & Design* **11 (9)** 3802-3812
[14] Shankar SS, Rai A, Ahmad A, Sastry M 2004 *Journal of Colloid and Interface Science.* **275(4)** 496-502
[15] Solomon SD, Bahadory M, Jeyarajasingam AV, Rutkowsky SA, Boritz C, Mulfinger L 2007 *Journal of Chemical Education* **84 (2)** 322-325
[16] Feldheim DL dan Foss CA 2002 *Metal Nanoparticles: Synthesis, Characterization, and Applications.* New York (US) : Marcel Dekker.
[17] Shi Q, Vitchuli N, Nowak J, Noar J, Caldwell JM, Breidt F, Bourham M, McCord M, Zhang X 2011 *Journal of Materials Chemistry* **21** 10330-10335
[18] Dankovich TA 2014 *Environmental Science Nano* **1** 367-378
[19] Meejoo S. Maneeprakorn W, Winotai P 2006 *Thermocimica Acta* **447(1)** 115-120
[20] Vukamonović M, Repnik U, Bergant TZ, Kostanjsek R, Skapin SD, Suvorov D 2015 *Biomaterials Science & Engineering* **1(10)** 935-946
[21] Mocanu A, Furtos G, Rapuntean S, Horovitsz O, Flore C, Garbo C, Danisteanu A, Rapuntean G, Prejmerean C, Cotisel MT 2014 *Applied Surface Science* **298** 225-235

IOP Conf. Series: Earth and Environmental Science **58** (2017) 012065 doi:10.1088/1755-1315/58/1/012065

The Flexible Carbon Activated Electrodes made from Coconut Shell Waste for Supercapacitor Application

E Taer[1*], W S Mustika[1*], Agustino[1], Fajarini[1], N Hidayu[1] and R Taslim[2]
[1]Department of Physics, Faculty of Mathematics and Natural Science, University of Riau, INDONESIA.
[2]Department of Industrial Engineering, Faculty of Science and Technology, State Islamic University of Sultan Syarif Kasim, INDONESIA.

Email: erman_taer@yahoo.co.id , widyasintamustika@ymail.com

Abstract. The flexible carbon activated electrode supercapacitor application has successfully prepared from coconut shell based carbon materials that produced by local farmers. The Increasing physical and electrochemical properties of coconut shell carbon has been done by chemical activation using KOH and microwave radiation assisted. A Flexible carbon activated electrodes was produced by by spraying method. Physical and electrochemical properties were tested are the specific surface area and specific capacitance. Extensive surface and optimum capacitance were obtained from a flexible carbon electrodes made from coconut shell based carbon is 194.429 m^2 / g and 10.55 F / g, respectively.

1. Introduction

Electrochemical double layer capacitors (EDLC) or also known as supercapacitors have become an object of research interest in the area study of energy storage device [1]. The electrodes in this device is the main component that determines the amount of energy and power that can be stored in a supercapacitor cell. Looking for a superior electrode is still in progress today, one way is to select the basic raw materials that can be developed into various forms and modifications pores for carbon electrodes. Carbon materials attract to be studied because the advantage of some properties such as the abundant of materials availability, cheap price, pores are easily controlled and stable in the physical and chemical characteristic [2]. Some carbon materials were produced from waste of agriculture. Coconut based waste materials, such as shell has been studied its compatibility as carbon electrode for supercapacitor application. Several previous studies have reviewed on several sides in the preparation of coconut shell as a supercapacitor electrode. Yang in 2010 showed that carbon activated materials can be made from the coconut shell with high surface area [3]. Geng in 2013 has reviewed the various activation agent to produce a high pore surface area activated carbon from coconut shell [4], Hu in 2014 [5] Jain 2014 [6] has reviewed the use of a polymer electrolyte to the coconut shell based electrodes for device of supercapacitor. Some studies that has been shown the coconut shell can be developed as a potential raw material in the production of supercapacitor electrodes.

The Improvement in energy storage capability and power of the supercapacitor and the excellence competitiveness with other energy storage device can also be done by adding dimension and primacy in the shape. Flexible supercapacitor cell will facilitate the use of supercapacitors in many appliances. One way of making the flexible supercapacitor cells can be done using the method of spraying the porous carbon material on a sheet of flexible media. This study will report the manufacture of flexible supercapacitor carbon electrodes from carbon material that produced by coconut farmers through the spraying technique. The production of activated carbon from coconut shell is done with by using chemical activation agent with potassium hydroxide and assisted with physical activation using microwave radiation.

2. Experiment

Coconut shell based carbon (CSC) is obtained from a merchant retail in Panam market, Pekanbaru, Riau Province, Indonesia. Carbon based on the coconut shell is produced by local farmers through conventional carbonization techniques. The process of production of activated carbon was begins with the destruction process of the CSC through manual milling techniques to produce small grain size of 0.5 mm. CSC granules was continued by ball milling process for 20 hours to produce a finer grain size is in the range of 50-100 mm. Furthermore, CSCs were activated by chemical activation using KOH as an activating agent with the composition 1: 3, 1: 4 and 1: 5 in weigh ratio. Differences in the composition of carbon and KOH generate different samples and labelled as A1, A2 and A3 for 1: 3, 1: 4 and 1; 5, respectively, the same procedure has been reported before [7, 8]. The next step of each sample was irradiated using microwaves radiated at a power of 600 watts for 20 minutes. Furthermore, sample A1, A2 and A3 that have been irradiated were labelled A1 + R, + R A2 and A3 + R. The last step in the preparation of activated carbon is washed using distilled water and dryed to obtain powder coconut shell activated carbon (ACSC).

A cell supercapacitor was constructed by using some of component such as: a pair of Stainless Steel current collector, a separator using whatman filter paper, a pair of ACSC electrodes, and 1 M of H_2SO_4 as an electrolyte. The process of production an activated carbon electrode includes several steps: first, active carbon powder soaked into 1 M H_2SO_4 solution for 2 x 24 hours. Further, the electrolyte solution and ACSC sprayed on one side of the separator with a distance of 15 cm and the number of sprays 3 times. Spraying carbon electrode was conducted at a pressure of 8 kg / cm^2. Carbon sheets are then dried at a temperature of 65°C for 5 minutes. The same process is subsequently repeated on the other side of the separator. The last stage, the electrode were dried again and flexible carbon electrode were obtained from coconut shell based ACSC. Furthermore, the electrode is ready for fabrication to be a supercapacitor cells for examination, like Figure 1.

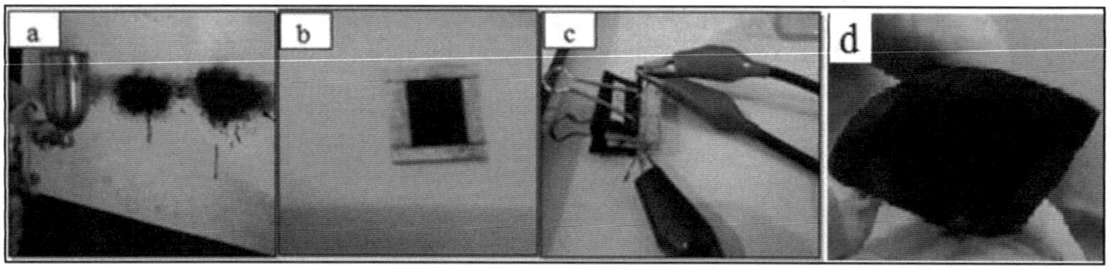

Figure 1. Fabrication of cell supercapacitor: a) process of spraying activated carbon, b) flexible electrode cell, c) flexible supercapacitor cells from coconut shell, d) bent electrode samples showed the flexible characteristic.

Physical properties of ACSC Flexible were performed on the surface morphology and surface area properties. The surface morphology was characterized using a Scanning Electron Microscope (SEM) using a Hitachi S-3400N with a magnification of 10.000 times. ACSC surface area measured by a N_2 gas absorption using the Instrument Nova Station B at a temperature of 77 K. Electrochemical properties supercapacitor cell were carriout by using Cyclic voltammetry (CV) used the handmade instrument UR RAD-ER 5841 which has been calibrated. The potential window for CV measurement of 0 V to 0.5 V with a scan rate of 1 mV / s and is controlled by software in C ++.

3. Result and Discussion

SEM micrograph images of each of the various samples are shown in Figure 2. SEM micrographs show the difference in colour of dark and light. The differences of brightness level of SEM morphological image may indicate differences in the nature of pores in the particles ACSC. Based on Figure 2, Figure 2b and 2 d which is a sample A3 and A3 + R have the highest brightness level and the lowest. Brightness level morphological differences have relevance to the surface area of the sample.

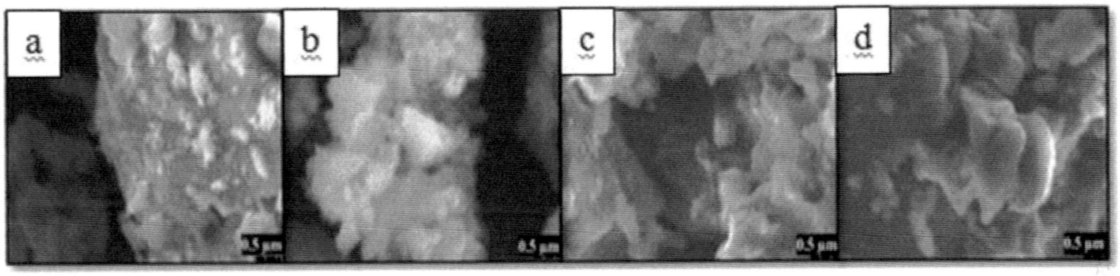

Figure 2. SEM image of coconut shell activated carbon a) A1, b) A3), c) A1+R, d) A3+R

The use of activators KOH on this research has resulted average pores in microporous [9], the pore size of the pore with diameter of ≤ 2 nm. The existence of these pores is indicated by the brightness level morphological SEM lighter. The increasing in the concentration of KOH can increase the number of micropores thus increasing the brightness of SEM micrographs. The power in microwave radiation was produce more light micrograph at A1 + R compared to A1. The process of radiation have resulted in enlarging existing pores and increase the formation of new microporous. On the other hand, the microwave radiation levels uniformly provide better surface regularity on the A1+R and A3 + R. However, the increase of KOH concentration in the samples A3 + R produces SEM micrograph darker. This is due to the formation of a new pore structure on A1 + R better than A3 + R. The increase in the concentration of KOH at A3 + R inhibits the formation of new porous structure by microwave.

CV measurement data obtained in the form of a change of current density vs voltage in cycle curve. The current density is expressed for the two states of charge and discharge condition. Sample data for the CV characteristics measurement are shown in Figure 2. The area formed by the charge and discharge currents have become a key factor in determining the nature of capacitance. The specific capacitation (Csp) of electrode can be calculated using the formula 1, where Ic, Id, s and m each are: cas current density, discharge current density , the scan rate and the mass of the electrode. The complete results of the charge, discharge for the entire electrode of supercapacitor cells is shown in Table 1. Table 1 shows that the electrode A1 + R has the best value of Ic-Id. Based on the formula 1 is clearly shown that the specific capacitance is affected by the Ic-Id and the mass of the electrode. Sample A1 + R shows the highest value of Ic-Id, this result can be attributed that these sample would yield the highest specific capacitance value. Instead A1 sample showed the lowest value of Ic-Id. These results suggest that the microwave radiation has changed the nature of the sample pore A1. The formation process flow occurs in a double layer supercapacitors caused by the process of forming a double layer of charge between ions and electrons into the electrode pores. These results clearly show that the samples A1 + R has a pore structure that is easy difused by ion so that the ion-electron pairs more formed.

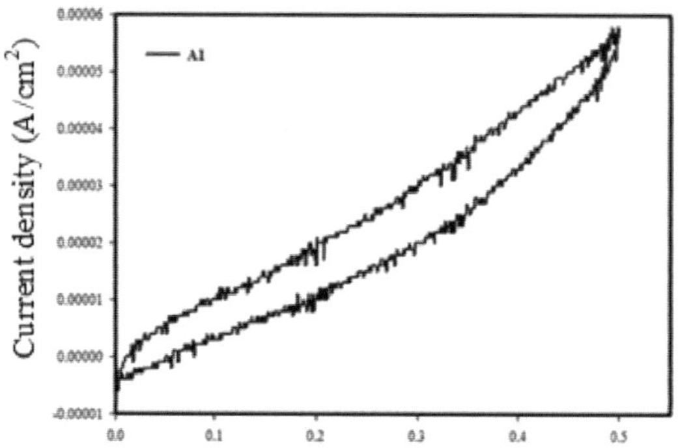

Figure 3. Cv measurement data for A1 sample.

Table 1. Data of CV measurement of coconut based activated carbon supercapacitor electrode.

Samples	Scan rate (V/s)	Mass (g)	Ic (A)	Id (A)	Ic-Id
A1	0.001	0.0165	0.000024	0.000015	0.000009
A2	0.001	0.0190	0.000025	0.000012	0.000013
A3	0.001	0.0175	0.000103	0.000061	0.000042
A1 + R	0.001	0.0165	0.000274	0.000100	0.000174
A2 + R	0.001	0.0180	0.000204	0.000087	0.000117
A3 + R	0.001	0.0170	0.000234	0.000162	0.000072

$$C_{sp} = \frac{(I_C - I_D)}{S\ x\ m} \qquad (1)$$

ACSC specific capacitance values correlated to surface area [10]. Specific capacitance value of the electrode and the BET surface area ACSC presented in Table 2. The surface area data of the sample A1 and A3 shows that the increase in carbon and KOH ration, KOH effect on increasing the BET surface area (SBET) [11]. The process of microwave radiation is also shown a role in the increasing in the surface area of the sample.

The increases in the concentration of KOH enhance the Csp and BET surface area on the samples of A1, A2 and A3. Microwave radiation for each various concentrations of KOH activator have been increased the Csp. The A1 sample that have been radiation treatment produce a new pore structure so that the surface area A1 + R larger than A1. Radiation treatment increases SBET [12]. On the other hand, the microwave radiation of 600 Watt to A3 causes a reduction in the BET surface area of the sample A3 + R. This situation is related to the A3 sample has a larger pore size, so the addition of radiation causes the pore size more increases, this situation is causing the decreasing in BET surface area. The Csp for A3 + R sample higher than A3 may influence by surface regularity A3 + R better than A3.

Table 2. The value of specific capacitance and BET surface area of coconut shell based activated carbon electrode.

Sampel	S_{BET} ($m^2.g^{-1}$)	C_{sp} ($F.g^{-1}$)
A1	153.815	0.54
A2	-	0.63
A3	200.695	2.40
A1 + R	194.429	10.55
A2 + R	-	6.50
A3 + R	82.755	4.24

Comparison of several flexible supercapacitor electrodes with different precursors are presented in Table 3. This data includes the method of manufacture and the specific capacitance of the flexible electrode. This table shows that the flexible carbon electrodes from carbonized coconut shell by conventionally carbonization process have relatively low capacitive properties than other raw materials. The abundant availability of coconut shell is the advantages of these samples so it certainly can be produced at a relatively low cost. Subsequent studies for physical properties of flexible supercapacitor electrode of coconut shell are expected to increase the electrochemical properties the electrode. Alternative improvement of physical and electrochemical properties still can be more improved such as modification in the carbonization process. Generally, this study has successfully demonstrated that the production of coconut shell based carbon by local farmers as potential materials to be developed as raw materials for a flexible carbon electrode in supercapacitor as an energy storage device.

Table 3. Different flexible supercapacitor electrode from several raw material.

Precursor	Method	Csp (F/g)	Reference
Polyacrylonitrile+ polyvinylpyrrolidone +SiO2	Co-electro spinning	242	[13]
Polyethersulfone	Magnetic stirring and ultrasonication	169.4	[14]
TPA + PAN + PVP	Electrospinning	175	[15]
Multi Wall Carbon Nanotubes	Spray method	145	[16]
Activated Carbon Cloth + Polyaniline	Electrodeposition	241	[17]
Carbon Cloth+MnO	Bonding Cu-Wire	425	[18]
Conventional Coconut Shell	Spray method	10.55	Recent study

4. Conclusions

This study has demonstrated the opportunities use of carbon from coconut shell as a flexible activated carbon electrode. The abundant availability, low priced and easily to be activated were good properties of coconut shell based carbon that can be relied upon. Finally, it can be concluded that this study has opened up space in the utilization of carbon material from coconut shell as a potential candidate in the production of activated carbon electrodes for supercapacitor application.

Acknowledgements

The authors express their gratitude for the funding from Kemenristek Dikti through a Competency Research Grant to Dr. Erman Taer, which titled: Penyediaan nano karbon sebagai inti elektroda komposit untuk aplikasi pada superkapasitor.

References

[1] Kötz R and Carlen M 2000 *Electrochimica Acta* **45** 2483-2498

[2] Taer E, Taslim R, Deraman M 2016 *AIP Conf. Proc* **1712** 050011

[3] Yang K, Penga J, Srinivasakannan C, Zhang L, Xia H, Duan X 2010 *Bioresource Technology* **101(15)** 6163–6169

[4] Geng X, Li L, Zhang M, An B, Zhu X 2013 *Journal of Environmental Sciences* **25(1)** S110-S117

[5] Hu C, Qu W, Rajagopalan R, Randall C 2014 *Journal of Power Sources* **272** 90-99

[6] Jain A, Tripathi SK 2014 *Materials Science and Engineering:B* **183** 54–60

[7] Taer E, Deraman M, Taslim R, Iwantono 2013 *AIP Conf. Proc* **1712** 33-37

[8] Taer E, Iwantono, Manik ST, Taslim R, Dahlan D, Deraman M 2014 *Advance Materials Research* **896** 179-182

[9] Taer E, Deraman M, Taslim R and Iwantono 2013 *AIP Conf. Proc.* **1554** 33-37

[10] Taer E, Deraman M, Talib IA, Awitdrus A, Hashmi SA, Umar AA 2011 *Int.J Electrochemical. Sci.* **6** 3301-3315

[11] Xiao Y, Long C, Zheng M-T, Dong H-W, Lei B-F, Zhang H-R, Liu Y-L 2014 *Chinese Chemical Letters* **25** 865-868

[12] Li W, Peng J, Zhang L, Yang K, Xia H, Zhang S, Guo SH 2009 *Waste Management* **29** 756-760

[13] Fan L, Yang L, Ni X, Han J, Guo R, Zhang C-J 2016 *Carbon* **107** 629-637

[14] Zhao X, Ran F, Shen K, Yang Y, Wu J, Niu X, Kong L, Kong L, Chen S 2016 *Journal of Power Source* **329** 104-114

[15] Cheng Y, Huong L, Xiao X, Yao B, Yuan L, Li T, Hu Z, Wang B, Wan J, Zhou J 2015 *Nano energy* **15** 66-74

[16] Huang C and Grant PS 2013 *Scientific Report* **3** 2393

[17] Zhong M, Song Y, Li Y, Ma C, Zhai X, Shi J, Guo Q, Liu L 2012 *Journal of power sources* **217** 6-12

[18] Chen Y-C, Hsu Y-K, Lin Y-Gu, Lin Y-K, Horng Y-Y, Chen L-C, Chen K-H 2011 *Electrochemical acta* **56** 7124-7130

Inhibition of α-Glucosidase, Total Phenolic Content and Flavonoid Content on Skin Fruit and Flesh Extracts of Some Varieties of Snake Fruits

E Rohaeti[1]*, M R Fauzi[1] and I Batubara[1,2]

[1]Department of Chemistry, Bogor Agriculture University, INDONESIA.
[2]Tropical Biopharmaca Research Center, Bogor Agricultural University, INDONESIA

Email : etirohaeti@apps.ipb.ac.id

Abstract. This study aimed to determine the antidiabetic activity of the skin fruit and flesh of snack fruit through α-glucosidase inhibition and correlated with total phenolic and flavonoid content as well as thin layer chromatography bio-autography. Seven varieties of varieties of skin and flesh of the fruits each extracted by maceration using ethanol 70%. The results show the highest power of the α -glucosidase inhibition obtained at Manonjaya skin extract with IC_{50} value of 17.9 μg/mL. The TLC pattern indicates the presence of four active spot on skin extract and two spots on flesh extracts on the use of solvent BuOH:HAc:water (6:2:2). The highest phenolic content obtained at skin fruit extract of Salak Mawar 186.15 ± 1.66 mg of gallic acid equivalents per gram extract. The highest total flavonoid content obtained in Salak Malaka skin fruit extract that is 7:43 ± 0:04 milli gram of quercetin equivalents

1. Introduction

Controlling blood sugar levels is one way to treat type 2 diabetes mellitus that can delay glucose absorption, ie by inhibiting the enzyme α-glucosidase [1]. This enzyme catalyzes carbohydrates into glucose in the intestinal. Inhibition of α glucosidase can reduce the digestion of complex carbohydrates and absorption of glucose so it can reduce the increasing blood glucose levels.

Indonesia has plants including fruits that can inhibit α-glucosidase. Snack fruits (*Salacca Reinw edulis*) is one of the tropical fruits which are abundant in Indonesia. In Indonesia, almost all parts of snack fruit can be used, such as the fruit used as food, while the seeds can be made into coffee[2]. Normally the skin fruit of snack fruit are not used anymore and became waste, but there are a small percentage of people using the skin fruit and believe it may help diabetic patient. In addition, the flesh fruits also have benefits as an antioxidant. Based on research Leontowicz *et al.* [3] showed that the flesh fruit part of snack fruits of Bangkok, Thailand has high antioxidant capacity. In addition, research of Ariviani and Parnanto [4] showed that the varieties of Salak pondoh, Salak Nglumut, and Salak Bali also has antioxidant capacity.

According Sahputra [5], skin fruit and flesh of snack fruits contain flavonoids, tannins, and alkaloid. The skin fruit and flesh extracts of snack fruits from Yogyakarta do not have α-glucosidase inhibitory activity, while skin fruit extract and flesh extract of snack fruits from Samarinda have inhibitory effect on α-glucosidase. So, it is interesting to find the activity of other variety of snack fruit on α-glucosidase inhibitory. The aims of this study is to determine the α-glucosidase inhibition activity of skin fruit extract and flesh extract of snack fruits from different varieties, determine the total phenolic and flavonoid content as well as its autobiography thin layer chromatography profile.

2. Materials and Methods

This study is covering sample preparation, extraction by maceration using ethanol 70%, total phenolic and flavonoid content, and profile of active components are determined by TLC.

2.1. Sample Preparation

Sample preparation is including sample collection from Bandung, Tasikmalaya, and Jonggol, West Java, Indonesia. All the snack fruit variety that collected is listed in Table 1. The skin fruit was separated with flesh and seed. The skin fruit and flesh then was drying with oven at 60°C.

Table 1. Snack fruit variety used in this study.

No	Lokal Name	Scientific Name	Collection Place
1	Salak Cililin	*Salacca zalacca* cf Cultivar 'Jawa Lokal'	Bandung, West Java
2	Salak Condet	*Salacca zalacca*	Mekarsari Fruit Park, Jonggol, West Java
3	Salak Malaka	*Salacca zalacca* cf Cultivar 'Manon Jaya'	Bandung, West Java
4	Salak Manonjaya	*Salacca zalacca* cf Cultivar 'Manon Jaya'	Tasikmalaya, West Java
5	Salak Mawar	*Salacca zalacca*	Mekarsari Fruit Park, Jonggol, West Java
6	Salak Pondoh	*Salacca zalacca* cf Cultivar 'Pondoh'	Tasikmalaya, West Java
7	Salak Pondoh Mekarsari	*Salacca zalacca* cf Cultivar 'Pondoh'	Mekarsari Fruit Park, Jonggol, West Java

2.2. Extraction

The dried skin fruit and flesh of snack fruits were extracted with 70% ethanol at 1: 5 by maceration for 24 hours. The extract has been obtained is then evaporated with a rotary evaporator at a temperature of 40-50 ° C in order to obtain the crude extract.

2.3. Total Phenolic Content Determination

Each extract about 25 mg diluted with methanol: water (1: 1) into a 25 ml flask. About 0.9 mL of standard solution (gallic acid) and extract was added by 4.5 mL reagent Folin-Ciocalteau, and shaken with a vortex. After 3 minutes, each solution was added with 3.6 mL of Na_2CO_3 7.5%, shaken and incubated for 1 hour. The absorbance of the standard solution and the sample was measured by UV-Vis spectrophotometer at a wavelength of 765 nm. Total phenol content of samples is determined using a regression equation of standard gallic acid.

2.4. Total Flavonoid Content Determination

Total flavonoid content is determined based on the method of Chang *et al.* [6] with modifications. A total of 10 μL of the extract solution put in a 96 wells microplate, then added with 60 μL ethanol, 10 μL of 10% aluminum chloride, 10 μL of 1M potassium acetate and 120 μL of distilled water. The mixture was incubated for 30 min at room temperature. The absorbance was measured at a wavelength of 415 nm with a microplate reader. Total flavonoid content is reported as Quercetin Equivalent (QE) from the calibration curve of quercetin as standard solution.

2.5. α-Glucosidase Inhibition Test
A total of 10 mL sample solution was added with 50 mL of phosphate buffer pH 7:00 and 25 mL solution of the substrate p-nitrophenyl-α-D-glucopyranose (pNPG), and then incubated at 37 ° C for 5 minutes. The mixture was added with 25 mL of α-glucosidase enzyme solution and incubated for another 15 minutes at 37 ° C. After completion of incubation, the mixture was added 100 mL Na_2CO_3. About 100 μL of the solution is put in a 96 wells microplate and the absorbance of reaction product (p-nitrophenol) was measured at 400 nm.

2.6. Thin Layer Chromatography Profile
Each extract was analysed by Thin Layer Chromatography on silica gel $G_{60}F_{254}$ and BuOH;HAc'water (6:2:2) as eluent. The spots were observed in the UV lamp at 254 and 366 nm. After the finger print of the sample at the two UV wave length were documented, then the TLC plate was sprayed with a solution of α-glucosidase enzyme and incubated for 60 min at room temperature. After incubation the plate sprayed with a solution of p-nitrophenyl α-D-glucopyranose. Spot yellow ribbon will appear after the plate stayed at room temperature. The spot showed that the tested sample is active as an inhibitor of α-glucosidase [7].

3. Results and Discussion
The samples used in the study are seven varieties of snackfruits namely Salak Cililin and Salak Malacca from Bandung, Salak Pondoh and Salak Manonjaya from Tasikmalaya, and Salak Pondoh, Salak Condet, and Salak Mawar from Mekarsari Fruit Garden (Table 1). The physical appearance of whole fruits can be seen in Figure 1. The different physical appearance could result in different chemical component on the flesh and skin fruit of snake fruit. The separated flesh and skin fruit of snack fruits then were extracted by maceration prior to examine the activity as α-glucosidase inhibition. Maceration at room temperature was used in this study because this method is using simple tools and compounds will not damage because it is not high temperature. The yield of extraction is shown in Table 2.

Figure 1. Physical appearance snack fruit used in this study.

The extraction method used to extract the skin and flesh of fruits is maceration. Maceration the extraction process without heating, i.e. immersing the samples using a solvent with stirring performed several times at room temperature. Usually extraction by maceration takes 3 days at the turn of the solvent every day. Extraction using maceration technique has the advantage, using simple tools and samples extracted will be safer because it does not require a high temperature so that the active compound contained in the sample was not damaged.

Based on total phenolic content in Table 2 shows that each extract of the skin fruit and flesh have different total phenolic content. It is influenced by differences in location and the variation of fruits. The skin fruit extract had higher total phenolic content compare with flesh extract. Among the skin fruit, the highest total phenolic content was found in Salak Mawar skin fruit extract, while the lowest total phenolic content was found in Salak Cililin skin fruit extract. Among the flesh extract, Salak Manonjaya had the highest total phenolic content and Salak Cililin had the lowest.

Total flavonoid content of all extract is shown in Table 2. It shows that each extract of the skin fruit and flesh have different total flavonoids content. Total flavonoid content on skin fruits extract is higher compare with the flesh extract. This phenomenon also appears in total phenolic content. It means the skin fruit is good to be used as source of secondary metabolite such as phenolic and flavonoid. Among the skin fruit, Salak Malaka skin fruit extract had the highest total flavonoids content, while Salak Cililin skin fruit extract had the lowest content. Among the flesh extract, the Salak Manonjaya flesh extract is the highest and the lowest present in Salak Mawar flesh extract.

Table 2. Yield, Phenolic, Flavonoid Contents and IC_{50} value of all extracts.

Snack Fruit Name	Part	Yield (%)	Phenolic content (mg gallic acid equivalen/g extract)	Flavonoid Content (mg quercetin equivalen/g extract)	α-glucosidase IC_{50} (ppm)
Cililin	Skin fruit	8.77	41.90 ± 2.53	2.92 ± 0.06	405.6
	Flesh	79.85	21.05 ± 0.00	0.71 ± 0.18	467.8
Condet	Skin fruit	8.61	141.24 ± 4.07	5.75 ± 0.30	337.5
	Flesh	76.28	21.69 ± 4.66	0.61 ± 0.00	417.9
Malaka	Skin fruit	9.09	132.79 ± 1.92	7.43 ± 0.04	437.5
	Flesh	61.86	61.82 ± 4.50	1.09 ± 0.05	507.1
Manonjaya	Skin fruit	12.12	136.63 ± 2.43	6.45 ± 0.44	17.9
	Flesh	63.92	90.91 ± 4.09	1.24 ± 0.08	502.9
Mawar	Skin fruit	8.92	186.15 ± 1.66	7.01 ± 0.97	27.7
	Flesh	46.98	54.74 ± 1.75	0.46 ± 0.00	274.2
Pondoh	Skin fruit	12.37	138.67 ± 7.23	6.75 ± 0.10	99.1
	Flesh	57.91	65.75 ± 6.37	1.00 ± 0.05	421.7
Pondoh Mekarsari	Skin fruit	10.00	123.85 ± 4.06	4.16 ± 0.20	201.9
	Flesh	93.47	64.99 ± 3.87	0.62 ± 0.00	539.1
Akarbosa (positive control)					6.8

Inhibition of α-glucosidase is expressed in IC_{50}, the concentration of the extract to inhibit 50% activity. The lower IC_{50} value, more potent the extract as α-glucosidase inhibitor. Akarbosa is used as positive control. Based on IC_{50} values in Table 2, the best IC_{50} values for extract obtained in the Salak Manonjaya skin fruit extract, followed by Salak Mawar skin fruit extract. Among the flesh extract, the best α-glucosidase inhibitor is found in Salak Mawar flesh extract. These results suggest that skin fruit extract and flesh extract of Salak Manonjaya and Salak Mawar are potent as α-glucosidase inhibitor. Salak Manonjaya skin fruit extract as the best α-glucosidase inhibitor had the highest phenolic and flavonoids content. Phytochemical compounds in plants have ability to inhibit the α-glucosidase, such as alkaloids, triterpenoids, flavonoids, and also phenolic [8-10].

Thin layer chromatography (TLC) - Bio-autography is a simple and rapid chromatographic method to separate and identify the active compounds. This technique uses direct observations on the spot which is formed after spraying with directly reagents [11]. TLC bio-autography have been developed on inhibiting enzyme activity such as tyrosinase inhibitors, acetylcholinesterase inhibitors, xanthin oxidase inhibitors, and glucosidase inhibitors. TLC bio-autography of α-glucosidase inhibitor is performed by spraying the plate with α-glucosidase enzyme and p-nitrophenyl α-D-glucopyranose as the substrate, the active spot is the spot with yellowish colour on the TLC plate.

The bioautogram of extract is shown in Figure 2 for the skin fruit extract and Figure 3 for the flesh extract. About 4 spots are active from skin fruit extract while on the flesh extract only two active spot. This proves that the skin fruits extract more active in inhibiting the α-glucosidase and also acts as an antidiabetic.

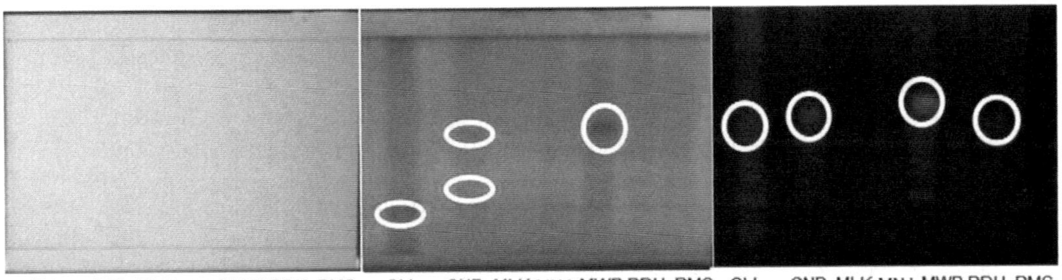

Figure 2. TLC Bioautogram α- Glucosidase of skin fruit extracts of Salak Cililin (CLL), Condet (CND), Malaka (MLK), Mawar (MWR), Pondoh (PDH), Pondoh Mekarsari (PMS).

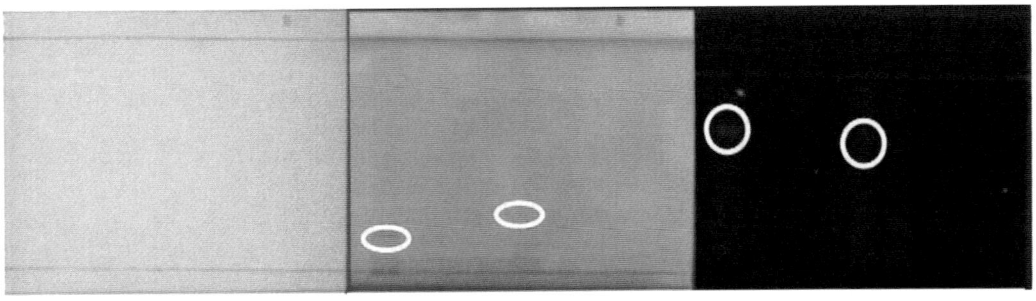

Figure 3. TLC Bioautogram α- Glucosidase of flesh extracts of Salak Cililin (CLL), Condet (CND), Malaka (MLK), Mawar (MWR), Pondoh (PDH), Pondoh Mekarsari (PMS).

4. Conclusion

The highest of total phenolic content of the extract were obtained from the skin fruit of Salak Mawar and the from the flesh of Salak Manonjaya, which each amount are respectively 186.15 and 90.91 equivalent gallic acid/ g extract. The highest of total flavonoid levels were obtained in the extract of skin fruit of Salak Malaka and in the extract of flesh of Salak Manonjaya which each amount are respectively 7.43 and 1.24 mg equivalent quercetin/g extract. The strongest Inhibition of α-glucosidase, stated by the IC50 value, was obtained in the extract of the skin fruits of Salak Manonjaya which amount is 17.9 μg/ mL. Based on the results of the TLC bioautografi inhibitor of α-glucosidase using a solvent which is mixed of butanol: glacial acetic acid: water (6: 2: 2) it was produced active spots for about 4 spots on the extract of skin fruits and another 2 spots on the extract of flesh.

Acknowledgments

The authors thanks to Taman Buah Mekarsari for the snack fruit collection used in this study.

References

[1] Yuhao Li 2005 *Journal of Ethnopharmacology* **99** 239-244

[2] Nazaruddin, Kristiawati 1992 18 *Varietas Salak* (Jakarta (ID): Penebar Swadaya)

[3] Leontowicz H, Leontowicz M, Drzewiecki J, Haruenkit R, Poovarodom S, Park YS, Jung ST, Kang S, Trakhtenberg S, Gorinstein S 2006 *European Food Research and Technology* **223** 697-703

[4] Ariviani S, Parnanto NHR 2013 *Agritechnology* **33(3)** 325-330
[5] Sahputra FM 2008 Potensi ekstrak kulit dan daging buah salak sebagai antidiabetes [Skripsi] (Bogor (ID): Bogor Agricultural University)
[6] Chang CC, Yang MH, Wen HM, Chern JC 2002 *Journal Food Drug Analytical* **10** 178-182
[7] Pires S, Hmicha B, Marston A, Hostettmann K 2009 *Phytochemical Analysis* **20** 511-515
[8] Patel MB, Mishra SM 2012 *Journal Functional Food* **4** 79-86
[9] Lai YC, Chen CK, Tsai SF, Lee SS 2012 *Phytochemistry* **74** 206-211
[10] Wang H, Du YJ, Song HC 2010 *Food Chemistry* **123** 6-13
[11] Macek K 2005 *Pharmaceutical Application of Thin Layer and Paper Chromatography* 3rd Ed (New York (US): Elsevier Publishing Company)

IOP Conf. Series: Earth and Environmental Science **58** (2017) 012067 doi:10.1088/1755-1315/58/1/012067

In Vivo Inhibition of Lipid Accumulation in *Caenorhabditis elegans*

Sulistiyani[1,2*], E P Purwakusumah[1,2] and D Andrianto[1]
[1] Departement of Biochemistry, Bogor Agricultural University, Indonesia
2 Tropical Biopharmaca Research Center, Bogor Agricultural University, Indonesia

Email: sulistiyani@apps.ipb.ac.id

Abstract. This is a preliminary research report on the use of *Caenorhabditis elegans* as a model to establish anti-obesity screening assay of the natural plant resources. Nematode *C. elegans* has been used as experimental animal model for understanding lipid accumulation. The objective of this research was to investigate the effect of selected plant extracts on lipid accumulation in *C. elegans*. Currently no report could be found regarding lipid accumulation in *C.elegans* treated with ethanolic leaf extracts of jabon merah (*Anthocephalus macrophyllus)*, jati belanda (*Guazuma ulmifolia)*, and Mindi (*Melia Azedarach)* plants. Lipid accumulation was determined qualitatively using lipid staining method and quantitatively by colorimetry using sulpho-phospho-vanillin reagent. Data showed that lipid accumulation was inhibited up to 72% by extract of *M. azedarach*, about 35% by both of *A. macrophyllus* and *G. ulmifolia* extracts, and up to 25% by orlistat (a synthetic slimming drug). Ethanolic extract of *A. macrophyllus*, *G. ulmifolia*, and *M. azedarach* leaves were shown to inhibit lipid accumulation in *C. elegans* and *M. azedarach* leaves extracts was the most effective inhibitor. *C.elegans* were shown to be an effective model for in vivo lipid accumulation mechanism and potential to be used as a rapid screening assay for bioactive compounds with lipid accumulation inhibitory activity.

1. Introduction

C. elegans is a nematode that has been used as experimental animal for studying lipid metabolism [1]. The lipid metabolism in this nematode is similar to that of mammals, but it lacks leptin-like molecules and their receptor, which regulates lipid storage through appetite control. Therefore, *C. elegans* model could provide information regarding lipid storage and mobilization without the influence of hormonal inhibition of food intake [2]. Lipid storage in *C. elegans* is regulated by *tub-1* gene. Mutation on this gene will result in excessive lipid accumulation as shown by a mutant strain RB1600 of *C. elegans*. Kim et al. [2] used this mutant strain for studying anti-obesity activity of taurine.

Both *Anthocephalus spp*.and *G. ulmifolia* have been used as slimming ingredients in traditional medicine of Indonesia [3,4]. Administration of *Guazuma ulmifolia* leaves ethanolic extract has been reported to reduce abdominal fat of Sprague Dawleyrats as much as 27.06% [5]. No reports regarding the effect of *Anthocephalus macrophylus* on lipid metabolism, only recent report by Acharyya et al. [6] that methanolic extract of the root of jabon putih (*Anthocephalus cadamba*) reduced low density lipoprotein (LDL) concentration in Wistar rats and increased the high density lipoprotein (HDL) concentrations. Similarly, methanolic leaf extract of *Melia azedarach* has been shown to posses antioxidant activity [7]. Very few reports was found on the lipid metabolism activity of *M. azedarach*.

IOP Conf. Series: Earth and Environmental Science **58** (2017) 012067 doi:10.1088/1755-1315/58/1/012067

According to Indian traditional medicine (Ayurveda), *M. azedarach* can stimulate the excretion of urine [8]. The water- and methanolic- extracts of M. azedarach was reported to reduce the body lipid mass of hyperlipidemic albino Wistar rats significantly [9].

2. Materials and Methods

Plant materials of *A. macrophyllus, G. ulmofolia,* and *M. Azedarach* were obtained from Tropical Biopharmaca Research Center (Trop BRC) Conservation and Cultivation Station at Cikabayan, IPB Dramaga Campus, Bogor. *C .elegans* strains of N2 (wild type) and RB1600 along with *Eschericia coli* strain OP50 were provided by the Caenorhabditis Genetics Center (CGC) at University of Minnesota, Minneapolis (USA). Orlistat (Xenecal®) , a commercially available slimming agent, was purchased from a general drugstore in Bogor. Culture media and chemicals used for nematode cultivation, maintenance, and experiments were Luria Bertani (LB) broth (Himedia), agar (Himedia), S Medium (NaCl, pepton, $CaCl_2$ 1M, MgSO4, and buffer KPO4 1M pH 6), M9 medium (KH2PO4, Na2HPO4, NaCl, MgSO4 1 M), cholesterol (Sigma), NaCl, pepton, CaCl2, MgSO4, bufer KPO4 1M pH 6, antifungal nistatin (Kimia Farma), and streptomycin (Kimia Farma). Reagents and other chemicals for sample extraction, nematode staining, and quantitative lipid determination were all purchased from Merck, Darmstad, Germany.

2.1. Preparation of Ethanolic Extracts

Dried leaves was prepared as reported by Sulistiyani et al. [10]. Samples were dried at $50^{o}C$ and then the dried leaves were mashed into 40 mesh. Ethanolic leaf extracts of these plants were prepared by maceration of each dried leaves with ethanol 96% (1:10) at room temperature for 2x24hours. The extracts were filtered and concentrated with rotary vacum evaporator at $50^{o}C$ to produce semi-dried crude ethanolic extracts.

2.2. Experimental Design

The wild type (N2) and the mutant (RB1600) strains of *C. elegans* in this research were used in 6 experimental condition groups: 1) normal condition (in standard media contained 5 ppm cholesterol), 2) high lipid control (in media contained 10ppm cholesterol), 3) high lipid with addition of orlistat; high lipid treatment with addition of : 4) *A. macrophyllus*; 5) *G. ulmifolia*; and 6) *M. azedarach* ethanolic extracts. Each treatment groups of *C. elegans* were incubated with 3 doses of extracts: 25, 50, 100 ppm (orlistat and *M. azedarach*) and 250, 500, 1000 ppm for *A. macrophyllus* and *G. ulmifolia*. All assays were performed in triplicates. Liquid cultures of *C. elegans* were incubated at $25^{o}C$ for 40 hours in each experimental medium until it reached life phase early L4. Additional 9.5 hour incubation were done to reach a complete L4 life phase before measurement of total lipid.

2.3. Culture of C. elegans and Extract Treatments

C.elegans were grown on Nematode Growth Medium (NGM) agar plates and cultivated according to Stiernagle [11] and were kept at $25^{o}C$. Nematodes were maintained on living *E.coli* strain OP50 as a food source which was seeded to the surface of NGM plates. The stock of *E.coli* was prepared by culturing it overnight in an LB medium until concentrated particles were formed. During each experiment, the nematodes were grown in liquid culture medium (S Medium) prepared as described [12 Lewis & Flemming 1995] that contained concentrated *E.coli* and were maintained at room temperature ($25^{o}C$). The orlistat- and extract-containing experimental media were prepared by adding indicated concentration of orlistat or ethanolic plant extract into high lipid (10 ppm cholesterol)-containing S Medium. Liquid cultures of C. elegans were treated with aliquots of experimental media in microplates kept in incubator for 12-40 hours at $37^{o}C$.

2.4 Qualitiative Analysis of Lipid Accumulation

Qualitative analysis of lipid accumulation in the nematodes was determined using lipid staining method of Kim *et al.*[2] For the staining, the nematodes were removed from the experimental cultures with ose to be placed on the preparative slides and starved for 6 hours. After they were washed in M9 media, the nematodes were fixed with 1% paraformamide 1% in M9. Then the preprative slides were rised with 95% ethanol. The staining were performed in 70% ethanol saturated with Sudan black B. The effect of treatment was studied via microscopy on stained nematodes.

2.5 Total Lipid Quantitative Analysis

For quantitative analysis, total lipid of the *C.elegans* was extracted using the method of Bligh and Dyer [13]. Nematodes were washed off the experimental media by gentle centrifugation to pellet worms, and washed twice in 0.9% NaCl, After incubation with in 0.9% NaCl for 20 minutes at room temperature, nematodes were pelleted by centrifugation and resuspended in 0.5 mL H_2O. The worm suspension were mixed with 3 mL of methanol:kloroform (2:1) solution and vortexed for 5 minutes and then 1 mL of chloroform and 1.8 mL of H_2O were added. The mixture was vortexed again for 5 minutes until separated into 2 phases, the upper phase was the polar part and the bottom phase was the organic part that contained the cellular lipid (triglycerides, membrane phospholipid, cholesterol etc.). After the upper phase was discarded, the remaining lipid-containing fraction was dried until all the solvent was evaporated. Then mixture solution of methanol:chloroform (1:1) 0.2 mL was added to the sample and ready for total lipid determination.

Total lipid concentrations were determined by colorimetry using sulpho-phospho-vanillin reagent as described by Knight et al. [14] based on the reaction between lipid carbonium ion and active carbonyl group of the phospho-vanilin forming cherry-red color product. Pure olive oil was used as standard and was dissolved in chloroform at concentrations of: 2.5, 5, 10, 25, 50, and 100 ppm. Samples and standards (0.2 mL) each was mixed gently with 1 ml concentrated sulfuric acid to hydrolize the lipid ester. These mixtures were heated in waterbath at $100^{o}C$ for 20 minutes under the acid/fume hood for a highly reactive lipid carbonium ion was formed in this reaction. Subsequently, reaction tubes were cooled down under running tap water for 5 minutes , and 2 mL of vanilin-fosfate was added prior to incubation at room temperature for 30 minutes. This reaction yielded cherry-red colored product which was stable within 60 minutes. The absorbance of standards and sample mixtures were measured at 525 nm using using a spectrophotometer.

3 Results and Discussions

3.1 Effect of High Lipid-Diet on Cellular Lipid of C. elegans

Experiments were conducted to examine the effect of medium with high cholesterol content on lipid accumulation in both the wild type N2 and RB 1600 mutant strains. The Rb1600 *C. elegans* strain has mutation in the tub-1 gene [2]. Functional loss of tub-1 in *C.elegans* leads to accumulation of lipids [15, 16]. Figure 1 shows that the lipid content of the wild type N2 strains (36 ppm) was slightly higher than that of the RB1600 strain (25 ppm) when incubated with medium contained standard amount of 5 ppm cholesterol. Incubation of the nematodes with higher concentration of cholesterol (10 ppm) in the medium increased the total cellular lipid in the wild type N2 as well as the RB1600 strains by 77% compared to those incubated with standard cholestesterol concentration. These quantitative data were supported by Sudan black staining , which showed no substantial difference appeared in the fat stores of both strains of nematodes incubated with a normal medium (data not shown). When cultured on a high lipid medium, however, the nematodes was seen to accumulate excess lipid as shown by the appearance of higher number of dark stained lipid droplets (Figure 2). Visually, the lipid droplets in the wild type N2 were more noticeable than the mutants (data not shown). In this study, the RB1600 mutants showed less response to high cholesterol regarding its cellular lipid accumulation. It is not clear whether this was due to experimental condition or other characteristic of this mutant. In *C elegans*, the gene products of *tub-1* are thought to manage lipid accumulation in synergistic manner with those of *kat-1*, which govern fatty acid beta-oxidation [17]. *kat-1* encodes a beta-oxidation enzyme, 3-ketoacyl-CoA and found to be expressed in intestine, body wall muscles, and in pharynx. The tub-1 mutants demonstrate an obese phenotype especially under an impaired kat-1 gene [2].

Incubation of the *C.elegans* with high cholesterol-content medium resulted in cholesterol accumulation in lipid droplets in the form of cholesteryl ester and other lipids. In *C. elegans*, cholesterol is converted to dafakronic acid which in turn will trigger translation of daf-12 gene. The gene product is fatty acyl reductase-1 (FARD-1) which will catalyze the fatty acid catabolism[18]. Thus, incubation with cholesterol will not increase the accumulation of triglyceride. Triglyceride could be accumulated by increasing the amount of bacteria (i.e. *E. coli* OP50) thus increasing the bacterial fat [19]. *C. elegans* does not have pathway for cholesterol de novo synthesis from acetate [19] thus increase in total cellular lipid following incubation with cholesterol represents the efficiency in cholesterol absorption by the nematodes. In other words, any reduction in the amount of cellular lipid in this study may be attributed particularly to inhibiton of cholesterol absorption.

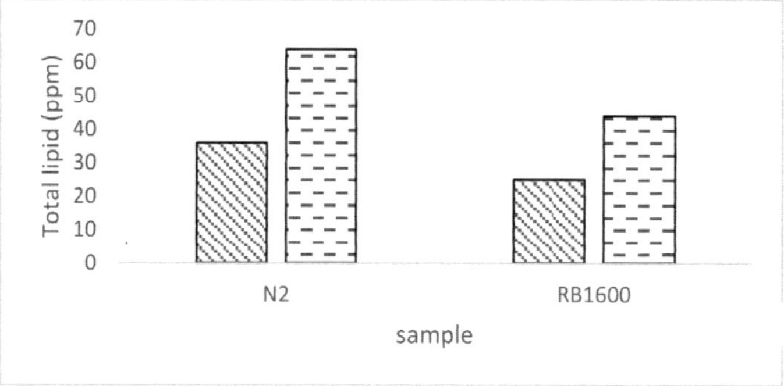

Figure 1. Lipid accumulation in wild-type N2 and mutant RB1600 *C.elegans* strains, each treated with medium contained normal cholesterol- (left bars) and high cholesterol- content (right bars) for 40 hours at 37°C.

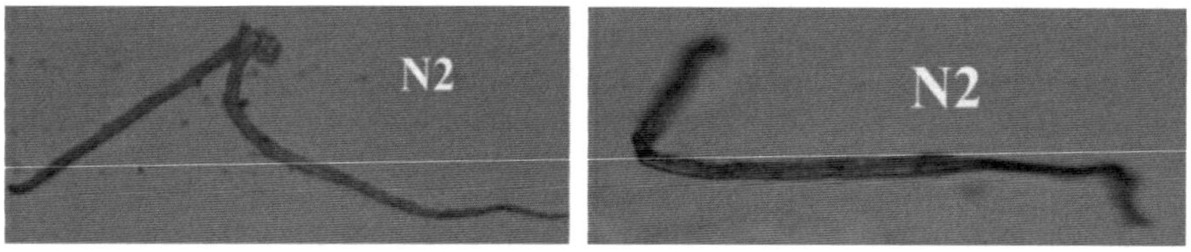

Figure 2. Sudan black staining of wild-type N2 *C.elegans* strains incubated in medium with normal cholesterol- (left) and high cholesterol-content (right) for 40 hours at 37°C.

3.2 Effect of Orlistat Treatment on Lipid Accumulation in Cholesterol-fed C. elegans.

Orlistat is a generic name for a commercial slimming agent [20]. Incubation of RB1600 mutant nematodes with orlistat at concentration of 25, 50, and 100 ppm cause reduction of total cellular lipid which shown as the appearance of dark spots of lipid droplets became less noticeable than the high lipid control (Figure 3 A). These observations are consistent with the quantitative data which show 33% to 70% reduction in total cellular lipid concentration compared to the cholesterol-fed control (Figure 3B). Similar effects of orlistat on the wild type N2 strains were observed, only in lesser degree. The data suggest that the mechanism of lipid stores in C. elegans of both strains was affected by orlistat. This is an important observation since orlistat often used as a positive control for body weight reducing (slimming) agent since its mechanism is to inhibit intestinal lipid absorption. Orlistat (tetrahydrolipstatin) is synthesized from lipase inhibitor derivative produced by Streptomyces toxytricini [21]. Orlistat inhibits lipid absorption by inhibition of lipase, including carboxyl ester lipase (pancreatic cholesterol esterase) in mammals [22]. Whereas C. elegans also has cholesterol esterase [23]. Thus, it is thought that orlistat inhibit the lipid (cholesterol) accumulation by inhibiton of lipase, particulaly cholesterol esterase.

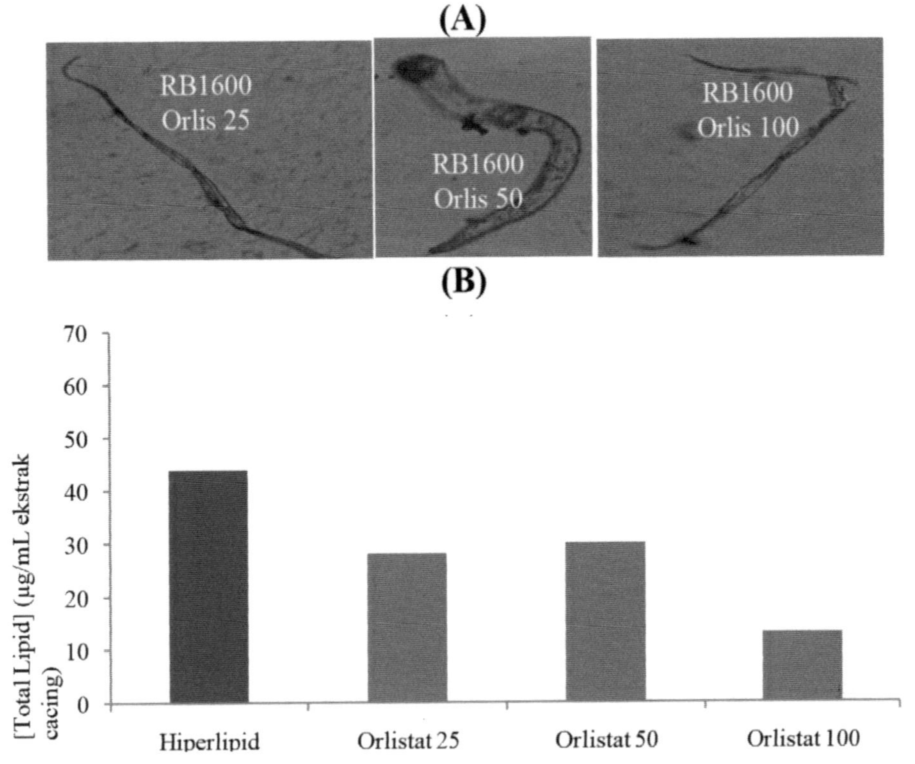

Figure 3. Lipid accumulation in *C.elegans* treated with orlistat: visually (A) and quantitatively (B)

3.3 Effect of Ethanolic Extracts of A. macrophyllus and G. ulmifolia on Lipid Accumulation in Cholesterol-fed C. elegans

Data showed that lipid accumulation was higher in cholesterol-fed (the high lipid) control N2 strain of *C.elegans* than those treated with *A.macrophyllus* or *G. ulmifolia* extracts at concentration of 250 ppm (Figure 4). Quantitative analysis showed that the lipid accumulation was inhibited about 35% by both *A. macrophyllus* and *G. ulmifolia* leaf extracts. This inhibition was higher than those of N2 treated by orlistat at concentration of 100 ppm which resulted only in 25% inhibition (data not shown). Sudan black staining indicated that the nematodes treated with *G.ulmifolia* showed no dark spots, while those treated with A. macrophyllus were seen to accumulate excess lipid as shown by the appearance of higher number of dark stained lipid droplets (Figure 4A). Visually, the lipid droplets in the high lipid control of wild type N2 were more noticeable than the others.

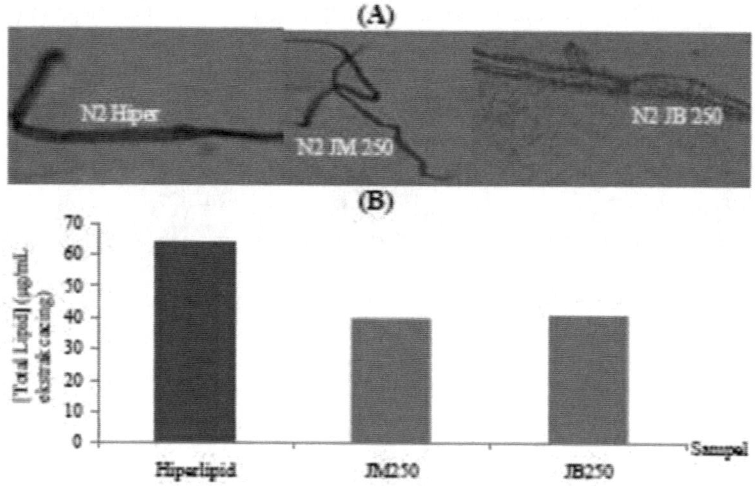

Figure 4. Lipid accumulation in wild type N2 of *C.elegans* treated with leaf extracts of *A.macrophyllus* (JM) and *G.ulmifolia* (JB) at concentration of 250 ppm each: visually (A) and quantitatively (B)

3.4 Effect of Ethanolic Extracts of M. azedarah *on Lipid Accumulation in Cholesterol-fed C. elegans*
Treatment with *M.azedarach* leaf extract resulted in much lower lipid accumulation, since this extracts were able to reduce lipid accumulation at concentrations of 25-50ppm as much as 72-78% (Figure 5). Sudan black staining also supported the reduction of lipid stores as showed by the lack of dark spots (data not shown).

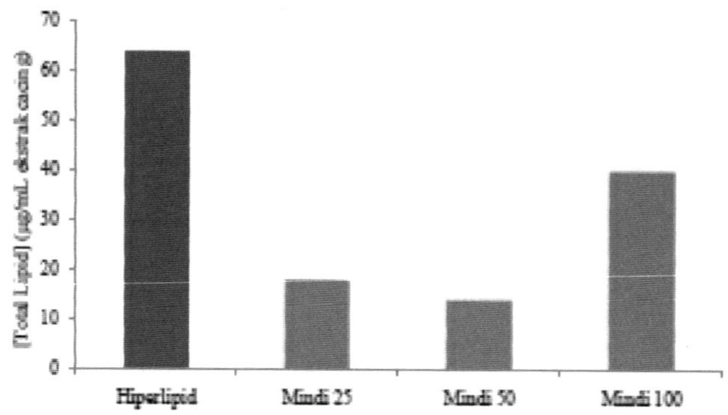

Figure 5. Lipid accumulation in *C.elegans* treated with M.*azedarach* leaf extract in various concentrations (ppm)

Taken together, these data showed that lipid (cholesterol) accumulation in *C. elegans* were inhibited by ethanolic extracts of *G. ulmifolia* and *A. macrophyllus* as much as 35% at concentration of 250 ppm, and by extracts of *M. azedarach* at much lower concentration (25 ppm) and stronger inhibition effect (72%). Recently, we reported the strong cytotoxicity effect of *M.azedarach* leaf extracts [24], which explained the small concentration it needed to exhibit its effect on the nematodes. The mechanisms by which these plant extracts exert their action may be mediated through their activity as lipase inhibitors. Our recent studies on ethanolic extracts of these plants indicated that all of them were potential lipase inhibitors as shown by their inhibitory activity against crude extract of fungal lipase [25]. Their inhibtory actvity at concentration of 100 ppm each was 59.7%, 61.9%, and 33,8% for *A. macrophyllus*, *G.ulmifolia*, and *M. azedarach*, respectively. In conclusion, cultures of *C. elegans* can be used as potential rapid screening assay for natural compounds with lipid reduction activities.

References

[1] Elle IC, Olsen LCB, Mosbech M, Rodkaer SV, Pultz D, Boelt SG, Fredens J, Sorensen P, Fargeman NJ 2008 *Lipid Insights* **1** 13-21
[2] Kim HM, Do CH, Lee DH 2010 *J Biomed Sci.* **17(1)** S33
[3] Mansur I, Tuheteru FD 2010 *Kayu Jabon.* Jakarta (ID): Penebar Swadaya Acharyya S, Dash GK, Abdullah MS 2013 Antihyperglycemic and antilipidemic activity of Anthocephalus cadamba (roxb.) Miq. roots. *Euro J Exp Bio.* **3(3)** 116-120
[4] Dalimartha S 2000 *Atlas Tumbuhan Obat Indonesia.* Jakarta (ID): Trubus Agriwidjaya Al-Suwailem AK, Tamimi AS, Al-Omar MA, Al-Suhibani MS 2006 Safety and mechanism of action of orlistat (tetrahydrolipstatin) as the first local antiobesity drug. *J Appl Sci Res.* **2(4)** 205-208
[5] Purnomo D 2008 Khasiat ramuan ekstrak daun jati belanda terhadap jumlah lemak abdomen tikus hiperlipidemia. [skripsi]. Bogor (ID): Institut Pertanian Bogor.
[6] Acharyya S, Dash GK, Abdullah MS 2013 *Euro J Exp Bio.* **3(3)**116-120
[7] Asadujjaman M, Saed A, Hossain M, Karmakar U 2013 *AJBIO* **1(2)** 118-122
[8] Azam MM, Mamun-Or-Rashid ANM, Towfique NM, Sen MK, Nasrin S 2013 *Am J BioSci.* **1(2)** 44-49
[9] Kumar VS, Sanghai DB, Rao MC, Shreedhara CS 2013 *Res J Pharm Tech.* **6(11)** 1195-1199
[10] Sulistiyani, Sari RK, Wahyuni WT 2014 Pengembangan produk antikolesterol dari tumbuhan hutan berbasis mekanisme penghambatan enzim HMG-CoA reduktase Laporan Akhir Tahun Penelitian Desentralisasi Baru IPB 2014
[11] Stiernagle T 2006 *Maintenance of C. elegans* Wormbook 1- 10 doi:10.1895/wormbook.1.101.1
[12] Lewis JA, Fleming JT 1995 In: *Methods in Cell Biology* **Vol. 48** HF Epstein and DC Shakes, eds. San Diego: Academic Press
[13] Bligh EG, Dyer WJ 1959 *Can J Biochem Physiol.* **37** 911-917
[14] Knight JA, Anderson S, Rawle JM 1972 *Clinical Chemistry* **18(3)** 199-202
[15] Ashrafi K, Chang FY, Watts JL *Nature* **421** 268-272
[16] Mukhopadyay A, Deplancke B, Walhout AJ 2005 *Cell Metab* **2** 35-42
[17] Mak HY, Nelson LS, Basson M 2006 *Nat Genet* **38** 363-368
[18] Antebi A 2015 *Nuclear receptor signal transduction in C. elegans* Wormbook 1- 49
[19] Braeckman BP, Houthoofd K, Vanfleteren JR 2009 *Intermediary Metabolism.* Wormbook. 1-24
[20] Mittendorfer B, Ostlund RE, Patterson BW, Klein S 2001 *Obes Res.* **9(10)** 599-604
[21] Al-Suwailem AK, Tamimi AS, Al-Omar MA, Al-Suhibani MS 2006 *J Appl Sci Res.* **2(4)** 205-208
[22] Wilding JPH 2008 *Pharmacotherapy of Obesity* Berlin (GE): Springer
[23] Merris M, Wang T, Soteropoulos P, Lenard J 2007 *J Lipid Res.* **48** 1159-1166
[24] Irshan Y 2015 Aktivitas sitotoksik dan antiproliferasi ekstrak etanol daun mindi (Melia azedarach) terhadap sel kanker paru manusia A549[skripsi] Bogor (ID): Institut Pertanian Bogor
[25] Sari GA 2016 Inhibisi ekstrak kasar lipase Aspergillus niger oleh ekstrak daun Guazoma ulmifolia Lamk, Melia azedarach, dan Anthocephalus macrophyllus secara in vitro[skripsi] Bogor (ID): Institut Pertanian Bogor

Isolation, Fractionation and Characterization of Catalase from *Neurospora crassa* (InaCC F226)

Suryani[1*], L Ambarsari[1] and E Lindawati[1]
[1]Department of Biochemistry, Faculty of Mathematics and Natural Sciences, Bogor Agricultural University, Indonesia

Email: ani3110@yahoo.com

Abstract. Catalase from Indigenous isolate *Neurospora crassa* InaCC F226 has been isolated, fractionated and characterized. Production of catalase by *Neurospora crassa* was done by using PDA medium (*Potato Dextrosa Agar*) and fractionated with ammonium sulphate with 20-80% saturation. Fraction 60% was optimum saturation of ammonium sulphate and had highest specific activity 3339.82 U/mg with purity 6.09 times, total protein 0.920 mg and yield 88.57%. The optimum pH and temperature for catalase activity were at 40^{0}C and pH 7.0, respectively. The metal ions that stimulated catalase activity acted were Ca^{2+}, Mn^{2+} and Zn^{2+}, and inhibitors were EDTA, Mg^{2+} and Cu^{2+}. Based on K_m and V_{max} values were 0.2384 mM and 13.3156 s/ mM.

1. Introduction

Catalase (E.C.1.11.1.6) is classified as oxidoreductase enzymes that contribute to defense cells and oxidative stress because its ability to decompose hydrogen peroxide (H_2O_2) into oxygen and water [8]. Catalase can be applied for food industry to detect level of calcium in milk and water [1], in the medical field for therapy acatalasemia [10], inhibition of tissue damage and tumor metastasis [16], development biosensor for detection of bacterial infection mastitis [9] and degradation of industrial waste containing H_2O_2..

Catalase also can be used as a fuel cell enzymatic or Enzymatic Fuel Cell (EFC) that use enzymes as biocatalysts to change biochemical energy into electrical energy directly [4]. In this present research, catalase was investigated and planned in order to utilize oxydoreductase as Enzymatic Fuel Cell.

Some fungi produced catalase such as *Neurospora crassa*, *Saccharomyces cerevisiae*, *Aspergillus niger*, *Pichia pastoris*, and *Septoria tritici* [2], while catalase-producing bacteria including *Bacillus subtilis* 168 [13], *Bacillus* sp. [11], and *Lactobacillus sakei* [2]. Catalase from microorganisms source have many advantages, such as catalytic activity and high specific activity, wide range activity for pH and temperature, and high stability to acids and alkaline [16]. However, the range of potential application is increasing rapidly.

Catalase from *Neurospora crassa* (InaCC F226) have not been yet optimized and characterized. Therefore, in the present work catalase was isolated, fractionated ammonium sulfate saturation level of 20-80%, optimized pH and temperature and determined the kinetic parameter in order to utilize oxydoreductase as Enzymatic Fuel Cell for future work.

2. Materials and Methods

2.1. Preparation of Crude Extract catalase from N. crassa [12].
Neurospora crassa (InaCC F226) isolate were obtained from Indonesian Culture Collection Indonesian Institute of Sciences have been transferred to PDA medium and incubated at 37^0C for 10 days until fungus konidium ready to be harvested. The cells were treated with 50 mM phosphate buffer pH 7.0 with volume 200 mL and 3 mL 75% H_2O_2 were added until no bubbles as a result of catalysis H_2O_2. Cells were homogenized by centrifugation at speed of 9500 rpm for 15 min at 4° C. Supernatant is used as crude extract of catalase.

2.2. Fractionation by Precipitation Ammonium Sulphate (NH4)$_2$SO$_4$)
Supernatant as crude catalase was precipitated with different concentration of ammonium sulfate at saturation 20, 40, 60, and 80%. The addition ammonium sulphate was done at 4°C and stirred until dissolved then centrifuged at 3000 rpm for 10 min at 4°C and stored overnight at 4° C. Pellet were resuspended in 50 mM phosphate buffer pH 7 of 1-2 mL.

2.3. Determination of Catalase Activity Assay [22]
The enzymes that obtained from each step of ammonium sulphate fractionation were incubated for 3 minutes at 35°C and measured the absorbance by using spectrophotometer at wavelength 240 nm. The catalase activity calculated by using coefficient of the half-life of enzymes was 43.6 M^{-1} cm^{-1}.

2.4. Determination of Protein Concentration by Bradford Method [6]
Bovine Serum Albumin (BSA) is used as standard protein for determination of protein concentration with the range 0020-0120 mg / mL. The enzyme that obtained from precipitation with 60% of ammonium sulfate and crude extract with total volume 100 mL were added with 5 ml of Bradford reagent and homogenized by vortex. After incubated 5 minutes, absorbance was measured at 595 nm wavelength.

2.5. Determination of Optimum Temperature and pH [12]
Optimum temperature and pH were determined by using the highest specific activity. Determination of optimum temperature were done at temperature range (20 - 60°C). Determination of optimum pH were done with different buffer started at pH 4.0- 6.0 (in 50 mM sodium acetate buffer), pH 7.0-10.0 (in 50 mM sodium phosphate buffer). All assays were done by using 0.1 mM H_2O_2 3.0 mL and 0.100 mL enzymes. Absorbance was measured by spectrophotometer at 240 nm.

2.6. Effect of Metal Ion on Catalase Activity [12]
Some compounds were used to examine the effect of metal ions consisting of EDTA, $MnCl_2$, $MgCl_2$, $CuSO_4$, $ZnCl_2$ and $CaCl_2$ with concentration of 5 mM. A total of 3.0 ml of 0.1 mM H_2O_2 was added with 50 mL of metal ion and incubated 2 minutes. Enzyme was added and incubated back for 3 minutes. Inactivating of enzyme was done by heating 90°C and measured absorbance at wavelength 240 nm.

2.7. Determination of Kinetics Parameter of Catalase (Km and Vmax) [12]
Kinetics parameters (Km and Vmax) of catalase were determinated with varying susbtrate H_2O_2 in the range 0.02 – 0. 24 mM. Decomposition of H_2O_2 levels were measured by using spectrophotometer at 240 nm. The apparent of km value and Vmax of catalase was estimated by analysis data of Michaelis-Menten /Lineweaver-Burk plots.

3. Results

3.1. Catalase Results Isolation and Purification

The specific activity of catalase was showed presented in Table 1. Based on the results, total protein and specific activity of the crude extract obtained at 6325 mg and 548.47 U/ mg. After fractionated with ammonium sulfate obtained the highest total protein in the fraction with 60% saturation was 0920 mg.

Table 1. Specific Activity of Catalase.

Fraction	Total Activity (U)	Total Protein (mg)	Specific Activity (U/mg)	Purification (fold)	Yield (%)
Crude extract	3469.11	6.325	548.47	1.00	100.00
Ammonium sulphate 60%	3072.64	0.920	3339.82	6.09	88.57

Total protein and the highest specific activity obtained after ammonium sulfate saturation at 60% were respectively 0.920 mg and 3339.82 U mg. The total protein and specific activity fraction of 60% was lower compared to [7], total protein and specific activity of catalase-1 isolated from *Neurospora crassa* 35% ammonium sulphate precipitation were 8.2 mg and 292,980.00 U/mg, respectively. According to [5], the lower of total protein and specific activity due to amino acids has many hydrophilic group (56.20%). It was required high concentration of ammonium sulphate to change electrostatic force that effect to low solubility of proteins in water and then protein will be precipitated. Beside that, enzyme has more hydrophilic amino acids that required higher salt concentration for precipitation.

The fraction 60% of ammonium sulphate precipitation with highest specific activity 3339.82 U/ mg was used to determine optimum pH and temperature to hydrolysis H_2O_2 as substrate. The enzyme activity will increase by increasing temperature up to optimum temperature. The increasing of temperature above optimum temperature will decrease enzyme activity. Enzyme activity increased with the range of temperature 20-40°C. Increasing of enzyme reaction will increase kinetic energy which accelerates vibrational motion, translational and rotational enzymes and substrates, thereby increasing intensity of collision between substrate and enzyme until the optimum temperature reached.

3.2. Optimum temperature and pH of catalase

Optimum temperature of catalase *N. crassa* (InaCC F226) was 40°C with relative activity reached 100.00%. According to [21], catalase of *Vibrio rumoiensis* had optimum temperature 40°C. At optimum temperature, collision between enzyme and substrate is most effective. Beside that, formation of complex between enzyme substrates are easily and also products formation will increase.

Increasing pH 4.0-7.0 will increase catalase activity and reached activity 100.00% at pH 7.0 (neutral) as shown at Figure 2 . Catalase activity decreased at range pH 7.0-10.0. Temperature and pH optimum catalase isolated from *N. crassa* (InaCC F226) was 40°C and pH 7.0, respectively.

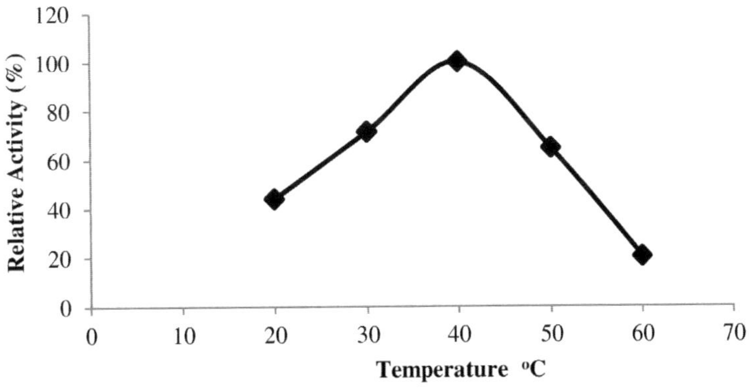

Figure 1. Relative activity (%) of catalase at different temperature.

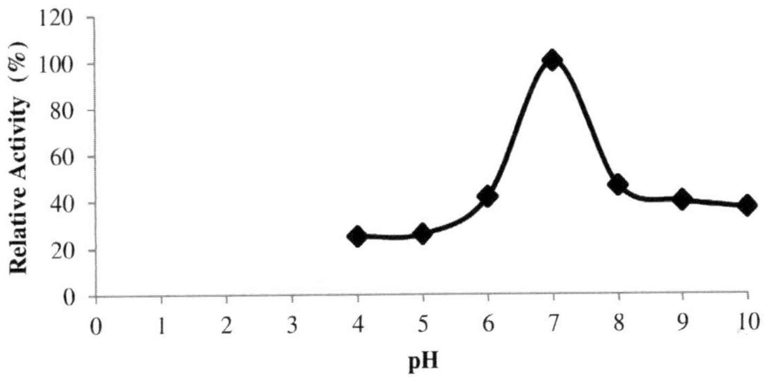

Figure 2. Relative activity (%) of catalase at different of pH.

One of factor that influence enzyme activity was pH, there was correlation between ionization of amino acids enzyme at active side. Optimum pH of *N. crassa* catalase (InaCC F226) was at pH 7.0 (neutral) with relative activity with 100.00%. Catalase from *Beauveria bassiana* has optimum at pH 7.0. At this optimum pH 7.0, active site of enzyme in maximum condition and most appropriate for binding to substrate and catalysis [18]. Catalase from *N. crassa* (InaCC F226) decreased relative activity in acidic and alkaline. pH of acids and bases will affect ionization changes and affect to catalytic ability that related to structure and enzyme active site [18].

3.3. Effect of Metal Ion on Catalase Activity

Some metal ions increased catalase activity were Ca^{2+}, Mn^{2+} and Zn^{2+} at concentration 5 mM. Ion Mn^{2+} and Zn^{2}+ increased relative activity up to 150.00% and 119.85%, and the highest activity catalase was obtained by adding Ca^{2+} with activity unit 266.63U/mL. Other research, [19] also reported that the addition of Ca^{2+} at level concentration of 0.5 mM can increase catalase activity by 111.2 % after purified by Sephadex G-75. Some metal ions gave inhibition effect of catalase activity i.e Cu^{2+}, Mg^{2+} and EDTA (Figure 3). EDTA and Mg^{2+} can reduced catalase activity became 86.03% and 77.94%, respectively. The highest of inhibition was Cu^{2+} with relative activity 44.78%.

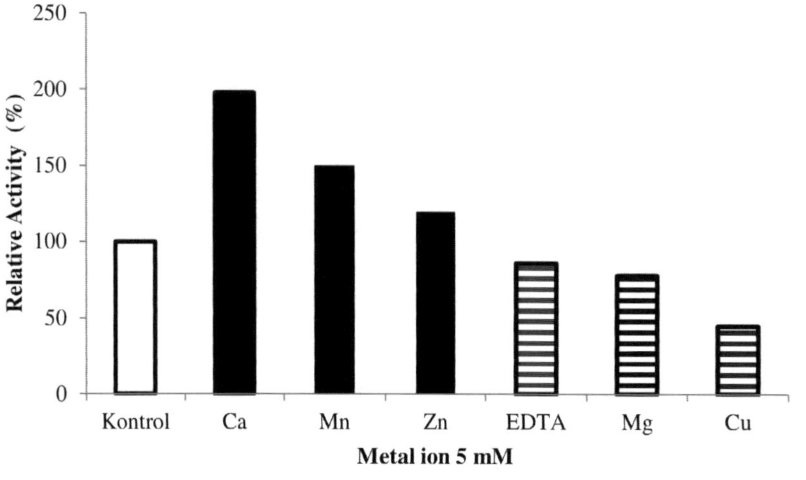

Figure 3. Relative activity of catalase at concentration 5 mM of metal ions.

Ions have ability to increase (activator) or decrease (inhibitors) enzyme activity after interaction with substrate [17]. Metal ions has important role for catalytic reaction by binding substrate to active site of enzyme, to stabilize active conformation or induce formation of binding or active site. In this present research, metal ions Ca^{2+}, Mn^{2+} and Zn^{2+} with concentration 5 mM had ability to increase catalase activity *N. crassa* (InaCC F226) and highest activity of catalase reached with Ca^{2+} as metal ion (relative activity 197.79%). Catalase has active site on the binding of Ca^{2+}, and supported for binding of to the substrate [14]. Mn^{2+} and Zn^{2+} increased catalase activity by enhancing stability of enzyme structure and also by changing conformation of catalase and enzyme active site of enzyme. Therefore, hydrolysis of substrate of catalase will increased [23].

Cu^{2+}, Mg^{2+} and concentration of 5 mM EDTA reduced catalase activity (Figure 3). EDTA inhibited catalase activity by chelating ions Fe^{3+} as structure component of catalase. Ion Mg^{2+} and Cu^{2+} inhibited catalase activity due to reduce reaction of oxidation reduction during decomposition of H_2O_2. Therefore, in this present research, metal ion Ca2+, Mn2+ and Zn2+ were the activator and metal ion Cu^{2+}, Mg^{2+} and EDTA were inhibitor of catalase activity from *N. crassa* (InaCC F226). The results of this research are in accordance with Kandukuri *et al.* that found metal ions Mn^{2+} and Ca^{2+} could stimulated catalase activity, while EDTA, Mg^{2+} and Cu^{2+} concentrations of 5 mM inhibited catalase activity from *Vigna mungo* [17]. Catalase had metal ion in active site. Addition of other metal ions at certain concentrations could affect stability of catalase structures or affect oxidation reduction reaction to active site enzyme, therefore addition of metal ion will affect activity.

3.4. Kinetics Parameter of Catalase

Kinetic properties of enzyme can be determined by maximum catalytic velocity (Vmax) and substrate concentration when kinetic rate reached half maximum (Km) [15]. Kinetics catalase *N. crassa* (InaCC F226) were determined through the Michaelis-Menten equation and Lineweaver Burk. Michaelis-Menten equation was used to determine concentration of H_2O_2 on the catalase activity. The rate of hydrogen peroxide decomposition as function of hydrogen peroxide concentration was given in Figure 4 Catalase activity at concentrations 0.02 up to 0.10 mM H_2O_2 increased significantly against substrate concentration, but at higher concentration 0.10 mM the increasing of enzyme activity was not significantly against substrate concentration.

Based on this curve, the increasing of catalase activity significantly different at concentrations of H_2O_2 from 0.02 up to 0.10 mM and started from 0.16 to 0.24 mM concentrations of H_2O_2 increasing in activity catalase are ramps (not significant). Based on Lineweaver Burk curve, value of Km and Vmax for catalase *N. crassa* (InaCC F226) obtained were 13.3156 0.2384 mM and s/mM, respectively (Figure 5).

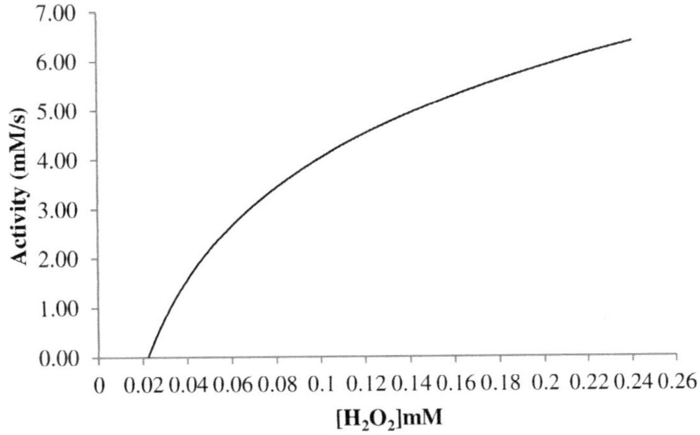

Figure 4. Michaelis Menten Curve : Correlation between Catalase Activity at Different Concentration of H_2O_2.

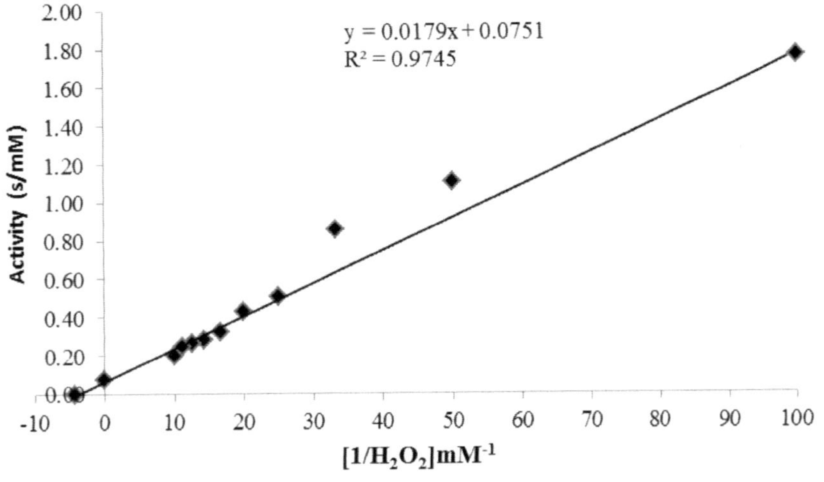

Figure 5. Lineweaver Burk curve : correlation between catalase activity at different concentration H_2O_2.

Compared to other research [7] that isolated *N.crassa,* Km value obtained from this study is lower. The lower of Km value was considered as high affinity of catalase from *N. Crassa* to H_2O_2 as the substrate. But, Vmax value obtained from this research is lower than the Vmax obtained from research [7].

4. Conclusions

Isolation of catalase from indigenous isolate *Neurospora crassa* (InaCC F226) by fractionation with 60% ammonium sulphate saturation showed highest specific activity. Optimum temperature and pH for catalase activity were 40°C and pH 7.0, respectively. The metal ions Ca^{2+} , Mn^{2+} and Zn^{2+} stimulated catalase activity , while EDTA, Cu^{2+} and Mg^{2+} inhibited activity of catalase. The value of Km for Catalase was 0.2384 and Vmax values was 13.3156 s mM / mM.

References

[1] Akyilmaz E, Kozgus O 2009 *J Food Chem.* **115(1)** 347–351
[2] An H, Zhou H, Hung Y, Wang G, Luan C, Mou J 2010 *J Mol Biotechnol.* **45** 155-160
[3] Anwar YAS 2006 Produksi dan karakterisasi enzim tanin asil hidrolase dari *Aspergillus niger* [*tesis*] Bogor (ID): Institut Pertanian Bogor
[4] Barton SC, Gallaway J, Atanassov P 2004 *J Chem Rev.* **104** 4867–4886
[5] Berg JM, Tymoczko JL, Sttyrer L 2002 *Biochemistry 5th Ed.* US: WH. Freeman & Company
[6] Bradford MM 1976 *J Analisys Biochem.* **72** 248–254
[7] Diaz A, Rangel P, Oca YM, Liedias F, Hansberg W 2001 *J Free Radic Biol Med.* **31(11)** 1323–1333
[8] Foyer CH, Noctor G 2000 *J New Phytol.* **146(3)** 359–388
[9] Futo P, Markus G, Kiss A, Adanyi N 2012 *J Electroanalysis* **24(1)** 107–113
[10] Goth L, Nagy T 2012 *J Arch Biochem Biophys.* **525** 195-200
[11] Hussein AA 2012 *J Int Res Biotechnol.* **3(10)** 207-214
[12] Kandukuri SS, Ayesha N, Shiva RS, Vijayalakshmi 2012 *J of Chromatography* **889** 50– 54
[13] Li J, Zhang Y, Chen H, Liu Y, Yang Y 2013 *J Biochem Eng.* **9** 72-83
[14] Manu UJS 2009 *J Food Chemistry* **114** 66–71
[15] Nelson DL, Cox MM 2008 *Principles of Biochemistry Fifth Edition.* United State of America (AS): Freeman and Company
[16] Nishikawa M, Hyoudou K, Kobayashi Y, Umeyama Y, Takakura Y, Mitsuru 2005 *J Control Release* **109(1)**101-107
[17] Page DS 1989 *Prinsip-prinsip Biokimia* Jakarta (ID): Erlangga
[18] Pedrini N, Juarez MP, Crespo R, Alaniz MJ 2006 *J Mycologia* **98(4)** 528-534
[19] Pugoh Santoso, Laksmi Ambarsari, Suryani, Yopi 2016 *Int J on Advanced Science, Engineering and Information Technology* **6(4)** 502-507
[20] Schliebs W, Wurtz C, Kunau WH, Veenhuis M, Rottensteiner HA 2006 *J Eukaryot Cell.* **5(9)** 1490–1502
[21] Sooch BS, Kauldhar BS, Puri M 2014 *J Biotechnology Advances* **9** 1-19
[22] Tijssen T 1985 *Elsevier* **173** 220
[23] Thys RCS, Brandelli A 2006 *J Appl. Microbiol.* **101** 1259-1268

IOP Conf. Series: Earth and Environmental Science **58** (2017) 012069 doi:10.1088/1755-1315/58/1/012069

Protein Characterization of Javan Cobra (*Naja sputatrix*) Venom Following Sun Exposure and Photo-Oxidation Treatment

Sulistiyani[1*], R S Biki[1] and D Andrianto[1]

[1]Departement of Biochemistry, Bogor Agricultural University, Indonesia

Email: sulistiyani@apps.ipb.ac.id

Abstract. Snake venom has always been known for its toxicity that can cause fatality, however, it is also one of the important biological resources to be used for disease treatment. In Indonesia, snake venom previously expose under the sun has been used for alternative treatment of some diseases such as dengue fever, atherosclerosis, cancer, and diabetes. There has been very little scientific evidence on the use of snake venom of Indonesia origin as well as its protein characteristic. Thus, the objective of this research is to characterize the protein content and the specific activity of the venom of Javan Cobra (*N.sputatrix*) when treated with sun exposure in comparison with photo-oxidation by ultraviolet. Qualitative analysis of protein contents was determined using sodium dodecyl sulfate polyacrylamide gel electrophoresis (SDS PAGE). The L-amino acid oxidase activity (LAAO) and the phospholipase A_2 (PLA_2) activities were determined using spectrophotometry. The venom's protein was separated into 5 main protein bands with molecular weight ranging from 14 to 108 kDa. A time course study showed that the venom lost 91% of its LAAO activity and 96% of PLA_2 activity after 6 hours of sun exposure. UV photo-oxidation carried out for 3 hours decreased 91% of LAAO activity, and almost diminished all of PLA_2 activity (99.8%). These findings suggest that the exposure of *N. sputatrix* venom under the sun and UV photo-oxidation decreased its toxicity as shown by the significant reduction of the enzymes activity, but did not affect the protein's integrity. Therefore, these approaches produced *N.sputatrix* venom with less toxicity but still withheld other characters of intact proteins.

1. Introduction

Snake venom has always been known for its toxicity that is potential for lethality, however, it is also one of the important biological resources to be used for disease treatment. The first use of snake venom in modern medicine has been reported in 1970 and has been approved by the US *Food and Drug Assosiation* (FDA) in 2004 as medicinal source for treatment of high blood pressure- and cardiovascular-related diseases [1]. Its applications ranging for treatment of heart disease, diabetes, cancer, as well as *Human Immunodeficiency Virus* (HIV) infection [2]. In China and India, snake venom has been used for treatment of stroke, ischemia, and haemorraghia [3]. Oral administration of snake venom had been showed to increase immune response [4]. It has been reported also that photo-oxidized venom was able to immunized rats against snake venom toxin [5]. Protein isolated from snake venom such as phospholipase A2 (PLA_2), L-amino acid oxidase (LAAO) and other proteins have been showed in vitro as potential cure for cancer through various mechanisms such as necrosis and apoptosis [6,7]. Snake venom also exhibited coagulation and fibrinolytic effects associated with hemostasis [2].

Several approaches could be used to reduce the toxic activity of the snake venom [5]. In India, photo-oxidation by ultraviolet was used to decrease the toxicity of snake venom, so the remaining effect is therapeutic activity [8,9]. The components of snake venom and its effects are varied and depending on the species, geographical location, climate, niche, sex, and age. Vyas et al. reported that the amounts of toxin being produced was decreased as the snake gets older [6]. In Indonesia, snake venom previously expose under the sun has been used for treatment of some diseases such as dengue fever, atherosclerosis, cancer, and diabetes. Particularly, javan cobra (*N. sputatrix*)'s venom pre-treated with sun exposure has been used for oral treatment of human cancer. Javan cobra, also known as *ular sendok*, is an Elapidae commonly found in Indonesia. There has been very limited information on the characteristics of the venom of this species of snake, especially the one that has been traditionally 'detoxified'. Thus, the objective of this research is to characterize the protein content and the specific activity of the venom of *N.sputatrix* when treated with sun exposure in comparison with photo-oxidation by ultraviolet. The venom's characterization includes: total protein and protein molecular weight determination, LAAO and PLA_2 enzymes activities, and anticoagulation activity. This research will provide preliminary scientific evidence on the application of snake venom as alternative therapy for infectious as well as non-infectious diseases.

2. Materials and Methods

The venom of *N.sputatrix* was obtained from The Indonesian Cobra Show (ICS), Bogor. Buffer solutions used in the experiments were as follows: Tris-HCl pH 7.4, Tris-HCl pH 6.8 and Tris-HCl pH 8.8. The standards solutions used in the experiments were *Bovine Serum Albumin* (BSA), L-leucin. and palmitic acid. The reagents and substrates used in enzymatic assays were Bradford reagent, ninhydrin, phenol-red, egg-yolk, and bovine whole blood. Supplies for running the SDS PAGE were acrylamide, bis-acrylamide, Sodium Dodecyl Sulphate (SDS), amonium persulphate (APS), tetramethyl-ethylenediamine (Temed), bromphenol blue, glicerol, mercaptoethanol, *Comassie Brilliant Blue*, methanol, glacial acetic acid, glicyine, dan unstained low molecular weight protein marker. All SDS PAGE materials were purchased from Thermo Scientific, USA. All chemicals for solutions such as NaOH, concentrated HCl, blue methylene, activated charcoal, DMSO, ethanol, sodium citrate, and $CaCl_2$ all were purchased from Merck, Darmstad, Germany.

2.1. Experimental Design

The experimentals approach of this research was to characterize several parameters of *N. sputatrix* venom activities when untreated or treated with either sun exposure or UV-photooxidation. Those parameters were: total protein content, protein molecular weight qualitative analysis, specific activity of LAAO and PLA_2 enzymes, and coagulation/anticoagulation activity of the venom.

2.2. Samples Preparation

Snake venom was milked from N. Sputatrix using a glass drinking cup, after which the collected venom was transported in a cool box (4-8°C) and kept at -20°C until used for experiment. Sun-exposed venom was prepared by placing the sample in a closed petri dish directly under the sun during the daylight. The sun exposure was carried out for: 2, 4, and 6 hours. The UV photooxidation was done following the modified method of Gawade et al.[9]. Briefly, snake venom sample was diluted (1:2) with Tris-HCl buffer pH 7.4 and 2 ml of 0.05 M methylene blue solution was added. This mixture was poured in a petri dish and was incubated in a 37°C waterbath under the UV lamp (Philips 25 Watt, germicidal 254 nm) at distance of 10 cm. The UV-photooxidation was carried out for 3 hours. Following the incubation, the UV-treated sample was transfered into a beaker glass and 0.5 gram of activated charcoal was added and mixed for 15 minutes. The mixture was filtered and the filtrate was kept in a vial at -20°C until experiment.

2.3. Qualitative Protein Analysis with SDS PAGE

Qualitative analysis of protein contents was determined using SDS PAGE with 13% polyacrylamide. Diluted samples of snake venom (2%,v/v) were added with mercaptoethanol-containing reducing sample buffer. The sample mixtures was heated in boiling water for 15 minutes. After cooled down, all samples were loaded into wells (10μl each) in the polyacrylamide. Protein molecular weight marker was loaded in a 5-μl volume.The SDS PAGE was run for 1 hour 30 minutes at 80 volts. The gel was stained with solution of Coomassie brilliant blue for 12 hours. Destaining of the gel was done to clear up the remaining stain by soaking the gel in boiling water until all stains in the background was gone and all markers clearly stained. The electrophoregram was analyzed using PhotocapMw application (Vilber Lourmat, versi 10.0)

2.4. Determination of Total Protein Content

Total protein concentration was determined quantitatively using Bradford method [10] Samples of snake venom in aliquotes of 50 μL were diluted to 100 mL with destilated H_2O. The BSA standard solutions at concentration of: 0,40, 80, 120, 160, and 200 μg/mL and diluted samples (100 μL) were mixed thoroughly with Bradford reagents and incubated at room temperature for 5 minutes. The absorbancy was measured using spectrophotometer at 615 nm wave-length.

2.5. Assay of L-Amino Acid Oxidase (LAAO) Acitivity

The LAAO activity was determined by ninhydrin test [11] The activity of the enzyme was determined by measuring the remaining amino acid substrate left after certain time of incubations. Samples of snake venom in 0.1M Tris-HCl Buffer pH 7.4 at concentration of 10 mg/mL (0.1 mL) were added to 4.9 mL of L-leucin solution in similar buffer (800 μg/mL) and the mixtures were incubated at 37°C for periods of: 0, 2.5, 5, and 10 minutes. The reactions were stopped by heating the mixtures in waterbath at 100 °C for 10 minutes. Subsequently , 4 mL of this mixtures was taken for amino acid concentration analysis. Amino acid standard curve was made using 4 ml each of L-leucine standard solutions at concentrations of: 0, 500, 600, 700, 800, 900, and 1000 μg/mL. All samples and standard solutions were added with 1 ml of ninhydrin reagen. All reaction tubes were sealed with aluminium foil and the mixtures were shaken well, after which all were put in 100°C waterbath for 15 minutes. After cooled down, all samples were measured their absorbance at 570 using spectrophotometer. All measurements were in triplicates.

2.6. Assay of Phospholipase A_2 (PLA_2) Activity

The PLA_2 activity was determined according to the modified method of de AraÚjo and Radvanyi [12]. The activity of the PLA_2 was determined by measuring the rate of product increment which was fatty acid that reduced the pH of the solution. The pH of the substrate, the fatty acid standard, and the phenol red indicator were all 7.4 at the beginning of the reaction. The phospholipid solution (2 ml) was incubated with the venom sample (1 mg/ml) at 37°C for 0, 20, 40, and 60 detik. The reations were stopped by incubating the mixtures in boiling water (water bath) for 10 minutes. The fatty acid contents in the reaction mixtures and standard solutions (3 ml each) were measured by addition of 0.5 ml of 0.1% phenol red. The fatty acid standard curve was made by using palmitic acid standard at concentrations of : 0, 30, 60, 90, 120, and 150 μg/mL. The absorbances were read using spectrophotometer at 540 nm. All assays were performed in triplicates.

2.7. Assay of Coagulant and Anticoagulant Activity [13]

The coagulant effect of N. Sputatrix venom was determined by recording the clotting times after addition of the venom to citrate-containing bovine plasma according to the method of Theakston & Reid [14]. Addition of 250 mM $CaCl_2$ alone to aliquots of plasma was used as a positive control. Bovine plasma was obtained from the whole blood by addition of 3.2% sodium citrate anticoagulant to the whole blood (1:9, v/v). The plasma was separated from the blood cell by centrifugation in low-speed centrifuge (Hettich Universal) at 3000 rpm. The supernatant was the plasma which contained fibrinogen and other inactivated coagulation factors. The plasma aliquots (3mL) to be assayed were prepared in reaction tubes and kept in 37°C waterbath. The venom at concentration of 1 mg/ml (0.2 mL) was mixed with the plasma and the recording of the clotting times was immediately begun. The anticoagulant activity was assayed using modified method of Suarez et al. [13] by incubating the citrated bovine plasma with the mixture of venom and 250 mM $CaCl_2$ and the clotting times recorded.

3. Results and Discussions

3.1. *Protein Molecular Weight Analysis*

Figure 1 shows that using SDS PAGE the venom's protein was separated into 5 main protein bands. PhotocapMw Application showed that the molecular weight of those bands were ranging from 15.5 to 107.9 kDa (Table 1). These observations are consistent with the ones reported by Suarez *et al.* (2016). The five observed protein bands were closely matched with phosphodiesterase, hyaluronidase, LAAO, prostatin-like protein, and PLA_2 reported in their works.There was no different between control venom and pre-treated venom samples. This observation indicates that the protein structure remained intact in the venom treated with either sun exposure or UV-photooxidation.

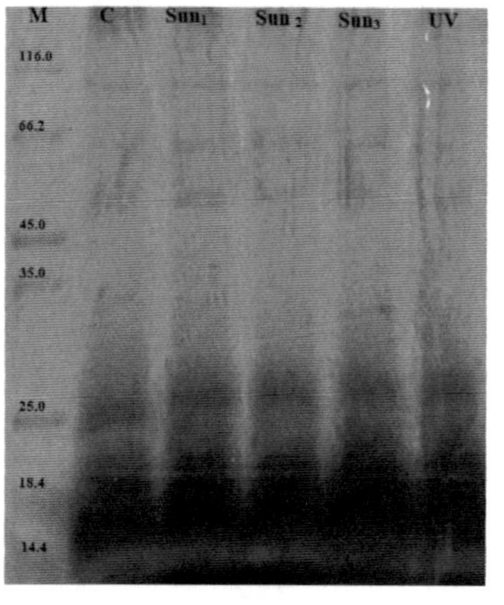

Figure 1. Electrophoregram of SDS PAGE of *N. sputatrix* venom. Lanes: C=control, untreated venom, Sun 1-3 = sun exposed (2,4,6hrs), UV=UV photo-oxidized (3 hrs)

Table 1. Molecular weight of *N. sputatrix* venom protein.

Molecular Weight (kDa)				
Control	Sun Exposure (2 hrs)	Sun exposure (4 hrs)	Sun exposure (6 hrs)	Photo-oxidation
102.9	102.9	106.1	102.9	107.8
64.9	64.5	65.9	65.2	63.8
53.9	54.3	55.1	55.5	54.7
25.3	25.7	26.2	26.6	27.2
14.9	15.5	15.7	15.8	15.6

3.2. *Total Protein Content*

Data in Table 2 shows that the protein content of the untreated venom was 20.46 % (w/v). The sun exposure for 6 hours increased protein content by 38.7%, which probably due to the concentrating effect of the heat. On the other hand, UV photo-oxidation for 3 hours decreased protein content by

22%. This reduction may be due to adsorption by activated charcoal. The data on protein content showed that the venom of *N.sputatrix* is a protein rich substance which account for approximately 90% of its total dry weight [15]. This protein content was used later to calculate the specific activity of the venom.

Table 2. Total protein content.

Samples Treatments	Protein content (% w/v)
Control	20.46 ± 0.43
Sun exposure, 2 hours	23.35 ± 0.16
Sun exposure, 4 hours	28.10 ± 0.16
Sun exposure, 6 hours	28.38 ± 0.16
Photo-oxidation	15.89 ± 0.00

3.3. L-Amino Acid Oxidase (LAAO) Activity

The LAAO specific activity of the control (untreated) venom of *N. sputatrix* was 0,229 μmol/min./mg (Figure 2). LAAOs (EC1.4.3.2) are hemorrhagic toxins present in snake venoms which are protein responsible for the yellowish color of the venom. The enzyme catalyzes oxidative deamination of L-amino acid to form ammonia, hydrogen peroxide, and ketoacid and commonly found in many kinds of snake venom. [16]. This enzyme has been demonstrated to alter platelet function [17]. A time course study on the effect of length of sun exposure showed that there was a significant decrease in LAAO enzyme activity during the first 2 hours of exposure and the enzyme lost 91% of its activity by 6 hours. Similar findings was observed for samples pre-treated with UV photo-oxidation for 3 hours which cause 91% reduction in the enzyme activity.

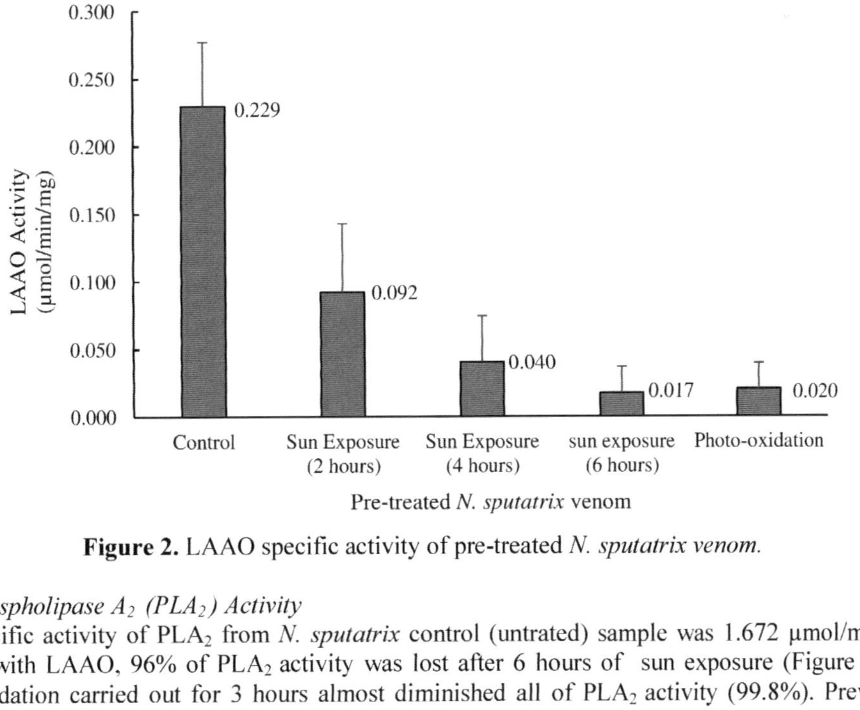

Figure 2. LAAO specific activity of pre-treated *N. sputatrix venom.*

3.4. Phospholipase A_2 (PLA_2) Activity

The specific activity of PLA_2 from *N. sputatrix* control (untrated) sample was 1.672 μmol/min./mg. Similar with LAAO, 96% of PLA_2 activity was lost after 6 hours of sun exposure (Figure 3). UV photooxidation carried out for 3 hours almost diminished all of PLA_2 activity (99.8%). Previously,

Sallau *et al.* [18] reported that PLA_2 activity of *Echis ocellatus* was as much as 2.39 μmol/min.t/mg. This enzyme is the most abundant enzymes found in snake venom and is responsible for the phospholipid degradation which eventualy damaged the cell membrane. This activity results in the venom's toxicity, particularly when it is entering the blood circulation. The PLA_2 catalyzes the hydrolysis of the fatty acid by attacking the ester bond at the *sn*-2 of the phospholipid molecule thus liberating free fatty acid and lysophospholipid [18]. Thus, the decrease in the activity of PLA_2 as well as that of LAAO following sun exposure and UV photo-oxidation indicates that the venom had lost its toxicity due to the treatments. Our observations are consistent with the works of Tejasen & Ottolenghi [19] which reported the reduction of snake venom toxicity following UV photo-oxidation as shown by the decrease in PLA_2 and phosphodiesterase activities.

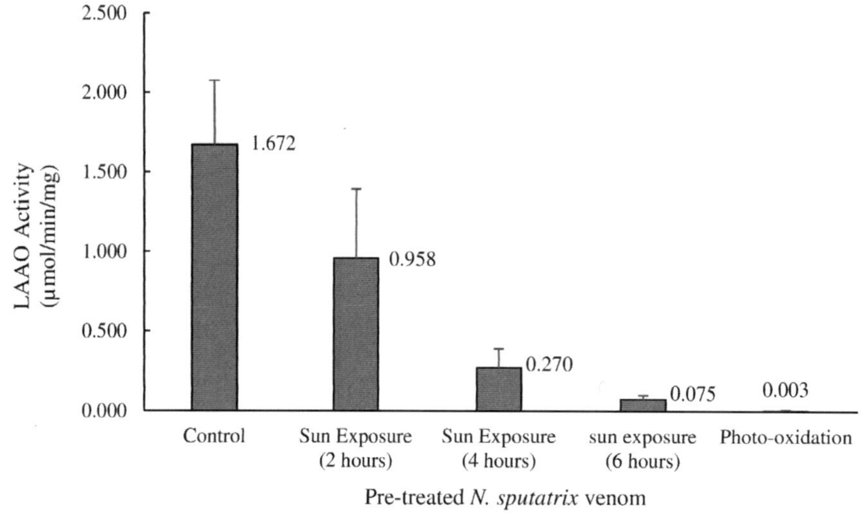

Figure 3. PLA_2 specific actifity of pre-treated *N. sputatrix* venom.

3.5. Coagulant and Anti-Coagulant Activity

As shown in Table 3, the positive control for coagulant activity was observed by addition of $CaCl_2$ solution into the citrate-containing plasma which will induce the plasma to clot. The calcium added into the citrated plasma stimulated the mechanisme of clotting. When the citrated plasma was mixed with *N. sputatrix* venom alone, no clotting was observed, indicating that the venom does not posses coagulant activity. This lack of coagulant activity were still exhibited regardless of the sun exposure or the UV-photo-oxidation pre-treatment of the venom, which suggest that the *N. Sputatrix* venom does not contain enzymes that clot fibrinogen, unlike the venoms of several snakes previously reported by Pirkle et al. [20]. The lack of coagulant activity of the venom of the Colombian coral snake *Micrurus dumerilii* has been reported recently [13]. On the other hand, when tested for anticoagulant activity, the untreated *N. sputatrix* venom exhibited positive result. The anticoagulant activity was also observed with the venom exposed to the sun for 2- and 4- hours. Interestingly, the 6 hour-sun exposed and photo-oxidized venoms were no longer demonstrated the anticoagulant activities. Zingali [21] reported several snake venom proteins which possessed anticoagulant activity, among others were PLA_2 and LAAO. The PLA_2s affect blood clotting through the hydrolysis of procoagulant phospholipid or by binding to them [22, 23]. The anticoagulant activity of the untreated venom shown in our present data is correlated with the highest specific activity showed in the LAAO and PLA_2 assay. This anticoagulant activities were no longer presence when the activity of both enzymes were diminished following 6-hour sun exposure or UV photo-oxidation.

Table 3. Plasma clotting time with the addition of pre-treated *N. sputatrix* venom.

Samples	Clotting time
Plasma +$CaCl_2$	Clots observed after 193 ± 7 sec.; coagulant effect
Plasma +Untreated venom	None (after 20 minutes observation); no coagulant effect
Plasma +Sun exposed (2 hrs) venom	None (idem)
Plasma +Sun exposed (4 hrs) venom	None (idem)
Plasma +Sun exposed (6 hrs) venom	None (idem)
Plasma +Photo-oxidized venom	None (idem)
Plasma +Untreated venom + $CaCl_2$	None (idem), anticoagulant effect
Plasma + Sun exposed (2 hrs) + $CaCl_2$	None (idem), anticoagulant effect
Plasma+Sun exposed (4 hrs) + $CaCl_2$	None (idem), anticoagulant effect
Plasma+Sun exposed (6 hrs) + $CaCl_2$	Clots observed after 307 ± 13 sec. ; no anticoagulation effect
Plasma+$CaCl_2$+Photo-oxidized venom	Clots observed after 227 ± 22 sec. ; no anticoagulant effect

These observations suggest that prolong (6 hours) sun-exposure and the UVphoto-oxidation were able to inactivate the LAAO and PLA_2 which are consistent with the venom detoxification [Tejasen &Ottolenghi]. The inactivation of those enzymes resulted in the loss of anticoagulant activity of *the N. sputatrix* venom. Nevertheless, the proteins contained in the venom remain intact under the prolong sun exposure as well as UV photo-oxidation. Taken together, these findings suggest that sun exposure and UVphoto-oxidation significantly decreased the activity of LAAO and PLA_2.of the *N.sputatrix* venom, but both treatments did not affect the physical characteristic of the venom's protein content. Further studies on the immunogenicity of the treated venom will provide insight to the traditional healing property of the snake venoms for various diseases.

References

[1] Mohney G 2013 Toxic medicine: How venom can heal [internet] (March 22, 2016) http://abcnews.go.com/Health/toxic-medicine-venom-lead-medical-breakthroughs/story?id=18558299
[2] Fatima LD and Fatah C 2014 *J Clin Toxicol* **4(2)** 1-9
[3] Koh DCI, Armugam A, Jeyaseelan K 2006 *Cell Mol Life Sci* **63** 3030-3041
[4] Anindyajati G, Budiono B, Joyosumarto R 2009 *Jurnal Matematika, Sains, dan Teknologi* **10(2)** 75-80
[5] Kocholaty W 1966 *Toxicon* **3** 175-186
[6] Vyas VK, Brahmbhatt K, Bhatt H, Parmar U *Asian Pac J Trop Biomed* **3(2)** 156-162
[7] Lee ML, Fung SY, Chung I, Pailoor J, Cheah SH, Tan NH 2014 *Int J Med Sci* **11** 593-601
[8] Mukund H, Gawade SP 2011 *J Venom Anim Toxins incl Trop Dis* **17(4)** 422-429
[9] Gawade SP, Tamboli MA, Thorat SS 2016 *Indian J Pharmaceutical Ed Res* **50** 123-129
[10] Bradford MM 1976 *Analytical Biochem.* **72** 248-254
[11] Yu Z, Wang Y, Zhou N, Zhao M, Qiu J, Lin J 2014 *Appl. Biochem. Biotechnol.* **174** 13-27
[12] Araújo AL, Radvanyi F 1987 *Toxicon* **25** 1181-1188
[13] Suarez PR, Nunez V, Fernandez J, Lomonte B *J Proteomics* **16** 2-52
[14] Theakston RDG, Reid HA 1983 *Bull World Health Org* **61** 949-956
[15] Aarti C and Kusro A 2013 *Int. J. Pure App. Biosci.* **1(6)** 24-29
[16] More SS, Kiran KM, Veena SM, Gadag JR 2010 *J Venom Anim Toxins incl Trop Dis* **16(1)** 60-75
[17] Sakurai Y, Takatsuka H, Yoshioka A, Matsui T, Suzuki M, Titani K, Fujimura Y 2001 *Toxicon* **39** 1827-1833

[18] Sallau AB, Ibrahim MA, Saluhu A, Patrick FU 2008 *African J Biochem Res* **2(4)** 98-101
[19] Tejasen P, Ottolenghi A 1970 *Toxicon* **8** 225-233
[20] Pirkle H, Theodor I, Miyada D, Dimmons G 1986 *J Biol. Chem*. **261** 8830-8835
[21] Zingali RB 2007 *Toxin Reviews* **26** 25-46
[22] Markland FS 1998 *Toxicon* **39** 371-375
[23] Mounier CM, Bon C, Kini RM 2001 *Haemostasis* **31** 279-287

The Phase Transformation and Crystal Structure Studies of Strontium Substituted Barium Monoferrite

A Mulyawan[1*], W A Adi[1], S Mustofa[1] and A Fisli[1]

[1] Center for Science and Tecnhnology of Advanced Materials, National Nuclear Energy Agency of Indonesia (BATAN), Indonesia.

Email: ademulyawan@batan.go.id

Abstract. Unlike other AFe_2O_4 ferrite materials, Barium Monoferrite ($BaFe_2O_4$) have an orthorhombic structure which is very interesting to further study the crystal structure and phase formation. In this study, Strontium substituted Barium Monoferrite in the form of $Ba_{(1-x)}Sr_{(x)}Fe_2O_4$ has successfully been synthesized through solid state reaction method which includes $BaCO_3$, $SrCO_3$, and Fe_2O_3 as starting materials. $Ba_{(1-x)}Sr_{(x)}Fe_2O_4$ was made by varying the dopant composition of Strontium (Sr^{2+}) from x = 0, 0.1, 0.3, and 0.5. Each composition was assisted by ethanol and continued to the milling process for 5 hours then followed by sintering process at 900 °C for 5 hours. The phase transformation was studied by using X-ray diffractometer (XRD) and Rietveld refinement using General Structure Analysis System (GSAS) also 3D crystal visualization using VESTA. Referring to the refinement results, a single phase of $BaFe_2O_4$ was formed in x = 0 and 0.1. The composition has orthorhombic structure, space group B *b*21*m*, and lattice parameters of a =19.0229, b =5.3814 c =8.4524 Å, $\alpha = \beta = \gamma = 90^o$ and a =18.9978, b =5.3802 c =8.4385 Å, $\alpha = \beta = \gamma = 90^o$ respectively. In the composition of x = 0.3 it was found that the phase of $BaSrFe_4O_8$ begin to form due to the overload expansion of the Sr^{2+} occupancy which made the distortion of the initial lattice parameters and finally in the x = 0.5 composition the single phase of $BaSrFe_4O_8$ was clearly formed. Energy Dispersive Spectroscopy (EDS) was used to confirm the change of the material structure by measuring the elemental compound composition ratio. The result of EDS spectra clearly exhibited the dominant elements were Barium (Ba), Strontium (Sr), Iron (Fe), and Oxygen (O) with the compound ratio (Atomic percentage and mass percentage) correspond to the $BaFe_2O_4$ and $BaSrFe_4O_8$ phase.

Keywords : Barium Monoferrite; Solid State Reaction; Strontium Substitution; Crystal Structure, Phase Transformation, Atomic Structure Simulation

1. Introduction

Ferrite magnetic materials is either in the form of soft magnetic or hard magnetic have been applied in such a wide technology application area. Spinel magnetic materials with the general formula $M^{2+}Fe_2O_4$ (divalent ion metal, M = Ni, Co, Mn, Zn, Mg, Fe, etc) were highly studied due to the unique magnetic and electric properties which very appropriate to be applied as sensor materials, nano devices, microwave devices,etc[1]

Unlike the usual $M^{2+}Fe_2O_4$ ferrite which exhibits spinel structure, Barium Monoferrite ($Ba^{2+}Fe_2O_4$) have an orthorhombic structure where the Fe^{3+} is bonded tetrahedronally with O^{2-} cations. These type ferrite material are less studied compared to the spinel ferrite type due to the assumption only as a coexisting phase into the M-type Barium Hexaferrite ($BaFe_{12}O_{19}$) along with the Fe_2O_3[2]. $BaFe_2O_4$ has several advantages compared to the spinel ferrite such as high capacity of magnetization, high coercivity, highly stable, high permittivity, low-band-gap ferrite material [3]. Due to the low band gap properties, Barium Monoferrite oftenly only used as an important candidate for visible light photo-catalysis material and ceramic pigment[4] (1). Another beneficial properties of $BaFe_2O_4$ such high permittivity are rarely studied. The development of microwave and radar absorber materials (M-RAM) are growing rapidly to be applied in the military application and information technology. The candidate material which would be applied as M-RAM materials should have high permitivity and permeability values [5]. The high permittivity of the $BaFe_2O_4$ will make this ferrite material as a good candidate to be applied as M-RAM materials. In this work, the low permeability properties of the $BaFe_2O_4$ will be improved by substituting the Sr^{3+} cation into the $BaFe_2O_4$ structure in the form of $Ba_{(1-x)}Sr_{(x)}Fe_2O_4$.

There are several methods that have been conducted to synthesize ferrite magnetic materials such as combustion method, solid state reaction method, glass crystallization method [6-8]. In this study, in order to improve the permeability value, the Sr^{3+} cations were substituted into $BaFe_2O_4$ by using solid state reaction method with High Energy Milling machine. This technique was used due to the simplicity and efficiency for preparing the fine powder of the $Ba_{(1-x)}Sr_{(x)}Fe_2O_4$ powder. In this regards, The aim of this study is to understand the mechanism of the phase transformation when the $BaFe_2O_4$ structure was substituted by Sr^{3+} cations which are never studied before, as far the literature research by the author.

2. Experimental Procedure

2.1 Sample preparation

The starting materials which employed to synthesize substituted Strontium to Barium Monoferrrite, in the form of $Ba_{(1-x)}Sr_{(x)}Fe_2O_4$, are Barium Carbonate powder ($BaCO_3$, Merck), Strontium Carbonate powder ($SrCO_3$, Merck), and Iron (III) Oxide powder (Fe_2O_3, Sigma Aldrich). All of the Barium Strontium Monoferrite (BSMFO) samples were made by varying the Strontium (Sr) substitution to Barium Monoferrite ($BaFe_2O_4$) in the range of x = 0, 0.1, 0.3, and 0.5 respectively, with the chemical reaction are shown as below:

$$(1\text{-}x)BaCO_3 + (x)SrCO_3 + Fe_2O_3 \rightarrow Ba(1\text{-}x)Sr(x)Fe_2O_4 + CO_2 \qquad (1)$$

The BSMFO compositions have been synthesized by using solid state reaction method. The synthesize process was started by mixing all the compound materials in stainless steel vial and assisted by ethanol then being milled for 5 hours. After the milling process, the raw BSMFO cmpositions were heated in the oven to evaporate the ethanol and followed by sintering process in furnace at 900 ^{0}C in ambient condition for 5 hours, all samples were finally grinded by using Agate mortar until fine powder of BSMFO were formed.

2.2 Characterization technique

In order to confirm the phase composition to all of the BSMFO samples, the X-ray diffraction pattern was carried out by using X-ray Diffractometer Panalytical Phillips with Cu Kα radiation (λ=0.15406 nm), with the range of the 2Θ angles started from 10^0 to 80^0 with the scan step 0.0263^0 per second. The elemental compound ratio (atomic percentage and mass percentage) was measured by using Energy Dispersive Spectroscopy (EDS) at 20 kV which attached in Scanning Electron Microscope (SEM) JEOL JSM-6510 LA.

3. Results and Discussion

The x-ray diffraction patterns of all BSMFO composition (x = 0, 0.1, 0.3, and 0.5) are shown in figure 1. In order to identify the phase that have been formed in the composition, all diffraction peaks were firstly confirmed by using Match! ver.3.1 software which reffered to the Crystallography Open Database (COD). The $BaFe_2O_4$ phase was identified by the COD number 96-430-9916 and the $BaSrFe_4O_8$ was identified by the COD number 96-210-6519. Referring to all difraction peaks, it was known that the single phase of $BaFe_2O_4$ has been formed in the composition of x = 0 and 0.1, the most intense peak of the $BaFe_2O_4$ phase was indexed to (212) crystal plane. In both composition the single phase formation of $BaFe_2O_4$ were formed without the14e55 presence of any secondary phases such as Fe_2O_3, Fe_3O_4, or even the residual of $BaCO_3$ and $SrCO_3$ which oftenly formed through the crystallization process of $BaFe_2O_4$. Silvana da dalt,et al also tried to synthesize $BaFe_2O_4$ by using combustion method which reported the presence of Fe_2O_3 and Fe_3O_4 as secondary phase until 1000 ^{0}C [4]. According to the inset figure 1.(b) and 1.(c), eventhough the single phase of $BaFe_2O_4$ was successfully formed in the composition of x = 0 and 0.1, the dominant diffraction pattern of x = 0.1 composition were slight changed, it happened due to the occupation of the Sr^{2+} to the initial lattice of $BaFe_2O_4$. This phenomenon became cleared when the Strontium substitution was increased in the composition of x = 0.3 in which the diffraction pattern clearly exhibited the phase transformation from $BaFe_2O_4$ to $BaSrFe_4O_8$. In the composition of x = 0.5 the crystal structure have completely transformed into a single phase of $BaSrFe_4O_8$, it has two dominant peaks which indexed to (102) and (110) crystal plane. The phase confirmation process then continued to the refinement process by using General Structure Analysis System (GSAS) software with Rietveld analyze method to further study the effect of the Strontium substitution to the initial lattice of $BaFe_2O_4$. The detailed refinement graphics, lattice parameters, density, volume of unit cell, and the phase fraction were shown in table 1 and figure 2.

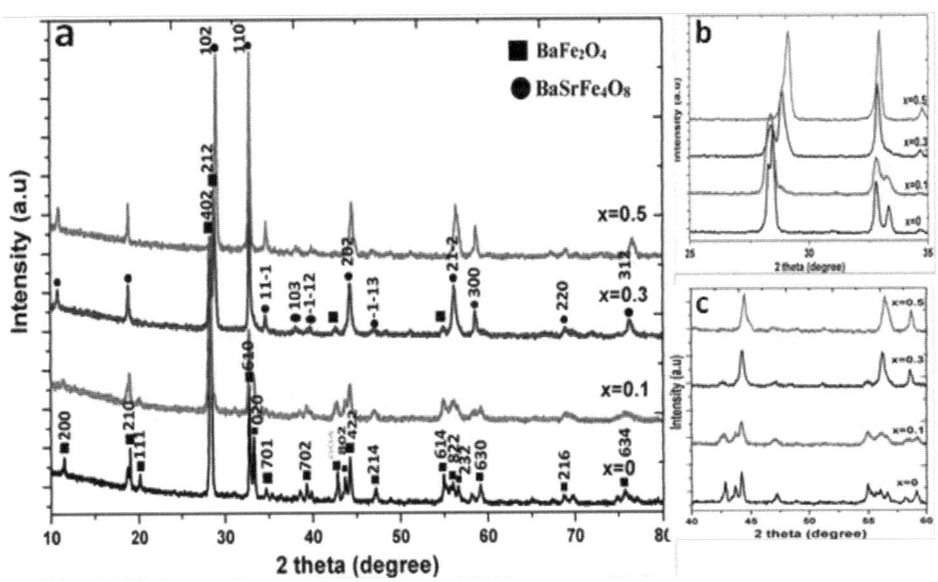

Figure 1. (a) X-ray diffraction pattern of all BSMFO compositions (Strontium substitution, x = 0, 0.1, 0.3, and 0.5), **(b)** inset figure of diffraction pattern from 2Θ = 25 to 35^0, **(c)** inset figure of diffraction pattern from 2Θ = 40 to 60^0.

Table 1. Refinement result from GSAS software for all of the BSMFO compositions (Strontium substitution, x = 0, 0.1, 0.3, and 0.5).

Sample (*x*)	Phase	Lattice parameter (Å) *a*	*b*	*c*	V (Å ^{3})	ρ (g/cm ^{3})	Fraction wt%	R_{wp} (%)	χ^2
0	$BaFe_2O_4$	19.023	5.382	8.452	865.279	4.806	100	4.900	1.292
0.1	$BaFe_2O_4$	18.964	5.381	8.465	863.744	4.738	100	4.510	1.846
0.3	$BaFe_2O_4$	18.383	5.360	8.443	852.434	2.575	28.882	4.900	2.403
	$BaSrFe_4O_8$	5.446	5.446	8.185	210.217	4.482	71.185		
0.5	$BaSrFe_4O_8$	5.445	5.445	8.102	207.998	4.733	100	4.500	2.300

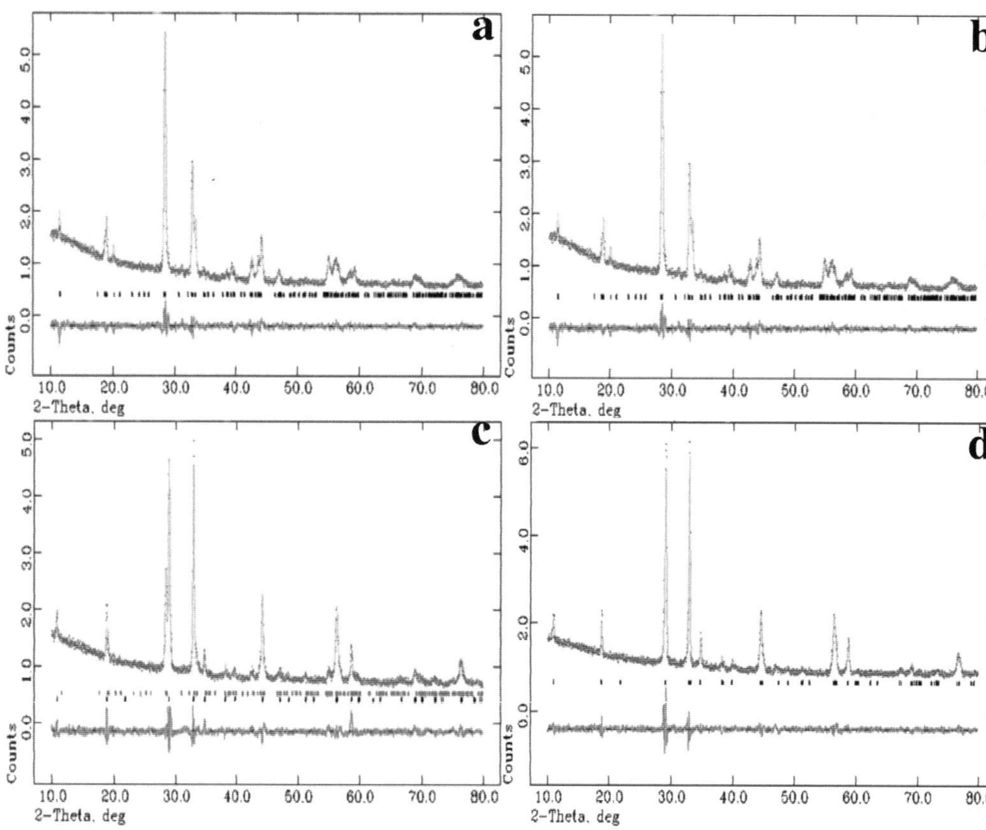

Figure 2. The refinement graphics to all of the BSMFO compositions as the output of the GSAS software **(a)** x = 0, **(b)** x = 0.1, **(c)** x = 0.3, and **(d)** x = 0.5

The obtained lattice parameters for both phase of $BaFe_2O_4$ and $BaSrFe_4O_8$ in all BSMFO composition showed a good suitability to the COD database, $BaFe_2O_4$ phase have the orthorhombic crystal structure with space group C m c 21 and detailed lattice parameter is a = 19.04±2^0, b = 5.3826±2^0, c = 8.4538±2^0 and α = β = γ = 90^0, in other side $BaSrFe_4O_8$ have the trigonal crystal structure with space group P -3 1 m and detailed the lattice parameter is a = b = 5.45±2^0, c = 8.10±2^0 and α = β = 90^0, γ =120^0. The lattice parameters and the 2Θ diffraction angles of this experiment also correspond to another experiment results[2,3].

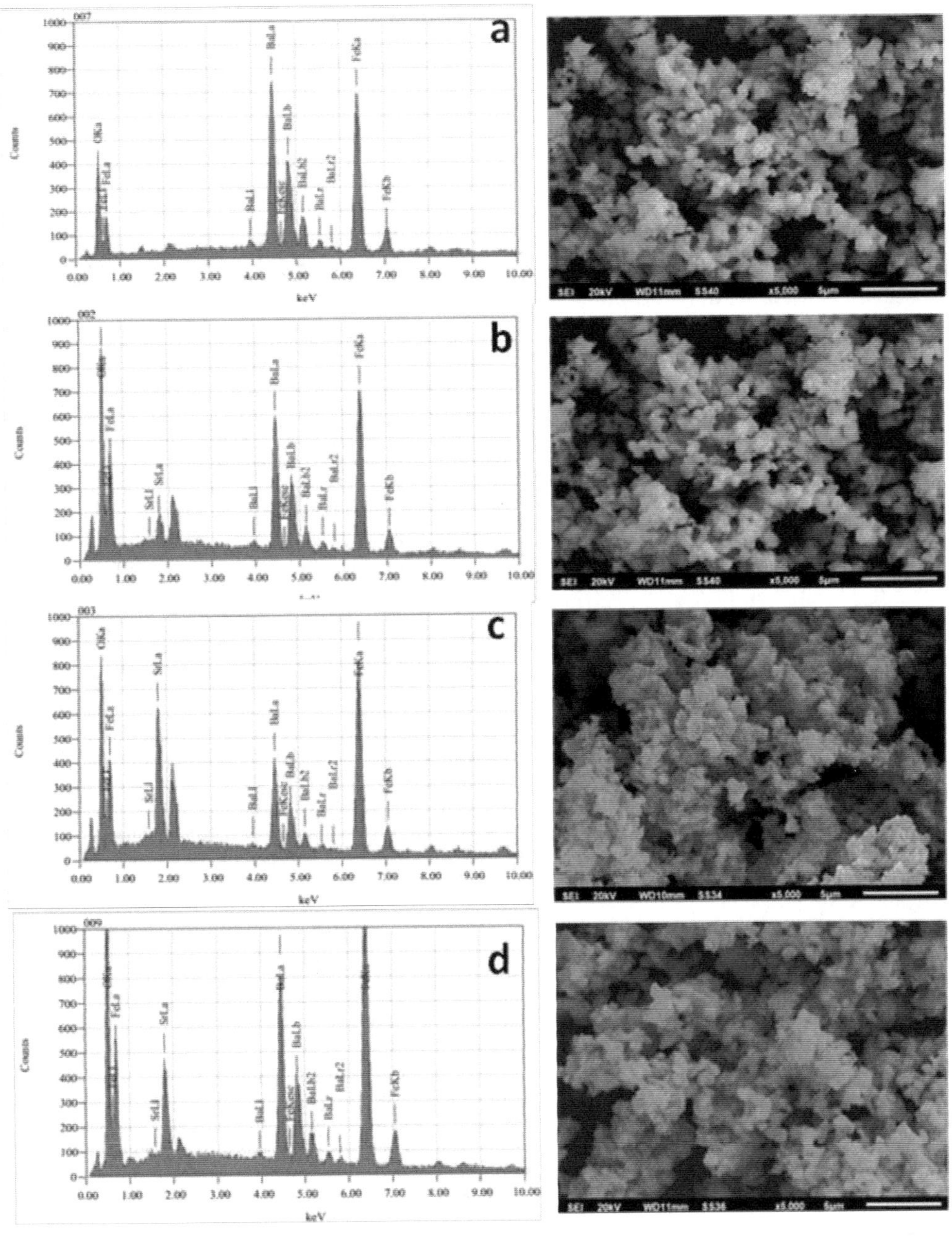

Figure 3. The Energy Dispersive Spectroscopy (EDS) and microstructure images of all BSMFO compositions; **(a)** x = 0, **(b)** x = 0.1, **(c)** x = 0.3, and **(d)** x = 0.5

As shown in table 1, there was a gradual decrease for lattice parameters of $BaFe_2O_4$ phase align with the increasing contain of Strontium (Sr^{2+}) which influenced the value of the cell volume and the density. This phenomenon happened due to the replacement of Ba^{2+} by the Sr^{2+} cation, Sr^{2+} have a lower ionic radius with the value of 2.15Å rather than Ba^{2+} cation with the value of 2.24Å. In the composition of x = 0.1 the Sr^{2+} cation were perfectly substitute the position of the Ba^{2+} and made the crystal structure to be a bit more shrinkage compared to the initial lattice in the composition of x = 0. The same conclusion was also gained to another related research by M.Ait Haddouch,et al who tried to synthesize the $Ba_{(1-x)}Sr_{(x)}WO_4$ ceramics who also found that the substitution of the Sr^{2+} will decrease the lattice parameters and influenced the cell volume [9]. In the composition of x = 0.3, the abundance of the Sr^{2+} created an overload expansion to the initial lattice which had an orthorhombic crystal structure, this made a great distortion in the lattice and finally created another phase; $BaSrFe_4O_8$ which had a trigonal crystal structure with weight fraction 28.882% for $BaFe_2O_4$ and 71.185% for $BaSrFe_4O_8$. The trigonal structure have been fully formed in the composition of x = 0.5 due to the value of lattice parameters database, it made the cell volume were slightly different compared to the composition of x=0.3 which still affected by the $BaFe_2O_4$.

The characterization process using Energy Dispersive Spectroscopy (EDS) is a powerful tool to determine the relative elemental compound ratio (in mass percentage and atomic percentage) which directly correspond to the phase that have been formed in the composition. The relative elemental compound ratio, microstructure images, and EDS graphics were shown in the table 1 and figure 3. Scanning electron microscope (SEM) images were taken at 5000x magnification. Referring to the SEM images from figure 3, it can be noticed that the particle have homogenous size distribution with low degree of agglomeration. The EDS spectra images of all BSMFO compo sitions also shown in figure 3. The result for all of the BSMFO compositions clearly exhibited the dominant elements of Barium (Ba), Strontium (Sr), Iron (Fe), and Oxygen (O). Referring to the atomic percentage in the table 2, the quantification ratio of each elemental compound showed a good agreement to the $BaFe_2O_4$ and $BaSrFe_4O_8$ phase with an error less than 0.4%. In the composition of x = 0, the quantification ratio between Ba : Fe were in the range of 1 : 2 which correspond to the phase of $BaFe_2O_4$. The same quantification ratio was also found in the composition of x = 0.1, which related to the phase of $BaFe_2O_4$, the amount of Sr which detected in this composition have confirmed that the Sr^{2+} were successfully replaced the Ba^{2+} cation in the lattice. The unusual quantification result in the composition of x = 0.3 were likely caused by the mixed composition between the $BaFe_2O_4$ and $BaSrFe_4O_8$.The quantification ratio in the composition of x = 0.5 between Ba : Sr : Fe showed a good agreement to the phase of $BaSrFe_4O_8$ with the value of 1 : 1 : 4. All of the quantification ratio was well correspond to the x-ray diffraction pattern.

Table 2. EDS results of all BSMFO compositions (Strontium substitution, x = 0, 0.1, 0.3, and 0.5).

Strontium substitution (x)	Elemental Compound							
	Atomic Percentage (%)				Mass Percentage (%)			
	Ba	Sr	Fe	O	Ba	Sr	Fe	O
0	22.46	-	41.03	36.52	51.75	-	38.44	9.80
0.1	12.08	1.98	27.47	58.47	38.57	4.03	35.65	21.74
0.3	10.37	4.09	28.33	57.21	33.28	8.38	36.96	21.38
0.5	7.08	7.28	27.19	58.45	23.93	15.70	37.36	23.01

 IOP Publishing
doi:10.1088/1755-1315/58/1/012070

Table 3. Selected inter-atomic distances (Å) for all BSMFO compositions (Strontium substitution, x = 0, 0.1, 0.3, and 0.5).

Composition	x = 0 (Å)	x = 0.1(Å)	x = 0.3(Å)	x = 0.5(Å)
Ba-O	2.7430	2.8163	2.8573	2.9307
Sr-O	-	-	2.60629	2.49926
Fe-O	1.8842	1.8791	1.8444	1.8136

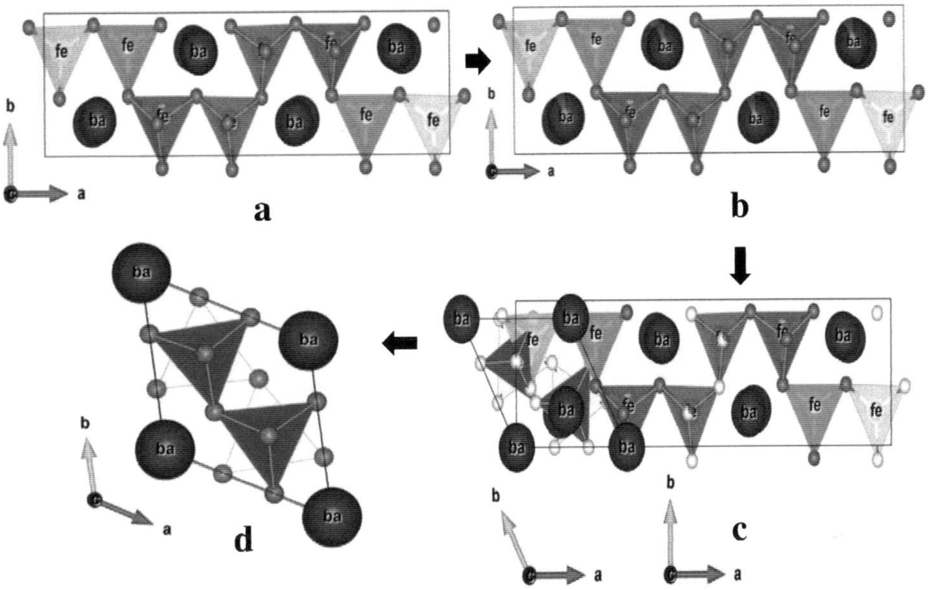

Figure 4. The gradual changes of the crystal structure from orthorhombic (space group C m c 21) into trigonal (space group P -3 1 m) in 2-dimensional form; **(a)** x = 0, **(b)** x = 0.1, **(c)** x = 0.3, and **(d)** x = 0.5

The visualization of the atomic structure in order to clarify the gradual changes of the crystal structure of each composition were carried out by using vesta software. The gradual changes were shown in the 2-dimensional figure (figure 4) and 3-dimensional form (figure 5). The orthorhombic unit cell of $BaFe_2O_4$ phase was outlined by the solid black lines in the composition of x = 0 and 0.1, which based on the X-ray diffraction pattern had a single phase of $BaFe_2O_4$. The orthorhombic unit cell was built by the two type of cations Ba^{2+} and Fe^{3+} (12 Ba atom and 16 Fe atom) and the Oxygen anions (36 atom). In the composition of x = 0, the Ba^{2+}, Fe^{3+}, and O^{2-} were located in the 4a and 8b sites in which the Fe atom were tetragonally bonded with the Oxygen as shown in the figure 4(a) and 5(a). The perfect substitution of the Sr^{2+} into the initial lattice of the was shown in the figure 4(b) and 5(b) represented by a slight green colour decorated the spherical Barium atom,. It can be noticed that the occupancy of the Sr^{2+} was not disturbing the form of the intial lattice however with decreasing the cation size (substituted by the Sr^{2+}), it affected the inter-atomic distances either for Ba-O or Sr-O in this composition which it became greater than the initial lattice in the composition of x = 0 (table 1). The overload expansion of Sr^{2+} in the composition of x = 0.3 which have made a great distortion in the initial lattice and transformed into $BaSrFe_4O_8$ was clearly shown in the figure 4(c) and 5(c) where the Sr^{2+} were entirely moved into the 1b site and made the distance even smaller. The greatest distance between Ba-O was known in the composition of x = 0.5 due to the great distortion which made the phase fully transformed into a complete phase of $BaSrFe_4O_8$ in which the Fe^{3+} still bonded tetragonally into the Oxygen anions but the the Ba^{2+} was moved into the corner of the site as shown in the figure 4(d) and 5(d).

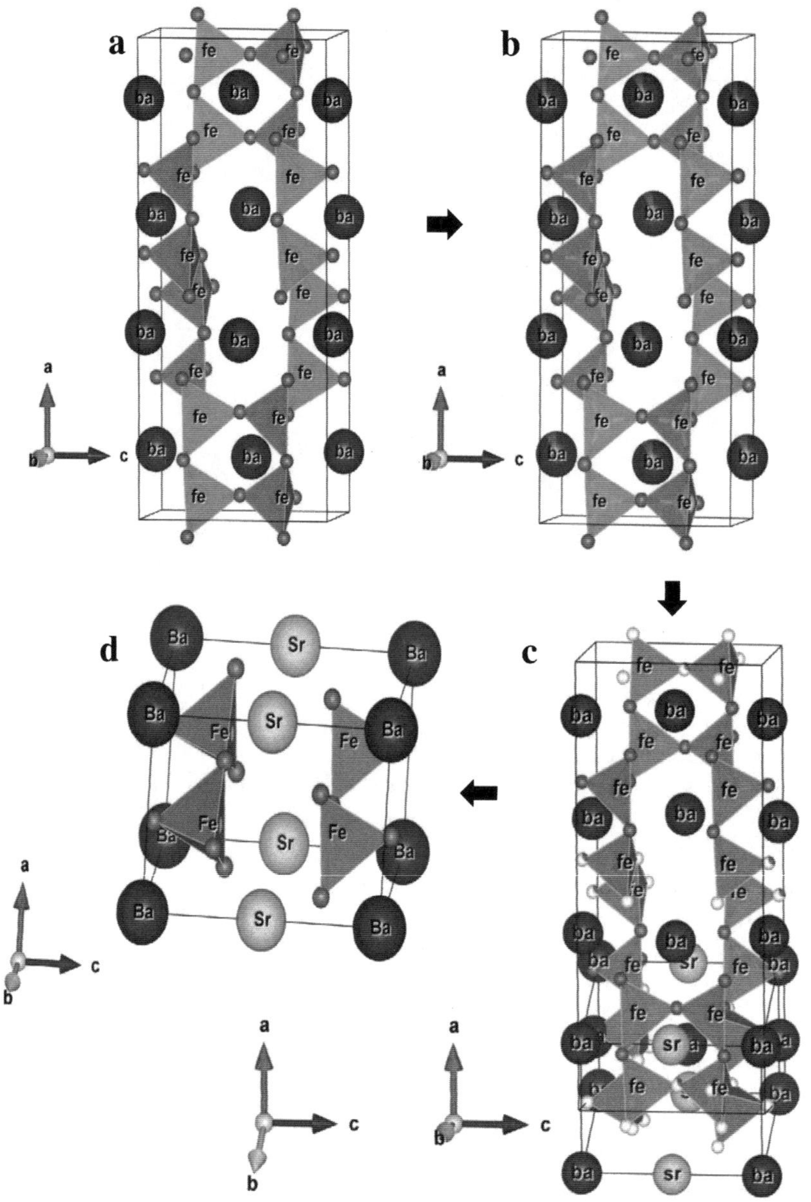

Figure 5. The gradual changes of the crystal structure from orthorhombic (space group C m c 21) into trigonal (space group P -3 1 m) in 3-dimensional form; **(a)** x = 0, **(b)** x = 0.1, **(c)** x = 0.3, and **(d)** x = 0.5

4. Conclusion

Referring to the the result of the X-ray diffraction pattern, it was known that the single phase of $BaFe_2O_4$ were successfully made by using solid state reaction method with High Energy Milling in the composition of x = 0 and 0.1. Due to the Rietveld refinements using GSAS software, it was observed that there was a gradual decrease for lattice parameters of $BaFe_2O_4$ phase align with the increasing contain of Strontium (Sr^{2+}) which influenced the value of the cell volume and the density. It was also found that in the composition of x 0.3 the presence of $BaSrFe_4O_8$ phase was begin to form due to to the overload expansion of the Sr^{3+} cations and fully formed in the composition of x = 0.5.

References

[1] Nejati K and Zabihi R 2012 *Chem Cent J* **6(1)** 23
[2] Candeia RA, Souza MAF, Bernardi MIB, Maestrelli SC 2006 *Ceramics International* 2–6
[3] Borse PH, Cho CR, Lim KT 2011 *Journal of the Korean Physical Society* **58(6)**1672–6
[4] Dalt S Da, Sousa BB, Alves AK, Bergmann CP 2011 *Material Research* **14(3)** 505–7
[5] Manaf A, Adi WA 2013 *J Sains Mater Indonesia* **14(2)** 136–41
[6] Ram M, Arya GS, Parmar K, Kotnala RK, Negi NS, Delhi N 2015 *Acta Physica Polonica A* **8(3)** 329–36
[7] Tudorache F, Popa PD, Brinza F, Tascu S 2012 *Acta Physica Polonica A* **121(1)** 95–7
[8] Akhtar MN, Yahya N, Hussain PB 2009 *IJBAS-IJENS* **9** 37–40
[9] Haddouch MA, Aharbil Y, Tamraoui Y 2015 *J. Mater. Environ. Sci.* **6(12)** 3403–15